Springer Series in Statistics

Advisors:
P. Bickel, P. Diggle, S. Fienberg, K. Krickeberg,
I. Olkin, N. Wermuth, S. Zeger

Springer
New York
Berlin
Heidelberg
Barcelona
Hong Kong
London
Milan
Paris
Singapore
Tokyo

Springer Series in Statistics

Andersen/Borgan/Gill/Keiding: Statistical Models Based on Counting Processes.
Andrews/Herzberg: Data: A Collection of Problems from Many Fields for the Student and Research Worker.
Anscombe: Computing in Statistical Science through APL.
Berger: Statistical Decision Theory and Bayesian Analysis, 2nd edition.
Bolfarine/Zacks: Prediction Theory for Finite Populations.
Borg/Groenen: Modern Multidimensional Scaling: Theory and Applications
Brémaud: Point Processes and Queues: Martingale Dynamics.
Brockwell/Davis: Time Series: Theory and Methods, 2nd edition.
Daley/Vere-Jones: An Introduction to the Theory of Point Processes.
Dzhaparidze: Parameter Estimation and Hypothesis Testing in Spectral Analysis of Stationary Time Series.
Fahrmeir/Tutz: Multivariate Statistical Modelling Based on Generalized Linear Models.
Farebrother: Fitting Linear Relationships: A History of the Calculus of Observations 1750 - 1900.
Farrell: Multivariate Calculation.
Federer: Statistical Design and Analysis for Intercropping Experiments, Volume I: Two Crops.
Federer: Statistical Design and Analysis for Intercropping Experiments, Volume II: Three or More Crops.
Fienberg/Hoaglin/Kruskal/Tanur (Eds.): A Statistical Model: Frederick Mosteller's Contributions to Statistics, Science and Public Policy.
Fisher/Sen: The Collected Works of Wassily Hoeffding.
Good: Permutation Tests: A Practical Guide to Resampling Methods for Testing Hypotheses.
Goodman/Kruskal: Measures of Association for Cross Classifications.
Gouriéroux: ARCH Models and Financial Applications.
Grandell: Aspects of Risk Theory.
Haberman: Advanced Statistics, Volume I: Description of Populations.
Hall: The Bootstrap and Edgeworth Expansion.
Härdle: Smoothing Techniques: With Implementation in S.
Hart: Nonparametric Smoothing and Lack-of-Fit Tests.
Hartigan: Bayes Theory.
Heyde: Quasi-Likelihood and its Application: A General Approach to Optimal Parameter Estimation.
Heyer: Theory of Statistical Experiments.
Huet/Bouvier/Gruet/Jolivet: Statistical Tools for Nonlinear Regression: A Practical Guide with S-PLUS Examples.
Jolliffe: Principal Component Analysis.
Kolen/Brennan: Test Equating: Methods and Practices.
Kotz/Johnson (Eds.): Breakthroughs in Statistics Volume I.

(continued after index)

Mike West
Jeff Harrison

Bayesian Forecasting and Dynamic Models

Second Edition

With 115 Illustrations

Springer

Mike West
Institute of Statistics and
Decision Science
Duke University
Durham, NC 27708-0251
USA

Jeff Harrison
Department of Statistics
University of Warwick
Coventry CV4 7AL
England

Library of Congress Cataloging-in-Publication Data
West, Mike, 1959-
Bayesian forecasting and dynamic models / Mike West, Jeff Harrison. — 2nd ed.
p. cm. — (Springer series in statistics)
Includes bibliographical references and index.
ISBN 0-387-94725-6 (hc : alk. paper)
1. Bayesian statistical decision theory. 2. Linear models (Statistics) I. Harrison, Jeff. II. Title. III. Series.
QA279.5.W47 1997
519.5′5—dc20 96-38166

Printed on acid-free paper.

Production managed by Robert Wexler; manufacturing supervised by Johanna Tschebull.
Photocomposed pages prepared from the authors' AMS-LaTeX file.
Printed and bound by Maple-Vail Book Manufacturing Group, York, PA.
Printed in the United States of America.

9 8 7 6 5 4 3 2 (Corrected second printing, 1999)

ISBN 0-387-94725-6 Springer-Verlag New York Berlin Heidelberg SPIN 10708511

Errata

West/Harrison: Bayesian Forecasting and Dynamic Models

Second Edition

0-387-94725-6

Chapter 1, page 31.

The following paragraph should appear at the end of Chapter 1, p. 31.

> The field continues to develop and flourish. As we approach the new millenium, we see exciting developments in new fields of application, and increasing sophistication in modelling developments and advanced computation. Interested readers might explore some recent studies in Aguilar and West (1998a,b), Aguilar, Huerta, Prado and West (1999), Cooper and Harrison (1997), and Prado, Krystal and West (1999), for example. Readers interested in contacting at least some of the more recently documented developments, publications and software, can explore the resources and links on the author web site indicated in the Preface.

Chapter 17, page 651.

The following paragraph should appear at the end of Chapter 17, p. 651.

> As an aside, note the more general Jordan forms for $\mathbf{G}$ matrices of non-observable models that appear in the proof of Theorem 5.2 (the requisite generalisations of the linear algebraic theory can be found, for example, in Theorem 8.5 of Nerig (1969).) In such cases any system matrix with one eigenvalue e of multiplicity n can be reduced to a form $\text{diag}(\mathbf{J}_{r_1}(e), \ldots, \mathbf{J}_{r_m}(e))$ with $r_1 + \ldots + r_m = n$ and where each $\mathbf{J}_{r_i}(e)$ matrix is a standard Jordan block. This might even be the diagonal case where each $r_i = 1$ when $\mathbf{G} = e\mathbf{I}$. This completes the general theory but is of little practical interest.

PREFACE

This text is concerned with Bayesian learning, inference and forecasting in dynamic environments. We describe the structure and theory of classes of dynamic models and their uses in forecasting and time series analysis.

The principles, models and methods of Bayesian forecasting and time series analysis have been developed extensively during the last thirty years. This development has involved thorough investigation of mathematical and statistical aspects of forecasting models and related techniques. With this has come experience with applications in a variety of areas in commercial, industrial, scientific, and socio-economic fields. Much of the technical development has been driven by the needs of forecasting practitioners and applied researchers. As a result, there now exists a relatively complete statistical and mathematical framework, presented and illustrated here. In writing and revising this book, our primary goals have been to present a reasonably comprehensive view of Bayesian ideas and methods in modelling and forecasting, particularly to provide a solid reference source for advanced university students and research workers.

In line with these goals, we present thorough discussion of mathematical and statistical features of Bayesian analyses of dynamic models, with many illustrations, examples and exercises. Much of the text will be accessible to advanced undergraduate and graduate/postgraduate students in statistics, mathematics and related fields. The book is suitable as a text for advanced courses in such disciplines. On the less mathematical side, we have attempted to include sufficient material covering practical problems, motivation, modelling and data analysis in order that the ideas and techniques of Bayesian forecasting be accessible to students, research workers and practitioners in business, economic and scientific disciplines.

Prerequisites for the technical material in the book include a knowledge of undergraduate calculus and linear algebra, and a working knowledge of probability and statistics such as provided in intermediate undergraduate statistics courses. This second edition includes many more exercises. These exercises are a mixture of drill, mathematical and statistical calculations, generalisations of text material and more practically orientated problems that will involve the use of computers and access to software. It is fair to say that much insight into the practical issues of model construction and usage can be gained by students involved in writing their own software, at least for the simpler models. Computer demonstrations, particularly using graphical displays, and use of suitable software by students, should be an integral part of any course on advanced statistical modelling and forecasting.

Since the first edition appeared in 1989, the field has experienced growth in research, both theoretical and methodological, as well as in developments in computation, especially via simulation methods, and in more diverse applications. The revision for this second edition involved updates and refinements of original material, together with additional material based on research and development in the field during the early 1990s. In terms of new or substantially revised material, we note novel theory and methodology of dynamic linear model (DLM) analyses, including developments in retrospective time series analysis (Section 4.8), model estimation and diagnostics (Section 4.9), and in the theory of limiting results in time series dynamic linear models (Section 5.5). New material has been added on stationary time series models (Sections 5.6 and 9.4), on important new methods of time series decompositions in the state-space framework (Sections 9.5, 9.6 and 15.3), on time-varying parameter autoregressive DLMs (Section 9.6), and on inference and application of autoregressive component DLMs (Section 15.3). New results and methods of model monitoring and assessment, developed from a Bayesian decision analytic viewpoint (Section 11.6), complement original material on intervention and model assessment. Substantial new material has been added on statistical computation and simulation methods for Bayesian analysis of non-linear models, including, in particular, a new chapter, Chapter 15, focussed mainly on Markov Chain Monte Carlo approaches in dynamic models. This rapidly growing area represents one of the currently critical research frontiers in statistics, and in time series modelling and analysis specifically. Throughout the book the new and revised material includes additional illustration and references, as well as theory and methods, and new exercises in several chapters.

Following the publication of the first edition, we developed a related text that discusses and illustrates application of a standard class of dynamic linear models, essentially those of Chapters 10 and 11 here. That 1994 text, *Applied Bayesian Forecasting and Time Series Analysis* by Andy Pole, Mike West and Jeff Harrison, includes an extensive guide to the use of the BATS software package that implements the model class. BATS, written in C by Andy Pole, and with support from Chris Pole, runs under Windows95 and DOS. BATS was developed from original versions in APL; many of the examples and graphs here were produced using the APL version. More examples appear in the 1994 text, which readers may find to be a useful adjunct to the current, more comprehensive reference text. S-Plus software for dynamic modelling, forecasting and retrospective analysis is available from the second author in collaboration with Robin Reed (Harrison and Reed 1996). Additional software for time-varying autoregressions and time series decompositions (developed in Section 9.5, 9.6 and 15.3) is available in Matlab and Fortran90/S-Plus code from the first author in collaboration with Raquel Prado. This can be found at the Duke web site, http://www.stat.duke.edu/.

The field continues to develop and flourish. Readers interested in keeping up with at least some of the post-publication developments of the authors and their coauthors can explore the resources and links at the Duke web site, as indicated above.

Thematically, material in the book can be loosely partitioned into four sets, each of three consecutive chapters, a final set of four chapters on more advanced topics, and an appendix.

A. Introduction

The first three chapters provide a broad introduction to the basic principles, modelling ideas and practice of Bayesian forecasting and dynamic models. In Chapter 1 we discuss general principles of modelling, learning and forecasting, and aspects of the role of forecasters within decision systems. Here we introduce basic elements of dynamic modelling and Bayesian forecasting. Chapter 2 is devoted to the simplest, and most widely used, dynamic model, known as the first-order polynomial model, or steady model. In this setting, the simplest mathematical framework, we introduce the approach to sequential learning and forecasting, describe important theoretical model features, consider practical issues of model choice and intervention, and relate the approach to well-known alternatives. Chapter 3 continues the introduction to dynamic modelling through simple, dynamic regression models. Readers will be familiar with standard regression concepts, so that the rather simple extension of straight line regression models to dynamic regression will be easily appreciated.

B. Dynamic linear model theory and structure

Chapters 4, 5 and 6 provide a comprehensive coverage of the theoretical structure of the class of *dynamic linear models* (DLMs) and Bayesian analyses within the class. Chapter 4 is key. Here we introduce the fundamental concepts, principles, general framework, definitions and notation, and fully develop the distribution theory associated with dynamic linear models. This includes complete and detailed descriptions of entire joint distributions relevant to sequential learning, forecasting and retrospective analysis. Chapter 5 is concerned with a special subclass of DLMs, referred to as *time series models,* that relate naturally to most existing methods for time series forecasting. In this second edition, a new elegant proof of variance convergence for constant DLMs is given, more convergence results are provided, and new material on stationary time series models is included. Chapter 6 focuses on two important aspects of model design and specification, namely component modelling and discounting.

C. Classes of dynamic models

Chapters 7, 8 and 9 describe in greater detail the structure of important special classes of dynamic models and their analyses. Chapter 7 is devoted to time series models for polynomial trends, particularly important cases being first-order polynomials of Chapter 2 and second-order polynomials, or linear trend models. Chapter 8 concerns dynamic linear models for seasonal time series, describing approaches through seasonal factor representations and harmonic models based on Fourier representations. Chapter 9 concerns relationships between time series modelled through dynamic regressions, extending Chapter 3, and extended models for transfer effects of independent variables. This second edition substantially expands discussion and development of classical ARMA and related models, and connections with dynamic linear modelling, in Chapter 9, including new and practically useful developments in time-varying parameter autoregressive models.

D. DLMs in practice, intervention and monitoring

Chapter 10 illustrates the application of standard classes of dynamic models for analysis and forecasting of time series with polynomial trends, seasonal and regression components. Also discussed are various practical model modifications and data analytic considerations. Chapter 11 focuses on intervention as a key feature of complete forecasting systems. We describe modes of subjective intervention in dynamic models, concepts and techniques of forecast model monitoring and assessment, and methods of feed-forward and feed-back control. The second edition expands this chapter with new material on model assessment based on Bayesian decision analysis, with consequent links to cusum methods of model monitoring. Chapter 12 is concerned with multi-process models by which a forecaster may combine several basic DLMs for a variety of purposes. These include model identification, approximation of more complex models, and modelling of highly irregular behaviour in time series, such as outlying observations and abrupt changes in pattern.

E. Advanced topics

Chapters 13, 14, 15 and 16 are concerned with more advanced and recently developed models. In Chapters 13 and 14 we consider approaches to learning and forecasting in dynamic, non-linear models, where the neat theory of linear models does not directly apply. Chapter 13 describes some standard methods of analytic and numerical approximations, and also some more advanced approaches based on numerical integration; this leads into new developments in Bayesian computation based on stochastic simulation. Chapter 14 develops non-normal models and explores methods and applications in the class of dynamic generalised linear models. Chapter 15, a completely new chapter for the second edition, is wholly concerned with dynamic model analysis via methods of stochastic simulation, discussing, in particular, recent advances in Gibbs sampling with application to autoregressive component DLMs. In Chapter 16 we return to primarily linear models but consider aspects of modelling and forecasting in multivariate settings.

Acknowledgements

Section 1.4 of Chapter 1 briefly reviews historical developments and influences on our own work. Over the years many people and organisations have contributed, directly and indirectly, to material presented in this book and to our own efforts in developing the material. Others have helped enormously in suggesting changes and corrections and in assisting with proof-reading the second edition. In particular, we would mention Colin Stevens, Roy Johnston, Alan Scott, Mike Quinn, and past and present research students Mike Green, Jim Smith, Rei Souza, Muhammed Akram, Jamal Ameen, Helio Migon, Tomek Brus, Dani Gamerman, Jose Quintana, Emanuel Barbosa, Klaus Vasconcellos, Parma Veerapen, Raquel Prado, Gabriel Huerta Gomez, Omar Aguilar Chavez, Heidi Ashih and Susan Paddock.

Special thanks go to Andy Pole, our colleague and co-author for almost a decade, and the leader in developing the BATS package. We would also like to thank all our colleagues, past and present, for their interest and inputs, in particular, Jim Smith, Ewart Shaw, Adrian Smith, Dennis Lindley, Tom Leonard, Peter Müller, Tony O'Hagan, and Robin Reed.

Among the companies who have supported the development of Bayesian forecasting we must single out Imperial Chemical Industries plc. The initial work on forecasting began there in 1957 and took its Bayesian flavour in the late 1960's. That company has continued to support development, particularly during the late 1980s, and we would particularly like to acknowledge the continued interest, help and encouragement of Richard Munton, Steve Smith and Mike Taylor. Other companies have supported our work and aided software development. Among these are British Gas Corporation, through the guiding efforts of Chris Burston and Paul Smith, Information Services International (formerly Mars Group Services), Millward Brown, Unilever, and IBM.

In addition, we acknowledge the support and facilities of our home departments and institutions, the Institute of Statistics and Decision Sciences at Duke University, and the Department of Statistics at the University of Warwick. Finally, research support received from the Science Research Council (U.K.) and the National Science Foundation (U.S.A.) has contributed, directly and indirectly, to various aspects of our work and hence to the development of this book.

Mike West & Jeff Harrison
1999

CONTENTS

CHAPTER 1

INTRODUCTION

1.1 MODELLING, LEARNING AND FORECASTING

1.1.1 Perspective

This book concerns modelling, learning, and forecasting. A basic view of scientific modelling is that a model is a *"simplified description of a system ... that assists calculations and predictions"* (Oxford English Dictionary). Broadly, a model is any descriptive, explanatory scheme that organises information and experiences, thus providing a medium for learning and forecasting. The prime reason for modelling is to provide efficient learning processes that will increase understanding and enable wise decisions.

In one way the whole operation of an organisation can be viewed as comprising a sequence of decisions based upon a continual stream of information. Consequently there is an accumulation of knowledge that, in principle, should lead to improved understanding and better decisions. Suitably formulated and utilised, descriptive models provide vehicles for such learning.

The foundation for learning is the Scientific Method. It is often assumed that scientific investigation and learning are concerned with the pursuit and identification of a single "true" model, but this is certainly not our position. Models do not represent truth. Rather they are ways of viewing a system, its problems and their contexts, that enable good decisions and enhance performance both in the short and long term. Within a model framework, the scientific learning process facilitates the routine, coherent processing of information, that leads to revised views about the future and hence to rational actions.

Since a model organises personal experiences and information, it is always a subjective picture anchored in the past. Consequently, a derived forecast, being a hypothesis, conjecture, extrapolation, or speculative view about something future, may well prove to be "far from the mark", resulting in a sizeable forecast error. But, it is exactly the forecast errors that stimulate learning and a good system will efficiently utilise them in order to improve performance through model enhancement. Forecasting systems often ignore the wider aspects of learning and are founded on myopic models incapable of development and only able to make overly restricted predictions.

One desirable property of a forecasting and learning system is that the way of viewing should not change radically too frequently; otherwise confidence is impaired, communication breaks down, and performance deteriorates. Hence the fundamentals of an operational model should remain constant for considerable periods of time, regular change only affecting small details. That is, there should be a routine way of learning during phases when predictions and decisions appear adequate, and an "exceptional" way when they seem unsatisfactory. The routine adjustment will generally be concerned with small improvements in the estimates of model quantities whereas the exceptions may involve basic model revision.

The principle by which such systems operate is that of *Management by Exception,* an important part of the Scientific Method. Essentially, information is routinely processed within an accepted conceptual and qualitative framework within which the major features of an operational model remain unchanged, minor quantitative modifications are accommodated, and the relative merits of any rival models noted. Exceptions arise in two main ways. The first occurs when non-routine expert information anticipates a future major change that will not properly be reflected by routine learning. The second occurs when performance monitoring, using quality control techniques applied to the forecast errors, identifies deficiencies, thus questioning model adequacy and prompting expert intervention. In either case, for effective communication, the model structure must be descriptive, robust and meaningful.

1.1.2 Model structure

Model structure is critical to performance. A good structure will provide model properties that include

- Description
- Control
- Robustness

Consequently we view the structuring of a model through a triple

$$\mathcal{M}: \qquad \{C, F, Q\}.$$

The component C describes the **conceptual** basis, F the model **form**, and Q the **quantified** form.

C : The concepts C provide an abstract view of a model. They may express decision centre objectives; scientific or socio-economic laws; behavioural characteristics etc. As such they are expected to be very durable and rarely changed. Further, at any moment in time, rival models representing alternative views may be founded upon the same conceptual base C (although this is not always the case).

F : The qualitative form F represents the conceptual in descriptive terms, selecting appropriate variables and defining relationships. For example a government may be seen as a decision centre wishing to retain power. This may be a part of a general conceptual view of an economic sector, which helps to express the type of decisions to be taken as circumstances evolve. At any time there may be choices about the way in which general objectives are accomplished. With one policy, the relevant control variables will be from set A, say, and at a time of policy revision, they may suddenly change to a set B. Form defines the relevant sets of variables and their relationships, perhaps in terms of algebraic, geometric, graphical, and flow sheet representations.

Q : Often many, if not all, rival models will have a common qualitative parametric form and differ only at the quantitative level Q, in the values given to the parameters. Then a single form is often durable for reasonable periods of time. It is at the quantitative level that frequent change occurs. Here the merits of rival parametric values are continually changing as new information is received. Generally these are small changes that are in accord with the uncertainty conditional upon the adequacy of both concept and form.

Description aims at providing meaning and explanation in an acceptable and communicative way. This is necessary for unifying all concerned with a decision process and its effects. It brings confidence from the fact that all are working and learning together with a well defined view. In particular, it encourages growth in understanding, participation, and progressive change. In most decision situations, anticipation of major change is the critical factor upon which the life of an organisation may depend. Hence it is vital to promote creative thinking about the basic concepts and form, and to improve intervention at all levels. Two important aspects of an effective description are parsimony and perspective. Parsimony means simplicity. It excludes the irrelevant and uses familiar canonical concepts, forms, and learning procedures, bringing all the power of past experience to bear. Perspective is concerned with the relative importance of the various model characteristics and what they do and do not affect.

Control is usually taken to mean actually influencing the behaviour of the system being modelled. Given such control there is great opportunity for effective learning by experimenting. In process control this may mean carrying out dynamic experimentation, as embodied in the principles and practice of evolutionary operation (Box and Draper 1969). All the principles and power of the statistical design of experiments can then be utilised. Another aspect of control occurs when the system being modelled cannot be controlled or can only be partially controlled. Then the decision makers may still have control in the sense of having freedom to respond wisely to predictions about systems that they cannot influence. For example a farmer who assesses the weather in order to make farming decisions cannot directly influence the weather but can utilise his forecast to control his actions, perhaps waiting until "the time is ripe". Currently there is much power in the hands of remote, centralised organisations that can bring disaster to those they control simply because of their selfish desires and ignorance of the systems they control. It is then vital that the true nature of the "controlled" system be understood along with the motivations and reactive responses of the power centre. One look at the past control of European agriculture and industry illustrates this point only too well.

Robustness is a key property for a learning system. Essentially the aim is to structure so that at "exceptional" times, intervention is efficiently and economically accomplished. This means only changing that which needs changing. Thus the objective is to extract the maximum from history so that all relevant information is retained whilst accommodating the new. The structure $\{C, F, Q\}$ provides a major source of robustness. It offers the opportunity of carrying out major changes at the quantitative level Q whilst retaining the model form F. It provides a way of retaining the conceptual C whilst drastically altering aspects of form F. However, it is also crucial to structure within each of these levels so that major changes affect only the relevant aspects and do not damage others. At the quantitative level, when operating within a parametrised form, modelling component features of a system through distinct though related *model components* is recommended. Each component describes a particular aspect such as price elasticity, the transfer response of an input flow change, a seasonal effect, or a trend. Then if intervention necessitates a major change concerning any particular component this can be accomplished without affecting other components.

1.1.3 The role of mathematics

The role of mathematics and statistics is as a language. It is a very powerful language since far-reaching implications can often be deduced from very simple statements. Nevertheless, the function of mathematics must be seen in perspective. It expresses a view in a way analogous to that of paint on canvas. In this sense it is only as good as the artist who uses the materials and the audience who see the result. Like all sources of power, mathematics can be well used or abused. Selective, marginal, and conditional interpretations are the key weapons of the deceiver. Any modeller, just like an artist, must of necessity select a view of the context under study and is thus, either innocently or deliberately, likely to mislead. Choosing an appropriate view is often very hard work. It may have nothing to do with mathematics, although of course it can involve some data analysis. Many modellers pay scant regard to this vital preliminary effort in their eagerness to play with computers and equations. Consequently, so many mathematical models are inappropriate and misleading. With today's computing power it is not far from the truth to say that if a system can be coherently described then it can be expressed mathematically and modelled. To summarise, our position is that modelling is an art; that the first task is to define objectives; the second to select a consistent view of the system; and only later, and if appropriate, to use a mathematical description.

1.1.4 Dynamic models

Learning is dynamic. At any particular time a model describes a routine way of viewing a context, with possible competing views described through alternative models. However, because of uncertainty, the routine view itself is likely to comprise a set of models. For example, consider the view that an output variable Y is related to an input variable X according to a parametrised form

$$Y = X\theta + \epsilon.$$

Here θ is an uncertain parameter and ϵ an uncertain, random error term. Further suppose that the forecaster's beliefs about the parameter θ are expressed through a probability distribution $P(\theta)$. Then the view may be described as comprising a set of models, one for each possible value of θ, each with measurable support $P(\cdot)$. This embodies one form of model uncertainty. The dynamic nature of processes and systems demands also that uncertainty due to the passage of time be recognised. Then, because the model form is only locally appropriate in time, it is necessary to routinely characterise θ as slowly evolving. Further, at some future time, the model form may change, possibly involving quite different input variables. Such typical applications require **dynamic models**, defined generally as *"sequences of sets of models"*.

At any given time, a dynamic model $\mathcal{M}$ will comprise member models M, with the forecaster's uncertainty described through a *prior* distribution $P(M)$, $(M \in \mathcal{M})$. In producing a forecast from the dynamic model for an output Y, each member model M will provide a conditional forecast in terms of a probability distribution $P(Y|M)$. In the above example M relates directly to the parametrisation θ and thus to the component Q of the model. This is typical. More widely, however, M may involve uncertain aspects of form F and even conceptual descriptions C. The forecast from the dynamic model $\mathcal{M}$ is then simply defined by the marginal probability distribution, namely

$$P(Y) = \int_{M \in \mathcal{M}} P(Y|M) dP(M).$$

Alternative dynamic models are rivals in the sense that they compete with the routine model for its prime position, thus providing a means of performing model monitoring and assessment. Although it is often the case, a dynamic model is not restricted to members having the same form and conceptual base; the framework is entirely general.

1.1.5 Routine learning

Bayesian methodology offers a comprehensive way of routine learning that is not dependent upon any particular assumptions. For simplicity consider a dynamic model $\mathcal{M}$ with member models M. Before receiving an observation of uncertain quantities Y, for each $M \in \mathcal{M}$, the forecaster has a *prior* probability $P(M)$ describing the relative uncertainty concerning the *truth* of M. Each member model M provides a means of forecasting Y through a conditional probability distribution $P(Y|M)$, which specifies the view about future possible values of Y, conditional upon the *truth* of that particular description M.

By the laws of probability, these two sets of probabilities combine to provide a joint probability distribution, written in terms of densities as

$$p(Y, M) = p(Y|M)p(M).$$

When Y is observed to take a value Y^*, say, the updated probability distribution for M given $Y = Y^*$ is defined by the conditional density

$$p(M|Y^*) \propto p(Y^*, M),$$

or equivalently

$$p(M|Y^*) \propto p(Y^*|M)p(M).$$

This means of updating or learning is often expressed as

$$\textbf{Posterior} \propto \textbf{Observed likelihood} \times \textbf{Prior.}$$

The proportionality constant is simply the normalising quantity $p(Y^*)$, the prior density for Y at the observed value Y^*, which ensures that the "posterior" density over $M \in \mathcal{M}$ is normalised to unit probability. Hence, for any observed value of Y, the Bayes' theorem representation is

$$p(M|Y) = p(Y|M)p(M)/p(Y), \qquad (M \in \mathcal{M}).$$

Given the dynamic model $\mathcal{M}$, all the routine information contained in the observation Y is expressed through the likelihoods $p(Y|M)$.

1.1.6 Model construction

Some forecasters approach model building in an unsatisfactory way, often resorting to processing historical output data by computer and accepting whatever mathematical expression emerges. Such people are always asking "Where do you get your models from?" The answer is very simple. We apply the natural scientific approach to model building.

As previously stated, the first step is to clarify objectives. If both macro and micro decisions are to be made using the model, think about the value of structuring the model hierarchically; initially modelling within hierarchical levels whilst not losing sight of the relationships between levels. The next step is to decide upon the conceptual basis, and then to consider significant factors and relationships. In particular, the aim is to explain as much of the significant variation as possible. This may involve identifying "sure thing" relations that arise because of accepted physical laws, relationships, constraints etc. Often "lead" variables are involved. For example, if you are predicting beef supply at the butchers, it is clear that there is a lag of three years from impregnation of a cow to the resulting meat reaching the shop. Within a "closed" community it is obvious that the decision point for expansion at retail level is three years previously. It is also very clear, from biological considerations, that "in order to expand you must first contract". That is, more heifers will have to be taken back into the breeding herd at the expense of next year's meat supply. Germinal decision points and these delayed response dynamics appear in many systems only to be completely ignored by many modellers and "fire brigade" decision makers. Hence the reason for so many problems that arise, not only in agriculture (Harrison and Quinn 1978), but in many other economic sectors.

Another key step is to assess the nature of the process being studied. Is it purposeful? It is no use studying the detailed flow of a river in forecasting its path; the main thing is to recognise that all rivers follow the principle of steepest descent. Similarly, when studying a process dependent upon other decision makers, it may be critical to assess their purposes, what effect the environment will have on them, how their perceptions match reality, and how their statements match their actions. In selecting model factors, control factors are to be prized.

When, and only when, any significant structure has been modelled should the modeller resort to time series. A time series model is essentially a confession of ignorance, generally describing situations statistically without relating them to explanatory variables. That is not to say that pure time series models are useless. For example, time series models involving polynomial trends and seasonal components may be very useful in, say, short-term sales forecasting, where experience, perhaps across many similar products has led to empirical growth laws and defensible arguments for seasonality. The danger arises when mathematical expressions, such as general stationary noise models, are adopted without any substantial foundation. Then historic peculiarities are likely to suggest totally inappropriate models.

The key message for the modeller is *"THINK, and do not sacrifice yourself to mathematical magic."* This is not to rule out exploratory data analysis. History and data may be used both to assess contending models and to stimulate creative thought, but data analysis should not replace preliminary contextual thinking, nor should it promote mathematical formulae that have no defensible explanations. All analytic methods have some contribution to offer, but they must be seen as servants of explanatory thought and not its usurper.

1.2 FORECAST AND DECISION SYSTEMS

1.2.1 Integration

As Dr Johnson said, *"People need to be reminded more often than they need to be instructed."* This is true in modelling, especially when dealing with open decision systems. There is much that we know or are able to recognise but that we are not anxious to see or practise. Good modelling demands hard thinking, and good forecasting requires an integrated view of the role of forecasting within decision systems.

The consequences of failing to harmonize forecasts and decisions is sharply, and rather humorously, exemplified in the following incident that occurred over thirty years ago in the brewery trade. At the end of the summer of 1965, one of us received a call from "Peter." Peter had just moved to a famous brewer and had been given the responsibility for short-term forecasting. Since he knew very little about this, he sought advice. Upon visiting, he said that his task was to produce beer forecasts for two weeks ahead. He did not appear to be enthralled about this, and when asked what decisions rested on the forecasts his reply was "I am not interested in the decisions. My job is simply to forecast." Pressed further, he said that the forecasts would be used in production planning. However, he would not see the importance of integrating forecasting and control. Prior to his visit, some of his data had been examined using the ICI MULDO package (Harrison and Scott 1965; Harrison 1965); the data appears in Figure 1.1.

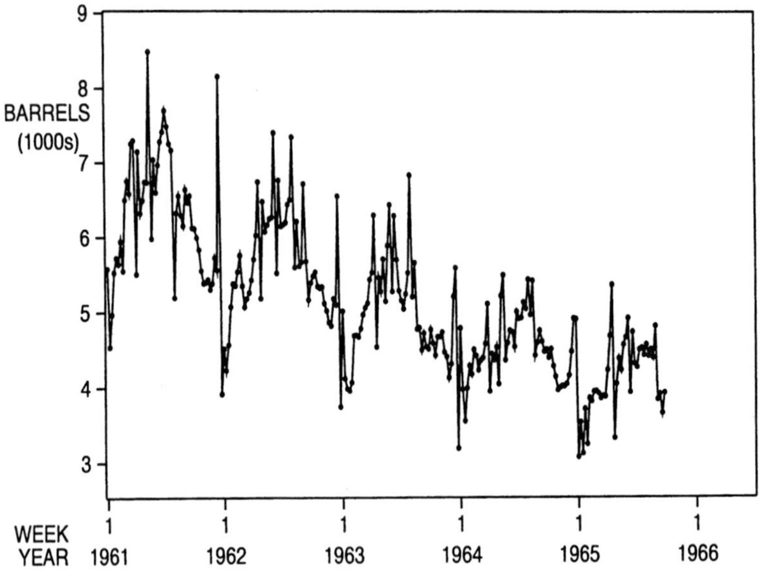

Figure 1.1 Barrels of beer (in thousands) sold in a sector of the UK market during 1961 to 1966

As is evident from the figure, the data showed a marked seasonal effect with a summer peak, a pronounced festive component, particularly at Christmas, and a declining trend. A few months later Peter telephoned to announce that he was in trouble. He did not volunteer the extent of his misfortunes. "No, I have not been following the methods you advised. I simply estimated based upon the last few weeks. This seemed to give as good forecasts as yours and to begin with I did fine." Six weeks later it become clear just how his misfortunes had grown. The story made the front page of a number of national daily newspapers, with the colourful *Daily Mirror* version reproduced below in Figure 1.2. The reader is left to speculate how a computer came to be blamed when apparently Peter had never had access to one.

Computer sends the beer for a Burton

It happened, of all places, in Burton-upon-Trent, the town made famous by beer.

First, they found they had **TOO MUCH** beer.

Thousands of gallons too much — all because a computer went wrong.

The computer over-estimated how much thirsty revellers could swallow over the Christmas and New Year holidays — and now the beer is too old for use.

Brewery officials in the Staffordshire beer "capital" ordered: "Down the drain with it ..."

Secret

The throwaway beer — more than 9,000 casks of best bitter and pale ale produced by the Bass-Worthington group — is worth about £100,000.

Tankers are pouring it down the drain at a secret spot.

But now the brewery officials are feeling sour once again ...

Some pubs in the town yesterday reported a beer **SHORTAGE**.

Production of fresh supplies has been disrupted — again because of the Christmas and New Year holidays.

A brewery spokesman said: "We were caught on the hop."

Figure 1.2 Caught on the hop. (*Daily Mirror*, January 12th 1966)

1.2.2 Choice of information

Often there is a choice concerning the observed information on which forecasts are based. Modellers are generally aware that they may need to allow for differences in numbers of working days, seasonal holidays, deflated prices, etc. However, there are other considerations. It is clear that in controlling a system, information that is relatively independent of the performance of that system is preferable to that which is dependent upon it. In many cases little thought is given to the matter or its consequences. As an example, in short-term sales forecasting for stock control and production planning, modellers may, without question, accept sales statistics as their routine observations. The real objective is to forecast customer requirements. Clearly, sales statistics represent what is sold. As such they reflect the ability of the system to meet requirements, and not necessarily the actual requirements themselves. The use of sales statistics may result in excessive variation, particularly when products compete for manufacturing capacity, as illustrated by the graph in Figure 1.3.

There can also be an insensitivity to increases in demand, particularly when manufacture is continuous. Other problems include recording delays, due to waiting for suitable transport, and the possibility of negative figures when returns occur. In general it is preferable to collect order or demand

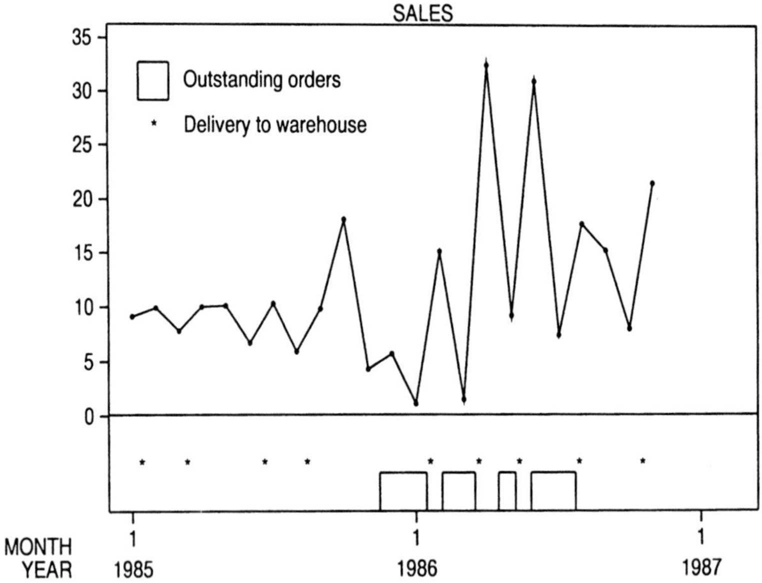

Figure 1.3 Sales variation induced by production system

statistics, although even these can mislead on requirements in periods of shortage when customers deliberately request more than they need, knowing that they will only receive a partial allocation. However, the short-term problem is then one of allocation.

Another example of care is when one of us correctly forecast a bumper linseed harvest, but then, without thinking, converted it into its oil equivalent. Unfortunately this proved to be a significant over-estimate of the available oil since crushing capacity was limited. So, although the abundant harvest materialised, the oil price did not fall nearly as far as predicted.

These examples suffice to remind modellers to consider both the direct relevance and the quality of their observations.

1.2.3 Prospective intervention

A major problem for many mathematically based learning systems has been that of accommodating subjective information. This is particularly important at times of major change. Consider the example of the history of UK sulphuric acid production. Production statistics from 1870 to 1987 are shown in Figure 1.4 (Source: *Monthly Digest of Statistics*).

Clearly there were major changes that, if not anticipated, would have been extremely costly to producers. The feature of these changes is the

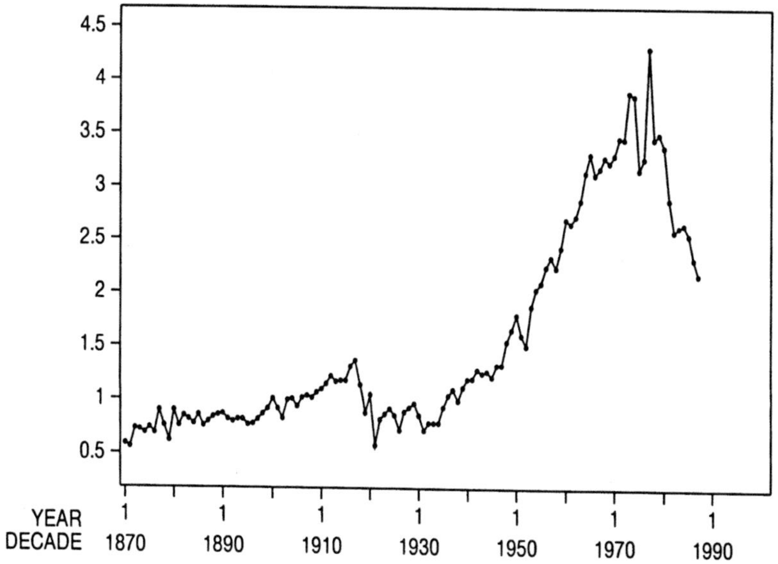

Figure 1.4 UK sulphuric acid production (tons$\times 10^6$)

sharp decline and demise of previously significant end uses due to technological and social change. The first end use to suffer was the Leblanc process for the production of soda. The ammonia-soda process uses no sulphuric acid and finally superseded Leblanc around the end of 1920. More recently, hydrochloric acid has challenged in steel pickling since it offers a faster process and is recoverable. The importance of an early awareness and assessment of the effect of these developments and consequent prospective intervention is self-evident (Harrison and Pearce 1972).

The vital role of a model structure facilitating expert intervention has already been stressed. Usually, at times of intervention, there is additional uncertainty with an accompanying sharp change in beliefs about the future. Being phrased in terms of meaningful probabilities, the Bayes' approach offers a natural way of accommodating such uncertain information, thus facilitating intervention. This is one of the prime reasons for its adoption as the routine learning method within management by exception systems.

As an example, consider short-term forecasting of the UK retail price index, (RPI) as in Figure 1.5, with a routine view based on local linear extrapolation. In June 1979, the chancellor's announcement that the value-added tax (VAT) was to be raised from 10% to 15% makes intervention essential to sustained short-term forecasting accuracy. Prior to this the July price level might have been assessed by a forecaster as 221.5 with an associated uncertainty represented by a standard deviation of 1. However, upon hearing of the change in VAT, the forecaster is likely to assess the effect as increasing the RPI by an extra 3.5%, leading to a revised estimate of 229.5, with the increased uncertainty being reflected by a standard deviation of 3. The actual mechanics of the change can, and should, be designed with the intervener in mind. Direct communication in terms of statistical terminology is not at all necessary, provided the chosen intervention method can be statistically interpreted. Figure 1.5 further illustrates the VAT incident, representing simple forecasts based on the foregoing intervention. The forecasts are one-step ahead, that for each monthly value of the RPI being made the previous month, and are plotted with associated uncertainties represented simply by intervals of one standard deviation (sd) either side of the point forecasts.

Some people object to intervention on the grounds of subjectivity. Let us be quite clear: as soon as anyone starts to model they are being subjective. Of course, any facility that enables subjective communication is open to abuse, but the answer is to monitor and control subjectivity rather than stupidly prohibiting it. This may be accomplished within the routine monitoring system or otherwise. It is important that such monitoring be performed as part of the learning process, for upon first encountering complete learning and forecasting systems, decision makers are tempted to intervene on ill-founded hunches, on situations that are already accommodated in the routine model, or simply due to wishful thinking.

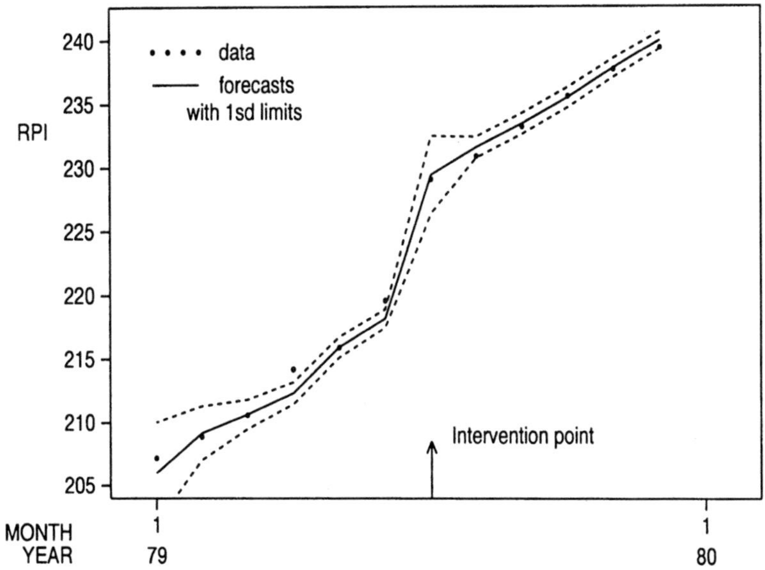

Figure 1.5 UK RPI forecasting with intervention

1.2.4 Monitoring

A continual assessment of the performance of any regularly used forecasting and decision system is vital to its effective use. Model monitoring is concerned with detecting inadequacies in the current routine model M, and, in particular, in signalling significant forecast errors caused by major unanticipated events. Practitioners will be familiar with such schemes as applied in quality control and inventory management. Of course it is hoped that expert intervention will anticipate relevant major events, but in the absence of such feed-forward information, monitoring systems continually assess the routine model, signalling doubtful performance based on the occurrence of unexpectedly large forecast errors. At such times, relevant explanatory external information may be sought or automatic procedures applied, that are designed to correct for specified types of change.

Within the Bayesian approach, the extent to which observations accord with predictions is measured through predictive probabilities, possibly with an associated specified loss function. Writing $P(Y|M)$ as the predictive probability distribution for a future quantity Y based on model M, model adequacy may be questioned if the eventually observed value $Y = y$ does not accord with $P(y|M)$. The initial task in designing any monitoring system is to decide what constitutes bad forecasting, that is, how to measure accord with forecast distributions in the context of decisions that are dependent upon these forecasts.

1.2.5 Retrospective assessment

In addition to learning and prediction, retrospection is often informative. Typically this is called a *"What happened?"* analysis. In the light of all the current information, the objective is to estimate what happened in the past in order to improve understanding and subsequent performance.

A typical marketing example occurs when a novel promotional campaign is undertaken. Initially, prospective intervention communicates the predicted effects, which are likely to involve increased sales estimates, and their associated uncertainties over the campaign period. This is necessary in order to ensure that appropriate product stocks are available. The progress of the campaign's effect will be suitably monitored. Then, some time after completion, the campaign will be retrospectively assessed in order to increase the store of market knowledge. This involves properly attributing effects to the many contributing sources of variation such as price, seasonality, competitive action, etc. Retrospective analysis is particularly useful for examining what happened at times of major change, especially where there is debate or ignorance about possible explanations. As such it is an integral part of a good learning system.

1.2.6 Utilities and decisions

A statistician, economist, or management scientist usually looks at a decision as comprising a forecast, or belief, and a utility, or reward, function. In a simple horse-racing situation, you may have beliefs about the relative merits of the runners that lead you to suppose that horse A is the most likely to win and horse B the least likely to win. However, the bookmaker's odds may be such that you decide to bet on horse B.

Formally, in a simple situation, you may have a view about the outcome of a future random quantity Y conditional on your decision a expressed through a forecast or probability distribution function $P(Y|a)$, possibly depending on a. A reward function $U(Y,a)$ expresses your gain or loss if outcome Y happens when you take decision a. Thus, for each decision a it is possible to calculate the merit as reflected in the expected reward

$$\mathrm{r}(a) = \int U(Y,a)dP(Y|a).$$

The optimal Bayes' decision is that a that maximises $r(a)$, as in De Groot (1971) and Berger (1985).

Accepting this formulation, it is clear that in general, for a given forecast distribution, provided the utility favours a particular decision strongly enough, that decision will be chosen. Also, for a given utility function, if the probability distribution or forecast relating to decisions sufficiently favours a particular decision, then that decision will be chosen.

A decision maker occupies a powerful situation and often has strong self interests that lead to utilities quite different from those of others. On the other hand, often the decision maker cannot openly express his selfish utilities and is therefore forced to articulate in terms of a utility function that is acceptable to others who may later authorise the decision or participate in implementation. In such a situation, forecasting offers a major source of power, which can be used to manipulate decisions. Furthermore, the extent of forecast manipulation in business, government, etc. should not be underestimated. As Bertrand Russell (1921) remarks, *"It becomes necessary to entertain whole systems of false beliefs in order to hide the nature of what is desired."* It is not uncommon for practitioners to be asked to undertake forecasting exercises where the directive is that a given decision should result. For example, "you must not show that this plant should not be built," and "we must show that television advertising does not influence the total alcoholic drinks market, but only the shares of competing brands." The former case involving extended capacity occurs regularly. On one occasion one of us was strongly reprimanded for carrying out a check that revealed that the current market of a recently formed organisation was over 40% less than the figures being used to justify extended capacity. The reasons for such exhortations are clear. Unless the plant is built, "I will be out of a job", "we banks cannot lend money", "unemployment statistics will increase", etc., so that many powerful groups have very strong and similar utility functions.

The key weapon of the manipulator is *selection.* This may be data selection, the selection of a particular set of circumstances to illustrate an argument, the reference to forecasts from other, similarly interested organisations, the selection of consultants and forecasters who need business, the selection of an inexperienced project team, etc.

There is, as readers will be well aware, a recent history in gross over-forecasting across a host of sectors by business and government, the following examples being typical.

EXAMPLE 1.1. Consider a forecast of UK consumption of low density polyethylene, denoted LDP, made in 1971 (Harrison and Pearce 1972). The history to that date is shown in Figure 1.6. Historically, trend curves backed by end use analysis had performed exceedingly well in estimating UK consumption and the predicted fast growth had supported past decisions to extend production capacity. In 1971, the same approach estimated a 1980 market of about 470 thousand long tons, which did not suit the utility of managers. The only new significant use was in plastic refuse sacks, and no end use argument could be proposed for increasing this forecast other than surprise or the imposition of unfounded growth constraints. The basic use of trend curves is as a control mechanism. Historically, various empirical rules have been derived such as "the percentage growth in demand decreases with time", "real production costs decrease with cumulative output', etc. These form a standard by which proposals can be judged.

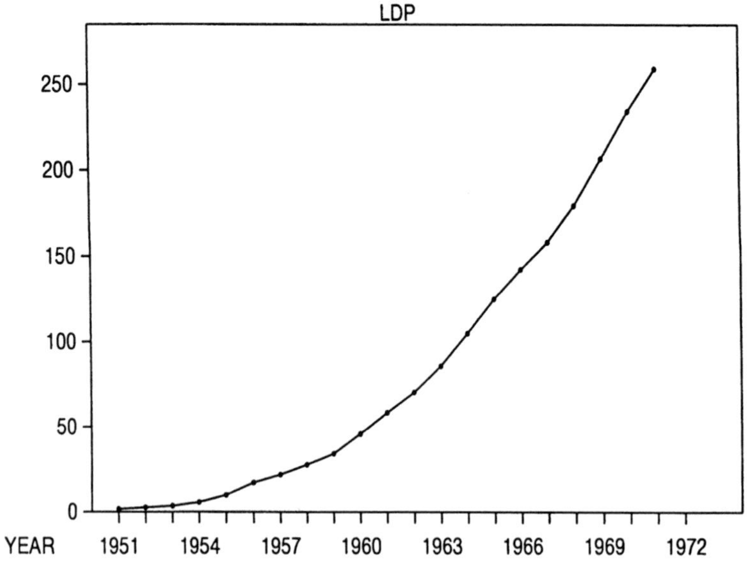

Figure 1.6 UK consumption of LDP (long tons$\times 10^6$)

Basically, if a case conforms to these rules, it may be safe to sanction extension proposals. If however, the case rests on a vastly different forecast then there must be a very sound and strong reason why the future will evolve in such an unexpected way. In the LDP situation the growth rate had already fallen to 10% and could be expected to fall further to about 6% by 1980. Thus, all empirical evidence pointed to a struggle for the market to exceed 500 thousand long tons by 1980. However, the suppliers of the raw material ethylene had produced an incredible 1980 forecast of 906 thousand long tons. Furthermore, the board of the plastics company for whom the forecasts were produced decided to plump for an unjustified figure of 860 thousand long tons. To come to fruition, both these forecasts needed an average growth rate of about 13% over the next ten years. To say the least, such a growth rate recovery would have been most remarkable, and any proposal based upon this would certainly have had to have been very well founded. The market researcher who did the end use analysis was retired. Extensions were sanctioned, and throughout the next decade, regretted. From the mid 1970s onwards capacity in Western Europe was generally between 30 and 40% over capacity and in 1979 estimated forecasts continually lowered, finally reaching the region of the 470 mark.

EXAMPLE 1.2. The above example is not mentioned as an isolated case but as typical of decision makers. For example, Figure 1.7 shows how

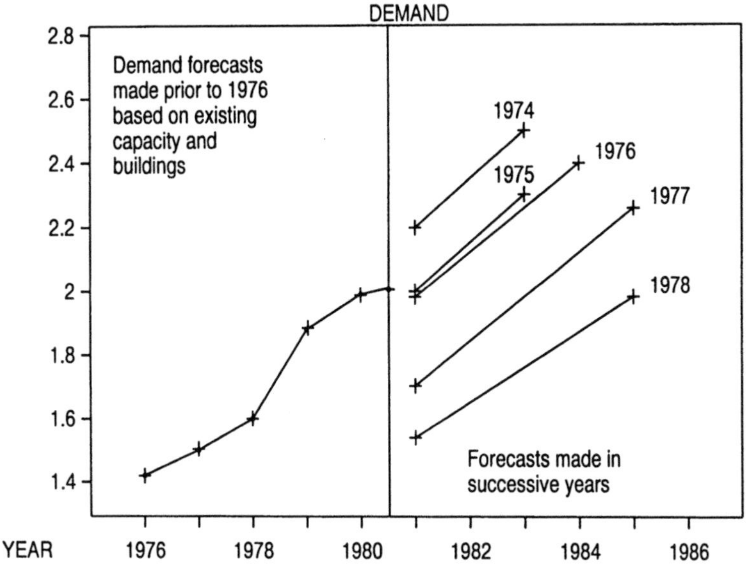

Figure 1.7 Official UK ethylene forecasts (tonnes$\times 10^6$)

official forecasts for ethylene production were continually reduced over the years throughout the 1970s (Source: *Financial Times*, August 9th 1978).

EXAMPLE 1.3. The profits figures and official forecasts shown in Figure 1.8, for a period ending around 1970, should cause no embarrassment now. Up to this time, this particular company had never under-forecast profits. Usually at a boom time, which then regularly occurred with the business cycle at about five-year intervals, the forecast profits over the next three "recession" years would be about fifty percent too high. It was said that even if improved forecasts could be obtained, they could not be issued and used because of the consequences in the "City". Hence totally inappropriate forecasts were adopted, resulting in great losses in efficiency as projects were cancelled, workers laid off, etc.

It is evident that the profits figures show a cyclic behaviour typical of the second-order dynamics generated by reactive control trying to force a system to violate its natural characteristics. Essentially, over this period, the UK government operated a deflation/reflation policy governed by the state of UK unemployment and trade balance; unions took advantage of boom periods to negotiate inappropriate wage rates and "golden" shifts; and companies took actions to expand activities and profits.

Figure 1.9 (from Harrison and Smith 1980) gives an insight into what was happening. In 1971, a dynamic systems model led to the forecast that the mid 1970s would experience the deepest depression since the 1930s.

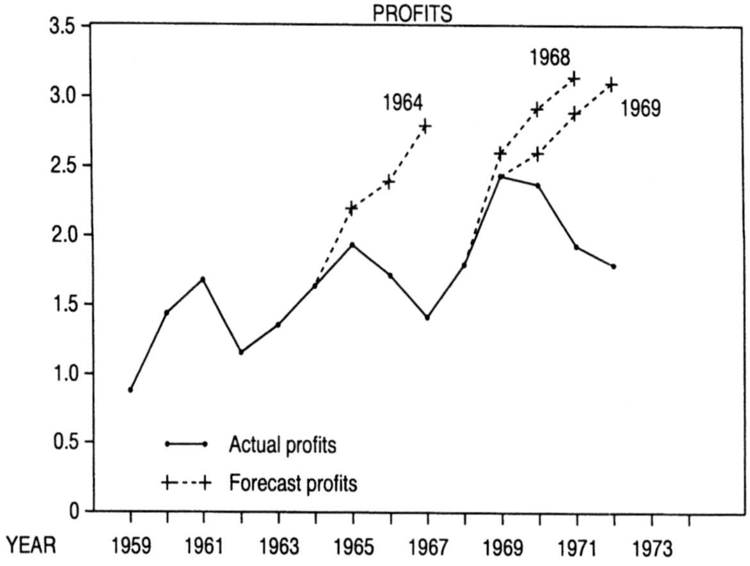

Figure 1.8 Profits and official forecasts (£10^9)

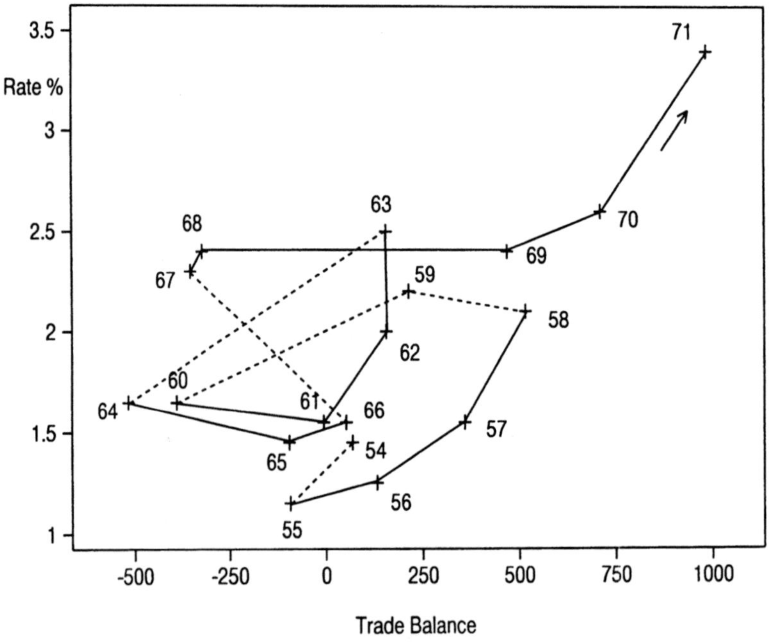

Figure 1.9 UK unemployment rate and trade balance

This soundly based forecast was received with both astonishment, abuse, and absolutely no rational counter arguments. So in that year, one of us wagered that UK unemployment would exceed 2 million before 1980. Although this figure was not officially achieved until later, in 1980 it was conceded that the bet had been successful, since in order to depress unemployment statistics, the government had introduced new compilation methods that omitted previously included groups. However, we refused this concession on the grounds that a professional forecaster should have anticipated manipulation of the figures.

1.2.7 Macro and micro forecasts

The production of the above profit figures is interesting and is illustrative of a major point. The process was conducted from the bottom up over thousands of products. Individual point forecasts would be produced, generally judged on their merits as sales targets and summed through a hierarchy into a single figure, that could be amended by the treasurer but was rarely altered in any significant way. The merits of this system are that it virtually guarantees that top management get a forecast that satisfies their desires and for which they can shelve a good measure of responsibility. The properties of this forecast system are that it focuses on point forecasts which are sales targets, nearly always exceeding the immediate previous achievement, and that it completely ignores the major factors causing profit variation.

This last point is critical for a forecaster. For example, in 1970 within this company there was much argument about the effect of the business cycle. One camp would argue that because it did not affect product A, or product B, or any other individual products, then it did not exist. Others would say that it quite clearly existed if one looked at the total business. The difference of opinion arises because of a lack of appreciation of the law of large numbers which one of us demonstrated easily as follows.

Suppose that, at time t, there are a thousand products with sales Y_i, modelled simply as

$$Y_i = \mu_i + B + \epsilon_i, \qquad i = 1, \dots, 1000,$$

where the μ_i's are known means, the ϵ_i's are independent error terms with zero means and variances 99, denoted by $\epsilon_i \sim [0, 99]$, and $B \sim [0; 1]$ is a common source of variation. Clearly, each Y_i has variance $V[Y_i] = 100$, with the negligible common factor B accounting for only 1% of the individual product variance, or 0.5% of the standard deviation.

Now consider the total sales S, given by

$$S = \mu + 1000B + \sum_{i=1}^{1000} \epsilon_i.$$

The variance of S is

$$V[S] = 10^6 V[B] + \sum_{i=1}^{1000} V[\epsilon_i] = 1099 \times 10^3.$$

Now the common factor B accounts for 91% of the total variance and over 95% of the total standard deviation. The example can be generalised but the key point is vitally clear, namely that factors that dominate a system at one level of aggregation may be insignificant at other levels and vice-versa (Green and Harrison 1972; Harrison 1985).

One important message to be drawn from this is that a model should focus on a limited range of questions and not try to answer both macro and micro questions unless it is particularly well structured hierarchically with models within models. On the other hand, forecasts from a range of models for parts, subtotals, totals, etc., often need to be consistent. Appreciating how to combine and constrain forecasts is a necessary skill for a forecaster. In the above example, stock control forecasts for the thousand individual products may be produced individually using a simple time series model. However, for macro stock decisions, a model of the total business needs to be made that can anticipate major swings in product demand which, in turn, can affect the total stock policy by their translation into action at the individual product level.

Unlike traditional forecasting methods, the Bayes' approach offers easy, natural ways of combining and deriving consistent forecasts.

1.3 BAYESIAN MODELLING AND FORECASTING

1.3.1 Preliminaries

This book is primarily concerned with modelling and forecasting single time series with attention focusing on the mathematical and statistical structure of classes of dynamic models and data analysis. Whilst of necessity the emphasis is on detailed features of particular models, readers should not lose sight of the wider considerations relevant to real-life modelling and forecasting as discussed above. This noted, a full understanding of detailed structure is necessary in order to appreciate the practical implications.

Bayesian statistics is founded on the fundamental premise that all uncertainties should be represented and measured by probabilities. Its justification lies in the formal, axiomatic development of the normative framework for rational, coherent, individual behaviour in the face of uncertainty. Individuals desiring to behave (at least in principle) in this way are led to act as if their uncertainties are represented using subjective probability. In addition to ensuring coherence, the Bayesian paradigm provides simple rules for the management of uncertainties, based on the laws of probability. Analyses of complex problems, with possibly many different but interacting sources of uncertainty, become problems of mathematical manipulation, and so are well-defined. The laws of probability then apply to produce probabilistic inferences about any quantity, or collection of quantities, of interest. In forecasting, such quantities may be future values of time series and model parameters; related forecasts being the model's predictive probability distributions. Thus Bayesian forecasting involves the provision of forecast information in terms of probability distributions that represent and summarise current uncertain knowledge and beliefs. *All* probabilities are subjective, the beliefs represented being those of an individual, the forecaster or modeller responsible for the provision of forecast information.

Throughout the book, many underlying principles and features of the Bayesian approach are identified and described in the contexts of various forecasting models and problems. Routine manipulation of collections of probability distributions to identify those relevant to forecasting and related inferences are performed repeatedly. This provides a useful introduction to general Bayesian ideas for the inexperienced reader, although it is no substitute for more substantial introductions such as can be found in the works of Lindley (1965), De Groot (1971), Berger (1985), O'Hagan (1994) and Bernado and Smith (1994).

1.3.2 Basic notation

Notation is introduced in the context of modelling a series of real-valued quantities observed over time, a generic value being denoted by Y. The time index t is used as a suffix for the time series, so that Y_t denotes the tth value of the series. Conventionally, observations begin at $t = 1$, the series developing as $Y_1, Y_2, \ldots$, or Y_t, $(t = 1, 2, \ldots,)$. There is no need for the series to be equally spaced in time although for convenience Y_t will often be referred to as the observation at time t. Random quantities and their outcomes or realised values are not distinguished. Thus, prior to observing the value of the series at time t, Y_t denotes the unknown, or, more appropriately, *uncertain* random quantity, which becomes known, or certain, when observed. The context and notations used in the specification of models and probability distributions make the distinctions as necessary.

Throughout the book all probability distributions are assumed discrete or continuous, having densities defined with respect to Lebesgue or counting measures as appropriate. Generally, continuous distributions with continuous densities are used, models being mainly based on standard and familiar parametric forms such as normal, Student T, and so forth. Fuller notational considerations are given in the first Section of Chapter 17, a general mathematical and statistical appendix to the book. Some basic notation is as follows. Density functions are denoted by $p(\cdot)$, and labelled by their arguments. Conditioning events, information sets and quantities are identified as necessary in the argument, following a vertical line. For example, the distribution of the random quantity Y conditional on an information set D has density $p(Y|D)$, and that given D plus additional information denoted by H is simply $p(Y|D, H)$ or $p(Y|H, D)$. The joint density of two random quantities Y and X given D is $p(Y, X|D)$ or $p(X, Y|D)$. Conditional upon X taking any specified value, whether hypothetical or actually observed, the conditional density of Y given both D and X is simply $p(Y|X, D)$ or $p(Y|D, X)$. Note here that the distinction between X as an uncertain random quantity and as a realised value is clear from the notation. In referring to the distributions of random quantities, this notation is used without the p. Thus we talk of the distributions of Y, $(Y|D)$, $(Y, X|D)$ and $(Y|X, D)$, for example. Expectations of functions of random quantities are denoted by $\mathrm{E}[\cdot]$; thus $\mathrm{E}[Y|D]$ is the conditional expectation, or mean, of $(Y|D)$. Variances and covariances are represented using $\mathrm{V}[\cdot]$ and $\mathrm{C}[\cdot,\cdot]$ respectively; thus $\mathrm{V}[Y|D]$ is the variance of $(Y|D)$ and $\mathrm{C}[Y, X|D]$ the covariance of $(Y, X|D)$.

Conventionally, both uppercase and lowercase roman characters are used for quantities that are either known or uncertain but observable. Thus, at any time point t, the observed values of the time series $Y_1, Y_2, \ldots, Y_t$ will usually be known, those of $Y_{t+1}, Y_{t+2}, \ldots$, being uncertain but potentially observable at some future time. Uncertain, unobservable random quantities are denoted by Greek characters, and referred to as unknown model parameters. For example, the mean and variance of a random quantity will be denoted by roman characters if they are known, but by Greek characters if they are unknown distributional parameters. In the latter case, the uncertain nature of the parameters will usually be made explicit by their inclusion in the conditioning of distributions. Thus $p(Y|\mu, D)$ is the density of Y given the information set D and in addition, the value of an uncertain random quantity μ. On the other hand, not all assumed quantities are recognised in the conditioning.

Notation for specific distributions is introduced and referenced as necessary. Distributions related to the multivariate normal play a central role throughout the book, so the notation is mentioned here with further details in Sections 17.2 and 17.3. In particular, $(Y|D) \sim \mathrm{N}[m, V]$ when $(Y|D)$ is normally distributed with known mean m and known variance V. Then $X = (Y - m)/\sqrt{V}$ has a standard normal distribution, $(X|D) \sim \mathrm{N}[0, 1]$. Similarly, $(Y|D) \sim \mathrm{T}_k[m, V]$ when $(Y|D)$ has a Student T distribution on k degrees of freedom, with mode m and scale V. If $k > 1$ then $\mathrm{E}[Y|D] = m$; if $k > 2$ then $\mathrm{V}[Y|D] = Vk/(k-2)$. $X = (Y-m)/\sqrt{V}$ has a standard T distribution, $(X|D) \sim \mathrm{T}_k[0, 1]$. Similar notation applies to other distributions specified by a small number of parameters. In addition, $(Y|D) \sim [m, V]$ signifies $\mathrm{E}[Y|D] = m$ and $\mathrm{V}[Y|D] = V$, whatever the distribution.

Vector and matrix quantities are denoted by bold typeface, whether Roman or Greek. Matrices are always uppercase; vectors may or may not be depending on context. The above notation for distributions has the following multivariate counterparts as detailed in Section 17.2: let $\mathbf{Y}$ be a vector of n random quantities having some joint distribution, $\mathbf{m}$ a known n-dimensional vector, and $\mathbf{V}$ a known $n \times n$ non-negative definite matrix. Then $(\mathbf{Y}|D) \sim \mathrm{N}[\mathbf{m}, \mathbf{V}]$ when $(\mathbf{Y}|D)$ has a multivariate normal distribution with known mean $\mathbf{m}$ and known variance matrix $\mathbf{V}$. Similarly, $(\mathbf{Y}|D) \sim \mathrm{T}_k[\mathbf{m}, \mathbf{V}]$ when $(\mathbf{Y}|D)$ has a multivariate Student T distribution on k degrees of freedom, with mode $\mathbf{m}$ and scale matrix $\mathbf{V}$. If $k > 1$ then $\mathrm{E}[\mathbf{Y}|D] = \mathbf{m}$; if $k > 2$ then $\mathrm{V}[\mathbf{Y}|D] = \mathbf{V}k/(k-2)$. Finally, $(\mathbf{Y}|D) \sim [\mathbf{m}, \mathbf{V}]$ when $\mathrm{E}[\mathbf{Y}|D] = \mathbf{m}$ and $\mathrm{V}[\mathbf{Y}|D] = \mathbf{V}$, with the distributional form otherwise unspecified.

1.3.3 Dynamic models

Mathematical and statistical modelling of time series processes is based on classes of *dynamic models*, the term *dynamic* relating to changes in such processes due to the passage of time as a fundamental motive force. The most widely known applied subclass is that of *normal dynamic linear models*, referred to simply as *dynamic linear models*, or *DLMs*, when the normality is understood. This class of models forms the basis for much of the development in the book.

The fundamental principles used by a Bayesian forecaster in structuring forecasting problems through dynamic models are discussed in detail in Section 4.1 of Chapter 4 and comprise

- parametric models with meaningful dynamic parameters;
- a probabilistic representation of information;
- a sequential model definition utilising conditional independence;
- robust conditionally independent model components;
- forecasts derived as probability distributions;
- a facility for incorporating expert information;
- model quality control.

A sequential model definition and structuring is natural in the time series context. As time evolves, information relevant to forecasting the future is received and may be used to revise the forecaster's views, whether this revision be at the quantitative, the form, or the conceptual level in the general model structure $\mathcal{M}$ of Section 1.1. The sequential approach focuses attention on statements about the future development of a time series conditional on existing information. Suppose, with no loss of generality, that the time origin $t = 0$ represents the current time, and that existing information available to, and recognised by, a forecaster is denoted by the

$$\textit{initial information set:} \qquad D_0.$$

This represents all the available relevant starting information that is used to form initial views about the future, including history and all defining model quantities. In forecasting ahead to any time $t > 0$, the primary objective is the calculation of the forecast distribution for $(Y_t|D_0)$. Similarly, at any time t, statements made concerning the future are conditional on the existing

$$\textit{information set at time } t: \qquad D_t.$$

Thus, statements made at time t about any interesting random quantities are based on D_t; in particular, forecasting ahead to time $s > t$ involves consideration of the forecast distribution for $(Y_s|D_t)$. As time evolves, so does the forecaster's information. Observing the value of Y_t at time t implies that D_t includes both the previous information set D_{t-1} and the observation Y_t. If this is all the relevant information, then $D_t = \{Y_t, D_{t-1}\}$, although often, as discussed earlier, further relevant information will be incorporated in revising or updating the forecaster's view of the future. Denoting all additional relevant information at time t by I_t leads to the

$$\textit{information updating:} \qquad D_t = \{I_t, D_{t-1}\}, \qquad\qquad (t = 1, 2, \dots).$$

The sequential focus is emphasised by describing the future development of the series via a probability distribution for $Y_t, Y_{t+1}, \dots$, conditional on past information D_{t-1}. Usually such a distribution depends upon defining parameters determining distributional forms and moments, functional relationships, and so forth. Focusing on one-step ahead, the beliefs of the forecaster are structured in terms of a parametric model,

$$p(Y_t \mid \boldsymbol{\theta}_t, D_{t-1}),$$

where $\boldsymbol{\theta}_t$ is a defining parameter vector at time t. This mathematical and statistical representation is the language providing communication between the forecaster, model and decision makers. As such, the parameters must represent meaningful constructs. Indexing $\boldsymbol{\theta}_t$ by t indicates that the parametrisation may be dynamic. In addition, although often the number and meaning of the elements of $\boldsymbol{\theta}_t$ will be stable, there are occasions on which $\boldsymbol{\theta}_t$ will be expanded, contracted or changed in meaning according to the forecaster's existing view of the time series. In particular, this is true of *open systems*, such as arise in typical social, economic, and biological environments, where influential factors affecting the time series process are themselves subject to variation based on the state of the system generating the process. In such cases, changes in $\boldsymbol{\theta}_t$ may be required to reflect system learning and the exercise of purposeful control. Such events, although recognisable when they happen, may be difficult to anticipate and so will not generally be included in the model until occurrence.

The model parameters $\boldsymbol{\theta}_t$ provide the means by which information relevant to forecasting the future is summarised and used in forming forecast distributions. The learning process sequentially revises the state of knowledge about such parameters. Probabilistic representation of all uncertain knowledge is the essence of the Bayesian approach, whether such knowledge relates to future, potentially observable quantities or unobservable model parameters. At time t, historical information D_{t-1} is summarised through a *prior* distribution for future model parameters: prior, that is, to observing Y_t, but of course *posterior* to the information set D_{t-1}. The prior density $p(\boldsymbol{\theta}_t \mid D_{t-1})$ and the posterior $p(\boldsymbol{\theta}_t \mid D_t)$ provide a concise, coherent and effective transfer of information on the time series process through time. In addition to processing the information deriving from observations and feeding it forward to forecast future development, this probabilistic encoding allows new information from external sources to be formally incorporated in the system. It also extends naturally to allow for expansion or contraction of the parameter vector in open systems, with varying degrees of uncertainty associated with the effects of such external interventions and changes. Further, inferences about system development and change are directly drawn from components of these distributions in a standard statistical manner.

Two very simple models, the subjects of Chapters 2 and 3, exemplify the class of dynamic models, giving concrete settings for the above general ideas. The first is an example of a particular univariate normal dynamic linear model which is briefly defined in order to introduce the quadruple notation used in those chapters.

Definition 1.1. For each t, the univariate, uniparameter normal dynamic linear model, represented by the quadruple $\{F_t, \lambda, V_t, W_t\}$, is defined by:

$$\begin{array}{lll} \text{Observation equation:} & Y_t = F_t\mu_t + \nu_t, & \nu_t \sim \mathrm{N}[0, V_t], \\ \text{System equation:} & \mu_t = \lambda\mu_{t-1} + \omega_t, & \omega_t \sim \mathrm{N}[0, W_t], \\ \text{Initial information:} & (\mu_0 \mid D_0) \sim \mathrm{N}[m_0, C_0], & \end{array}$$

where the error sequences ν_t and ω_t are independent, and mutually independent. In addition, they are independent of $(\mu_0 \mid D_0)$. The values of the variance sequences V_t and W_t may be unknown, but the constant λ and relevant values of the sequence F_t are known.

EXAMPLE 1.4. An archetype statistical model assumes that observations are independent and identically normally distributed, denoted by $(Y_t|\mu) \sim \mathrm{N}[\mu, V]$, $(t = 1, 2, \dots)$, This is the trivial DLM $\{1, 1, V, 0\}$. Changes over time in the mean, and sometimes the variance, of this model may be unavoidable features when observations are made on a process or system that is itself continually evolving. Such changes are usually gradual, reflecting continuous slow changes in environmental conditions. However, changes are occasionally more abrupt, often responses to significant shifts in major influential factors. For example, a normal model may be assumed as a suitable representation of the random variation in "steady" consumer demand for a product, but the *level* of demand will rarely remain absolutely constant over time. A simple extension of the archetype incorporating a time varying mean provides a considerable degree of flexibility. Then, subscripting μ by t, $(Y_t|\mu_t) \sim \mathrm{N}[\mu_t, V]$, or

$$Y_t = \mu_t + \nu_t, \quad \text{and} \quad \nu_t \sim \mathrm{N}[0, V],$$

where μ_t represents the level of the series at time t, and ν_t random, *observational* error or noise about the underlying level.

One of the simplest, non-degenerate ways in which the level can be dynamically modelled is as a *random walk*. In such a case, time evolution of the level, defining the process model, is represented as

$$\mu_t = \mu_{t-1} + \omega_t, \quad \text{and} \quad \omega_t \sim \mathrm{N}[0, W],$$

where ω_t represents purely random, unpredictable changes in level between time $t-1$ and t. This is the standard DLM $\{1, 1, V, W\}$. Chapter 2 is devoted to the study of such models, which find wide application in short-term forecasting. In product demand forecasting and inventory management, for instance, underlying levels may be assumed to be *locally* constant, but will be expected to change significantly over longer periods of time. The zero-mean and independence assumptions for the ω_t series are consistent with a view that this longer-term variation cannot be systematically predicted, being described as purely stochastic.

With reference to the earlier general discussion, if the variances V and W are known, then the model parameter $\boldsymbol{\theta}_t$ at time t represents the uncertain level μ_t alone. Otherwise, $\boldsymbol{\theta}_t$ may include uncertain variances for either or both of ν_t and ω_t. With known variances, their values are included in the initial information set D_0, so with parameter $\boldsymbol{\theta}_t = \mu_t$ for all t, $p(Y_t|\boldsymbol{\theta}_t, D_{t-1}) = p(Y_t|\mu_t, D_0)$ is the density of $(Y_t|\mu_t, D_0) \sim \mathrm{N}[\mu_t, V]$. The historical information set D_{t-1}, which includes all past values of the Y series, leads to the prior distribution for $\boldsymbol{\theta}_t = \mu_t$ given D_{t-1}. In demand forecasting, for example, historical values of the Y series may lead the forecaster to believe that next month's demand level, μ_t, is most likely to be in the region of about 250, but is unlikely to be below 230 or to exceed 270. One possible representation of this prior view is that

$$(\mu_t|D_{t-1}) \sim \mathrm{N}[250, 100],$$

having an expected and most likely value of 250, and about 95% probability of lying between 230 and 270. This sort of model structuring is developed extensively in Chapter 2.

EXAMPLE 1.5. Regression modelling is central to much of statistical practice, being concerned with the construction of a quantitative description of relationships between observables, such as between two time series. Consider a second time series represented by observations X_t, $(t = 1, 2, \ldots)$, observed contemporaneously with Y_t. Regression modelling often focuses on the extent to which changes in the mean μ_t of Y_t can be explained through X_t, and possibly past values X_s for $s < t$. Common terminology refers to Y_t as the response or dependent variable series and X_t as the regressor or independent variable series. Then μ_t is the mean response, related to the regressor variable through a mean response function $\mu_t = r_t(X_t, X_{t-1}, \ldots)$ defining the regression. For example, a simple linear model for the effect of the current X_t on the current mean is $\mu_t = \alpha + \beta X_t$, where the defining parameters α and β take suitable values. Models of this sort find wide use in prediction, interpolation, estimation, and control contexts. Construction of models in practice is guided by certain objectives specific to the problem area, such as short-term forecasting of the response series. For some such purposes, simple static linear models may well be satisfactory *locally*, but as in the previous example, are unlikely to adequately describe the global relationship, i.e. as time evolves *and* as X_t varies. Flexibility in modelling such changes can be provided simply by allowing for the possibility of time variation in the coefficients, so that

$$\mu_t = \alpha_t + \beta_t X_t.$$

Thus, although the *form* of regression model is linear in X_t for all t, the quantified model may have different defining parameters at different times. This distinction, between an appropriate *local model form* and an appropriate *quantified local model*, is critical and fundamental in dynamic modelling. Often the values of the independent variable X_t change only slowly in time and an appropriate local model description is that above, $\mu_t = \alpha_t + \beta_t X_t$, where the parameters vary only slightly from one time point to the next. As in the previous example, this may be modelled using random walk type evolutions for the defining parameters, such as

$$\alpha_t = \alpha_{t-1} + \delta\alpha_t,$$
$$\beta_t = \beta_{t-1} + \delta\beta_t,$$

where $\delta\alpha_t$ and $\delta\beta_t$ are zero-mean error terms. These evolution equations express the concept of *local constancy* of the parameters, subject to variation controlled and modelled through the distribution of the evolution error terms $\delta\alpha_t$ and $\delta\beta_t$. Small degrees of variation here imply a stable linear regression function over time, larger values modelling greater volatility and suggesting caution in extrapolating or forecasting too far ahead in time based on the current quantified linear model. The usual static regression model, so commonly used, is obviously obtained as a special case of this dynamic regression when both evolution errors are identically zero for all time.

Finally, the primary goals of the forecaster are attained by directly applying probability laws. The above components provide one representation of a joint distribution for the observations and parameters, namely

$$p(Y_t, \boldsymbol{\theta}_t \mid D_{t-1}) = p(Y_t \mid \boldsymbol{\theta}_t, D_{t-1})\, p(\boldsymbol{\theta}_t \mid D_{t-1}),$$

from which the relevant one-step forecast may be deduced as the marginal

$$p(Y_t \mid D_{t-1}) = \int p(Y_t, \boldsymbol{\theta}_t \mid D_{t-1}) d\boldsymbol{\theta}_t.$$

Inference for the future Y_t is simply a standard statistical problem of summarising the forecast distribution and a coherent optimal decision policy may be derived with the introduction of a utility function.

1.4 HISTORICAL PERSPECTIVE AND BIBLIOGRAPHIC COMMENTS

Our approach to modelling and forecasting synthesises concepts, models and methods whose development has been influenced by work in many fields. It would be impossible to fully document historical influences. As already discussed, much is involved in modelling that cannot be routinely described using formal, mathematical structures, particularly in the stages of model formulation, choice and criticism. However, in line with the fundamental concepts of scientific method, we identify the Bayesian approach as the framework for routine learning and organisation of uncertain knowledge within complete forecasting systems. Over the last fifty years there has been rapidly increasing support for the Bayesian approach as a means of scientific learning and decision making, with notable recent acceptance by practitioners driven by the need to adopt the common-sense principles on which it is founded. Axiomatic foundations notwithstanding, decision makers find it natural to phrase beliefs as normed or probability measures, as do academics, though some prefer not to recognise the fact. This seems to have been done ever since gambling started — and what else is decision making except compulsory gambling? We all face "one-off" decisions that have to be made with no chance of repeating the experience, so the value of classical statistics is very limited for decision makers.

Important influences on current Bayesian thinking and practice may be found, in particular, in the books of Box and Tiao (1973), De Finetti (1974/75), De Groot (1971), Jeffreys (1961), Lindley (1965), and Savage (1954). More recent recent contributions include Berger (1985), O'Hagan (1994) and Bernardo and Smith (1994), which include useful bibliographies.

Concerning approaches to practical forecasting, little had been done in industry and government in the way of mathematical or socio-economic modelling before the arrival of computers in the 1950s. Exponentially weighted moving averages, EWMAs, and Holt's (1957) linear growth and seasonal model began to find use in the mid to late 1950s in forecasting for stock control and production planning, with one of us (PJH) involved in pioneering their use at Imperial Chemical Industries Ltd., (ICI). In this company, Holt's method became established, and apart from taking over the routine of forecasting, led to great improvements in the accuracy of forecasts. In fact, the methods were used in conjunction with interventions so that a complete forecasting system was in operation at that time. This involved the product sales control department in adjusting the computer forecasts whenever they felt it necessary, though initially these adjustments did not always improve the forecasts, probably because hunches and wishful thinking appeared to them as definite information. Methods developed then are still in use now in sales forecasting and stock control.

The work of Brown (1959, 1962), promoting the use of discount techniques in forecasting, was and still is a major influence for practitioners. A parsimonious concept anchored to the fundamental familiar principle of discounting is very attractive. Work at ICI at the start of the 1960s used the principle of discounting in models for series with trend and seasonality, but applied two discount factors, one for trend and one for seasonality. As far as we know, this was the first development of multiple discounting, incorporated in two forecasting programmes, Seatrend and Doubts, in 1963, presented at the Royal Statistical Society's annual conference held at Cardiff in 1964 and later published in Harrison (1965). However, what will not be known by a literature search is what really happened in development. The society had asked for a paper to be circulated in advance, and this presented the multiple discount methods but concluded that, because of simplicity, it was preferable to use a single discount factor as in Brown's exponentially weighted regression (EWR) approach. However, because of the difficulties in programming, numerical comparisons were unavailable at the time of writing. Almost as soon as the paper was sent to the RSS the results were obtained, and surprisingly indicated the enormous improvements to be obtained using multiple discounting. This led to a reversal of the issued conclusions through a quickly written seven page addendum that was circulated at the conference. One reader wrote "the paper read like a thriller: the rise and fall of Brown". However, as is usual with research, the published paper shows no trace of how the work actually progressed. Basically, the conclusion was that different model components need discounting at different rates, a view that now dominates in structuring models using discounting ideas in Bayesian forecasting.

Further developments at ICI were described in a paper presented to the Operational Research Society Conference in 1965, the manuscript never being published but available in its original form as a Warwick Research Report (Harrison and Scott 1965). The contents will surprise many people; they include discussion of complete forecasting systems emphasising intervention and monitoring and present several specific dynamic models. Some of the modelling work was based on developments by Muth (1960), Nerlove and Wage (1964), Thiel and Wage (1964), and Whittle (1965), though a major concern was with more general models, with an emphasis on sensitivity to departures from optimal adaptive parameters, partially motivated by unfounded, inexperienced opposition to discounting from certain academics. The monitoring work in that paper was largely based on the backward cusum controller developed initially in 1961 from cusums in quality control and published in Harrison and Davies (1964). Much of this followed developments by Page (1954), Barnard (1959), the work of ICI through Woodward and Goldsmith (1964), and of British Nylon Spinners, later to become part of ICI, through Ewan and Kemp (1960).

So by 1964, simple, parametrised, structural models were in use, employing double discounting and emphasising the approach through complete forecasting systems operating according to the principle of management by exception, enabling prospective and retrospective interventions. By the end of the 1960s, the basic models had been extended and generalised to a wide class of dynamic linear models (though not then referred to as such), and to multi-process models that were to model outliers and sudden changes in structure. Some indication of the extent of the developments, both theoretical and practical, was then reported in Harrison and Stevens (1971). Implemented forecasting systems based on this took full advantage of the intervention facilities of the Bayesian approach and also performed automatic model discrimination using multi-process models. Important also for the use of subjective and quantitative information is the 1970 application in the mail order business reported rather later in Green and Harrison (1973).

At about the same time, in 1969, it became clear that some of the mathematical models were similar to those used in engineering control. It is now well known that in normal DLMs with known variances, the recurrence relationships for sequential updating of posterior distributions are essentially equivalent to the Kalman filter equations, based on the early work of Kalman (1960, 1963) in engineering control, using a minimum variance approach. It was clearly not, as many people appear to believe, that Bayesian forecasting is founded upon Kalman filtering (see Harrison and Stevens 1976a, and discussion; and reply to the discussion by Davis of West, Harrison and Migon 1985). To say that "Bayesian forecasting is Kalman Filtering" is akin to saying that statistical inference is least squares.

The late 1970s and 1980s have seen much further development and application of Bayesian forecasting models and methods. Notable amongst these are procedures for variance learning (West 1982; Smith and West 1983), discounting (Ameen and Harrison 1984, 1985a and b), monitoring and intervention (West 1986; West and Harrison 1986; West, Harrison and Pole 1987; Harrison and Veerapen 1994; Pole, West and Harrison 1994), non-normal and non-linear model structures (Souza 1981; Smith 1979; Migon 1984; Migon and Harrison 1985; West, Harrison and Migon 1985), reference analyses (Pole and West 1987), and many others. Since the revolution in computational statistics, beginning in 1990, major new directions in research, model development and application have been opened up, involving analysis using simulation methods. Much of the dynamic modelling activity is discussed in Chapter 15, a new chapter in the second edition, with references to the key areas and recent developments.

Computing developments have led to wider usage, easing communication with less technically orientated practitioners. We do hold the view that modelling is an art, and particularly so is Bayesian forecasting. Any software package is just that: a package of specific, selected facilities, and the limitations of such software are too easily seen by some as the limitations of the whole approach. Indeed, the early Bayesian forecasting package SHAFT, produced in the 1970s by Colin Stevens, led to a widespread view that Bayesian forecasting was the single model discussed in Harrison and Stevens (1971). This is like defining an animal as a fox. However, it is now abundantly clear that to treat the armoury of the Bayesian approach as a paint box without even a numbered canvas is not going to get many pictures painted. Some results of our approaches to the problem are the software packages mentioned in the Preface.

The above discussion concentrates closely on what we identify as direct influences on Bayesian forecasting and dynamic modelling as we present the subject in this book. There has, of course, been considerable development of dynamic modelling and related forecasting techniques outside the Bayesian framework, particularly in control engineering. Good references include Anderson and Moore (1979), Jazwinski (1970), Sage and Melsa (1971), for example. Details of work by statisticians, econometricians and others may be found, with useful references, in Duncan and Horne (1972), Harvey (1981), Spall (1988), Thiel (1971), and Young (1984), for example.

CHAPTER 2

INTRODUCTION TO THE DLM: THE FIRST-ORDER POLYNOMIAL MODEL

2.1 INTRODUCTION

Many important concepts and features of dynamic linear models appear in the simplest and most widely used **first-order polynomial model.** Consequently it offers an excellent introduction to DLMs and is examined in detail in this chapter. This DLM is the simple, yet non-trivial, time series model in which the observation series Y_t is represented as

$$Y_t = \mu_t + \nu_t, \qquad \nu_t \sim \mathrm{N}[0, V_t]$$

where μ_t is the **level** of the series at time t and ν_t is the **observational error**. The time evolution of the level is then modelled as a simple random walk, or *locally constant mean,*

$$\mu_t = \mu_{t-1} + \omega_t, \qquad \omega_t \sim \mathrm{N}[0, W_t],$$

with **evolution error** ω_t.

The observational and evolution error sequences comprise internally and mutually independent normal random variables. So, for all t and all s with $t \neq s$, ν_t and ν_s are independent, ω_t and ω_s are independent, and ν_t and ω_s are independent. To begin it is also assumed that the variances V_t and W_t are known for each time t. The foregoing observational and evolution equations may also be expressed for each $t = 1, 2, \ldots,$ as

$$(Y_t \mid \mu_t) \sim \mathrm{N}[\mu_t, V_t],$$
$$(\mu_t \mid \mu_{t-1}) \sim \mathrm{N}[\mu_{t-1}, W_t].$$

Figures 2.1 (a) and (b) show time graphs $\{Y_t, \mu_t, t\}$ of such series. In each the starting value is $\mu_0 = 25$, and the observation variance $V_t = 1$ is constant. The evolution variances are also constant, with $W = 0.05$ in (a) and $W = 0.5$ in (b). Thus in (a) the evolution variance, $W = V/20$, is small compared to the observational variance leading to a typical locally constant level, whereas in (b) the variance, $W = V/2$, is ten times larger, resulting in much greater variation in the level μ_t.

This model is used effectively in numerous applications, particularly in short-term forecasting for production planning and stock control. For example, in modelling market demand for a product, μ_t represents *true* underlying market demand at time t with ν_t describing random fluctuation, which arises in the actual placement of customer orders, about this level. *Locally* in time, that is, a few periods forwards or backwards, the underlying demand μ_t is characterised as roughly constant. Significant changes over longer periods of time are expected, but the zero-mean and independent nature of the ω_t series imply that the modeller does not wish to anticipate

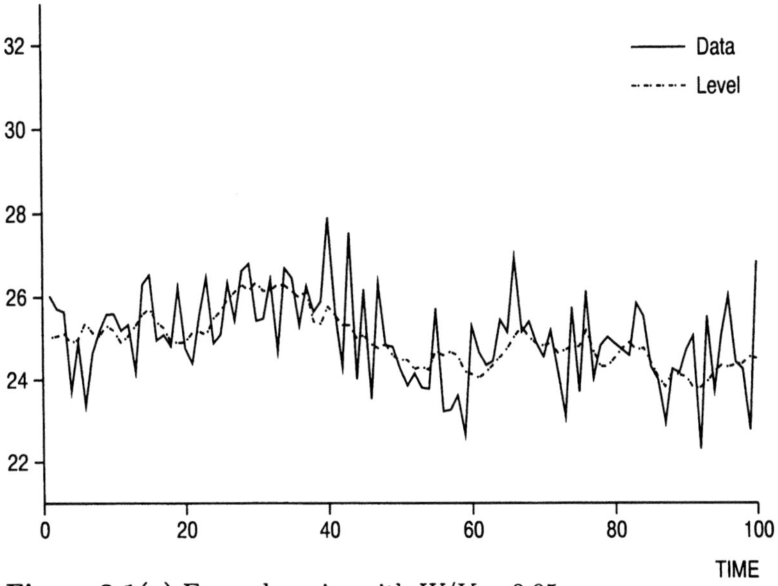

Figure 2.1(a) Example series with $W/V = 0.05$

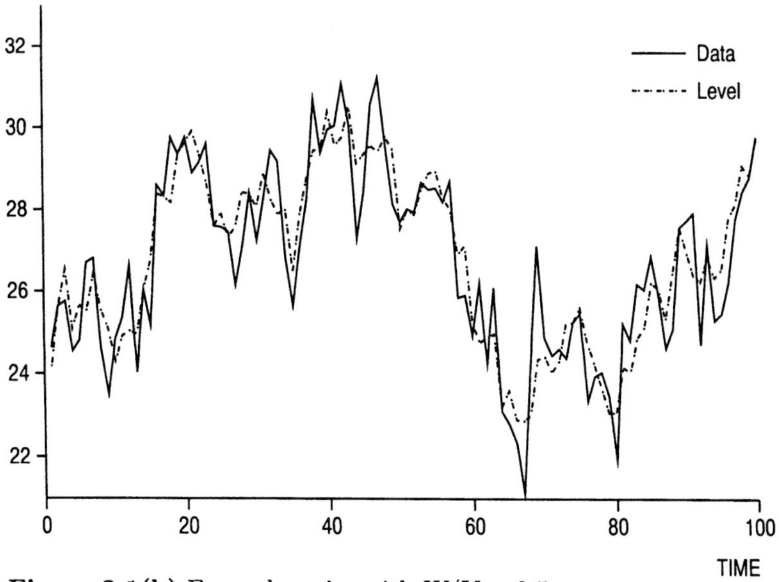

Figure 2.1(b) Example series with $W/V = 0.5$

the form of this longer term variation, merely describing it as a purely stochastic process. It is useful to think of μ_t as a smooth function of time $\mu(t)$ with an associated Taylor series representation

$$\mu(t + \delta t) = \mu(t) + \text{higher-order terms},$$

with the model simply describing the higher-order terms as zero-mean noise. This is the genesis of the first-order polynomial DLM: the level model is a locally constant (first-order polynomial) proxy for the underlying evolution. Sometimes, although a little misleadingly, the DLM has been referred to as a *steady* model. A guide as to the model's suitability for particular applications is that, upon forecasting k–steps ahead from time t, the expected value of the series conditional on the current level is just

$$\mathrm{E}[Y_{t+k} \mid \mu_t] = \mathrm{E}[\mu_{t+k} \mid \mu_t] = \mu_t.$$

At time t, given the existing information D_t, the forecaster's posterior distribution for μ_t will have a mean m_t depending on past data, so that the **forecast function** $f_t(\cdot)$ is constant, being given by

$$f_t(k) = \mathrm{E}[Y_{t+k} \mid D_t] = \mathrm{E}[\mu_t \mid D_t] = m_t,$$

for all $k > 0$. Consequently, this DLM is useful only for short-term application, and particularly in cases in which the observation variation, as measured by V_t, is considerably greater than the systematic level variation, measured by W_t. Such design considerations are discussed later in Section 2.3.

2.2 THE DLM AND RECURRENCE RELATIONSHIPS

2.2.1 Definition

In accord with Definition 1.1, for each time t this model is characterised by the quadruple $\{1, 1, V_t, W_t\}$ and formally defined as follows:

Definition 2.1. For each t, the DLM $\{1, 1, V_t, W_t\}$ is defined by

Observation equation:	$Y_t = \mu_t + \nu_t,$	$\nu_t \sim \mathrm{N}[0, V_t],$
System equation:	$\mu_t = \mu_{t-1} + \omega_t,$	$\omega_t \sim \mathrm{N}[0, W_t],$
Initial information:	$(\mu_0 \mid D_0) \sim \mathrm{N}[m_0, C_0],$	

where the error sequences ν_t and ω_t are internally independent, mutually independent, and independent of $(\mu_0 | D_0)$.

Initial information is the probabilistic representation of the forecaster's beliefs about the level μ_0 at time $t = 0$. The mean m_0 is a point estimate of this level, and the variance C_0 measures the associated uncertainty. Each information set D_v comprises all the information available at time v, including D_0, the values of the variances $\{V_t, W_t : t > 0\}$, and the values of the observations $Y_v, Y_{v-1}, \ldots, Y_1$. Thus, the only new information becoming available at any time t is the observed value Y_t, so that $D_t = \{Y_t, D_{t-1}\}$.

2.2.2 Updating equations

The following theorem provides the key probability distributions necessary for effective forecasting, control, and learning.

Theorem 2.1. *In the DLM of Definition 2.1 the one-step forecast and level posterior distributions for any time $t > 0$ can be obtained sequentially as follows:*

(a) Posterior for μ_{t-1} :

$$(\mu_{t-1} \mid D_{t-1}) \sim \mathrm{N}[m_{t-1}, C_{t-1}].$$

(b) Prior for μ_t :

$$(\mu_t \mid D_{t-1}) \sim \mathrm{N}[m_{t-1}, R_t],$$

$$\text{where } R_t = C_{t-1} + W_t.$$

(c) 1-step forecast:

$$(Y_t \mid D_{t-1}) \sim \mathrm{N}[f_t, Q_t],$$

$$\text{where } f_t = m_{t-1} \text{ and } Q_t = R_t + V_t.$$

(d) Posterior for μ_t :

$$(\mu_t \mid D_t) \sim \mathrm{N}[m_t, C_t],$$

$$\text{with } m_t = m_{t-1} + A_t e_t \text{ and } C_t = A_t V_t,$$

$$\text{where } A_t = R_t / Q_t \text{ , and } e_t = Y_t - f_t.$$

Proof. Two methods of deriving (d) will be used, each instructive and illuminating in its own right. The first, most important, and generally applicable method employs Bayes' theorem and standard Bayesian calculations. The second method, appropriate for all DLMs, provides an elegant derivation using the additivity, linearity, and distributional closure properties of normal linear models. Although this latter method extends to all normal DLMs, the former is required as a general approach for non-normal models. Both methods use standard results associated with normal distributions and Bayesian normal procedures as detailed in some generality in Section 17.2.

Proof is by induction. Assume the truth of the distribution in (a). Then conditional on D_{t-1}, μ_t is the sum of two independent normal random quantities μ_{t-1} and ω_t, and so is itself normal. The mean and variance are obtained by adding means and variances of the summands, leading to (b):

$$(\mu_t \mid D_{t-1}) \sim \mathrm{N}[m_{t-1}, R_t], \qquad \text{where } R_t = C_{t-1} + W_t.$$

Similarly, conditional upon D_{t-1}, Y_t is the sum of the independent normal quantities μ_t and ν_t and so is normal, leading to (c):

$$(Y_t \mid D_{t-1}) \sim \mathrm{N}[m_{t-1}, Q_t], \qquad \text{where } Q_t = R_t + V_t.$$

As mentioned above, (d) is derived twice, using two different techniques:

(1) **Updating via Bayes' Theorem**
The Bayesian method is general, applying to all models, no matter what the distributional assumptions. The observation equation provides the observational density function

$$p(Y_t \mid \mu_t, D_{t-1}) = (2\pi V_t)^{-1/2} \exp[-(Y_t - \mu_t)^2/(2V_t)].$$

From (b), the prior for μ_t given D_{t-1} has probability density function (pdf)

$$p(\mu_t \mid D_{t-1}) = (2\pi R_t)^{-1/2} \exp[-(\mu_t - m_{t-1})^2/(2R_t)].$$

On observing Y_t, the *likelihood* for μ_t is proportional to the observed density viewed as a function of μ_t. So with $D_t = \{D_{t-1}, Y_t\}$, from Bayes' theorem, the posterior for μ_t is

$$p(\mu_t \mid D_t) = p(\mu_t \mid D_{t-1})p(Y_t \mid \mu_t, D_{t-1})/p(Y_t \mid D_{t-1}).$$

Concentrating on this as a function of μ_t alone, ignoring multiplicative factors depending on the now known value Y_t and other constants, leads to the proportional form of Bayes' theorem,

$$\begin{aligned} p(\mu_t \mid D_t) &\propto p(\mu_t \mid D_{t-1})p(Y_t \mid \mu_t, D_{t-1}) \\ &\propto \exp[-(\mu_t - m_{t-1})^2/2R_t - (Y_t - \mu_t)^2/2V_t]. \end{aligned}$$

In applications of Bayes' theorem the natural logarithmic scale provides simpler additive expressions. So with differing constants k_i,

$$\begin{aligned} 2\ln[p(\mu_t \mid D_t)] &= k_1 - (\mu_t - m_{t-1})^2 R_t^{-1} - (Y_t - \mu_t)^2 V_t^{-1}, \\ &= k_2 - (R_t^{-1} + V_t^{-1})\mu_t^2 + 2(R_t^{-1} m_{t-1} + V_t^{-1} Y_t)\mu_t, \\ &= k_3 - (\mu_t - m_t)^2 C_t^{-1}, \end{aligned}$$

where, with $A_t = R_t/Q_t = R_t/(R_t + V_t)$, and $e_t = Y_t - m_{t-1}$,

$$C_t = 1/[R_t^{-1} + V_t^{-1}] = R_t V_t/Q_t = A_t V_t$$

and

$$m_t = C_t(m_{t-1}/R_t + Y_t/V_t) = m_{t-1} + A_t e_t.$$

Upon exponentiating, $p(\mu_t \mid D_t) \propto \exp[-(\mu_t - m_t)^2/2C_t]$, so that $(\mu_t \mid D_t) \sim \mathrm{N}[m_t, C_t]$ as stated in (d).

(2) **Proof based on standard normal theory.**
Within this special normal, linear framework, a more specific proof based on the bivariate normal distribution is derived by

(i) calculating the joint distribution of $(Y_t, \mu_t \mid D_{t-1})$, and

(ii) deducing the conditional distribution $(\mu_t \mid Y_t, D_{t-1})$.

Any linear function of Y_t and μ_t is a linear combination of the independent normal quantities ν_t, ω_t, and μ_{t-1}, and so, conditional on

D_{t-1}, is normally distributed. Then by definition, $(Y_t, \mu_t \mid D_{t-1})$ is bivariate normal. To identify the mean vector and variance matrix, note that $(\mu_t \mid D_{t-1}) \sim \mathrm{N}[m_{t-1}, R_t]$ and $(Y_t \mid D_{t-1}) \sim \mathrm{N}[f_t, Q_t]$ as the marginal distributions in (b) and (c). Using additivity and the independence of μ_t and ν_t, the remaining covariance is

$$\mathrm{C}[Y_t, \mu_t \mid D_{t-1}] = \mathrm{C}[\mu_t + \nu_t, \mu_t \mid D_{t-1}] = \mathrm{V}[\mu_t \mid D_{t-1}] = R_t.$$

Hence the joint distribution is

$$\left(\begin{matrix} Y_t \\ \mu_t \end{matrix} \middle| D_{t-1}\right) \sim \mathrm{N}\left[\begin{pmatrix} m_{t-1} \\ m_{t-1} \end{pmatrix}, \begin{pmatrix} Q_t & R_t \\ R_t & R_t \end{pmatrix}\right].$$

General multivariate normal theory (Section 17.2) may be applied to obtain the required distribution conditional on Y_t, but the particular characteristics of the bivariate normal are used here. The correlation $\rho_t = R_t/(R_t Q_t)^{1/2}$ is clearly positive, with $\rho_t^2 = R_t/Q_t = A_t$. So using the referenced material,

$$(\mu_t \mid Y_t, D_{t-1}) \sim \mathrm{N}[m_t, C_t],$$

with

$$m_t = m_{t-1} + \rho_t^2 (Y_t - m_{t-1})$$

and

$$C_t = (1 - \rho_t^2) R_t = R_t V_t / Q_t = A_t V_t.$$

In this case A_t can be interpreted as both the squared correlation and the regression coefficient of μ_t on Y_t.

The result (d) has been established in two ways conditional upon (a). The complete proof follows by induction since (a) is true for $t = 1$ directly from Definition 2.1.

$\diamond$

Some discussion of the various elements of the distributions is in order. First, e_t is the one step ahead forecast error, the difference between the observed value Y_t and its expectation f_t. Second, A_t is the prior regression coefficient of μ_t upon Y_t and, in this particular case, is the square of their correlation coefficient; clearly $0 \le A_t \le 1$. The results are computationally simple and elegant due to the use of normal distributions for each model component. Using these results, sequential updating and revision of forecasts is direct. It is worth noting that an alternative representation for m_t is

$$m_t = A_t Y_t + (1 - A_t) m_{t-1},$$

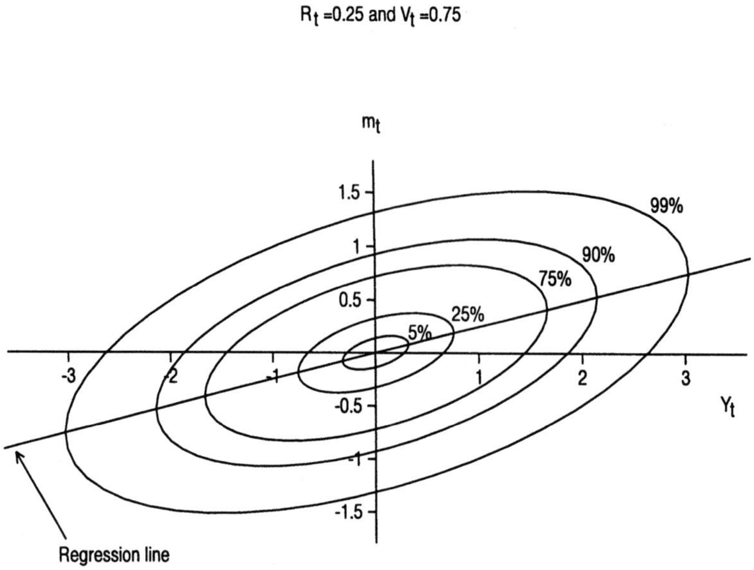

Figure 2.2 Probability contours of a bivariate normal

showing that m_t is a weighted average of the prior level estimate m_{t-1} and the observation Y_t. The **adaptive coefficient** A_t, or weight, defining this combination lies between 0 and 1, being closer to 0 when $R_t < V_t$ so that the prior distribution is more concentrated than the likelihood, and being closer to 1 when the prior is more diffuse, or less informative, than the likelihood. In addition, the posterior is less diffuse than the prior since $C_t < R_t$, representing an increase in information about μ_t due to the additional observation Y_t.

A simple case with $m_{t-1} = 0$, $R_t = 0.25$ and $V_t = 0.75$ is illustrated in Figure 2.2 which provides a contour plot of the joint density. Here $A_t = 0.25$, and the plotted regression line, $m_t = Y_t/4$, simply expresses m_t as a function of Y_t.

2.2.3 Forecast distributions

At time t, the two main forecast distributions are the marginals $(Y_{t+k} \mid D_t)$ and $(X_t(k) \mid D_t)$, where $X_t(k) = Y_{t+1} + Y_{t+2} + \cdots + Y_{t+k}$, for $k > 0$. The former is the **k-step ahead forecast** and the latter the **k-step lead time forecast**.

Theorem 2.2. *For $k > 0$, the following distributions exist:*

(a) *k-step ahead:* $(Y_{t+k} \mid D_t) \sim \mathrm{N}[m_t, Q_t(k)]$,

(b) *k-step lead-time:* $(X_t(k) \mid D_t) \sim \mathrm{N}[km_t, L_t(k)]$,

where

$$Q_t(k) = C_t + \sum_{j=1}^{k} W_{t+j} + V_{t+k}$$

and

$$L_t(k) = k^2 C_t + \sum_{j=1}^{k} V_{t+j} + \sum_{j=1}^{k} j^2 W_{t+k+1-j}.$$

Proof. From the evolution equation for μ_t and the observational equation for Y_t, for $k \geq 1$,

$$\mu_{t+k} = \mu_t + \sum_{j=1}^{k} \omega_{t+j},$$

$$Y_{t+k} = \mu_t + \sum_{j=1}^{k} \omega_{t+j} + \nu_{t+k}.$$

Since all terms are normal and mutually independent, $(Y_{t+k} \mid D_t)$ is normal and the mean and variance follow directly. For the lead time, note that

$$X_t(k) = k\mu_t + \sum_{j=1}^{k} j\omega_{t+k+1-j} + \sum_{j=1}^{k} \nu_{t+j},$$

which is clearly normal with mean $k\mu_t$. Using the independence structure of the error terms,

$$\mathrm{V}[X_t(k) \mid D_t] = k^2 C_t + \sum_{j=1}^{k} j^2 W_{t+k+1-j} + \sum_{j=1}^{k} V_{t+j},$$

and the stated form of $L_t(k)$ follows.

◇

2.3 THE CONSTANT MODEL

2.3.1 Introduction

The special case in which the observational and evolution variances are constant in time is referred to as a **constant** model. It is characterised by the quadruple $\{1, 1, V, W\}$ and defined as

$$\begin{aligned} &\text{Observation equation:} && Y_t = \mu_t + \nu_t, && \nu_t \sim \mathrm{N}[0, V], \\ &\text{System equation:} && \mu_t = \mu_{t-1} + \omega_t, && \omega_t \sim \mathrm{N}[0, W], \\ &\text{Initial information:} && (\mu_0 \mid D_0) \sim \mathrm{N}[m_0, C_0]. && \end{aligned}$$

The important, positive constant $r = W/V$ relates to the engineering concept of a *signal-to-noise* ratio, measuring the sustained system variance relative to the ephemeral observation variance. Since the information sets $D_t = \{Y_t, D_{t-1}\}$ contain no information external to the time series, the model is called **closed** (to external information). Although apparently restricted, the closed, constant, first-order polynomial model is of practical and theoretical interest, allowing the derivation of important limiting results, which illuminate the structure of more general DLMs and relate directly to classical time series models and popular point forecasting methods.

EXAMPLE 2.1. A pharmaceutical company markets KURIT, an ethical drug, which currently sells an average of 100 units per month. Medical advice leads to a change in drug formulation that is expected to result in wider market demand for the product. It is agreed that from January, $t = 1$, the new formulation with new packaging will replace the current product, but the price and brand name KURIT remains unchanged. In order to plan production, stocks and raw material supplies, short-term forecasts of future demand are required. The drug is used regularly by individual patients, so that demand tends to be locally constant in time. So a constant first-order polynomial DLM is adopted for the total monthly sales. Sales fluctuations and observational variation about demand level are expected to considerably exceed month-to-month variation in the demand level, so that W is small compared to V. In accord with this, the constant DLM $\{1, 1, 100, 5\}$, which operated successfully on the old formulation, is retained for the new formulation.

In December, $t = 0$, the expert market view for the new product is that demand is most likely to have expanded by about 30%, to 130 units per month. It is believed that demand is unlikely to have fallen by more than 10 units or to have increased by more than 70. This range of 80 units is taken as representing 4 standard deviations for μ_0. Hence the initial view of the company prior to launch is described by $m_0 = 130$ and $C_0 = 400$, so that

$$(\mu_0 \mid D_0) \sim \mathrm{N}[130, 400].$$

Table 2.1. KURIT example

Month	Forecast Distribution		Adaptive Coeff.	Datum	Error	Posterior Information	
t	Q_t	f_t	A_t	Y_t	e_t	m_t	C_t
0						130.0	400
1	505	130.0	0.80	150	20.0	146.0	80
2	185	146.0	0.46	136	−10.0	141.4	46
3	151	141.4	0.34	143	1.6	141.9	34
4	139	141.9	0.28	154	12.1	145.3	28
5	133	145.3	0.25	135	−10.3	142.6	25
6	130	142.6	0.23	148	5.3	143.9	23
7	128	143.9	0.22	128	−15.9	140.4	22
8	127	140.4	0.21	149	8.6	142.2	21
9	126	142.2	0.21	146	3.8	143.0	20
10	125	143.0	0.20				

Consequently, the operational routine model for sales Y_t in month t is

$$Y_t = \mu_t + \nu_t, \qquad \nu_t \sim \mathrm{N}[0, 100],$$

$$\mu_t = \mu_{t-1} + \omega_t, \qquad \omega_t \sim \mathrm{N}[0, 5],$$

$$(\mu_0 \mid D_0) \sim \mathrm{N}[130, 400].$$

Here $r = 0.05$, a low signal-to-noise ratio typical in this sort of application.

Observations over the next few months and the various components of the one-step forecasting and updating recurrence relationships are given in Table 2.1. Figure 2.3a provides a time plot of the observations and one-step forecasts, and Figure 2.4a is a time plot of the adaptive coefficient A_t, both plots extending to September, $t = 9$, for Table 2.1. Initially, at $t = 0$, the company's prior view of market demand is vitally important in making decisions about production and stocks. Subsequently, however, the value of this particular subjective prior diminishes rapidly as data is received. For example, the adaptive coefficient A_1 takes the value 0.8, so that

$$m_1 = m_0 + 0.8e_1 = (4Y_1 + m_0)/5.$$

Thus, the January observation is given 4 times the weight of the prior mean m_0 in calculating the posterior mean m_1. At $t = 2$, $A_2 = 0.46$ and

$$m_2 = m_1 + 0.46e_2 = 0.46Y_2 + 0.43Y_1 + 0.11m_0,$$

so that Y_2 is also relatively highly weighted and m_0 contributes only 11% of the weight of information incorporated in m_2. As t increases, A_t appears to decay rapidly to a limiting value near 0.2. In fact, the next section shows that A_t does converge to exactly $A = 0.2$ in this example. Finally, the coefficient of m_0 in m_t is simply $(1 - A_t)(1 - A_{t-1}) \dots (1 - A_1)$, so that m_0 contributes only 1% of the information used to calculate m_{10}, and as t increases the relevance of this subjective prior decays to zero.

2.3.2 Intervention to incorporate external information

Model closure is unwise: it precludes the use of extra external information, which is essential in coping with exceptional circumstances, and it is just not acceptable in applied dynamic systems. One of the major advantages of the Bayesian approach lies in the ease with which subjective information and model/data information combine. In the example, at $t = 9$, the level posterior and one-step ahead forecast distributions are

$$(\mu_9 \mid D_9) \sim \mathrm{N}[143, 20],$$
$$(Y_{10} \mid D_9) \sim \mathrm{N}[143, 125].$$

Suppose that information is now received concerning the pending withdrawal from the market of a major competitive drug BURNIT due to suspected side effects. This will occur at $t = 10$, when patients who were prescribed BURNIT will switch to a competitor. This information is received at $t = 9$ and is denoted by S_9. It is known that BURNIT currently accounts for roughly 50% of the market, which leads the company to estimate a 100% increase in KURIT demand, $\mathrm{E}[\mu_{10} \mid D_9, S_9] = 286$. However, uncertainty about this figure is high, with estimated increased demand ranging within the company's marketing department from a pessimistic value of 80 units to an optimistic 200. After discussion, it is agreed to model the change in demand at time 10 as

$$(\omega_{10} \mid D_9, S_9) \sim \mathrm{N}[143, 900],$$

leading to the revised one-step ahead forecast distributions

$$(\mu_{10} \mid D_9, S_9) \sim \mathrm{N}[286, 920]$$

and

$$(Y_{10} \mid D_9, S_9) \sim \mathrm{N}[286,\ 1020].$$

Consequently, A_{10} increases from 0.2 to 0.9, providing much faster adaptation to the immediately forthcoming data than would happen without intervention. Observing $Y_{10} = 326$ implies $e_{10} = 40$ and

$$(\mu_{10} \mid D_{10}) \sim \mathrm{N}[322, 90].$$

Note that the conditioning information set here is $D_{10} = \{Y_{10}, D_9, S_9\}$. Figures 2.3b and 2.4b show the continued time graphs for the following six months, from $t = 10$ to $t = 15$.

When unexpected and relevant external information of this sort becomes available, intervention is of paramount importance for good decision making. From this example, it is clearly unsatisfactory to confine intervention

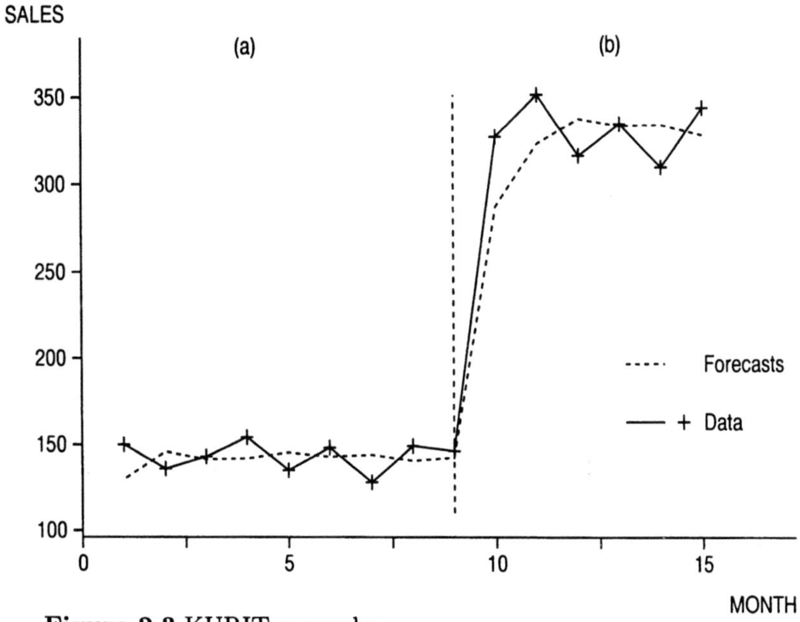

Figure 2.3 KURIT examples

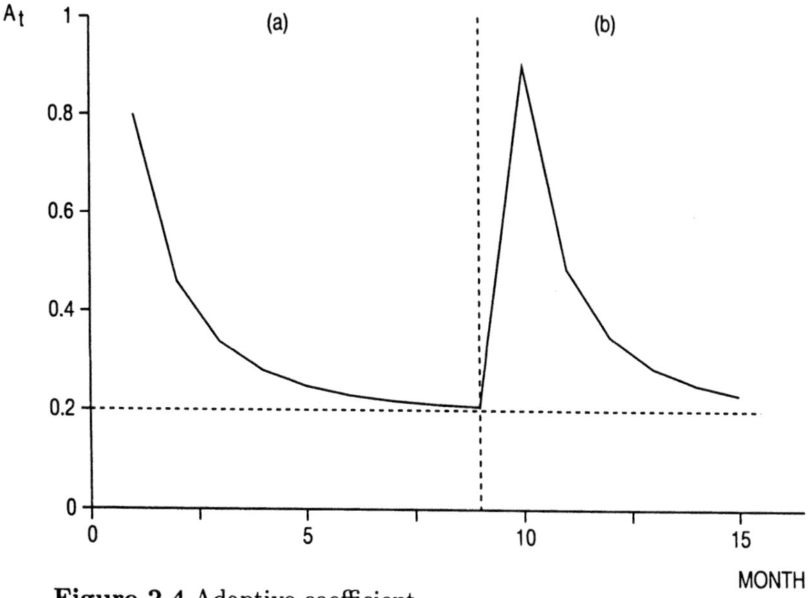

Figure 2.4 Adaptive coefficient

to an alteration of the mean estimate of demand, although this is unfortunately an approach adopted by some forecasters. The associated large variance reflects the true increase in uncertainty, leads to less weight being associated with data prior to the change, adapts more quickly to the immediately forthcoming forecast errors, and results in more reliable forecasts.

2.3.3 Limiting behaviour and convergence

In the closed model, the rate of adaptation to new data, as measured by the adaptive coefficient A_t, rapidly converges to a constant value as follows.

Theorem 2.3. *Define $r = W/V$. As $t \to \infty$, $A_t \to A$ and $C_t \to C = AV$, where*

$$\lim_{t\to\infty} A_t = A = \frac{r}{2}\left(\sqrt{1+\frac{4}{r}} - 1\right).$$

Proof. Since $0 < A_t < 1$ and $C_t = A_t V$, it follows that C_t is bounded, with

$$0 < C_t \leq V, \qquad \text{for all } t.$$

So, using the recursions $C_t^{-1} = R_t^{-1} + V^{-1}$ and $R_t = C_{t-1} + W$,

$$C_t^{-1} - C_{t-1}^{-1} = R_t^{-1} - R_{t-1}^{-1} = K_t(C_{t-1}^{-1} - C_{t-2}^{-1}),$$

where $K_t = C_{t-1}C_{t-2}/(R_t R_{t-1}) > 0$. So C_t is a monotonic and bounded sequence and its limit, say C, exists. Consequently, R_t converges to $R = C + W$. Also, using $C_t = R_t V/(R_t + V)$, it follows that $C = RV/(R+V)$, implying

$$C^2 + CW - VW = 0.$$

This quadratic has just one positive root,

$$C = rV(\sqrt{1+4/r} - 1)/2.$$

Since $C_t = A_t V$, then A_t converges to $A = C/V$, which is the required function of r. A useful inversion of this relationship leads to

$$r = A^2/(1-A).$$

◇

A is a function of $r = W/V$ alone, the relationship being illustrated in the following table of values of $1/r$ and A:

$1/r$	9900	380	90	20	6	0.75	0.01
A	0.01	0.05	0.10	0.20	0.33	0.67	0.99

Summarising the above results and some easily deduced consequences we have:

(i) $A_t \to A = r(\sqrt{1+4/r} - 1)/2$ with $r = A^2/(1-A)$,

(ii) $Q_t \to Q = V/(1-A)$,

(iii) $C_t \to C = AV$,

(iv) $R_t \to R = AQ$,

(v) $W = A^2Q$ and $V = (1-A)Q$.

C_t is monotonic increasing/decreasing according to whether the initial variance C_0 is less/greater than the limiting value C. Similar comments apply to the form of convergence of the sequence A_t.

An exact expression for A_t is derived in Harrison (1985a), the details of the proof given here in the Appendix to this chapter, Section 2.7. In summary, the result is as follows: Define $\delta = 1 - A$. Clearly $0 < \delta < 1$, and for large t, $m_t \approx \delta m_{t-1} + (1-\delta)Y_t$. Given C_0, the initial adaptive coefficient is $A_1 = (C_0 + W)/(C_0 + W + V)$ and

$$A_t = A\left[\frac{(1-\delta^{2t-2})A + (\delta + \delta^{2t-2})A_1}{(1+\delta^{2t-1})A + (\delta - \delta^{2t-1})A_1}\right],$$

from which the following are deduced:

(a) In the case of a very vague initial prior in which C_0^{-1} is close to zero, $A_1 \approx 1$, and A_t is monotonically decreasing with

$$A_t \approx A(1+\delta^{2t-1})/(1-\delta^{2t}),$$

or, writing $\delta_t = 1 - A_t$,

$$\delta_t \approx \delta(1-\delta^{2t-2})/(1-\delta^{2t}).$$

(b) At the other extreme, when the initial prior is very precise with C_0 close to zero, then $A_1 \approx 0$ and A_t is monotonically increasing with

$$A_t \approx A(1-\delta^{2t-2})/(1+\delta^{2t-1}),$$

or

$$\delta_t \approx \delta(1+\delta^{2t-3})/(1+\delta^{2t-1}).$$

(c) Convergence is exponentially fast, being monotonically increasing for $A_1 < A$, and monotonically decreasing for $A_1 > A$.

2.3.4 General comments

The case $A = 1$ corresponds to $V = 0$, when $A_t = 1$ for $t \geq 2$. This is implied in practical cases by $r \to \infty$ and reduces the model to a pure random walk, $Y_t = Y_{t-1} + \omega_t$. Then $m_t = Y_t$ and the model is of little use for prediction. Often, when applied to series such as daily share, stock and commodity prices, the appropriate values of V in a first-order model may appear to be close to zero or very small compared to the systematic variance W. In other words the DLM appears to imply that all movement between daily prices may be attributed to movement in the underlying price level μ_t. This feature has led many economic forecasters to conclude that such series are purely random and that no forecast of future prices better than $(Y_{t+k} \mid D_t) \sim \mathrm{N}[Y_t, kW]$ is available. Such conclusions are quite erroneous and reflect myopic modelling; clearly the model is restricted and inappropriate as far as forecasting is concerned. It is rather like looking at a cathedral through a microscope and concluding that it has no discernable form. In other words, other models and ways of seeing are required.

The major applications of the constant model are in short-term forecasting and control, when the main benefit is derived from data smoothing. Denoting the forecasting horizon by L sampling periods, it is advisable to choose a sampling interval such that $1 \leq L \leq 4$ and $20 \leq 1/r \leq 1000$. In the limit as $t \to \infty$, the point predictor is $m_t = AY_t + (1 - A)m_{t-1}$. Clearly, if $A = 1$ then $m_t = Y_t$ and any changes in Y_t are fully reflected in the predictor. On the other hand, if A is close to zero, then $m_t \approx m_{t-1}$ and none of the changes in the series will be captured by the predictor. The larger the value of A, the more sensitive is the predictor to the latest values of the observation series. So there is a dilemma arising from the conflict between sensitivity and robustness requirements. Note that $Y_t = \mu_{t-1} + \omega_t + \nu_t$. On the one hand, it is desirable to have a large value of A so that any sustained changes that occur through the signal ω_t are fully incorporated in the estimate m_t; on the other hand, a small value is desirable in order that the corrupting random noise, ν_t, be excluded from the estimate. The selected value of A reflects the relative expected variation in ω_t and ν_t through the ratio r. Applications should always employ the *Principle of Management by Exception,* embedding the routine model within a complete forecasting system. So r is chosen to give good routine forecasts and treats occasional sharp level changes and maverick observations as exceptions, which will either be anticipated by experts or signaled by the forecast monitoring system and referred to experts. Relevant expert subjective information is combined with data as illustrated in the previous "Kurit" example. Further details of such important topics are left until later chapters.

2.3.5 Limiting predictors and alternative methods

In the formal sense that $\lim_{t\to\infty}(m_t - m_{t-1} - Ae_t) = 0$, the limiting one-step ahead point forecast, $f_{t+1} = m_t = \mathrm{E}[Y_{t+1} \mid D_t]$, may be written

$$m_t = (1 - A)m_{t-1} + AY_t = m_{t-1} + Ae_t.$$

Several commonly used point forecasting methods, and the Box–Jenkins ARIMA(0,1,1) predictor, adopt this *limiting* forecast function of the constant first-order polynomial DLM:

(a) **Holt's point predictor**
Holt (1957) introduced a widely used point forecasting method,

$$f_t(k) = M_t, \quad \text{where} \quad M_t = \alpha Y_t + (1-\alpha)M_{t-1},$$

which is equivalent to the asymptotic form of m_t with $\alpha = A$. For finite t, and at times of intervention, Holt's method is insufficiently responsive, the most recent observation always receiving the same weight, as is seen on writing $\delta = 1 - A$, when

$$M_t = AY_t + \sum_{j=1}^{t-1} \delta^j Y_{t-j} + \delta^t M_0.$$

(b) **Exponentially weighted moving averages (EWMA)**
For a parameter $0 < \delta < 1$, the EWMA of a sequence $Y_t, \ldots, Y_1$, is

$$M_t = \frac{(1-\delta)}{(1-\delta^t)} \sum_{j=0}^{t-1} \delta^j Y_{t-j}.$$

Practitioners usually apply and refer to an EWMA in its limiting form

$$M_t = (1-\delta)Y_t + \delta M_{t-1},$$

which, with $\delta = 1 - A$, is identical to Holt's predictor. Formally, given m_0 in the closed, constant DLM, an EWMA M_0, and any $\epsilon > 0$, we have

$$\lim_{t \to \infty} \Pr(|m_t - M_t| > \epsilon) = 0.$$

(c) **Brown's exponentially weighted regression (EWR)**
Brown's (1962) forecast function for a locally constant mean of an infinite data set $Y_t, \ldots, Y_0, \ldots$, is $f_t(k) = \hat{\mu}_t$, where, for a given discount factor $0 < \delta < 1$, the EWR estimate $\hat{\mu}_t = \mu$ minimizes the discounted sum of squares

$$S_t(\mu) = \sum_{j=0}^{\infty} \delta^j (Y_{t-j} - \mu)^2.$$

It is easily shown that $\hat{\mu}_t$ is unique and equal to the EWMA M_t.

In each of the three foregoing cases, the classical point predictors use a forecast function equivalent to the limiting forecast function of the closed, constant DLM. Unlike the DLM however, these forecasting methods are ad hoc devices that, although simple to apply, have no coherent basis. With little relevant data their limiting forms are inappropriate, they are difficult to interpret, they have none of the DLM flexibility, and with the exception of EWR, they do not easily generalize.

(d) **An alternative ARIMA(0,1,1) model representation**
The predictors of Box and Jenkins (1976) founded on ARIMA models are very popular. Their widely applied ARIMA(0,1,1) predictor produces identical point predictions to those of Holt and Brown, so corresponding to the limiting predictor of a closed, constant, first-order polynomial DLM. This is demonstrated as follows:

Given a series Y_t generated by the DLM $\{1, 1, V, W\}$, for $t > 1$,

$$Y_t - Y_{t-1} = \nu_t - \nu_{t-1} + \omega_t.$$

So the first difference of the observation series may be represented as an ARIMA(0,1,1) process

$$Y_t - Y_{t-1} = a_t - \delta a_{t-1}, \qquad a_t \sim \mathrm{N}[0, Q],$$

where $\delta = 1 - A$ and the a_t are independent random variables.

From the two relationships $m_{t-1} = Y_t - e_t$ and $m_t = m_{t-1} + A_t e_t$,

$$Y_t - Y_{t-1} = e_t - (1 - A_{t-1})e_{t-1},$$

so that

$$\lim_{t\to\infty} [Y_t - Y_{t-1} - e_t + \delta e_{t-1}] = 0.$$

The Box–Jenkins ARIMA(0,1,1) predictor is

$$Y_t - Y_{t-1} = e_t - \delta e_{t-1},$$

thus assuming the limit form and replacing the unknown random variable a_t by the observed one step ahead forecast error e_t. It can be shown that as t increases, $(e_t - a_t)$ converges in probability to zero. But clearly, as happens with little data or at times of intervention, if Q_t differs from Q, equating e_t and a_t is quite inappropriate.

2.3.6 *Forecast distributions*

At time t, the k-step ahead and k-lead time forecast distributions are special cases of those in Theorem 2.2. Using that theorem,

(a) *k-step ahead:* $(Y_{t+k} \mid D_t) \sim \mathrm{N}[m_t, Q_t(k)]$,

(b) *k-step lead time:* $(X_t(k) \mid D_t) \sim \mathrm{N}[km_t, L_t(k)]$,

where

$$Q_t(k) = C_t + kW + V$$

and

$$L_t(k) = k^2 C_t + kV + k(k+1)(2k+1)W/6.$$

It is of interest to note that the lead time forecast *coefficient of variation*, defined as $L_t(k)^{1/2}/(km_t)$, is monotonically decreasing for k less than or equal to the integer part of $k_0 = \sqrt{(0.5 + 3/r)}$, and monotonically increasing for $k > k_0$. Note that k_0 is independent of C_t, being a function only of r. For example, a value of $r = 0.05$, corresponding to $A = 0.2$, implies that $k_0 = 8$. Then, with the limiting value of variance $C = 0.2V$ substituted for C_t, it follows that $L_t(8)/(64L_t(1)) = (0.62)^2$, so that the limiting coefficient of variation for the one-step ahead forecast is 61% greater than that for the 8-step lead time forecast.

2.4 SPECIFICATION OF EVOLUTION VARIANCE W_t

The forecasting performance of a first-order polynomial DLM $\{1, 1, V_t, W_t\}$, constant or otherwise, depends heavily on choosing appropriate values for the variances, V_t and W_t. The problem of the former variance is considered in Section 2.5 below; here the choice of the latter is examined.

2.4.1 Robustness to values of W in the constant model

Suppose that a forecaster applies the constant model with observational variance V and signal to noise ratio r when the data are truly generated by a model with values V_0 and r_0. Of course this example is purely hypothetical since no mathematical model can exactly represent a "true" process, but it provides important insight into the questions of robustness and choice of variances. Convergence is fast, so consider the limiting form of the model,

$$Y_t - Y_{t-1} = e_t - \delta e_{t-1},$$

where $\delta = 1 - A$ and the forecaster assumes $(e_t \mid D_{t-1}) \sim \mathrm{N}[0, Q]$ with $Q = V/\delta$. The true limiting model is such that

$$Y_t - Y_{t-1} = a_t - \delta_0 a_{t-1},$$

where $\delta_0 = 1 - A_0$, and, in fact, $(a_t \mid D_{t-1}) \sim \mathrm{N}[0, Q_0]$ with $Q_0 = V_0/\delta_0$. Equating the two expressions gives $e_t - \delta e_{t-1} = a_t - \delta_0 a_{t-1}$, so that

$$e_t = a_t + (\delta - \delta_0) \sum_{j=0}^{t-2} \delta^j a_{t-j-1}.$$

Given the independence of the true error sequence a_t, it follows that the true initial prior mean of the model errors is $\mathrm{E}[e_t \mid D_0] = 0$, and as $t \to \infty$, the corresponding limiting variances are

$$\mathrm{V}[e_t \mid D_0] = Q_0[1 + (\delta - \delta_0)^2/(1 - \delta^2)].$$

It is immediately apparent that a gross penalty is incurred if the value of r, and hence W, is too low. As $r \to 0$, then $\delta \to 1$, and the true variance of the above error sequence becomes infinite, whereas the forecaster uses the

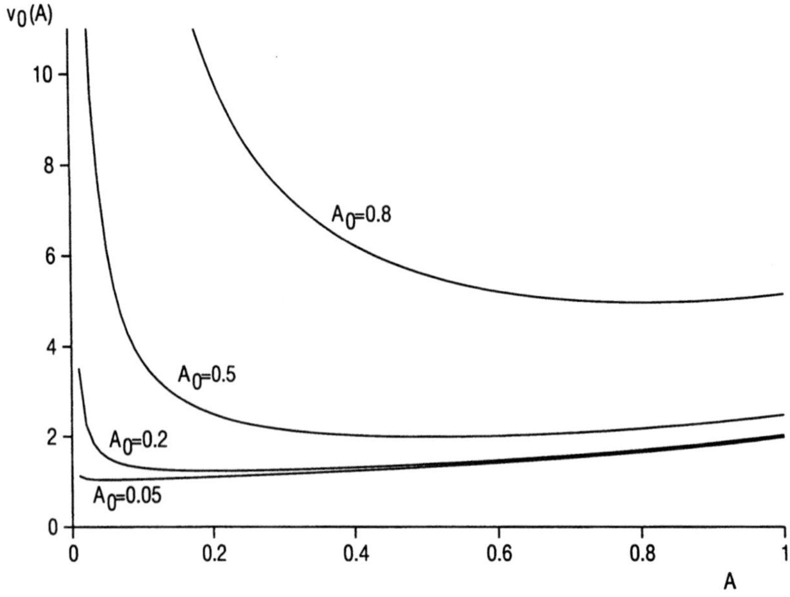

Figure 2.5 Limiting error variance with misspecified A

value V from the model! At the other extreme, as $r \to \infty$, then $\delta \to 0$, and the true variance tends to $Q_0(1+\delta_0^2)$. Hence, in an unmonitored system, choosing r too large is preferable to choosing it too small, over adaptation limiting the margin for error. The case $r = 0$ implies a constant level $\mu_t = \mu$, a model that has been much used in practice but whose pitfalls are now clear. Figure 2.5 illustrates this feature by plotting the function $\mathrm{v}_0(A) = [1+(\delta-\delta_0)^2/(1-\delta^2)]/\delta_0$ as a function of the assumed value of $A = 1-\delta$ for various true values $A_0 = 0.05, 0.2, 0.5, 0.8$.

Additionally, misspecification leads to correlated errors. Clearly, the true covariance between the model forecast errors e_t and e_{t-k} is

$$\mathrm{C}[e_t, e_{t-k} \mid D_0] = Q_0(\delta-\delta_0)\delta^{k-1}(1-\delta\delta_0)/(1-\delta^2),$$

giving correlations

$$C_0(k) = (\delta-\delta_0)(1-\delta\delta_0)\delta^{k-1}/(1-2\delta\delta_0+\delta_0^2),$$

instead of zero as assumed in the model. If $\delta > \delta_0$, so that $A < A_0$, the e_t are positively correlated; otherwise they are negatively correlated. In the limit, as $\delta_0 \to 1$, so that μ_t is actually constant,

$$0 > C_0(k) = -\delta^{k-1}(1-\delta)/2.$$

At the other extreme, as $\delta_0 \to 0$, so that Y_t is a pure random walk,

$$0 < C_0(k) = \delta^k.$$

Thus an under-adaptive model ($\delta > \delta_0$) leads to positively correlated forecast errors, and an over-adaptive model ($\delta < \delta_0$) leads to negatively correlated errors. Further details and discussion appear in Harrison (1967) and Roberts and Harrison (1984).

2.4.2 Discount factors as an aid to choosing W_t

By definition, for the constant model, $R_t = C_{t-1} + W$, and in the limit, $R = C + W = C/(1-A)$. Thus, $W = AC/(1-A)$, so that W is a fixed proportion of C. This is a natural way of thinking about the system variance: between observations, the addition of the error ω_t leads to an additive increase of $W = 100A/(1-A)\%$ of the initial uncertainty C. Since $\delta = 1-A$, it follows that $R = C/\delta$. Choice of δ by reference to the limiting rate of adaptation to data then guides the choice of W. For $A = 0.1$, $\delta = 0.9$ and $W \approx 0.11C$; this increases to $0.25C$ for $\delta = 0.8$. Bearing in mind that in this DLM the limiting behaviour is rapidly achieved, it is convenient and natural to adopt a constant *rate* of increase of uncertainty, or decay of information, for all t rather than just in the limiting case. Thus, for a given **discount factor** δ, typically between 0.8 and 1, choosing

$$W_t = C_{t-1}(1-\delta)/\delta$$

for each t implies

$$R_t = C_{t-1}/\delta.$$

This DLM is not a constant model but quickly converges to the constant DLM $\{1, 1, V, rV\}$ with $r = (1-\delta)^2/\delta$, as is easily seen upon noting that

$$C_t^{-1} = V^{-1} + R_t^{-1} = V^{-1} + \delta C_{t-1}^{-1} = V^{-1}[1 + \delta + \cdots + \delta^{t-1}] + \delta^t C_0^{-1},$$

so that the limiting value of C_t is $C = (1-\delta)V = AV$.

Further discussion of this and more general **discount** models appears in later chapters. Here the concept is introduced as a simple and natural way of structuring and assigning values to the W_t sequence in the first-order polynomial model. Note that it applies directly to the general model with variances V_t and W_t since W_t, as defined above, depends only on the values of C_{t-1} and δ, which are known at time $t-1$. Furthermore, with the assumption of the known values of the V_t sequence in the initial information set D_0, W_t is also known at $t = 0$ and is calculated from the above recursion as a function of δ, C_0, and $V_{t-1}, \ldots, V_1$.

2.5 UNKNOWN OBSERVATIONAL VARIANCES

2.5.1 Introduction

Typically, the observational variance sequence V_t will not be precisely known, although forecasters will usually have prior beliefs about particular features of the sequence. Examples include $V_t = V$ as in the constant model but with V unknown; weighted variances $V_t = Vk_t$ where the weight sequence k_t is known but V is unknown; $V_t = Vk_t(\mu_t)$ where $k_t(\cdot)$ is a possibly time dependent variance function of the mean, such as the common power function $k_t(\mu_t) = \mu_t^p$ for some power p, and V is unknown; etc. In each case, the forecaster may also suspect that the nominally constant scale factor V does, in fact, vary slowly in time, reflecting natural dynamics of the time series observational process, as well as allowing for errors of approximation in the model. All these cases are of practical importance and are developed further in later chapters. Here only the case of the first-order polynomial model, in which $V_t = V$ is an unknown constant, is considered. Often working with the reciprocal of V, or the **precision** $\phi = 1/V$, clarifies and simplifies proceedings. Refer to Section 17.3 for a wider theoretical discussion of normal linear models with unknown variance.

A simple, closed-form Bayesian analysis of the DLM with unknown, constant variance V is available if a particular structure is imposed on the W_t sequence and on the initial prior for μ_0. This structure enables a *conjugate* sequential updating procedure for V, equivalently for ϕ, in addition to that for μ_t. To motivate this structure, recall that in the constant model it is natural to specify the evolution variance as a multiple of the observational variance for all time. Applying this to the general model suggests that at each time t, for known V,

$$(\omega_t \mid V, D_{t-1}) \sim \mathrm{N}[0, VW_t^*].$$

Thus, W_t^* is the variance of ω_t if $V = 1$, or equivalently $\phi = 1$. With this structure, suppose that C_0 is very large compared to V. In particular, as C_0 increases, $C_1 \approx V$, and for $t > 1$, C_t is given by V multiplied by that value obtained from the standard model in which $V = 1$. Indeed, this always happens eventually, whatever the value of C_0, since the effect of the initial prior rapidly decays. So the form of the C_t sequence as deduced from the data leads to scaling by V just as with W_t^* above. In particular, if C_0 is scaled by V directly, then so is C_t for all t. This feature is the final component of the required model structure for the following closed, conjugate analysis.

2.5.2 The case of a constant unknown variance

The analysis for V follows standard Bayesian theory as detailed in Section 17.3. The conjugate analysis is based on gamma distributions for ϕ, and thus inverse gamma distributions for V, for all time, and derives from the following model definition.

Definition 2.2. For each t, the model is defined by

Observation equation:	$Y_t = \mu_t + \nu_t,$	$\nu_t \sim \mathrm{N}[0, V],$
System equation:	$\mu_t = \mu_{t-1} + \omega_t,$	$\omega_t \sim \mathrm{N}[0, VW_t^*],$
Initial information:	$(\mu_0 \mid D_0, V) \sim \mathrm{N}[m_0, VC_0^*],$	
	$(\phi \mid D_0) \sim \mathrm{G}[n_0/2, d_0/2],$	

for some known m_0, C_0^*, W_t^*, n_0 and d_0. In addition, the usual independence assumptions of Definition 2.1 hold, now conditional on V.

The final component is a gamma prior with density

$$p(\phi \mid D_0) = \frac{(d_0/2)^{n_0/2}}{\Gamma(n_0/2)} \phi^{n_0/2-1} \mathrm{e}^{-\phi d_0/2}, \qquad \phi > 0.$$

Ignoring the normalisation constant,

$$p(\phi \mid D_0) \propto \phi^{n_0/2-1} \mathrm{e}^{-\phi d_0/2}.$$

The mean of this prior is

$$\mathrm{E}[\phi \mid D_0] = \frac{n_0}{d_0} = \frac{1}{S_0},$$

where S_0 is a prior point estimate of the observational variance $V = \phi^{-1}$. Alternative expressions of the prior are that when n_0 is integral, the multiple d_0 of ϕ has a chi-square distribution with n_0 degrees of freedom,

$$(d_0\phi \mid D_0) \sim \chi^2_{n_0},$$

or that $(V/d_0 \mid D_0)$ has an inverse chi-squared distribution with the same degrees of freedom. Notice that in specifying the prior, the forecaster must choose the prior estimate S_0 and the associated degrees of freedom n_0, in addition to m_0 and C_0^*. The starred variances C_0^* and the sequence W_0^*, assumed known, are multiplied by V to provide the actual variances when V is known. Thus they are referred to as *observation-scale free variances*, explicitly recognising that they are independent of the scale of the observations determined by V.

The results obtained below are based on Student T distributions for the level parameters and forecasts. Again details are given in Section 17.3 but for convenience, a short summary is now given. A real random quantity μ has a T distribution with n degrees of freedom, mode m and scale C, denoted by

$$\mu \sim \mathrm{T}_n[m, C],$$

if and only if the density of μ is given by

$$p(\mu) \propto \left[n + \frac{(\mu - m)^2}{C}\right]^{-(n+1)/2}.$$

The quantity $(\mu - m)/\sqrt{C}$ has a standard T distribution on n degrees of freedom, so

$$\frac{\mu - m}{\sqrt{C}} \sim \mathrm{T}_n[0, 1].$$

As n increases, this distribution converges to the standard normal distribution, and the notation is chosen by analogy with this case. The analogy is important; the usual, known variance model is based on normal distributions that are replaced by T forms when V is unknown. Note that $\mathrm{E}[\mu] = m$ when $n > 1$ and $\mathrm{V}[\mu] = Cn/(n-2)$ when $n > 2$, so $\mathrm{V}[\mu] \approx C$ for large n.

From the model definition, unconditionally (with respect to V),

$$(\mu_0 \mid D_o) \sim \mathrm{T}_{n_0}[m_0, C_0],$$

where $C_0 = S_0 C_0^*$ is the scale of the marginal prior T distribution. In specifying the initial prior, the forecaster specifies

(i) the distribution of V via n_0 and d_0, and

(ii) the distribution of μ_0 via m_0 and C_0.

These may be considered separately. Note that from the T prior, if $n_0 > 2$, $\mathrm{V}[\mu_0 \mid D_0] = C_0 n_0/(n_0 - 2)$, and that for large n_0, $V \approx S_0$ so that

$$(\mu_0 | D_0) \approx \mathrm{N}[m_0, C_0].$$

Theorem 2.4. *With the above model, the following distributional results obtain at each time $t \geq 1$:*

(a) *Conditional on V:*
Define $R_t^* = C_{t-1}^* + W_t^*$, $f_t = m_{t-1}$, $Q_t^* = R_t^* + 1$, $e_t = Y_t - f_t$ and $A_t = R_t^*/Q_t^*$. Then

$$\begin{aligned}(\mu_{t-1} \mid D_{t-1}, V) &\sim \mathrm{N}[m_{t-1}, VC_{t-1}^*],\\ (\mu_t \mid D_{t-1}, V) &\sim \mathrm{N}[m_{t-1}, VR_t^*],\\ (Y_t \mid D_{t-1}, V) &\sim \mathrm{N}[f_t, VQ_t^*],\\ (\mu_t \mid D_t, V) &\sim \mathrm{N}[m_t, VC_t^*],\end{aligned}$$

with

$$m_t = m_{t-1} + A_t e_t \quad \text{and} \quad C_t^* = R_t^* - A_t^2 Q_t^* = A_t.$$

(b) *For the precision $\phi = V^{-1}$:*

$$\begin{aligned}(\phi \mid D_{t-1}) &\sim \mathrm{G}[n_{t-1}/2, d_{t-1}/2],\\ (\phi \mid D_t) &\sim \mathrm{G}[n_t/2, d_t/2],\end{aligned}$$

where

$$n_t = n_{t-1} + 1 \quad \text{and} \quad d_t = d_{t-1} + e_t^2/Q_t^*.$$

(c) *Unconditional on V :*
Define $S_{t-1} = d_{t-1}/n_{t-1}$, $C_{t-1} = S_{t-1}C_{t-1}^*$, $R_t = S_{t-1}R_t^*$, $Q_t = S_{t-1}Q_t^*$, $C_t = S_tC_t^*$, *and* $S_t = d_t/n_t$. *Then*

$$\begin{aligned}(\mu_{t-1} \mid D_{t-1}) &\sim \mathrm{T}_{n_{t-1}}[m_{t-1}, C_{t-1}],\\ (\mu_t \mid D_{t-1}) &\sim \mathrm{T}_{n_{t-1}}[m_{t-1}, R_t],\\ (Y_t \mid D_{t-1}) &\sim \mathrm{T}_{n_{t-1}}[f_t, Q_t],\\ (\mu_t \mid D_t) &\sim \mathrm{T}_{n_t}[m_t, C_t],\end{aligned}$$

(d) *Operational definition of the updating equations:*
Defining $Q_t = R_t + S_{t-1}$ *and* $A_t = R_t/Q_t$,

$$\begin{aligned}m_t &= m_{t-1} + A_te_t,\\ C_t &= (S_t/S_{t-1})\,[R_t - A_t^2Q_t] = A_tS_t,\\ n_t &= n_{t-1} + 1,\\ d_t &= d_{t-1} + S_{t-1}e_t^2/Q_t,\\ S_t &= d_t/n_t.\end{aligned}$$

Proof. The results in (a) are the known variance results of Theorem 2.1. The remainder of the proof is by induction. From (a)

$$p(Y_t \mid D_{t-1}, \phi) \propto \phi^{\frac{1}{2}} \exp(-\phi e_t^2/2Q_t^*).$$

Now, by Bayes' Theorem, the posterior for ϕ is

$$p(\phi \mid D_t) \propto p(\phi \mid D_{t-1})\, p(Y_t \mid D_{t-1}, \phi).$$

Using the prior from (b) and the above likelihood,

$$p(\phi \mid D_t) \propto \phi^{(n_{t-1}+1)/2-1} \exp[-(d_{t-1} + e_t^2/Q_t^*)\phi/2],$$

thereby establishing $(\phi \mid D_t) \sim \mathrm{G}[n_t/2, d_t/2]$ as in (b) with updated parameters $\{n_t, d_t\}$ as in (d).

Results (c) follow directly from the normal/gamma/T theory mentioned earlier, and as reviewed in detail in Sections 17.3.1 and 17.3.2: simply integrate the conditional normal distributions in (a) with respect to the appropriate prior/posterior gamma distribution for ϕ in (b). Noting that the results are true for $t = 1$, the inductive proof is complete.

$\diamond$

This theorem provides the key results. At time t, the prior mean of ϕ is $\mathrm{E}[\phi \mid D_{t-1}] = n_{t-1}/d_{t-1} = 1/S_{t-1}$, where $S_{t-1} = d_{t-1}/n_{t-1}$ is a prior point estimate of $V = 1/\phi$. The posterior estimate is $S_t = d_t/n_t$. The updating equations for the parameters defining the T prior/posterior and forecast distributions are essentially the same as the standard, known variance equations with the estimate S_{t-1} appearing as the variance. The only differences are the scaling, S_t/S_{t-1}, in deriving C_t and the scaling of e_t^2/Q_t, by S_{t-1}, in calculating d_t, both to correct for the revised estimate of V. Equations (d) may be used in practice, the starred, scale free versions appearing only to communicate the theoretical structure.

The estimate S_t of the variance V can be written in the recursive form

$$S_t = S_{t-1} + (e_t^2/Q_t - 1)S_{t-1}/n_t.$$

For a fixed h, $S_t = S_{t-1} + h(e_t^2 - S_{t-1})$ was recursively used by some early EWMA systems as an estimate of the *one-step forecast* variance rather than of the *observational* variance V, an important distinction. Such an estimate is obviously suspect, particularly when t and/or n_t are small. The prediction variance is actually given, from the Student T forecast distribution, as $\mathrm{V}[Y_{t+1} \mid D_t] = Q_{t+1}n_t/(n_t - 2)$, when $n_t > 2$. For large n_t, this variance is approximately Q_{t+1}. If the model is constant, then $Q_{t+1} \approx S_t/\delta$ as t and n_t increase, so that the one-step variance is approximately S_t/δ. In this special case, as $t \to \infty$, the *limiting* prediction variance may be written in the above mentioned ad hoc form, as

$$\mathrm{V}[Y_{t+1} \mid D_t] \approx \mathrm{V}[Y_t \mid D_{t-1}] + (\delta/n_t)(e_t^2 - \mathrm{V}[Y_t \mid D_{t-1}]).$$

2.5.3 Summary

Key aspects of model structure, updating and forecasting are summarised in the table below (continued on the following page).

First-Order Polynomial DLM, with Constant Variance V		
Observation:	$Y_t = \mu_t + \nu_t,$	$\nu_t \sim \mathrm{N}[0, V],$
System:	$\mu_t = \mu_{t-1} + \omega_t,$	$\omega_t \sim \mathrm{T}_{n_{t-1}}[0, W_t].$
Information:	$(\mu_{t-1} \mid D_{t-1}) \sim \mathrm{T}_{n_{t-1}}[m_{t-1}, C_{t-1}],$	
	$(\phi \mid D_{t-1}) \sim \mathrm{G}\left[\frac{n_{t-1}}{2}, \frac{n_{t-1}S_{t-1}}{2}\right].$	
Forecast:	$(\mu_t \mid D_{t-1}) \sim \mathrm{T}_{n_{t-1}}[m_{t-1}, R_t],$	
	$(Y_t \mid D_{t-1}) \sim \mathrm{T}_{n_{t-1}}[f_t, Q_t],$	
with $f_t = m_{t-1},$	$R_t = C_{t-1} + W_t,$	$Q_t = R_t + S_{t-1}.$

Updating Recurrence Relationships
$(\mu_t \mid D_t) \sim \mathrm{T}_{n_t}[m_t, C_t],$ with $m_t = m_{t-1} + A_t e_t,$ $C_t = A_t S_t.$ $(\phi \mid D_t) \sim \mathrm{G}\left[\frac{n_t}{2}, \frac{n_t S_t}{2}\right],$ with $n_t = n_{t-1} + 1,$ $S_t = S_{t-1} + \frac{S_{t-1}}{n_t}\left(\frac{e_t^2}{Q_t} - 1\right),$ where $e_t = Y_t - f_t,$ and $A_t = R_t/Q_t.$
k-Step Forecast Distributions
$(Y_{t+k} \mid D_t) \sim \mathrm{T}_{n_t}[m_t, Q_t(k)],$ $(X_t(k) \mid D_t) \sim \mathrm{T}_{n_t}[km_t, L_t(k)],$ with $Q_t(k) = C_t + \sum_{j=1}^{k} W_{t+j} + S_t,$ and $L_t(k) = k^2 C_t + \sum_{j=1}^{k} j^2 W_{t+k+1-j} + kS_t,$ where for $j > 0$ and scale free variances $W_{t+j}^*,$ $W_{t+j} = S_t W_{t+j}^*.$

2.6 ILLUSTRATION

The series of Table 2.2 represents the first differences of the logged monthly USA/UK exchange rate \$/£ from January 1975 to July 1984. From the time plot in Figure 2.6 it is evident that there was considerable short-term variation about a changing level.

The data are examined using a first-order polynomial DLM simply to demonstrate analyses using several closed DLMs that differ only in the values of their evolution variances $\{W_t\}$. As recommended in Section 2.4, these variances are specified by a discount factor δ, the four models examined having discount values of 0.7, 0.8, 0.9 and 1.0, the last corresponding to the degenerate static model with $W_t = 0$, characterising the observations as a simple normal random sample. In each case the initial prior distribution is defined by $m_0 = 0, C_0 = 1, n_0 = 1$ and $d_0 = 0.01$. This vague, uninformative, joint prior specification implies that given D_0, μ_0 lies between -0.1 and 0.1 with probability 0.5, and between -0.63 and

Table 2.2. USA/UK exchange rate index (×100)

Year	Month (Jan – Jun) & (Jul – Dec)					
75	1.35	1.00	−1.96	−2.17	−1.78	−4.21
	−3.30	−1.43	−1.35	−0.34	−1.38	0.30
76	−0.10	−4.13	−5.12	−2.13	−1.17	−1.24
	1.01	−3.02	−5.40	−0.12	2.47	2.06
77	−0.18	0.29	0.23	0.00	0.06	0.17
	0.98	0.17	1.59	2.62	1.96	4.28
78	0.26	−1.66	−3.03	−1.80	1.04	3.06
	2.50	0.87	2.42	−2.37	1.22	1.05
79	−0.05	1.68	1.70	−0.73	2.59	6.77
	−0.98	−1.71	−2.53	−0.61	3.14	2.96
80	1.01	−3.69	0.45	3.89	1.38	1.57
	−0.08	1.30	0.62	−0.87	−2.11	2.48
81	−4.73	−2.70	−2.45	−4.17	−5.76	−5.09
	−2.92	−0.22	1.42	3.26	0.05	−0.95
82	−2.14	−2.19	−1.96	2.18	−2.97	−1.89
	0.12	−0.76	−0.94	−3.90	−0.86	−2.88
83	−2.58	−2.78	3.30	2.06	−1.54	−1.30
	−1.78	−0.13	−0.20	−1.35	−2.82	−1.97
84	2.25	1.17	−2.29	−2.49	−0.87	−4.15
	−0.53					

0.63 with probability 0.9. The data and sequences of one-step point forecasts from the two models with $\delta = 0.8$ and $\delta = 1.0$ appear in Figure 2.7. As expected, the degree of adaptation to new data increases as δ decreases, leading to more erratic forecast sequences. To compare the models, and hence the suitability of the different discount factors, Table 2.3 displays various summary quantities.

The first two are commonly used measures of forecast accuracy, namely the mean absolute deviation, MAD $= \sum_{t=1}^{115} |e_t|/115$, and mean square error, MSE $= \sum_{t=1}^{115} e_t^2/115$, for the entire series. The third summary is based on the observed predictive density

$$p(Y_{115}, Y_{114}, \ldots, Y_1 \mid D_0) = \prod_{t=1}^{t=115} p(Y_t \mid D_{t-1}),$$

the product of the sequence of one-step forecast densities evaluated at the actual observation, and provides a measure of predictive performance of the model that is actually a likelihood for δ, since the DLMs differ only

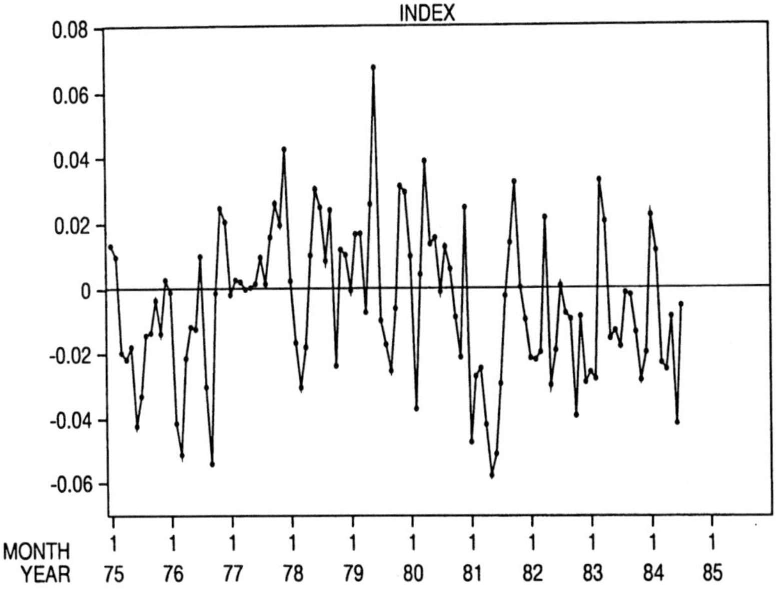

Figure 2.6 USA/UK exchange rate index

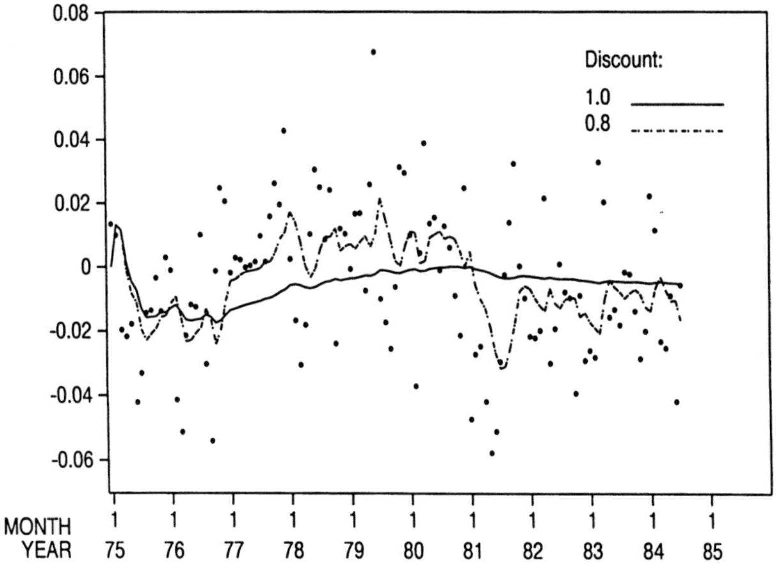

Figure 2.7 USA/UK exchange rate index and forecasts

Table 2.3. Exchange rate Index example

δ	MAD	$\sqrt{\text{MSE}}$	LLR	90% Interval		SD
1.0	0.019	0.024	0.00	−0.009	−0.001	0.025
0.9	0.018	0.022	3.62	−0.024	0.000	0.023
0.8	0.018	0.022	2.89	−0.030	0.002	0.022
0.7	0.018	0.023	0.96	−0.035	0.003	0.021

with respect to their known discount factors. Viewing δ as an uncertain quantity, we could now introduce a prior distribution and hence calculate a posterior for δ. However, for illustrative purposes, just the above likelihoods are examined to obtain a rough data-based guide. For convenience a log scale is used and the measures defined relative to the model with $\delta = 1$. Thus, LLR is the log-likelihood ratio for the values 0.9, 0.8, 0.7 relative to 1.0, larger values indicating a higher degree of support from the data.

From Table 2.3, the MSE and MAD measures indicate better predictive performance for δ between 0.8 and 0.9 than otherwise, the latter more in favour of the smaller value 0.8. The LLR measure, however, favours 0.9 due to the fact that it takes into account the variances of forecast distributions that are ignored by the MAD and MSE. In this measure, there is a balance between forecast accuracy as measured by the e_t sequence and forecast precision as measured by the spread of the predictive distributions. The value of 0.8 for δ leads to more diffuse distributions than that of 0.9, which counts against the model in the LLR measure of predictive performance. In particular, note that by comparison with the others on all three measures of performance, the static model is clearly unsatisfactory.

Further information provided in the table indicates final values of some of the interesting quantities. In particular 90% posterior probability intervals for the final level μ_{115} based on the posterior T distributions are given. The estimated standard deviation SD $= \sqrt{S_{115}}$ is also quoted. Note that as expected, smaller values of δ lead to

(i) faster decay of information about the level between observations and so wider posterior intervals;

(ii) smaller estimates of observational variance, as indicated by the standard deviations.

In this example the differences in the estimates of the observation variance V are not large. In some applications, as illustrated in the next chapter, the observational variance can be markedly over-estimated by models with discount factors too close to 1, leading to much more diffuse forecast distributions than those from models with suitably lower discount values.

2.7 APPENDIX

The exact expression for A_t in the constant model is derived. In the closed, constant model $C_t = A_t V$, $R_t = A_{t-1}V + W$, and $Q_t = R_t + V$. So, with $\lim_{t\to\infty} A_t = A$, $\delta = 1 - A$, and $r = W/V$,

$$A_{t+1} = \frac{C_t + W}{C_t + W + V} = \frac{A_t + r}{A_t + r + 1},$$

with limiting value

$$A = \frac{A + r}{A + r + 1}.$$

Define $u_t = 1/(A_t - A)$ for each t, and note that by subtraction,

$$u_{t+1} = u_t(A_t + r + 1)(A + r + 1).$$

Now, $r = A^2/\delta$, so $A + r + 1 = 1/\delta$ and $A_t + r + 1 = 1/u_t + 1/\delta$, whereupon $\delta^2 u_{t+1} = u_t + \delta$, and so

$$u_t = \frac{\delta(1 - \delta^{2(t-1)}) + u_1(1 - \delta^2)}{(1 - \delta^2)\delta^{2(t-1)}}.$$

Note that $1 - \delta^2 = A(1 + \delta)$ and substitute for $u_1 = 1/(A_1 - A)$ to get

$$A_t - A = \frac{A(A_1 - A)(1 + \delta)\delta^{2(t-1)}}{(1 + \delta^{2t-1})A + (\delta - \delta^{2t-1})A_1}.$$

After rearrangement, the general solution is

$$A_t = A\frac{(1 - \delta^{2t-2})A + (\delta + \delta^{2t-2})A_1}{(1 + \delta^{2t-1})A + (\delta - \delta^{2t-1})A_1}.$$

2.8 EXERCISES

Unless stated otherwise, the exercises relate to the first-order polynomial DLM $\{1, 1, V_t, W_t\}$ with known variances $\{V_t, W_t\}$ and/or discount factor δ and with $D_t = \{Y_t, D_{t-1}\}$:

$$\begin{aligned} Y_t &= \mu_t + \nu_t & \nu_t &\sim \mathrm{N}[0, V_t], \\ \mu_t &= \mu_{t-1} + \omega_t & \omega_t &\sim \mathrm{N}[0, W_t], \\ (\mu_{t-1} \mid D_{t-1}) &\sim \mathrm{N}[m_{t-1}, C_{t-1}]. \end{aligned}$$

(1) Write a computer program to graph 100 simulated observations from the DLM $\{1, 1, 1, W\}$ starting with $\mu_0 = 25$. Simulate several series for each value of $W = 0.05$ and 0.5. From these simulations, become familiar with the forms of behaviour such series can display.

(2) For the DLM $\{1, 1, V_t, W_t\}$ show that
   (a) the posterior precision of $(\mu_t|D_t)$ is the sum of the prior precision of $(\mu_t|D_{t-1})$ and the observation precision of $(Y_t|\mu_t)$,

namely

$$C_t^{-1} = R_t^{-1} + V_t^{-1};$$

(b) the posterior mean of $(\mu_t|D_t)$ is a weighted average of the sum of the prior mean $\mathrm{E}[\mu_t|D_{t-1}]$ and the observation Y_t with weights proportional to the precisions R_t^{-1} and V_t^{-1}, namely

$$m_t = C_t(R_t^{-1}m_{t-1} + V_t^{-1}Y_t).$$

(3) Consider the DLM $\{1, 1, V_t, W_t\}$ extended so that $\nu_t \sim \mathrm{N}[\bar{v}_t, V_t]$ and $\omega_t \sim \mathrm{N}[\bar{w}_t, W_t]$ may have non-zero means. Obtain the recurrence relations for $\{m_t, C_t\}$
 (a) using Bayes' theorem;
 (b) deriving the joint distribution $(\mu_t, Y_t \mid D_{t-1})$ and using normal theory to obtain the appropriate conditional distribution.

(4) Show that the static DLM $\{1, 1, V, 0\}$, is equivalent to the model

$$\begin{aligned}(Y_t|\mu) &\sim \mathrm{N}[\mu, V],\\ (\mu|D_0) &\sim \mathrm{N}[m_0, C_0].\end{aligned}$$

Now suppose that C_0 is very large relative to V, so that $VC_0^{-1} \approx 0$. Show that

(a) $$m_1 \approx Y_1 \quad \text{and} \quad C_1 \approx V;$$

(b) $$m_t \approx \frac{1}{t}\sum_{j=1}^{t} Y_j \quad \text{and} \quad C_t \approx \frac{V}{t}.$$

Comment on these results in relation to classical estimates.

(5) For the constant DLM $\{1, 1, 100, 4\}$, if $(\mu_t|D_t) \sim N[200, 20]$, what are your forecasts of

$$(a)\ (Y_{t+4}|D_t), \quad (b)\ (Y_{t+1} + Y_{t+2}|D_t), \quad (c)\ (Y_{t+3} + Y_{t+4}|D_t)\ ?$$

(6) Suppose that Y_t is a missing observation, so that $D_t = D_{t-1}$. Given $(\mu_{t-1}|D_{t-1}) \sim \mathrm{N}[m_{t-1}, C_{t-1}]$, obtain the distributions of

$$(\mu_t|D_t) \quad \text{and} \quad (Y_{t+1}|D_t).$$

Do this for the constant DLM $\{1, 1, 100, 4\}$ when

$$(\mu_{t-1}|D_{t-1}) \sim \mathrm{N}[200, 40].$$

(7) Bearing in mind the previous question, suggest a method for coping with outliers and general maverick observations with respect to subsequent forecasts.

(8) For the DLM $\{1, 1, V_t, W_t\}$, with $(\mu_t|D_{t-1}) \sim \mathrm{N}[m_{t-1}, R_t]$,
 (a) obtain the joint distribution of $(\nu_t, Y_t|D_{t-1})$.
 (b) Hence prove that the posterior distribution for ν_t is

$$(\nu_t|D_t) \sim \mathrm{N}[(1 - A_t)e_t,\ A_tV_t].$$

(c) Could you have deduced (b) immediately from $(\mu_t|D_t)$?

(9) It is often of interest to perform a retrospective analysis that looks back in time to make inferences about historical levels of a time series based on all the current data. As a simple case, consider inferences about μ_{t-1} based on $D_t = \{Y_t, D_{t-1}\}$.

(a) Obtain the joint distribution $(\mu_{t-1}, Y_t|D_{t-1})$;

(b) hence with $B_{t-1} = C_{t-1}/R_t$ deduce that

$$(\mu_{t-1}|D_t) \sim \mathrm{N}[a_t(-1), R_t(-1)],$$

where

$$a_t(-1) = m_{t-1} + B_{t-1}(m_t - m_{t-1})$$

and

$$R_t(-1) = C_{t-1} - B_{t-1}^2(R_t - C_t).$$

(c) Write these equations for the discount DLM of Section 2.4.2.

(10) For the constant DLM $\{1, 1, V, W\}$, $(\mu_{t-1}|D_{t-1}) \sim \mathrm{N}[m_{t-1}, C_{t-1}]$, suppose that the data recording procedure at times t and $t+1$ is such that Y_t and Y_{t+1} cannot be separately observed, but $X = Y_t + Y_{t+1}$ is observed at $t+1$. Hence $D_t = D_{t-1}$ and $D_{t+1} = \{D_{t-1}, X\}$.

(a) Obtain the distributions of $(X|D_{t-1})$ and $(\mu_{t+1}|D_{t+1})$.

(b) Generalise this result to the case

$$X = \sum_{v=0}^{k} Y_{t+v} \quad \text{and} \quad D_{t+k} = \{X, D_{t-1}\}.$$

(c) For integers j and k such that $0 \le j < j + k$, find the forecast distribution of $\sum_{v=j}^{j+k} Y_{t+v}$ given D_{t-1}.

(11) There is a maxim, "When in doubt about a parameter value err on the side of more uncertainty." To investigate this, repeat the exercise of Example 2.1 using in turn the following prior settings:

(a) $(\mu_0|D_0) \sim \mathrm{N}[650, 100000]$;

(b) $(\mu_0|D_0) \sim \mathrm{N}[130, 4]$;

(c) $(\mu_0|D_0) \sim \mathrm{N}[11, 1]$;

In particular, examine the time graphs of $\{A_t\}$, $\{f_t, f_t \pm Q_t^{1/2}, Y_t\}$, and of $\{m_t, Y_t\}$. What conclusions do you draw? We once designed a more general forecasting system which the customer tried to break by setting priors with silly prior means m_0 and large variances C_0. He drew the conclusion that the system was so robust it could not be broken. How would you show that it could be broken if it were not protected by a monitoring system?

(12) Another maxim is, "In a complete forecast system higher rather than lower values of the discount factor are to be preferred." In-

vestigate this by redoing Example 2.1 using the prior $(\mu_0|D_0) \sim \mathrm{N}[130, 400]$ but employing the discount DLM so that $R_t = C_{t-1}/\delta$. Use in turn the discount factors $\delta = 0.8, 1.0$ and 0.01. In particular, examine time graphs of the $\{f_t, Y_t\}$ in each case. What conclusions do you draw? Do you see any mimicry? Too many systems fall between two stools in trying to select adaptive/discount factors that will not overly respond to random fluctuations yet will quickly adapt to major changes; the result is an unsatisfactory compromise. A complete forecasting system generally chooses high discount factors, usually $0.8 \leq \delta < 0.99$, to capture the routine system movements but relies on a monitoring system to signal major changes that need to be brought to the notice of decision makers and that require expert intervention.

(13) In the constant DLM $\{1, 1, V, W\}$, verify the limiting identities

$$R = AV/(1-A), \quad Q = V/(1-A), \quad W = A^2 Q.$$

(14) In the closed, constant DLM with limiting values A, C, R, etc., prove that the sequence C_t decreases/increases as t increases according to whether C_0 is greater/less than the limiting value C. Show that the sequence A_t behaves similarly.

(15) Discount weighted regression applied to a locally constant process estimates the current level at time t as that value M_t of μ that given $Y_1, \ldots, Y_t$, minimises the discounted sum of squares

$$S_t(\mu) = \sum_{j=0}^{t-1} \delta^j (Y_{t-j} - \mu)^2.$$

(a) Prove that M_t is a discount weighted average of the t observations

$$M_t = \frac{1-\delta}{1-\delta^t} \sum_{v=0}^{t-1} \delta^v Y_{t-v}.$$

(b) Show that writing $e_t = Y_t - M_{t-1}$, neat recurrence forms are

$$M_t = \frac{1-\delta}{1-\delta^t} Y_t + \frac{1-\delta^{t-1}}{1-\delta^t} \delta M_{t-1}$$

and

$$M_t = M_{t-1} + \frac{1-\delta}{1-\delta^t} e_t.$$

(c) Show that as $t \to \infty$ the limiting form of this recurrence relationship is that of Brown's method of EWR, Section 2.3.5(c),

$$M_t = \delta M_{t-1} + (1-\delta) Y_t = M_{t-1} + (1-\delta) e_t.$$

(16) In the context of question (16) on DWR, note that as $t \to \infty$,

$$V[e_t|D_{t-1}] \to Q \quad \text{and} \quad (Y_{t+1} - Y_t - e_{t+1} + \delta e_t) \to 0.$$

This suggests that the process can be modelled as

$$Y_{t+1} - Y_t = a_{t+1} - \delta a_t,$$

where $a_t \sim N[0, Q]$ are independent random variables. Then an estimate of Q given $Y_{t+1}, \ldots, Y_1$ is

$$\hat{Q}(t+1) = \frac{1}{t}\sum_{v=1}^{t}\frac{(Y_{v+1} - Y_v)^2}{1+\delta^2}.$$

(a) Do you consider this a reasonable point estimate of Q?
(b) Show that

$$\hat{Q}(t+1) = \hat{Q}(t) + \frac{1}{t}\left\{\frac{(y_{t+1} - y_t)^2}{1+\delta^2} - \hat{Q}(t)\right\},$$

and that a reasonable point estimate of $V[Y_t|D_{t-1}]$ is

$$\hat{Q}_t = \left\{\delta + \frac{(1-\delta)(1-\delta^t)}{(1-\delta^{t-1})^2}\right\}\hat{Q}(t-1),$$

with $t-1$ degrees of freedom.

(17) In the $\{1, 1, V, W_t\}$ discount DLM with constant discount factor δ, suppose that C_0 is very large relative to V. Show that

(a) $$C_t \approx V(1-\delta)/(1-\delta^t), \qquad \text{for all } t \geq 1;$$

(b) $$m_t \approx \frac{(1-\delta)}{(1-\delta^t)}\sum_{j=0}^{t-1}\delta^j Y_{t-j},$$

(c) $$m_t \approx \frac{1-\delta}{1-\delta^t}Y_t + \frac{1-\delta^{t-1}}{1-\delta^t}\delta m_{t-1},$$

(d) $$m_t \approx m_{t-1} + \frac{1-\delta}{1-\delta^t}e_t.$$

(e) Compare these results with those of the relevant DWR approach in question (16) above. What do you conclude? What do you think about applying that variance estimate $\hat{Q}_t$ of Q, from question (16), to this DLM? If you do adopt the method, what is the corresponding point estimate of V?

(18) In the constant DLM $\{1, 1, V, W\}$, show that $R_t = C_{t-1}/\delta_t$, where δ_t lies between 0 and 1. Thus, the constant DLM updating equations are equivalent to those in a discount DLM with discount factors δ_t changing over time. Find the limiting value of δ_t as t increases, and verify that δ_t increases/decreases with t according to whether the initial variance C_0 lies below/above the limiting value C.

(19) Consider the lead time forecast variance $L_t(k)$ in Section 2.3.6.
  (a) Show that the value of k minimising the lead time coefficient of variation is independent of C_t. What is this value when $V = 97$ and $W = 6$?
  (b) Supposing that $C_t = C$, the limiting value, show that the corresponding value of $L_t(k)/V$ depends only on k and $r = W/V$. For each value of $r = 0.05$ and $r = 0.2$, plot the ratio $L_t(k)/V$ as a function of k over $k = 1, \ldots, 20$. Comment on the form of the plots and the differences between the two cases.

(20) Become familiar with just how heavy-tailed Student T distributions with small and moderate degrees of freedom are relative to normal distributions. To do this graph the distribution using an appropriate computer package and find the upper 90%, 95%, 97.5% and 99% points of the $\mathrm{T}_n[0,1]$ distribution for n =2, 5, 10 and 20 degrees of freedom, comparing these with those of the $\mathrm{N}[0,1]$ distribution. Statistical tables can also be used (Lindley and Scott, 1984, p45).

(21) Perform analyses of the USA/UK exchange rate index series along the lines of those in Section 2.6, one for each value of the discount factor $\delta = 0.6, 0.65, \ldots, 0.95, 1$. Relative to the DLM with $\delta = 1$, plot the MSE, MAD and LLR measures as functions of δ. Comment on these plots. Sensitivity analyses explore how inferences change with respect to model assumptions. At $t = 115$, explore how sensitive this model is to values of δ in terms of inferences about the final level μ_{115}, the variance V and the next observation Y_{116}.

(22) In the DLM $\{1, 1, 1, W\}$, define $Z_t = Y_{t+1} - Y_t$. Show that for integer k such that $|k| > 1$,

$$\mathrm{E}[Z_t] = 0, \quad \mathrm{V}[Z_t] = 2 + W, \quad \mathrm{C}[Z_t, Z_{t-1}] = -1$$

and $\mathrm{C}[Z_t, Z_{t+k}] = 0$. Based upon $n+1$ observations $(Y_1, \ldots, Y_{n+1})$, giving the n values $(Z_1, \ldots, Z_n)$, the usual sample estimate of the autocorrelation coefficient of lag k, $\mathrm{C}[Z_t, Z_{t+k}]/\mathrm{V}[Z_t]$, is

$$r_k = \sum_{i=1}^{n-k} Z_{i+k} Z_i \Big/ \sum_{i=1}^{n} Z_i^2.$$

Using the computer program of question 1, generate 100 values of z_i and plot the sample autocorrelation graph $\{r_k, k : \ k = 0, \ldots, 12\}$ for $W = 0.05$ and also $W = 0.5$. Assuming the model true, the prior marginal distribution of r_k, for every $|k| > 1$, is roughly $\mathrm{N}\left[0, 1/\sqrt{n}\right]$. Do the data support or contradict the model? This is an approach used in identifying the constant DLM and an ARIMA(0,1,1) model.

Supposing the more general DLM $\{1, 1, V_t, W_t\}$, show that again $\mathrm{C}[Z_t, Z_{t+k}] = 0$ for all $|k| > 1$, so the graph $\{r_k, k : \ k > 1\}$ is expected to look exactly the same. Note also that if V_t/W_t is constant, the whole graph $\{r_k, k\}$ is expected to look exactly the same. What is r_k now measuring and what are the implications for identifying the constant DLM and the ARIMA(0,1,1)?

(23) Suppose an observation series $\{Y_t\}$ is generated by the constant DLM $\{1, 1, V^*, W^*\}$. We can write $Y_t - Y_{t-1} = a_t - \delta^* a_{t-1}$ where $a_t \sim \mathrm{N}[0, Q^*]$ are independent random variables and Q^* is the associated limiting one-step forecast variance. In order to investigate robustness, suppose a non-optimal DLM $\{1, 1, V, W\}$ is employed, so that in the limit, $Y_t - Y_{t-1} = e_t - \delta e_{t-1}$ where the errors will have a larger variance Q and no longer be independent. Show that for integer k such that $|k| \geq 1$,

$$Q = \mathrm{V}[e_t] = [1 + (\delta - \delta^*)^2/(1 - \delta^2)]Q^*$$

and

$$\mathrm{C}(k) = \mathrm{C}[e_{t+k}, e_t] = \delta^{|k|-1} Q^* (\delta - \delta^*)(1 - \delta\delta^*)/(1 - \delta^2).$$

Examine graphs of $\{\delta,\ Q/Q^*\}$ and of $\{\delta,\ \mathrm{C}(1)/Q\}$ for the typical practical cases $\delta^* = 0.9$, $\delta^* = 0.8$ and for the atypical case $\delta^* = 0.5$.

CHAPTER 3

INTRODUCTION TO THE DLM: THE DYNAMIC REGRESSION MODEL

3.1 INTRODUCTION

In this chapter some basic concepts underlying the general DLM theory are introduced and developed in the context of dynamic linear regression. The general multiple regression model is discussed, but details of analysis and examples are considered only for the very special case of straight line regression through the origin. Although seemingly trivial, this particular case effectively illustrates the important messages without the technical complications of larger and more practically important models.

Regression modelling concerns the construction of a mathematical and statistical description of the effect of *independent* or *regressor* variables on the *response* time series Y_t. Considering a single such variable, represented by a time series of observations X_t, regression modelling often begins by relating the mean response function μ_t of the original series to X_t, and possibly X_s, for $s < t$, via a particular *regression function*. For example, a simple linear model for the effect of the current X_t on the current mean is

$$\mu_t = \alpha + \beta X_t,$$

where the defining parameters α and β take suitable values. Models of this sort may be used in a variety of prediction, interpolation, estimation and control contexts, such as

(i) Prediction using a *lead* variable, or indicator: For example, for month t, μ_t is the current underlying monthly demand for roofing tiles, Y_t the corresponding observed demand, and X_t the number of new housing starts made nine months previously, in month $t-9$;
(ii) Prediction using a *proxy* variable: For example, in predicting population growth, $X_t = t$ is time itself;
(iii) Control using a *control variable:* For example, the temperature level μ_t of water from a shower can be related to the setting X_t of the tap mixing incoming hot and cold water flows;
(iv) Interpolation: Suppose that the response Y_t represents a measurement of the latitude of a satellite at time t, when interest lies in describing and estimating the trajectory up to that time from $t = 0$, described by $\mu_t = \alpha + \beta t$.

In each of the above cases both Y_t and X_t are scale measurements. However, practically important and interesting models often include *categorical* regressor variables that classify the response into groups according to type, presence or absence of a control, etc. Classificatory regressor variables such as these are treated just like measured values of scale variables. In some

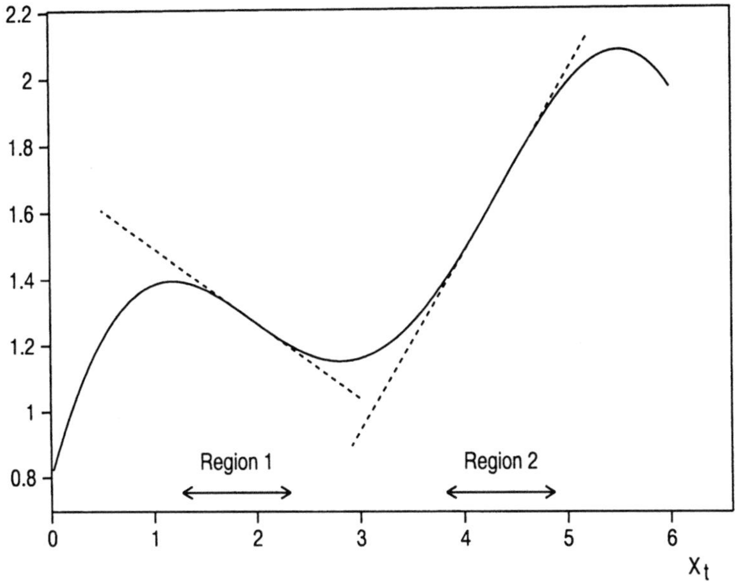

Figure 3.1 Local linearity of μ_t as a function of X_t

applications the Y_t series will also be a series of discrete, categorical variables, so that normal models are clearly inappropriate. Non-normal models for this sort of problem are developed in a later chapter.

In practice, model construction is guided by specified objectives. Rarely is a modeller seeking to establish an all embracing model purporting to represent a "true" relationship between the response and regressor. Rather, the model is a way of looking at the problem that is required to capture those features of importance in answering specific questions about the relationship. A frequent objective, for example, is short-term forecasting of the response series.

Suppose that there really is an underlying, unknown and complex relationship $f(\mu_t, X_t, t) = c$ between the level of the series μ_t, the regressor variable X_t, and time itself. If the modeller believes that this relationship is sufficiently *smooth* and well-behaved locally as a function of both X_t and t, then for short-term prediction, that is, *local* inference, a local approximating model of the form $\mu_t = \alpha_t + \beta_t X_t$ may well satisfy the specific objectives. Note the word *form* and the t index of the coefficients α and β. The form of a linear model may adequately describe the *qualitative* local characteristics of the relationship for all X_t and t, but the *quantification* of this form may well have to change according to the locality. For illustration suppose that the true relationship between the mean response and the regressor is as given in Figure 3.1 for all t.

It can be seen that the general form $\mu = \alpha + \beta X$ is always locally appropriate as represented by the tangent at X. However, in region 1, β must be negative, whereas in region 2 it is clearly positive. Similarly, the intercept coefficient α differs markedly between the two regions.

The distinction between an appropriate local model form and an appropriate quantified local model is a critical, fundamental concept in dynamic modelling. Often in practice, the values of the independent variable X_t change rather slowly in time, so that an appropriate local model description is that above, namely

$$\mu_t = \alpha_t + \beta_t X_t,$$

where the parameter values vary only slightly from one time point to the next. In modelling this quantitative variation, the modeller develops a dynamic model. For example, a simple important representation of slowly evolving parameters is that of a random walk, with

$$\mathrm{E}\left[\alpha_t \mid \alpha_{t-1}, \beta_{t-1}\right] = \alpha_{t-1}$$

and

$$\mathrm{E}\left[\beta_t \mid \alpha_{t-1}, \beta_{t-1}\right] = \beta_{t-1},$$

together with a measure of the variance associated with these changes. This random walk model is a characteristic of the regression DLM.

Modellers must continually be aware that model building is always a selective process. Often there are several, if not many, independent variables that may be considered as useful and important predictors of Y_t. The modeller typically identifies just a few of these variables that are judged to be of greatest importance. Those that are judged unimportant, or indeed, of which the modeller is not conscious, are omitted from the model, with the result that their effects are either carried via those regressors in the model or, commonly, lumped together into error terms with some broad statistical description. It is important that this selectivity be identified as potentially contributing to inadequacies in the chosen model.

As a simple example suppose that Y_t is the percentage yield of a chemical process that operates under different temperature and pressure controls. Over the operating region, the mean yield μ_t is related to temperature X_t and pressure Z_t according to

$$\mu_t = 80 - (X_t - 100)^2 - 2(X_t - 100)(Z_t - 2) - 10(Z_t - 2)^2. \tag{3.1}$$

This relationship, shown in Figure 3.2, is typical of the elliptical nature of a yield response function of temperature and pressure in the region of the optimum operating conditions. In this case, mean yield is maximised at $X_t = 100$ and $Z_t = 2$.

Consider now two chemical plants running the same process. The first operates with constant pressure $Z_t = 1$ for which, from above,

$$\mu_t = 70 - (X_t - 100)^2 + 2(X_t - 100), \tag{3.2}$$

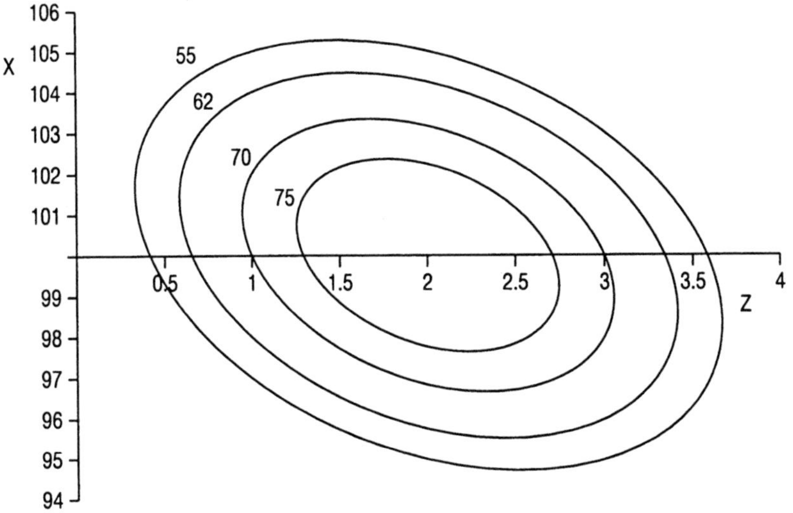

Figure 3.2 Contours of quadratic mean response function

so that the effect of raising temperature from 98° to 102° is to *increase* the mean yield by about 13% from 62 to 70. The second plant operates with constant pressure $Z_t = 3$, so that

$$\mu_t = 70 - (X_t - 100)^2 - 2(X_t - 100), \tag{3.3}$$

and here the effect of the above temperature increase is to *decrease* the mean yield by about 11% from 70 to 62! Suppose that each plant currently operates with $X_t = 100$, so that they both have a mean yield of 70%. Ignoring the difference in pressure settings will lead to a sharp difference of opinion between plant managers on the effect of raising temperature; one will claim it to be beneficial, the other detrimental. Consideration of pressure as a contributory factor clearly identifies the source of confusion and conflict.

The general point of this example is that all models are conditional, although the conditions under which they are constructed are often *not* explicit. Thus, quite frequently, various conflicting models may be proposed for equivalent situations, each supported by empirical evidence and associated statistical tests. The truth may be that although each model is *conditionally* correct, the conditioning may be so restrictive as to render the models practically useless.

Further study of the example provides valuable insight and a pointer to possible useful modification of simple conditional models. Consider the operating model for the first plant (3.2). If, due to uncontrollable circumstances, pressure begins to increase, moving from $Z_t = 1$ to $Z_t = 3$, then the model becomes invalid and practically misleading. In this and more complex situations, and particularly in modelling open systems, the lack of awareness of the variation in, and the interaction with, excluded or unidentified variables causes confusion and misleading inferences. Simple modifications of conditional models to provide a small degree of flexibility of response to changing external conditions and variables are possible. In the example, the operating model may be rewritten as

$$\mu_t = \alpha(Z_t) + \beta(Z_t)X_t + \gamma X_t^2,$$

where the coefficients $\alpha(\cdot)$ and $\beta(\cdot)$ (and in more complex situations γ too) are functions of pressure Z_t. If it is assumed that pressure changes only slowly in time, then the concept of local modelling described earlier suggests that to account for some of the variability due to the unidentified pressure variable (and possibly others too), a simple local model would be that above with coefficients $\alpha(\cdot)$ and $\beta(\cdot)$ replaced by time-varying quantities, to give

$$\mu_t = \alpha_t + \beta_t X_t + \gamma X_t^2.$$

A simple dynamic model for the coefficients, such as a random walk, will now provide a means of responding and adapting to changes in underlying conditions and related variables.

It is worth exploring the distinction between the qualitative, that is, the local model form, and the quantitative by reference to Taylor series expansions. Temporarily suppressing the dependence on time, suppose there exists some unknown, complex, but smooth underlying relationship between the mean response function μ and an independent variable X, of the form $\mu = f(X, Z)$, where Z represents a set of possibly many related but omitted variables. For any given Z, the functional dependence on X can be represented locally in a neighbourhood of any point X_0 by the form

$$\mu = f(X_0, Z) + \frac{\partial f(X_0, Z)}{\partial X_0}(X - X_0),$$

or

$$\mu = \alpha_0(Z) + \beta_0(Z)X.$$

In considering the use of a linear regression on X for the response function over time, it is clear that two factors are important in assessing the worth of such a model:

(a) the adequacy of the linear approximation as a function of X for any given Z; and
(b) the implicit assumption of constancy as a function of Z.

Situations in which X and Z vary slowly over time, or in which Z varies slowly and the response is close to linear in X for given Z, provide the most satisfactory cases. Here the response function can be adequately represented as

$$\mu_t = \alpha_t + \beta_t X_t,$$

with the coefficients changing slowly over time according to the simple random walk

$$\begin{pmatrix}\alpha_t\\ \beta_t\end{pmatrix} = \begin{pmatrix}\alpha_{t-1}\\ \beta_{t-1}\end{pmatrix} + \omega_t,$$

where ω_t is a zero-mean random vector. This expresses the concept of *local constancy* of the parameters, subject to variation controlled by the variance matrix of ω_t, say $V[\omega_t] = \mathbf{W}_t$. Clearly, small values of $\mathbf{W}_t$ imply a stable linear function over time, larger values leading to greater volatility and suggesting caution in extrapolating or forecasting too far ahead in time based on the current quantified linear model. The common static regression model is obviously the special case of this dynamic regression in which $\omega_t = \mathbf{0}$ for all t.

The foregoing discussion has identified some of the potential dangers in employing simple static models and also reveals reasons why they often prove inadequate in practice. A final point concerns the suggestion that the above type of dynamic model is likely to break down if there happens to be a large, abrupt change in either X or Z, or both. In the chemical plant example, a sudden change in pressure from 1 to 3 leads to an enormous shift in the locally appropriate pair of coefficients α and β, with the latter even reversing its sign. However, in spite of this marked quantitative change, it is still the case that the qualitative form is locally durable with respect to large changes in either X or Z. Furthermore, the changes in parameter values may be estimated and adapted to by introducing larger variances, $\mathbf{W}_t$, for abrupt change points. In later chapters models that can cope with discontinuous changes in an otherwise smooth process are a prominent feature of our practical approach.

3.2 THE MULTIPLE REGRESSION DLM

For reference, before considering the case of a single regressor variable in detail, the structure of the general dynamic regression model is specified. Suppose that n regressor variables are identified and labelled $X_1, \ldots, X_n$. The value of the ith variable X_i at time t is denoted by X_{ti}, with the convention that a constant term is represented by $X_{t1} = 1$, for all t. The regression DLM is now defined.

Definition 3.1. For each t, the model is defined by

Observation equation:	$Y_t = \mathbf{F}'_t\,\boldsymbol{\theta}_t + \nu_t\,,$	$\nu_t \sim \mathrm{N}[0, V_t],$
System equation:	$\boldsymbol{\theta}_t = \boldsymbol{\theta}_{t-1} + \boldsymbol{\omega}_t\,,$	$\boldsymbol{\omega}_t \sim \mathrm{N}[\mathbf{0}, \mathbf{W}_t],$

where $\mathbf{F}_t = (X_{t1}, \ldots, X_{tn})'$ is the **regression vector,** $\boldsymbol{\theta}_t$ is the $n \times 1$ **regression parameter vector**, and $\mathbf{W}_t$ is the **evolution variance matrix** for $\boldsymbol{\theta}_t$.

The standard static regression model has the above form with $\mathbf{W}_t = \mathbf{0}$ for all t, so that $\boldsymbol{\theta}_t = \boldsymbol{\theta}$ is constant in time. Following the introductory discussion, the regression DLM assumes that the linear regression form is only locally appropriate in time, with the regression parameter vector varying according to a random walk. The evolution error term $\boldsymbol{\omega}_t$ describes the changes in the elements of the parameter vector between times $t-1$ and t. The zero mean vector reflects the belief that $\boldsymbol{\theta}_t$ is *expected* to be constant over the interval, whilst the variance matrix $\mathbf{W}_t$ governs the extent of the movements in $\boldsymbol{\theta}_t$ and hence the extent of the time period over which the assumption of local constancy is reasonable. Finally, the error sequences ν_t and $\boldsymbol{\omega}_t$ are each assumed to be independent sequences. Additionally, ν_t is independent of $\boldsymbol{\omega}_s$ for all t and s.

An important special case obtains when $n = 1$ and a constant term is included in the model. The result is a straight line regression on $X = X_t$ specified by $\mathbf{F}_t = (1, X_t)'$ and $\boldsymbol{\theta}_t = (\alpha_t, \beta_t)'$. Then

$$\begin{aligned} Y_t &= \alpha_t + \beta_t X_t + \nu_t, \qquad \nu_t \sim \mathrm{N}[0, V_t], \\ \alpha_t &= \alpha_{t-1} + \omega_{t1}, \\ \beta_t &= \beta_{t-1} + \omega_{t2}, \end{aligned}$$

where $\boldsymbol{\omega}_t = (\omega_{t1}, \omega_{t2})' \sim \mathrm{N}[0, \mathbf{W}_t]$.

3.3 DYNAMIC STRAIGHT LINE THROUGH THE ORIGIN

3.3.1 Introduction and definition

For illustrative purposes the simple dynamic straight line through the origin is considered. Formally, this is a special case of the straight line model for which $\alpha_t = 0$ and $\boldsymbol{\theta}_t = \theta_t = \beta_t$ for each t. Thus it is assumed that a straight line passing through the origin models the local relationship, but that in different localities the appropriate slope values differ. For illustration, three data sets are now given and examined later in Sections 3.3.4 and 3.4.2.

In Table 3.1, for $t = 1970$ to 1982, the response series Y_t is the USA total annual milk production, and F_t the total number of milk cows, with θ_t representing the average annual milk output per cow in year t. The model is not of primary interest for forecasting the Y_t series but rather for assessing changing productivity over time.

Table 3.2 concerns a leading indicator that is used to forecast annual product sales. Here Y_t is the change in annual sales between years $t-1$ and t, and F_t is the lead indicator measured as the change in an industrial production index between the years $t-2$ and $t-1$.

Table 3.1. USA annual milk production and milk cows

Year t	1970	1971	1972	1973	1974	1975	1976
Y_t: Milk (lbs $\times 10^9$)	117.0	118.6	120.0	115.5	115.6	115.4	120.2
F_t: Cows $\times 10^6$	12.0	11.8	11.7	11.4	11.2	11.1	11.0
Year t	1977	1978	1979	1980	1981	1982	
Y_t: Milk (lbs $\times 10^9$)	122.7	121.5	123.4	128.5	130.0	135.8	
F_t: Cows $\times 10^6$	11.0	10.8	10.7	10.8	10.9	11.0	

Table 3.2. Change in annual sales and industrial production

Year t	1	2	3	4	5	6	7	8	9	10	11	12	13	14
Y_t	12	11	9	5	3	0	-5	-7	-6	-3	7	10	13	12
F_t	4	4	3	2	1	-1	-3	-4	-3	-1	2	3	4	4

Table 3.3. Company sales/ Total market data

	Company Sales Y				Total Market F			
	Quarter				Quarter			
Year	1	2	3	4	1	2	3	4
1975	71.2	52.7	44.0	64.5	161.7	126.4	105.5	150.7
1976	70.2	52.3	45.2	66.8	162.1	124.2	107.2	156.0
1977	72.4	55.1	48.9	64.8	165.8	130.8	114.3	152.4
1978	73.3	56.5	50.0	66.8	166.7	132.8	115.8	155.6
1979	80.2	58.8	51.1	67.9	183.0	138.3	119.1	157.3
1980	73.8	55.9	49.8	66.6	169.1	128.6	112.2	149.5
1981	70.0	54.8	48.7	67.7	156.9	123.4	108.8	153.3
1982	70.4	52.7	49.1	64.8	158.3	119.5	107.7	145.0
1983	70.0	55.3	50.1	65.6	155.3	123.1	109.2	144.8
1984	72.7	55.2	51.5	66.2	160.6	119.1	109.5	144.8
1985	75.5	58.5			165.8	127.4		

The final data set, in Table 3.3, displays a company's quarterly sales, Y_t in standardised units, relative to the total market sales F_t, over the years 1975 to mid-1985. Primary interest lies in assessing the way their relationship has changed over time, and in forecasting one year ahead. A major feature of this data set is the marked annual seasonal pattern exhibited by each

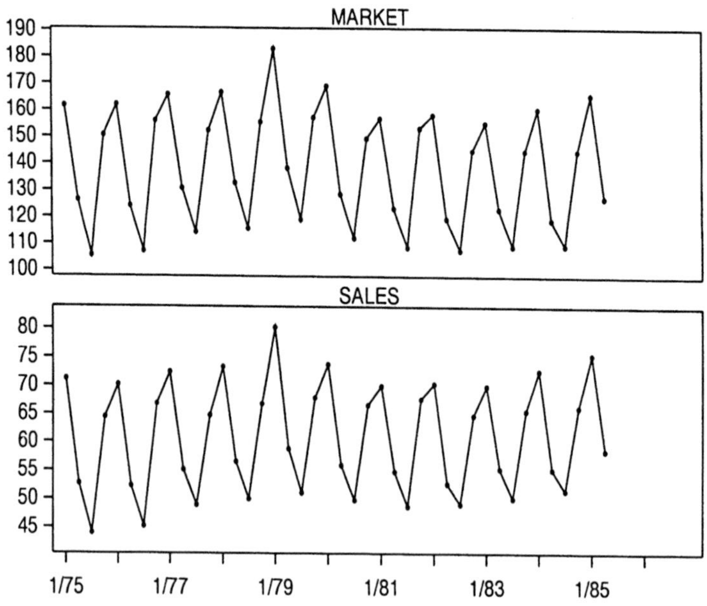

Figure 3.3 Company sales and total market series

of the series over the year. The two series are plotted over time in Figure 3.3. A simple scatter plot of Y_t versus F_t, given in Figure 3.4, removes this seasonality and seems to support a simple, essentially static straight line regression with, from the nature of the data, zero origin and with θ_t representing the market share as a proportion of the total market.

Definition 3.2. The model form is a special case of Definitions 1.1 and 3.1, being characterised by the quadruple $\{F_t, 1, V_t, W_t\}$ as

$$\begin{aligned} &\text{Observation equation:} && Y_t = F_t\theta_t + \nu_t, && \nu_t \sim \mathrm{N}[0, V_t], \\ &\text{System equation:} && \theta_t = \theta_{t-1} + \omega_t, && \omega_t \sim \mathrm{N}[0, W_t], \\ &\text{Initial information:} && (\theta_0 \mid D_0) \sim \mathrm{N}[m_0, C_0], && \end{aligned}$$

for some mean m_0 and variances C_0, V_t and W_t.

The sequential model description for the series requires that the defining quantities at time t be known at that time. Similarly, when forecasting more than one step ahead to time $t+k$ at time t, the corresponding quantities F_{t+k}, V_{t+k}, and W_{t+k} must belong to the current information set D_t. In general for this chapter, and unless otherwise specified, it will be assumed that the information set D_0 contains all the future values of F_t, V_t, and W_t, so that $D_t = \{Y_t, D_{t-1}\}$ for each t. Finally there is the usual assumption of mutual independence of the error sequences and D_0.

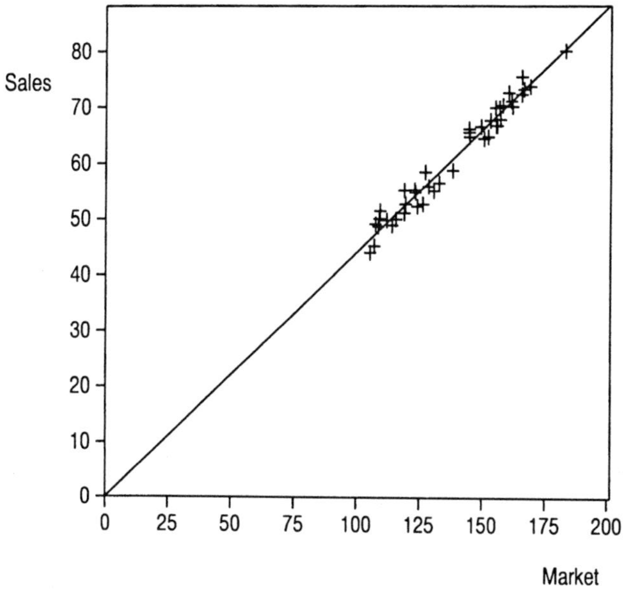

Figure 3.4 Company sales versus total market sales

3.3.2 Updating and forecasting equations

Theorem 3.1. *One-step forecast and posterior distributions are given, for each t, as follows:*

(a) Posterior for θ_{t-1} :
$$(\theta_{t-1} \mid D_{t-1}) \sim \mathrm{N}[m_{t-1}, C_{t-1}],$$
for some mean m_{t-1} and variance C_{t-1}.

(b) Prior for θ_t :
$$(\theta_t \mid D_{t-1}) \sim \mathrm{N}[m_{t-1}, R_t],$$
where $R_t = C_{t-1} + W_t$.

(c) 1-step forecast :
$$(Y_t \mid D_{t-1}) \sim \mathrm{N}[f_t, Q_t],$$
where $f_t = F_t m_{t-1}$ and $Q_t = F_t^2 R_t + V_t$.

(d) Posterior for θ_t :
$$(\theta_t \mid D_t) \sim \mathrm{N}[m_t, C_t],$$
with $m_t = m_{t-1} + A_t e_t$ and $C_t = R_t V_t / Q_t$,

where $A_t = R_t F_t / Q_t$ and $e_t = Y_t - f_t$.

Proof. The proof is by induction, following that of Theorem 2.1. At any time $t > 1$ assume that (a) is true, noting that this is the case at $t = 1$.

Writing Y_t and θ_t in terms of linear functions of θ_{t-1}, ν_t and ω_t, it follows that Y_t and θ_t have a bivariate normal distribution conditional on D_{t-1}, with means and variances as stated above, establishing (b) and (c). For the covariance,

$$\begin{aligned}C[\theta_t, Y_t|D_{t-1}] &= C[\theta_t, F_t\theta_t + \nu_t|D_{t-1}] = C[\theta_t, F_t\theta_t|D_{t-1}] + C[\theta_t, \nu_t|D_{t-1}]\\ &= V[\theta_t|D_{t-1}]F_t + 0 = R_tF_t,\end{aligned}$$

so that

$$\left(\begin{matrix}\theta_t\\ Y_t\end{matrix}\middle| D_{t-1}\right) \sim N\left[\begin{pmatrix}m_{t-1}\\ f_t\end{pmatrix}, \begin{pmatrix}R_t & R_tF_t\\ F_tR_t & Q_t\end{pmatrix}\right].$$

The regression coefficient of θ_t on Y_t is then $A_t = R_tF_t/Q_t$. Hence, using normal theory from Section 17.2,

$$(\theta_t \mid Y_t, D_{t-1}) \sim N[m_t, C_t],$$

where

$$m_t = m_{t-1} + A_t(Y_t - f_t)$$

and

$$C_t = R_t - (R_tF_t)^2/Q_t.$$

This latter equation reduces to $C_t = R_tV_t/Q_t$ and (d) follows.

◇

For forecasting at time t, the forecaster requires the k-step ahead marginal distributions, $p(Y_{t+k} \mid D_t)$, which are as follows.

Theorem 3.2. *For $k > 0$, the k-step ahead forecast distributions are*

$$(Y_{t+k} \mid D_t) \sim N[f_t(k), Q_t(k)]$$

and

$$(\theta_{t+k} \mid D_t) \sim N[m_t, R_t(k)],$$

where

$$f_t(k) = F_{t+k}m_t,$$

$$R_t(k) = C_t + \sum_{r=1}^{k} W_{t+r},$$

and

$$Q_t(k) = F_{t+k}^2R_t(k) + V_{t+k}.$$

Proof. From the evolution equation for θ_t, for $k \geq 1$,

$$\theta_{t+k} = \theta_t + \sum_{r=1}^{k} \omega_{t+r} ,$$

which, together with the observational equation, gives

$$Y_{t+k} = F_{t+k}\theta_t + F_{t+k} \sum_{r=1}^{k} \omega_{t+r} + \nu_{t+k}.$$

Since all terms are normal and mutually independent, $(Y_{t+k} \mid D_t)$ is normal and the mean and variance follow directly as the sums of the means and variances respectively.

◇

3.3.3. General comments

The following points are noteworthy.

(i) The posterior mean m_t is obtained by correcting the prior mean m_{t-1} with a term proportional to the forecast error e_t. The coefficient $A_t = R_t F_t / Q_t$ scales the correction term according to the relative precisions of the prior and likelihood, as measured by R_t/Q_t, and by the regressor value F_t. So the correction always takes the sign of F_t and may be unbounded.

(ii) The posterior precision C_t^{-1} is

$$C_t^{-1} = Q_t(R_t V_t)^{-1} = R_t^{-1} + F_t^2 V_t^{-1},$$

so that for $F_t \neq 0$, it always exceeds the prior precision R_t^{-1}. Thus, the posterior for θ_t is never more diffuse than the prior. Further, the precision increases with $|F_t|$. If, however, $F_t = 0$, then Y_t provides no information on θ_t, and $C_t = R_t$. If $F_t = 0$ for a sequence of observations then the sequence C_t continues to grow by the addition of further W_t terms, reflecting an increasingly diffuse posterior. Thus, although there may exist an appropriate regression relationship changing in time, information relevant to this relationship is not forthcoming. Although seemingly trivial, this point is vital when considering the case of multiple regression. The chemical plant operation of Section 3.1 is a case in point. In view of concern about falling short of set production targets, plant managers are usually wary of varying operating conditions away from standard, well-used control conditions. In such cases, the absence of planned variation in operating conditions means that no new information is obtained about the effect on chemical yield of changing these conditions. As described in Section 3.1, the yield relationships with controlling factors change in time and thus actual optimum operating conditions can move away from those initially identified. The lack of incoming information about the effects of changing variables, such as temperature, together with a decreasing precision associated with the model parameters, means that the actual operating conditions will become far from optimal. The need for a continual flow of information on the effects of changes in operating conditions was recognised by G.E.P. Box when at Imperial Chemical Industries. This led to the development of Evolutionary Operation, which advocates continued small variation in conditions near the currently identified optimum so that movements away from this can be identified (Box and Draper 1969).

(iii) Consider the special case of constant variances, $V_t = V$ and $W_t = W$ for each t. In general, the sequence C_t is neither monotonic nor convergent. However, in the degenerate constant model given by $F_t = F \neq 0$, the corresponding model for the scaled series Y_t/F is a constant, first-order polynomial model with observational variance V/F^2. It follows from Theorem 2.3 that

$$\lim_{t\to\infty} A_t = A(F)$$

and

$$\lim_{t\to\infty} C_t = C(F),$$

where, with $r(F) = WF^2/V$,

$$A(F) = r(F)[\sqrt{1+4/r(F)} - 1]/2$$

and

$$C(F) = A(F)V/F^2.$$

Consequently, in the general case of bounded regressor values, where say $a < F_t < b$ for all t, as t increases, A_t will lie in the interval $[A(a), A(b)]$ and C_t in the interval $[C(c), C(d)]$, where $c = \max(|a|, |b|)$ and $d = \min\{|u| : u \in [a, b]\}$.

(iv) For the static model in which $W_t = 0$ for all t, θ is constant, so that

$$C_t^{-1} = C_0^{-1} + \sum_{r=1}^{t} F_r^2 V_r^{-1}$$

and

$$m_t = C_t C_0^{-1} m_0 + C_t \sum_{r=1}^{t} F_r^2 V_r^{-1} Y_r.$$

This is the standard posterior distribution derived from a non-sequential analysis of the constant regression model. In the particular cases of either large t or relatively little prior knowledge, as represented by small C_0^{-1}, m_t is approximately the usual maximum likelihood point estimate of θ, and C_t the associated variance.

3.3.4 Illustrations

The model is now illustrated by application to the first two data sets introduced above. The objective is simply to demonstrate the analysis and highlight practically interesting features. Reasonably uninformative priors are chosen at $t = 0$ for the purpose of illustration.

EXAMPLE 3.1. For convenience the years of this data set, in Table 3.1, are renumbered 1 to 13. The constant variance model with $V = 1$ and $W = 0.05$ is applied. θ_t may be interpreted as the level, in thousands of pounds, of milk per cow for year t. Initially we set $m_0 = 10$ and $C_0 = 100$, representing a high degree of uncertainty about θ_0. Table 3.4a gives the values of m_t, C_t and A_t calculated sequentially according to Theorem 3.1 (d).

The m_t series is seen to increase except at $t = 4$. This reflects increasing efficiency in general dairy production over time through better management and improving breeds of cow. In practice, a modeller would wish to incorporate this feature by modelling "growth" in θ_t, the current, constant model obviously being deficient. However, even though it is unsatisfactory for long-term prediction, the dynamic nature of this simple model does lead to reasonable short-term forecasts. For example, at $t = 11$ and $t = 12$, the one-step ahead forecasts are $(Y_{12} \mid D_{11}) \sim \mathrm{N}[129.2, 7.8]$ and $(Y_{13} \mid D_{12}) \sim \mathrm{N}[131.1, 7.9]$. The actual observations $Y_{12} = 130.0$ and $Y_{13} = 135.8$ are well within acceptable forecast limits. By comparison, Table 3.4b gives results from the standard static model with $W_t = 0$. It is clear that forecasts from this model are totally unsatisfactory, with, for example, $(Y_{12} \mid D_{11}) \sim \mathrm{N}[116.1, 1.09]$ and $(Y_{13} \mid D_{12}) \sim \mathrm{N}[118.2, 1.08]$. The beneficial effects of assuming a model form holding only locally rather than globally are clearly highlighted. The dynamic assumption leads to greater robustness and helps to compensate for model inadequacies that at first may not be anticipated or noticed. Since the independent variable F_t varies between 10.7 and 12.0, it is not surprising that in the dynamic model the values of C_t and A_t settle down to vary within narrow regions. By contrast, in the static model they both decay to zero, so that the model responds less and less to the most recent data points and movements in milk productivity.

EXAMPLE 3.2. Leading indicator. The sales data in Table 3.2 are analysed with the constant model in which $V = 1$ and $W = 0.01$. The initial, relatively diffuse prior sets $m_0 = 2$ and $C_0 = 0.81$. Table 3.5 gives the values of m_t, C_t and A_t. Neither the variance nor the adaptive coefficient are monotonic, and there is no convergence. Here F_t takes negative and positive values, and A_t takes the sign of the current regressor value. When the

Table 3.4. Analyses for milk data

			(a) Dynamic model			(b) Static model		
t	F_t	Y_t	m_t	$1000C_t$	$100A_t$	m_t	$1000C_t$	$100A_t$
1	12.0	117.0	9.75	6.94	8.3	9.75	6.94	8.3
2	11.8	118.6	10.01	6.38	7.5	9.90	3.53	4.2
3	11.7	120.0	10.23	6.47	7.6	10.01	2.38	2.8
4	11.4	115.6	10.14	6.77	7.7	10.04	1.82	2.1
5	11.2	115.6	10.30	6.99	7.8	10.09	1.48	1.7
6	11.1	115.4	10.38	7.10	7.9	10.14	1.25	1.4
7	11.0	120.2	10.86	7.22	7.9	10.24	1.09	1.2
8	11.0	122.7	11.12	7.22	8.1	10.35	0.96	1.1
9	10.8	121.5	11.23	7.46	8.1	10.44	0.86	0.9
10	10.7	123.4	11.49	7.58	8.1	10.54	0.79	0.8
11	10.8	128.5	11.85	7.46	8.1	10.65	0.72	0.8
12	10.9	130.0	11.92	7.34	8.0	10.75	0.66	0.7
13	11.0	135.8	12.29	7.22	7.9	10.87	0.61	0.7

latter is near zero, C_t tends to increase, since the information provided by Y_t is not compensating for the increased uncertainty about the regression in the movement from θ_{t-1} to θ_t. On the other hand, when F_t increases in absolute value, the observations are very informative, leading to decreases in C_t and increases in the absolute value of A_t.

Looking at m_t, it begins in the region of 3, then drops near to 2 before rising again to 3, apparently completing a cycle. Since F_t varies in a similar manner, there is a suggestion that a large part of this variation in m_t could be accounted for by relating it to F_t, and a more elaborate model is worth considering. A simple, tentative example would be to extend the regression to include a quadratic term in F_t, thus implying a multiple regression rather than the simple straight line here.

One further point that may be queried is that this simple model deals with *changes* in both response and regressor variables, rather than their original values. Typically this will lead to time series that are not conditionally independent, as the model implies. If, for example, $Y_t = U_t - U_{t-1}$, then Y_t and Y_{t-1} have U_{t-1} in common and so will be dependent. The reason why many difference models are used in classical forecasting approaches is that there is a desire to use static, stationary models that would be more reasonable for the differenced series Y_t than for U_t, on the basis that dynamic changes in the latter that cannot be handled by static models can be partially eliminated by differencing. Although this reasoning may be validated in some cases, it is much better and sounder to develop dynamic, stochastic, explanatory models directly for the undifferenced data U_t.

Table 3.5. Analysis of annual sales data

t	1	2	3	4	5	6	7
F_t	4	4	3	2	1	−1	−3
Y_t	12	11	9	5	3	0	−5
m_t	2.93	2.84	2.88	2.83	2.84	2.69	2.33
$100C_t$	5.81	3.26	3.08	3.51	4.31	5.04	3.92
$10A_t$	2.32	1.30	0.92	0.70	0.43	−0.50	−1.17
t	8	9	10	11	12	13	14
F_t	−4	−3	−1	2	3	4	4
Y_t	−7	−6	−3	7	10	13	12
m_t	2.08	2.06	2.09	2.31	2.63	2.88	2.93
$100C_t$	2.75	2.80	3.67	3.93	3.42	2.59	2.28
$10A_t$	−0.84	−0.84	−0.37	0.79	1.02	1.03	0.91

3.4. MODEL VARIANCES AND SUMMARY

3.4.1 Summary of updating and forecasting equations

Estimation of the observational variance and assignment of values to the evolution variance series are problems discussed in detail for the general DLM in later chapters. With a constant observational variance, the analysis is a minor generalisation of that for the first-order polynomial DLM as developed in Section 2.5. The corresponding results are summarised in the following table. The observational variance, if assumed unknown but constant, is estimated using the fully conjugate Bayesian analysis based on gamma prior/posterior distributions for the precision parameter.

From the DLM observation equation it is clear that W_t is not invariant to the scale of measurement of the independent variable F_t. Because of this, and the fact that the amount of information about θ_t conveyed by an observation varies with $|F_t|$, there is difficulty in assigning suitable values to the evolution variance sequence. If as in Section 2.4.2, a discount approach is used with discount factor $0 < \delta < 1$, then $W_t = C_{t-1}(\delta^{-1}-1)$ is naturally defined as a multiple of the variance C_{t-1}, so that $R_t = C_t/\delta$.

Section 4.5 deals with the general DLM of which this is a particular case, and the reader is referred to that section for further details, which are easily translated for this simple DLM.

The summary table is completely analogous to that of Section 2.5.3 for the first-order polynomial model; n_t is the degree-of-freedom parameter for the posterior distribution of the scale ϕ, increasing by one for each observation, and S_t is the point estimate of V.

Regression DLM with Zero Origin and Constant Variance

Observation: $Y_t = F_t\theta_t + \nu_t, \qquad \nu_t \sim \mathrm{N}[0, V],$

System: $\theta_t = \theta_{t-1} + \omega_t, \qquad \omega_t \sim \mathrm{T}_{n_{t-1}}[0, W_t].$

Information:

$$(\theta_{t-1} \mid D_{t-1}) \sim \mathrm{T}_{n_{t-1}}[m_{t-1}, C_{t-1}],$$
$$(\phi \mid D_{t-1}) \sim \mathrm{G}\left[\tfrac{n_{t-1}}{2}, \tfrac{n_{t-1}S_{t-1}}{2}\right].$$

Forecast:

$$(\theta_t \mid D_{t-1}) \sim \mathrm{T}_{n_{t-1}}[m_{t-1}, R_t],$$
$$(Y_t \mid D_{t-1}) \sim \mathrm{T}_{n_{t-1}}[f_t, Q_t],$$

$$R_t = C_{t-1} + W_t, \quad f_t = F_t m_{t-1}, \quad \text{and} \quad Q_t = F_t^2 R_t + S_{t-1}.$$

Updating Recurrence Relationships

Writing $e_t = Y_t - f_t$, and $A_t = F_t R_t / Q_t$,

(i) $(\phi \mid D_t) \sim \mathrm{G}\left[\tfrac{n_t}{2}, \tfrac{n_t S_t}{2}\right],$

$$n_t = n_{t-1} + 1,$$
$$S_t = S_{t-1} + (e_t^2/Q_t - 1)S_{t-1}/n_t.$$

(ii) $(\theta_t \mid D_t) \sim \mathrm{T}_{n_t}[m_t, C_t],$

$$m_t = m_{t-1} + A_t e_t,$$
$$C_t = R_t S_t / Q_t.$$

3.4.2 Example

The company sales/total market data set of Table 3.3 and Figure 3.3 is used to illustrate the sequential analysis and also to highlight some further practical points. Sales are plotted against total market data in Figure 3.4, demonstrating an apparently stable linear relationship. θ_t is the expected market share of the company in month t, the prior estimate of θ_0 at $t = 0$ being $m_0 = 0.45$, in line with pre-1975 information, and the associated uncertainty being represented by $C_0 = 0.0025$. A loose prior for the observational precision is assigned by setting $n_0 = S_0 = 1$. This implies that $(\theta_0 \mid D_0) \sim \mathrm{T}_1[0.45, 0.0025]$, with, in particular, 90% prior probability that θ_0 lies between 0.13 and 0.77, symmetrically about the mode 0.45.

It is of interest to consider several analyses of the data using this model and initial prior, differing only through the value of the discount factor δ. Consider first an analysis with $\delta = 0.6$ corresponding to a view that θ_t

may vary markedly over time. Given the initial prior, sequential updating and one-step ahead forecasting proceed directly. Figure 3.5 displays a plot over time of the raw one-step ahead forecast errors e_t. Note that there is a preponderance of positive errors during the latter half of the time period, years 1980 onwards, indicating that the model is generally under-forecasting the data series Y_t during that time.

Figure 3.6 provides a plot over time of the on-line estimated value m_t of the regression parameter θ_t, referred to as the estimated *trajectory* of the coefficient. This is the solid line in the figure. An indication of uncertainty about the value at time t is indicated by the dashed lines symmetrically located either side of the estimate. These limits provide 90% posterior probability intervals (in fact, HPD intervals) for the corresponding values of θ_t, calculated from the Student T posterior distributions of $(\theta_t|D_t)$. It is clear that θ_t drifts upwards over time, particularly over the last six or seven years of data. The model, though not predicting such positive drift, adapts as data are processed, sequentially adjusting the posterior to favour higher values consistent with the data. The fact that the model is very adaptive (with δ scandalously low at 0.6), leads to the resulting marked degree of inferred change in θ_t. However, since the model implies that θ_t undergoes a random walk, the positive drift cannot be anticipated and so, generally, under-forecasting results as evidenced in the forecast errors in Figure 3.5.

Consider now an analysis with $\delta = 1$, so the model is a static regression with constant $\theta_t = \theta_0$. The sequentially calculated one-step ahead errors e_t from this analysis appear in Figure 3.7. The preponderance of positive errors is evident as with the previous analysis, but the effect is profound. From 1980 onwards, all the errors are positive, and tending to increase, indicating continual deterioration in model adequacy. Figure 3.8 displays the on-line estimates m_t with 90% HPD intervals. Again the model is adapting to higher values of θ_t as time progresses, but the rate of adaptation is much lower than the previous, very adaptive model with $\delta = 0.6$. This under-adaptation leads to increasingly poor forecasts as time progresses.

Obviously, the regression model as specified is deficient for this data set; a revised model anticipating drift in θ_t is desirable. Despite this, the model with $\delta = 0.6$ adapts sufficiently to new data that the forecast errors, though tending to be positive, are relatively small. By comparison, the static regression model with $\delta = 1$ performs extremely poorly in one-step ahead forecasting due to its poor adaptability. Thus, interpreting Figure 3.4 as suggesting a simple static regression is certain to mislead. This plot masks the effects of time on the relationship between the two series. Many readers will immediately note that a simple plot of the observed market shares Y_t/F_t (empirical estimates of the θ_t values) over time indicates the increasing nature of θ_t and suggest a more appropriate model form; Figure 3.9 provides such a plot, the ratio values appearing as crosses. Also plotted are the sequences of on-line estimates m_t from each of the two analyses described above; the solid line is that from the adaptive analysis with

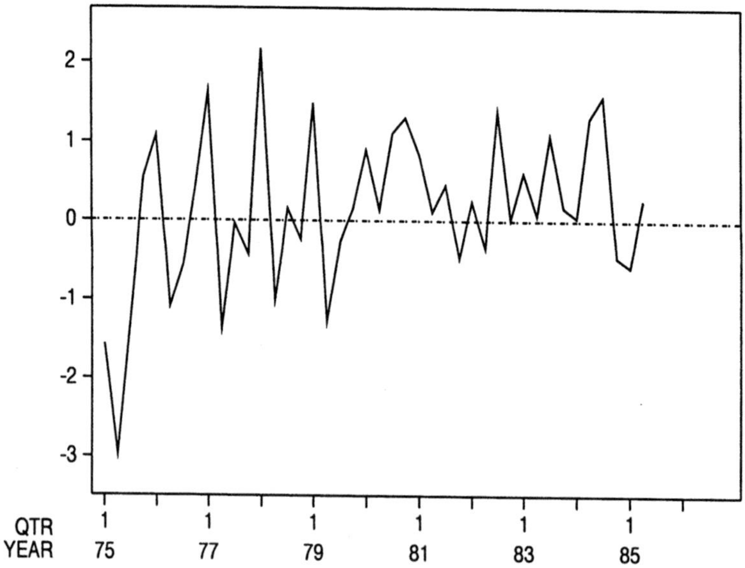

Figure 3.5 One-step ahead forecast errors : $\delta = 0.6$

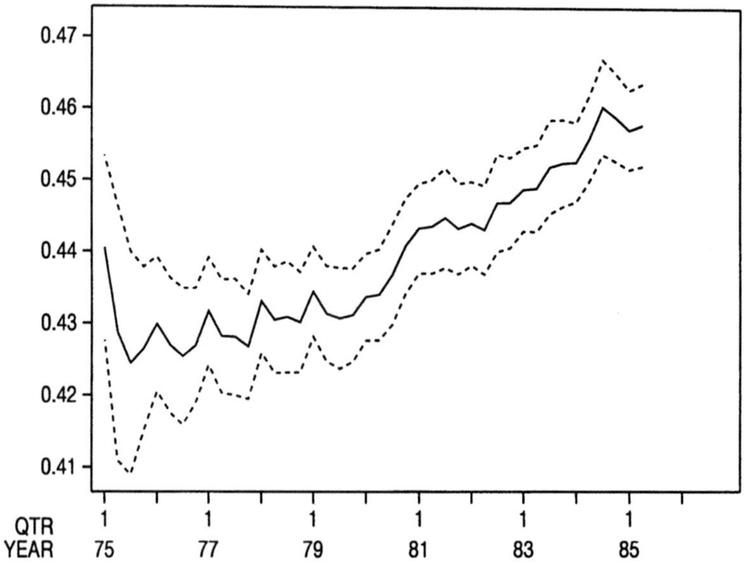

Figure 3.6 On-line estimated trajectory of θ_t: $\delta = 0.6$

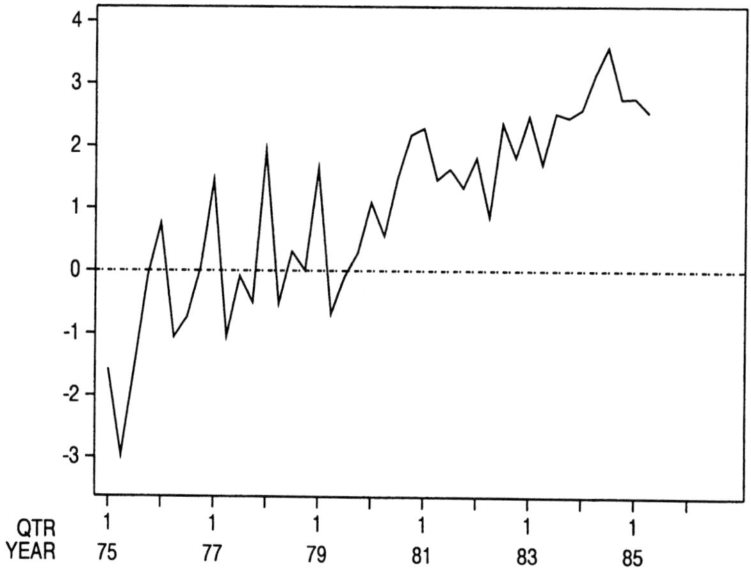

Figure 3.7 One-step ahead forecast errors : $\delta = 1.0$

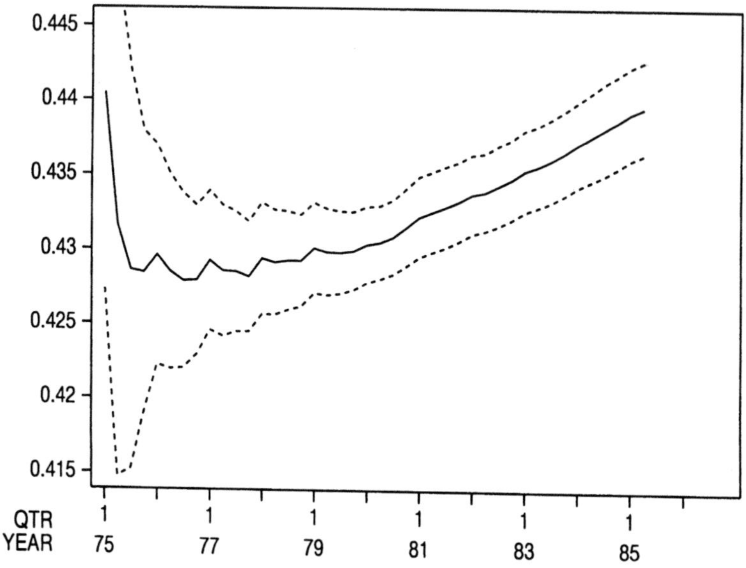

Figure 3.8 On-line estimated trajectory of θ_t: $\delta = 1.0$

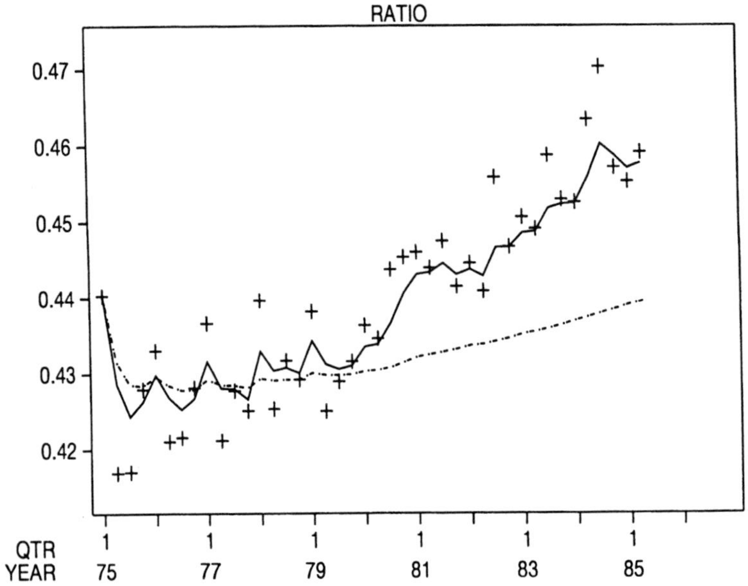

Figure 3.9 Ratio Y_t/F_t versus time t, with on-line estimates

$\delta = 0.6$, the dashed line from the static model. It is evident that the adaptive model tracks changes in θ_t rather well.

To further compare the two analyses, details are given below. Tables 3.6 and 3.7 display the data and some components of the prior/posterior distributions from the two analyses for the final 12 quarters, $t = 31, \ldots, 42$.

As noted above, the increasing market share is identified most clearly by the model with the lower discount of 0.6, where m_t approaches 0.46 at the end of the data, though it is not continually increasing. The one-step forecasts are good, although the increasing θ_t leads to a dominance of positive forecast errors e_t, albeit very small compared to their distributions. Note that $n_t = t + 1$, so that the Student T forecast distributions are close to normality and Q_t is then roughly the variance associated with e_t. In contrast, the static regression model with discount 1.0 is far less adaptive, the m_t sequence slowly increasing to 0.44 at the end of the data. This is well below the more reasonable values near 0.46, as is evident from the sustained, and significant, under-forecasting of the Y_t series. The key point here is that, although the model lacks a growth term and so is really inappropriate, the lower discount analysis adapts well and the resulting short-term forecasts are acceptable. The standard static model is extremely poor by comparison, and the message extends to more general models, and regressions in particular.

Table 3.6. Sales data example: discount factor = 0.6

t	F_t	Y_t	f_t	$Q_t^{1/2}$	$S_t^{1/2}$	m_t	$C_t^{1/2}$
31	107.7	49.1	47.71	0.94	0.80	0.447	0.0040
32	145.0	64.8	64.79	1.10	0.79	0.447	0.0037
33	155.3	70.0	69.39	1.09	0.78	0.449	0.0035
34	123.1	55.3	55.23	0.96	0.77	0.449	0.0036
35	109.2	50.1	49.02	0.92	0.77	0.452	0.0039
36	144.8	65.6	65.43	1.06	0.76	0.452	0.0036
37	160.6	72.7	72.66	1.07	0.75	0.453	0.0033
38	119.1	55.2	53.90	0.91	0.76	0.456	0.0036
39	109.5	51.5	49.93	0.92	0.78	0.460	0.0039
40	144.8	66.2	66.65	1.07	0.77	0.459	0.0037
41	165.8	75.5	76.08	1.10	0.77	0.457	0.0033
42	127.4	58.5	58.23	0.94	0.76	0.458	0.0034

Table 3.7. Sales data example: discount factor = 1.0

t	F_t	Y_t	f_t	$Q_t^{1/2}$	$S_t^{1/2}$	m_t	$C_t^{1/2}$
31	107.7	49.1	46.74	1.19	1.23	0.434	0.0016
32	145.0	64.8	62.98	1.26	1.25	0.435	0.0016
33	155.3	70.0	67.52	1.28	1.30	0.435	0.0016
34	123.1	55.3	53.59	1.32	1.32	0.436	0.0016
35	109.2	50.1	47.58	1.33	1.36	0.436	0.0016
36	144.8	65.6	63.14	1.38	1.40	0.437	0.0017
37	160.6	72.7	70.11	1.43	1.44	0.437	0.0017
38	119.1	55.2	52.06	1.46	1.51	0.438	0.0017
39	109.5	51.5	47.92	1.52	1.59	0.438	0.0018
40	144.8	66.2	63.44	1.62	1.63	0.439	0.0018
41	165.8	75.5	72.73	1.66	1.66	0.439	0.0019
42	127.4	58.5	55.96	1.68	1.69	0.440	0.0019

Note further that the more adaptive model allows for a much greater decay of information about θ_t over time, and this results in a larger posterior variance. At the end of the data, for example, the posterior standard deviation of θ_{42} in the adaptive model is almost 80% greater than that in the static model. This is a large difference due to a very small discount at 0.6. Typically, if the form of the model is reliable over a reasonable period,

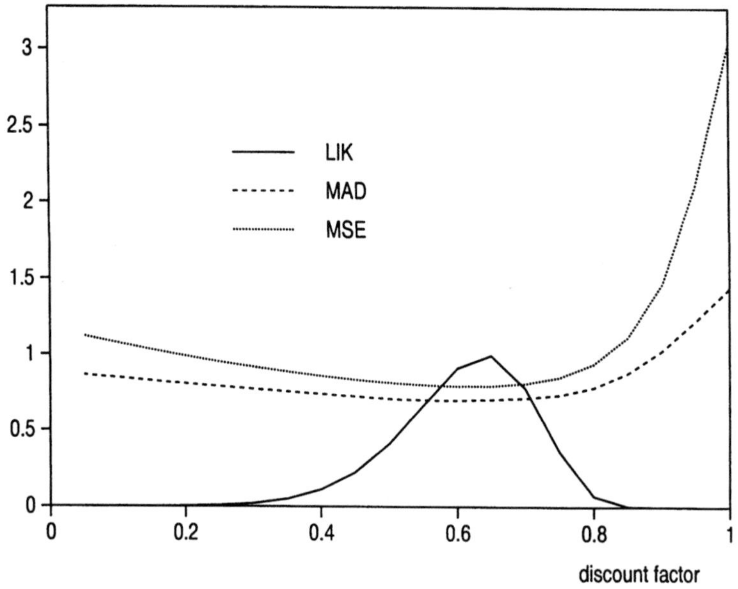

Figure 3.10 MSE, MAD and LIK measures as functions of δ

as is clearly not the case here, then discount factors for regression will exceed 0.9. Concerning the static model, note that in addition to very poor forecasts, the posteriors for θ_t are overly precise, being highly concentrated about a mode that is far from suitable as an estimate of θ_t!

The adaptive model correctly attributes a high degree of variation in the Y_t series to movement in θ_t, and so much less than the static model to observational variation about level. The final estimate of observational standard deviation in the adaptive model is 0.76 compared with 1.69 in the static case. Further, note that the one-step forecast variances in the adaptive model are much smaller than those in the static model. The final one-step forecast standard deviation in the former is 0.94 compared to 1.68 in the latter. This is due to the point above; the observational variance is heavily over-estimated in the static model. Thus, in addition to providing much more accurate point forecasts than the static model, the adaptive model produces much more concentrated forecast distributions. By any measures of comparative performance, the adaptive model is much better than the standard static model.

Similar analyses with different values of δ can be assessed and compared using the MSE, MAD and LLR (log likelihood ratio) criteria as demonstrated in the example of Section 2.7. Each of these measures is calculated from analyses with $\delta = 0.05, 0.1, \ldots, 0.95, 1.0$. MSE and MAD measures are graphed as functions of δ over this range in Figure 3.10.

Also plotted is the actual model likelihood LIK=exp(LLR). This figure clearly indicates that a static model, $\delta = 1.0$, is highly inappropriate. Both MSE and MAD measures, as functions of δ, decay rapidly from their maxima at $\delta = 1.0$ to minima near $\delta = 0.6$, thereafter rising only slightly as δ approaches 0. The curves are flat between $\delta = 0.4$ and $\delta = 0.8$, indicating the usual marked robustness to particular values within a suitable range. The likelihood function LIK for δ peaks between 0.6 and 0.7, being negligible above $\delta = 0.85$. Highly adaptive models with δ less than 0.5 have low likelihood, being penalised since the corresponding one-step forecast distributions are very diffuse.

3.5 EXERCISES

Unless otherwise stated these exercises refer to a univariate time series $\{Y_t\}$ modelled by the closed regression DLM $\{F_t, 1, V_t, W_t\}$ of Definition 3.2 with known variances and/or known discount factor δ as follows:

Observation equation:	$Y_t = F_t\theta_t + \nu_t,$	$\nu_t \sim \mathrm{N}[0, V_t],$
System equation:	$\theta_t = \theta_{t-1} + \omega_t,$	$\omega_t \sim \mathrm{N}[0, W_t],$
Initial prior:	$(\theta_0 \mid D_0) \sim \mathrm{N}[m_0, C_0].$	

(1) In the DLM $\{F_t, 1, 100, 0\}$ suppose that the sequence $\theta_t = \theta$ is a precisely known constant but that the regressor variable sequence F_t is not controllable. You model F_t as a sequence of independent normal random quantities, $F_t \sim \mathrm{N}[0, 400]$. Given D_{t-1}, answer the following questions.
  (a) Prove that Y_t and F_t have a bivariate normal distribution and identify its mean vector and variance matrix.
  (b) What is the correlation between Y_t and F_t?
  (c) What is the regression coefficient of F_t on Y_t?
  (d) What is the posterior distribution of $(F_t|Y_t, D_{t-1})$?

(2) In the previous question suppose that the sequence θ_t is known but not constant. You also adopt a random walk model for F_t, so that $F_t = F_{t-1} + \epsilon_t$ with independent $\epsilon_t \sim \mathrm{N}[0, U]$. Show that your overall model is equivalent to the simple regression DLM $\{\theta_t, 1, 100, U\}$ with regressor variable θ_t and parameter F_t.

(3) In the DLM $\{F_t, 1, V_t, W_t\}$, suppose that $F_t \neq 0$ for all t.
  (a) Show that the series $X_t = Y_t/F_t$ follows a first-order polynomial DLM and identify it fully.
  (b) Verify that the updating equations for the regression DLM can be deduced from those of the first-order polynomial DLM.
  (c) What happens if $F_t = 0$?

(4) One measure of the predictability of Y_t at time $t-1$ is the modulus of the reciprocal of the coefficient of variation, given by $|f_t/Q_t^{1/2}|$.

Explore this measure as a function of $F_t \in [-100, 100]$ for each of the cases $R_t = 0,\ 10,\ 20, 50$, when $m_t = 10$, and $V_t = 100$.

(5) Suppose that V_t is a known function of a control variable F_t. In particular, let $V_t = V(a + |F_t|^p)$ for known quantities V, a and p.
 (a) How should F_t be chosen in order to maximise the posterior precision C_t^{-1} subject to $|F_t| < k$ for some $k > 0$? What is the optimal design value of F_t in the case $a = 8$, $p = 3$ and $k = 10$?
 (b) How does this problem change when V is unknown and is estimated from the data?

(6) For the discount regression DLM in which V_t is known for all t and $R_t = C_{t-1}/\delta$, show that the updating equations can be written as

$$m_t = C_t \delta^t C_0^{-1} m_0 + C_t \sum_{j=0}^{t-1} \delta^j F_{t-j} V_{t-j}^{-1} Y_{t-j}$$

and

$$C_t^{-1} = \delta^t C_0^{-1} + \sum_{j=0}^{t-1} \delta^j F_{t-j}^2 V_{t-j}^{-1}.$$

Deduce that as $t \to \infty$, $C_t^{-1} \to 0$ and

$$m_t \to \sum_{j=0}^{t-1} \delta^j F_{t-j} V_{t-j}^{-1} Y_{t-j} \Big/ \sum_{j=0}^{t-1} \delta^j F_{t-j}^2 V_{t-j}^{-1}.$$

(7) Consider discount weighted regression applied to the estimation of a parameter θ_t by the value m_t. In DWR, the estimate m_t is chosen to minimise the discounted sum of squares

$$S(\theta) = \sum_{j=0}^{t-1} \delta^j \left(Y_{t-j} - F_{t-j}\theta\right)^2,$$

where all quantities other than θ are known.
 (a) Show that

$$m_t = \sum_{j=0}^{t-1} \delta^j F_{t-j} Y_{t-j} \Big/ \sum_{j=0}^{t-1} \delta^j F_{t-j}^2.$$

 (b) Generalising (a), suppose that

$$Y_t = F_t\theta + \nu_t, \qquad \nu_t \sim \mathrm{N}[0, V_t],$$

and that m_t is more appropriately chosen to minimise

$$S(\theta) = \sum_{j=0}^{t-1} \delta^j V_{t-j}^{-1} \left(Y_{t-j} - F_{t-j}\theta\right)^2.$$

Show that

$$m_t = \sum_{j=0}^{t-1} \delta^j F_{t-j} V_{t-j}^{-1} Y_{t-j} \Big/ \sum_{j=0}^{t-1} \delta^j F_{t-j}^2 V_{t-j}^{-1}.$$

(c) Compare these results with those of the previous question to see that the estimates correspond to those from the discount DLM with the uninformative prior $C_0^{-1} = 0$.

(8) Suppose that $V_t = Vk_t$, where $V = 1/\phi$ is unknown and k_t is a known variance multiplier. Show how the analysis summarised in the table in Section 3.4.1 is modified.

(9) Consider the simple regression DLM $\{(-1)^t k, 1, V, W\}$, in which $k > 0$ is a known constant.

(a) By reference to the first-order polynomial constant DLM convergence results or otherwise, prove that $\lim_{t\to\infty} C_t = C$ exists. Obtain C and the limiting values of Q_t and $|A_t|$.

(b) Treating the limiting value C as a function of k, verify that it is equal to W when $k = \sqrt{V/2W}$.

(10) Consider the company sales/total market series in the example of Section 3.4.2. Perform similar analyses of this data using the same DLM but varying the discount factor over the range $0.6, 0.65, \ldots, 1$. Explore the sensitivity to inferences about the time trajectory of θ_t as the discount factor varies in the following ways:

(a) Plot m_t versus t, with intervals based on $C_t^{1/2}$ to represent uncertainty, for each value of δ and comment on differences with respect to δ.

(b) Compare the final estimates of observational variance S_{42} as δ varies. Do the same for prediction variances Q_{42}. Discuss the patterns of behaviour.

(c) Use MSE, MAD and LLR measures to assess the predictive performance of the models relative to the static model defined by $\delta = 1$.

(11) Consider a *retrospective analysis* in which inferences are made about historical parametric values based on the current data. In particular, this question concerns inferences about θ_{t-1} given D_t for the DLM $\{F_t, 1, V_t, W_t\}$ with known variances.

(a) Use the system equation directly to show that

$$\mathrm{C}[\theta_t, \theta_{t-1} | D_{t-1}] = B_{t-1} \mathrm{V}[\theta_t | D_{t-1}],$$

for some B_{t-1} lying between 0 and 1, and identify B_{t-1}.

(b) Deduce that

$$\mathrm{C}[\theta_{t-1}, Y_t | D_{t-1}] = B_{t-1} \mathrm{C}[\theta_t, Y_t | D_{t-1}].$$

(c) Hence identify the moments of the joint normal distribution of $(\theta_{t-1}, \theta_t, Y_t | D_{t-1})$, and from this, those of the conditional distribution of $(\theta_{t-1}|D_t)$ (by conditioning on Y_t in addition to D_{t-1}). Verify that the regression coefficient of θ_{t-1} on Y_t is $B_{t-1}A_t$, where A_t is the usual regression coefficient (adaptive coefficient) of θ_t on Y_t given D_{t-1}.

(d) Deduce that $(\theta_{t-1}|D_t)$ is normal with moments that can be written as

$$\mathrm{E}[\theta_{t-1}|D_t] = m_{t-1} + B_{t-1}(\mathrm{E}[\theta_t|D_t] - \mathrm{E}[\theta_t|D_{t-1}])$$

and

$$\mathrm{V}[\theta_{t-1}|D_t] = C_{t-1} - B_{t-1}^2(\mathrm{V}[\theta_t|D_{t-1}] - \mathrm{V}[\theta_t|D_t]).$$

(12) Generalise the results of the previous exercise to allow retrospection back over time for more than one step, calculating the distribution of $(\theta_{t-k}|D_t)$ for any k, $(0 \le k \le t)$. Do this as follows:

(a) Using the observation and evolution equations directly, show that for any $r \ge 1$,

$$\mathrm{C}[\theta_{t-k}, Y_{t-k+r}|D_{t-k}] = B_{t-k}\mathrm{C}[\theta_{t-k+1}, Y_{t-k+r}|D_{t-k}],$$

where for any s, $B_s = C_s/R_{s+1}$ lies between 0 and 1.

(b) Writing $\mathbf{X}_t(k) = (Y_{t-k+1}, \dots, Y_t)'$, deduce from (a) that

$$\mathrm{C}[\theta_{t-k}, \mathbf{X}_t(k)|D_{t-k}] = B_{t-k}\mathrm{C}[\theta_{t-k+1}, \mathbf{X}_t(k)|D_{t-k}].$$

(c) Hence identify the moments of the joint normal distribution of

$$(\theta_{t-k}, \theta_{t-k+1}, \mathbf{X}_t(k)|D_{t-k}),$$

and from this those of the conditional distributions of $(\theta_{t-k}|D_t)$ and $(\theta_{t-k+1}|D_t)$ (by conditioning on $\mathbf{X}_t(k)$ in addition to D_{t-k} and noting that $D_t = \{\mathbf{X}_t(k), D_{t-k}\}$). Using (b), verify that the regression coefficient vector of θ_{t-k} on $\mathbf{X}_t(k)$ is B_{t-k} times that of θ_{t-k+1} on $\mathbf{X}_t(k)$.

(d) Deduce that $(\theta_{t-k}|D_t)$ is normal with moments that can be written as

$$\begin{aligned}\mathrm{E}[\theta_{t-k}|D_t] =& m_{t-k}+\\ & B_{t-k}(\mathrm{E}[\theta_{t-k+1}|D_t] - \mathrm{E}[\theta_{t-k+1}|D_{t-k}])\end{aligned}$$

and

$$\begin{aligned}\mathrm{V}[\theta_{t-k}|D_t] =& C_{t-k}-\\ & B_{t-k}^2(\mathrm{V}[\theta_{t-k+1}|D_{t-k}] - \mathrm{V}[\theta_{t-k+1}|D_t]).\end{aligned}$$

(e) Let the above moments be denoted by $a_t(-k)$ and $R_t(-k)$, so that $(\theta_{t-k}|D_t) \sim \mathrm{N}[a_t(-k), R_t(-k)]$. Verify that the above, retrospective updating equations provide these moments backwards over time for $k = t-1, t-2, \dots, 0$ via

$$a_t(-k) = m_{t-k} + B_{t-k}[a_t(-k+1) - a_{t-k+1}]$$

and

$$R_t(-k) = C_{t-k} - B_{t-k}^2[R_t(-k+1) - R_{t-k+1}]$$

with $a_s = m_{s-1}$ and $R_s = C_{s-1}+W_s$ for all s, and initial values $a_t(0) = m_t$ and $R_t(0) = C_t$.

(13) In the last two questions, employing the discount regression model $\{F_t, 1, V, W_t\}$, where $W_t = C_{t-1}(\delta^{-1} - 1)$, show that for $k \geq 0$, $B_{t-1} = \delta$,

$$a_t(-k) = a_{t-1}(-k+1) + \delta^k A_t e_t$$

and

$$R_t(-k) = R_{t-1}(-k+1) - \delta^{2k} A_t^2 Q_t.$$

This provides for neat and simple updating of the retrospective means and variances.

(14) Suppose the yield Y_t of the tth batch of a manufacturing plant is truly represented by

$$Y_t = 70 - (X_t - 3)^2 + \eta_t, \qquad \eta_t \sim \mathrm{N}[0, V],$$

$$X_t \sim \{F_t, 1, V, W\}, \qquad (\theta_0|D_0) \sim \mathrm{N}[1, V].$$

Initially, the setting $F_1 = 3$ is optimal in the sense of maximising the expected yield.

(a) If F_t is kept constant at 3, or if from any other specified time it is kept constant at its then perceived optimal value, what are the consequences?

(b) Plant managers have production targets to meet and dislike changing operating conditions, fearing a drop in yield. If you were the production director would you approve this attitude or would you introduce a policy encouraging plant managers to make regular small experimental variations about the then perceived optimal value of F_t?

(15) The following data set refers to an internationally famous canned product. The objective is to establish a relationship between market share and price in order to make short-term pricing decisions. The observation series Y_t is the percentage market share for quarter t minus 42%, and F_t is a linear function of the real price.

Qtr. t	1	2	3	4	5	6
Y_t	0.45	0.83	1.45	0.88	-1.43	-1.50
F_t	-0.50	-1.30	-1.50	-0.84	-0.65	-1.19
Qtr t	**7**	**8**	**9**	**10**	**11**	**12**
Y_t	-2.33	-0.78	0.58	1.10	?	?
F_t	2.12	0.46	-0.63	-1.22	-2.00	2.00

Adopt the simple discount regression DLM $\{F_t, 1, V, W_t\}$ with $\delta = 0.975$.

(a) Carry out sequential forecasting with known variance $V = 0.2$ and $\theta_0 \sim \mathrm{N}[0, 1]$. Either by hand or computer prepare a calculation table that produces $R_t, A_t, f_t, Q_t, e_t, m_t$ and C_t for each t. What are your final inferences about the price elasticity θ_{10}? What is your forecast for the market share in the next two quarters?

(b) Repeat the analysis and inference when V is an unknown constant variance starting with $n_0 = 1$ and $S_0 = 0.2$.

CHAPTER 4

THE DYNAMIC LINEAR MODEL

4.1 OVERVIEW

The first-order polynomial and simple regression models of the preceding two chapters illustrate many basic concepts and important features of the general class of normal dynamic linear models, referred to as dynamic linear models (DLMs) when the normality is understood. This class is described and analysed here, providing a basis for the special cases and generalisations that follow in later chapters. The principles employed in Bayesian forecasting and dynamic modelling involve

(i) parametric models with meaningful dynamic parameters;
(ii) a probabilistic representation of information about parameters;
(iii) a sequential model definition utilising conditional independence;
(iv) robust conditionally independent model components;
(v) forecasts derived as probability distributions;
(vi) a facility for incorporating expert information;
(vii) model quality control.

Suppose interest lies in a scalar series Y_t and that at time $t-1$ the current information set is D_{t-1}. The first step in the Bayesian approach is to examine the forecasting context and to select a meaningful parametrisation, $\boldsymbol{\theta}_{t-1}$, such that all the historical information relevant to predicting future observations is contained in the information about $\boldsymbol{\theta}_{t-1}$. In particular the modeller represents this relevant information in terms of the probability distribution $(\boldsymbol{\theta}_{t-1} \mid D_{t-1})$. In statistical terms, given D_{t-1}, $(\boldsymbol{\theta}_{t-1} \mid D_{t-1})$ is sufficient for predicting the future. The parameter together with this probability distribution defines how the modeller views the context at time $t-1$. Clearly, the parameters must be meaningful to decision makers who use the forecasts and who also supply occasional expert information. Indexing $\boldsymbol{\theta}_t$ by t indicates that the parametrisation is dynamic. In addition, although often the number and meaning of the elements of $\boldsymbol{\theta}_t$ will be stable, there are occasions when $\boldsymbol{\theta}_t$ will be expanded, contracted or changed in meaning according to the forecaster's existing view of the time series. This is particularly so with *open systems*, such as typically arise in social, economic and biological environments, where influential factors affecting the time series process are themselves subject to variation based on the state of the system generating the process. In such cases, changes in $\boldsymbol{\theta}_t$ may be required to reflect system learning and the exercise of purposeful control. Such events, although recognizable when they happen, may be difficult to identify initially and so will not typically be included in the model until occurrence.

The next modelling step is that of relating the current information to the future so that predictive distributions such as $(Y_{t+k} \mid D_{t-1})$ can be derived. This is accomplished by specifying a sequential parametric relation $(\boldsymbol{\theta}_t \mid \boldsymbol{\theta}_{t-1}, D_{t-1})$ together with an observation relation $(Y_t \mid \boldsymbol{\theta}_t, D_{t-1})$. In combination with $(\boldsymbol{\theta}_{t-1} \mid D_{t-1})$ these distributions enable the derivation of a full joint forecast distribution. The crucial structural property enabling effective dynamic modelling is conditional independence, most strikingly conveyed by the graph in Figure 4.1. The key feature is that given the present, the future is independent of the past. In particular, at time t, given $\boldsymbol{\theta}_t$, the past, present, and future are mutually independent. Also, given just D_t, all the information concerning the future is contained in the posterior parametric distribution $(\boldsymbol{\theta}_t \mid D_t)$. Further, if this distribution is normal, $\mathrm{N}[\mathbf{m}_t, \mathbf{C}_t]$, then given D_t, the pair $\{\mathbf{m}_t, \mathbf{C}_t\}$ contains all the relevant information about the future, so that in the usual statistical sense, given D_t, $\{\mathbf{m}_t, \mathbf{C}_t\}$ is sufficient for $\{Y_{t+1}, \boldsymbol{\theta}_{t+1}, \ldots, Y_{t+k}, \boldsymbol{\theta}_{t+k}\}$.

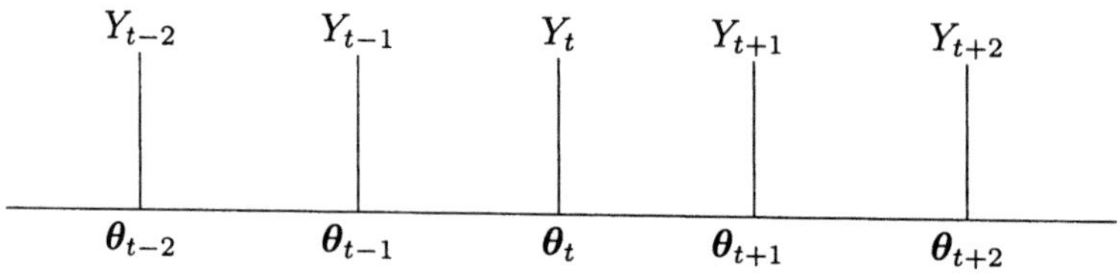

Figure 4.1 The DLM conditional independence structure

Conditional independence also features strongly in initial model building and in choosing an appropriate parametrisation. For example, the linear superposition principle states that any linear combination of deterministic linear models is a linear model. This extends to a normal linear superposition principle:

Any linear combination of independent normal DLMs is a normal DLM.

The case of a two-component DLM, with $\boldsymbol{\theta}'_t = (\boldsymbol{\theta}'_{t1}, \boldsymbol{\theta}'_{t2})$, is graphed in Figure 4.2. Here, conditional upon $\boldsymbol{\theta}_t$, the two series of components $\{\boldsymbol{\theta}_{t+i,1},\ i > 0\}$ and $\{\boldsymbol{\theta}_{t+i,2},\ i > 0\}$ evolve independently. The important consequence is that in most practical cases, a DLM can be decomposed into a linear combination of simple canonical DLMs, the good news being that a modeller only needs to master a very few simple DLMs in order to become proficient at model building. Another advantage of the component structure is robustness. If the DLM breaks down or expert intervention takes place, only the affected component needs to be serviced. All the information on other components is retained.

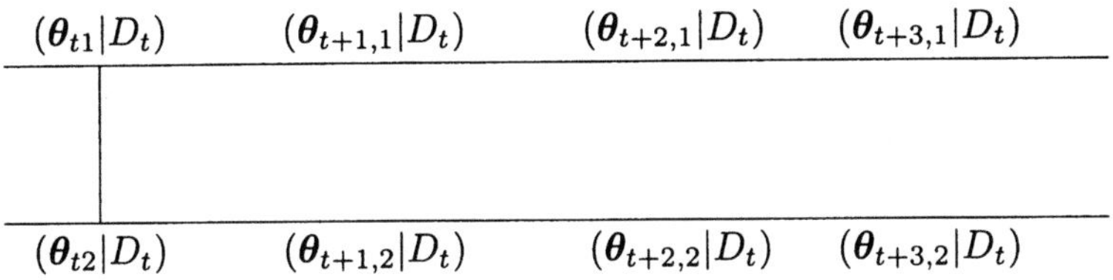

Figure 4.2 The parametric conditional independence structure

Forecasts are obtained by applying probability laws so that the joint distribution for the observations and parameters at time t may be derived via

$$p(Y_t, \boldsymbol{\theta}_t \mid D_{t-1}) = p(Y_t \mid \boldsymbol{\theta}_t, D_{t-1})\, p(\boldsymbol{\theta}_t \mid D_{t-1}).$$

The one-step forecast is simply the marginal distribution $(Y_t \mid D_{t-1})$, and the posterior, $(\boldsymbol{\theta}_t \mid D_t)$, is the conditional distribution $(\boldsymbol{\theta}_t \mid Y_t, D_{t-1})$. Inferences and decisions follow standard Bayesian procedures.

Operational Bayesian models specifically aim to incorporate information from any relevant source, including subjective expert views, leading to amended and updated model structures. The probabilistic formulation easily assimilates such information, naturally accommodating specified changes and associated uncertainties. Further, it offers simple procedures for combining forecasts and producing forecasts subject to specified constraints. This is particularly important for an organisation that requires consistency of forecasts at all levels and needs to exercise overall control by constraining detailed operations in line with macro directives. For example, the forecasts and stock control of many thousands of items must be consistent with forecasts of total demand and a constraint on total stock expenditure.

In practice, Bayesian dynamic models operate in accordance with the principle of *Management by Exception.* This involves the "routine" use of a proposed model unless exceptional circumstances arise. Such exceptions occur in two main distinct ways. The first is when relevant expert information from a source external to the system is received. Examples in consumer demand forecasting include information on patent expiry, licensing changes, new regulatory legislation, strikes, supply difficulties, forthcoming spot orders, and so on. This type of information, usually feed-forward and anticipatory, is naturally included in the existing system by formally combining it with existing probabilistic information. By contrast, the second type of exception is feedback, which occurs when a monitoring system used to assess the qualitative performance of the routine model signals a significant inadequacy. These monitoring systems are similar to those used in manufacturing quality control and are usually based on the recent forecast errors. The reaction to such signals may be relevant expert explanatory information, that is accommodated as above, or the introduction of automatic default procedures, that model specific types of inadequacy and are designed to distinguish and correct for model deficiencies.

4.2 DEFINITIONS AND NOTATION

For reference, the general normal DLM is defined for a vector observation $\mathbf{Y}_t$, rather than the more usual scalar special case. However, much of the discussion, in this chapter and elsewhere, is restricted to the scalar case. Let $\mathbf{Y}_t$ be an $(r \times 1)$ vector observation on the time series over times $t = 1, 2, \ldots$, following the model now defined.

Definition 4.1. The general normal dynamic linear model (DLM) is characterised by a set of *quadruples*

$$\{\mathbf{F}, \mathbf{G}, \mathbf{V}, \mathbf{W}\}_t = \{\mathbf{F}_t, \mathbf{G}_t, \mathbf{V}_t, \mathbf{W}_t\}$$

for each time t, where

(a) $\mathbf{F}_t$ is a known $(n \times r)$ matrix;
(b) $\mathbf{G}_t$ is a known $(n \times n)$ matrix;
(c) $\mathbf{V}_t$ is a known $(r \times r)$ variance matrix;
(d) $\mathbf{W}_t$ is a known $(n \times n)$ variance matrix.

This quadruple defines the model relating $\mathbf{Y}_t$ to the $(n \times 1)$ parameter vector $\boldsymbol{\theta}_t$ at time t, and the $\boldsymbol{\theta}_t$ sequence through time, via the sequentially specified distributions

$$(\mathbf{Y}_t \mid \boldsymbol{\theta}_t) \sim \mathrm{N}[\mathbf{F}_t'\boldsymbol{\theta}_t, \mathbf{V}_t] \tag{4.1a}$$

and

$$(\boldsymbol{\theta}_t \mid \boldsymbol{\theta}_{t-1}) \sim \mathrm{N}[\mathbf{G}_t\boldsymbol{\theta}_{t-1}, \mathbf{W}_t]. \tag{4.1b}$$

Equations (4.1) are also implicitly conditional on D_{t-1}, the information set available prior to time t. In particular, this includes the values of the defining variances $\mathbf{V}_t$ and $\mathbf{W}_t$ and the past observations $\mathbf{Y}_{t-1}, \mathbf{Y}_{t-2}, \ldots$, as well as the initial information set D_0. For notational simplicity, D_{t-1} is not explicitly recognized in the conditioning in equations (4.1), but it should be remembered that it is always conditioned upon.

An alternative representation of these defining equations is

$$\mathbf{Y}_t = \mathbf{F}_t'\boldsymbol{\theta}_t + \boldsymbol{\nu}_t, \qquad \boldsymbol{\nu}_t \sim \mathrm{N}[\mathbf{0}, \mathbf{V}_t], \tag{4.2a}$$

and

$$\boldsymbol{\theta}_t = \mathbf{G}_t\boldsymbol{\theta}_{t-1} + \boldsymbol{\omega}_t, \qquad \boldsymbol{\omega}_t \sim \mathrm{N}[\mathbf{0}, \mathbf{W}_t]. \tag{4.2b}$$

The error sequences $\boldsymbol{\nu}_t$ and $\boldsymbol{\omega}_t$ are internally and mutually independent. Equation (4.2a) is the **observation equation** for the model, defining the sampling distribution for $\mathbf{Y}_t$ conditional on the quantity $\boldsymbol{\theta}_t$. The conditional independence structure of Figure 4.1 applies. So, given $\boldsymbol{\theta}_t$, $\mathbf{Y}_t$ is independent of the all other observations and parameter values; and in general, given the present, the future is independent of the past. This equation relates the $\mathbf{Y}_t$ to $\boldsymbol{\theta}_t$ via a dynamic linear regression with a multivariate normal error structure having known, though possibly time varying, observational variance matrix $\mathbf{V}_t$. For time t

(e) $\mathbf{F}_t$ is the design matrix of known values of independent variables;
(f) $\boldsymbol{\theta}_t$ is the state, or system, vector:
(g) $\boldsymbol{\mu}_t = \mathbf{F}_t'\boldsymbol{\theta}_t$ is the mean response, or level;
(h) $\boldsymbol{\nu}_t$ is the observational error.

Equation (4.2b) is the **evolution, state** or **system equation**, defining the time evolution of the state vector. The conditional independence property shows a one-step Markov evolution so that, given $\boldsymbol{\theta}_{t-1}$ and the known values of $\mathbf{G}_t$ and $\mathbf{W}_t$, $\boldsymbol{\theta}_t$ is independent of D_{t-1}. That is, given $\boldsymbol{\theta}_{t-1}$, the distribution of $\boldsymbol{\theta}_t$ is fully determined independently of values of $\mathbf{Y}_{t-1}$ and all the state vectors and observations prior to time $t-1$. The deterministic component of the evolution is the transition from state $\boldsymbol{\theta}_{t-1}$ to $\mathbf{G}_t\boldsymbol{\theta}_{t-1}$, a simple linear transformation of $\boldsymbol{\theta}_{t-1}$. The evolution is completed with the addition of the random vector $\boldsymbol{\omega}_t$. At time t,

(i) $\mathbf{G}_t$ is the evolution, system, transfer or state matrix;
(j) $\boldsymbol{\omega}_t$ is the system, or evolution, error with evolution variance $\mathbf{W}_t$.

Finally, note that the defining quadruple, assumed known throughout, does not appear in the conditioning of the distributions. For notational clarity, the convention followed throughout the book is that in general, known quantities will not be made explicit in conditioning distributions. Some further related discussion appears below in Section 4.3.

Definition 4.2. Of special interest are the following two subsets of the general class of DLMs.

(i) If the pair $\{\mathbf{F}, \mathbf{G}\}_t$ is constant for all t then the model is referred to as a **time series DLM**, or **TSDLM**.

(ii) A TSDLM whose observation and evolution variances are constant for all t is referred to as a **constant DLM**.

Thus, a constant DLM is characterised by a single quadruple

$$\{\mathbf{F}, \mathbf{G}, \mathbf{V}, \mathbf{W}\}.$$

It will be seen that this important subset of DLMs includes essentially all classical linear time series models.

The general univariate DLM is defined by Definition 4.1 with $r = 1$ and is therefore characterised by a quadruple

$$\{\mathbf{F}_t, \mathbf{G}_t, V_t, \mathbf{W}_t\},$$

leading to

$$(Y_t \mid \boldsymbol{\theta}_t) \sim \mathrm{N}[\mathbf{F}_t'\boldsymbol{\theta}_t, V_t]$$

and

$$(\boldsymbol{\theta}_t \mid \boldsymbol{\theta}_{t-1}) \sim \mathrm{N}[\mathbf{G}_t\boldsymbol{\theta}_{t-1}, \mathbf{W}_t].$$

These equations, together with the initial prior at time 0, provide the full definition as follows.

Definition 4.3. For each t, the general univariate DLM is defined by:

Observation equation:	$Y_t = \mathbf{F}_t'\boldsymbol{\theta}_t + \nu_t,$	$\nu_t \sim \mathrm{N}[0, V_t],$
System equation:	$\boldsymbol{\theta}_t = \mathbf{G}_t\boldsymbol{\theta}_{t-1} + \boldsymbol{\omega}_t,$	$\boldsymbol{\omega}_t \sim \mathrm{N}[\mathbf{0}, \mathbf{W}_t],$
Initial information:	$(\boldsymbol{\theta}_0 \mid D_0) \sim \mathrm{N}[\mathbf{m}_0, \mathbf{C}_0],$	

for some prior moments $\mathbf{m}_0$ and $\mathbf{C}_0$. The observational and evolution error sequences are assumed to be internally and mutually independent, and are independent of $(\boldsymbol{\theta}_0 \mid D_0)$.

Some comments about slightly different model definitions are in order. First note that with the initial prior specified for time 0, this definition applies in particular when the data $Y_1, Y_2, \ldots$ represent the continuation of a previously observed series, the time origin $t = 0$ just being an arbitrary label. In such cases, the initial prior is viewed as sufficiently summarising the information from the past, $\boldsymbol{\theta}_0$ having the concrete interpretation of the final state vector for the historical data. Otherwise, $\boldsymbol{\theta}_0$ has no such interpretation and the model may be equivalently initialised by specifying a normal prior, $(\boldsymbol{\theta}_1|D_0)$, for the first state vector.

Secondly, apparently more general models could be obtained by allowing the error sequences $\{\nu_t\}$ and $\{\boldsymbol{\omega}_t\}$ to be both autocorrelated and cross-correlated, and some definitions of dynamic linear models would allow for this structure. However, it is always possible to reformulate such a correlated model in terms of one that satisfies the independence assumptions. Thus, nothing is lost by imposing this restriction that leads to the simplest and most easily analysed mathematical form. Further, the independence model is more meaningful and natural. The ν_t error is simply a random perturbation in the measurement process that affects the observation Y_t but has no further influence on the series. By contrast, $\boldsymbol{\omega}_t$ influences the development of the system into the future. The independence assumption clearly separates these two sources of stochastic input and clarifies their roles.

The model may be simply generalised to allow for known, non-zero means for either of the noise terms ν_t or ω_t. In addition, some or all of the noise components can be assumed known by taking the appropriate variances (and covariances) to be zero. These features are not central to the model theory but do appear in particular models and are discussed in later chapters as required.

4.3 UPDATING EQUATIONS: THE UNIVARIATE DLM

Consider univariate DLMs that are closed to external information at times $t \geq 1$, so that given initial prior information D_0 at $t = 0$, at any future time t the available information set is simply

$$D_t = \{Y_t, D_{t-1}\},$$

where Y_t is the observed value of the series at time t. To formally incorporate the known values of the defining quadruples $\{\mathbf{F}, \mathbf{G}, V, \mathbf{W}\}_t$ for each t, it is assumed that D_0 includes these values. This convention is purely for notational convenience and economy in explanation since only those values that are to be used in calculating required forecast distributions need to be known at any particular time.

The central characteristic of the normal model is that at any time, existing information about the system is represented and sufficiently summarised by the posterior distribution for the current state vector. The key results, that have trivial extension to the more general multivariate DLM, are as follows.

Theorem 4.1. *In the univariate DLM of Definition 4.3, one-step forecast and posterior distributions are given, for each t, as follows:*

(a) *Posterior at $t-1$:*
For some mean $\mathbf{m}_{t-1}$ and variance matrix $\mathbf{C}_{t-1}$,

$$(\boldsymbol{\theta}_{t-1} \mid D_{t-1}) \sim \mathrm{N}[\mathbf{m}_{t-1}, \mathbf{C}_{t-1}].$$

(b) *Prior at t:*

$$(\boldsymbol{\theta}_t \mid D_{t-1}) \sim \mathrm{N}[\mathbf{a}_t, \mathbf{R}_t],$$

where

$$\mathbf{a}_t = \mathbf{G}_t\mathbf{m}_{t-1} \qquad \text{and} \qquad \mathbf{R}_t = \mathbf{G}_t\mathbf{C}_{t-1}\mathbf{G}_t' + \mathbf{W}_t.$$

(c) *One-step forecast:*

$$(Y_t \mid D_{t-1}) \sim \mathrm{N}[f_t, Q_t],$$

where

$$f_t = \mathbf{F}_t'\mathbf{a}_t \qquad \text{and} \qquad Q_t = \mathbf{F}_t'\mathbf{R}_t\mathbf{F}_t + V_t.$$

(d) *Posterior at t:*

$$(\boldsymbol{\theta}_t \mid D_t) \sim \mathrm{N}[\mathbf{m}_t, \mathbf{C}_t],$$

with

$$\mathbf{m}_t = \mathbf{a}_t + \mathbf{A}_t e_t \qquad \textit{and} \qquad \mathbf{C}_t = \mathbf{R}_t - \mathbf{A}_t Q_t \mathbf{A}_t',$$

where

$$\mathbf{A}_t = \mathbf{R}_t \mathbf{F}_t Q_t^{-1} \qquad \textit{and} \qquad e_t = Y_t - f_t.$$

Proof. The proof is by induction using the multivariate normal distribution theory of Section 17.2. Suppose (a) true. Two proofs of (d) are given. The first important proof utilises the normality of all distributions. The second general Bayes' procedure is applicable to any distributions.

Note that sometimes the expressions used in the proof may look a little odd. The reason for this is so that both the statement of the theorem and the proof are exactly valid for the multivariate case, in which $\mathbf{Y}_t$ is a vector.

First, with terms defined in the theorem statement, establish the joint distribution:

$$\begin{pmatrix} \boldsymbol{\theta}_t \\ Y_t \end{pmatrix} \Bigg| D_{t-1} \Bigg) \sim \mathrm{N}\left[\begin{pmatrix} \mathbf{a}_t \\ f_t \end{pmatrix}, \begin{pmatrix} \mathbf{R}_t & \mathbf{A}_t Q_t \\ Q_t \mathbf{A}_t' & Q_t \end{pmatrix}\right].$$

The following results are derived from basic facts concerning means and variance matrices of linear functions of normal random vectors.

(a) The system equation $\boldsymbol{\theta}_t = \mathbf{G}_t \boldsymbol{\theta}_{t-1} + \boldsymbol{\omega}_t$, $\boldsymbol{\omega}_t \sim \mathrm{N}[\mathbf{0}, \mathbf{W}_t]$, and the prior $(\boldsymbol{\theta}_{t-1} \mid D_{t-1}) \sim \mathrm{N}[\mathbf{m}_{t-1}, \mathbf{C}_{t-1}]$ lead to $(\boldsymbol{\theta}_t \mid D_{t-1}) \sim \mathrm{N}[\mathbf{a}_t, \mathbf{R}_t]$.

(b) The observation equation $Y_t = \mathbf{F}_t' \boldsymbol{\theta}_t + \nu_t$, $\nu_t \sim \mathrm{N}[0, V_t]$, and the prior of (a) lead to $(Y_t | D_{t-1}) \sim \mathrm{N}[\mathbf{F}_t' \mathbf{a}_t, \mathbf{F}_t' \mathbf{R}_t \mathbf{F}_t + V_t]$ and prove that the joint distribution of Y_t and $\boldsymbol{\theta}_t$ is normal.

(c) The joint distribution is established upon noting the covariance

$$\begin{aligned} \mathrm{C}[\boldsymbol{\theta}_t, Y_t \mid D_{t-1}] &= \mathrm{C}[\boldsymbol{\theta}_t, \mathbf{F}_t' \boldsymbol{\theta}_t + \nu_t \mid D_{t-1}] \\ &= \mathrm{V}[\boldsymbol{\theta}_t \mid D_{t-1}] \mathbf{F}_t + \mathbf{0} = \mathbf{R}_t \mathbf{F}_t = \mathbf{A}_t Q_t. \end{aligned}$$

(d) *Proof of (d) using normal theory.*
The regression vector of $\boldsymbol{\theta}_t$ on Y_t is $\mathbf{A}_t$. The standard normal theory of Section 17.2 immediately supplies the required conditional distribution $(\boldsymbol{\theta}_t \mid Y_t, D_{t-1})$.

(d) *Proof of (d) using Bayes' theorem.*
Bayes' theorem implies that as a function of $\boldsymbol{\theta}_t$,

$$p(\boldsymbol{\theta}_t \mid D_t) \propto p(\boldsymbol{\theta}_t \mid D_{t-1})\, p(Y_t \mid \boldsymbol{\theta}_t).$$

The second term is derived from the observation equation as

$$p(Y_t \mid \boldsymbol{\theta}_t) \propto \exp\{-(Y_t - \mathbf{F}_t' \boldsymbol{\theta}_t)' V_t^{-1} (Y_t - \mathbf{F}_t' \boldsymbol{\theta}_t)/2\},$$

and the first from $(\boldsymbol{\theta}_t \mid D_{t-1}) \sim \mathrm{N}[\mathbf{a}_t, \mathbf{R}_t]$ as

$$p(\boldsymbol{\theta}_t \mid D_{t-1}) \propto \exp\left\{-(\boldsymbol{\theta}_t - \mathbf{a}_t)'\mathbf{R}_t^{-1}(\boldsymbol{\theta}_t - \mathbf{a}_t)/2\right\}.$$

Taking natural logarithms and multiplying by -2,

$$\begin{aligned}-2\ln[p(\boldsymbol{\theta}_t \mid D_t)] =&(\boldsymbol{\theta}_t - \mathbf{a}_t)'\mathbf{R}_t^{-1}(\boldsymbol{\theta}_t - \mathbf{a}_t)\\ &+ (Y_t - \mathbf{F}_t'\boldsymbol{\theta}_t)'V_t^{-1}(Y_t - \mathbf{F}_t'\boldsymbol{\theta}_t) + \text{constant},\end{aligned}$$

the constant not involving $\boldsymbol{\theta}_t$. This quadratic function of $\boldsymbol{\theta}_t$ can be expanded and rearranged with a new constant as

$$\boldsymbol{\theta}_t'(\mathbf{R}_t^{-1} + \mathbf{F}_t V_t^{-1}\mathbf{F}_t')\boldsymbol{\theta}_t - 2\boldsymbol{\theta}_t'(\mathbf{R}_t^{-1}\mathbf{a}_t + \mathbf{F}_t V_t^{-1} Y_t) + \text{constant}.$$

With $\mathbf{C}_t$ as in the theorem statement,

$$(\mathbf{R}_t^{-1} + \mathbf{F}_t V_t^{-1}\mathbf{F}_t')\mathbf{C}_t = \mathbf{I},$$

the $n \times n$ identity matrix, so that

$$\mathbf{R}_t^{-1} + \mathbf{F}_t V_t^{-1}\mathbf{F}_t' = \mathbf{C}_t^{-1}.$$

With $\mathbf{m}_t$ as in the theorem statement

$$\mathbf{C}_t^{-1}\mathbf{m}_t = \mathbf{R}_t^{-1}\mathbf{a}_t + \mathbf{F}_t V_t^{-1} Y_t.$$

Consequently, with differing constants,

$$\begin{aligned}-2\ln[p(\boldsymbol{\theta}_t \mid D_t)] &= \boldsymbol{\theta}_t'\mathbf{C}_t^{-1}\boldsymbol{\theta}_t - 2\boldsymbol{\theta}_t'\mathbf{C}_t^{-1}\mathbf{m}_t + \text{constant}\\ &= (\boldsymbol{\theta}_t - \mathbf{m}_t)'\mathbf{C}_t^{-1}(\boldsymbol{\theta}_t - \mathbf{m}_t) + \text{constant}.\end{aligned}$$

Upon exponentiating, $(\boldsymbol{\theta}_t|D_t) \sim \mathrm{N}[\mathbf{m}_t, \mathbf{C}_t]$, since

$$p(\boldsymbol{\theta}_t \mid D_t) \propto \exp\{-(\boldsymbol{\theta}_t - \mathbf{m}_t)'\mathbf{C}_t^{-1}(\boldsymbol{\theta}_t - \mathbf{m}_t)/2\}, \qquad \boldsymbol{\theta}_t \in \mathbb{R}^n.$$

◇

Key Identities

The identities below have been produced in deriving the above results. Note that they are valid for the multivariate case with **I** replacing 1 throughout.

(a) $\mathbf{A}_t = \mathbf{R}_t\mathbf{F}_t Q_t^{-1} = \mathbf{C}_t\mathbf{F}_t V_t^{-1}$;

(b) $\mathbf{C}_t = \mathbf{R}_t - \mathbf{A}_t Q_t \mathbf{A}_t' = \mathbf{R}_t(\mathbf{I} - \mathbf{F}_t\mathbf{A}_t')$;

(c) $\mathbf{C}_t^{-1} = \mathbf{R}_t^{-1} + \mathbf{F}_t V_t^{-1}\mathbf{F}_t'$;

(d) $Q_t = (1 - \mathbf{F}_t'\mathbf{A}_t)^{-1} V_t$;

(e) $\mathbf{F}_t'\mathbf{A}_t = 1 - V_t Q_t^{-1}$.

By way of terminology, e_t is the one-step forecast error and $\mathbf{A}_t$ the adaptive vector at time t.

As noted earlier, the model definition may be marginally extended to incorporate known, non-zero means $\mathrm{E}[\nu_t]$ for the observational noise and

$\mathrm{E}[\boldsymbol{\omega}_t]$ for the evolution noise. It is trivial, and left to the reader, to verify the extensions, namely that the above results apply with the modification $\mathbf{a}_t = \mathbf{G}_t\mathbf{m}_{t-1} + \mathrm{E}[\boldsymbol{\omega}_t]$ and $f_t = \mathbf{F}_t'\mathbf{a}_t + \mathrm{E}[\nu_t]$.

4.4. FORECAST DISTRIBUTIONS

Definition 4.4. The **forecast function** $f_t(k)$, at any time t, is defined for all integers $k \geq 0$ as

$$f_t(k) = \mathrm{E}[\mu_{t+k}|D_t] = \mathrm{E}[\mathbf{F}_{t+k}'\boldsymbol{\theta}_{t+k} \mid D_t],$$

where

$$\mu_v = \mathbf{F}_v'\boldsymbol{\theta}_v$$

is the **mean response function** for any time $v \geq 0$.

For k strictly greater than 0, the forecast function provides the expected values of future observations given current information,

$$f_t(k) = \mathrm{E}[Y_{t+k} \mid D_t], \qquad \text{for } \ k \geq 1.$$

However, for completeness, the definition is given in terms of the expected values of the mean response μ_{t+k} rather than $Y_{t+k} = \mu_{t+k} + \nu_{t+k}$, so including a posterior point estimate of the current level of the series, namely $f_t(0) = \mathrm{E}[\mu_t \mid D_t]$. The forecast function is of major importance in designing DLMs as will be evident in future chapters.

The following results provide the full forecast distributions. The forecast functions are central components.

Theorem 4.2. *For $0 \leq j < k$, at each time t the future joint distribution is normal and defined by the following covariances and k-step marginal distributions:*

(a) *State distribution:* $\quad (\boldsymbol{\theta}_{t+k} \mid D_t) \sim \mathrm{N}[\mathbf{a}_t(k), \mathbf{R}_t(k)],$

(b) *Forecast distribution :* $\quad (Y_{t+k} \mid D_t) \sim \mathrm{N}[f_t(k), Q_t(k)],$

(c) *State covariances:* $\quad \mathrm{C}[\boldsymbol{\theta}_{t+k}, \boldsymbol{\theta}_{t+j} \mid D_t] = \mathbf{C}_t(k,j),$

(d) *Obsn. covariances:* $\quad \mathrm{C}[Y_{t+k}, Y_{t+j} \mid D_t] = \mathbf{F}_{t+k}'\mathbf{C}_t(k,j)\mathbf{F}_{t+j},$

(e) *Other covariances:* $\quad \mathrm{C}[\boldsymbol{\theta}_{t+k}, Y_{t+j} \mid D_t] = \mathbf{C}_t(k,j)\mathbf{F}_{t+j},$

$\quad \mathrm{C}[Y_{t+k}, \boldsymbol{\theta}_{t+j} \mid D_t] = \mathbf{F}_{t+k}'\mathbf{C}_t(k,j),$

where

$$f_t(k) = \mathbf{F}_{t+k}'\mathbf{a}_t(k) \quad \text{and} \quad Q_t(k) = \mathbf{F}_{t+k}'\mathbf{R}_t(k)\mathbf{F}_{t+k} + V_{t+k},$$

that may be recursively calculated using

$$\mathbf{a}_t(k) = \mathbf{G}_{t+k}\mathbf{a}_t(k-1),$$
$$\mathbf{R}_t(k) = \mathbf{G}_{t+k}\mathbf{R}_t(k-1)\mathbf{G}'_{t+k} + \mathbf{W}_{t+k},$$
$$\mathbf{C}_t(k,j) = \mathbf{G}_{t+k}\mathbf{C}_t(k-1,j), \;\; k = j+1, \ldots,$$

together with starting values $\mathbf{a}_t(0) = \mathbf{m}_t$, $\mathbf{R}_t(0) = \mathbf{C}_t$ *and* $\mathbf{C}_t(j,j) = \mathbf{R}_t(j)$.

Proof. Define the $n \times n$ matrices $\mathbf{H}_{t+k}(r) = \mathbf{G}_{t+k}\mathbf{G}_{t+k-1}\ldots\mathbf{G}_{t+k-r+1}$ for all t and integer $r \leq k$, with $\mathbf{H}_{t+k}(0) = \mathbf{I}$. From repeated application of the state evolution equation,

$$\boldsymbol{\theta}_{t+k} = \mathbf{H}_{t+k}(k)\boldsymbol{\theta}_t + \sum_{r=1}^{k} \mathbf{H}_{t+k}(k-r)\boldsymbol{\omega}_{t+r}.$$

Thus, by linearity and independence of the normal summands,

$$(\boldsymbol{\theta}_{t+k} \mid D_t) \sim \mathrm{N}[\mathbf{a}_t(k), \mathbf{R}_t(k)],$$

where $\mathbf{a}_t(k) = \mathbf{H}_{t+k}(k)\mathbf{m}_t = \mathbf{G}_{t+k}\mathbf{a}_t(k-1)$ and

$$\begin{aligned}\mathbf{R}_t(k) &= \mathbf{H}_{t+k}(k)\mathbf{C}_t\mathbf{H}_{t+k}(k)' + \sum_{r=1}^{k} \mathbf{H}_{t+k}(k-r)\mathbf{W}_{t+r}\mathbf{H}_{t+k}(k-r)' \\ &= \mathbf{G}_{t+k}\mathbf{R}_t(k-1)\mathbf{G}'_{t+k} + \mathbf{W}_{t+k},\end{aligned}$$

with starting values $\mathbf{a}_t(0) = \mathbf{m}_t$, and $\mathbf{R}_t(0) = \mathbf{C}_t$. This establishes (a).

Using the observation equation at time $t+k$ the forecast distribution (b) is deduced as $(Y_{t+k} \mid D_t) \sim \mathrm{N}[f_t(k), Q_t(k)]$, where

$$f_t(k) = \mathbf{F}'_{t+k}\mathbf{a}_t(k) \quad \text{and} \quad Q_t(k) = \mathbf{F}'_{t+k}\mathbf{R}_t(k)\mathbf{F}_{t+k} + V_{t+k}.$$

The covariances are easily obtained using the conditional independence structure. For example, given D_t and $j < k$,

$$\begin{aligned}\mathrm{C}[\boldsymbol{\theta}_{t+k}, \boldsymbol{\theta}_{t+j}] &= \mathrm{C}[\mathbf{G}_{t+k}\boldsymbol{\theta}_{t+k-1} + \boldsymbol{\omega}_{t+k}, \boldsymbol{\theta}_{t+j}] = \mathbf{G}_{t+k}\mathrm{C}[\boldsymbol{\theta}_{t+k-1}, \boldsymbol{\theta}_{t+j}] \\ &= \mathbf{H}_{t+k}(k-j)\mathrm{C}[\boldsymbol{\theta}_{t+j}, \boldsymbol{\theta}_{t+j}] = \mathbf{H}_{t+k}(k-j)\mathbf{R}_t(j),\end{aligned}$$

as required.

Similar derivations of the other covariances are left to the reader. Since any linear function of the future Ys and $\boldsymbol{\theta}$s is normal it follows that their joint distribution is multivariate normal.

◇

In the special cases when $\mathbf{G}_t$ and/or $\mathbf{F}_t$ are constant, the above results simplify considerably and are so important that they are now detailed.

Corollary 4.1. *If the evolution matrix $\mathbf{G}_t = \mathbf{G}$ is constant for all t, then for $k \geq 0$,*

$$\mathbf{a}_t(k) = \mathbf{G}^k \mathbf{m}_t \quad \text{and} \quad f_t(k) = \mathbf{F}'_{t+k}\mathbf{G}^k \mathbf{m}_t.$$

If additionally $\mathbf{F}_t = \mathbf{F}$ for all t, so that the model is a TSDLM, then the forecast function has the form $f_t(k) = \mathbf{F}'\mathbf{G}^k\mathbf{m}_t$.

In the latter case, the potential form of the forecast function, as a function of the step ahead k, is entirely determined by the powers of the system matrix $\mathbf{G}$. This is a fundamental guiding feature in time series model design, investigated extensively in Chapters 5 and 6.

Corollary 4.2. *If the evolution matrix $\mathbf{G}_t = \mathbf{G}$ is constant for all t, then for $k, v \geq 0$,*

$$\mathbf{R}_t(k) = \mathbf{G}^k \mathbf{C}_t \mathbf{G}^k + \sum_{i=0}^{k-1} \mathbf{G}^i \mathbf{W}_{t+k-i} \mathbf{G}'^i$$

and $\mathbf{C}_t(k+v,k) = \mathbf{G}^v \mathbf{R}_t(k)$. If in addition, $\mathbf{F}_t = \mathbf{F}$ for all t, then $\mathbf{Q}_t(k) = \mathbf{F}'\mathbf{R}_t(k)\mathbf{F} + V_{t+k}$ and $\mathrm{C}[Y_{t+k+v}, Y_{t+k}] = \mathbf{F}'\mathbf{C}_t(k+v,k)\mathbf{F}$.

4.5 OBSERVATIONAL VARIANCES

So far, the defining quadruples of the univariate DLM have been assumed known for all time. Generally, the regression vectors $\mathbf{F}_t$ and the evolution matrices $\mathbf{G}_t$ are defined by the modeller in accordance with model design principles discussed in the next two chapters. The evolution variance matrix is also chosen by the modeller, usually applying the discount principle explored in Chapter 6. However the remaining element of each quadruple, the observational variance V_t, is often unknown, and large relative to the system variance $\mathbf{W}_t$. Thus, being the major source of forecasting uncertainty, appropriate Bayesian learning procedures for unknown observational variances are presented. In this chapter attention is restricted to the special case of an unknown constant variance, $V_t = V$ for all t. As in earlier chapters, its reciprocal, the observation precision, is represented by $\phi = 1/V$. Generalisations to the important cases of both stochastically changing variances and variance laws are introduced later, where the basic analysis of this section is appropriately modified.

Working in terms of the unknown precision parameter $\phi = V^{-1}$, a fully conjugate Bayesian analysis, corresponding to that introduced in Sections 2.5 and 3.4, is now developed, the key feature being that in defining the DLM, all variances and covariances are scaled by V.

Definition 4.5. For each t, writing $\phi = V^{-1}$, the DLM is defined by

Observation equation:	$Y_t = \mathbf{F}_t'\boldsymbol{\theta}_t + \nu_t,$	$\nu_t \sim \mathrm{N}[0, V],$
System equation:	$\boldsymbol{\theta}_t = \mathbf{G}_t\boldsymbol{\theta}_{t-1} + \boldsymbol{\omega}_t,$	$\boldsymbol{\omega}_t \sim \mathrm{N}[0,\ V\mathbf{W}_t^*],$
Initial information:	$(\boldsymbol{\theta}_0 \mid D_0, \phi) \sim \mathrm{N}[\mathbf{m}_0,\ V\mathbf{C}_0^*],$	
	$(\phi \mid D_0) \sim \mathrm{G}\left[\frac{n_0}{2}, \frac{n_0 S_0}{2}\right].$	

The initial quantities $\mathbf{m}_0$, $\mathbf{C}_0^*$, n_0, and S_0 are specified, as are the matrices $\{\mathbf{F}_t, \mathbf{G}_t, \mathbf{W}_t^*\}$. Notice that all variances and covariances have V as a multiplier, or scale factor, providing a *scale-free* model in terms of the starred *scale-free* variances $\mathbf{C}_0^*$ and $\mathbf{W}_t^*$. No generality is lost by this. For V fixed, the model coincides with the original Definition 4.3 with the scale factor V simply being absorbed into these matrices.

Conditionally on V, or equivalently ϕ, being known, the usual conditional independence assumptions hold. As in chapters 2 and 3, $\mathrm{E}[\phi \mid D_0] = 1/S_0$, where S_0 is a prior point estimate of the observational variance V.

The results of Theorem 4.3 and the summary Section 4.6 are based on multivariate T distributions for the state vector at all times. Details are given in Section 17.3.3 but a short summary is in order here. By analogy with that for normal distributions, the notation for Student T distributions is extended to multivariate cases. In particular, if the $n \times 1$ random vector $\boldsymbol{\theta}$ has a multivariate T distribution with h degrees of freedom, mode $\mathbf{m}$ and positive definite scale matrix $\mathbf{C}$, then the density is

$$p(\boldsymbol{\theta}) \propto \{h + (\boldsymbol{\theta} - \mathbf{m})'\mathbf{C}^{-1}(\boldsymbol{\theta} - \mathbf{m})\}^{-(n+h)/2}.$$

Following the univariate case, the notation is simply $\boldsymbol{\theta} \sim \mathrm{T}_h[\mathbf{m}, \mathbf{C}]$ with $\mathrm{E}[\boldsymbol{\theta}] = \mathbf{m}$ if $h > 1$ and $\mathrm{V}[\boldsymbol{\theta}] = \mathbf{C}h/(h-2)$ if $h > 2$. As $h \to \infty$ the distribution converges to the normal $\boldsymbol{\theta} \sim \mathrm{N}[\mathbf{m}, \mathbf{C}]$.

Theorem 4.3. *With the above DLM, the following distributional results obtain at each time $t \geq 1$.*

(a) *Conditional on V:*

$$(\boldsymbol{\theta}_{t-1} \mid D_{t-1}, V) \sim \mathrm{N}[\mathbf{m}_{t-1}, V\mathbf{C}_{t-1}^*],$$
$$(\boldsymbol{\theta}_t \mid D_{t-1}, V) \sim \mathrm{N}[\mathbf{a}_t, V\mathbf{R}_t^*],$$
$$(Y_t \mid D_{t-1}, V) \sim \mathrm{N}[f_t, VQ_t^*],$$
$$(\boldsymbol{\theta}_t \mid D_t, V) \sim \mathrm{N}[\mathbf{m}_t, V\mathbf{C}_t^*],$$

with $\mathbf{a}_t = \mathbf{G}_t\mathbf{m}_{t-1}$, $\quad \mathbf{R}_t^* = \mathbf{G}_t\mathbf{C}_{t-1}^*\mathbf{G}_t' + \mathbf{W}_t^*$,

$f_t = \mathbf{F}_t'\mathbf{a}_t$, $\quad Q_t^* = 1 + \mathbf{F}_t'\mathbf{R}_t^*\mathbf{F}_t$,

$e_t = Y_t - f_t$, $\quad \mathbf{A}_t = \mathbf{R}_t^*\mathbf{F}_t/Q_t^*$,

and $\mathbf{m}_t = \mathbf{a}_t + \mathbf{A}_t e_t$, $\quad \mathbf{C}_t^* = \mathbf{R}_t^* - \mathbf{A}_t\mathbf{A}_t'Q_t^*$.

(b) *For the precision* $\phi = V^{-1}$*:*

$$(\phi \mid D_{t-1}) \sim \mathrm{G}\left[\frac{n_{t-1}}{2}, \frac{n_{t-1}S_{t-1}}{2}\right],$$

$$(\phi \mid D_t) \sim \mathrm{G}\left[\frac{n_t}{2}, \frac{n_t S_t}{2}\right],$$

where $n_t = n_{t-1} + 1$ *and* $n_t S_t = n_{t-1}S_{t-1} + e_t^2/Q_t^*$.

(c) *Unconditional on* V *:*

$$(\boldsymbol{\theta}_{t-1} \mid D_{t-1}) \sim \mathrm{T}_{n_{t-1}}[\mathbf{m}_{t-1}, \mathbf{C}_{t-1}],$$

$$(\boldsymbol{\theta}_t \mid D_{t-1}) \sim \mathrm{T}_{n_{t-1}}[\mathbf{a}_t, \mathbf{R}_t],$$

$$(Y_t \mid D_{t-1}) \sim \mathrm{T}_{n_{t-1}}[f_t, Q_t],$$

$$(\boldsymbol{\theta}_t \mid D_t) \sim \mathrm{T}_{n_t}[\mathbf{m}_t, \mathbf{C}_t],$$

where $\mathbf{R}_t = S_{t-1}\mathbf{R}_t^*, \quad Q_t = S_{t-1}Q_t^*$ *and* $\mathbf{C}_t = S_t\mathbf{C}_t^*$.

(d) *Operational definition of updating equations:*
With $Q_t = \mathbf{F}_t'\mathbf{R}_t\mathbf{F}_t + S_{t-1}$ *and* $\mathbf{A}_t = \mathbf{R}_t\mathbf{F}_t/Q_t$,

$$n_t = n_{t-1} + 1 \quad \text{and} \quad S_t = S_{t-1} + \frac{S_{t-1}}{n_t}\left(\frac{e_t^2}{Q_t} - 1\right),$$

$$\mathbf{m}_t = \mathbf{a}_t + \mathbf{A}_t e_t \quad \text{and} \quad \mathbf{C}_t = \frac{S_t}{S_{t-1}}\left(\mathbf{R}_t - \mathbf{A}_t\mathbf{A}_t'Q_t\right).$$

Proof. Given the model definition, the results in (a) follow directly from Theorem 4.1. They are simply the known variance results. The rest of the proof is by induction, using standard normal/gamma results as detailed in Section 17.3. Assume that the prior for the precision ϕ in (b) is true. Then writing $d_t = n_t S_t$ for all t, we have $(\phi \mid D_{t-1}) \sim \mathrm{G}[n_{t-1}/2, d_{t-1}/2]$ with density function

$$p(\phi \mid D_{t-1}) \propto \phi^{n_{t-1}/2-1}\exp(-d_{t-1}\phi/2)$$

for $\phi > 0$. From (a), we have

$$p(Y_t \mid D_{t-1}, \phi) \propto \phi^{\frac{1}{2}}\exp(-e_t^2\phi/2Q_t^*),$$

so that by Bayes' theorem, the posterior for ϕ is simply

$$p(\phi \mid D_t) \propto p(\phi \mid D_{t-1})\, p(Y_t \mid D_{t-1}, \phi),$$

or just

$$p(\phi \mid D_t) \propto \phi^{n_t/2-1}\exp(-d_t\phi/2),$$

where $n_t = n_{t-1} + 1$ and $d_t = d_{t-1} + e_t^2/Q_t^*$. As $S_t = d_t/n_t$, this establishes (b). Using results in Section 17.3, if the $n \times 1$ random vector $\boldsymbol{\theta}$ has distribution $(\boldsymbol{\theta} \mid \phi) \sim \mathrm{N}[\mathbf{m}, \mathbf{C}^*/\phi]$ with $\phi \sim \mathrm{G}[n/2, nS/2]$, then, unconditionally, $\boldsymbol{\theta} \sim \mathrm{T}_n[\mathbf{m}, \mathbf{C}^* S]$. So the results in (c) follow by marginalisation of the distributions in (a) with respect to the appropriate prior/posterior gamma distribution for ϕ. The summary updating equations in (d) simply follow from those in (c), the difference being that the variances and covariances in the T distributions (all now unstarred) include the relevant estimate of V. Noting that the results are true for $t = 1$ completes the inductive proof.

◇

This theorem provides key results. At time t, the prior mean of ϕ is $\mathrm{E}[\phi \mid D_{t-1}] = n_{t-1}/d_{t-1} = 1/S_{t-1}$, where $S_{t-1} = d_{t-1}/n_{t-1}$ is a prior point estimate of $V = 1/\phi$. Similarly, the posterior estimate is $S_t = d_t/n_t$. The updating equations for the parameters defining the T prior/posterior and forecast distributions are essentially the same as the standard, known variance equations with the estimate S_{t-1} appearing as the variance. The only difference lies in the scaling by S_t/S_{t-1} in the update for $\mathbf{C}_t$ to correct for the updated estimate of V. Equations (d) are used in practice, the starred, scale-free versions appearing only to communicate the theory.

4.6 SUMMARY

For the univariate DLM the above results are tabulated here (and continued on the following page) for easy reference.

Univariate DLM: unknown, constant variance $V = \phi^{-1}$		
Observation:	$Y_t = \mathbf{F}_t'\boldsymbol{\theta}_t + \nu_t,$	$\nu_t \sim \mathrm{N}[0, V],$
System:	$\boldsymbol{\theta}_t = \mathbf{G}_t\boldsymbol{\theta}_{t-1} + \boldsymbol{\omega}_t,$	$\boldsymbol{\omega}_t \sim \mathrm{T}_{n_{t-1}}[\mathbf{0}, \mathbf{W}_t].$
Information:	$(\boldsymbol{\theta}_{t-1} \mid D_{t-1}) \sim \mathrm{T}_{n_{t-1}}[\mathbf{m}_{t-1},\ \mathbf{C}_{t-1}],$	
	$(\phi \mid D_{t-1}) \sim \mathrm{G}\left[\frac{n_{t-1}}{2},\ \frac{n_{t-1}S_{t-1}}{2}\right].$	
Forecast:	$(Y_t \mid D_{t-1}) \sim \mathrm{T}_{n_{t-1}}[f_t,\ Q_t],$	
	$(\boldsymbol{\theta}_t \mid D_{t-1}) \sim \mathrm{T}_{n_{t-1}}[\mathbf{a}_t,\ \mathbf{R}_t],$	
where	$\mathbf{R}_t = \mathbf{G}_t\mathbf{C}_{t-1}\mathbf{G}_t' + \mathbf{W}_t,$	$\mathbf{a}_t = \mathbf{G}_t\mathbf{m}_{t-1},$
	$Q_t = \mathbf{F}_t'\mathbf{R}_t\mathbf{F}_t + S_{t-1},$	$f_t = \mathbf{F}_t'\mathbf{a}_t.$

Updating Recurrence Relationships
$(\phi \mid D_t) \sim \mathrm{G}\left[\frac{n_t}{2},\ \frac{n_t S_t}{2}\right],$ $(\boldsymbol{\theta}_t \mid D_t) \sim \mathrm{T}_{n_t}[\mathbf{m}_t,\ \mathbf{C}_t],$ with $e_t = Y_t - f_t$ and $\mathbf{A}_t = \mathbf{R}_t\mathbf{F}_t/Q_t,$ $n_t = n_{t-1} + 1,$ $S_t = S_{t-1} + \frac{S_{t-1}}{n_t}\left(\frac{e_t^2}{Q_t} - 1\right),$ $\mathbf{m}_t = \mathbf{a}_t + \mathbf{A}_t e_t,$ $\mathbf{C}_t = \frac{S_t}{S_{t-1}}(\mathbf{R}_t - \mathbf{A}_t\mathbf{A}_t' Q_t).$
Forecast Distributions $k \geq 1$
$(\boldsymbol{\theta}_{t+k} \mid D_t) \sim \mathrm{T}_{n_t}[\mathbf{a}_t(k),\ \mathbf{R}_t(k)],$ $(Y_{t+k} \mid D_t) \sim \mathrm{T}_{n_t}[f_t(k),\ Q_t(k)].$ The moments are as defined in Theorem 4.2   but with V_t replaced by the estimate S_t.

4.7 FILTERING RECURRENCES

Our attention has so far focussed on the future, the sequential updating being geared to producing and revising forecasts. However, there is often great interest in looking back in time in order to get a clearer picture of what happened. Interest now lies in inference about past state vectors $\boldsymbol{\theta}_t, \boldsymbol{\theta}_{t-1}, \ldots$. There is no difficulty in obtaining retrospective marginal distributions, such as $(\boldsymbol{\theta}_{t-k} \mid D_t)$, based upon data before, at, and after time $t-k$. Nor is there any difficulty in obtaining a full joint retrospective distribution for any set of past parameters. Both this and Section 4.8 concentrate on the derivation of such retrospective distributions and sequential updating procedures.

Retrospective analysis sets out to answer, "What Happened?" Such questions provide information that is likely to improve understanding and future decision making. For example, in manufacture it is of interest to examine the *underlying* quality $\{\mu_{t-k}, k = 0, 1, \ldots, t\}$ in order to detect unexplained changes or drifts, the object being to interpret these and thus improve future control. In many economic series, such as employment figures, retail prices etc., there is great interest both in seasonal patterns and in historical *deseasonalised* figures, the latter being used to show trends and the *real* state of affairs. Other examples occur when an organisation changes its policy or a company carries out an unusual advertising campaign. Initially there may be great uncertainty about the consequences. This uncertainty can be formulated as a prior probability distribution. Then, as subsequent data are received, the distribution relating to the period of change is updated giving a increasingly sharp picture of the effect and perhaps its cause. In turn, this may lead to subsequent policy changes and model revision. Consequently, retrospective analysis contributes vitally to growth in understanding, model development, and performance.

The use of recent data to revise inferences about previous values of the state vector is called **filtering**, this information being filtered back to previous time points. The distribution of $(\boldsymbol{\theta}_{t-k} \mid D_t)$, for $k \geq 1$, is called the k-step **filtered** distribution for the state vector at time t, analogous to the k-step ahead forecast distribution. A related concept is that of **smoothing** a time series. The retrospective estimation of the historical development of a time series mean response function μ_t using the filtered distributions $(\mu_{t-k} \mid D_t)$ for $k \geq 1$ is called **smoothing** the series.

At any given time t, the filtered distributions may be derived recursively backwards in time using relationships, proven below, that are similar in structure to the standard sequential updating equations. For $k \geq 1$ the definition of the k-step ahead state forecast distributions with moments $\mathbf{a}_t(k)$ and $\mathbf{R}_t(k)$ is extended to negative arguments $\mathbf{a}_t(-k)$ and $\mathbf{R}_t(-k)$ and the following results are proved using Bayes Theorem. Section 4.8 gives an alternative proof of the full retrospective distribution using powerful conditional independence results, that have wide application.

Theorem 4.4. *In the univariate DLM $\{\mathbf{F}_t, \mathbf{G}_t, V_t, \mathbf{W}_t\}$, for all t, define*

$$\mathbf{B}_t = \mathbf{C}_t\mathbf{G}'_{t+1}\mathbf{R}^{-1}_{t+1}.$$

Then for all k, $(1 \leq k < t)$, the filtered marginal distributions are

$$(\boldsymbol{\theta}_{t-k} \mid D_t) \sim \mathrm{N}[\mathbf{a}_t(-k), \mathbf{R}_t(-k)]$$

where

$$\mathbf{a}_t(-k) = \mathbf{m}_{t-k} + \mathbf{B}_{t-k}[\mathbf{a}_t(-k+1) - \mathbf{a}_{t-k+1}]$$

and

$$\mathbf{R}_t(-k) = \mathbf{C}_{t-k} + \mathbf{B}_{t-k}[\mathbf{R}_t(-k+1) - \mathbf{R}_{t-k+1}]\mathbf{B}'_{t-k}$$

with starting values

$$\mathbf{a}_t(0) = \mathbf{m}_t \quad \textit{and} \quad \mathbf{R}_t(0) = \mathbf{C}_t,$$

and where as usual,

$$\mathbf{a}_{t-k}(1) = \mathbf{a}_{t-k+1} \quad \textit{and} \quad \mathbf{R}_{t-k}(1) = \mathbf{R}_{t-k+1}.$$

Proof. The filtered densities are defined recursively via

$$p(\boldsymbol{\theta}_{t-k} \mid D_t) = \int p(\boldsymbol{\theta}_{t-k} \mid \boldsymbol{\theta}_{t-k+1}, D_t)\, p(\boldsymbol{\theta}_{t-k+1} \mid D_t)\, d\boldsymbol{\theta}_{t-k+1}, \tag{4.3}$$

suggesting proof by induction. Assume the theorem true for $k-1$, so that it applies to the second term in the integrand of (4.3), so

$$(\boldsymbol{\theta}_{t-k+1} \mid D_t) \sim \mathrm{N}[\mathbf{a}_t(-k+1), \mathbf{R}_t(-k+1)].$$

Using Bayes' theorem, the first integrand term is

$$p(\boldsymbol{\theta}_{t-k} \mid \boldsymbol{\theta}_{t-k+1}, D_t) = \frac{p(\boldsymbol{\theta}_{t-k} \mid \boldsymbol{\theta}_{t-k+1}, D_{t-k})\, p(\mathbf{Y} \mid \boldsymbol{\theta}_{t-k}, \boldsymbol{\theta}_{t-k+1}, D_{t-k})}{p(\mathbf{Y} \mid \boldsymbol{\theta}_{t-k+1}, D_{t-k})},$$

where $\mathbf{Y} = \{Y_{t-k+1}, \dots, Y_t\}$. Now, given $\boldsymbol{\theta}_{t-k+1}$, $\mathbf{Y}$ is independent of the previous value $\boldsymbol{\theta}_{t-k}$ so that the two terms $p(\mathbf{Y} \mid \cdot)$ cancel. By Bayes' theorem, the remaining term is

$$p(\boldsymbol{\theta}_{t-k} \mid \boldsymbol{\theta}_{t-k+1}, D_{t-k}) \propto p(\boldsymbol{\theta}_{t-k} \mid D_{t-k})\, p(\boldsymbol{\theta}_{t-k+1} \mid \boldsymbol{\theta}_{t-k}, D_{t-k}). \tag{4.4}$$

Now,

$$(\boldsymbol{\theta}_{t-k+1} \mid \boldsymbol{\theta}_{t-k}, D_{t-k}) \sim \mathrm{N}[\mathbf{G}_{t-k+1}\boldsymbol{\theta}_{t-k}, \mathbf{W}_{t-k+1}]$$

and

$$(\boldsymbol{\theta}_{t-k} \mid D_{t-k}) \sim \mathrm{N}[\mathbf{m}_{t-k}, \mathbf{C}_{t-k}]$$

define the joint distribution $p(\boldsymbol{\theta}_{t-k}, \boldsymbol{\theta}_{t-k+1} \mid D_{t-k})$. From this we obtain (4.4) as

$$(\boldsymbol{\theta}_{t-k} \mid \boldsymbol{\theta}_{t-k+1}, D_{t-k}) \sim \mathrm{N}[\mathbf{h}_t(k), \mathbf{H}_t(k)], \tag{4.5}$$

where

$$\mathbf{h}_t(k) = \mathbf{m}_{t-k} + \mathbf{C}_{t-k}\mathbf{G}'_{t-k+1}\mathbf{R}^{-1}_{t-k+1}[\boldsymbol{\theta}_{t-k+1} - \mathbf{a}_{t-k+1}]$$

and

$$\mathbf{H}_t(k) = \mathbf{C}_{t-k} - \mathbf{C}_{t-k}\mathbf{G}'_{t-k+1}\mathbf{R}^{-1}_{t-k+1}\mathbf{G}_{t-k+1}\mathbf{C}_{t-k}.$$

Since $\mathbf{B}_t = \mathbf{C}_t\mathbf{G}'_{t+1}\mathbf{R}^{-1}_{t+1}$, it follows that

$$\mathbf{h}_t(k) = \mathbf{m}_{t-k} + \mathbf{B}_{t-k}[\boldsymbol{\theta}_{t-k+1} - \mathbf{a}_{t-k+1}]$$

and

$$\mathbf{H}_t(k) = \mathbf{C}_{t-k} - \mathbf{B}_{t-k}\mathbf{R}_{t-k+1}\mathbf{B}'_{t-k}.$$

Returning to (4.3), the required density $p(\boldsymbol{\theta}_{t-k} \mid D_t)$ is the expectation of (4.5) with respect to $(\boldsymbol{\theta}_{t-k+1} \mid D_t)$. This was earlier postulated to have the form stated in the Theorem and so we directly deduce that

$$(\boldsymbol{\theta}_{t-k} \mid D_t) \sim \mathrm{N}[\mathbf{a}_t(-k), \mathbf{R}_t(-k)]$$

where

$$\mathbf{a}_t(-k) = \mathrm{E}[\mathbf{h}_t(k) \mid D_t]$$

and

$$\mathbf{R}_t(-k) = \mathrm{E}[\mathbf{H}_t(k) \mid D_t] + \mathrm{V}[\mathbf{h}_t(k) \mid D_t].$$

The moments here are the values stated in the Theorem, namely

$$\mathbf{a}_t(-k) = \mathbf{m}_{t-k} + \mathbf{B}_{t-k}[\mathbf{a}_t(-k+1) - \mathbf{a}_{t-k+1}]$$

and

$$\mathbf{R}_t(-k) = \mathbf{C}_{t-k} - \mathbf{B}_{t-k}[\mathbf{R}_{t-k+1} - \mathbf{R}_t(-k+1)]\mathbf{B}'_{t-k}.$$

The theorem is completed by induction, since it is true for the case $k = 1$ with

$$\mathbf{a}_t(-k+1) = \mathbf{a}_t(0) = \mathbf{m}_t$$

and

$$\mathbf{R}_t(-k+1) = \mathbf{R}_t(0) = \mathbf{C}_t.$$

◇

Corollary 4.3. *The case of an unknown constant variance* $V_t = V = \phi^{-1}$ *: If the conjugate analysis of Section 4.5 is applied, then*

$$(\boldsymbol{\theta}_{t-k} \mid D_t) \sim \mathrm{T}_{n_t}[\mathbf{a}_t(-k),\ (S_t/S_{t-k})\mathbf{R}_t(-k)].$$

Note that as with the sequential updating equations of Section 4.6, a change of scale is implied when the unknown observational variance is estimated by the conjugate normal/gamma procedure.

Corollary 4.4. *The corresponding smoothed distributions for the mean response of the series, when* V *is unknown, are*

$$(\mu_{t-k} \mid D_t) \sim \mathrm{T}_{n_t}[f_t(-k),\ (S_t/S_{t-k})\mathbf{F}'_{t-k}\mathbf{R}_t(-k)\mathbf{F}_{t-k}],$$

where extending the forecast function notation to negative arguments,

$$f_t(-k) = \mathbf{F}'_{t-k}\mathbf{a}_t(-k).$$

4.8 RETROSPECTIVE ANALYSIS

4.8.1 Introduction

Some powerful and general conditional independence results (Harrison and Veerapen 1993), are used here to develop further aspects of the theoretical structure of DLM distribution theory useful in restrospection. Some of the supporting theory from this reference appears in Theorems 4.16 and 4.17, together with Corollary 4.10, in the Appendix, Section 4.11. Although phrased in terms of normal distributions, these results hold under weaker assumptions and are important when considering general problems regarding the incorporation and deletion of information in the form of linear functions of observations, expert opinion, or external forecasts. Theorem 4.17 immediately allows the derivation of the entire historical joint parameter distribution $(\boldsymbol{\theta}_1, \ldots, \boldsymbol{\theta}_t | D_t)$ together with two useful sets of recurrence relations: the one already derived in Section 4.7 and the other being important when continual retrospection is required for a specified period, perhaps relating to a policy change or other significant event. Finally, recurrences are given for the dual problem of deleting previously incorporated data. This is required when assessing the influence of observations, for jackknife analysis, and especially in model assessment when calculating distributions such as $(Y_{t-k} \mid D_t - Y_{t-k})$, i.e., that for Y_{t-k} based upon all the data D_t except Y_{t-k} itself. This enables the associated retrospective jackknife forecast residuals to be assessed as in Harrison and West (1991).

4.8.2 The DLM retrospective parametric distribution

The results of Appendix 4.11 are applied here to the multivariate DLM $\{\mathbf{F}_t, \mathbf{G}_t, \mathbf{V}_t, \mathbf{W}_t\}$ with parameter $\boldsymbol{\theta}_t$. The conditional independence structure follows that of Figure 4.3. The notation of Section 4.7 for the multivariate DLM is adopted, with $\mathbf{B}_{t-k} = \mathbf{C}_{t-k}\mathbf{G}'_{t-k+1}\mathbf{R}^{-1}_{t-k+1}$ for all $t = 1, 2, \ldots$ and k such that $1 \leq k \leq t$.

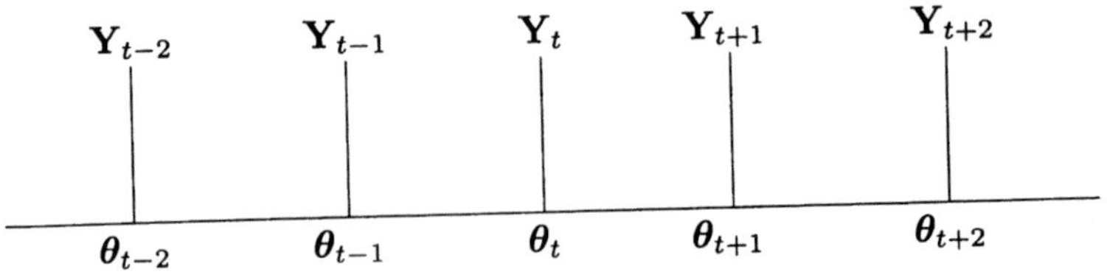

Figure 4.3 The DLM conditional independence structure

Definition 4.6. Given D_t, for any integers $k \geq j \geq 0$, define the regression matrix of $\boldsymbol{\theta}_{t-k}$ on $\boldsymbol{\theta}_{t-j}$ as $\mathbf{A}_{t-k,t-j}$.

Then $\mathbf{A}_{t-k,t-k} = \mathbf{I}$ and for any $k > 0$,

$$\mathbf{A}_{t-k,t-k+1} = \mathbf{B}_{t-k} \quad \text{and} \quad \mathbf{A}_{t-k,t} = \prod_{v=t-k}^{t-1} \mathbf{B}_v.$$

These follow from Corollary 4.10 in conjunction with

$$\begin{aligned} \mathrm{C}[\boldsymbol{\theta}_{t-k}, \boldsymbol{\theta}_{t-k+1}|D_{t-k}] &= \mathrm{C}[\boldsymbol{\theta}_{t-k}, \mathbf{G}_{t-k+1}\boldsymbol{\theta}_{t-k} + \boldsymbol{\omega}_{t-k+1}|D_{t-k}] \\ &= \mathbf{C}_{t-k}\mathbf{G}'_{t-k+1} = \mathbf{B}_{t-k}\mathbf{R}_{t-k+1}. \end{aligned}$$

Further, $\boldsymbol{\theta}_t$ and Y_{t+1} are conditionally independent given $\boldsymbol{\theta}_{t+1}$, written as $\boldsymbol{\theta}_t \perp\!\!\!\perp \mathbf{Y}_{t+1}|\boldsymbol{\theta}_{t+1}$. It follows that conditional on D_{t-1}, the regression matrix of $\boldsymbol{\theta}_{t-k}$ on $\mathbf{Y}_t$ is $\mathbf{A}_{t-k,t}\mathbf{A}_t$, where as usual, $\mathbf{A}_t$ is the regression matrix of $\boldsymbol{\theta}_t$ on $\mathbf{Y}_t$ given D_{t-1}. Retrospective recurrence relationships follow immediately from Theorem 4.17, the proof being trivial and left to the reader.

Theorem 4.5. *Given D_t the joint distribution of the historical parameters, $(\boldsymbol{\theta}_1, \ldots, \boldsymbol{\theta}_t|D_t)$ is defined by their marginal distributions and covariances, that may be calculated recurrently in either of the following ways:*

Retrospective Parametric Distribution
$(\boldsymbol{\theta}_{t-k}|D_t) \sim \mathrm{N}[\mathbf{a}_t(-k),\ \mathbf{R}_t(-k)], \qquad k \geq 0,$
$\mathrm{C}[\boldsymbol{\theta}_{t-k-j}, \boldsymbol{\theta}_{t-k}|D_t] = \mathbf{A}_{t-k-j,t-k}\mathbf{R}_t(-k), \qquad j \geq 0.$
Retrospective Recurrence Relations
With $\mathbf{B}_v = \mathbf{C}_v\mathbf{G}'_{v+1}\mathbf{R}^{-1}_{v+1}$ *and* $\mathbf{A}_{t-k,t} = \prod_{v=t-k}^{t-1}\mathbf{B}_v$,
(i) $\mathbf{a}_t(-k) = \mathbf{a}_{t-1}(-k+1) + \mathbf{A}_{t-k,t}\mathbf{A}_t\mathbf{e}_t,$
$\mathbf{R}_t(-k) = \mathbf{R}_{t-1}(-k+1) - \mathbf{A}_{t-k,t}\mathbf{A}_t\mathbf{Q}_t\mathbf{A}'_t\mathbf{A}'_{t-k,t},$
$\mathbf{A}_{t-k,t} = \mathbf{A}_{t-k,t-1}\mathbf{B}_{t-1}.$
(ii) $\mathbf{a}_t(-k) = \mathbf{m}_{t-k} + \mathbf{B}_{t-k}[\mathbf{a}_t(-k+1) - \mathbf{a}_{t-k+1}],$
$\mathbf{R}_t(-k) = \mathbf{C}_{t-k} + \mathbf{B}_{t-k}[\mathbf{R}_t(-k+1) - \mathbf{R}_{t-k+1}]\mathbf{B}'_{t-k}.$

Equations (i) in the above table are particularly useful for continually updating the whole history or just that part in which interest lies, such as a time of special promotion, policy change, or major external event. Notice that the recurrence relations of Theorem 4.1(d) are given by the special case

$k = 0$. Equations (ii), essentially the same as derived in Theorem 4.4, are useful when retrospection is only required at a particular time t, as is usual when a batch of data is to be analysed. Then starting with $\mathbf{a}_t(0) = \mathbf{m}_t$ and $\mathbf{R}_t(0) = \mathbf{C}_t$, we apply the equations sequentially backwards over time.

These results are conditional upon known variances. For the multivariate model $\{\mathbf{F}_t,\ \mathbf{G}_t,\ \mathbf{V}_t/\phi,\ \mathbf{W}_t/\phi\}$, where just the scale factor ϕ is unknown with mean $\mathrm{E}[\phi|D_t] = 1/S_t$, Corollaries 4.3 and 4.10 are directly applicable. Then for $k \geq 0$,

$$(\boldsymbol{\theta}_{t-k} \mid D_t) \sim \mathrm{T}_{n_t}[\mathbf{a}_t(-k),\ (S_t/S_{t-k})\mathbf{R}_t(-k)],$$

and for $j, k \geq 0$,

$$\mathrm{C}[\boldsymbol{\theta}_{t-k-j}, \boldsymbol{\theta}_{t-k}|D_t] = \mathbf{A}_{t-k-j,t-k}\mathrm{V}[\boldsymbol{\theta}_{t-k} \mid D_t].$$

4.8.3 Deleting observations

The following theorem provides a general result that leads to the theory for revising distributions when information is deleted. The $r \times 1$ vector $\mathbf{Y}$ is any set of observations and subjective information, $\mathbf{F}$ is a known matrix, $\mathbf{Z}$ any vector of observations and states, and $\phi = 1/V$ is a scalar precision. Let the following be proper distributions:

$$\begin{aligned}(\mathbf{Y} \mid \mathbf{Z}, \phi) &\sim \mathrm{N}[\mathbf{F}'\mathbf{Z},\ \mathbf{R}_{y|z}V],\\ (\mathbf{Z} \mid \mathbf{Y}, \phi) &\sim \mathrm{N}[\mathbf{a}_{z|y},\ \mathbf{R}_{z|y}V],\\ (\phi \mid \mathbf{Y}) &\sim \mathrm{G}\left[\frac{n_y}{2},\ \frac{n_y S_y}{2}\right],\end{aligned}$$

$$\left(\begin{matrix}\mathbf{Z}\\ \mathbf{Y}\end{matrix}\,\middle|\,\phi\right) \sim \mathrm{N}\left[\begin{pmatrix}\mathbf{a}_z\\ \mathbf{F}'\mathbf{a}_z\end{pmatrix}, \begin{pmatrix}\mathbf{R}_z & \mathbf{R}_z\mathbf{F}\\ \mathbf{F}'\mathbf{R}_z & \mathbf{R}_y\end{pmatrix}V\right],$$

where $\mathbf{R}_y = \mathbf{F}'\mathbf{R}_z\mathbf{F} + \mathbf{R}_{y|z}$.

Theorem 4.6. *Given the above distributional structure and notation, define*

$$\mathbf{d} = \mathbf{Y} - \mathbf{F}'\mathbf{a}_{z|y}, \quad \mathbf{R}_d = \mathbf{R}_{y|z} - \mathbf{F}'\mathbf{R}_{z|y}\mathbf{F}, \quad \text{and} \quad \mathbf{A}_{zd} = \mathbf{R}_{z|y}\mathbf{F}\mathbf{R}_d^{-1}.$$

The following results hold:

(1) *The leverage* $\mathbf{A}$, *of* $\mathbf{Y}$ *on* $\mathbf{Z}$, *is calculable as* $\mathbf{A} = \mathbf{R}_{z|y}\mathbf{F}\mathbf{R}_{y|z}^{-1}$.

(2) *Deleting* $\mathbf{Y}$ *we have*

$$(\mathbf{Z} \mid \phi) \sim \mathrm{N}[\mathbf{a}_z, \mathbf{R}_z V] \quad \text{and} \quad \phi \sim \mathrm{G}[n/2, nS/2],$$

where

$$\mathbf{a}_z = \mathbf{a}_{z|y} - \mathbf{A}_{zd}\mathbf{d},$$
$$\mathbf{R}_z = \mathbf{R}_{z|y} + \mathbf{A}_{zd}\mathbf{R}_d\mathbf{A}'_{zd},$$
$$n = n_y - r,$$

and

$$nS = n_y S_y - \mathbf{d}'\mathbf{R}_d^{-1}\mathbf{d}.$$

(3) *The jackknife forecast for* $\mathbf{Y}$ *is*

$$\mathbf{Y} \sim \mathrm{T}_n[\mathbf{F}'\mathbf{a}_z,\ \mathbf{R}_y S],$$

where $\mathbf{R}_y = \mathbf{F}'\mathbf{R}_z\mathbf{F} + \mathbf{R}_{y|z}$.

Proof. The proof follows Harrison and Veerapen (1993). Notice first that $p(\mathbf{Z} \mid \mathbf{Y}, \phi) \equiv p(\mathbf{Z} \mid \mathbf{d}, \phi)$ and

$$\begin{pmatrix} \mathbf{Z} \\ \mathbf{d} \end{pmatrix} \Big| \phi \sim \mathrm{N}\left[\begin{pmatrix} \mathbf{a}_z \\ \mathbf{0} \end{pmatrix}, \begin{pmatrix} \mathbf{R}_z & \mathbf{A}_{zd}\mathbf{R}_d \\ \mathbf{R}_d\mathbf{A}'_{zd} & \mathbf{R}_d \end{pmatrix} V\right].$$

From the usual Bayes' updating for $(\mathbf{Z} \mid \mathbf{Y}, \phi)$ we see that

$$\mathbf{R}_{z|y}^{-1} = \mathbf{R}_z^{-1} + \mathbf{F}\mathbf{R}_{y|z}^{-1}\mathbf{F}'$$

and

$$\mathbf{R}_{z|y}^{-1}\mathbf{a}_{z|y} = \mathbf{R}_z^{-1}\mathbf{a}_z + \mathbf{F}\mathbf{R}_{y|z}^{-1}\mathbf{Y}.$$

From the latter equation, the leverage is $\mathbf{A} = \mathbf{R}_{z|y}\mathbf{F}\mathbf{R}_{y|z}^{-1}$, proving (1). The variance result for $\mathbf{R}_z$ in (2) follows from the precision equation above, i.e., the identity

$$\mathbf{R}_z^{-1} = \mathbf{R}_{z|y}^{-1} - \mathbf{F}\mathbf{R}_{y|z}^{-1}\mathbf{F}'$$

implies that

$$\mathbf{R}_z = \mathbf{R}_{z|y} + \mathbf{A}_{zd}\mathbf{R}_d\mathbf{A}'_{zd}.$$

The result for the mean in (2) follows similarly, i.e., the identity

$$\mathbf{R}_z^{-1}\mathbf{a}_z = (\mathbf{R}_z^{-1} + \mathbf{F}\mathbf{R}_{y|z}^{-1}\mathbf{F}')\mathbf{a}_{z|y} - \mathbf{F}\mathbf{R}_{y|z}^{-1}\mathbf{Y}$$

implies that

$$\mathbf{a}_z = \mathbf{a}_{z|y} - \mathbf{R}_z\mathbf{F}\mathbf{R}_{y|z}^{-1}(\mathbf{Y} - \mathbf{F}'\mathbf{a}_{z|y}) = \mathbf{a}_{z|y} - \mathbf{A}_{zd}\mathbf{d}.$$

Write the jackknife residual as $\mathbf{e} = \mathbf{Y} - \mathbf{F}'\mathbf{a}_z$. Then

$$\mathbf{e} = \mathbf{Y} - \mathbf{F}'\mathbf{a}_{z|y} + \mathbf{F}'(\mathbf{a}_{z|y} - \mathbf{a}_z) = \mathbf{d} + \mathbf{F}'\mathbf{A}\mathbf{e},$$

which leads to

$$\mathbf{d} = (\mathbf{I}_r - \mathbf{F}'\mathbf{A})\mathbf{e} = \mathbf{R}_{y|z}\mathbf{R}_y^{-1}\mathbf{e}.$$

Also,

$$\mathbf{d} = \mathbf{Y} - \mathbf{F}'\mathbf{a}_z - \mathbf{F}'\mathbf{A}_{zd}\mathbf{d}$$

implies

$$\mathbf{e} = (\mathbf{I}_r + \mathbf{F}'\mathbf{R}_{z|y}\mathbf{F}\mathbf{R}_d^{-1})\mathbf{d} = \mathbf{R}_{y|z}\mathbf{R}_d^{-1}\mathbf{d}.$$

Now, using the identities $n_y = n + r$, $\mathbf{d} = \mathbf{R}_{y|z}\mathbf{R}_y^{-1}\mathbf{e}$ and $\mathbf{e} = \mathbf{R}_{y|z}\mathbf{R}_d^{-1}\mathbf{d}$, we see that

$$n_y S_y = nS + \mathbf{e}'\mathbf{R}_y^{-1}\mathbf{e} = nS + \mathbf{d}'\mathbf{R}_{y|z}^{-1}\mathbf{R}_{y|z}\mathbf{R}_d^{-1}\mathbf{d} = nS + \mathbf{d}'\mathbf{R}_d^{-1}\mathbf{d}.$$

Then $nS = n_y S_y - \mathbf{d}'\mathbf{R}_d^{-1}\mathbf{d}$ and $n = n_y - r$, completing (2). The jackknife forecast of (3) now follows immediately.

◇

4.8.4 Deleting observations in the DLM

The following results apply to the multivariate DLM

$$\{\mathbf{F}_t,\ \mathbf{G}_t,\ \mathbf{V}_t/\phi,\ \mathbf{W}_t/\phi\},$$

where the variance sequences $\{\mathbf{V}_t\}$ and $\{\mathbf{W}_t\}$ are known but the scale parameter ϕ is unknown. Consequently, the univariate DLM with unknown constant variance $V = \phi^{-1}$ is the special case $\mathbf{V}_t = 1$, as in Harrison and West (1991).

Write $D_t(-k) = D_t - \mathbf{Y}_{t-k}$ to be the current information except for the observed value of $\mathbf{Y}_{t-k}$. Given D_t, define the following retrospective quantities relating to time $t-k$:

$$\begin{aligned}
\mathbf{e}_t(-k) &= \mathbf{Y}_{t-k} - \mathbf{F}'_{t-k}\mathbf{a}_t(-k),\\
\mathbf{Q}_t(-k) &= \mathbf{V}_{t-k} - \mathbf{F}'_{t-k}\mathbf{R}_t(-k)\mathbf{F}_{t-k},\\
\mathbf{A}_t(-k) &= \mathbf{R}_t(-k)\mathbf{F}_{t-k}\{\mathbf{Q}_t(-k)\}^{-1}.
\end{aligned}$$

Theorem 4.7. *Deleting the observation $\mathbf{Y}_{t-k}$, the marginal distributions are*

$$(\boldsymbol{\theta}_{t-k}|D_t(-k)) \sim \mathrm{T}_n[\mathbf{a}_{t,k},\ \mathbf{R}_{t,k}]$$

and

$$(\phi|D_t(-k)) \sim \mathrm{G}\left[\frac{n}{2},\ \frac{nS_{t,k}}{2}\right],$$

where

$$\begin{aligned}
\mathbf{a}_{t,k} &= \mathbf{a}_t(-k) - \mathbf{A}_t(-k)\mathbf{e}_t(-k),\\
\mathbf{R}_{t,k} &= \mathbf{R}_t(-k) + \mathbf{A}_t(-k)\mathbf{Q}_t(-k)\mathbf{A}'_t(-k),\\
n &= n_t - r,
\end{aligned}$$

and

$$S_{t,k} = S_t + \frac{1}{n}\left[r - \mathbf{e}_t'(-k)\{\mathbf{Q}_t(-k)\}^{-1}\mathbf{e}_t(-k)\right].$$

Proof. This is a just special case of Theorem 4.6, the correspondence being $\mathbf{Z} = \boldsymbol{\theta}_{t-k}$, $\mathbf{Y} = \mathbf{Y}_{t-k}$, $\mathbf{R}_d = \mathbf{Q}_t(-k)$, $\mathbf{R}_{y|z} = V_{t-k}$, $\mathbf{d} = \mathbf{e}_t(-k)$ and $\mathbf{A}_{zd} = \mathbf{A}_t(-k)$, with the joint distribution of $(\mathbf{Z}, \mathbf{Y})$ being (implicitly) conditioned on $D_t(-k)$.

◇

Definition 4.7. The leverage of the observation $\mathbf{Y}_{t-k}$ is the regression matrix of $\boldsymbol{\theta}_{t-k}$ on $\mathbf{Y}_{t-k}$ given $D_t(-k)$, namely

$$\mathbf{R}_{t,k}\mathbf{F}_{t-k}(\mathbf{F}_{t-k}'\mathbf{R}_{t,k}\mathbf{F}_{t-k} + \mathbf{V}_{t-k})^{-1}.$$

This leverage measures the influence that the individual observation $\mathbf{Y}_{t-k}$ has on the parametric mean $\boldsymbol{\theta}_{t-k}$.

Definition 4.8. The jackknife forecast for $\mathbf{Y}_{t-k}$ is the distribution

$$(\mathbf{Y}_{t-k}|D_t(-k)) \sim \mathrm{T}_n[\mathbf{f}_{t,k},\ \mathbf{Q}_{t,k}],$$

where

$$\mathbf{f}_{t,k} = \mathbf{F}_{t-k}'\mathbf{a}_{t,k} \quad \text{and} \quad \mathbf{Q}_{t,k} = \mathbf{F}_{t-k}'\mathbf{R}_{t,k}\mathbf{F}_{t-k} + \mathbf{V}_{t-k}S_{tk}.$$

This is the forecast for $\mathbf{Y}_{t-k}$ given all the information except for $\mathbf{Y}_{t-k}$ itself. A time graph of these forecasts and observations is informative.

Definition 4.9. The jackknife residual is

$$\mathbf{e}_{t,k} = \mathbf{Y}_{t-k} - \mathbf{f}_{t,k}.$$

The set of standardised jackknife residuals is useful in assessing model adequacy, outlying observations and influential data points.

Theorem 4.8. *Define* $\mathbf{A}_{t-k,t-j} = \mathbf{R}_t(-k)\mathbf{A}_{t-j,t-k}'\{\mathbf{R}_t(-j)\}^{-1}$. *Deleting the observation* $\mathbf{Y}_{t-k}$, *the revised retrospective parametric marginal distributions are*

$$(\boldsymbol{\theta}_{t-j}|D_t(-k)) \sim \mathrm{T}_n[\mathbf{a}_{t,k}(-j),\ \mathbf{R}_{t,k}(-j)], \qquad k < j,$$

where

$$\mathbf{a}_{t,k}(-j) = \mathbf{a}_t(-j) + \mathbf{A}_{t-k,t-j}[\mathbf{a}_{t,k} - \mathbf{a}_t(-k)]$$

and

$$\mathbf{R}_{t,k}(-j) = \mathbf{R}_t(-j) + \mathbf{A}_{t-k,t-j}[\mathbf{R}_{t,k} - \mathbf{R}_t(-k)]\mathbf{A}_{t-k,t-j}',$$

which for $k > j$ may be calculated recurrently using relationships

$$\mathbf{a}_{t,k}(-j) = \mathbf{a}_t(-j) + \mathbf{B}_{t-j}[\mathbf{a}_{t,k}(-j+1) - \mathbf{a}_t(-j+1)]$$

and

$$\mathbf{R}_{t,k}(-j) = \mathbf{R}_t(-j) + \mathbf{B}_{t-j}[\mathbf{R}_{t,k}(-j+1) - \mathbf{R}_t(-j+1)]\mathbf{B}'_{t-j}.$$

The proofs of these results with extensions to stochastic variances and eliminating blocks of data are given in Harrison and Veerapen (1993).

4.9 LINEAR BAYES' OPTIMALITY

4.9.1 Introduction

Theorem 4.1 provides the key updating equations for dynamic models assuming normality of the observational and evolution error sequences and the prior at $t = 0$. However, the recurrences for $\mathbf{m}_t$ and $\mathbf{C}_t$ may also be derived using approaches that do not invoke the normal assumption, since they possess strong optimality properties that are derived when the distributions are only partially specified in terms of means and variances.

Sections 4.9.2 and 4.9.3 describe the decision theoretically based linear Bayes' estimation procedure and its application to DLMs. Section 4.9.4 gives an alternative precise probabilistic derivation for the recurrences using what we term the weak Bayes' approach. The idea here is that $\phi(\boldsymbol{\theta}_t, Y_t)$, a function of the parameter $\boldsymbol{\theta}_t$ and the observation Y_t, is modelled as independent of Y_t, so that upon observing the value $Y_t = y$, the posterior distribution $\phi(\boldsymbol{\theta}_t, y)$ is identical to the prior distribution $\phi(\boldsymbol{\theta}_t, Y_t)$.

Non-Bayesian techniques of minimum variance/least squares estimation will be familiar to some readers versed in Kalman filtering and are adopted by many authors as a basis for the recurrence relations; see, for example, Anderson and Moore (1979), Harvey (1981), Jazwinski (1970), Kalman (1960, 1963), and Sage and Melsa (1971). Section 4.9.5 briefly discusses the relationship between these and Bayes' methods.

Both linear Bayes' and weak Bayes' approaches assume the standard DLM but drop the normality assumptions. Thus, in the univariate case the model equations become

$$\begin{aligned} Y_t &= \mathbf{F}'_t\boldsymbol{\theta}_t + \nu_t, & \nu_t &\sim [0, V_t], \\ \boldsymbol{\theta}_t &= \mathbf{G}_t\boldsymbol{\theta}_{t-1} + \boldsymbol{\omega}_t, & \boldsymbol{\omega}_t &\sim [\mathbf{0}, \mathbf{W}_t], \end{aligned} \tag{4.6}$$

$$(\boldsymbol{\theta}_0 \mid D_0) \sim [\mathbf{m}_0, \mathbf{C}_0].$$

Provided they are consistent with their defined first and second-order moments all the distributions are now free to take any form whatsoever.

4.9.2. Linear Bayes' Estimation

Linear Bayes' estimation is now detailed in a general setting in which inferences are to be made about the n-vector $\boldsymbol{\theta}$ based upon a p-vector observation $\mathbf{Y}$, given their joint distribution and employing the following decision-theoretic framework. Further discussion of the principles, theory, and applications can be found in Hartigan (1969), and Goldstein (1976).

Let $\mathbf{d}$ be any estimate of $\boldsymbol{\theta}$, and suppose that accuracy in estimation is measured by a loss function $\mathrm{L}(\boldsymbol{\theta}, \mathbf{d})$. Then the estimate $\mathbf{d} = \mathbf{m} = \mathbf{m}(\mathbf{Y})$ is optimal with respect to the loss function if the posterior risk, or expected loss, function $r(\mathbf{d}) = \mathrm{E}[\mathrm{L}(\boldsymbol{\theta}, \mathbf{d}) \mid \mathbf{Y}]$ is minimised as a function of $\mathbf{d}$ when $\mathbf{d} = \mathbf{m}$. In particular, if the loss function is quadratic,

$$\mathrm{L}(\boldsymbol{\theta}, \mathbf{d}) = (\boldsymbol{\theta} - \mathbf{d})'(\boldsymbol{\theta} - \mathbf{d}) = \text{ trace } (\boldsymbol{\theta} - \mathbf{d})(\boldsymbol{\theta} - \mathbf{d})',$$

then the posterior risk is minimised at $\mathbf{m} = \mathrm{E}[\boldsymbol{\theta} \mid \mathbf{Y}]$, the posterior mean, and the minimum risk is the trace of the posterior variance matrix, equivalently the sum of posterior variances of the elements of $\boldsymbol{\theta}$, namely, $r(\mathbf{m}) = \mathrm{trace}\mathrm{V}[\boldsymbol{\theta} \mid \mathbf{Y}]$. This is a standard result in Bayesian decision theory; see, for example, De Groot (1971), or Berger (1985).

However, suppose now that the decision maker has only partially specified the joint distribution of $\boldsymbol{\theta}$ and $\mathbf{Y}$, providing just the joint first- and second-order moments, the mean and variance matrix

$$\begin{pmatrix} \boldsymbol{\theta} \\ \mathbf{Y} \end{pmatrix} \sim \left[\begin{pmatrix} \mathbf{a} \\ \mathbf{f} \end{pmatrix}, \begin{pmatrix} \mathbf{R} & \mathbf{AQ} \\ \mathbf{QA}' & \mathbf{Q} \end{pmatrix} \right]. \tag{4.7}$$

With a quadratic loss function, or indeed any other, this specification does not provide enough information to identify the optimal estimate, nor the posterior mean, nor the posterior variance. They are, in fact, undefined. The LBE method side-steps this problem, providing an alternative estimate that may be viewed as an approximation to the optimal procedure. Since the posterior risk function cannot be calculated, the overall risk

$$r(\mathbf{d}) = \mathrm{trace}\, \mathrm{E}[(\boldsymbol{\theta} - \mathbf{d})(\boldsymbol{\theta} - \mathbf{d})'], \tag{4.8}$$

unconditional on $\mathbf{Y}$, is used instead. Furthermore, estimates $\mathbf{d} = \mathbf{d}(\mathbf{Y})$ of $\boldsymbol{\theta}$ are restricted to linear functions of $\mathbf{Y}$, of the form

$$\mathbf{d}(\mathbf{Y}) = \mathbf{h} + \mathbf{HY}, \tag{4.9}$$

for some $n \times 1$ vector $\mathbf{h}$ and $n \times p$ matrix $\mathbf{H}$. Clearly, this may be viewed as a standard application of linear regression; the unknown regression function $\mathrm{E}[\boldsymbol{\theta} \mid \mathbf{Y}]$ is approximated by the linear model above.

Definition 4.10. **A linear Bayes' estimate (LBE)** of $\boldsymbol{\theta}$ is a linear form (4.9) that is optimal in the sense of minimising the overall risk (4.8).

The main result is as follows.

Theorem 4.9. *In the above framework, the unique LBE of* $\boldsymbol{\theta}$ *is*

$$\mathbf{m} = \mathbf{a} + \mathbf{A}(\mathbf{Y} - \mathbf{f}).$$

The associated **risk matrix (RM)**, *given by*

$$\mathbf{C} = \mathbf{R} - \mathbf{AQA}',$$

is the value of $\mathrm{E}[(\boldsymbol{\theta} - \mathbf{m})(\boldsymbol{\theta} - \mathbf{m})']$, *so that the minimum risk is simply* $r(\mathbf{m}) = \mathrm{trace}(\mathbf{C})$.

Proof. For $\mathbf{d} = \mathbf{h} + \mathbf{HY}$, define $\mathbf{R}(\mathbf{d}) = \mathrm{E}[(\boldsymbol{\theta} - \mathbf{d})(\boldsymbol{\theta} - \mathbf{d})']$. Then

$$\mathbf{R}(\mathbf{d}) = \mathbf{R} + \mathbf{HQH}' - \mathbf{AQH}' - \mathbf{HQA}' + (\mathbf{a} - \mathbf{h} - \mathbf{Hf})(\mathbf{a} - \mathbf{h} - \mathbf{Hf})',$$

that may be written, using the identity

$$(\mathbf{H} - \mathbf{A})\mathbf{Q}(\mathbf{H} - \mathbf{A})' = \mathbf{AQA}' + \mathbf{HQH}' - \mathbf{AQH}' - \mathbf{HQA}',$$

as

$$\mathbf{R}(\mathbf{d}) = \mathbf{R} - \mathbf{AQA}' + (\mathbf{H} - \mathbf{A})\mathbf{Q}(\mathbf{H} - \mathbf{A})' + (\mathbf{a} - \mathbf{h} - \mathbf{Hf})(\mathbf{a} - \mathbf{h} - \mathbf{Hf})'.$$

Hence the risk $r(\mathbf{d}) = \mathrm{trace}\mathbf{R}(\mathbf{d})$ is the sum of three terms:

(1) $\mathrm{trace}(\mathbf{R} - \mathbf{AQA}')$, that is independent of $\mathbf{d}$;
(2) $\mathrm{trace}(\mathbf{H} - \mathbf{A})\mathbf{Q}(\mathbf{H} - \mathbf{A})'$, that has a minimum value of 0 at $\mathbf{H} = \mathbf{A}$; and
(3) $(\mathbf{a} - \mathbf{h} - \mathbf{Hf})'(\mathbf{a} - \mathbf{h} - \mathbf{Hf})$, that has a minimum value of 0 at $\mathbf{h} + \mathbf{Hf} = \mathbf{a}$.

Thus $r(\mathbf{d})$ is minimised at $\mathbf{H} = \mathbf{A}$ and $\mathbf{h} = \mathbf{a} - \mathbf{Af}$, so as required,

$$\mathbf{d}(\mathbf{Y}) = \mathbf{a} + \mathbf{A}(\mathbf{Y} - \mathbf{f}) = \mathbf{m}.$$

Also, at $\mathbf{d} = \mathbf{m}$, the risk matrix is, as required,

$$\mathrm{E}[(\boldsymbol{\theta} - \mathbf{m})(\boldsymbol{\theta} - \mathbf{m})'] = \mathbf{R} - \mathbf{AQA}' = \mathbf{C}.$$

◇

Corollary 4.5. *If* $\boldsymbol{\theta}_t$ *is any subvector of* $\boldsymbol{\theta}$, *then the LBE of* $\boldsymbol{\theta}_t$ *and the associated RM from the marginal distribution of* $\mathbf{Y}$ *and* $\boldsymbol{\theta}_t$ *coincide with the relevant marginal terms of* $\mathbf{m}$ *and* $\mathbf{C}$.

Corollary 4.6. $\mathbf{m}$ *and* $\mathbf{C}$ *are equivalent to the posterior mean and variance matrix of* $(\boldsymbol{\theta} \mid \mathbf{Y})$ *in the case of joint normality of* $\mathbf{Y}$ *and* $\boldsymbol{\theta}$.

The use and interpretation of $\mathbf{m}$ and $\mathbf{C}$ is as approximations to posterior moments. Given the restriction to linear estimates $\mathbf{d}$ in $\mathbf{Y}$, $\mathbf{m}$ is sometimes called the **linear posterior mean** and $\mathbf{C}$ the **linear posterior variance** of $(\boldsymbol{\theta} \mid \mathbf{Y})$. This terminology is due to Hartigan (1969).

4.9.3. Linear Bayes' estimation in the DLM

Consider the DLM specified by equations (4.6) without any further distributional assumptions. The observational and evolution error sequences, assumed independent in the normal framework, may now be dependent, so long as they remain uncorrelated. Although not generally true, under normality assumptions this implies independence. Suppose, in addition, the initial prior is partially specified in terms of moments as

$$(\boldsymbol{\theta}_0 \mid D_0) \sim [\mathbf{m}_0, \mathbf{C}_0],$$

being uncorrelated with the error sequences. Finally, $D_t = \{D_{t-1}, Y_t\}$, with D_0 containing the known values of the error variance sequences.

Theorem 4.10. *The moments* $\mathbf{m}_t$ *and* $\mathbf{C}_t$ *as defined in the normal DLM of Theorem 4.1 are the linear posterior mean and variance matrix of* $(\boldsymbol{\theta}_t \mid D_t)$.

Proof. For any t, let $\mathbf{Y}$ and $\boldsymbol{\theta}$ be defined respectively as the $t \times 1$ and $n(t+1) \times 1$ vectors

$$\mathbf{Y} = (Y_t, Y_{t-1}, \ldots, Y_1)'$$

and

$$\boldsymbol{\theta} = (\boldsymbol{\theta}_t', \boldsymbol{\theta}_{t-1}', \ldots, \boldsymbol{\theta}_0')'.$$

The linear structure of the DLM implies that the first- and second-order moments of the initial forecast distribution for $\mathbf{Y}$ and $\boldsymbol{\theta}$ are then

$$\left(\begin{matrix}\boldsymbol{\theta}\\ \mathbf{Y}\end{matrix}\middle| D_0\right) \sim \left[\begin{pmatrix}\mathbf{a}\\ \mathbf{f}\end{pmatrix}, \begin{pmatrix}\mathbf{R} & \mathbf{AQ}\\ \mathbf{QA}' & \mathbf{Q}\end{pmatrix}\right],$$

where the component means, variances and covariances are precisely those in the special case of normality. The actual values of these moments are not important here; the feature of interest is that they are the same, whatever the full joint distribution may be.

From Corollary 4.5, the first $n \times 1$ subvector $\boldsymbol{\theta}_t$ of $\boldsymbol{\theta}$ has LBE given by the corresponding subvector of the LBE $\mathbf{m}$ of $\boldsymbol{\theta}$. But from Corollary 4.6, $\mathbf{m}$ is just the posterior mean for $\boldsymbol{\theta}$ in the normal case, so that the required subvector is $\mathbf{m}_t$. Similarly, the corresponding RM is the submatrix $\mathbf{C}_t$.

◇

Corollary 4.7. *For integers* $k \geq 1$, *the normal DLM moments*

$$[f_t(k), Q_t(k)] \qquad \text{and} \qquad [\mathbf{a}_t(k), \mathbf{R}_t(k)]$$

of the k-step ahead forecast distributions and the corresponding moments

$$[f_t(-k), \mathbf{F}_{t-k}'\mathbf{R}_t(-k)\mathbf{F}_{t-k}] \qquad \text{and} \qquad [\mathbf{a}_t(-k), \mathbf{R}_t(-k)]$$

of the k-step smoothed and filtered distributions are also the linear posterior means and variances of the corresponding random quantities.

These results provide one justification for the use of the sequential updating, forecasting and filtering recurrences outside the strict assumptions of normality. They have been presented in the case of a known variance DLM, but it should be noted that as in the normal model, the recurrences apply when the observational variances are known up to a constant scale factor, when all variances are simply scaled by this factor as in Section 4.5. Estimation of the unknown scale then proceeds using the updated estimates S_t of Section 4.5. This procedure may also be justified within a modified linear Bayes' framework although this is not detailed here.

4.9.4. Weak Bayes' estimation in the DLM

Weak Bayes' estimation makes probabilistic statements that facilitate recurrences for various distributional characteristics without recourse to loss functions. The framework was introduced in Harrison (1996), and is illustrated here in an application to DLMs.

Consider vectors $\mathbf{Y}$ and $\boldsymbol{\theta}$, jointly distributed with first- and second-order moments

$$\begin{pmatrix}\boldsymbol{\theta}\\ \mathbf{Y}\end{pmatrix} \sim \left[\begin{pmatrix}\mathbf{a}\\ \mathbf{f}\end{pmatrix}, \begin{pmatrix}\mathbf{R} & \mathbf{AQ}\\ \mathbf{QA}' & \mathbf{Q}\end{pmatrix}\right].$$

Then, with a transformation matrix

$$\mathbf{L} = \begin{pmatrix}\mathbf{I} & -\mathbf{A}\\ \mathbf{0} & \mathbf{I}\end{pmatrix},$$

the transformed vector

$$\begin{pmatrix}\boldsymbol{\theta} - \mathbf{AY}\\ \mathbf{Y}\end{pmatrix} = \mathbf{L}\begin{pmatrix}\boldsymbol{\theta}\\ \mathbf{Y}\end{pmatrix}$$

has moment structure

$$\begin{pmatrix}\boldsymbol{\theta} - \mathbf{AY}\\ \mathbf{Y}\end{pmatrix} \sim \left[\begin{pmatrix}\mathbf{a} - \mathbf{Af}\\ \mathbf{f}\end{pmatrix}, \begin{pmatrix}\mathbf{R} - \mathbf{AQA}' & \mathbf{0}\\ \mathbf{0}' & \mathbf{Q}\end{pmatrix}\right].$$

Now, practical modelling is usually probabilistically local: that is, a modeller is usually only concerned that the model is, in some sense, "true in probability", especially if model monitoring and assessment procedures operate. Hence it is often satisfactory to postulate that, and to act as if, *the posterior mean and variance of* $\boldsymbol{\theta} - \mathbf{AY}$ *are independent of the observed value of* $\mathbf{Y}$. This statement is, of course, precisely and theoretically true if the joint distribution is normal. Otherwise, it is often essentially assumed as a practical proposition, and acted upon in updating the distribution of $\boldsymbol{\theta}$ on observing unexceptional values of $\mathbf{Y}$. Hence, in many applications it is safe for routine operation, exceptional observations being those that will trigger the associated model monitoring procedure and prompt expert intervention. The immediate consequence is that given an observed value $\mathbf{Y} = \mathbf{y}$, say, the posterior mean and variance of $(\boldsymbol{\theta} - \mathbf{A}\mathbf{y})$ equal the prior mean and variance of $(\boldsymbol{\theta} - \mathbf{A}\mathbf{Y})$; this is true for any observed value $\mathbf{y}$. As a result,

$$(\boldsymbol{\theta} \mid \mathbf{Y}) \sim [\mathbf{m},\ \mathbf{C}]$$

where

$$\mathbf{m} = \mathbf{a} + \mathbf{A}\mathbf{e}$$

and

$$\mathbf{C} = \mathbf{R} - \mathbf{A}\mathbf{Q}\mathbf{A}'$$

with $\mathbf{e} = \mathbf{Y} - \mathbf{f}$. These recurrence relationships are identical to those of linear Bayes' but are based upon a precise modelling assumption rather than a loss function approach. This weak approach easily extends to the case of the unknown variance, so that without specifying the usual gamma distributions, the two recurrence relations for n_t and S_t are identical to those in Section 4.6. Details of this and the general approach are given in Harrison (1996).

4.9.5. Minimum linear least squares estimates

It is worth pointing out that essentially minimum linear least squares methodology depends upon the following simple theorem.

Theorem 4.11. *Let $\boldsymbol{\theta}$ and $\mathbf{Y}$ be random vectors, as above, with joint first- and second-order moment structure*

$$\begin{pmatrix} \boldsymbol{\theta} \\ \mathbf{Y} \end{pmatrix} \sim \left[\begin{pmatrix} \mathbf{a} \\ \mathbf{f} \end{pmatrix}, \begin{pmatrix} \mathbf{R} & \mathbf{A}\mathbf{Q} \\ \mathbf{Q}\mathbf{A}' & \mathbf{Q} \end{pmatrix} \right].$$

Then for any conformable vector $\mathbf{h}$ and matrix $\mathbf{H}$, and for all conformable vectors $\mathbf{l}$,

$$\begin{aligned} &\mathrm{E}[\mathbf{l}(\boldsymbol{\theta} - \mathbf{a} - \mathbf{A}\mathbf{Y} + \mathbf{A}\mathbf{f})(\boldsymbol{\theta} - \mathbf{a} - \mathbf{A}\mathbf{Y} + \mathbf{A}\mathbf{f})'\mathbf{l}'] \\ &\qquad \leq \mathrm{E}[\mathbf{l}(\boldsymbol{\theta} - \mathbf{h} - \mathbf{H}\mathbf{Y})(\boldsymbol{\theta} - \mathbf{h} - \mathbf{H}\mathbf{Y})'\mathbf{l}']. \end{aligned}$$

Proof. From Section 4.9.4,

$$\mathrm{E}[\boldsymbol{\theta} - \mathbf{a} - \mathbf{A}\mathbf{Y} + \mathbf{A}\mathbf{f}] = \mathbf{0}$$

and

$$\mathrm{C}[\boldsymbol{\theta} - \mathbf{A}\mathbf{Y}, \mathbf{Y}] = \mathbf{0}$$

so that

$$\begin{aligned}
\mathrm{E}[\mathbf{l}(\boldsymbol{\theta} - \mathbf{h} - \mathbf{HY})(\boldsymbol{\theta} - \mathbf{h} - \mathbf{HY})'\mathbf{l}'] \\
&\geq \mathrm{V}[\mathbf{l}(\boldsymbol{\theta} - \mathbf{HY})] \\
&= \mathrm{V}[\mathbf{l}(\boldsymbol{\theta} - \mathbf{AY}) + \mathbf{l}(\mathbf{A} - \mathbf{H})\mathbf{Y}] \\
&= \mathrm{V}[\mathbf{l}(\boldsymbol{\theta} - \mathbf{AY})] + \mathrm{V}[\mathbf{l}(\mathbf{A} - \mathbf{H})\mathbf{Y}] \\
&\geq \mathrm{V}[\mathbf{l}(\boldsymbol{\theta} - \mathbf{AY})] \\
&= \mathrm{E}[\mathbf{l}(\boldsymbol{\theta} - \mathbf{a} - \mathbf{AY} + \mathbf{Af})(\boldsymbol{\theta} - \mathbf{a} - \mathbf{AY} + \mathbf{Af})'\mathbf{l}'].
\end{aligned}$$

◇

So, among all linear functions $\phi(\mathbf{Y}) = \mathbf{h} + \mathbf{HY}$, the LBE $\mathbf{a} + \mathbf{A}(\mathbf{Y} - \mathbf{f})$ minimizes the expected value of the variance $\mathrm{V}[\boldsymbol{\theta} - \phi(\mathbf{Y})]$ and also the corresponding squared deviation,

A priori, as in Section 4.9.4, it is certainly the case that in this least squares/minimum variance framework, $\boldsymbol{\theta} - \mathbf{a} - \mathbf{A}(\mathbf{Y} - \mathbf{f}) \sim [\mathbf{0},\ \mathbf{R} - \mathbf{AQA}']$. However, this gives no route to deducing the posterior parameter distribution $(\boldsymbol{\theta}|\mathbf{Y})$, nor even just its future forecast means and variances. The same criticism applies to pure likelihood methods. Logical progress requires further modelling statements at least equivalent to the weak Bayes' statement, whereupon the posterior moments emerge as $[\mathbf{m}, \mathbf{C}]$.

4.10 REFERENCE ANALYSIS OF THE DLM†

4.10.1 Introductory comments

The specification of proper, possibly highly informative, prior distributions to initialise models is beneficial if relevant, easily articulated prior information exists. Whilst informative priors and the incorporation of expert information is central to Bayesian forecasting, there is an important role for reference analyses, that use standard vague or uninformative priors (Bernardo 1979) to initialise models without further inputs from the user.

Such analyses are developed in this section. The reference analysis of a DLM (including learning about unknown observational variances) has a structure differing from standard analysis that is of theoretical and practical interest. A reference prior based analysis provides a reference level, or benchmark, against which alternative analyses may be compared. In particular, a retrospective reference analysis applied to historic data familiarises a modeller with the past data and context, quickly leading to the development of a refined model. Further, the implications for inference of various informative priors may be gauged and highlighted by comparison with a reference analysis. We have also found that users unfamiliar with

---

†This section may be omitted on a first reading

the full complexities of a proper Bayesian analysis appreciate the reference facility, which offers an easily applied default analysis, enabling them to use the methods immediately, without any demanding prerequisites, and to rapidly gain experience and understanding through practice.

In spite of the extensive research into the development of reference priors, there is no unique representation of a state of complete "ignorance" within the Bayesian framework. However, there has emerged a consensus on the problem of normal, linear regression models, and in this area there is what may be termed a *standard* analysis, the implications of the standard reference prior being well investigated and understood as, for example, in Box and Tiao (1973), and De Groot (1970). Since the DLM has the same basic linear regression structure, this standard reference prior is now adopted.

The relevant theoretical results for the DLM are stated here and some are proven. All the results in this section, together with full proofs, further discussion and related theory, are to be found in Pole and West (1989).

4.10.2 Sequential updating equations in reference analysis

The results in this section, summarised in Theorem 4.12 below, are general, applying to all univariate DLMs in which the system evolution variance matrices $\mathbf{W}_t$ are non-singular for all t. The cases of observational variances either known for all time or unknown but constant are considered together. Section 4.10.3 develops similar ideas for the important special *static* case of $\mathbf{W}_t = \mathbf{0}$.

DLM reference analysis is based on one of the following reference prior specifications for time $t = 1$ (e.g., Box and Tiao 1973):

(1) For the DLM of Definition 4.3, with known observational variances, the reference initial prior specification is defined via

$$p(\boldsymbol{\theta}_1|D_0) \propto \textit{constant}.$$

(2) For the DLM of Definition 4.5, with $V_t = V$ unknown, the reference initial prior specification is defined via

$$p(\boldsymbol{\theta}_1, V|D_0) \propto V^{-1}.$$

Definitions 4.11. In the DLMs of Definition 4.3 (V known) and of Definition 4.5 (V unknown), sequentially define the following quantities:

$$\begin{aligned}
\mathbf{H}_t &= \mathbf{W}_t^{-1} - \mathbf{W}_t^{-1}\mathbf{G}_t\mathbf{P}_t^{-1}\mathbf{G}_t'\mathbf{W}_t^{-1},\\
\mathbf{P}_t &= \mathbf{G}_t'\mathbf{W}_t^{-1}\mathbf{G}_t + \mathbf{K}_{t-1},\\
\mathbf{h}_t &= \mathbf{W}_t^{-1}\mathbf{G}_t\mathbf{P}_t^{-1}\mathbf{k}_{t-1},
\end{aligned}$$

$$\text{and if } V_t = V \text{ is unknown,} \quad \begin{cases} \mathbf{K}_t &= \mathbf{H}_t + \mathbf{F}_t\mathbf{F}_t',\\ \mathbf{k}_t &= \mathbf{h}_t + \mathbf{F}_t Y_t, \end{cases}$$

$$\text{but if } V_t = V \text{ is known,} \qquad \begin{cases} \mathbf{K}_t &= \mathbf{H}_t + \mathbf{F}_t\mathbf{F}_t'/V_t, \\ \mathbf{k}_t &= \mathbf{h}_t + \mathbf{F}_t Y_t/V_t, \end{cases}$$

both having initial values $\mathbf{H}_1 = \mathbf{0}$ and $\mathbf{h}_1 = \mathbf{0}$.

For unknown $V_t = V$ in the above definitions, $\mathbf{W}_t$ is replaced by the scale-free matrix $\mathbf{W}_t^*$. Also in this case, define

$$\begin{aligned} \gamma_t &= \gamma_{t-1} + 1, \\ \lambda_t &= \delta_{t-1} - \mathbf{k}_{t-1}'\mathbf{P}_t^{-1}\mathbf{k}_{t-1}, \\ \delta_t &= \lambda_t + Y_t^2, \end{aligned}$$

with initial values $\lambda_1 = 0$ and $\gamma_0 = 0$.

Theorem 4.12.

(1) **Case of variance known**
In the model of Definition 4.3, with the reference prior

$$p(\boldsymbol{\theta}_1|D_0) \propto \textit{constant},$$

and with defining parameters as given in Definition 4.12, the prior and posterior distributions of the state vector at time t are given by the (possibly improper) probability density functions

$$p(\boldsymbol{\theta}_t|D_{t-1}) \propto \exp\{-\tfrac{1}{2}(\boldsymbol{\theta}_t'\mathbf{H}_t\boldsymbol{\theta}_t - 2\boldsymbol{\theta}_t'\mathbf{h}_t)\}$$

and

$$p(\boldsymbol{\theta}_t|D_t) \propto \exp\{-\tfrac{1}{2}(\boldsymbol{\theta}_t'\mathbf{K}_t\boldsymbol{\theta}_t - 2\boldsymbol{\theta}_t'\mathbf{k}_t)\}.$$

(2) **Case of variance unknown**
In the DLM of Definition 4.5, with

$$p(\boldsymbol{\theta}_1, V|D_t) \propto V^{-1},$$

the joint prior and posterior distributions of the state vector and the observation variance at time $t = 1, 2, \ldots$ are given by

$$p(\boldsymbol{\theta}_t, V|D_{t-1}) \propto \\ V^{-\left(1+\frac{\gamma_{t-1}}{2}\right)} \exp\{-\tfrac{1}{2}V^{-1}(\boldsymbol{\theta}_t'\mathbf{H}_t\boldsymbol{\theta}_t - 2\boldsymbol{\theta}_t'\mathbf{h}_t + \lambda_t)\}$$

and

$$p(\boldsymbol{\theta}_t, V|D_t) \propto \\ V^{-\left(1+\frac{\gamma_t}{2}\right)} \exp\{-\tfrac{1}{2}V^{-1}(\boldsymbol{\theta}_t'\mathbf{K}_t\boldsymbol{\theta}_t - 2\boldsymbol{\theta}_t'\mathbf{k}_t + \delta_t)\}.$$

Proof. The proof of the results in the variance unknown case are given, (the variance known case being left to the reader). So, $(\boldsymbol{\omega}_t|V) \sim \mathrm{N}[\mathbf{0}, V\mathbf{W}_t^*]$ and $\mathbf{W}_t^*$ replaces $\mathbf{W}_t$ in Definition 4.11.

The proof is by induction. Assume the prior for $(\boldsymbol{\theta}_t, V|D_{t-1})$ as in (2) above,

$$p(\boldsymbol{\theta}_t, V|D_{t-1}) \propto V^{-(1+\frac{\gamma_{t-1}}{2})}\exp\{-\tfrac{1}{2}V^{-1}(\boldsymbol{\theta}_t'\mathbf{H}_t\boldsymbol{\theta}_t - 2\boldsymbol{\theta}_t'\mathbf{h}_t + \lambda_t)\}.$$

The likelihood from the observation Y_t is

$$p(Y_t|\boldsymbol{\theta}_t, V, D_{t-1}) \propto V^{-\frac{1}{2}}\exp\{-\tfrac{1}{2}V^{-1}(Y_t - \mathbf{F}_t'\boldsymbol{\theta}_t)^2\}.$$

So the posterior for time t is of the form stated in the theorem.

Consider now the implied prior for $t+1$, specifically

$$\begin{aligned} p(\boldsymbol{\theta}_{t+1}, V|D_t) &= \int p(\boldsymbol{\theta}_{t+1}, V|\boldsymbol{\theta}_t, D_t)p(\boldsymbol{\theta}_t|D_t)d\boldsymbol{\theta}_t \\ &= \int p(\boldsymbol{\theta}_{t+1}|\boldsymbol{\theta}_t, V, D_t)p(\boldsymbol{\theta}_t, V|D_t)d\boldsymbol{\theta}_t. \end{aligned}$$

The first term in the integral is the normal $\mathrm{N}[\mathbf{G}_{t+1}\boldsymbol{\theta}_t, V\mathbf{W}_{t+1}^*]$ pdf, so that

$$\begin{aligned} p(\boldsymbol{\theta}_{t+1}, &V|D_t) \propto \\ &\int V^{-\frac{n}{2}}\exp\{-\tfrac{1}{2}V^{-1}(\boldsymbol{\theta}_{t+1} - \mathbf{G}_{t+1}\boldsymbol{\theta}_t)'\mathbf{W}_{t+1}^{*\,-1}(\boldsymbol{\theta}_{t+1} - \mathbf{G}_{t+1}\boldsymbol{\theta}_t)\} \\ &\qquad \times V^{-(1+\frac{\gamma_t}{2})}\exp\{-\tfrac{1}{2}V^{-1}(\boldsymbol{\theta}_t'\mathbf{K}_t\boldsymbol{\theta}_t - 2\boldsymbol{\theta}_t'\mathbf{k}_t + \delta_t)\}d\boldsymbol{\theta}_t, \end{aligned}$$

which reduces to

$$V^{-(1+\frac{\gamma_t+n}{2})}\int \exp\{-\tfrac{1}{2}V^{-1}[(\boldsymbol{\theta}_t - \boldsymbol{\alpha}_{t+1})'\mathbf{P}_{t+1}(\boldsymbol{\theta}_t - \boldsymbol{\alpha}_{t+1}) + \mathbf{R}_{t+1}]\}d\boldsymbol{\theta}_t,$$

where

$$\begin{aligned} \mathbf{P}_{t+1} &= \mathbf{K}_t + \mathbf{G}_{t+1}'\mathbf{W}_{t+1}^{*\,-1}\mathbf{G}_{t+1}, \\ \boldsymbol{\alpha}_{t+1} &= \mathbf{P}_{t+1}^{-1}(\mathbf{k}_t + \mathbf{G}_{t+1}'\mathbf{W}_{t+1}^{*\,-1}\boldsymbol{\theta}_{t+1}), \\ \mathbf{R}_{t+1} &= \boldsymbol{\theta}_{t+1}'\mathbf{W}_{t+1}^{*\,-1}\boldsymbol{\theta}_{t+1} + \delta_t - \boldsymbol{\alpha}_{t+1}'\mathbf{P}_{t+1}\boldsymbol{\alpha}_{t+1}. \end{aligned}$$

It easily follows that

$$p(\boldsymbol{\theta}_{t+1}, V|D_t) \propto V^{-(1+\frac{\gamma_t}{2})}\exp\{-\tfrac{1}{2}V^{-1}\mathbf{R}_{t+1}\}.$$

Expanding $\mathbf{R}_{t+1}$ gives

$$\mathbf{R}_{t+1} = \boldsymbol{\theta}_{t+1}'\mathbf{H}_{t+1}\boldsymbol{\theta}_{t+1} - 2\boldsymbol{\theta}_{t+1}'\mathbf{h}_{t+1} + \lambda_{t+1},$$

where $\mathbf{H}_{t+1}$, $\mathbf{h}_{t+1}$, and λ_{t+1} are defined as stated in the theorem. It remains to validate the theorem for $t=1$. Setting $\mathbf{H}_1 = 0$, $\mathbf{h}_1 = 0$, $\lambda_1 = 0$ and $\gamma_0 = 0$ provides a direct validation.

◇

Recall that the model has n parameters in the state vector at each time t and so, starting from the reference prior, the posteriors will be improper until at least time n in the observation known case, time $n+1$ in the observation unknown case. After sufficient observations have been processed the improper posterior distributions characterising the model become proper. Although the above recursions remain valid it is more usual to revert to the standard forms involving directly updating the state vector posterior mean and covariances since these do not require any matrix inversions. The number of observations required to achieve proper distributions depends on the form of the model and the data. As mentioned above, the least number that will suffice is the number of unknown parameters in the model, including 1 for the observational variance if unknown. This requires that there are no missing data in these first observations and also that there are no problems of collinearity if the model includes regressors. In practice, more than the minimum will rarely be required. For generality, however, define

$$[n] = \min\{t : \text{posterior distributions are proper}\}.$$

The necessary relationships between the quantities defining the posterior distributions as represented in Theorem 4.12 and those in the original representation are easily obtained as follows, the proof being left as an exercise for the reader.

Corollary 4.8.

(1) **Case of variance known**
For $t \geq [n]$, the posterior distribution of $(\boldsymbol{\theta}_t|D_t)$ is as in Section 4.3, with

$$\mathbf{C}_t = \mathbf{K}_t^{-1} \qquad \textit{and} \qquad \mathbf{m}_t = \mathbf{K}_t^{-1}\mathbf{k}_t.$$

(2) **Case of variance unknown**
For $t \geq [n]$, the posterior distribution of $(\boldsymbol{\theta}_t, V|D_t)$ is as in Section 4.6, with

$$\mathbf{C}_t = S_t\mathbf{K}_t^{-1} \qquad \textit{and} \qquad \mathbf{m}_t = \mathbf{K}_t^{-1}\mathbf{k}_t,$$

where $S_t = d_t/n_t$ as usual, with $n_t = \gamma_t - n$ and $d_t = \delta_t - \mathbf{k}_t'\mathbf{m}_t$. In the usual case that $[n] = n+1$, then $n_{n+1} = 1$ and it is easily shown that $d_{n+1} = S_{n+1} = e_{n+1}^2/Q_{n+1}^*$.

4.10.3 Important special case of $\mathbf{W}_t = \mathbf{0}$

Consider now the case of models with deterministic evolution equations, that is, those in which $\mathbf{W}_t = \mathbf{W}_t^* = \mathbf{0}$. Whilst of interest from the point of view of theoretical completion and generality, this special case is discussed primarily for practical reasons. The basic motivation derives from the need to specify $\mathbf{W}_t$ in the recursions detailed above. As has been mentioned elsewhere this problem has led to much resistance to the use of these models by practitioners and the introduction of discount techniques in the usual conjugate prior analysis. Unfortunately, these methods do not apply in the reference analysis for $t < [n]$ because the posterior covariances do not then exist. Hence, an alternative approach is required, and the practical use of zero covariances initially, $\mathbf{W}_t = \mathbf{0}$ for $t = 1, 2, \dots, [n]$ is recommended. The rationale behind this is as follows.

In the reference analysis with $n+1$ model parameters (including V), we need $[n]$ (at least $n+1$) observations to obtain a fully specified joint posterior distribution: one observation for each parameter. More generally, at time $t = [n]$ we have essentially only one observation's worth of information for each parameter. Thus, it is *impossible* to detect or estimate any changes in parameters during the first $(n+1)$ observation time points over which the reference analysis is performed. Consequently, use of non-zero $\mathbf{W}_t$ matrices is irrelevant since they basically allow for changes that cannot be estimated, and so we lose nothing by setting them to zero for $t = 1, 2, \dots, [n]$. At time $t = [n]$, the posteriors are fully specified, and future parametric changes can be identified. Thus, at this time, we revert to a full dynamic model with suitable, non-zero evolution covariance matrices.

Theorem 4.13. *In the framework of Theorem 4.12, suppose that $\mathbf{G}_t$ is non-singular and $\mathbf{W}_t = \mathbf{W}_t^* = \mathbf{0}$. Then the prior and posterior distributions of $\boldsymbol{\theta}_t$ and V have the forms of Theorem 4.12 with recursions defined as follows:*

(1) **Case of variance known**

$$\begin{aligned} \mathbf{H}_t &= \mathbf{G}_t^{-1\prime}\mathbf{K}_{t-1}\mathbf{G}_t^{-1}, \\ \mathbf{h}_t &= \mathbf{G}_t^{-1\prime}\mathbf{k}_{t-1}. \end{aligned}$$

(2) **Case of variance unknown**

$$\begin{aligned} \mathbf{H}_t &= \mathbf{G}_t^{-1\prime}\mathbf{K}_{t-1}\mathbf{G}_t^{-1}, \\ \mathbf{h}_t &= \mathbf{G}_t^{-1\prime}\mathbf{k}_{t-1}, \\ \lambda_t &= \delta_{t-1}. \end{aligned}$$

Proof.

(1) **Case of variance known**

Again the proof is inductive. Suppose first that $p(\boldsymbol{\theta}_{t-1}|D_{t-1})$ has the stated form. Then the system equation is $\boldsymbol{\theta}_t = \mathbf{G}_t\boldsymbol{\theta}_{t-1}$, and this may be inverted when $\mathbf{G}_t$ is non-singular so that $\boldsymbol{\theta}_{t-1} = \mathbf{G}_t^{-1}\boldsymbol{\theta}_t$. Applying this linear transformation, that has a constant Jacobian, to $p(\boldsymbol{\theta}_{t-1}|D_{t-1})$, immediately results in the prior

$$p(\boldsymbol{\theta}_t|D_{t-1}) \propto \exp\{-\tfrac{1}{2}(\boldsymbol{\theta}_t'\mathbf{G}_t^{-1\prime}\mathbf{K}_{t-1}\mathbf{G}_t^{-1}\boldsymbol{\theta}_t - 2\boldsymbol{\theta}_t'\mathbf{G}_t^{-1\prime}\mathbf{k}_{t-1})\}.$$

Hence, multiplying by the likelihood, the posterior $p(\boldsymbol{\theta}_t|D_t)$ is

$$\propto \exp\{-\tfrac{1}{2}[\boldsymbol{\theta}_t'\mathbf{H}_t\boldsymbol{\theta}_t - 2\boldsymbol{\theta}_t'\mathbf{h}_t + V_t^{-1}(Y_t - \mathbf{F}_t'\boldsymbol{\theta}_t)'(Y_t - \mathbf{F}_t'\boldsymbol{\theta}_t)]\}$$
$$\propto \exp\{-\tfrac{1}{2}(\boldsymbol{\theta}_t'\mathbf{K}_t\boldsymbol{\theta}_t - 2\boldsymbol{\theta}_t'\mathbf{k}_t)\}.$$

Initially, for $t = 1$,

$$p(\boldsymbol{\theta}_1|D_1) \propto \exp\{-\tfrac{1}{2}V^{-1}(\boldsymbol{\theta}_1'\mathbf{F}_1\mathbf{F}_1'\boldsymbol{\theta}_1 - \boldsymbol{\theta}_1'\mathbf{F}_1 Y_1)\},$$

and the result follows by induction.

(2) **Case of variance unknown**
The proof follows as a simple extension of that for part (1) and is left as an exercise.

◇

The updating equations derived in Theorem 4.13 are of key practical importance. The assumption that the system matrices $\mathbf{G}_t$ are non-singular is obviously crucial to the results. In practical models this is typically satisfied. This is true, in particular, of the important class of time series models. The following section covers more cases when this is not so.

4.10.4 Filtering

Filtering in the case of reference initial priors uses exactly the same results as in the conjugate (proper) priors case for times $t \geq [n]$ since all the distributions in this range are proper. In particular, the usual recursions of Section 4.7 are valid. However, for $t < [n]$ these recursions do not apply since the on-line posterior means and covariances required do not exist. The following theorem provides the solutions for this case.

Theorem 4.14. *In the framework of Theorem 4.12, the filtered distributions in the DLM for times $t - r$, $r = 0, 1, 2, ..., [n] - 1$ are defined as follows.*

(1) **Case of variance known**

$$p(\boldsymbol{\theta}_{t-r}|D_t) \propto \exp\{-\tfrac{1}{2}[\boldsymbol{\theta}_{t-r}'\mathbf{K}_t(-r)\boldsymbol{\theta}_{t-r} - 2\boldsymbol{\theta}_{t-r}'\mathbf{k}_t(-r)]\}.$$

(2) **Case of variance unknown**

$$p(\boldsymbol{\theta}_{t-r}, V|D_t) \propto$$
$$V^{-(1+\frac{t}{2})}\exp\{-\tfrac{1}{2}V^{-1}[\boldsymbol{\theta}_{t-r}'\mathbf{K}_t(-r)\boldsymbol{\theta}_{t-r} - 2\boldsymbol{\theta}_{t-r}'\mathbf{k}_t(-r) + \delta_t(-r)]\}$$

where the defining quantities are calculated recursively backwards in time according to

$$\begin{aligned}
\mathbf{K}_t(-r) &= \mathbf{G}'_{t-r+1}\mathbf{W}^{-1}_{t-r+1}\mathbf{G}_{t-r+1} \\
&\quad - \mathbf{G}'_{t-r+1}\mathbf{W}^{-1}_{t-r+1}\mathbf{P}^{-1}_t(-r+1)\mathbf{W}^{-1}_{t-r+1}\mathbf{G}_{t-r+1} + \mathbf{K}_{t-r}, \\
\mathbf{P}_t(-r+1) &= \mathbf{W}^{-1}_{t-r+1} + \mathbf{K}_t(-r+1) - \mathbf{H}_{t-r+1}, \\
\mathbf{k}_t(-r) &= \mathbf{k}_{t-r} + \mathbf{G}'_{t-r+1}\mathbf{W}^{-1}_{t-r+1}\mathbf{P}^{-1}_t(-r+1)[\mathbf{k}_t(-r+1) - \mathbf{h}_{t-r+1}], \\
\delta_t(-r) &= \delta_t(-r+1) - \lambda_{t-r+1} \\
&\quad - [\mathbf{k}_t(-r+1) - \mathbf{h}_{t-r+1}]'\mathbf{P}^{-1}_t(-r+1)[\mathbf{k}_t(-r+1) - \mathbf{h}_{t-r+1}]
\end{aligned}$$

and $\mathbf{H}_t$, $\mathbf{h}_t$, $\mathbf{K}_t$, $\mathbf{k}_t$, λ_t, δ_t *are as in Definition 4.11. Again note that, in the case of* V *unknown,* $\mathbf{W}^*_t$ *replaces* $\mathbf{W}_t$ *throughout, for all* t*. Starting values for these recursions are* $\mathbf{K}_t(0) = \mathbf{K}_t$, $\mathbf{k}_t(0) = \mathbf{k}_t$ *and* $\delta_t(0) = \delta_t$.

Proof. An exercise for the reader. Complete proofs of these results appear in Pole and West (1989a).

◇

For theoretical completion, note that the recursions in Theorem 4.14 also apply for $t > [n]$. Although the standard filtering equations should be used, those of the theorem provide the relevant distributions, as follows.

Corollary 4.9. *For* $t > [n]$*, the distributions defined in Theorem 4.14 are proper, as given in Section 4.7, with*

(1) **Case of variance known**

$$(\boldsymbol{\theta}_{t-r}|D_t) \sim \mathrm{N}[\mathbf{a}_t(-r), \mathbf{R}_t(-r)],$$

where $\mathbf{a}_t(-r) = \mathbf{K}_t(-r)^{-1}\mathbf{k}_t(-r)$ *and* $\mathbf{R}_t(-r) = \mathbf{K}_t(-r)^{-1}$.

(2) **Case of variance unknown**

$$(\boldsymbol{\theta}_{t-r}|D_t) \sim \mathrm{T}_{\gamma_t-[n]}[\mathbf{a}_t(-r), \mathbf{R}_t(-r)],$$

where $\mathbf{a}_t(-r) = \mathbf{K}_t(-r)^{-1}\mathbf{k}_t(-r)$ *and* $\mathbf{R}_t(-r) = S_t\mathbf{K}_t(-r)^{-1}$.

Theorem 4.15. *In the case of zero evolution disturbance variances as in Theorem 4.13, the results of Theorem 4.14 are still valid but with the following changes to the recursions:*

$$\begin{aligned}
\mathbf{K}_t(-r) &= \mathbf{G}'_{t-r+1}\mathbf{K}_t(-r+1)\mathbf{G}_{t-r+1}, \\
\mathbf{k}_t(-r) &= \mathbf{G}'_{t-r+1}\mathbf{k}_t(-r+1), \\
\delta_t(-r) &= \delta_t(-r+1).
\end{aligned}$$

Proof. Again left as an exercise.

◇

4.11 APPENDIX: CONDITIONAL INDEPENDENCE

Definition 4.12. Random vectors $\mathbf{X}$ and $\mathbf{U}$ are conditionally independent given the random vector $\mathbf{Z}$, written $\mathbf{X} \perp\!\!\!\perp \mathbf{U}|\mathbf{Z}$, if

$$p(\mathbf{X}|\mathbf{Z}) = p(\mathbf{X}|\mathbf{Z},\mathbf{U})$$

and

$$p(\mathbf{U}|\mathbf{Z}) = p(\mathbf{U}|\mathbf{Z},\mathbf{X})$$

for all $\mathbf{X}, \mathbf{Z}$ and $\mathbf{U}$.

The pictorial representation of $\mathbf{X} \perp\!\!\!\perp \mathbf{U}|\mathbf{Z}$ is

$$\mathbf{X} \text{ ——— } \mathbf{Z} \text{ ——— } \mathbf{U}$$

Under the full joint distribution $p(\mathbf{X},\mathbf{Z},\mathbf{U})$, based upon whatever implicit prior information is assumed, then (a) the relevant information concerning $\mathbf{X}$ supplied by $(\mathbf{Z},\mathbf{U})$ is provided by $\mathbf{Z}$ alone, and, similarly, (b) the relevant information about $\mathbf{U}$ supplied by $(\mathbf{Z},\mathbf{X})$ is provided by $\mathbf{Z}$ alone.

Theorem 4.16. *If the random vectors $\mathbf{X}$, $\mathbf{Z}$ and $\mathbf{U}$ have a joint normal distribution such that $\mathbf{X} \perp\!\!\!\perp \mathbf{U}|\mathbf{Z}$, then the regression matrix $\mathbf{A}_{xu}$, of $\mathbf{X}$ on $\mathbf{U}$, is the product of the regression matrix $\mathbf{A}_{xz}$, of $\mathbf{X}$ on $\mathbf{Z}$, and the regression matrix $\mathbf{A}_{zu}$, of $\mathbf{Z}$ on $\mathbf{U}$. Thus*

$$\begin{pmatrix} \mathbf{X} \\ \mathbf{Z} \\ \mathbf{U} \end{pmatrix} \sim \mathrm{N}\left[\begin{pmatrix} \mu_x \\ \mu_z \\ \mu_u \end{pmatrix} ; \begin{pmatrix} \mathbf{R}_x & \mathbf{A}_{xz}\mathbf{R}_z & \mathbf{A}_{xu}\mathbf{R}_u \\ \mathbf{R}_z\mathbf{A}'_{xz} & \mathbf{R}_z & \mathbf{A}_{zu}\mathbf{R}_u \\ \mathbf{R}_u\mathbf{A}'_{xu} & \mathbf{R}_u\mathbf{A}'_{zu} & \mathbf{R}_u \end{pmatrix}\right],$$

with

$$\mathbf{A}_{xu} = \mathbf{A}_{xz}\mathbf{A}_{zu} \; .$$

Proof. Conditional independence implies that

$$\begin{aligned} 0 = \mathrm{C}[\mathbf{X},\mathbf{U}|\mathbf{Z}] &= \mathrm{C}[\mathbf{X},\mathbf{U}] - \mathrm{C}[\mathbf{X},\mathbf{Z}]\,\{\mathrm{V}[\mathbf{Z}]\}^{-1}\,\mathrm{C}[\mathbf{Z},\mathbf{U}] \\ &= \mathrm{C}[\mathbf{X},\mathbf{U}] - \mathbf{A}_{xz}\mathbf{R}_z\mathbf{R}_z^{-1}\mathbf{A}_{zu}\mathbf{R}_u = \mathrm{C}[\mathbf{X},\mathbf{U}] - \mathbf{A}_{xz}\mathbf{A}_{zu}\mathbf{R}_u \; . \end{aligned}$$

◇

Corollary 4.10. *Suppose* $\mathbf{X}_i$, $(i = 1, \ldots, n)$, *are normal random vectors such that for all* i *and for all* $1 < j < k < n$, $\mathbf{X}_i \perp\!\!\!\perp \mathbf{X}_{i+k} | \mathbf{X}_{i+j}$. *Pictorially,*

$$\mathbf{X}_1 \text{ --- } \mathbf{X}_2 - \cdots - \mathbf{X}_i - \cdots - \mathbf{X}_{i+j} - \cdots - \mathbf{X}_{i+k} - \cdots - \mathbf{X}_{n-1} \text{ --- } \mathbf{X}_n$$

Then, writing the regression matrix of $\mathbf{X}_i$ *on* $\mathbf{X}_{i+1}$ *as* $\mathbf{A}_{i,i+1}$, *the regression matrix of* $\mathbf{X}_1$ *on* $\mathbf{X}_n$ *is*

$$\mathbf{A}_{1,n} = \prod_{i=1}^{n-1} \mathbf{A}_{i,i+1} \,.$$

Theorem 4.17. *Under the conditions of Theorem 4.16,*

$$\begin{pmatrix} \mathbf{X} \\ \mathbf{Z} \end{pmatrix} \Bigg| \mathbf{U} \sim \mathrm{N}\left[\begin{pmatrix} \boldsymbol{\mu}_{x|u} \\ \boldsymbol{\mu}_{z|u} \end{pmatrix} ; \begin{pmatrix} \mathbf{R}_{x|u} & \mathbf{A}_{xz}\mathbf{R}_{z|u} \\ \mathbf{R}_{z|u}\mathbf{A}'_{xz} & \mathbf{R}_{z|u} \end{pmatrix} \right],$$

and so

(i) the regression matrix of $\mathbf{X}$ *on* $\mathbf{Z}$ *remains unchanged:* $\mathbf{A}_{xz|u} = \mathbf{A}_{xz}$,

(ii) the distribution of $(\mathbf{X}|\mathbf{U}) \sim \mathrm{N}[\boldsymbol{\mu}_{x|u}, \mathbf{R}_{x|u}]$ *may be calculated from the distribution* $(\mathbf{Z}|\mathbf{U}) \sim \mathrm{N}[\boldsymbol{\mu}_{z|u}, \mathbf{R}_{z|u}]$ *via the equations*

$$\boldsymbol{\mu}_{x|u} = \boldsymbol{\mu}_x + \mathbf{A}_{xz}(\boldsymbol{\mu}_{z|u} - \boldsymbol{\mu}_z)$$

and

$$\mathbf{R}_{x|u} = \mathbf{R}_x + \mathbf{A}_{xz}\left(\mathbf{R}_{z|u} - \mathbf{R}_z\right)\mathbf{A}'_{xz}.$$

Proof. Apply conditional normal theory results to the distribution of $(\mathbf{X}, \mathbf{Z}, \mathbf{U})$ to deduce the following identities, and so prove the results:

(i)
$$\begin{aligned} \mathrm{C}[\mathbf{X}, \mathbf{Z}|\mathbf{U}] &= \mathbf{A}_{xz}\mathbf{R}_z - \mathbf{A}_{xu}\mathbf{R}_u\mathbf{A}'_{zu} \\ &= \mathbf{A}_{xz}\mathbf{R}_z - \mathbf{A}_{xz}\mathbf{A}_{zu}\mathbf{R}_u\mathbf{A}'_{zu} = \mathbf{A}_{xz}\mathbf{R}_{z|u}. \end{aligned}$$

(iia)
$$\begin{aligned} \boldsymbol{\mu}_{z|u} &= \boldsymbol{\mu}_z + \mathbf{A}_{zu}(\mathbf{U} - \boldsymbol{\mu}_u) \Longrightarrow \mathbf{A}_{zu}(\mathbf{U} - \boldsymbol{\mu}_u) = \boldsymbol{\mu}_{z|u} - \boldsymbol{\mu}_z \\ \Longrightarrow \boldsymbol{\mu}_{x|u} &= \boldsymbol{\mu}_x + \mathbf{A}_{xz}\mathbf{A}_{zu}(\mathbf{U} - \boldsymbol{\mu}_u) = \boldsymbol{\mu}_x + \mathbf{A}_{xz}(\boldsymbol{\mu}_{z|u} - \boldsymbol{\mu}_z). \end{aligned}$$

(iib)
$$\begin{aligned} \mathbf{R}_{z|u} &= \mathbf{R}_z - \mathbf{A}_{zu}\mathbf{R}_u\mathbf{A}'_{zu} \Longrightarrow \mathbf{A}_{zu}\mathbf{R}_u\mathbf{A}'_{zu} = \mathbf{R}_z - \mathbf{R}_{z|u} \\ \Longrightarrow \mathbf{R}_{x|u} &= \mathbf{R}_x - \mathbf{A}_{xz}\mathbf{A}_{zu}\mathbf{R}_u\mathbf{A}'_{zu}\mathbf{A}'_{xz} \\ &= \mathbf{R}_x - \mathbf{A}_{xz}(\mathbf{R}_z - \mathbf{R}_{z|u})\mathbf{A}'_{xz}. \end{aligned}$$

◇

4.12 EXERCISES

Unless otherwise stated, the questions below concern a standard, univariate DLM with known variances, i.e., $\{\mathbf{F}_t, \mathbf{G}_t, V_t, \mathbf{W}_t\}$ defined by

$$\begin{aligned} Y_t &= \mathbf{F}_t'\boldsymbol{\theta}_t + \nu_t, & \nu_t &\sim \mathrm{N}[0, V_t], \\ \boldsymbol{\theta}_t &= \mathbf{G}_t\boldsymbol{\theta}_{t-1} + \boldsymbol{\omega}_t, & \boldsymbol{\omega}_t &\sim \mathrm{N}[\mathbf{0}, \mathbf{W}_t], \end{aligned}$$

$$(\boldsymbol{\theta}_{t-1} \mid D_{t-1}) \sim \mathrm{N}[\mathbf{m}_{t-1}, \mathbf{C}_{t-1}].$$

(1) Consider the DLM $\{\mathbf{F}_t, \mathbf{G}, V_t, \mathbf{W}_t\}$.
   (a) If $\mathbf{G}$ is of full rank, prove that the DLM can be reparametrised to the DLM
$$\left\{ \begin{pmatrix} \mathbf{F}_t \\ 1 \end{pmatrix}, \begin{pmatrix} \mathbf{G} & \mathbf{0} \\ \mathbf{0}' & 0 \end{pmatrix}, 0, \begin{pmatrix} \mathbf{W}_t & \mathbf{0} \\ \mathbf{0}' & V_t \end{pmatrix} \right\}.$$
   (b) Now show how to accommodate the observation variance in the system equation when $\mathbf{G}$ is singular.

(2) Consider the constant DLM $\{\mathbf{F}, \mathbf{G}, V, \mathbf{W}\}$, generalised so that
$$\nu_t \sim \mathrm{N}[\bar{v}, V] \quad \text{and} \quad \boldsymbol{\omega}_t \sim \mathrm{N}[\bar{\boldsymbol{w}}, \mathbf{W}].$$
Show that $(\boldsymbol{\theta}_t|D_t) \sim \mathrm{N}[\mathbf{m}_t, \mathbf{C}_t]$ and derive recurrence relationships for $\mathbf{m}_t$ and $\mathbf{C}_t$
   (a) by using Bayes' theorem for updating, and
   (b) by deriving the joint distribution of $(\boldsymbol{\theta}_t, Y_t \mid D_{t-1})$ and using normal theory to obtain the conditional probability distribution.
   (c) What is the forecast function $f_t(k)$?
   (d) How do the results generalise to the DLM $\{\mathbf{F}, \mathbf{G}, V, \mathbf{W}\}_t$ with
$$\nu_t \sim \mathrm{N}[\bar{v}_t, V_t] \quad \text{and} \quad \boldsymbol{\omega}_t \sim \mathrm{N}[\bar{\boldsymbol{w}}_t, \mathbf{W}_t]?$$

(3) For the constant DLM $\{\mathbf{F}, \mathbf{G}, V, \mathbf{W}\}$, given $(\boldsymbol{\theta}_t \mid D_t) \sim \mathrm{N}[\mathbf{m}_t, \mathbf{C}_t]$, obtain
   (a) the k-step ahead forecast distribution $p(Y_{t+k} \mid D_t)$;
   (b) the k-step lead-time forecast distribution $p(X_{t,k}|D_t)$ where
$$X_{t,k} = \sum_{r=1}^{k} Y_{t+r}.$$

(4) For the univariate DLM define $b_t = V_t/Q_t$.
   (a) Show that given D_t, the posterior distribution for the mean response $\mu_t = \mathbf{F}_t'\boldsymbol{\theta}_t$ is
$$(\mu_t|D_t) \sim \mathrm{N}[f_t(0), Q_t(0)],$$
where
$$\mathrm{E}[\mu_t|D_t] = f_t(0) = \mathbf{F}_t'\mathbf{m}_t$$

and

$$\mathrm{V}[\mu_t|D_t] = Q_t(0) = \mathbf{F}_t'\mathbf{C}_t\mathbf{F}_t.$$

Use the recurrence equations for $\mathbf{m}_t$ and $\mathbf{C}_t$ to show that for some appropriate scalar A_t that you should define,

(b) $\mathrm{E}[\mu_t|D_t]$ can be updated using either the equation

$$\mathrm{E}[\mu_t|D_t] = \mathrm{E}[\mu_t|D_{t-1}] + A_t e_t$$

or

$$f_t(0) = A_t Y_t + (1 - A_t) f_t,$$

interpreting $f_t(0)$ as a weighted average of two estimates of μ_t;

(c) $\mathrm{V}[\mu_t|D_t]$ can be updated using either the equation

$$\mathrm{V}[\mu_t|D_t] = (1 - A_t)\mathrm{V}[\mu_t|D_{t-1}]$$

or

$$\mathrm{V}[\mu_t|D_t] = Q_t(0) = A_t V_t.$$

(5) Write $\mathbf{H}_t = \mathbf{W}_t - \mathbf{W}_t\mathbf{F}_t\mathbf{F}_t'\mathbf{W}_t/Q_t$, $\mathbf{L}_t = (1 - A_t)\mathbf{W}_t\mathbf{F}_t$, and $A_t = 1 - V_t/Q_t$. Prove that the posterior distribution of the observation and evolution errors is

$$\begin{pmatrix} \nu_t \\ \omega_t \end{pmatrix} \Bigg| D_t \sim \mathrm{N}\left[\begin{pmatrix} 1 - A_t \\ \mathbf{W}_t\mathbf{F}_t/Q_t \end{pmatrix} e_t,\ \begin{pmatrix} A_t V_t & -\mathbf{L}_t' \\ -\mathbf{L}_t & \mathbf{H}_t \end{pmatrix}\right].$$

(6) Consider the DLM $\{\mathbf{F}_t, \mathbf{G}_t, V_t, V_t\mathbf{W}_t^*\}$ with unknown variances V_t, but in which the observational errors are heteroscedastic, so that

$$\nu_t \sim \mathrm{N}[0, k_t V],$$

where $V = \phi^{-1}$ and k_t is a known, positive variance multiplier. Also,

$$(\phi \mid D_{t-1}) \sim \mathrm{G}[n_{t-1}/2,\ n_{t-1}S_{t-1}/2].$$

(a) What is the posterior $(\phi \mid D_t)$?

(b) How are the summary results of Section 4.6 affected?

(7) Consider the closed, constant DLM $\{1, \lambda, V, W\}$ with λ, V and W known, and $|\lambda| < 1$.

(a) Obtain the k-step forecast distribution $(Y_{t+k}|D_t)$ as a function of m_t, C_t and λ.

(b) Show that as $k \to \infty$, $(Y_{t+k}|D_t)$ converges in distribution to

$$\mathrm{N}\left[0,\ V + W/(1 - \lambda^2)\right].$$

(c) Obtain the joint forecast distribution for $(Y_{t+1}, Y_{t+2}, Y_{t+3})$.

(d) Obtain the k-step lead-time forecast distribution $p(X_{t,k}|D_t)$ where $X_{t,k} = \sum_{r=1}^{k} Y_{t+r}$.

(8) Generalise Theorem 4.1 to a DLM whose observation and evolution noise are instantaneously correlated. Specifically, suppose that $C[\omega_t, \nu_t] = \mathbf{c}_t$, a known n-vector of covariance terms, but that all other assumptions remain valid. Show now that Theorem 4.1 applies with the modifications

$$Q_t = (\mathbf{F}_t'\mathbf{R}_t\mathbf{F}_t + V_t) + 2\mathbf{F}_t'\mathbf{c}_t$$

and

$$\mathbf{A}_t = (\mathbf{R}_t\mathbf{F}_t + \mathbf{c}_t)/Q_t.$$

If $\mathbf{G}$ is of full rank, show how to reformulate the DLM in the standard form so the observation and evolution noise are uncorrelated.

(9) Given D_t, your posterior distribution is such that

$$(\boldsymbol{\theta}_t|D_t) \sim \mathrm{N}[\mathbf{m}_t, \mathbf{C}_t] \quad \text{and} \quad (\boldsymbol{\theta}_{t-k}|D_t) \sim \mathrm{N}[\mathbf{a}_t(-k), \mathbf{R}_t(-k)].$$

The regression matrix of $\boldsymbol{\theta}_{t-k}$ on $\boldsymbol{\theta}_t$ is $\mathbf{A}_{t-k,t}$. You are now about to receive additional information $\mathbf{Z}$, that might be an external forecast, expert opinion, or more observations. If

$$(\boldsymbol{\theta}_1, \ldots, \boldsymbol{\theta}_{t-1}) \perp\!\!\!\perp \mathbf{Z}|\boldsymbol{\theta}_t,$$

and given the information $\mathbf{Z}$, your revised distribution is

$$(\boldsymbol{\theta}_t|D_t, \mathbf{Z}) \sim \mathrm{N}[\mathbf{m}_t + \boldsymbol{\epsilon}, \mathbf{C}_t - \mathbf{6}],$$

what is your revised distribution for $(\boldsymbol{\theta}_{t-k}|D_t, \mathbf{Z})$?

(10) Prove the retrospective results, sets (i) and (ii) of Theorem 4.5, using the conditional independence structure of the DLM and the conditional independence results of the Appendix, Section 4.11.

(11) With discount factor δ, the discount regression DLM $\{\mathbf{F}_t, \mathbf{I}, \mathbf{V}, \mathbf{W}_t\}$ is such that $\mathbf{I}$ is the identity matrix and $\mathbf{W}_t = \mathbf{C}_{t-1}(1-\delta)/\delta$. Given D_t, and with integer $k > 0$, show that
  (a) $\mathbf{R}_t = \mathbf{C}_{t-1}/\delta$;
  (b) the regression matrix of $\boldsymbol{\theta}_{t-k}$ on $\boldsymbol{\theta}_{t-k+1}$ is $\mathbf{B}_{t-k} = \delta\mathbf{I}$;
  (c) the regression matrix of $\boldsymbol{\theta}_{t-k}$ on $\boldsymbol{\theta}_t$ is $\mathbf{A}_{t-k,t} = \delta^k\mathbf{I}$;
  (d) the filtering recurrences of Theorem 4.5 simplify to

$$\begin{aligned}
\text{(i)} \qquad \mathbf{a}_t(-k) &= \mathbf{a}_{t-1}(-k+1) + \delta^k\mathbf{A}_t\mathbf{e}_t, \\
\mathbf{R}_t(-k) &= \mathbf{R}_{t-1}(-k+1) - \delta^{2k}\mathbf{A}_t\mathbf{Q}_t\mathbf{A}_t', \\
\text{(ii)} \qquad \mathbf{a}_t(-k) &= \mathbf{m}_{t-k} + \delta[\mathbf{a}_t(-k+1) - \mathbf{a}_{t-k+1}], \\
\mathbf{R}_t(-k) &= \mathbf{C}_{t-k} + \delta^2[\mathbf{R}_t(-k+1) - \mathbf{R}_{t-k+1}].
\end{aligned}$$

(12) Generalise Theorem 4.1 to the multivariate DLM $\{\mathbf{F}_t, \mathbf{G}_t, \mathbf{V}_t, \mathbf{W}_t\}$.

(13) Verify the filtering and smoothing results in Corollaries 4.3 and 4.4 relating to the DLM with an unknown constant observational variance V.

(14) Prove the reference prior results stated in part (1) of Theorem 4.12, that relate to the case of a known observational variance V.

(15) Prove the reference prior results stated in part (2) of Theorem 4.13, relating to the case of an unknown observational variance V.

(16) Prove the results stated in Corollary 4.8, providing the conversion from reference analysis updating to the usual recurrence equations when posteriors become proper.

(17) Consider the first-order, polynomial model $\{1, 1, V, W\}$, with $n = 1$ and $\boldsymbol{\theta}_t = \mu_t$, in the reference analysis updating of Section 4.10.
   (a) Assume that V is known. Using the known variance results of Theorem 4.12 and Corollary 4.8, show that $(\mu_1|D_1) \sim \mathrm{N}[Y_1, V]$.
   (b) Assume that $V = 1/\phi$ is unknown, so that the results of Theorem 4.13 and Corollary 4.8 apply. Show that the posterior for μ_t and V becomes proper and of standard form at $t = 2$, and identify the defining quantities m_2, C_2, n_2 and S_2.
   (c) In (b), show directly how the results simplify in the case $W = 0$.

(18) Consider the 2-dimensional model

$$\left\{\begin{pmatrix}1\\0\end{pmatrix}, \begin{pmatrix}1 & 1\\0 & 1\end{pmatrix}, V, \mathbf{0}\right\}$$

assuming V to be known. Apply Theorem 4.13 and Corollary 4.8 to deduce the posterior for $(\boldsymbol{\theta}_2|D_2)$.

(19) Consider the discount DLM $\{\mathbf{F}_t, \mathbf{I}, V, \ V\mathbf{W}_t^*\}$ with unknown but constant variance V. The discount factor is $0 < \delta \leq 1$, so that $W_t^* = C_{t-1}^*(1-\delta)/\delta$, and $\mathbf{R}_t^{*-1} = \delta \mathbf{C}_{t-1}^{*\ -1}$. Initially, ignorance is formulated so the prior precision of $\boldsymbol{\theta}_0$ is $\mathbf{C}_0^{-1} = \mathbf{0}$, the prior degrees of freedom for $\phi = 1/V$ are $n_0 = 0$, $\mathbf{m}_0 = \mathbf{0}$ and $S_0 = 0$. Use the notation of Theorem 4.3 so that starred variances are conditioned on $V = 1$.
   (a) Prove that $\mathbf{C}_1^{*-1} = \mathbf{F}_1\mathbf{F}_1'$.
   (b) Prove that

$$\mathbf{C}_t^{*-1} = \sum_{v=0}^{t-1} \delta^v \mathbf{F}_{t-v}\mathbf{F}_{t-v}'.$$

   (c) Suppose that rank $\mathbf{C}_t^{*-1} =$ rank $\mathbf{C}_{t-1}^{*-1} = r_{t-1}$. Show that the relationship between Y_t and the information D_{t-1} can be modelled by a DLM $\{\tilde{\mathbf{F}}_t, \mathbf{I}, V, \ V\tilde{\mathbf{W}}_t\}$ with a parameter of dimension r_{t-1} that, conditional upon V, has a proper distribution. Consequently, show that a conditional forecast exists such that

$$(Y_t|D_{t-1}, V) \sim \mathrm{N}[f_t, VQ_t^*],$$

and that given Y_t, the variance can be updated according to

$$n_t = n_{t-1} + 1,$$
$$S_t = S_{t-1} + (e_t^2/Q_t^* - S_{t-1})/n_t,$$

with $n_1 = 1$ and $S_1 = e_1^2$. Show further that if $n_{t-1} > 0$, the unconditional forecast

$$(Y_t|D_{t-1}) \sim \mathrm{T}_{n_{t-1}}[f_t, Q_t]$$

exists. Suggest a method of obtaining $\mathbf{m}_t$ for this reduced DLM.

(d) Now suppose rank $\mathbf{C}_t^{*-1} = 1 +$ rank $\mathbf{C}_{t-1}^{*-1}$. Show that no forecast of $(Y_t|D_{t-1}, V)$ is possible. However, given Y_t, show that although $\{n_t, S_t\} = \{n_{t-1}, S_{t-1}\}$, the dimension of the design space for which forecasts can now be derived is increased by 1.

(e) Collinearity can be a real problem. For example, in the simplest regression discount DLM, a price variable used as a regressor may be held constant for quite a long period before being changed. Obviously, information is being gathered, so that useful conditional forecasts can be made based upon this price. However, forecasts conditional on other prices cannot be made until after a price change has been experienced. In general, at time $t-1$, the forecast design space is spanned by $\mathbf{F}_1, \ldots, \mathbf{F}_{t-1}$. Construct an algorithm to provide a reference analysis accommodating collinearity that at time $t-1$, enables forecasts and monitoring for time t, whenever $\mathbf{F}_t$ is contained in the current forecast design space and $n_{t-1} \geq 1$.

(f) Generalise this approach beyond discount regression DLMs. Some relevant work is presented in Vasconcellos (1992).

CHAPTER 5

UNIVARIATE TIME SERIES DLM THEORY

5.1 UNIVARIATE TIME SERIES DLMS

As introduced in Definition 4.2 of the previous chapter, the class of univariate time series DLMs, or TSDLMs, is defined by quadruples

$$\{\mathbf{F}, \mathbf{G}, V_t, \mathbf{W}_t\},$$

for any V_t and $\mathbf{W}_t$. We often use the shorthand notation

$$\{\mathbf{F}, \mathbf{G}, \cdot, \cdot\}.$$

This chapter explores the theoretical structure of this important class of models. Much of classical time series analysis concerns itself with models of *stationary* processes (Box and Jenkins 1976), otherwise referred to as processes exhibiting *stationarity*. It will be shown that such models can be formulated as constant TSDLMs, namely as special DLMs for which the whole quadruple $\{\mathbf{F}, \mathbf{G}, V, \mathbf{W}\}$ is constant for all t. In practice, this constancy is usually a restrictive assumption, particularly since the variances V_t and $\mathbf{W}_t$ often vary both randomly and as a function of the level of the series. Consequently, we do not restrict attention to DLMs with constant quadruples, but consider processes that cannot typically be reduced to stationarity.

The mean response function $\mu_{t+k} = \mathrm{E}[Y_{t+k} \mid \boldsymbol{\theta}_{t+k}] = \mathbf{F}'\boldsymbol{\theta}_{t+k}$ and the forecast function $f_t(k) = \mathrm{E}[\mu_{t+k} \mid D_t] = \mathbf{F}'\mathbf{G}^k\mathbf{m}_t$ are of particular interest. The structure of the forecast function is central in describing the implications of a given DLM and in designing DLMs consistent with a forecaster's view of the future development of a series. A further ingredient in this design activity is the concept of *observability* of DLMs, which is where this chapter starts.

5.2 OBSERVABILITY

5.2.1 Introduction and definition

Observability is a fundamental concept in linear systems theory that has ramifications for TSDLMs. Primarily, it relates to the DLM parametrisation $\boldsymbol{\theta}_t$. As an introduction, consider the evolution error-free case with $\mathbf{W}_t = \mathbf{0}$ for all t, so that $\boldsymbol{\theta}_t = \mathbf{G}\boldsymbol{\theta}_{t-1}$ and $\mu_{t+k} = \mathbf{F}'\mathbf{G}^k\boldsymbol{\theta}_t$. The state vector, comprising n elements, is chosen to reflect identifiable features of the time series. Accordingly, values of the mean response over time provide information on the state vector. Clearly, at least n distinct values of the mean response are required for complete identification, with parametric parsimony suggesting that no more than n are necessary. The n distinct values starting at t, denoted by $\boldsymbol{\mu}_t = (\mu_t, \mu_{t+1}, \ldots, \mu_{t+n-1})'$, are related to the state vector via

$$\boldsymbol{\mu}_t = \mathbf{T}\boldsymbol{\theta}_t,$$

where $\mathbf{T}$ is the $n \times n$ observability matrix

$$\mathbf{T} = \begin{pmatrix} \mathbf{F}' \\ \mathbf{F}'\mathbf{G} \\ \vdots \\ \mathbf{F}'\mathbf{G}^{n-1} \end{pmatrix}. \tag{5.1}$$

Thus, in order to precisely determine the state vector from the necessary minimum of n consecutive values of the mean response function, $\mathbf{T}$ must be of full rank. Then

$$\boldsymbol{\theta}_t = \mathbf{T}^{-1}\boldsymbol{\mu}_t.$$

These ideas of parametric economy and identifiability in the case of a deterministic evolution motivate the formal definition of observability in the stochastic case.

Definition 5.1. Any TSDLM $\{\mathbf{F}, \mathbf{G}, \cdot, \cdot\}$ is **observable** if and only if the $n \times n$ **observability matrix T** in (5.1) has full rank n.

5.2.2 Examples

EXAMPLE 5.1. The model

$$\left\{\begin{pmatrix} 1 \\ 0 \end{pmatrix}, \begin{pmatrix} 1 & 1 \\ 0 & 1 \end{pmatrix}, \cdot, \cdot\right\}$$

is observable since $\mathbf{T}$ is of rank 2, with

$$\mathbf{T} = \begin{pmatrix} 1 & 0 \\ 1 & 1 \end{pmatrix}.$$

If the system variance $\mathbf{W}_t = \mathbf{0}$, then $\boldsymbol{\theta}_t = \mathbf{T}^{-1}\boldsymbol{\mu}_t$ where $\boldsymbol{\mu}_t' = (\mu_t, \mu_{t+1})$. Thus, $\boldsymbol{\theta}_t' = (\mu_t, \mu_{t+1} - \mu_t)$ comprises $\theta_{t1} = \mu_t$, the current level of the series, and $\theta_{t2} = \mu_{t+1} - \mu_t$ the growth in level between times t and $t+1$.

EXAMPLE 5.2. The model

$$\left\{\begin{pmatrix} 1 \\ 1 \end{pmatrix}, \begin{pmatrix} 1 & 0 \\ 0 & 1 \end{pmatrix}, \cdot, \cdot\right\}$$

is unobservable since $\mathbf{T}$ only has rank 1, with

$$\mathbf{T} = \begin{pmatrix} 1 & 1 \\ 1 & 1 \end{pmatrix}.$$

To see what is happening, write $\boldsymbol{\theta}_t' = (\theta_{t1}, \theta_{t2})$, $\boldsymbol{\omega}_t' = (\omega_{t1}, \omega_{t2})$, and the observation and evolution equations as

$$\begin{aligned} Y_t &= \theta_{t1} + \theta_{t2} + \nu_t, \\ \theta_{t1} &= \theta_{t-1,1} + \omega_{t1}, \\ \theta_{t2} &= \theta_{t-1,2} + \omega_{t2}. \end{aligned}$$

Noting that $\mu_t = \theta_{t1} + \theta_{t2}$ and defining $\psi_t = \theta_{t1} - \theta_{t2}$, the model becomes

$$\begin{aligned} Y_t &= \mu_t + \nu_t, \\ \mu_t &= \mu_{t-1} + \delta_{t1}, \\ \psi_t &= \psi_{t-1} + \delta_{t2}, \end{aligned}$$

where $\delta_{t1} = \omega_{t1} + \omega_{t2}$ and $\delta_{t2} = \omega_{t1} - \omega_{t2}$. The first two equations completely define the DLM as a first-order polynomial DLM. The random quantity ψ_t has no influence on the mean response, and in this sense is redundant. The model is overparametrised.

Example 5.2 illustrates the general result that if $\mathbf{T}$ has rank $n-r$ for some r, $(1 \le r < n)$, the DLM can be reparametrised, by a linear transformation of the state vector, to an observable DLM of dimension $n-r$ that has the same mean response function.

EXAMPLE 5.3. Consider the reparametrised model in Example 5.2 and suppose that

$$\left(\begin{matrix}\mu_0 \\ \psi_0\end{matrix}\middle| D_0\right) \sim \mathrm{N}\left[\begin{pmatrix} m_{\mu,0} \\ m_{\psi,0}\end{pmatrix}, \begin{pmatrix} C_{\mu,0} & 0 \\ 0 & C_{\psi,0}\end{pmatrix}\right],$$

with, for all t,

$$\mathbf{W}_t = \begin{pmatrix} W_\mu & 0 \\ 0 & W_\psi \end{pmatrix}.$$

In this case there is no correlation between Y_t and ψ_t, so the observations provide no information about the latter. Applying the updating equations leads to $\mathbf{C}_t$ diagonal for all t and the marginal posterior

$$(\psi_t \mid D_t) \sim \mathrm{N}[m_{\psi,0},\ C_{\psi,0} + tW_\psi],$$

with increasingly large variance.

EXAMPLE 5.4. Example 5.3 suggests that in the case of unobservability, the observations provide no information about some of the state parameters, or some linear functions of them. In that specific case, this is always true unless $\mathbf{W}_t$ has non-zero off-diagonal elements introducing correlations between the state vector elements. For example, consider the unobservable model

$$\left\{\begin{pmatrix}1 \\ 0\end{pmatrix}, \begin{pmatrix}1 & 0 \\ 0 & 0\end{pmatrix}, 2, \begin{pmatrix}1 & U \\ U & W\end{pmatrix}\right\}.$$

Clearly θ_{t2} is redundant, an equivalent mean response function being provided by the reduced model $\{1, 1, 2, 1\}$. This is a first-order polynomial model with, from Chapter 2, limiting distribution $(\theta_{t1} \mid D_t) \sim \mathrm{N}[m_{t1}, 1]$ as $t \to \infty$. In the 2-dimensional model, however, it may be verified that the updating equations lead to the limiting form

$$(\theta_{t2} \mid D_t) \sim \mathrm{N}[Ue_t/4,\ W - U^2/4],$$

as $t \to \infty$. Thus, unless $U = 0$, Y_t is informative about the parameter that, for forecasting purposes, is redundant.

EXAMPLE 5.5. In a deterministic model $\{\mathbf{F}, \mathbf{G}, 0, \mathbf{0}\}$, $Y_{t+k} = \mu_{t+k}$ for all k, so that defining

$$\mathbf{Y}_t = (Y_t, Y_{t+1}, \ldots, Y_{t+n-1})'$$

we have $\mathbf{Y}_t = \boldsymbol{\mu}_t = \mathbf{T}\boldsymbol{\theta}_t$. In the observable case it follows that

$$\boldsymbol{\theta}_t = \mathbf{T}^{-1}\mathbf{Y}_t.$$

Hence the values of the state vector elements are precisely determined, or observed, as linear functions of any n values of the series. If, however, $\mathbf{T}$ has rank $n-r$, then every Y_t is a linear combination of any $n-r$ observations, say $Y_1, Y_2, \ldots, Y_{n-r}$, and $\boldsymbol{\theta}_t$ is *not* fully determinable from the observations. This corresponds closely to the concept of observability as applied in linear systems theory and related fields.

5.2.3 *Observability and the forecast function*

For a TSDLM with $f_t(k) = \mathbf{F}'\mathbf{G}^k\mathbf{m}_t$, for $k \geq 0$,

$$\mathbf{T}\mathbf{m}_t = \begin{pmatrix} f_t(0) \\ f_t(1) \\ \vdots \\ f_t(n-1) \end{pmatrix}$$

and

$$\mathbf{T}\mathbf{G}^k\mathbf{m}_t = \begin{pmatrix} f_t(k) \\ f_t(k+1) \\ \vdots \\ f_t(n+k-1) \end{pmatrix}.$$

So observability implies that the first n consecutive terms of the forecast function are linearly independent, and vice versa. Any further values of the forecast function are then linear combinations of the first n values. If, however, $\mathbf{T}$ has rank $n-r$, then all values of the forecast function are linear functions of the first $n-r$.

EXAMPLE 5.6. The model

$$\left\{\begin{pmatrix}1\\0\end{pmatrix},\begin{pmatrix}1&1\\0&1\end{pmatrix},\dots\right\}$$

has forecast function

$$f_t(k) = m_{t1} + km_{t2} = f_t(0) + k[f_t(1) - f_t(0)],$$

a polynomial of order 2, a straight line.

EXAMPLE 5.7. The model

$$\left\{\begin{pmatrix}0\\1\end{pmatrix},\begin{pmatrix}1&1\\0&1\end{pmatrix},\dots\right\}$$

is unobservable, with $\mathbf{T}$ of rank 1, and the forecast function

$$f_t(k) = m_{t2} = f_t(0)$$

is reduced to that of a first-order polynomial model.

5.2.4 Constrained observability

Because of indeterminism, parametric redundancy and the reducibility of unobservable to observable models, from now on it will be assumed that working TSDLMs are observable unless otherwise stated. In some cases models with a singular observability matrix are employed provided that they are subject to additional structure leading to observability in a wider sense.

For example, consider the deterministic model

$$\left\{\begin{pmatrix}1\\1\\0\end{pmatrix},\begin{pmatrix}1&0&0\\0&0&1\\0&1&0\end{pmatrix},0,\mathbf{0}\right\}$$

having an observability matrix $\mathbf{T}$ of rank 2, where

$$\mathbf{T} = \begin{pmatrix}1&1&0\\1&0&1\\1&1&0\end{pmatrix}.$$

This model is unobservable. However, if θ_{t2} and θ_{t3} represent the effects of a factor variable, say a seasonal cycle of periodicity 2, then for all t, the modeller will typically apply a *constraint* of the form

$$\theta_{t2} + \theta_{t3} = 0.$$

Now, for all $t \geq 2$, $\boldsymbol{\theta}_t$ is determined from the mean response function and constraint via

$$\begin{aligned}\mu_t &= \theta_{t1} + \theta_{t2},\\ \mu_{t+1} &= \theta_{t1} + \theta_{t3},\\ 0 &= \theta_{t2} + \theta_{t3},\end{aligned}$$

since then

$$\theta_{t1} = (\mu_t + \mu_{t+1})/2$$

and

$$\theta_{t2} = -\theta_{t3} = (\mu_t - \mu_{t+1})/2.$$

Such models for *effects* are common in statistics, relating to classifying factors such as seasonal period, blocking variables, treatment regimes and so on. Rather different constraints arise in some applications, such as in studies of compositional data, where perhaps some of the elements of the state vector represent proportions that must sum to 1, again implying a constraint on a linear function of $\boldsymbol{\theta}_t$.

Clearly the above unobservable model can be reduced to the observable model

$$\left\{ \begin{pmatrix} 1 \\ 1 \end{pmatrix}, \begin{pmatrix} 1 & 0 \\ 0 & -1 \end{pmatrix}, 0, \mathbf{0} \right\},$$

producing the same forecast function. However, for practical interpretation and communication, it is may be desirable to retain the full unobservable DLM. Hence a wider definition of observability is required to cover models subject to linear constraints.

Definition 5.2. Suppose the unobservable model $\{\mathbf{F}, \mathbf{G}, V_t, \mathbf{W}_t\}$ of dimension n is subject to constraints on the state vector of the form

$$\mathbf{C}\boldsymbol{\theta}_t = \mathbf{c},$$

for some known, constant matrix $\mathbf{C}$ and vector $\mathbf{c}$. Then the DLM is said to be **constrained observable** if and only if the **extended observability matrix**

$$\begin{pmatrix} \mathbf{T} \\ \mathbf{C} \end{pmatrix}$$

has full rank n.

5.3 SIMILAR AND EQUIVALENT MODELS

5.3.1 Introduction

The concept of observability allows a modeller to restrict attention to a subclass of DLMs that are parsimoniously parametrised whilst providing the full range of forecast functions. However, this subclass is still large, and any given form of forecast function may typically be derived from many observable models. In designing DLMs, further guidelines are needed in order to identify small, practically meaningful collections of suitable models and usually a preferred single DLM for any given forecast function. The two key concepts are **similarity** and **equivalence** of TSDLMs. Similarity identifies and groups together all observable models consistent with any given forecast function. Two such models are called **similar models.** Equivalence strengthens this relationship by requiring that in addition to having the same *qualitative form* of forecast function, the **quantitative** specification of the full forecast distributions for future observations be precisely the same. Any two models producing the same forecasts are called **equivalent models.** Similarity and equivalence are essentially related to the **reparametrisation** of a model via a linear map of the state vector. This is particularly useful in model building when a simpler, identified, **canonical** model may be reparametrised to provide an equivalent model that is operationally more meaningful, efficient and easily understood.

5.3.2 Similar models

Consider two observable TSDLMs, M and M_1, characterised by quadruples

$$\begin{aligned} M: &\qquad \{\mathbf{F}, \mathbf{G}, V_t, \mathbf{W}_t\}, \\ M_1: &\qquad \{\mathbf{F}_1, \mathbf{G}_1, V_{1t}, \mathbf{W}_{1t}\}, \end{aligned}$$

having forecast functions $f_t(k)$ and $f_{1t}(k)$ respectively. The formal definition of similarity is as follows.

Definition 5.3. M and M_1 are **similar** models, denoted by $M \sim M_1$, if and only if the system matrices $\mathbf{G}$ and $\mathbf{G}_1$ have identical eigenvalues.

The implications of similarity are best appreciated in the special case when the system matrices both have n *distinct* eigenvalues $\lambda_1, \ldots, \lambda_n$. Here $\mathbf{G}$ is diagonalisable. If $\boldsymbol{\Lambda} = \text{diag}(\lambda_1, \ldots, \lambda_n)$, then there exists a non-singular $n \times n$ matrix $\mathbf{E}$ such that for all $k \geq 0$,

$$\mathbf{G} = \mathbf{E}\boldsymbol{\Lambda}\mathbf{E}^{-1} \qquad \text{and} \qquad \mathbf{G}^k = \mathbf{E}\boldsymbol{\Lambda}^k\mathbf{E}^{-1}.$$

By definition, the forecast function of M is

$$f_t(k) = \mathbf{F}'\mathbf{G}^k\mathbf{m}_t = \mathbf{F}'\mathbf{E}\boldsymbol{\Lambda}^k\mathbf{E}^{-1}\mathbf{m}_t = \sum_{r=1}^{n} a_{tr}\lambda_r^k,$$

for some coefficients $a_{t1}, \ldots, a_{tn}$ that do not depend on k. Since $M \sim M_1$, the forecast function for M_1 takes the similar form

$$f_{1t}(k) = \sum_{r=1}^{n} b_{tr}\lambda_r^k,$$

for some coefficients $b_{t1}, \ldots, b_{tn}$ not involving $\boldsymbol{\Lambda}$. So, as functions of the step-ahead integer k, M and M_1 have forecast functions of precisely the same *algebraic form*. This is the key to understanding similarity. If a modeller has a specific forecast function form in mind, then for forecasting purposes, any two similar models are qualitatively identical. Although the above example concerns the special case of distinct eigenvalues, it is *always* the case that two observable models have the same form of forecast function if and only if they are similar. Alternatively, M and M_1 are similar models if and only if the system matrices $\mathbf{G}$ and $\mathbf{G}_1$ are *similar matrices*, so that for some non-singular $n \times n$ **similarity matrix H**,

$$\mathbf{G} = \mathbf{H}\mathbf{G}_1\mathbf{H}^{-1}.$$

Hence, similarity of observable models is defined via similarity of system matrices. Further discussion of these points follows in Section 5.4, where the eigenstructure of system matrices is thoroughly explored.

5.3.3 Equivalent models and reparametrisation

Represent the form of the observable DLM $M_1 = \{\mathbf{F}_1, \mathbf{G}_1, V_{1t}, \mathbf{W}_{1t}\}$ as

$$Y_t = \mathbf{F}_1'\boldsymbol{\theta}_{1t} + \nu_{1t}, \qquad \nu_{1t} \sim \mathrm{N}[0, V_{1t}],$$

$$\boldsymbol{\theta}_{1t} = \mathbf{G}_1\boldsymbol{\theta}_{1t-1} + \boldsymbol{\omega}_{1t}, \qquad \boldsymbol{\omega}_{1t} \sim \mathrm{N}[\mathbf{0}, \mathbf{W}_{1t}].$$

Given any $n \times n$ nonsingular matrix $\mathbf{H}$, M_1 may be **reparametrised** by linearly transforming the state vector $\boldsymbol{\theta}_{1t}$ so that for all t,

$$\boldsymbol{\theta}_t = \mathbf{H}\boldsymbol{\theta}_{1t} \qquad (5.2)$$

and

$$\boldsymbol{\theta}_{1t} = \mathbf{H}^{-1}\boldsymbol{\theta}_t.$$

Then

$$Y_t = \mathbf{F}_1'\mathbf{H}^{-1}\boldsymbol{\theta}_t + \nu_{1t},$$
$$\boldsymbol{\theta}_t = \mathbf{H}\mathbf{G}_1\mathbf{H}^{-1}\boldsymbol{\theta}_{t-1} + \mathbf{H}\boldsymbol{\omega}_{1t}.$$

Defining $\mathbf{F}$ and $\mathbf{G}$ via the equations

$$\begin{aligned} \mathbf{F}' &= \mathbf{F}_1'\mathbf{H}^{-1}, \\ \mathbf{G} &= \mathbf{H}\mathbf{G}_1\mathbf{H}^{-1}, \end{aligned} \qquad (5.3)$$

and $\mathbf{m}_t = \mathbf{H}\mathbf{m}_{1t}$, we can write

$$f_{1t}(k) = \mathbf{F}_1'\mathbf{G}_1^k\mathbf{m}_{1t} = \mathbf{F}_1'\mathbf{H}^{-1}\mathbf{H}\mathbf{G}_1^k\mathbf{H}^{-1}\mathbf{H}\mathbf{m}_{1t} = \mathbf{F}'\mathbf{G}^k\mathbf{m}_t.$$

It follows that *any* model $M = \{\mathbf{F}, \mathbf{G}, \cdot, \cdot\}$, with $\mathbf{F}$ and $\mathbf{G}$ given by (5.3) for some $\mathbf{H}$, is similar to M_1. Further, the matrix $\mathbf{H}$ is defined as follows.

Theorem 5.1. *If for some nonsingular matrix* $\mathbf{H}$, $M = \{\mathbf{F}, \mathbf{G}, \cdot, \cdot\}$ *and* $M_1 = \{\mathbf{F}_1, \mathbf{G}_1, \cdot, \cdot\}$ *have respective observability matrices* $\mathbf{T}$ *and* $\mathbf{T}_1$ *and are such that* $\mathbf{F}' = \mathbf{F}_1'\mathbf{H}^{-1}$ *and* $\mathbf{G} = \mathbf{H}\mathbf{G}_1\mathbf{H}^{-1}$, *then,*

$$(i) \quad M \sim M_1 \qquad \text{and} \qquad (ii) \quad \mathbf{H} = \mathbf{T}^{-1}\mathbf{T}_1.$$

Proof. (i) follows from the definition of similarity since the system matrices are similar, having similarity matrix $\mathbf{H}$.

(ii) follows from the definition of $\mathbf{T}$ and $\mathbf{T}_1$, since

$$\mathbf{T}_1 = \mathbf{T}_1\mathbf{H}^{-1}\mathbf{H} = \mathbf{T}\mathbf{H}, \quad \text{so that} \quad \mathbf{H} = \mathbf{T}^{-1}\mathbf{T}_1.$$

◇

$\mathbf{H}$ is called the **similarity matrix** of the transformation from M_1 to M, since it is the similarity matrix transforming $\mathbf{G}_1$ to $\mathbf{G}$.

EXAMPLE 5.8. Let

$$\mathbf{F} = \begin{pmatrix} 1 \\ 0 \end{pmatrix}, \qquad \mathbf{G} = \begin{pmatrix} 1 & 1 \\ 0 & 1 \end{pmatrix},$$

$$\mathbf{F}_1 = \begin{pmatrix} -6 \\ 5 \end{pmatrix}, \qquad \mathbf{G}_1 = \begin{pmatrix} 9 & 16 \\ -4 & -7 \end{pmatrix}.$$

M and M_1 are observable, with

$$\mathbf{T} = \begin{pmatrix} 1 & 0 \\ 1 & 1 \end{pmatrix}, \qquad \mathbf{T}_1 = \begin{pmatrix} -6 & 5 \\ -74 & -131 \end{pmatrix}.$$

$\mathbf{G}$ and $\mathbf{G}_1$ each have a single eigenvalue 1 of multiplicity 2, so that $M \sim M_1$. Also,

$$\mathbf{H} = \mathbf{T}^{-1}\mathbf{T}_1 = \begin{pmatrix} -6 & 5 \\ -68 & -136 \end{pmatrix},$$

and it is easily verified that $\mathbf{F}_1 = \mathbf{H}'\mathbf{F}$ and $\mathbf{G}_1 = \mathbf{H}^{-1}\mathbf{G}\mathbf{H}$.

The forecast function has the form

$$f_t(k) = f_t(0) + k[f_t(1) - f_t(0)],$$

although the precise numerical values may differ between models.

Further features of the reparametrisation defined by (5.2) to (5.3) are as follows.

(a) The defining state vector $\boldsymbol{\theta}_t$ of M may be obtained as a linear transformation of $\boldsymbol{\theta}_{1t}$ in M_1 via

$$\boldsymbol{\theta}_t = \mathbf{H}\boldsymbol{\theta}_{1t},$$

and vice versa. One model may thus be reparametrised to obtain the other as far as the structural components $\mathbf{F}$, $\mathbf{G}$ and $\boldsymbol{\theta}_t$ are concerned.

(b) The full defining quadruple of M : $\{\mathbf{F}, \mathbf{G}, V_t, \mathbf{W}_t\}$ is obtained via this reparametrisation from that of M_1 : $\{\mathbf{F}_1, \mathbf{G}_1, V_{1t}, \mathbf{W}_{1t}\}$ if in addition to (a), the variances are related via

$$V_t = V_{1t} \quad \text{and} \quad \mathbf{W}_t = \mathbf{H}\mathbf{W}_{1t}\mathbf{H}'.$$

(c) If in addition, $(\boldsymbol{\theta}_t \mid D_t) \sim \mathrm{N}[\mathbf{m}_t, \mathbf{C}_t]$ and $(\boldsymbol{\theta}_{1t} \mid D_t) \sim \mathrm{N}[\mathbf{m}_{1t}, \mathbf{C}_{1t}]$ are related via

$$\mathbf{m}_t = \mathbf{H}\mathbf{m}_{1t}, \qquad \mathbf{C}_t = \mathbf{H}\mathbf{C}_{1t}\mathbf{H}',$$

then the entire quantitative specification of M is obtained from that of M_1 by the reparametrisation. In particular, this is true if these relationships hold between the initial priors at $t = 0$.

These final comments motivate the concept of model equivalence.

Definition 5.4. Consider two similar TSDLMs M and M_1 with similarity matrix $\mathbf{H} = \mathbf{T}^{-1}\mathbf{T}_1$. Suppose $M = \{\mathbf{F}, \mathbf{G}, V_t, \mathbf{W}_t\}$ with initial moments $(\mathbf{m}_0, \mathbf{C}_0)$, and $M_1 = \{\mathbf{F}_1, \mathbf{G}_1, V_{1t}, \mathbf{W}_{1t}\}$, with initial moments $(\mathbf{m}_{1,0}, \mathbf{C}_{1,0})$. Then M and M_1 are said to be **equivalent**, denoted by $M \equiv M_1$, if

$$V_t = V_{1t} \quad \text{and} \quad \mathbf{W}_t = \mathbf{H}\mathbf{W}_{1t}\mathbf{H}'$$

for all t, with

$$\mathbf{m}_0 = \mathbf{H}\mathbf{m}_{1,0} \quad \text{and} \quad \mathbf{C}_0 = \mathbf{H}\mathbf{C}_{1,0}\mathbf{H}'.$$

EXAMPLE 5.9. Consider the models

$$M: \quad \left\{ \begin{pmatrix} 1 \\ 0 \end{pmatrix}, \begin{pmatrix} \lambda & 1 \\ 0 & \rho \end{pmatrix}, V, \begin{pmatrix} 2 & \rho-\lambda \\ \rho-\lambda & (\rho-\lambda)^2 \end{pmatrix} \right\},$$

$$M_1: \quad \left\{ \begin{pmatrix} 1 \\ 1 \end{pmatrix}, \begin{pmatrix} \lambda & 0 \\ 0 & \rho \end{pmatrix}, V, \begin{pmatrix} 1 & 0 \\ 0 & 1 \end{pmatrix} \right\},$$

where λ and ρ are real and distinct. Then

$$\mathbf{T} = \begin{pmatrix} 1 & 0 \\ \lambda & 1 \end{pmatrix}, \qquad \mathbf{T}_1 = \begin{pmatrix} 1 & 1 \\ \lambda & \rho \end{pmatrix},$$

so that both models are observable. Thus, since both system matrices have identical eigenvalues λ and ρ, $M \sim M_1$. Further,

$$\mathbf{H} = \begin{pmatrix} 1 & 1 \\ 0 & \rho-\lambda \end{pmatrix},$$

and it can be verified that $\mathbf{F}' = \mathbf{F}_1'\mathbf{H}^{-1}$ and $\mathbf{W} = \mathbf{H}\mathbf{W}_1\mathbf{H}'$. So, if the initial priors conform, in the sense described in (c) above, then $M \equiv M_1$. Finally, notice that if $\rho = \lambda$, the models cannot even be similar, let alone equivalent, since although M is still observable, M_1 is not.

5.3.4. General equivalence

The definition of equivalence is a key concept in DLM design. However, as Harrison and Akram (1983) point out, it fails to apply in degenerate cases where two similar models that produce precisely the same forecast distributions cannot be linked by a linear transformation. The anomaly arises for $n \geq 2$ when there is an uncountable set of models that produce exactly the same forecast distributions but differ from M only through the system variance matrix $\mathbf{W}_t$. This reveals a fundamental ambiguity concerning $\mathbf{W}_t$ and the interpretation of $\boldsymbol{\theta}_t$, indicating that the modeller needs to impose some restraining structure, such as canonical component modelling and discounting as in Chapter 6.

For the general case, with M defined by the quadruple $\{\mathbf{F}, \mathbf{G}, V_t, \mathbf{W}_t\}$, let $\boldsymbol{\epsilon}_t$ be an independent sequence of random $n-$vectors with

$$(\boldsymbol{\epsilon}_t \mid D_{t-1}) \sim \mathrm{N}[\mathbf{0}, \mathbf{U}_t].$$

If for each t, $\mathbf{U}_t$ is a non-negative definite matrix such that $\mathbf{F}'\mathbf{U}_t\mathbf{F} = 0$, then $\mathbf{F}'\boldsymbol{\epsilon}_t = 0$ with probability one for all t. Consider now the model defined by adding the term $\mathbf{G}\boldsymbol{\epsilon}_{t-1}$ to the system evolution equation M so

$$\begin{aligned} Y_t &= \mathbf{F}'\boldsymbol{\theta}_t + \nu_t, \\ \boldsymbol{\theta}_t &= \mathbf{G}\boldsymbol{\theta}_{t-1} + (\boldsymbol{\omega}_t + \mathbf{G}\boldsymbol{\epsilon}_{t-1}). \end{aligned}$$

Defining the new state vector $\boldsymbol{\psi}_t$ via

$$\boldsymbol{\psi}_t = \boldsymbol{\theta}_t + \boldsymbol{\epsilon}_t,$$

and remembering that $\mathbf{F}'\boldsymbol{\epsilon}_t = 0$, we have

$$\begin{aligned} Y_t &= \mathbf{F}'\boldsymbol{\psi}_t + \nu_t, \\ \boldsymbol{\psi}_t &= \mathbf{G}\boldsymbol{\psi}_{t-1} + \boldsymbol{\omega}_{1t}, \end{aligned}$$

where $\boldsymbol{\omega}_{1t} = \boldsymbol{\omega}_t + \boldsymbol{\epsilon}_t$ is an independent error sequence with variance

$$\mathbf{W}_{1t} = \mathbf{W}_t + \mathbf{U}_t.$$

The *stochastic* shift from $\boldsymbol{\theta}_t$ to $\boldsymbol{\psi}_t$ transforms M to the model M_1 given by $\{\mathbf{F}, \mathbf{G}, V_t, \mathbf{W}_{1t}\}$. Thus, for *any* specified $\mathbf{W}_{1t}$, the model $\{\mathbf{F}, \mathbf{G}, V_t, \mathbf{W}_{1t}\}$ can be written in an uncountable number of ways by choosing $\mathbf{W}_t$ and $\mathbf{U}_t$ to give $\mathbf{W}_{1t}$ as above. Clearly, although reparametrisations that change the parametric interpretation are involved, they are stochastic and cannot be expressed as a deterministic linear transformation of the state vector.

EXAMPLE 5.10. To show that this can in fact be done, consider the model with $\mathbf{F}' = (1, 1)$ and $\mathbf{G} = \mathbf{I}$, the 2×2 identity matrix. In order to satisfy $\mathbf{F}'\boldsymbol{\epsilon}_t = 0$ for all t, we require $\boldsymbol{\epsilon}_t$ be of the form $\boldsymbol{\epsilon}_t = (\epsilon_{t1}, -\epsilon_{t1})'$. Then, with $\epsilon_{t1} \sim \mathrm{N}[0, U_t]$ for some $U_t > 0$,

$$\mathbf{U}_t = U_t \begin{pmatrix} 1 & -1 \\ -1 & 1 \end{pmatrix}.$$

Thus, given any variance matrix

$$\mathbf{W}_{1t} = \begin{pmatrix} W_{t1} & W_{t3} \\ W_{t3} & W_{t2} \end{pmatrix},$$

to obtain $\mathbf{W}_{1t} = \mathbf{W}_t + \mathbf{U}_t$ we simply choose $\mathbf{W}_t = \mathbf{W}_{1t} - \mathbf{U}_t$ and this is a valid variance matrix whenever

$$0 < U_t < (W_{t1}W_{t2} - W_{t3}^2)/(W_{t1} + W_{t2} + 2W_{t3}).$$

This can always be satisfied since, as $\mathbf{W}_{1t}$ is positive definite, the numerator of this upper bound is the determinant $|\mathbf{W}_{1t}| > 0$ and the denominator is the positive term $\mathbf{F}'\mathbf{W}_{1t}\mathbf{F}$.

These stochastic shift models seem to suggest that any specified model can be transformed to one with a simple *diagonal* evolution variance matrix. This is a very appealing simplification and in a large number of practical cases can be done. However, this is not always possible, as in Example 5.10, where $\mathbf{W}_t$ will be diagonal only when $W_{t3} = -U_t$, which is never possible if $W_{t3} > 0$.

This discussion motivates the following definition.

Definition 5.5. The models M and M_1, that are either observable or constrained observable, are said to be **generally equivalent** if and only if they produce exactly the same forecast distributions.

5.4 CANONICAL MODELS

5.4.1 Introduction

Similarity groups together observable DLMs with similar forecast functions. Within each such group, we identify particular models with specific, simple structures that provide *canonical* DLMs consistent with the required forecast function form. Similarity is related to the eigenstructure of system evolution matrices, and this section explores the various possible eigenvalue configurations that arise. In relation to simple, canonical matrices that are similar to any given $\mathbf{G}$, the focus is naturally on diagonal, block diagonal and **Jordan forms**. Supporting material on the linear algebra associated with these matrices appears in Section 17.4.

To begin, consider the simple case in which the $n \times n$ system matrix has a single real eigenvalue of multiplicity n, when a Jordan block system matrix is fundamental.

Definition 5.6. The $n \times n$ **Jordan block** is defined, for real or complex λ, as the $n \times n$ upper diagonal matrix

$$\mathbf{J}_n(\lambda) = \begin{pmatrix} \lambda & 1 & 0 & 0 & \dots & 0 \\ 0 & \lambda & 1 & 0 & \dots & 0 \\ 0 & 0 & \lambda & 1 & \dots & 0 \\ \vdots & \vdots & \vdots & \vdots & \ddots & \vdots \\ 0 & 0 & 0 & 0 & \dots & 1 \\ 0 & 0 & 0 & 0 & \dots & \lambda \end{pmatrix}.$$

Thus, the diagonal elements are all equal to λ, those on the super-diagonal are 1, and the remaining elements are 0.

5.4.2 System matrix with one real eigenvalue

Suppose that $\mathbf{G}$ is one of the uncountable number of matrices having a single real eigenvalue λ of multiplicity $n > 1$. The simplest example is $\mathbf{G} = \lambda\mathbf{I}$, a multiple of the $n \times n$ identity matrix. The first result, of great importance, shows that the class of observable models is restricted to the subset whose system matrices are similar to the Jordan block $\mathbf{J}_n(\lambda)$.

Theorem 5.2. *If* $\mathbf{G}$ *has one eigenvalue* λ *of multiplicity* n *but is not similar to* $\mathbf{J}_n(\lambda)$, *then any TSDLM* $\{\mathbf{F}, \mathbf{G}, \cdot, \cdot\}$ *is unobservable.*

Proof. The proof uses properties of similar matrices and Jordan forms for which reference may be made to Section 17.4.3. Since $\mathbf{G}$ is not similar to the Jordan block, it must be similar to a Jordan form

$$\mathbf{J}_s = \text{block diag}\,[\mathbf{J}_{n_1}(\lambda), \mathbf{J}_{n_2}(\lambda), \ldots, \mathbf{J}_{n_s}(\lambda)],$$

for some $s \geq 2$, $n_1 + n_2 + \cdots + n_s = n$, and $n_r \geq 1$, $(r = 1, \ldots, s)$. For each r, let $\mathbf{f}_r$ be any n_r-dimensional vector, and define $\mathbf{F}_s$ via $\mathbf{F}'_s = (\mathbf{f}'_1, \mathbf{f}'_2, \ldots, \mathbf{f}'_s)$. Then the observability matrix $\mathbf{T}$ of any model with regression vector $\mathbf{F}_s$ and system matrix $\mathbf{J}_s$ has rows

$$\mathbf{t}'_{k+1} = \mathbf{F}'_s \mathbf{J}^k_s(\lambda) = [\mathbf{f}'_1 \mathbf{J}^k_{n_1}(\lambda), \ldots, \mathbf{f}'_s \mathbf{J}^k_{n_s}(\lambda)], \qquad (k = 0, \ldots, n-1).$$

Define $m = \max\{n_1, \ldots n_s\}$, so that $m \leq n - s + 1$. Using the referenced appendix, it follows that for $k > m$,

$$\sum_{r=1}^{k} \binom{k}{r} (-\lambda)^r \mathbf{t}_{k-r+1} = \mathbf{0}.$$

This implies that $\mathbf{T}$ is of less than full rank, having at most $m < n$ linearly independent rows, and that the DLM is not observable.

$\diamond$

Corollary 5.1. *If* $\mathbf{G}$ *and* $\mathbf{G}_1$ *each have a single eigenvalue* λ *of multiplicity* n *and for some* $\mathbf{F}$ *and* $\mathbf{F}_1$, *the models* $\{\mathbf{F}, \mathbf{G}, \cdot, \cdot\}$ *and* $\{\mathbf{F}_1, \mathbf{G}_1, \cdot, \cdot\}$ *are observable, then* $\mathbf{G}$ *is similar to* $\mathbf{G}_1$ *and the models are similar.*

These results identify the class of observable DLMs with a single multiple eigenvalue λ as those observable models whose system matrix is similar to the canonical Jordan block $\mathbf{J}_n(\lambda)$. The class is uncountable even if $n = 1$. The following result identifies further structure that leads to the identification of a unique canonical model within the class.

Theorem 5.3. *Any TSDLM* $\{\mathbf{F}, \mathbf{J}_n(\lambda), \cdot, \cdot\}$ *is observable if and only if the first element of* $\mathbf{F}$ *is non-zero.*

Proof. Let $\mathbf{F}' = (f_1, \ldots, f_n)$ and the rows of $\mathbf{T}$ be

$$\mathbf{t}'_{r+1} = \mathbf{F}'\mathbf{J}_n^r(\lambda), \qquad (r = 0, \ldots, n-1).$$

Let $\mathbf{A}$ be the $n \times n$ matrix whose rows are

$$\mathbf{a}'_{k+1} = \sum_{r=0}^{k} \binom{k}{r} (-\lambda)^{k-r} \mathbf{t}'_{r+1}, \qquad (k = 0, \ldots, n-1).$$

Then

$$\mathbf{a}'_{k+1} = \mathbf{F}'[\mathbf{J}_n(\lambda) - \lambda \mathbf{I}_n]^k = (0, \ldots, 0, f_1, \ldots, f_{n-k}),$$

having k leading zeros. Thus $\mathbf{A}$ is an upper triangular matrix with leading diagonal $(f_1, \ldots, f_1)'$ with determinant f_1^n, and is non-singular if and only if $f_1 \neq 0$. The rows of $\mathbf{A}$ are constructed as linearly independent linear combinations of those of $\mathbf{T}$. So $\mathbf{A}$ and $\mathbf{T}$ have the same rank and the result follows.

◇

Notice that even the DLM with $\lambda = 0$ is observable since $\mathbf{J}_n(0)$ has rank $n-1$. Thus, although the system matrix of an observable DLM must be of at least rank $n-1$, it is not necessary that it be of full rank n.

Within the above class of similar observable models the simplest DLM is $\{\mathbf{E}_n, \mathbf{J}_n(\lambda), \cdot, \cdot\}$, for which $\mathbf{E}'_n = (1, 0, \ldots, 0)$, the only non-zero element of which is the leading 1. This specific form is adopted as the basic model with a single real eigenvalue. The $\mathbf{E}_n$ notation is used throughout the book.

Definition 5.7. Let $\mathbf{M} = \{\mathbf{F}, \mathbf{G}, V_t, \mathbf{W}_t\}$ be any observable TSDLM in which the system matrix $\mathbf{G}$ has a single real eigenvalue λ of multiplicity n. Let $\mathbf{T}$ be the observability matrix of this model and define

$$\mathbf{E}_n = (1, 0, \ldots, 0)'.$$

Then

(i) any model $\mathbf{M}_1 = \{\mathbf{E}_n, \mathbf{J}_n(\lambda), \cdot, \cdot\}$ with observability matrix $\mathbf{T}_1$ is defined as a **canonical similar model;** and

(ii) the model $\mathbf{M}_0 = \{\mathbf{E}_n, \mathbf{J}_n(\lambda), V_t, \mathbf{H}\mathbf{W}_t\mathbf{H}'\}$, where $\mathbf{H} = \mathbf{T}_0^{-1}\mathbf{T}$, is defined as the **canonical equivalent model**, so long as the initial priors are related as in Definition 5.4.

5.4.3 Multiple real eigenvalues

Suppose $\mathbf{G}$ has s distinct eigenvalues $\lambda_1, \ldots, \lambda_s$ with λ_i having multiplicity $r_i \geq 1$, so that $n = r_1 + \cdots + r_s$. Again using Section 17.4.3 it follows that $\mathbf{G}$ is similar to the block diagonal **Jordan form** -matrix

$$\mathbf{J} = \text{block diag}\,[\mathbf{J}_{r_1}(\lambda_1), \ldots, \mathbf{J}_{r_s}(\lambda_s)],$$

defined by the superposition of s Jordan blocks, one for each eigenvalue and having dimension given by the multiplicity of that eigenvalue. In such cases a generalisation of Theorem 5.3 shows that the DLM is observable if and only if it is similar to models of the form

$$\{\mathbf{E}, \mathbf{J}, \cdot, \cdot\},$$

where

$$\mathbf{E}' = (\mathbf{E}'_{r_1}, \ldots, \mathbf{E}'_{r_s})$$

is constructed as the corresponding catenation of s vectors of the form

$$\mathbf{E}'_r = (1, 0, \ldots, 0)$$

of dimension r, for $r = r_1, \ldots, r_s$. The forms of $\mathbf{E}$ and $\mathbf{J}$ provide the algebraically simplest similar models and are adopted as canonical.

Definition 5.8. Let $\mathbf{M} = \{\mathbf{F}, \mathbf{G}, V_t, \mathbf{W}_t\}$ be any observable TSDLM in which the system matrix $\mathbf{G}$ has s distinct real eigenvalues $\lambda_1, \ldots, \lambda_s$ with multiplicities $r_1, \ldots, r_s$ respectively. Let $\mathbf{T}$ be the observability matrix of this model and define

$$\mathbf{E} = (\mathbf{E}'_{r_1}, \ldots, \mathbf{E}'_{r_s})'$$

and

$$\mathbf{J} = \text{block diag}[\mathbf{J}_{r_1}(\lambda_1), \ldots, \mathbf{J}_{r_s}(\lambda_s)].$$

Then

(i) any model $\mathbf{M}_1 = \{\mathbf{E}, \mathbf{J}, \cdot, \cdot\}$ with observability matrix $\mathbf{T}_1$ is defined as a **canonical similar model**; and
(ii) the model $\mathbf{M}_0 = \{\mathbf{E}, \mathbf{J}, V_t, \mathbf{H}\mathbf{W}_t\mathbf{H}'\}$, where $\mathbf{H} = \mathbf{T}_0^{-1}\mathbf{T}$, is defined as the **canonical equivalent model**, so long as the initial priors are related as in Definition 5.4.

5.4.4 Complex eigenvalues when $n = 2$

Suppose that $n = 2$ and the 2×2 system matrix $\mathbf{G}$ has complex eigenvalues. $\mathbf{G}$ is real valued so the eigenvalues are a complex conjugate pair of the form

$$\lambda_1 = \lambda e^{i\omega} \qquad \text{and} \qquad \lambda_2 = \lambda e^{-i\omega},$$

for some real λ and ω, i being the imaginary square root of -1. Thus, $\mathbf{G}$ is similar to diag(λ_1, λ_2). As in the case of distinct real eigenvalues of Section 5.4.3, the model is similar to any DLM $\{(1,1)', \text{diag}(\lambda_1, \lambda_2), \cdot, \cdot\}$. However, this canonical similar model is not used since it results in a complex parametrisation, horrifying to practitioners and clearly to be avoided.

Instead, a *real canonical form* of $\mathbf{G}$ and the associated similar, **real** DLMs are identified. To proceed, note that, again following Section 17.4.3, if

$$\mathbf{H} = \begin{pmatrix} 1 & 1 \\ i & -i \end{pmatrix},$$

then

$$\mathbf{H}\begin{pmatrix} \lambda_1 & 0 \\ 0 & \lambda_2 \end{pmatrix}\mathbf{H}^{-1} = \lambda\begin{pmatrix} \cos(\omega) & \sin(\omega) \\ -\sin(\omega) & \cos(\omega) \end{pmatrix}$$

and

$$\mathbf{H}'\begin{pmatrix} 1 \\ 0 \end{pmatrix} = \begin{pmatrix} 1 \\ 1 \end{pmatrix},$$

from which it follows that the model is similar to any model with regression vector $(1,0)'$ and system matrix with the above cos/sin form.

Definition 5.9. Let the observable TSDLM $\{\mathbf{F}, \mathbf{G}, V_t, \mathbf{W}_t\}$ with observability matrix $\mathbf{T}$ be any 2-dimensional model in which the system matrix $\mathbf{G}$ has a pair of distinct, complex conjugate eigenvalues

$$\lambda_1 = \lambda e^{i\omega} \qquad \text{and} \qquad \lambda_2 = \lambda e^{-i\omega},$$

for real, non-zero λ and ω. Define

$$\mathbf{J}_2(\lambda, \omega) = \lambda\begin{pmatrix} \cos(\omega) & \sin(\omega) \\ -\sin(\omega) & \cos(\omega) \end{pmatrix}.$$

Then

(i) any model $\mathbf{M}_1 = \{\mathbf{E}_2, \mathbf{J}_2(\lambda,\omega), \cdot, \cdot\}$ with observability matrix $\mathbf{T}_1$ is defined as a **real canonical similar model;** and
(ii) the model $\mathbf{M}_0 = \{\mathbf{E}_2, \mathbf{J}_2(\lambda,\omega), V_t, \mathbf{H}\mathbf{W}_t\mathbf{H}'\}$, where $\mathbf{H} = \mathbf{T}_0^{-1}\mathbf{T}$, is defined as the **real canonical equivalent model**, so long as the initial priors are related as in Definition 5.4.

It is easily checked that the observability matrix of M_0 is simply

$$\mathbf{T}_0 = \begin{pmatrix} 1 & 0 \\ \lambda\cos(\omega) & \lambda\sin(\omega) \end{pmatrix}.$$

5.4.5 Multiple complex eigenvalues†

Again following Section 17.4, we directly define the real canonical models for cases in which the system matrix has multiple complex eigenvalues. Although rare in practice, there are instances in which such models may be used, and these canonical forms provide the simplest construction. Since

† Sections 5.4.5 and 5.4.6 are of rather theoretical interest and may be omitted without loss on a first reading.

complex eigenvalues must occur in conjugate pairs and $\mathbf{G}$ is real valued, the model dimension n must be even.

Definition 5.10. Let $\{\mathbf{F}, \mathbf{G}, V_t, \mathbf{W}_t\}$ be any observable TSDLM of dimension $n = 2v$, where v is some positive integer. Suppose the system matrix $\mathbf{G}$ has v multiples of a pair of distinct complex conjugate eigenvalues

$$\lambda_1 = \lambda e^{i\omega} \qquad \text{and} \qquad \lambda_2 = \lambda e^{-i\omega},$$

for real, non-zero λ and ω. Let $\mathbf{T}$ be the $n \times n$ observability matrix of this model and define

$$\mathbf{J}_{2,v}(\lambda,\omega) = \begin{pmatrix} \mathbf{J}_2(\lambda,\omega) & \mathbf{I} & \mathbf{0} & \dots & \mathbf{0} \\ \mathbf{0} & \mathbf{J}_2(\lambda,\omega) & \mathbf{I} & \dots & \mathbf{0} \\ \vdots & \vdots & \vdots & \ddots & \vdots \\ \mathbf{0} & \mathbf{0} & \mathbf{0} & \dots & \mathbf{I} \\ \mathbf{0} & \mathbf{0} & \mathbf{0} & \dots & \mathbf{J}_2(\lambda,\omega) \end{pmatrix}.$$

Thus, $\mathbf{J}_{2,v}(\lambda,\omega)$ is a $2v \times 2v$ block matrix comprising 2×2 submatrices. The v diagonal blocks are the basic cos/sin 2×2 blocks $\mathbf{J}_2(\lambda,\omega)$, the super-diagonal blocks are the 2×2 identity $\mathbf{I}$, and the remainder are zero blocks. Finally, define the $2v \times 1$ vector $\mathbf{E}_{2,v}$ via

$$\mathbf{E}_{2,v} = (\mathbf{E}_2', \dots, \mathbf{E}_2')'.$$

Then

(i) any model $\mathbf{M}_1 = \{\mathbf{E}_{2,v}, \mathbf{J}_{2,v}(\lambda,\omega), \cdot, \cdot\}$ with observability matrix $\mathbf{T}_1$ is defined as a **real canonical similar model;** and
(ii) the model $\mathbf{M}_0 = \{\mathbf{E}_{2,v}, \mathbf{J}_{2,v}(\lambda,\omega), V_t, \mathbf{H}\mathbf{W}_t\mathbf{H}'\}$, where $\mathbf{H} = \mathbf{T}_0^{-1}\mathbf{T}$, is defined as the **real canonical equivalent model**, so long as the initial priors are related as in Definition 5.4.

Note in particular the special case of Section 5.4.4 when $v = 1$, $\mathbf{E}_{2,1} = \mathbf{E}_2$, and $\mathbf{J}_{2,1}(\lambda,\omega) = \mathbf{J}_2(\lambda,\omega)$.

5.4.6 General case

In the most general case, $\mathbf{G}$ has s real and distinct eigenvalues $\lambda_1, \dots, \lambda_s$ of multiplicities $r_1, \dots, r_s$ respectively, and v pairs of complex conjugate eigenvalues

$$\lambda_{s+k} e^{i\omega_k} \qquad \text{and} \qquad \lambda_{s+k} e^{-i\omega_k}, \qquad\qquad (k = 1, \dots, v),$$

for some real, distinct $\lambda_{s+1}, \dots, \lambda_{s+v}$, and some real, distinct $\omega_1, \dots, \omega_v$, with the kth pair having multiplicity r_{s+k}, $(k = 1, \dots, v)$, respectively. Note that the dimension of the model is now

$$n = \sum_{k=1}^{s} r_k + 2\sum_{k=1}^{v} r_{s+k}.$$

The simplest real similar model is based on a system matrix formed by the superposition of $s+v$ diagonal blocks each corresponding to the canonical form for an individual real and/or pair of complex conjugate eigenvalues. Again the reader is referred to Section 17.4.

Definition 5.11. Let $\mathbf{M} = \{\mathbf{F}, \mathbf{G}, V_t, \mathbf{W}_t\}$ be any n-dimensional observable TSDLM with the eigenvalue structure as detailed above and observability matrix $\mathbf{T}$. Define the $n \times n$ block diagonal matrix $\mathbf{J}$ as

$$\mathbf{J} = \text{block diag}\,[\mathbf{J}_{r_1}(\lambda_1), \mathbf{J}_{r_2}(\lambda_2), \ldots, \mathbf{J}_{r_s}(\lambda_s);$$
$$\mathbf{J}_{2,r_{s+1}}(\lambda_{s+1}, \omega_1), \mathbf{J}_{2,r_{s+2}}(\lambda_{s+2}, \omega_2), \ldots, \mathbf{J}_{2,r_{s+v}}(\lambda_{s+v}, \omega_v)],$$

and the $n \times 1$ vector $\mathbf{E}$ as

$$\mathbf{E} = (\mathbf{E}'_{r_1}, \mathbf{E}'_{r_2}, \ldots, \mathbf{E}'_{r_s}; \mathbf{E}'_{2,r_{s+1}}, \mathbf{E}'_{2,r_{s+2}}, \ldots, \mathbf{E}'_{2,r_{s+v}})'.$$

Then

(i) any model $\mathbf{M}_1 = \{\mathbf{E}, \mathbf{J}, \cdot, \cdot\}$ with observability matrix $\mathbf{T}_1$ is defined as a **real canonical similar model;** and
(ii) the model $\mathbf{M}_0 = \{\mathbf{E}, \mathbf{J}, V_t, \mathbf{H}\mathbf{W}_t\mathbf{H}'\}$, where $\mathbf{H} = \mathbf{T}_0^{-1}\mathbf{T}$, is defined as the **real canonical equivalent model**, so long as the initial priors are related as in Definition 5.4.

This general canonical form is constructed from the simpler canonical models for each of the real and complex pairs of eigenvalues. The individual system matrices are simply superposed to form an overall block diagonal $\mathbf{J}$, and the corresponding individual regression vectors are catenated in the same order to provide the general vector $\mathbf{E}$. This construction of a very general model from the component building blocks provided by simpler models is a key concept in model design, the subject of the next chapter.

5.5 LIMITING RESULTS FOR CONSTANT MODELS

5.5.1 Introduction

A feature of constant DLMs is that variances converge to limiting values, often rapidly. In this section, we give general convergence results based on an approach that depends only on the existence of the first two moments and not on assumptions of normal distributions. This follows Harrison (1997). Related though less general results can be found in Anderson and Moore (1979). The convergence of variances, and consequently of the adaptive vectors $\mathbf{A}_t$, reveals the relationship between constant DLMs and classical point forecasting methods that generally adopt the limiting recurrence relationship for the parametric mean $\mathbf{m}_t$ and the limiting one-step forecast variance Q as discussed in Section 5.5.6.

Consider the univariate constant observable DLM $\{\mathbf{F}, \mathbf{G}, V, \mathbf{W}\}$, with observability matrix $\mathbf{T}$. Using the notation of Section 4.2, the initial information is $(\boldsymbol{\theta}_0|D_0) \sim \mathrm{N}[\mathbf{m}_0, \mathbf{C}_0]$, and thereafter the information set at time t is $D_t = \{Y_t, D_{t-1}\}$. For $p = 0, 1, \ldots, t-1$, let $D_{p,t} = \{\boldsymbol{\theta}_p, Y_{p+1}, \ldots, Y_t\}$. Define

$$\mathbf{Y}_t = (Y_{t-n+1}, \ldots, Y_t)',$$
$$\psi_{t-n+1} = \nu_{t-n+1},$$
$$\psi_{t-n+1+i} = \nu_{t-n+1+i} + \mathbf{F}' \sum_{j=0}^{i-1} \mathbf{G}^j \boldsymbol{\omega}_{t-n+1+i-j},$$
$$\boldsymbol{\psi}_t = (\psi_{t-n+1}, \ldots, \psi_t)',$$

and

$$\boldsymbol{\epsilon}_t = \sum_{i=0}^{n-2} \mathbf{G}^i \boldsymbol{\omega}_{t-i} - \mathbf{G}^{n-1}\mathbf{T}^{-1}\boldsymbol{\psi}_t.$$

Thus, $\boldsymbol{\epsilon}_t$ is a linear function of $\boldsymbol{\psi}_t$ and $\{\boldsymbol{\omega}_{t-n+2}, \ldots, \boldsymbol{\omega}_t\}$, so that for all $t \geq n$, $\mathrm{V}[\boldsymbol{\epsilon}_t] = \mathbf{S}$ is a finite constant variance matrix.

From the observation and system equations $\mathbf{Y}_t = \mathbf{T}\boldsymbol{\theta}_{t-n+1} + \boldsymbol{\psi}_t$ and

$$\boldsymbol{\theta}_t = \mathbf{G}^{n-1}\boldsymbol{\theta}_{t-n+1} + \sum_{i=0}^{n-2} \mathbf{G}^i \boldsymbol{\omega}_{t-i},$$

so that

$$\boldsymbol{\theta}_t = \mathbf{G}^{n-1}\mathbf{T}^{-1}\mathbf{Y}_t + \boldsymbol{\epsilon}_t. \tag{5.4}$$

5.5.2 Key preliminaries

Definition 5.12. For two general $n \times n$ finite variance matrices $\mathbf{M}$ and $\mathbf{S}$, write $\mathbf{M} \leq \mathbf{S}$ to signify that $\mathbf{l}'\mathbf{M}\mathbf{l} \leq \mathbf{l}'\mathbf{S}\mathbf{l}$, for all n-vectors $\mathbf{l}$. In such cases, we say that $\mathbf{M}$ is bounded above by $\mathbf{S}$, and that $\mathbf{S}$ is bounded below by $\mathbf{M}$.

Convergence: All variance matrices are bounded below by $\mathbf{0}$. It follows that if $\{\mathbf{M}_t\}$ is a sequence of finite variance matrices that is bounded above by $\mathbf{S}$, and is such that for all t, either *(i)* $\mathbf{M}_{t+1} \leq \mathbf{M}_t$ or *(ii)* $\mathbf{M}_{t+1} \geq \mathbf{M}_t$, then $\lim_{t\to\infty} \mathbf{M}_t = \mathbf{M}$ exists.

Theorem 5.4. *For all $t \geq n$, the variance sequence $\{\mathbf{C}_t\}$ is bounded above and below, with $\mathbf{0} \leq \mathbf{C}_t \leq \mathbf{S}$.*

Proof. $\mathbf{C}_t$ is independent of the values of $\{Y_t\}$, and from (5.4),

$$\mathbf{C}_t \leq \mathrm{V}[\boldsymbol{\theta}_n | \mathbf{G}^{n-1}\mathbf{T}^{-1}\mathbf{Y}_n] = \mathrm{V}[\boldsymbol{\epsilon}_t] = \mathbf{S}.$$

◇

Key results. For $p = 0, 1, \dots, t-1$, we know that

$$\mathrm{V}[\boldsymbol{\theta}|D_t] = \mathrm{E}[\mathrm{V}[\boldsymbol{\theta}|D_{p,t}, D_t]|D_t] + \mathrm{V}[\mathrm{E}[\boldsymbol{\theta}|D_{p,t}, D_t]|D_t]. \tag{5.5}$$

From the recurrence relations, $\mathbf{m}_t$ is a linear function of $(Y_t, \dots, Y_{p+1}, \mathbf{m}_p)$. Given $D_{p,t}$, $\boldsymbol{\theta}_p$ is precisely known, so for appropriate known $n \times 1$ vectors $\mathbf{b}_{t-p,0}, \dots, \mathbf{b}_{t-p,t-p+1}$ and $n \times n$ matrix $\mathbf{B}_{t-p}$,

$$\mathrm{E}[\boldsymbol{\theta}_t|D_{p,t}, D_t] = \mathrm{E}[\boldsymbol{\theta}_t|D_{p,t}] = \sum_{i=0}^{t-p+1} \mathbf{b}_{t-p,i} Y_{t-i} + \mathbf{B}_{t-p}\boldsymbol{\theta}_p. \tag{5.6}$$

Also, it is clear that

$$\mathrm{V}[\boldsymbol{\theta}_p|D_t] \geq \mathrm{V}[\boldsymbol{\theta}_p|D_{0,t}] \geq \mathrm{V}[\boldsymbol{\theta}_p|\boldsymbol{\theta}_{p+1}, Y_p, \boldsymbol{\theta}_{p-1}] > \mathbf{0}, \tag{5.7}$$

based on the definitions of the respective conditioning information sets and on the fact that additional information increases precision, or decreases variance, in this context of multivariate normal distributions.

These key results underlie the proofs of the following general results.

5.5.3 The convergence theorem

Theorem 5.5. *For any observable constant DLM, the limiting variance*

$$\lim_{t\to\infty} \mathbf{C}_t = \mathbf{C}$$

exists and is independent of the initial information D_0.

Proof. Throughout, p is a given integer in the range $0 \leq p < t$. Write $\mathrm{V}[\boldsymbol{\theta}_t|D_{0,t}] = \mathbf{C}_t^*$. Note that the sequence $\{\mathbf{C}_t^*\}$ is independent of the actual values comprising $D_{0,t}$, and so, using (5.6),

$$\mathrm{E}[\mathrm{V}[\boldsymbol{\theta}_t|D_{p,t}]|D_t] = \mathrm{V}[\boldsymbol{\theta}_t|D_{p,t}] = \mathrm{V}[\boldsymbol{\theta}_{t-p}|D_{0,t-p}] = \mathbf{C}_{t-p}^*$$

and

$$\mathrm{E}[\boldsymbol{\theta}_t|D_t, \boldsymbol{\theta}_p] = \mathrm{E}[\boldsymbol{\theta}_t|D_{p,t}] = \sum_{i=0}^{t-p+1} \mathbf{b}_{t-p,i} Y_{t-i} + \mathbf{B}_{t-p}\boldsymbol{\theta}_p.$$

Now proceed as follows:

(i) First prove the theorem for any DLM when $D_0 = D_{0,0}$ precisely specifies $\boldsymbol{\theta}_0$, so that $\mathbf{C}_0 = \mathbf{C}_0^* = \mathbf{0}$. Using (5.5),

$$\begin{aligned}\mathbf{C}_t^* &= \mathrm{E}[\mathrm{V}[\boldsymbol{\theta}_t|D_{p,t}]|D_{0,t}] + \mathrm{V}[\mathrm{E}[\boldsymbol{\theta}_t|D_{p,t}]|D_{0,t}] \\ &= \mathbf{C}_{t-p}^* + \mathrm{V}[\mathbf{B}_{t-p}\boldsymbol{\theta}_p|D_{0,t}] \geq \mathbf{C}_{t-p}^*.\end{aligned}$$

Monotonicity and $\mathbf{0} \leq \mathbf{C}_t \leq \mathbf{S}$ prove that $\lim_{t\to\infty} \mathbf{C}_t^* = \mathbf{C}^*$ exists. Further,

$$\lim_{t\to\infty} \mathrm{V}[\mathbf{B}_{t-p}\boldsymbol{\theta}_p|D_{0,t}] = \mathbf{0}. \tag{5.8}$$

(ii) We now give the proof for any prior D_0 when $V\mathbf{W} > \mathbf{0}$. Using (5.8) and (5.7),

$$\mathrm{V}[\boldsymbol{\theta}_p|D_{0,t}] \geq \mathrm{V}[\boldsymbol{\theta}_p|\boldsymbol{\theta}_{p+1}, Y_p, \boldsymbol{\theta}_{p-1}] > \mathbf{0},$$

so that

$$\lim_{t\to\infty} \mathbf{B}_t = \mathbf{0}. \tag{5.9}$$

Consequently, $\lim_{t\to\infty} \mathbf{C}_t = \mathbf{C}^*$; this follows by employing (5.5), (5.6) and (5.9) to show that

$$\mathrm{V}[\boldsymbol{\theta}_t|D_t] = \mathrm{E}[\mathrm{V}[\boldsymbol{\theta}_t|D_{p,t}]|D_t] + \mathrm{V}[\mathrm{E}[\boldsymbol{\theta}_t|D_{p,t}]|D_t]$$

or

$$\mathbf{C}_t = \mathbf{C}^*_{t-p} + \mathrm{V}[\mathbf{B}_{t-p}\boldsymbol{\theta}_p|D_t] \longrightarrow \mathbf{C}^*.$$

(iii) The remaining DLMs are wholly or partially deterministic and/or static with $V\mathbf{W} \not> \mathbf{0}$. With identity matrix $\mathbf{I}$, consider the subset of DLMs defined by $\{\mathbf{F}, \mathbf{G}, V + x, \mathbf{W} + x\mathbf{I});\ D_0 : 0 \leq x < 1\}$. For any given x let the variance sequence be $\{\mathbf{C}_t(x)\}$. Then
(a) $\lim_{t\to\infty} \mathbf{C}^*_t(0) = \mathbf{C}^*(0)$ exists, from (i).
(b) $\lim_{t\to\infty} \mathbf{C}_t(x) = \mathbf{C}^*(x)$, for all $0 < x < 1$, from (ii).
(c) $\mathbf{C}_t(x)$ is bounded, continuous with x, monotonic in x.
Hence as $t \to \infty$, $\mathbf{C}_t(x)$ tends (converges) uniformly to $\mathbf{C}(x)$, and

$$\begin{aligned}\lim_{t\to\infty} \mathbf{C}_t(0) &= \lim_{t\to\infty}\lim_{x\to 0} \mathbf{C}_t(x)\\ &= \lim_{x\to 0}\lim_{t\to\infty} \mathbf{C}_t(x) = \lim_{x\to 0} \mathbf{C}^*(x) = \mathbf{C}^*(0).\end{aligned}$$

◇

Corollary 5.2.

$$\lim_{t\to\infty} \mathbf{R}_t = \mathbf{R} = \mathbf{GCG}' + \mathbf{W},$$

$$\lim_{t\to\infty} Q_t = Q = \mathbf{F}'\mathbf{RF} + V,$$

$$\lim_{t\to\infty} \mathbf{A}_t = \mathbf{A} = \mathbf{RF}/Q.$$

5.5.4 Multivariate models

Consider the multivariate constant DLM of Definition 4.1. Assume the error terms have known means $\mathrm{E}[\boldsymbol{\nu}_t] = \bar{\boldsymbol{\nu}}_t$ and $\mathrm{E}[\boldsymbol{\omega}_t] = \bar{\boldsymbol{\omega}}_t$, and known variances, as usual.

Definition 5.13. The multivariate DLM $\{\mathbf{F}, \mathbf{G}, \cdot, \cdot\}$ is observable if and only if the $nr \times n$ observability matrix $\mathbf{T}$ is of full rank n. That is, there exists a vector $\mathbf{l} \in \mathbb{R}^r$ such that the univariate DLM $\{\mathbf{l}'\mathbf{F}, \mathbf{G}, \cdot, \cdot\}$ for the univariate time series $\{\mathbf{l}'\mathbf{Y}_t\}$ is observable.

The extension to known means $\bar{\boldsymbol{\nu}}_t$ and $\bar{\boldsymbol{\omega}}_t$ does not affect covariance measures. Considering the univariate series $\mathbf{l}'(\mathbf{Y}_t - \bar{\boldsymbol{\nu}}_t - \mathbf{F}'\bar{\boldsymbol{\omega}}_t)$ shows $\mathbf{C}_t$ to be bounded. Then, with $\mathbf{Y}_t - \bar{\boldsymbol{\nu}}_t - \mathbf{F}'\bar{\boldsymbol{\omega}}_t$ for Y_t, the theorem takes exactly the same course, with the result that

$$\lim_{t\to\infty} \mathbf{C}_t = \mathbf{C} = \mathbf{C}^*,$$

$$\lim_{t\to\infty} \mathbf{R}_t = \mathbf{R} = \mathbf{G}\mathbf{C}\mathbf{G}' + \mathbf{W},$$

$$\lim_{t\to\infty} \mathbf{Q}_t = \mathbf{Q} = \mathbf{F}'\mathbf{R}\mathbf{F} + \mathbf{V},$$

$$\lim_{t\to\infty} \mathbf{A}_t = \mathbf{A} = \mathbf{R}\mathbf{F}\mathbf{Q}^{-1}.$$

5.5.5 Convergence for a non-observable DLM

Although a sufficient condition, observability is not a necessary condition for convergence of the variance sequence $\{\mathbf{C}_t\}$. Convergence may occur for non-observable constant DLMs, such as constrained observable models. A simple example is a DLM comprising a level parameter in addition to parameters representing each of the four quarterly seasonal factors. The observability matrix $\mathbf{T}$ is singular, but with the additional constraint that the sum of the seasonal effects is zero, $\mathbf{C}_t$ converges.

A stationary constant DLM has all the eigenvalues of $\mathbf{G}$ inside the unit circle (see Definition 5.16). So, no matter what the rank of $\mathbf{T}$, $\mathbf{C}_t$ is bounded and convergence is assured even in the most trivial case in which $\mathbf{F} = \mathbf{0}$.

Consider the general constant DLM that, without loss of generality, can be expressed in partitioned form as

$$\mathbf{F} = \begin{pmatrix} \mathbf{F}_1 \\ \mathbf{F}_2 \end{pmatrix} \quad \text{and} \quad \mathbf{G} = \begin{bmatrix} \mathbf{G}_1 & \mathbf{0} \\ \mathbf{0} & \mathbf{G}_2 \end{bmatrix},$$

where the eigenvalues of $\mathbf{G}_1$ lie on or outside the unit circle but those of $\mathbf{G}_2$ lie inside the unit circle. Convergence for the whole DLM occurs if the DLM $\{\mathbf{F}_1, \mathbf{G}_1, \cdot, \cdot\}$ is observable or constrained observable.

In all these cases it is easily shown that no matter what the initial prior, $\{\mathbf{C}_t,\ t \geq n\}$ is bounded above. Once this is demonstrated, the convergence proof is exactly the same as for observable DLMs.

5.5.6 Further limit results

For the univariate, observable and constant DLM, the limiting form of the updating equation for $\mathbf{m}_t$ is given by

$$\mathbf{m}_t \approx \mathbf{G}\mathbf{m}_{t-1} + \mathbf{A}e_t = \mathbf{H}\mathbf{m}_{t-1} + \mathbf{A}Y_t,$$

with $\mathbf{H} = (\mathbf{I} - \mathbf{A}\mathbf{F}')\mathbf{G}$. For a proper DLM with $\mathbf{W} > \mathbf{0}$, $\mathbf{H} = \mathbf{C}\mathbf{R}^{-1}\mathbf{G}$ and a limiting representation of the observation series in terms of the one-step forecast errors follows. Let B be the *backshift* operator, so for any time series X_t, $BX_t = X_{t-1}$ and $B^p X_t = X_{t-p}$.

Theorem 5.6. *In the univariate constant DLM* $\{\mathbf{F}, \mathbf{G}, V, \mathbf{W}\}$ *denote the eigenvalues of* $\mathbf{G}$ *by* λ_i, *and those of* $\mathbf{H}$ *by* ρ_i, $(i = 1, \ldots, n)$. *Then*

$$\lim_{t\to\infty} \left\{ \prod_{i=1}^{n} (1 - \lambda_i B) Y_t - \prod_{i=1}^{n} (1 - \rho_i B) e_t \right\} = 0. \tag{5.10}$$

Proof. Following Harrison and Akram (1983) and Ameen and Harrison (1985), let $\mathbf{P}_1(B)$ and $\mathbf{P}_2(B)$ be row vectors and $P_3(B)$ and $P_4(B)$ be scalars, all with elements that are polynomials in B of order not exceeding $n-1$. Employ the Cayley-Hamilton theorem (Section 17.4.2). From $Y_{t+1} = \mathbf{F}'\mathbf{G}\mathbf{m}_t + e_{t+1}$ and $\mathbf{m}_t = \mathbf{G}\mathbf{m}_{t-1} + \mathbf{A}_t e_t$ it follows that

$$\prod_{i=1}^{n} (1 - \lambda_i B) Y_{t+1} = \prod_{i=1}^{n} (1 - \lambda_i B) e_{t+1} + \mathbf{P}_1(B) \mathbf{A}_t e_t. \tag{5.11}$$

And, from the Bayes' proof of the recurrences in Theorem 4.1,

$$\lim_{t\to\infty} (\mathbf{m}_t - \mathbf{H}\mathbf{m}_{t-1} - \mathbf{A}Y_t) = \mathbf{0}.$$

Define $\mathbf{M}_0 = \mathbf{m}_0$, $\mathbf{M}_t = \mathbf{H}\mathbf{M}_{t-1} + \mathbf{A}Y_t$, and $X_{t+1} = \mathbf{F}'\mathbf{G}\mathbf{M}_t + e_{t+1}$, so that, again employing Cayley-Hamilton,

$$\prod_{i=1}^{n} (1 - \rho_i B) e_{t+1} = \prod_{i=1}^{n} (1 - \rho_i B) X_{t+1} + \mathbf{P}_2(B) \mathbf{A}_t Y_t. \tag{5.12}$$

Note that $\lim_{t\to\infty} \mathbf{M}_t = \mathbf{m}_t$ and $\lim_{t\to\infty} X_t = Y_t$, so from (5.11) and (5.12),

$$\lim_{t\to\infty} \left\{ \prod_{i=1}^{n} (1 - \lambda_i B) Y_t - [1 + BP_3(B)] e_t \right\} = 0$$

and

$$\lim_{t\to\infty} \left\{ \prod_{i=1}^{n} (1 - \rho_i B) e_t - [1 + BP_4(B)] Y_t \right\} = 0.$$

The order of the polynomials in B correspond, and since the equations are true for all allowable values (λ_i, ρ_i), $(i = 1, \ldots, n)$, the coefficients of each B^i can be equated to complete the proof.

◇

It should be noted that Theorems 5.5 and 5.6 are based solely on the forms of the updating equations in TSDLMs, with absolutely no assumptions about a "true" data generating process or about normality. In addition, they apply even if the observational variance V is unknown and subject to the usual variance learning. In such cases, the posterior distribution for V will concentrate about its mode as t increases, asymptotically degenerating, and the model therefore converges to a known variance model.

From Theorem 5.6, the limiting representation of the observation series in terms of forecast errors is given by

$$Y_t = \sum_{j=1}^{n} \alpha_j Y_{t-j} + e_t + \sum_{j=1}^{n} \beta_j e_{t-j}, \tag{5.13}$$

with coefficients given by

$$\alpha_1 = \sum_{i=1}^{n} \lambda_i, \quad \alpha_2 = -\sum_{i=1}^{n} \sum_{k=i+1}^{n} \lambda_i \lambda_k, \quad \alpha_n = (-1)^n \lambda_1 \lambda_2 \ldots \lambda_n,$$

$$\beta_1 = -\sum_{i=1}^{n} \rho_i, \quad \beta_2 = \sum_{i=1}^{n} \sum_{k=i+1}^{n} \rho_i \rho_k, \quad \beta_n = (-1)^{n+1} \rho_1 \rho_2 \ldots \rho_n.$$

This representation provides a link with familiar ARIMA predictors of Box and Jenkins (1976), and with alternative methods including exponentially weighted regression, or exponential smoothing (McKenzie 1976). The following comments on the relationship with ARIMA modelling are pertinent.

(1) Suppose that p of the eigenvalues of $\mathbf{G}$ satisfy $0 < \lambda < 1$, d are equal to 1, and $n-p-d$ are zero. Suppose also that q of the eigenvalues of $\mathbf{H}$ satisfy $0 < \rho < 1$ with the remainder being zero. Then equation (5.13) is an ARIMA(p, d, q) **predictor**, whether or not the errors $\{e_t\}$ are uncorrelated.

(2) The ARIMA predictor is a limiting result in the DLM and therefore primarily of theoretical interest. In practice the use of non-constant variances and interventions will mean that the limiting forms are rarely utilised.

Theorem 5.7. *In the univariate constant DLM $\{\mathbf{F}, \mathbf{G}, V, \mathbf{W}\}$ denote the eigenvalues of* $\mathbf{G}$ *by λ_i and those of* $\mathbf{H}$ *by ρ_i, $(i = 1, \ldots, n)$. Then if the series $\{Y_t\}$ is* **truly** *generated by this DLM, it can be represented as*

$$\prod_{i=1}^{n}(1 - \lambda_i B) Y_t = \prod_{i=1}^{n}(1 - \rho_i B) a_t, \tag{5.14}$$

where $a_t \sim N[0, Q]$ are uncorrelated random variables.

Proof. Only a sketch of the proof is given. Apply the Cayley-Hamilton theorem to show that

$$\prod_{i=1}^{n}(1-\lambda_i B)Y_t = \phi(\{\nu_t, \boldsymbol{\omega}_t\}, \ldots, \{\nu_{t-n}, \boldsymbol{\omega}_{t-n}\}),$$

where ϕ is a linear function of the random vectors $\{\boldsymbol{\omega}_t, \ldots, \boldsymbol{\omega}_{t-n+1}\}$ and random variables $\{\nu_t, \ldots, \nu_{t-n}\}$, and is independent of t. Following Harrison (1967), represent ϕ in MA(q) process form, as

$$\phi(\boldsymbol{\omega}_t, \ldots, \boldsymbol{\omega}_{t-n+1}, \nu_t \ldots, \nu_{t-n}) = a_t + \sum_{i=1}^{q} \psi_i a_{t-i}.$$

Finally, the previous limiting results and Theorem 5.6 identify the roots of this MA process as the ρ_i. Also, $\lim_{t\to\infty}(a_t - e_t) = 0$.

◇

From (5.14) it may be thought that Y_t must follow an ARIMA(p,d,q) process. The truth is, however, more general; the eigenvalues of $\mathbf{G}$ may take any values, so the framework encompasses explosive processes as well as processes with unit eigenvalues. In non-explosive cases, we will have some p of the eigenvalues satisfying $0 < |\lambda_i| < 1$, a further d such that $\lambda_i = 1$, with the remaining λ_i being zero, together with, typically, some number q of the ρ_k such that $0 < |\rho_k| < 1$ and the rest being zero. In these cases, Y_t can indeed be represented in the form of an ARIMA(p,d,q) process. Thus all ARIMA(p,d,q) processes can be represented by a member of a subclass of constant TSDLM's $\{\mathbf{F}, \mathbf{G}, V, \mathbf{W}\}$ with $n = \max\{p+d, q\}$ parameters. The corresponding ARIMA point predictors operate with limiting forms and consequently impose unnecessary restrictions in their application.

5.5.7 Retrospective limit results

Referring back to the retrospective results of Sections 4.7 and 4.8, notice that for any fixed integer $k > 0$,

$$\lim_{t\to\infty} \mathbf{B}_{t-k} = \mathbf{B} = \mathbf{C}\mathbf{G}'\mathbf{R}^{-1},$$

$$\lim_{t\to\infty} \mathbf{A}_{t-k,t} = \mathbf{C}\mathbf{H}'^{k-1}\mathbf{G}'\mathbf{R}^{-1},$$

so the limiting retrospective distribution can be obtained as follows.

Theorem 5.8. *Given any fixed integer k, as $t \to \infty$, the variance of the historical parameters $(\boldsymbol{\theta}_{t-k}, \ldots, \boldsymbol{\theta}_t | D_t)$ converges to a limit with elements*

$$\lim_{t\to\infty} \mathrm{C}[\boldsymbol{\theta}_{t-i-j}, \boldsymbol{\theta}_{t-i}|D_t] = \mathbf{C}[-(i+j), -i] = \mathbf{B}^j\mathbf{R}(-i)$$

for all i, j such that $i, j \geq 0$ and $i + j \leq k$, with $\mathbf{B} = \mathbf{C}\mathbf{G}'\mathbf{R}^{-1}$ and where the $\mathbf{R}(-k)$ may be recursively calculated according to

$$\mathbf{R}(-k) = \mathbf{C} + \mathbf{B}[\mathbf{R}(-k+1) - \mathbf{R}]\mathbf{B}',$$

with initial value $\mathbf{R}(0) = \mathbf{R}$.

With $\mathbf{a}_t(0) = \mathbf{m}_t$ and $\mathbf{a}_{t-1}(1) = \mathbf{a}_t$, the limiting form of the recursive equations for revising retrospective means may be written in two ways, namely

$$\begin{aligned}\mathbf{a}_t(-k) &= \mathbf{a}_{t-1}(-k+1) + \mathbf{B}^k\mathbf{A}e_t \\ &= \mathbf{m}_{t-k} + \mathbf{B}[\mathbf{a}_t(-k+1) - \mathbf{a}_{t-k+1}].\end{aligned}$$

The proof is straightforward, using the filtering and retrospective results of Sections 4.7 and 4.8, and is left to the reader.

5.5.8 Discount TSDLM limit results

Discount models are fully discussed later in Section 6.3, but are previewed here to tie in with the limiting theory above. In a single discount TSDLM $\{\mathbf{F}, \mathbf{G}, V, \mathbf{W}_t\}$, $\mathbf{W}_t$ is defined as $\mathbf{W}_t = (1-\delta)\mathbf{G}\mathbf{C}_{t-1}\mathbf{G}'/\delta$ for some discount factor δ such that $0 < \delta < \min\{1, \lambda_1^2, \dots \lambda_n^2\}$, where the λ_i are the eigenvalues of $\mathbf{G}$. In cases in which $\mathbf{G}$ is of full rank, the resulting form of the updating equations relates closely to discount weighted regression estimation and is of much interest.

Since $\mathbf{R}_t = \mathbf{G}\mathbf{C}_{t-1}\mathbf{G}'/\delta$ in this model, it follows that

$$\mathbf{C}_t^{-1} = \delta\mathbf{G}^{-1'}\mathbf{C}_{t-1}^{-1}\mathbf{G}^{-1} + V^{-1}\mathbf{F}\mathbf{F}',$$

so that

$$\lim_{t\to\infty} \mathbf{C}_t^{-1} = \mathbf{C}^{-1} = V^{-1}\sum_{v=0}^{\infty}\delta^v\mathbf{G}^{-v'}\mathbf{F}\mathbf{F}'\mathbf{G}^{-v}.$$

Clearly $\lim_{t\to\infty}\mathbf{W}_t = \mathbf{W}$, and the DLM converges rapidly to a constant DLM. Further, since $\mathbf{H} = \delta\mathbf{C}\mathbf{G}^{-1'}\mathbf{C}^{-1}$, the eigenvalues of $\mathbf{H}$ are simply $\rho_i = \delta/\lambda_i$. In (5.13) this leads to

$$\lim_{t\to\infty}\left\{\prod_{i=1}^{n}(1-\lambda_i B)Y_t - \prod_{i=1}^{n}\left(1 - \frac{\delta}{\lambda_i}B\right)e_t\right\} = 0. \tag{5.15}$$

Further,

$$\mathbf{B} = \mathbf{C}\mathbf{G}'\mathbf{R}^{-1} = \delta\mathbf{G}^{-1},$$

which simplifies computations in the limiting retrospective updating equations.

5.6 STATIONARITY

5.6.1 Introduction

Historically, stationarity has been a dominant concept throughout time series analysis, and it is found useful in modelling closed systems. Purely stationary models are, however, of limited value in modelling open systems that necessarily involve interventions and model modifications to adapt to changing circumstances in the forecasting environment. However, models that involve stationary component sub-models, and inherently non-stationary, time-varying extensions of traditional stationary models, are valuable in a variety of contexts. The definitions and elements of the theory of stationary processes are introduced here.

Definition 5.14. A random time series, or RTS, Z_t is

(1) an ordered finite or infinite set of random variables indexed by consecutive integers;
(2) (strictly) **stationary** if for any given integer n, the distribution function of any random vector $\mathbf{Z}_{t,n+1} = (Z_t, \dots, Z_{t+n})'$ is independent of the time t;
(3) **weakly stationary** if both $\mathrm{E}[\mathbf{Z}_{t,n+1}]$ and $\mathrm{V}[\mathbf{Z}_{t,n+1}]$ are independent of time t;
(4) **Gaussian (normal) stationary** if it is weakly stationary and the distribution of every $\mathbf{Z}_{t,n+1}$ is normal.

Definition 5.15. Given a weakly stationary RTS Z_t,

(1) $\mathrm{C}[Z_t, Z_{t+k}] = \gamma_k$ is the autocovariance at lag k;
(2) $\rho_k = \gamma_k/\gamma_0$ is the theoretical autocorrelation at lag k;
(3) The graph $\{k, \gamma_k\}$ is the theoretical autocorrelation function (ACF);
(4) The autocovariance generating function (ACGF) is

$$\gamma(B) = \sum_{v=-\infty}^{\infty} \gamma_{|v|} B^v,$$

for real arguments B such that $|B| < 1$; the definition is also useful, in a formal sense, when B is taken as the backshift operator.

Stationarity imposes a strong structure, with weak stationarity implying a common mean, $\mathrm{E}[Z_t] = \mu$ for all t, and the lag-dependent covariance structure $\mathrm{C}[X_t, X_{t+s}] = \gamma_s$ for all t and s. In a Gaussian process, these first and second-order moments are those of the implied multivariate normal distribution for any subset of the Z_t, and so completely characterise the RTS. Given the mean μ and the autocovariances $\gamma_0, \gamma_1, \dots,$ we use the notation

$$Z \sim \mathrm{S}[\mu; \gamma(B)],$$

and in the Gaussian case,

$$Z \sim \mathrm{GS}[\mu; \gamma(B)].$$

EXAMPLE 5.11. Consider a random time series generated according to $Z_t = 2a_t - a_{t-1}$, where $a_t \sim \mathrm{N}[0,1]$ independently. Based only on this specification, the RTS is Gaussian stationary, with the entire joint distribution defined by

$$\mu = 0, \quad \gamma_0 = 5, \quad \gamma_1 = -2, \quad \text{and} \quad \gamma_k = 0, \text{ for all } k > 1.$$

Thus,

$$Z_t \sim \mathrm{GS}[0;\ -2B^{-1} + 5 - 2B].$$

Note, however, that conditioning on, say, $Z_1 = 1$, implies that the RTS is no longer stationary, although the RTS $\{Z_t : t \geq 3\}$ is.

EXAMPLE 5.12. Consider an infinite random time series generated according to $Z_t - \lambda Z_{t-1} = a_t$, where $a_t \sim \mathrm{N}[\alpha, \sigma^2]$ independently. Based only on this specification, the RTS is Gaussian stationary if and only if $|\lambda| < 1$, and in that case the entire joint distribution is defined by

$$\mu = \alpha/(1-\lambda) \quad \text{and} \quad \gamma_k = \lambda^k \sigma^2/(1-\lambda^2), \quad \text{for all } k \geq 0,$$

so that

$$Z_t \sim \mathrm{GS}[\alpha/(1-\lambda);\ \sigma^2(1-\lambda B)^{-1}(1-\lambda B^{-1})^{-1}].$$

Again, note that observing Z_1 implies that the RTS is no longer stationary, nor, for any $k > 0$, is any subsequence $\{Z_t : t \geq k\}$ for $k > 0$.

5.6.2 Stationary DLMs

It is evident from the preceding examples that if interest lies in predicting a series Y_t generated by a DLM, then given D_t the future series $\{Y_{t+i} : i > 0\}$ will not be stationary except in trivial cases. If $\mathbf{W}_t > \mathbf{0}$, this future series will be stationary if and only if *(a)* the model is equivalent to an observable **constant** DLM $\{\mathbf{F}, \mathbf{G}, V, \mathbf{W}\}$; *(b)* all the eigenvalues of $\mathbf{G}$ satisfy $|\lambda_i| < 1$; and *(c)* $(\boldsymbol{\theta}_{t+1}|D_t) \sim [\mathbf{0}, \mathbf{R}]$, where $\mathbf{R} = \mathbf{GRG}' + \mathbf{W}$. Though the third condition here has zero probability of being true, it is useful to relax it and define a DLM to be (qualitatively) stationary based only upon the first two conditions.

Definition 5.16. A DLM $\{\mathbf{F}_t, \mathbf{G}_t, V_t, \mathbf{W}_t\}$ is a (zero mean) **weakly stationary DLM** if and only if

(1) it is equivalent to an observable constant DLM $\{\mathbf{F}, \mathbf{G}, V, \mathbf{W}\}$;
(2) the eigenvalues of $\mathbf{G}$ lie inside the unit circle, i.e., $|\lambda_i| < 1$, for $i = 1, \ldots, n$.

Naturally, all such normal DLMs are termed Gaussian stationary DLMs.

5.6.3 Identification

For a specified RTS Y_t generated by an observable TSDLM $\{\mathbf{F}, \mathbf{G}, V_t, \mathbf{W}_t\}$ together with the initial prior based on D_0, the entire joint distribution of all linear functions of the RTS is completely defined. Hence, in principle, given a subsequent series of observations, it is a straight-forward matter to compare the actual sampling and theoretical distributions. In particular, writing the eigenvalues of $\mathbf{G}$ as $\lambda_1, \ldots, \lambda_n$ and defining

$$Z_t = \prod_{i=1}^{n}(1 - \lambda_i B)Y_{t+n},$$

we know that for all $k > 0$ and $j > n$,

$$E[Z_{t+k}|D_t] = 0 \quad \text{and} \quad \mathrm{C}[Z_{t+k}, Z_{t+k+j}] = 0,$$

providing one basis for assessing the adequacy of the model and facilitating model identification.

Further, if the TSDLM is a constant DLM and the only eigenvalues of $\mathbf{G}$ that lie on or outside the unit circle are $\lambda_1, \ldots, \lambda_r$, then the implied DLM generating

$$U_t = \prod_{i=1}^{r}(1 - \lambda_i B)Y_{t+r}$$

is a stationary DLM that can be modelled as an ARMA process as discussed in Chapter 9.

This provides the basis of classical identification methods that examine linear functions of the observed series Y_t, seeking a parsimonious function U_t that appears weakly stationary. The emphasis is on the associated sample means and covariance structure of the U_t series either directly through the ACGF or its Fourier transform, the spectrum. The latter is particularly useful when the eigenvalues are complex, so that either the whole series or major components may follow a mixture of damped cosine waves of differing frequencies.

5.7 EXERCISES

(1) Determine whether, and if so under what conditions, the following TSDLMs $\{\mathbf{F}, \mathbf{G}, \cdot, \cdot\}$ are observable.

(a) $\{\mathbf{F}, \mathbf{G}\} = \{1, \lambda\}$ for some given real λ.

(b) With given real λ_1 and λ_2,

$$\mathbf{F} = \begin{pmatrix} 1 \\ 1 \end{pmatrix}, \qquad \mathbf{G} = \begin{pmatrix} \lambda_1 & 0 \\ 0 & \lambda_2 \end{pmatrix}.$$

(c)

$$\mathbf{F} = \begin{pmatrix} 1 \\ 0 \end{pmatrix}, \qquad \mathbf{G} = \begin{pmatrix} 0 & 1 \\ 0 & 0 \end{pmatrix}.$$

(d)

$$\mathbf{F} = \begin{pmatrix} 1 \\ 1 \\ 1 \end{pmatrix}, \qquad \mathbf{G} = \begin{pmatrix} 4 & -1 & 2 \\ 3 & 9 & 3 \\ 1 & 5 & 5 \end{pmatrix}.$$

(e) For a given real ω,

$$\mathbf{F} = \begin{pmatrix} 0 \\ 1 \end{pmatrix}, \qquad \mathbf{G} = \mathbf{J}_2(1, \omega) = \begin{pmatrix} \cos(\omega) & \sin(\omega) \\ -\sin(\omega) & \cos(\omega) \end{pmatrix}.$$

(f) $\{\mathbf{F}, \mathbf{G}\} = \{\mathbf{E}_n, \mathbf{J}_n(\lambda)\}$ for $n \geq 2$.

(2) Give an example of an observable, n-dimensional DLM whose system matrix $\mathbf{G}$ is of rank $n-1$. Show that a necessary but not sufficient condition for a TSDLM to be observable is that the rank of the $n \times n$ system matrix $\mathbf{G}$ is at least $n-1$.

(3) Consider the constant DLM

$$\left\{ \begin{pmatrix} 1 \\ 1 \end{pmatrix}, \begin{pmatrix} 1 & 0 \\ 0 & 1 \end{pmatrix}, 80, \begin{pmatrix} 2 & 1 \\ 1 & 2 \end{pmatrix} \right\}$$

with parameter vector $\boldsymbol{\theta}_t = (\theta'_{t1}, \theta'_{t2})'$.

(a) Is the DLM observable?

(b) Write down the observation and system equations.

(c) What is $\lim_{t \to \infty} \mathrm{V}[\theta_{t1} + \theta_{t2} | D_t]$?

(d) What is $\lim_{t \to \infty} \mathrm{V}[\theta_{t1} - \theta_{t2} | D_t]$?

(e) Provide an observable DLM to represent the series Y_t.

(4) Consider the constant DLM

$$\left\{ \begin{pmatrix} 1 \\ 0 \end{pmatrix}, \begin{pmatrix} 1 & 0 \\ 0 & 0 \end{pmatrix}, 5, \begin{pmatrix} 1 & 1 \\ 1 & 4 \end{pmatrix} \right\}$$

with parameter vector $\boldsymbol{\theta}_t = (\theta'_{t1}, \theta'_{t2})'$.

(a) Is the DLM observable?

(b) Write down the observation and system equations.

(c) Show that although the DLM is unobservable, the series Y_t can be represented by a DLM with only one parameter and that this is equivalent to a first-order polynomial DLM.

(d) Define this DLM $\{1, 1, V, W\}$.

(5) Obtain the algebraic form of the forecast function for each of the following TSDLMs $\{\mathbf{F}, \mathbf{G}, \cdot, \cdot\}$, with a real parametrisation, noting whether or not the models are observable.

(a) $\{\mathbf{F}, \mathbf{G}\} = \{1, \lambda\}$.
Investigate all the possible cases, i.e., $\lambda < -1$, $\lambda = -1$, $-1 < \lambda < 0$, $\lambda = 0$, $0 < \lambda < 1$, $\lambda = 1$ and $\lambda > 1$.

(b) $\mathbf{F}' = (1, 0, 0)$ and $\mathbf{G} = \mathbf{J}_3(\lambda)$ for $0 < \lambda < 1$.
Examine the form of $f_t(k)$ as a function of k, determining, in particular, the turning points of the forecast function.

(c) $\mathbf{F}' = (1, 0, 1, 0, 1)$ and

$$\mathbf{G} = \text{block diag}\left\{ \begin{pmatrix} 1 & 1 \\ 0 & 1 \end{pmatrix}, \lambda \begin{pmatrix} \cos(\omega) & \sin(\omega) \\ -\sin(\omega) & \cos(\omega) \end{pmatrix}, \phi \right\},$$

with $\lambda > 0$, $1 > \phi > 0$ and ω not an integer multiple of π.

(6) Consider a TSDLM with

$$\mathbf{F} = \begin{pmatrix} 1 \\ 0 \\ 0 \end{pmatrix}, \qquad \mathbf{G} = \begin{pmatrix} 0 & 1 & 0 \\ 0 & 0 & 1 \\ 1 & 0 & 0 \end{pmatrix}.$$

(a) Show that for all positive integers k and n, the forecast function

$$f_t(k) = f_t(k + 3n),$$

and so is *cyclical* of *period* 3.

(b) Show that the model is observable and transform it to canonical form.

(7) Generalise the previous example to models that are cyclical of period $n > 1$, having $\mathbf{F} = \mathbf{E}_n = (1, 0, \ldots, 0)'$ and

$$\mathbf{G} = \begin{pmatrix} \mathbf{0} & \mathbf{I} \\ 1 & \mathbf{0}' \end{pmatrix},$$

where $\mathbf{I}$ is the $(n-1) \times (n-1)$ identity matrix. Distinguish the cases of even and odd values of n.

(8) For some integer $n > 1$, suppose that $\mathbf{F}' = (1, \mathbf{E}_n')$ and

$$\mathbf{G} = \begin{pmatrix} 1 & 0 & \mathbf{0}' \\ \mathbf{0} & \mathbf{0} & \mathbf{I} \\ 0 & 1 & \mathbf{0}' \end{pmatrix},$$

where $\mathbf{I}$ is the $(n-1) \times (n-1)$ identity matrix.

(a) Show that the model is unobservable.

(b) Let $\boldsymbol{\phi}_t$ denote the final n elements of the state vector at time t, so that $\boldsymbol{\theta}_t' = (\theta_t, \boldsymbol{\phi}_t')$, say. If $\mathbf{1}'\boldsymbol{\phi}_t = 1$ show that the model is constrained observable.

(9) Transform the following observable models to canonical forms, identifying the corresponding similarity matrices $\mathbf{H}$ (Theorem 5.1).

(a) For some given real and distinct λ_1 and λ_2,

$$\mathbf{F} = \begin{pmatrix} 1 \\ 1 \end{pmatrix}, \qquad \mathbf{G} = \begin{pmatrix} \lambda_1 & 0 \\ 0 & \lambda_2 \end{pmatrix}.$$

(b)
$$\mathbf{F} = \begin{pmatrix} 1 \\ 1 \\ 1 \end{pmatrix}, \qquad \mathbf{G} = \begin{pmatrix} 1 & 1 & 1 \\ 0 & 1 & 1 \\ 0 & 0 & 0.5 \end{pmatrix}.$$

(c) Given ω is not an integer multiple of π,

$$\mathbf{F} = \begin{pmatrix} 1 \\ 1 \end{pmatrix}, \qquad \mathbf{G} = \begin{pmatrix} 0.5e^{i\omega} & 0 \\ 0 & 0.5e^{-i\omega} \end{pmatrix}.$$

(10) Consider any two TSDLMs M and M_1 characterised by quadruples

$$M_0 : \quad \{\mathbf{E}_3, \mathbf{J}_3(1), V_t, \mathbf{W}_{0,t}\},$$

$$M_1 : \quad \left\{ \begin{pmatrix} 1 \\ 0 \\ 0 \end{pmatrix}, \begin{pmatrix} 1 & 1 & 1 \\ 0 & 1 & 1 \\ 0 & 0 & 1 \end{pmatrix}, 1000, \mathbf{W} \right\}.$$

(a) Calculate the observability matrices $\mathbf{T}$ and $\mathbf{T}_1$ for M and M_1 and deduce that both models are observable.

(b) Show that the DLMs are similar, and that the similarity matrix $\mathbf{H} = \mathbf{T}^{-1}\mathbf{T}_1$ is given by

$$\mathbf{H} = \begin{pmatrix} 1 & 0 & 0 \\ 0 & 1 & 1 \\ 0 & 0 & 1 \end{pmatrix}.$$

(c) Identify the common form of the forecast function, and interpret the meaning of the state parameters in each model.

(d) If

$$\mathbf{W} = \begin{pmatrix} 100 & 0 & 0 \\ 0 & 9 & -1 \\ 0 & -1 & 1 \end{pmatrix}$$

and

$$(\boldsymbol{\theta}_0 | D_0, M_1) \sim \mathrm{N}\left[\begin{pmatrix} 100 \\ 3 \\ 1 \end{pmatrix}, 4\mathbf{W} \right],$$

under what conditions are M and M_1 equivalent?

(11) Consider the model $\{1, \lambda, V, W\}$ where λ is real. By applying Theorems 5.5 and 5.6 or otherwise, obtain
  (a) the limiting values of C, R, A and Q;
  (b) the limiting representation of Y_t as a linear function of Y_{t-1}, e_t and e_{t-1}.
  (c) Distinguish between, and comment upon, the cases $|\lambda| < 1$, $\lambda = 1$, and $|\lambda| > 1$.

(12) In Theorem 5.6, the eigenvalues of $\mathbf{H} = (\mathbf{I} - \mathbf{A}\mathbf{F}')\mathbf{G}$ determine the limiting representation of Y_t. Verify the identity $\mathbf{H} = \mathbf{C}\mathbf{R}^{-1}\mathbf{G}$.

(13) Given $|\lambda| \leq 1$ and $VW > 0$, obtain the limiting representations of the observation series Y_t in terms of past observations and one-step forecast errors e_t for the DLM

$$\left\{ \begin{pmatrix} 1 \\ 0 \end{pmatrix}, \begin{pmatrix} 1 & \lambda \\ 0 & \lambda \end{pmatrix}, V, V \begin{pmatrix} U + \lambda^2 W & \lambda^2 W \\ \lambda^2 W & \lambda^2 W \end{pmatrix} \right\}.$$

(14) Consider the constant model $\{1, 1, V, W\}$. Given an integer $k > 0$, obtain the limiting *retrospective* variance

$$\lim_{t\to\infty} \mathrm{V}[(\mu_t, \ldots, \mu_{t-k})'|D_t].$$

(15) If Y_t is a stationary random time series (RTS), show that for all integers k and times t, $\mathrm{C}[Y_t, Y_{t+k}] = \mathrm{C}[Y_t, Y_{t-k}]$.

(16) The *backshift operator* B operates on a time index t such that $B^k y_t = y_{t-k}$ and $B^k f(t) = f(t-k)$ for time series y_t and functions $f(t)$, and for all integers k. Write the following expressions in the form $\phi(B)Y_t = \theta(B)e_t$, where $\phi(B)$ and $\theta(B)$ are polynomials in B :
  (a) $Y_{t+1} - Y_t = e_t$;
  (b) $Y_{t+1} - Y_t = e_{t-2}$;
  (c) $Y_t - Y_{t-2} - e_t + 0.5e_{t-1} = 0$;
  (d) $Y_t - Y_{t-1} - Y_{t-12} + Y_{t-13} = e_t - 0.2e_{t-1} - 0.5e_{t-12} + 0.1e_{t-13}$;
  (e) $\sum_{v=0}^{n} \binom{n}{v} (-1)^v Y_{t-v} = \sum_{v=0}^{n} \binom{n}{v} (-\alpha)^v e_{t-v}$.

(17) Let $\ldots, y_{-1}, y_0, y_1, \ldots$ and $\ldots, a_{-1}, a_0, a_1, \ldots$ be infinite sequences, the latter bounded. Suppose that $|\alpha| < 1$ so that $\lim_{m\to\infty} \alpha^m y_k = 0$ for all integers k.
  (a) If $y_t - \alpha y_{t-1} = (1 - \alpha B)y_t = a_t$, prove that the inverted expression

$$y_t = (1 - \alpha B)^{-1} a_t = \sum_{i=0}^{\infty} (\alpha B)^i a_t = \sum_{i=0}^{\infty} \alpha^i a_{t-i}$$

  is valid and meaningful.
  (b) Prove by induction that the expression $\prod_{i=1}^{n} (1 - \alpha_i B) y_t = a_t$ is invertible to $y_t = \prod_{i=1}^{n} (1 - \alpha_i B)^{-1} a_t$ under the same conditions. This proves the important result that $\phi(B) y_t = a_t$

is invertible to $y_t = \phi(B)^{-1}a_t$ if and only if the roots of the equation $\phi(B) = 0$ all lie outside the unit circle (i.e., have modulus greater than 1).

(18) Given $e_t \sim \mathrm{N}[0, \sigma^2]$ independently, state which of the following RTSs are stationary and for these, derive the corresponding ACFs.
(a) $Y_t = e_t + 2e_{t-1}$.
(b) $Y_t + 2Y_{t-1} = e_t$.
(c) $Y_t = \mu_t$ where $\mu_t = g\mu_{t-1} + e_t$ and $|g| < 1$.
(d) $Y_t = \sum_{i=0}^{m} e_{t-i}/(m+1)$.

(19) In studying stock market prices, chartists often take moving averages of prices as indicators without being aware that such averaging usually introduces correlation, and that this can mislead through the resultant spurious patterns. To investigate this, let $Y_i \sim \mathrm{N}[\mu, \sigma^2]$ independently, and $X_t = \sum_{i=1}^{n} w_i Y_{t-i+1}$, where the w_i's are known constants such that $\sum_{i=1}^{n} w_i = 1$. That is, X_t is computed as a moving average of the values of Y_t. Find the induced correlation by deriving the ACF of X_t in the two cases
(a) $w_i = 1/n$, so that X_t is an arithmetic average,
(b) $w_i = (1-\beta)\beta^{i-1}/(1-\beta^n)$, so that X_t is a truncated EWMA.

(20) Suppose $X \sim \mathrm{S}[\mu_x; \gamma_x(B)]$ and $Y \sim \mathrm{S}[\mu_y; \gamma_y(B)]$ and that X_t is independent of Y_s for all t and s. Prove the following important theoretical results.

(a) If $Z_t = X_t + Y_t$ then $Z \sim S[\mu_x + \mu_y; \gamma_x(B) + \gamma_y(B)]$.

(b) For any real numbers l_1 and l_2, if $Z_t = l_1 X_t + l_2 Y_t$ then
$$Z \sim \mathrm{S}[l_1\mu_x + l_2\mu_y; l_1^2\gamma_x(B) + l_2^2\gamma_y(B)].$$

(c) If $Z_t = X_t - \alpha X_{t-1} = (1-\alpha B)X_t$ then
$$Z \sim \mathrm{S}[(1-\alpha)\mu_x; (1-\alpha B)(1-\alpha B^{-1})\gamma_x(B)].$$

(d) If $Z_t = \phi(B)X_t$, where $\phi(B)$ is a finite polynomial in B, then
$$Z \sim \mathrm{S}[\phi(1)\mu_x; \phi(B)\phi(B^{-1})\gamma_x(B)].$$

(21) $\{Y_t\}$ is a univariate random time series.
(a) You have $n+2$ observations $\mathbf{Y} = (Y_1, \ldots, Y_{n+2})$ available to test your theory that Y_t follows the NDLM
$$\left\{ \begin{pmatrix} 1 \\ 0 \end{pmatrix}, \begin{pmatrix} 1 & 1 \\ 0 & 1 \end{pmatrix}, 100, \begin{pmatrix} 5 & 0 \\ 0 & 1 \end{pmatrix} \right\}.$$
Obtain the theoretical distribution of an appropriate derived series and describe how this enables you to examine the adequacy of your proposed model.
(b) Suppose that the RTS $\{Y_t\}$ is such that the derived series
$$Z_t = Y_{t+1} - 0.9Y_t \sim \mathrm{GS}[0; -90B^{-1} + 200 - 90B] .$$

Construct an appropriate single-parameter constant DLM for $\{Y_t\}$, precisely quantifying the quadruple $\{F, G, V, W\}$.

(22) Let the RTS $\{Y_t\}$ be generated by the observable constant DLM $\{\mathbf{F}, \mathbf{G}, V, \mathbf{W}\}$, where $\mathbf{G}$ has eignevalues λ_i such that $|\lambda_i| < 1$ for $i = 1, \dots, n-r$, and $|\lambda_i| \geq 1$ for $i = n-r+1, \dots, n$. Show that the derived RTS

$$Z_t = \prod_{i=n-r+1}^{n} (1 - \lambda_i B) Y_{t+r}$$

can be appropriately modelled by the stationary observable DLM

$$\left\{ \begin{pmatrix} \mathbf{E}_{n-r} \\ \mathbf{E}_r \end{pmatrix}, \begin{pmatrix} \mathbf{G}^* & \mathbf{0} \\ \mathbf{0}' & \mathbf{J}_r(0) \end{pmatrix}, 0, \mathbf{W}^* \right\},$$

where $\mathbf{G}^*$ has eigenvalues $\lambda_1, \dots, \lambda_{n-r}$.

CHAPTER 6

MODEL SPECIFICATION AND DESIGN

6.1 BASIC FORECAST FUNCTIONS

Central to DLM specification and design is the development of appropriate form and structure of the forecast function, from Definition 4.4,

$$f_t(k) = \mathrm{E}[\mu_{t+k}|D_t] = \mathrm{E}[\mathbf{F}'_{t+k}\boldsymbol{\theta}_{t+k}|D_t],$$

for all $t, k > 0$. This defines both the qualitative form and the forecaster's numerical specification of the expected development of the time series. Consequently, it is of fundamental importance to the design and construction of appropriate DLMs. This chapter begins with a discussion of forecast functions derived from the various TSDLMs of the previous chapter. Together with complementary forecast (transfer) functions for the effects of independent variables, these provide the basis for designing all practically important dynamic linear models.

6.1.1 Real Jordan block system matrices

The simplest observable class of DLMs comprises those for which the system matrices each have a single real eigenvalue.

Theorem 6.1. *For real λ, the forecast function $f_t(k)$ $(k \geq 0)$ of any canonical model $\{\mathbf{E}_n, \mathbf{J}_n(\lambda), \cdot, \cdot\}$, and hence of any similar model, takes the following form:*

(1) *If, as with most practical models, $\lambda \neq 0$, then*

$$f_t(k) = \lambda^k \sum_{r=0}^{n-1} a_{tr} k^r,$$

where $a_{t0}, \ldots, a_{t,n-1}$ are linear functions of $\mathbf{m}_t = (m_{t1}, \ldots, m_{tn})'$, but are independent of k.

(2) *In the irregular case $\lambda = 0$, then*

$$f_t(k) = m_{t,k+1}, \qquad (0 \leq k < n),$$

$$f_t(k) = 0, \qquad (k \geq n).$$

Proof. Given $\mathrm{E}[\boldsymbol{\theta}_t \mid D_t] = \mathbf{m}_t$, by definition,

$$f_t(k) = \mathbf{E}'_n \mathbf{J}_n(\lambda)^k \mathbf{m}_t.$$

Using Section 17.4.3, when λ is non-zero,

$$f_t(k) = \lambda^k \sum_{r=0}^{n-1} a_{tr} k^r,$$

where the coefficients $a_{t0}, \dots, a_{t,n-1}$ depend on $\mathbf{m}_t$ and λ but not on k. It follows that any similar model has a forecast function of the same qualitative form in k, and any equivalent model has the identical quantified forecast function.

If $\lambda = 0$ then the only non-zero elements of $\mathbf{J}_n(\lambda)^k$ are those of the kth super-diagonal, which comprises unit elements. Hence $f_t(k) = m_{t,k+1}$, the $(k+1)^{st}$ element of $\mathbf{m}_t$, when $0 \le k < n$. For $k \ge n$, $\mathbf{J}_n(\lambda)^k = \mathbf{0}$ and the result follows.

$\diamond$

For the important practical cases, $\lambda \neq 0$, $f_t(k)$ has the form of the kth power of the eigenvalue λ multiplying a polynomial of order n in the step ahead index k.

EXAMPLE 6.1. Consider the case $n = 1$, so that the canonical model is $\{1, \lambda, \cdot, \cdot\}$, with scalars $\boldsymbol{\theta}_t = \mu_t$ and $\mathbf{m}_t = m_t$. Then

$$f_t(k) = m_t \lambda^k.$$

This special case is important since it illustrates the nature of the contribution of a single eigenvalue of multiplicity one to any observable DLM. The value of λ clearly determines the behaviour of the forecast function. The various possible cases, illustrated in Figure 6.1 with $m_t = 1$, are described.

(a) $\lambda = 0$.
Here $f_t(0) = m_t$, and for $k > 0$, $f_t(k) = 0$. The model is simply $Y_t = \omega_t + \nu_t$ with $m_t = \mathrm{E}[\omega_t | D_t]$.

(b) $\lambda = 1$.
Here $f_t(k) = m_t$ for all $k \ge 0$. This is the first-order polynomial DLM of Chapter 2.

(c) $0 < \lambda < 1$.
Here $f_t(k) = \lambda^k m_t$ decays to zero exponentially in k.

(d) $-1 < \lambda < 0$.
Here $f_t(k) = \lambda^k m_t$ oscillates between positive and negative values, exponentially decaying to zero in k.

(e) $\lambda = -1$.
Here $f_t(k) = (-1)^k m_t$ oscillates, taking the values m_t and $-m_t$ alternately. This is the forecast function of a **Nyquist harmonic**, and appears in models for cyclical or seasonal series in Chapter 8.

(f) $\lambda > 1$.
Here $f_t(k) = \lambda^k m_t$, and this explodes exponentially, and monotonically, to ∞ if $m_t > 0$, and to $-\infty$ if $m_t < 0$.

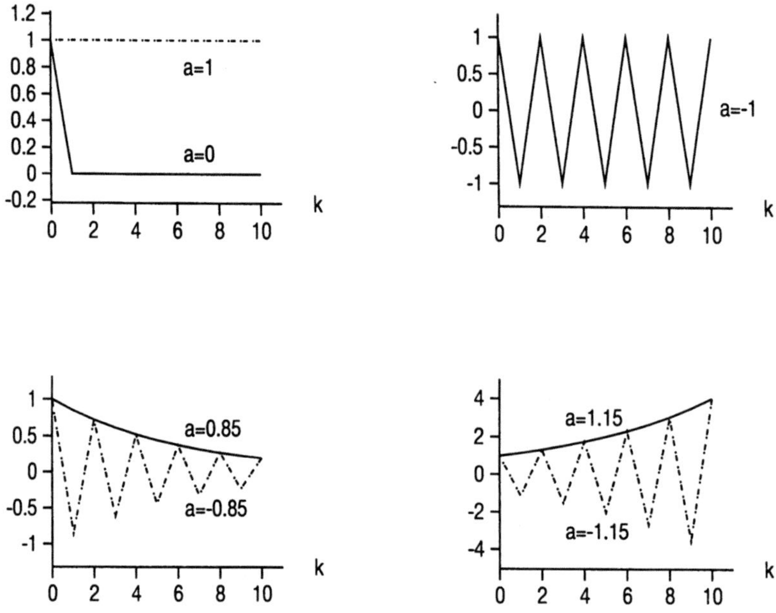

Figure 6.1 $f_t(k) = \lambda^k$ for various values of $\lambda = a$

(g) $\lambda < -1$.
Here $f_t(k) = \lambda^k m_t$ oscillates explosively when $m_t \neq 0$.

EXAMPLE 6.2. When $n = 2$ the canonical model has the form

$$\left\{ \begin{pmatrix} 1 \\ 0 \end{pmatrix}, \begin{pmatrix} \lambda & 1 \\ 0 & \lambda \end{pmatrix}, \cdot, \cdot \right\}.$$

With state vector $\boldsymbol{\theta}_t = (\theta_{t1}, \theta_{t2})'$ and evolution error $\boldsymbol{\omega}_t = (\omega_{t1}, \omega_{t2})'$ we have

$$\begin{aligned} Y_t &= \theta_{t1} + \nu_t, \\ \theta_{t1} &= \lambda\theta_{t-1,1} + \theta_{t-1,2} + \omega_{t1}, \\ \theta_{t2} &= \lambda\theta_{t-1,2} + \omega_{t2}. \end{aligned}$$

Writing $\mathbf{m}_t = (m_{t1}, m_{t2})'$, for $k \geq 0$ and $\lambda \neq 0$, we have

$$f_t(k) = (m_{t1} + k m_{t2}/\lambda)\,\lambda^k.$$

The various possible cases, determined by the value of λ, are described below and illustrated in Figure 6.2 with $m_{t1} = 1$ and $m_{t2} = 0.25$.

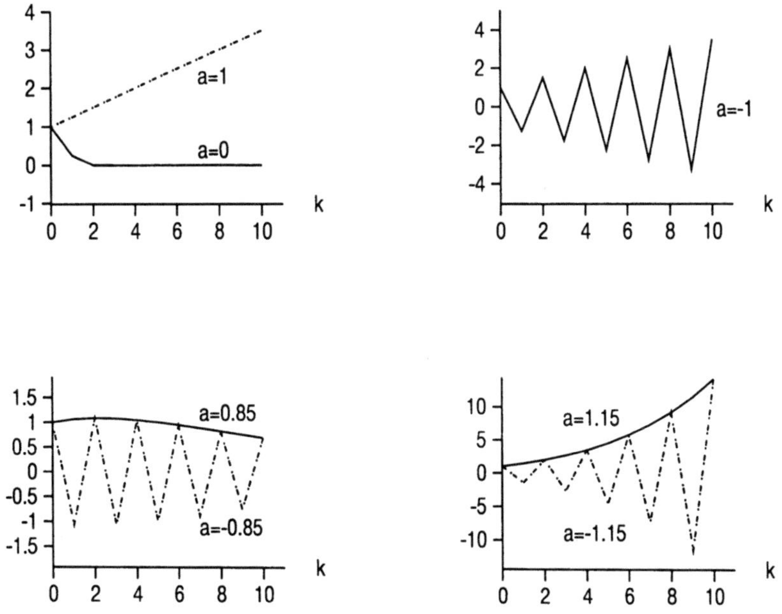

Figure 6.2 $f_t(k) = (1 + 0.25k/\lambda)\lambda^k$ for various values of $\lambda = a$

(a) $\lambda = 0$.
Here $f_t(0) = m_{t1}$, $f_t(1) = m_{t2}$ and $f_t(k) = 0$ for $k > 1$. The model is simply $Y_t = \omega_{t1} + \omega_{t-1,2} + \nu_t$. For a constant DLM, the Y_t series can be expressed as a moving-average process of order 1, MA(1).

(b) $\lambda = 1$.
Here $f_t(k) = m_{t1} + km_{t2}$ for all $k \geq 0$. The forecast function is a straight line, or polynomial of order 2. DLMs with this form of forecast function are extremely important in short-term forecasting where the model represents a "locally linear" development of the mean response function over time.

(c) $0 < \lambda < 1$.
Here $f_t(k) = (m_{t1} + ka_{t2})\lambda^k$, where $a_{t2} = m_{t2}/\lambda$, eventually decays exponentially to zero with k. The initial behaviour depends on the actual values of m_{t1} and m_{t2}. The forecast function converges to 0, possibly via an extremum.

(d) $-1 < \lambda < 0$.
Here $f_t(k) = (m_{t1} + ka_{t2})\lambda^k$ oscillates between positive and negative values of the case (c).

(e) $\lambda = -1$.
Here $f_t(k) = (-1)^k(m_{t1} - km_{t2})$ oscillates between a monotonic

series of points on each of two straight lines. This case is of restricted practical interest.

(f) $\lambda > 1$.
Here $f_t(k) = (m_{t1} + km_{t2})\lambda^k$ explodes to $\pm\infty$ according to the sign of m_{t2} or of m_{t1} if $m_{t2} = 0$.

(g) $\lambda < -1$.
Here $f_t(k)$ oscillates explosively between positive and negative values.

EXAMPLE 6.3. For any n, if $\lambda = 1$,

$$f_t(k) = a_{t0} + a_{t1}k + a_{t2}k^2 + \cdots + a_{t,n-1}k^{n-1},$$

which is a polynomial model of order $n-1$. For all n, these models provide the important class of **polynomial DLMs**: the expected behaviour of the series over the future period of interest is a polynomial of order $n-1$. Typically, this *local* description can be seen as a Taylor series approximation, using polynomial forms of low order 1, 2 or 3, say, to an unknown but essentially smooth mean response function. Chapter 2 was devoted to the case $n = 1$. Chapter 7 describes the general case, with particular attention devoted to linear growth models corresponding to the case $n = 2$.

EXAMPLE 6.4. In the special case of $\lambda = 0$, the model $\{\mathbf{E}_n, \mathbf{J}_n(0), V_t, \mathbf{W}_t\}$ has the form

$$\begin{aligned} Y_t &= \theta_{t1} + \nu_t, \\ \theta_{tr} &= \theta_{t-1,r+1} + \omega_{tr}, \qquad (r = 1, \ldots, n-1), \\ \theta_{tn} &= \omega_{tn}, \end{aligned}$$

so that

$$Y_t = \nu_t + \sum_{r=1}^{n} \omega_{t+1-r,r}.$$

From Theorem 6.1 with $\mathbf{m}_t = (m_{t1}, \ldots, m_{tn})'$,

$$f_t(k) = \begin{cases} \mathrm{E}[\theta_{t+k,1} \mid D_t] = m_{t,k+1}, & \text{for } 0 \le k < n; \\ 0, & \text{for } k \ge n. \end{cases}$$

For the first n steps, $k = 0, 1, \ldots, n-1$, the forecast function takes irregular values and thereafter is zero.

Note that an equivalent model is $\{\mathbf{E}_n, \mathbf{J}_n(0), 0, \mathbf{W}_{1t}\}$, where

$$\mathbf{W}_{1t} = \mathbf{W}_t + \begin{bmatrix} V_t & \mathbf{0}' \\ \mathbf{0} & \mathbf{0}_{n-1} \end{bmatrix},$$

with $\mathbf{0}_r$ being the $r \times r$ zero matrix. Thus, whenever zero eigenvalues occur, V_t can be set to zero by suitably amending $\mathbf{W}_t$. This is true even with the null parametric DLM $\{0, 0, V, 0\}$, which is equivalent to the single parameter DLM $\{1, 0, 0, V\}$. The incorporation of the observation noise ν_t into the system noise can be useful, particularly for handling "one-off" events such as promotional campaigns (Harrison 1988), and is standard in our research software.

If the model is constant with $V_t = V$ and $\mathbf{W}_t = \mathbf{W}$ for all t, then Y_t has a moving average MA$(n-1)$ representation

$$Y_t = \sum_{r=0}^{n-1} \psi_r \epsilon_{t-r},$$

where $\epsilon_t \sim \mathrm{N}[0, 1]$, $(t = 1, 2, \dots)$ is a sequence of independent random quantities. This representation may be useful to those readers familiar with standard linear, stationary time series modelling (e.g., Box and Jenkins 1976). Note that it is derived as the very special case of zero eigenvalues.

6.1.2 Single complex block system matrices

System matrices with complex eigenvalues lead to sinusoidal components in the forecast function. The simplest case, that of a single sine/cosine wave, corresponds to a pair of complex conjugate eigenvalues with $n = 2$.

Theorem 6.2. *In the 2-dimensional real canonical model*

$$\{\mathbf{E}_2, \mathbf{J}_2(\lambda, \omega), V_t, \mathbf{W}_t\}$$

with $\lambda \neq 0$, $0 < \omega < 2\pi$, m_{t1} *and* m_{t2} *all real, the forecast function is*

$$f_t(k) = [m_{t1}\cos(k\omega) + m_{t2}\sin(k\omega)]\,\lambda^k.$$

Proof. By induction, and using standard trigonometric identities, it is easily shown that for all integers k,

$$\mathbf{J}_2(\lambda, \omega)^k = \lambda^k \begin{pmatrix} \cos(k\omega) & \sin(k\omega) \\ -\sin(k\omega) & \cos(k\omega) \end{pmatrix} = \mathbf{J}_2(\lambda^k, k\omega).$$

Thus, with $\mathbf{m}_t = (m_{t1}, m_{t2})'$,

$$f_t(k) = \mathbf{E}_2'\mathbf{J}_2(\lambda, \omega)^k \mathbf{m}_t = [m_{t1}\cos(k\omega) + m_{t2}\sin(k\omega)]\,\lambda^k.$$

◇

An alternative expression for the forecast function is

$$f_t(k) = \lambda^k r_t \cos(k\omega + \phi_t),$$

where

(a) $r_t^2 = m_{t1}^2 + m_{t2}^2$, and $r_t > 0$ is the **amplitude** of the **periodic**, or **harmonic**, component $m_{t1}\cos(k\omega) + m_{t2}\sin(k\omega)$;

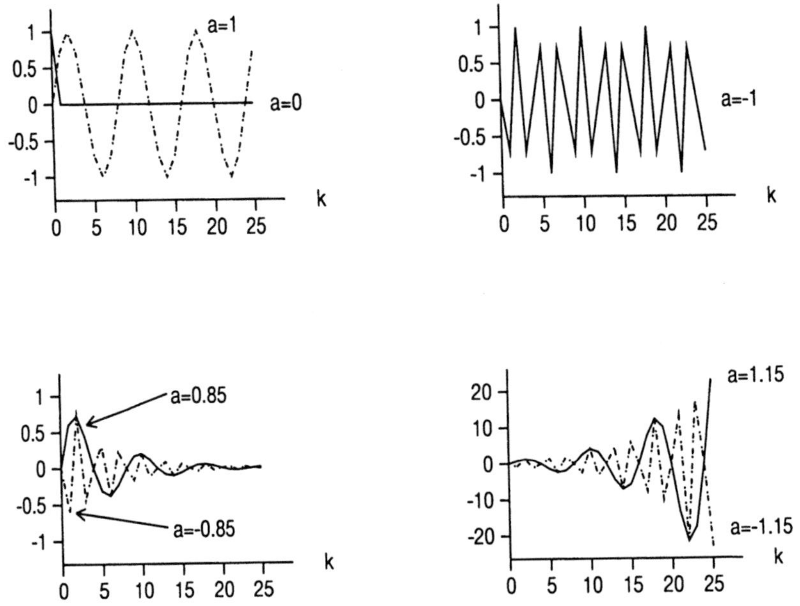

Figure 6.3 $f_t(k) = \lambda^k \cos(\pi k/4)$ for various values of $\lambda = a$

(b) $\phi_t = \arctan(-m_{t2}/m_{t1})$ is the **phase-angle**, or just the **phase**, of the periodic component; and

(c) ω is the **frequency** of the periodic component defining the **period**, $p = 2\pi/\omega$, over which the harmonic completes a full cycle; this follows since for all integers $h \geq 0$,

$$f_t(k) = f_t(k + 2\pi h/\omega).$$

The forecast function has the form of a sine/cosine wave modified by the multiplicative term λ^k. This latter term may *dampen* or *explode* the periodic component. If $|\lambda| < 1$, the sinusoidal form is dampened, decaying asymptotically to zero; if $|\lambda| > 1$ it is exploded, diverging as k increases. Negative values of λ lead to the forecast function oscillating between positive and negative values for consecutive values of k whilst either decaying to zero or diverging. Figure 6.3 illustrates the various possibilities for the particular coefficients $m_{t1} = 1$, $m_{t2} = 0$, and frequency $\omega = \pi/4$ or period $p = 8$. Of greatest practical importance are the cases with $0 < \lambda \leq 1$. In particular, $\lambda = 1$ leads to a pure cosine wave of frequency ω, a basic building block of seasonal time series models.

6.1.3 Models with multiple complex eigenvalues[†]

Essentially, models with the basic periodic forecast functions discussed in the previous section provide all the cyclical/seasonal behaviour associated with practical DLMs. Only rarely will a model with multiple complex eigenvalues be required. For completeness, the forecast functions of such models are now discussed, although most practitioners may safely ignore this section.

Refer to Definition 5.10 for the specification of the real canonical model in the case when a system matrix has a pair of complex conjugate eigenvalues $\lambda e^{i\omega}$ and $\lambda e^{-i\omega}$ with multiplicity v.

Theorem 6.3. *In the real canonical Jordan form model*

$$\{\mathbf{E}_{2,v}, \mathbf{J}_{2,v}(\lambda, \omega), V_t, \mathbf{W}_t\},$$

with λ, $\lambda\omega \neq 0$ and coefficients a_{tj} and b_{tj}, $(j = 0, \ldots, v-1)$ all real, the forecast function is

$$f_t(k) = \lambda^k \cos(k\omega) \sum_{j=0}^{v-1} a_{tj} k^j \ + \ \lambda^k \sin(k\omega) \sum_{j=0}^{v-1} b_{tj} k^j.$$

For amplitudes r_{tj} and phase angles ϕ_{tj}, $(j = 0, \ldots, v-1)$, a neater expression is

$$f_t(k) = \lambda^k \sum_{j=0}^{v-1} r_{tj} k^j \cos(k\omega + \phi_{tj}).$$

Proof. The proof, an exercise in linear algebra, is left to the reader.

◇

As mentioned above, this case is rarely used in practice. Of some interest are the particular models in which $\lambda = 1$ and $\phi_{tj} = 0$, $(j = 0, \ldots, v-1)$, so that the forecast function represents a cosine wave whose amplitude is a polynomial in k. For example, with $\lambda = 1$, $v = 2$, and $\phi_{tj} = 0$ for each j, the forecast function is

$$f_t(k) = (r_{t0} + r_{t1}k)\cos(k\omega).$$

Such forms might be useful in representing cyclic patterns for which the amplitude is increasing linearly.

[†]This section is of rather theoretical interest and may be omitted without loss on a first reading.

6.2 SPECIFICATION OF $\mathbf{F}_t$ AND $\mathbf{G}_t$

In applications, models are usually constructed by combining two or more component DLMs, each of which captures an individual feature of the real series under study. The construction of complex DLMs from component DLMs is referred to as **superposition**, and the reverse process, that of identifying components of a given model, as **decomposition**. These two important modelling concepts are now discussed in detail.

6.2.1 Superposition

Some examples introduce the basic ideas.

EXAMPLE 6.5. Consider the two special models, M_1, of dimension n, and M_2, of dimension 1, specified by quadruples

$$M_1: \quad \{\mathbf{F}, \mathbf{G}, 0, \mathbf{W}\} \qquad \text{and} \qquad M_2: \quad \{0, 0, V, 0\}.$$

Let M_1 have state vector $\boldsymbol{\theta}_t$ and generate a series Y_{1t}. Then with complete certainty, $Y_{1t} = \mu_t = \mathbf{F}'\boldsymbol{\theta}_t$. Model M_2 has no state vector, and generates an independent noise series, Y_{2t} according to $Y_{2t} = \nu_t \sim \mathrm{N}[0, V]$ independently. The composite series $Y_t = Y_{1t} + Y_{2t}$ then follows the DLM $\{\mathbf{F}, \mathbf{G}, V, \mathbf{W}\}$. In adding the two series, the new series created follows a more complex DLM defined by combining the quadruples in a particular way. This is a very simple example of superposition.

EXAMPLE 6.6. Consider two purely deterministic models M_1 and M_2, with state vectors $\boldsymbol{\theta}_{1t}$ and $\boldsymbol{\theta}_{2t}$ respectively, defined via

$$\begin{aligned} M_1: &\quad \{\mathbf{F}_1, \mathbf{G}_1, 0, \mathbf{0}\}; \\ M_2: &\quad \{\mathbf{F}_2, \mathbf{G}_2, 0, \mathbf{0}\}. \end{aligned}$$

Adding the observations generated by these two models produces a series with state vector $\boldsymbol{\theta}_t$, and generated by the DLM $\{\mathbf{F}, \mathbf{G}, 0, \mathbf{0}\}$ where

$$\boldsymbol{\theta}_t = \begin{pmatrix} \boldsymbol{\theta}_{1t} \\ \boldsymbol{\theta}_{2t} \end{pmatrix}, \quad \mathbf{F} = \begin{pmatrix} \mathbf{F}_1 \\ \mathbf{F}_2 \end{pmatrix} \quad \text{and} \quad \mathbf{G} = \begin{bmatrix} \mathbf{G}_1 & \mathbf{0} \\ \mathbf{0} & \mathbf{G}_2 \end{bmatrix}.$$

Extending the notation in an obvious way, the sum of the observation series from any h deterministic models $M_1, \ldots, M_h$ follows the DLM $\{\mathbf{F}, \mathbf{G}, 0, \mathbf{0}\}$ with

$$\mathbf{F}' = (\mathbf{F}'_1, \cdots, \mathbf{F}'_h)$$

and

$$\mathbf{G} = \text{block diag}[\mathbf{G}_1, \ldots, \mathbf{G}_h].$$

These examples illustrate the construction of DLMs from a collection of component DLMs by *superposition* of the corresponding state vectors, regression vectors and system matrices. As an aside, note that if the two component state vectors have common elements, the model formed by this superposition may be reparametrised to one of lower dimension. In general, we have the following result.

Theorem 6.4. *Consider h time series Y_{it} generated by DLMs*

$$M_i: \quad \{\mathbf{F}_{it}, \mathbf{G}_{it}, V_{it}, \mathbf{W}_{it}\}$$

for $i = 1, \ldots, h$. In M_i, the state vector $\boldsymbol{\theta}_{it}$ is of dimension n_i, and the observation and evolution error series are respectively ν_{it} and $\boldsymbol{\omega}_{it}$. The state vectors are distinct, and for all distinct $i \neq j$, the series ν_{it} and $\boldsymbol{\omega}_{it}$ are mutually independent of the series ν_{jt} and $\boldsymbol{\omega}_{jt}$.

Then the series

$$Y_t = \sum_{i=1}^{h} Y_{it}$$

follows the n-dimensional DLM $\{\mathbf{F}_t, \mathbf{G}_t, V_t, \mathbf{W}_t\}$ where $n = n_1 + \cdots + n_h$ and the state vector $\boldsymbol{\theta}_t$ and quadruple are given by

$$\boldsymbol{\theta}_t = \begin{pmatrix} \boldsymbol{\theta}_{1t} \\ \vdots \\ \boldsymbol{\theta}_{ht} \end{pmatrix}, \qquad \mathbf{F}_t = \begin{pmatrix} \mathbf{F}_{1t} \\ \vdots \\ \mathbf{F}_{ht} \end{pmatrix},$$

$$\mathbf{G}_t = \text{block diag}[\mathbf{G}_{1t}, \ldots, \mathbf{G}_{ht}],$$
$$\mathbf{W}_t = \text{block diag}[\mathbf{W}_{1t}, \ldots, \mathbf{W}_{ht}],$$

and

$$V_t = \sum_{i=1}^{h} V_{it}.$$

Proof. Summing the individual independent normal series,

$$Y_t = \mathbf{F}_t' \boldsymbol{\theta}_t + \nu_t, \qquad \nu_t = \sum_{i=1}^{h} \nu_{it} \sim \mathrm{N}[0, V_t].$$

With $\boldsymbol{\omega}_t' = (\boldsymbol{\omega}_{1t}', \ldots, \boldsymbol{\omega}_{ht}')$, noting that for all $i \neq j$, $\boldsymbol{\omega}_{it}$ and $\boldsymbol{\omega}_{jt}$ are independent,

$$\boldsymbol{\theta}_t = \mathbf{G}_t \boldsymbol{\theta}_{t-1} + \boldsymbol{\omega}_t, \qquad \boldsymbol{\omega}_t \sim \mathrm{N}[\mathbf{0}, \mathbf{W}].$$

The proof is completed on noting that the series $\{\omega_t\}$ is independent of the series $\{\nu_t\}$.

◇

This obvious, but highly important, result is termed the **Principle of Superposition.** It simply states that the linear combination of series generated by independent DLMs follows a DLM that is defined via the superposition of the corresponding model components. The principle depends upon the additivity properties associated with linear normal models. The strict model independence of the theorem is not crucial. A more general superposition requirement is that $\nu_{1t}, \ldots, \nu_{ht}$ and $\omega_{1t}, \ldots, \omega_{ht}$ each have a joint normal distribution and that the two series $\{\nu_t\}$ and $\{\omega_t\}$ are internally and mutually independent. Marginal normality of the terms within each model does not necessarily imply joint normality across models, although the practical circumstances in which joint normality is violated are rare and of little importance. So for practical purposes, a working superposition principle is that

A linear combination of DLMs is a DLM.

Usually, practical utilisation of the superposition principle naturally and appropriately adopts the independence assumptions of the theorem. Design implications of superposition do not depend on the independence structure since the additivity property is sufficient to determine the following forecast function result. The trivial proof is left to the reader.

Theorem 6.5. *Consider the models in Theorem 6.4 where the series $\{\nu_t\}$ and $\{\omega_t\}$ are internally and mutually independent series but where $(\nu_{1t}, \ldots, \nu_{ht})$ and $(\omega_{1t}, \ldots, \omega_{ht})$ each have a general joint normal distribution. Denote the forecast function form of M_i by $f_{it}(k)$. Then the forecast function form for the Y_t series generated by the superposition of the h component models is given by*

$$f_t(k) = \sum_{i=1}^{h} f_{it}(k).$$

Note that this is a qualitative statement regarding the forecast function form. If the conditions of Theorem 6.4 hold, it is also quantitatively true.

6.2.2 Decomposition and model design

The practical value of the superposition principle lies in the construction of models for complex problems by combining simpler components for easily identified features of the process. The employed technique is the reverse of superposition, namely the **decomposition** of models with complex forecast functions into simple, canonical components. These canonical components are few in number, familiar, easily understood, and allow a modeller to structure complex problems component by component. Superposition then provides the overall model for the series simply by aggregating the individual building blocks.

The starting point for model design is the form of the forecast function. Given this, observable models are constructed by identifying the component canonical forms. This effectively solves the design problem so far as choice of regression vector $\mathbf{F}_t$ and system matrix $\mathbf{G}_t$ is concerned. The practically important canonical components are related to forecast functions as follows.

(1) Suppose that for all $t \geq 0$ the forecast function has the form

$$f_t(k) = \lambda^k \sum_{r=0}^{n-1} a_{tr} k^r$$

for some given real $\lambda \neq 0$, integer $n \geq 1$, and real coefficients $a_{t0}, \ldots, a_{tn-1}$ not depending on k. From Theorem 6.1 the canonical model is immediately identified as

$$\{\mathbf{E}_n, \mathbf{J}_n(\lambda), \cdot, \cdot\}.$$

An observable TSDLM has the required forecast function form if and only if it is similar to this canonical model.

(2) The generalisation to several real eigenvalues is as follows. Suppose that the desired forecast function must have the full form

$$f_t(k) = \sum_{i=1}^{s} \left[\lambda_i^k \sum_{r=0}^{n_i-1} a_{tr}(i) k^r \right] = \sum_{i=1}^{s} f_{it}(k),$$

where $s > 1$ is integral, $\lambda_1, \ldots, \lambda_s \neq 0$ are real and distinct, and the real coefficients $a_{tr}(i)$ do not depend on k. As in (1) above, $f_{it}(k)$ is the forecast function of any observable model similar to the canonical form

$$\{\mathbf{E}_{n_i}, \mathbf{J}_{n_i}(\lambda_i), \cdot, \cdot\},$$

for $i = 1, \ldots, s$. Applying Theorem 6.5, the required forecast function form is provided by any observable model that is similar to $\{\mathbf{E}, \mathbf{J}, ., .\}$, where

$$\mathbf{E}' = (\mathbf{E}'_{n_1}, \ldots, \mathbf{E}'_{n_s})$$

and

$$\mathbf{J} \ = \text{block diag}[\mathbf{J}_{n_1}(\lambda_1), \ldots, \mathbf{J}_{n_s}(\lambda_s)].$$

(3) Suppose that for all $t \geq 0$, the forecast function has the form

$$f_t(k) = \lambda^k \sum_{r=0}^{v-1} a_{tr} k^r \cos(k\omega + \phi_{tr}),$$

for some real $\lambda \neq 0$, $0 < \omega < 2\pi$, integer $v > 0$, and real coefficients a_{tr} and ϕ_{tr} not depending on k. From Theorem 6.3, it follows that this form of forecast function is provided by any model similar to the real canonical form

$$\{\mathbf{E}_{2,v}, \mathbf{J}_{2,v}(\lambda, \omega), \cdot, \cdot\}.$$

(4) **General TSDLM**

For some non-negative integers s and v such that $s + v > 0$, the forecast function $f_t(k)$ of an observable time series model has the following general form:

$$f_t(k) = \sum_{i=1}^{s+v} f_{it}(k),$$

where:

- for $i = 1, \ldots, s$,

$$f_{it}(k) = \lambda_i^k \sum_{r=0}^{n_i-1} a_{tr}(i) k^r,$$

- for $i = s+1, \ldots, s+v$,

$$f_{it}(k) = \lambda_i^k \sum_{r=0}^{n_i-1} a_{tr}(i) k^r \cos[k\omega_i + \phi_{tr}(i)].$$

For each i and r, the integer $n_i \geq 1$ and the real, non-zero quantities λ_i, $0 < \omega_i < 2\pi$, $a_{tr}(i)$ and $\phi_{tr}(i)$ do not depend on k.

Following Theorem 6.5 and using the results of Section 5.4.6, this forecast function form is provided by any TSDLM similar to the real canonical model of Definition 5.11. This includes all real, non-zero eigenvalues of the system matrix for $i = 1, \ldots, s$, and complex pairs for $i = s+1, \ldots, s+v$. The most general model would also allow for zero eigenvalues, as in Example 6.4, adding a component

$$\{\mathbf{E}_{n_{s+v+1}}, \mathbf{J}_{n_{s+v+1}}(0), \cdot, \cdot\},$$

with forecast function

$$f_{s+v+1,t}(k) = \begin{cases} b_{tk}, & \text{for } 0 \leq k < n_{s+v+1}; \\ 0, & \text{for } k \geq n_{s+v+1}, \end{cases}$$

where n_{s+v+1} is the multiplicity of the zero eigenvalue and the b_{tk} are known constants.

(5) The above cases cover the forms of forecast function encountered in TSDLMs. Regression components for independent variables are rather simple in form. Suppose a related regressor variable gives rise to a time series X_t, with X_{t+k} known at t for $k \geq 0$. The

X_t may be raw or transformed values of a related series, or values filtered through a known, possibly non-linear, transfer function to provide a constructed *effect* variable that depends on past, or lagged, values of the related series. For example, the superposition of the first-order polynomial DLM $\{1, 1, \cdot, \cdot\}$ and the simple regression DLM $\{X_t, 1, \cdot, \cdot\}$ gives a model $\{(1, X_t)', \mathbf{I}, \cdot, \cdot\}$ with $f_t(k) = m_{t1} + m_{t2}X_{t+k}$.

Generalising to multiple linear regression DLMs, consider a collection of h possible regressor variables $X_{1t}, \ldots, X_{ht}$. By superposition of the corresponding h simple models and a first-order polynomial, a multiple regression DLM is obtained, namely

$$\{(1, X_{1t}, \ldots, X_{ht})', \mathbf{I}, \cdot, \cdot\}$$

with

$$f_t(k) = m_{t1} + \sum_{v=1}^{h} m_{t,v+1} X_{v,t+k}.$$

EXAMPLE 6.7. For the pure polynomial forecast function

$$f_t(k) = \sum_{r=0}^{n-1} a_{tr} k^r,$$

a unit eigenvalue of multiplicity n is required and the canonical model is

$$\{\mathbf{E}_n, \mathbf{J}_n(1), \cdot, \cdot\}.$$

Any similar model is called an **$\mathbf{n}^{th}$-order polynomial DLM.**

EXAMPLE 6.8. Suppose a modeller requires a forecast function that represents a single persistent harmonic oscillation of period p about a linear trend. Such forms are fundamental in short-term forecasting of seasonal series. From Example 6.7 with $n = 2$, the linear trend canonical component is

$$\{\mathbf{E}_2, \mathbf{J}_2(1), \cdot, \cdot\}.$$

From Theorem 6.2, the canonical DLM for the persistent cyclical term with frequency $\omega = 2\pi/p$ is

$$\{\mathbf{E}_2, \mathbf{J}_2(1, \omega), \cdot, \cdot\}.$$

The superposition of the two provides the required DLM

$$\left\{\begin{pmatrix} \mathbf{E}_2 \\ \mathbf{E}_2 \end{pmatrix}, \begin{pmatrix} \mathbf{J}_2(1) & \mathbf{0} \\ \mathbf{0} & \mathbf{J}_2(1,\omega) \end{pmatrix}, \cdot, \cdot\right\}.$$

This is referred to as a **second-order polynomial/seasonal** model in which the seasonal pattern has the form of a simple cosine wave.

EXAMPLE 6.9. Suppose the required forecast function is that of Example 6.8 but with the additional demand that

(i) the seasonal pattern is more complex, being modelled by adding another harmonic of frequency $\omega^* = 2\pi/p^*$, and

(ii) at any time t the forecast function converges geometrically to the second-order seasonal forecast function at a rate λ^k, where $0 < \lambda < 1$.

Two extra components are required: another canonical cyclic component and the canonical DLM $\{1, \lambda, \cdot, \cdot\}$. Employing superposition, an appropriate canonical DLM is

$$\left\{ \begin{pmatrix} \mathbf{E}_2 \\ 1 \\ \mathbf{E}_2 \\ \mathbf{E}_2 \end{pmatrix}, \begin{pmatrix} \mathbf{J}_2(1) & \mathbf{0} & \mathbf{0} & \mathbf{0} \\ \mathbf{0} & \lambda & \mathbf{0} & \mathbf{0} \\ \mathbf{0} & \mathbf{0} & \mathbf{J}_2(1,\omega) & \mathbf{0} \\ \mathbf{0} & \mathbf{0} & \mathbf{0} & \mathbf{J}_2(1,\omega^*) \end{pmatrix}, \cdot, \cdot \right\}.$$

In each of these examples the canonical models, corresponding to a stated forecast function, have been derived. If required, the canonical model may be transformed to a preferred similar model, by reparametrisation or by time shifts. The next two examples illustrate the reverse problem of finding the forecast function corresponding to a given TSDLM. Observability may be checked directly by examining the observability matrix $\mathbf{T}$. Then the eigenvalues of the system matrix and their multiplicity are identified, providing the forecast function, which is simply the sum of the forecast functions associated with each distinct real eigenvalue and each pair of distinct complex conjugate eigenvalues.

EXAMPLE 6.10. Consider the TSDLM

$$\left\{ \begin{pmatrix} 1 \\ 0 \\ 0 \\ 0 \end{pmatrix}, \begin{pmatrix} 1 & 1 & 0 & 0 \\ 0 & 0 & 1 & 0 \\ 0 & 0 & 0 & 1 \\ 0 & 1 & 0 & 0 \end{pmatrix}, \cdot, \cdot \right\}.$$

The DLM is observable since $\mathbf{T}$ is a lower triangular matrix of unit elements. The eigenvalues of $\mathbf{G}$ are the solutions of

$$0 = (\lambda - 1)(\lambda^3 - 1) = (\lambda - 1)^2(\lambda - e^{2\pi i/3})(\lambda - e^{-2\pi i/3}),$$

giving a real eigenvalue of 1 with multiplicity 2 and a complex conjugate pair $e^{\pm 2\pi i/3}$. So the canonical form of the DLM is that of Example 6.8 with $\omega = 2\pi/3$, namely

$$\left\{ \begin{pmatrix} \mathbf{E}_2 \\ \mathbf{E}_2 \end{pmatrix}, \begin{pmatrix} \mathbf{J}_2(1) & \mathbf{0} \\ \mathbf{0} & \mathbf{J}_2(1,\omega) \end{pmatrix}, \cdot, \cdot \right\},$$

for which the forecast function is

$$f_t(k) = a_{t0} + a_{t1}k + a_{t2}\cos(2\pi k/3) + a_{t2}\sin(2\pi k/3).$$

EXAMPLE 6.11. Now consider the TSDLM $\{\mathbf{E}_n, \mathbf{P}_n, \cdot, \cdot\}$, where $n = 2q+1$ for some integer $q > 0$, $\mathbf{I}$ is the $2q \times 2q$ identity matrix and

$$\mathbf{P}_n = \begin{pmatrix} \mathbf{0} & \mathbf{I} \\ 1 & \mathbf{0}' \end{pmatrix}.$$

The DLM is observable since $\mathbf{T}$ is the identity matrix. The eigenvalues, being the solutions of

$$0 = \lambda^n - 1,$$

are the nth roots of 1, namely $e^{iv\omega}$ for $v = 0, \ldots, n-1$, with $\omega = 2\pi/n$. So $\mathbf{G}$ has an eigenvalue 1 and q distinct pairs of complex conjugates $e^{\pm iv\omega}$, for $v = 1, \ldots, q$. The canonical DLM is

$$\{(1, \mathbf{E}_2', \ldots, \mathbf{E}_2')', \text{block diag}\,[1, \mathbf{J}_2(1,\omega), \mathbf{J}_2(1,2\omega), \ldots, \mathbf{J}_2(1,q\omega)]\,, \cdot, \cdot\},$$

with forecast function

$$f_t(k) = a_{t0} + \sum_{v=1}^{q} r_{tv} \cos(v\omega k + \phi_{tv}),$$

comprising the sum of the forecast functions of a first-order polynomial and the q harmonics, or cosine waves, the latter being called the **full seasonal effects** model of period n in this case of odd n.

6.3 DISCOUNT FACTORS AND COMPONENT MODEL SPECIFICATION

6.3.1 Component models

The above design principles lead naturally to DLM structures in **block** or **component** form. The system matrix is block diagonal with individual sub-matrices providing contributions from simple component models. The regression vector is partitioned into the catenation of corresponding sub-vectors. To complete the model specification, three further components are required namely the sequence of state evolution variance matrices $\mathbf{W}_t$, $(t = 1, \ldots)$; the observational variance sequence V_t, $(t = 1, \ldots)$; and the initial prior distribution for the state vector and the observational error variance, given D_0. Estimation of the constant observational variance has already been considered in Chapter 4, and in Chapter 10 it is generalised to stochastic and time dependent cases. The initial prior settings, and related questions concerning representation of subjective information of the forecaster in terms of probability distributions, are also covered extensively in later chapters and application-specific contexts. A general point here is that the component structure of DLMs typically leads to these initial priors being specified in terms of a collection of priors, one for each of the sub-vectors of $\boldsymbol{\theta}_0$ corresponding to the individual component models, with independence between components. This section concentrates on the specification of the sequence of evolution variance matrices $\mathbf{W}_t$, referring throughout to the general model of Definition 4.3, with minor, purely technical changes to cover the case of an unknown observational variance as in Definition 4.5.

The specification of the structure and magnitude of $\mathbf{W}_t$ is crucially important for successful modelling and forecasting. The values control the extent of the stochastic variation in the evolution of the model and hence determine the stability over time. In the system equation, $\mathbf{W}_t$ leads to an increase in uncertainty, or equivalently a loss of information, about the state vector between times $t-1$ and t. More precisely, consider the sequential information updating equations summarised in Section 4.6. At time $t-1$, the posterior for the current state vector has variance $\mathrm{V}[\boldsymbol{\theta}_{t-1} \mid D_{t-1}] = \mathbf{C}_{t-1}$, which, via the evolution equation, leads to a prior variance for $\boldsymbol{\theta}_t$ given by $\mathrm{V}[\boldsymbol{\theta}_t \mid D_{t-1}] = \mathbf{G}_t\mathbf{C}_{t-1}\mathbf{G}_t' + \mathbf{W}_t$. Let $\mathbf{P}_t$ denote the first term, that is,

$$\mathbf{P}_t = \mathbf{G}_t\mathbf{C}_{t-1}\mathbf{G}_t' = \mathrm{V}[\mathbf{G}_t\boldsymbol{\theta}_{t-1} \mid D_{t-1}].$$

$\mathbf{P}_t$ may be viewed as the appropriate prior variance in the standard DLM $\{\mathbf{F}_t, \mathbf{G}_t, V_t, \mathbf{0}\}$ with no evolution error at time t, and is the required prior variance corresponding to an ideal, stable state vector with no stochastic changes. In this DLM, with $\mathbf{W}_t = \mathbf{0}$, the system equation $\boldsymbol{\theta}_t = \mathbf{G}\boldsymbol{\theta}_{t-1}$ is postulated as *globally* true, whereas the dynamic modeller considers it only a *locally* appropriate description. That is, as discussed in Section 3.1, the form of system equation is treated as globally applicable, but the quantities defining this form are only locally apposite, being modelled, in routine application, as changing slowly in random fashion. Consequently, the system variance matrix $\mathbf{W}_t$ communicates how durable the model is. If $\mathbf{W}_t = \mathbf{0}$, the system model is globally reliable, whereas as $\mathbf{W}_t \to \infty$, the system model, and consequently the DLM itself, becomes totally unreliable and useless. It may also be said that $\mathbf{W}_t$ measures how quickly the value of the current information D_t decays with k, as k-step ahead predictions are made. So adding the evolution error $\boldsymbol{\omega}_t$ to $\mathbf{G}_t\boldsymbol{\theta}_{t-1}$ truly captures the modeller's view of the relationship between the state vectors $\boldsymbol{\theta}_{t-1}$ and $\boldsymbol{\theta}_t$. Given D_{t-1}, the effect is to increase the uncertainty from the *ideal* $\mathbf{P}_t$ to the realistic $\mathbf{R}_t = \mathbf{P}_t + \mathbf{W}_t$.

There are, however, a number of practical drawbacks associated with a system variance matrix $\mathbf{W}_t$:

(a) it is not invariant to the measurement scale of regressor variables as specified in $\mathbf{F}_t$;

(b) it is ambiguous: as shown in Section 5.3.4, if $n \geq 2$ there exists an uncountable number of equivalent *time shifted* DLM's, differing only in terms of their operational $\mathbf{W}_t$'s;

(c) effect components (e.g., treatment, block or seasonal) must satisfy constraints (e.g., sum to 0) and this demands that $\mathbf{W}_t$ satisfies corresponding requirements (e.g., every row and column sums to 0);

(d) the local durability of the DLM may vary with time (think of the local validity of a small Taylor series expansion) so generally there will not be an *optimal* value of $\mathbf{W}_t$ suitable for all times;

(e) most people have great difficulty in directly quantifying the variance and covariance elements, with the result that these are often grossly misspecified.

Consequently, practitioners require a better way of viewing the system evolution. One answer lies in discounting, which, being easy to apply and understand, overcomes the above difficulties. By definition, a discount factor δ satisfies the condition $0 < \delta \leq 1$. Usually δ is strictly less than 1, but the unit value is retained as a possibility since it relates to static models.

Chapters 2 and 3 introduced the idea of discounting for a single parameter. Now consider it for a canonical component model such as the second-order polynomial DLM $\{\mathbf{E}_2, \mathbf{J}_2(1), V, \mathbf{W}_t\}$. At any time t, given $(\boldsymbol{\theta}_{t-1}|D_{t-1}) \sim \mathrm{N}[\mathbf{m}_{t-1}, \mathbf{C}_{t-1}]$, the precision associated with $\boldsymbol{\theta}_{t-1}$ is $\mathbf{C}_{t-1}^{-1}$ and that of $\mathbf{G}\boldsymbol{\theta}_{t-1}$ is $\mathbf{P}_t^{-1}$. The latter represents precision associated with $\boldsymbol{\theta}_t$ were there to be no stochastic change at time t, so that the model is, in this sense, more "globally" durable. As the model is only locally appropriate, then the actual precision $\mathbf{R}_t^{-1}$ is reduced relative to $\mathbf{P}_t^{-1}$. The discount concept defines this decreased precision directly, via $\delta\mathbf{P}_t^{-1}$ or simply a proportion δ of the globally durable precision. The implied variance is

$$\mathrm{V}[\boldsymbol{\theta}_t|D_{t-1}] = \mathbf{R}_t = \frac{1}{\delta}\mathbf{P}_t.$$

This immediately leads to an identification of $\mathbf{W}_t$, since

$$\mathbf{R}_t = \mathbf{P}_t + \mathbf{W}_t,$$

so that

$$\mathbf{W}_t = \frac{1-\delta}{\delta}\mathbf{P}_t.$$

Furthermore, given δ and $\mathbf{C}_0$, the whole series $\{\mathbf{W}_t\}$ is identified. Note that both $\mathbf{R}_t$ and $\mathbf{W}_t$ have precisely the same internal correlation structure as $\mathbf{P}_t$. So the above drawbacks (a) to (e) are overcome; δ is invariant to scale changes in $\mathbf{F}_t$ and to parametric transformations. If effect constraints are initially satisfied by $\mathbf{C}_0$, then using the discount approach, they are satisfied by $\mathbf{R}_t$ and $\mathbf{W}_t$. The local durability of the model is easily controlled through the discount factor, which, if required, may be changed through time. Finally, there are few problems in selecting a discount factor; for polynomial, seasonal and regression components, δ will lie in $(0, 1]$ and is typically in the range $[0.9, 0.99]$ for routine analysis. The discount approach is parsimonious. Admittedly this means that discount models comprise a subset of DLMs, but very little is lost in terms of a potentially improved description while much is gained from the parsimony and simplicity of concept. For example, considering the second-order polynomial DLM, Harrison (1967) showed that the maximum loss in one step ahead prediction for typical general settings is an increase in standard deviation of less than 1%. Obviously, δ depends upon the sampling interval, but in most applications this and the model are chosen with respect to specific objectives; thus, if, for example, a low discount factor is applied, it generally signals an inadequate model and the need to obtain an improvement. The higher the discount factor the more durable the model, so the aim is to develop a model with a high discount factor provided it does not impair performance. Routine forecasting will reflect desirable stability while associated monitoring procedures will be responsible for signaling unusual events, sudden instabilities, and deteriorations in forecast performance that have not been anticipated by expert intervention.

The magnitude of variances and covariances is controlled by the discount factor in just the same way as described for the scalar case in Chapter 2. The implication is that information decays at the same rate for each of the elements of the state vector. This is particularly appropriate when the entire state vector is viewed as subject to change at a constant rate, without reference to components. This is often a suitable assumption in practice. Note, however, that the discount approach is not appropriate for the unusual case of a precisely known parameter θ_0 for which $C_0 = 0$ but $R_1 \neq 0$.

6.3.2 Component discounting

In developing early discount methods for trend/seasonal models, Harrison (1965) showed that single discount models are not always advisable. The point is that the trend and seasonal components often require different discount factors. This can arise when the seasonal characterisation is more durable than that of the trend, or if many more parameters are needed to specify seasonality. Recall the conditional independence structure of Figure 4.2, indicating model components evolving independently over time. For a DLM comprising the superposition of several components, the idea of one discount factor for each component is suggested. This raises the question as to what defines a model component. A TSDLM might be regarded as comprising r different components, one for each of the distinct real eigenvalues and one for each pair of complex conjugate eigenvalues of $\mathbf{G}$. Thus, in the model of Example 6.11, $r = q+1$ components are possible. However, if this model is being used as a first-order polynomial/seasonal model, most practitioners will prefer to model it as two operating components: the trend component, corresponding to the real unit eigenvalue, and the seasonal effect component, corresponding to all the complex eigenvalues. Then it is natural to associate one discount factor with trend and another with seasonality. By contrast, the model of Example 6.11 applied to a series arising in a physical sciences context will usually be viewed as comprising several components based on the collection of harmonics. The lower frequency harmonics may be much more durable and represent physical structure in the underlying process, and so require a higher discount factor than that appropriate for the high frequency harmonics, which usually reflect interference and extraneous noise. The low order harmonics may then be modelled individually as sub-model components, or grouped together as one component but separate from the higher frequency noise. Similar comments apply to regression models, as discussed in Section 6.2.2 (5). Here the practitioner may group a number of independent regressor variables together and treat them as a separate operating component. These comments should be borne in mind throughout; when we refer to a model component, we are usually talking about an operationally defined component or a sub-model.

As in Theorem 6.4, consider a DLM comprising the superposition of $h \geq 1$ sub-models M_i with state vectors $\boldsymbol{\theta}_{it}$, evolution errors $\boldsymbol{\omega}_{it}$, and of dimensions n_i, where $\sum_{i=1}^{h} n_i = n$. For each $i = 1, \dots, h$, write

$$M_i : \quad \{\mathbf{F}_{it}, \mathbf{G}_{it}, V_{it}, \mathbf{W}_{it}\}.$$

The DLM is thus specified by the state vector $\boldsymbol{\theta}_t$ and quadruples

$$\{\mathbf{F}_t, \mathbf{G}_t, V_t, \mathbf{W}_t\},$$

where

$$\mathbf{F}_t = \begin{pmatrix} \mathbf{F}_{1t} \\ \mathbf{F}_{2t} \\ \cdot \\ \cdot \\ \mathbf{F}_{ht} \end{pmatrix}, \quad \mathbf{G}_t = \begin{pmatrix} \mathbf{G}_{1t} & \mathbf{0} & \mathbf{0} & \dots & \mathbf{0} \\ \mathbf{0} & \mathbf{G}_{2t} & \mathbf{0} & \dots & \mathbf{0} \\ \mathbf{0} & \mathbf{0} & \mathbf{G}_{3t} & \dots & \mathbf{0} \\ \vdots & \vdots & \vdots & \ddots & \vdots \\ \mathbf{0} & \mathbf{0} & \mathbf{0} & \dots & \mathbf{G}_{ht} \end{pmatrix},$$

$$\boldsymbol{\theta}_t = \begin{pmatrix} \boldsymbol{\theta}_{1t} \\ \boldsymbol{\theta}_{2t} \\ \cdot \\ \cdot \\ \boldsymbol{\theta}_{ht} \end{pmatrix}, \quad \mathbf{W}_t = \begin{pmatrix} \mathbf{W}_{1t} & \mathbf{0} & \mathbf{0} & \dots & \mathbf{0} \\ \mathbf{0} & \mathbf{W}_{2t} & \mathbf{0} & \dots & \mathbf{0} \\ \mathbf{0} & \mathbf{0} & \mathbf{W}_{3t} & \dots & \mathbf{0} \\ \vdots & \vdots & \vdots & \ddots & \vdots \\ \mathbf{0} & \mathbf{0} & \mathbf{0} & \dots & \mathbf{W}_{ht} \end{pmatrix}.$$

At time t, the variance matrix

$$\mathbf{P}_t = \mathrm{V}[\mathbf{G}_t\boldsymbol{\theta}_{t-1} \mid D_{t-1}] = \mathbf{G}_t\mathbf{C}_{t-1}\mathbf{G}_t'$$

represents uncertainty about $\mathbf{G}_t\boldsymbol{\theta}_t$ *before* the addition of the evolution noise. Denote the diagonal block corresponding to the ith sub-model by $\mathbf{P}_{it}$, where

$$\mathbf{P}_{it} = \mathrm{V}[\mathbf{G}_{it}\boldsymbol{\theta}_{i,t-1} \mid D_{t-1}], \qquad (i = 1, \dots, h).$$

Although $\mathbf{P}_t$ will not generally be a block diagonal matrix, the block components $\mathbf{P}_{it}$ individually measure information about the sub-model state vectors. Now, adding the evolution noise $\boldsymbol{\omega}_t$ with block diagonal variance matrix $\mathbf{W}_t$ above, the prior variance matrix $\mathbf{R}_t$ for $\boldsymbol{\theta}_t$ has off-diagonal blocks identical to those of $\mathbf{P}_t$, but diagonal blocks

$$\mathbf{R}_{it} = \mathbf{P}_{it} + \mathbf{W}_{it}, \qquad (i = 1, \ldots, h).$$

The discount concept now applies naturally to component sub-models, as follows.

Definition 6.1. In the above framework, let $\delta_1, \ldots, \delta_h$ be any h discount factors, $(0 < \delta_i \leq 1;\ i = 1, \ldots, h)$, with δ_i being the discount factor associated with the component model M_i. Suppose that the component evolution variance matrices $\mathbf{W}_{it}$ are defined as in Section 6.3.1 above, via

$$\mathbf{W}_{it} = \frac{1 - \delta_i}{\delta_i}\mathbf{P}_{it}, \qquad (i = 1, \ldots, h).$$

Then the model is referred to as a **component discount DLM**.

The effect of component discounting is to model the decay in value of the current information at a possibly different rate for each component model. The modeller chooses the discount factors, some of which may, of course, be equal, to reflect belief about the durability, or stability, over time of the individual component models. Note that from an operational point of view in updating, the evolution from $\mathbf{P}_t$ to $\mathbf{R}_t$ need not make reference to the constructed $\mathbf{W}_t$ sequence. It is simply achieved by taking the component covariances as unchanged and dividing the block diagonal elements by the appropriate discount factors, so that for each i,

$$\mathbf{R}_{it} = \frac{1}{\delta_i}\mathbf{P}_{it}.$$

Block discounting is our recommended approach to structuring the evolution variance sequence in almost all applications. The approach is parsimonious, naturally interpretable, and robust. Sometimes a single discount factor applied to an entire model viewed as a single component will be adequate, but the flexibility remains to model up to n separate components, each with individual, though not necessarily distinct, discount factors. Importantly, the derived $\mathbf{W}_t$ matrix is naturally scaled, the discount factors being dimensionless quantities on a standardised scale. With or without variance learning, the discount construction applies directly. The following section describes some practical features of the use of component discount models. The basic ideas underlying multiple discounting have a long history, starting with Harrison (1965), but this specific approach was introduced in Ameen and Harrison (1985), described and developed in practical detail in Harrison and West (1986, 1987), Harrison (1988), and implemented in the BATS package of West, Harrison and Pole (1987) and Pole, West and Harrison (1994). Some theoretical variations are considered in Section 6.4 below, although they are of restricted practical interest.

6.3.3 Practical discount strategy

Discounting should be viewed as an elegant way of coping with the system evolution variance series $\mathbf{W}_t$. Of course, as far as *one-step ahead* forecasts are concerned, there is no need to refer to $\mathbf{W}_t$ explicitly since $\mathbf{R}_t = \mathbf{P}_t/\delta$. Looking further ahead than this single time point, it is *not* the case that repeat application with the same discount factor will produce the relevant sequence of variance matrices. For example, with a single component model having discount factor δ, repeated application would lead to the use of δ^k as a discount factor k-steps ahead with $\mathbf{R}_t(k) = \mathbf{G}^k\mathbf{C}_t\mathbf{G}'^k/\delta^k$. This implies an *exponential* decay in information, and this is not strictly consistent with the DLM, in which the information decays *arithmetically* through the addition of future evolution error variance matrices. Hence, though perfectly coherent one-step ahead, the discount approach must be applied with thought in extrapolating ahead (and also, therefore, when encountering missing values in the time series). Ameen and Harrison (1985) discuss this point, and use one-step discounting from $t = 0$ to determine the implied sequence $\mathbf{W}_t$ for all future times t. This is possible, since given the other model components, these matrices are simply functions of quantities assumed known initially. It can be seen that this is also possible in models where the observational variance is being estimated.

Since $|\mathbf{W}_t|$ is usually small relative to V_t, an alternative, more flexible, less computationally demanding, practical approach is suggested in Harrison and West (1986). This simply assumes that the one-step ahead evolution variance matrix is appropriate for extrapolation into the future, determining a *constant* step-ahead variance matrix. The resulting discount procedure is then as follows.

(1) Given $(\boldsymbol{\theta}_t|D_t)$, calculate $\mathbf{W}_{t+1} = \mathbf{P}_{t+1}(1-\delta)/\delta$.

(2) In forecasting k-steps ahead, adopt the conditionally constant variance

$$\mathrm{V}[\boldsymbol{\omega}_{t+k}|D_t] = \mathbf{W}_t(k) = \mathbf{W}_{t+1}, \qquad (k = 1, \ldots).$$

Thus, step-ahead forecast distributions will be based on the addition of evolution errors with the same variance matrix $\mathbf{W}_{t+1}$ for all k.

(3) The observation Y_{t+1} allows the posterior $(\boldsymbol{\theta}_{t+1}|D_{t+1})$ to be derived from which $\mathbf{P}_{t+2}$ and thus $\mathbf{W}_{t+2}$ are deduced. Thus forecasting ahead from time $t+1$, we have

$$\mathrm{V}[\boldsymbol{\omega}_{t+k}|D_{t+1}] = \mathbf{W}_{t+1}(k) = \mathbf{W}_{t+2}, \qquad (k = 1, \ldots).$$

(4) Proceed in this manner at time $t+2$, and so on.

The computational simplicity of this strategy is evident; at any time, a single evolution variance matrix is calculated and used k-steps ahead for any desired k. Note an important modification of the standard DLM analysis. Hitherto, the evolution errors were assumed to have variance matrices known for all time, and also independent of the history of the series. With the discount strategy this assumption has been weakened and modified to allow the variance matrices in the future to depend on the current state of information. Mathematically, the assumption that for any $k = 1, \dots$, $V[\boldsymbol{\omega}_{t+k}|D_t] = V[\boldsymbol{\omega}_{t+k}|D_0]$, has been revised; it is now the case that $V[\boldsymbol{\omega}_{t+k}|D_t] = \mathbf{W}_t(k)$ depends on t in addition to $t+k$. For example, at time t, the 2-step ahead variance matrix is

$$V[\boldsymbol{\omega}_{t+2}|D_t] = \mathbf{W}_t(2) = \mathbf{W}_{t+1}.$$

Obtaining a further observation, this is revised to

$$V[\boldsymbol{\omega}_{t+2}|D_{t+1}] = \mathbf{W}_{t+1}(1) = \mathbf{W}_{t+2}.$$

This modification is straightforward, and has no complicating consequences in practice. In updating and retrospection, the coherent value $\mathbf{W}_t$ as derived based on D_{t-1} is used in the relevant updating and filtering equations. As time progresses, the future evolution variances are revised, a process that is interpretable as a sequence of successive interventions.

6.4 FURTHER COMMENTS ON DISCOUNT MODELS†

From an applied viewpoint, the above framework provides a complete operational approach to structuring the evolution variance matrices of all DLMs. The use of single discount ideas to structure forecasting models based on TSDLMs is discussed in Brown (1962), Harrison (1965), Godolphin and Harrison (1975), and Harrison and Akram (1983). The first extension to multiple discount factors is to be found in Harrison (1965) and is discussed in Whittle (1965). The general extension to multiple discount factors for components described above is generally appropriate outside the restricted class of TSDLMs. Applications can be found in Ameen and Harrison (1985), West and Harrison (1986), Harrison and West (1986, 1987), with implementation in West, Harrison and Pole (1987) and in Pole, West, and Harrison (1994). It can be seen that these discount factors play a role analogous to those used in non-Bayesian point forecasting methods, in particular to exponential smoothing techniques (Ledolter and Abraham 1983, Chapters 3 and 4, for example), providing interpretation and meaning within the DLM framework. Some further theoretical discussion of discount models in general is now given.

Questions arise concerning the limiting behaviour of discount TSDLMs with constant triples $\{\mathbf{F}, \mathbf{G}, \mathbf{V}\}$. Consider the case of a single discount model in which for the specified and constant discount factor δ, it follows that

$$\mathbf{C}_t^{-1} = \delta \mathbf{P}_t^{-1} + \mathbf{F}\mathbf{V}^{-1}\mathbf{F}'.$$

---

†This section is of theoretical interest only, and may be omitted without loss on a first reading.

In cases in which $\mathbf{G}$ is non-singular, we then have

$$\lim_{t\to\infty} \mathbf{C}_t^{-1} = \sum_{v=0}^{\infty} \delta^v \mathbf{G}'^{-v}\mathbf{F}V^{-1}\mathbf{F}'\mathbf{G}^{-v}.$$

Based on this representation, Ameen and Harrison (1983) prove various limiting results. In particular, the following result, taken from that reference, is key.

Theorem 6.6. *Consider the canonical TSDLM with* $\mathbf{F} = (1, \ldots, 1)'$ *and* $\mathbf{G} = diag\,(\lambda_1, \ldots, \lambda_n)$, *where the* λ_i *are distinct, real or complex. Suppose the single discount strategy is applied with a discount factor* δ, *and define* $u_i = \delta^{1/2}/\lambda_i$ *for each* i. *Then if* $\delta < \min\{|\lambda_i^2|,\ i = 1, \ldots, n\}$, *the following limits exist:*

(1) $\lim_{t\to\infty} Q_t = V\prod_{i=1}^n u_i^{-2}$.

(2) $\lim_{t\to\infty} \mathbf{A}_t = (A_1, \ldots, A_n)'$, *where for* $i = 1, \ldots, n$,

$$A_i = (1 - u_i^2)\prod_{j\neq i}(1 - u_i u_j)/(1 - u_i/u_j).$$

(3) $\lim_{t\to\infty} \mathbf{C}_t^{-1} = \mathbf{K}V^{-1}$ *where* $\mathbf{K}$ *has elements* $K_{ij} = 1/(1 - u_i u_j)$ *for* $i, j = 1, \ldots, n$.

Some features of this kind of result are discussed in specific models in later Chapters, in particular Chapter 7, and have already been noted in the first-order polynomial model in Chapter 2. In practice, for more complex models using multiple discount factors applied to components as in the preceding sections, the updating equations are observed to converge to stable, limiting forms, although theoretical results for such models are (at time of writing) unavailable. It is conjectured that in any closed model $\{\mathbf{F}, \mathbf{G}, V, \mathbf{W}_t\}$ with $\mathbf{W}_t$ structured in block diagonal form as in Definition 6.1, the updating equations have stable limiting forms, with $\mathbf{C}_t$, $\mathbf{R}_t$, $\mathbf{A}_t$ and Q_t converging (rapidly) to finite limits $\mathbf{C}$, $\mathbf{R}$, $\mathbf{A}$ and Q. If this is so, then $\mathbf{W}_t$ also has a stable limiting form being based on $\mathbf{C}$ and the fixed discount factors. Thus, in the limiting forms, component discount TSDLMs are essentially standard DLMs with constant, block diagonal evolution variance matrices. The limiting representations of the observation series in generalised ARIMA form are then deducible from Section 5.5.

Some support for this conjectured limiting behaviour is provided in the work of Ameen and Harrison (1985) when considering alternatives to multiple discounting. These alternatives are very similar to component discounting, defining matrices $\mathbf{W}_t$ based on possibly several discount factors. Consider the n-dimensional DLM $\{\mathbf{F}_t, \mathbf{G}_t, V_t, \mathbf{W}_t\}$. Their approach involves the $n \times n$ diagonal discount matrix $\boldsymbol{\Delta}$, defined by

$$\boldsymbol{\Delta} = \text{diag}(\delta_1^{-1/2}, \ldots, \delta_n^{-1/2}).$$

Then, given the posterior variance matrix $\mathbf{C}_{t-1}$ at time $t-1$, two possible alternatives to component discounting define the prior variance matrix $\mathbf{R}_t$ at time t by either of the forms

$$\text{(a)} \quad \mathbf{R}_t = \boldsymbol{\Delta}\mathbf{G}_t\mathbf{C}_{t-1}\mathbf{G}_t'\boldsymbol{\Delta},$$
$$\text{(b)} \quad \mathbf{R}_t = \mathbf{G}_t\boldsymbol{\Delta}\mathbf{C}_{t-1}\boldsymbol{\Delta}\mathbf{G}_t'.$$

The above reference discusses scheme (a) and derives theoretical results about the stable limiting forms of updating equations when $\mathbf{F}$, $\mathbf{G}$ and V are constant over time. In particular, limiting results for $\mathbf{C}_t$, $\mathbf{A}_t$, Q_t and $\mathbf{W}_t$ show that convergence is typically rapid, so that the limiting form of the model is that of a standard, constant TSDLM $\{\mathbf{F}, \mathbf{G}, V, \mathbf{W}\}$. Following this, Theorem 5.6 applies to deliver the limiting representation of the observation and error series as

$$\lim_{t\to\infty} \left\{ \prod_{i=1}^{n}(1-\lambda_i B)Y_t - \prod_{i=1}^{n}(1-\rho_i B)e_t \right\} = 0,$$

where $\rho_1, \ldots, \rho_n$ are the eigenvalues of $(\mathbf{I} - \mathbf{AF}')\mathbf{G}$.

These models provide Bayesian analogues of standard point forecasting techniques based on the use of multiple discount factors, such as multiple exponential smoothing (Abraham and Ledolter 1983, Chapter 7; McKenzie 1974, 1976). Using either (a) or (b), note the following.

(1) Each method implies $\mathbf{W}_t = \mathbf{R}_t - \mathbf{G}_t\mathbf{C}_{t-1}\mathbf{G}_t'$. While obviously symmetric, and usually positive definite, it is theoretically possible that this matrix is not positive definite, and therefore not a valid evolution variance matrix. Component discounting has no such drawback.

(2) Unlike component discounting, $\mathbf{W}_t$ will not generally be block diagonal. Consequently, the desirable conditional independence structure of Figure 4.2 is lost.

(3) If the discount factors coincide, $\delta_i = \delta$ for $i = 1, \ldots, n$, then in both cases, $\mathbf{R}_t = \mathbf{G}_t\mathbf{C}_{t-1}\mathbf{G}_t'/\delta$ and the model is in fact a single component discount DLM.

6.5 *EXERCISES*

(1) Construct observable DLMs $\{\mathbf{F}, \mathbf{G}, \cdot, \cdot\}$ with forecast functions of the following forms:
   (a) $f_t(k) = a_{t1}\lambda_1^k + a_{t2}\lambda_2^k$ where $\lambda_1\lambda_2 \neq 0$ and $\lambda_1 \neq \lambda_2$.
   (b) $f_t(k) = a_{t1} + a_{t2}k + a_{t3}k^2 + a_{t4}k^3$.
   (c) $f_t(k) = a_{tj}$, where $j = k|4$, $(j = 1, \ldots, 4)$.
   (d) $f_t(k) = a_{tk}$ for $k = 0, 1, 2$ and 3, but $f_t(k) = 0$ for $k > 3$.
   (e) $f_t(k) = a_{t1} + a_{t2}k + a_{t3}\lambda^k + a_{t4}\lambda^k\cos(k\omega) + a_{t5}\lambda^k\sin(k\omega)$ for some λ and ω, where ω is not an integer multiple of π.

(f) $f_t(k) = a_{t1} + a_{t2}k + a_{t3}\lambda_1^k + a_{t4}\lambda_2^k \cos(k\omega) + a_{t5}\lambda_2^k \sin(k\omega)$ for some λ_1, λ_2 and ω, where ω is not an integer multiple of π.

(2) Construct an observable DLM for which the forecast function comprises a quadratic polynomial about which there is additive quarterly seasonal variation.

(3) Design an observable DLM $\{\mathbf{F}, \mathbf{G}, \cdot, \cdot\}$ such that the resultant forecast function has the form

$$f_t(k) = \sum_{v=1}^{3} f_{vt}(k),$$

where, with $a_{t1}, \ldots, a_{t6}$ known at time t,

$$\begin{aligned} f_{1t}(k) &= (a_{t1} + a_{t2}k)\lambda^k, \\ f_{2t}(k) &= a_{t3}\cos(k\pi/2) + a_{t4}\sin(k\pi/2), \\ f_{3t}(0) &= a_{t5}, \\ f_{3t}(1) &= a_{t6}, \end{aligned}$$

and

$$f_{3t}(k) = 0, \qquad k \geq 2.$$

(4) You construct a canonical DLM

$$\left\{ \begin{pmatrix} 1 \\ 1 \end{pmatrix}, \begin{pmatrix} 1 & 0 \\ 0 & \lambda \end{pmatrix}, \cdot, \cdot \right\}$$

with $\lambda = 0.9$, state vector $\boldsymbol{\theta}_t = (\theta_{1t}, \theta_{2t})'$, and initial prior

$$(\boldsymbol{\theta}_0 \mid D_0) \sim \mathrm{N}\left[\begin{pmatrix} 100 \\ -50 \end{pmatrix}, \begin{pmatrix} 100 & 0 \\ 0 & 25 \end{pmatrix} \right].$$

(a) Obtain the forecast function, showing that it has the form of a modified exponential.

(b) Interpret the parameters. For operation you wish to obtain an equivalent DLM

$$\left\{ \begin{pmatrix} 1 \\ 0 \end{pmatrix}, \begin{pmatrix} 1 & 1 \\ 0 & \lambda \end{pmatrix}, \cdot, \cdot \right\}$$

with parameter $\boldsymbol{\phi}_t = (\phi_{1t}, \phi_{2t})' = \mathbf{H}\boldsymbol{\theta}_t$.

(c) Obtain the matrix $\mathbf{H}$ and the prior $(\boldsymbol{\phi}_0 \mid D_0)$.

(d) Interpret the new parameters.

(5) Let Y_t be a Gaussian time series such that $Y_t - 0.9Y_{t-1} = a_t - 0.72a_{t-1}$, where $a_t \sim \mathrm{N}[0, 1]$ independently. Represent this process as a TSDLM.

(6) Consider three independent random time series X_t, Z_t and U_t generated by the following processes, in which B is the backward shift

operator:

$$(1-B)^2X_t = (1-0.95B)^2a_t, \qquad a_t \sim \mathrm{N}[0,\sigma_a^2],$$

$$(1-\sqrt{3}B+B^2)Z_t = (1-0.9\sqrt{3}B+0.81B^2)b_t, \qquad b_t \sim \mathrm{N}[0,\sigma_b^2],$$

$$U_t = (1-0.5B)c_t, \qquad c_t \sim \mathrm{N}[0,\sigma_c^2].$$

Let $Y_t = X_t+Z_t+U_t$. Construct a canonical observable pair $\{\mathbf{F},\mathbf{G}\}$ of a constant TSDLM for Y_t.

(7) Many observable discount TSDLMs $\{\mathbf{F},\mathbf{G},V,\mathbf{W}_t\}$, such as polynomial/seasonal models, are such that the eigenvalues λ_i of $\mathbf{G}$ lie on the unit circle, i.e., $|\lambda_i| = 1$. Consider a constant single discount model with discount factor $0 < \delta < 1$. As usual, the model has dimension n and $\mu_t = \mathbf{F}'\boldsymbol{\theta}_t$.

(a) Using the result of Theorem 6.6 or otherwise, prove that

$$\lim_{t\to\infty} Q_t = V/\delta^n.$$

(b) The total adaptation at time t may be defined as A_t where

$$\mathrm{E}[\mu_t|D_t] = \mathrm{E}[\mu_t|D_{t-1}] + A_t e_t.$$

Prove that

$$\lim_{t\to\infty} A_t = 1-\delta^n.$$

(c) Results (a) and (b) are important, giving insight into how a single discount factor is intimately related to the dimension of the DLM in terms of the number of elements, n, in the parameter vector. Think about this and its implications.

(8) In a general DLM, suppose that $\mathbf{G}_t$ is non-singular for all t. Suppose further that $\mathbf{W}_t$ is defined using a single discount factor δ so that $\mathbf{R}_t = \mathbf{G}_t\mathbf{C}_{t-1}\mathbf{G}_t'/\delta$ for all t, $(0<\delta<1)$.

(a) Show that $\mathbf{B}_{t-k} = \mathbf{C}_{t-k}\mathbf{G}'_{t-k+1}\mathbf{R}^{-1}_{t-k+1} = \delta\mathbf{G}^{-1}_{t-k+1}$ for any $k>0$.

(b) Hence show that the filtering recurrence equations in Theorem 4.4 simplify to

$$\mathbf{a}_t(-k) = (1-\delta)\mathbf{m}_{t-k} + \delta\mathbf{G}^{-1}_{t-k+1}\mathbf{a}_t(-k+1)$$

and

$$\mathbf{R}_t(-k) = (1-\delta)\mathbf{C}_{t-k} + \delta^2\mathbf{G}^{-1}_{t-k+1}\mathbf{R}_t(-k+1)(\mathbf{G}'_{t-k+1})^{-1}.$$

(b) Comment on the forms of these equations, with particular reference to the implied computational demands relative to those of the original, general recurrences.

(9) In the framework of the previous example, suppose that $\mathbf{G}_t = \mathbf{I}$, the $n\times n$ identity matrix, for all t, so that the model is a multiple

regression DLM as in Definition 3.1 of, or the important first-order polynomial DLM.

(a) Show that $\mathbf{B}_{t-k} = \delta\mathbf{I}$ and write down the simplified filtering equations corresponding to Theorem 4.5 (i) and (ii).

(b) Show also that

$$\mathbf{a}_t(-k) = (1-\delta)\sum_{v=0}^{k-1}\delta^v\mathbf{m}_{t-k+v} + \delta^k\mathbf{m}_t$$

and

$$\mathbf{R}_t(-k) = (1-\delta)\sum_{v=0}^{k-1}\delta^{2v}\mathbf{C}_{t-k+v} + \delta^{2k}\mathbf{C}_t.$$

(c) Now, for the following question parts, concentrate on the constant first-order polynomial DLM $\{1, 1, V, W\}$. For any fixed $k > 0$ show that

$$\lim_{t\to\infty} R_t(-k) = C(1+\delta^{2k+1})/(1+\delta),$$

where

$$C = \left(\sqrt{1+4V/W} - 1\right)W/2.$$

(d) In the framework of (c) above, show that with $D_t(-k) = D_t - Y_{t+k}$, the jackknife distribution of Theorem 4.9 is

$$(\mu_{t-k} \mid D_t(-k)) \sim \mathrm{N}[a_{t,k}, R_{t,k}],$$

where writing $A_t(-k) = R_t(-k)/(V - R_t(-k))$,

$$\begin{aligned} a_{t,k} &= a_t(-k) - A_t(-k)(Y_{t-k} - a_t(-k)),\\ R_{t,k} &= A_t(-k)V. \end{aligned}$$

Further, prove that

$$\lim_{t,k\to\infty} A_t(-k) = (1-\delta)/2\delta.$$

(10) Consider the discount DLM $\{1, \lambda, 1, W_t\}$ with discount factor δ so that $W_t = \lambda^2 C_{t-1}(\delta^{-1} - 1)$.

(a) Prove that if $0 < \delta < \lambda^2$, then

$$\lim_{t\to\infty} C_t = C = 1 - \delta/\lambda^2,$$

and obtain the corresponding limiting values of R_t, Q_t and A_t.

(b) What is the limiting value of C_t when $\delta \geq \lambda^2$?

(c) Using $m_t = \lambda m_{t-1} + Ae_t$, the limiting form of the updating equation, show that

$$Y_t - \lambda Y_{t-1} = e_t - (1 - \delta/\lambda)e_{t-1}.$$

Compare this result with the corresponding limiting result in the model $\{1, \lambda, 1, W\}$ obtained by applying Theorem 5.6.

(d) Consider the implications of these results for setting component discount factors in relationship to component eigenvalues.

(11) Consider the model

$$\left\{\begin{pmatrix}1\\1\end{pmatrix}, \begin{pmatrix}\lambda_1 & 0\\ 0 & \lambda_2\end{pmatrix}, 1, \mathbf{W}_t\right\}$$

for any distinct, non-zero values λ_1 and λ_2. Suppose a single discount model so that $\mathbf{W}_t = \mathbf{G}\mathbf{C}_{t-1}\mathbf{G}'(\delta^{-1} - 1)$. Let $\mathbf{K}_t = \mathbf{C}_t^{-1}$ for all t.

(a) Show that the updating equation for $\mathbf{C}_t$ can be written in terms of precision matrices as

$$\mathbf{K}_t = \delta\mathbf{G}^{-1}\mathbf{K}_{t-1}\mathbf{G}^{-1} + \mathbf{F}\mathbf{F}'.$$

(b) Writing

$$\mathbf{K}_t = \begin{pmatrix}K_{t1} & K_{t3}\\ K_{t3} & K_{t2}\end{pmatrix},$$

deduce the recurrence equations

$$\begin{aligned} K_{t1} &= 1 + \delta K_{t-1,1}/\lambda_1^2, \\ K_{t2} &= 1 + \delta K_{t-1,2}/\lambda_2^2, \\ K_{t3} &= 1 + \delta K_{t-1,3}/\lambda_1\lambda_2. \end{aligned}$$

(c) Deduce that as t increases, $\mathbf{K}_t$ converges to a limit if $\delta < \min\{\lambda_1^2, \lambda_2^2\}$.

(d) Assuming this to hold, deduce expressions for the elements of the limiting matrix, $\mathbf{K}$, say, as functions of δ, λ_1 and λ_2. Deduce the limiting variance matrix $\mathbf{C} = \mathbf{K}^{-1}$.

(e) Suppose that $\delta = 0.7$, $\lambda_1 = 0.8$ and $\lambda_2 = 0.9$. Calculate $\mathbf{K}$ and deduce the limiting values of $\mathbf{C}_t$, $\mathbf{R}_t$, Q_t and $\mathbf{A}_t$.

(12) Show that the results of the previous example apply in the case of complex eigenvalues $e^{\pm i\omega}$ for some real ω that is not an integer multiple of π.

(13) Consider a single discount normal DLM $\{\mathbf{F}, \mathbf{G}, V, \mathbf{W}_t\}$ in which by definition, $\mathbf{W}_t$ is implicitly defined by

$$\mathbf{R}_t = \delta^{-1}\mathbf{G}\mathbf{C}_{t-1}\mathbf{G}' = \mathbf{G}\mathbf{C}_{t-1}\mathbf{G}' + \mathbf{W}_t.$$

Show that the relationship between the distributions $p(\boldsymbol{\theta}_t \mid D_{t-1})$ and $p(\boldsymbol{\theta}_{t-1} \mid D_{t-1})$ is exactly as determined by application of the following "power discount" procedure.

Write $f_{t-1}(\cdot)$ for the density function of $(\boldsymbol{\theta}_{t-1}|D_{t-1})$, and *define* the density function $p(\boldsymbol{\theta}_t|D_{t-1})$ as $c_t f_{t-1}(\cdot)^{\delta}$ for an appropriate normalising constant c_t. This "power discount" procedure may be extended to non-normal, first-order polynomial dynamic models as in Smith (1979). See the final two exercises of Section 14.6 for key examples of this.

(14) Discount weighted regression (Ameen and Harrison 1984), adopts a forecast function $f_t(k) = \mathbf{F}'_{t+k}\mathbf{m}_t$, where $\mathbf{m}_t$ is that value of $\boldsymbol{\theta}$ minimising the discounted sum of squares

$$S_\delta(\boldsymbol{\theta}) = \sum_{i=0}^{t-1} \delta^i (Y_{t-i} - \mathbf{F}'_{t-i}\boldsymbol{\theta})^2$$

for some given discount factor δ. Define the n-vector,

$$\mathbf{h}_t = \sum_{i=0}^{t-1} \delta^i \mathbf{F}_{t-i} Y_{t-i} = \mathbf{F}_t Y_t + \delta \mathbf{h}_{t-1},$$

and the n-square matrix

$$\mathbf{H}_t = \sum_{i=0}^{t-1} \delta^i \mathbf{F}_{t-i}\mathbf{F}'_{t-i} = \mathbf{F}_t\mathbf{F}'_t + \delta \mathbf{H}_{t-1}.$$

Write $e_t = Y_t - \mathbf{F}'_t\mathbf{m}_{t-1}$, and assuming $t \geq n$ and that $\mathbf{H}_t$ is of full rank, $\mathbf{C}_t = \mathbf{H}_t^{-1}$.

(a) Show that

$$\frac{\partial S}{\partial \boldsymbol{\theta}} = -2(\mathbf{h}_t - \mathbf{H}_t\theta),$$

$$\frac{\partial^2 S}{\partial \boldsymbol{\theta} \partial \boldsymbol{\theta}'} = 2\mathbf{H}_t,$$

and so deduce

$$\mathbf{m}_t = \mathbf{H}_t^{-1}\mathbf{h}_t = \mathbf{C}_t\mathbf{h}_t.$$

(b) Writing $\mathbf{R}_t = \mathbf{C}_{t-1}/\delta$ and $\mathbf{A}_t = \mathbf{H}_t^{-1}\mathbf{F}_t$, show that

$$\mathbf{m}_t = \mathbf{m}_{t-1} + \mathbf{A}_t e_t.$$

(c) Show that

$$\mathbf{C}_t = (\mathbf{R}_t^{-1} + \mathbf{F}_t\mathbf{F}'_t)^{-1} = \mathbf{R}_t - \mathbf{A}_t Q_t \mathbf{A}'_t$$

and

$$Q_t = 1 + \mathbf{F}'_t\mathbf{R}_t\mathbf{F}_t.$$

(d) Compare these results with the recurrence relations for the single discount DLM $\{\mathbf{F}_t, \mathbf{I}, V, \mathbf{W}_t\}$ and draw your conclusion. Note that ordinary linear regression is the case $\delta = 1$.

CHAPTER 7

POLYNOMIAL TREND MODELS

7.1 INTRODUCTION

Polynomial models find wide use in time series and forecasting as they do in other branches of applied statistics, such as static regression and experimental design. In time series these models prove useful in describing trends that are generally viewed as smooth developments over time. Relative to the sampling interval of the series and the required forecast horizons, such trends are usually well approximated by low-order polynomial functions of time. Indeed, a first- or second-order polynomial component DLM is often quite adequate for short-term forecasting, either on its own or in combination with seasonal, regression and other components. Chapter 2 introduced the first-order polynomial model, which although very simple, is applied more than any other DLM. Next in practical importance is the second-order DLM, sometimes referred to as the linear growth model, which is the subject matter of much of this chapter. Higher-order polynomial models are also discussed, though it is rare to find applications employing polynomial DLMs of order greater than three (corresponding to quadratic growth). The structure of polynomial models is discussed in Harrison (1965, 1967), and theoretical aspects explored in Godolphin and Harrison (1975). See also Abraham and Ledolter (1983, chapter 3).

Polynomial DLMs are a subset of the class of time series DLMs, or TSDLMs, defined in Chapter 4 as those models whose $n \times 1$ regression vector $\mathbf{F}$ and $n \times n$ system matrix $\mathbf{G}$ are constant for all time.

Definition 7.1. Any observable TSDLM that for all $t \geq 0$ has a forecast function of the form

$$f_t(k) = a_{t0} + a_{t1}k + \cdots + a_{t,n-1}k^{n-1}\,, \qquad k \geq 0,$$

is defined as an $\mathbf{n^{th}}$**-order polynomial DLM**.

From Section 5.3, it follows that the system matrix $\mathbf{G}$ of an nth-order polynomial DLM has a single unit eigenvalue of multiplicity n. It is stressed that with this definition, $\mathbf{G}$ has no other eigenvalues, in particular, no zero eigenvalues. Consequently, following Section 5.4,

(1) A DLM is an nth-order polynomial model if and only if it is similar to the canonical model

$$\{\mathbf{E}_n, \mathbf{J}_n(1), \cdot, \cdot\}.$$

(2) Any DLM equivalent to the constant model $\{\mathbf{E}_n, \mathbf{J}_n(1), V, \mathbf{W}\}$ is a constant n^{th}-order polynomial DLM.

This chapter concentrates on the highly important second-order model, elaborating its structure and properties in isolation from other components.

The primary objectives are to familiarise the reader with the nature of the sequential updating equations, as was done in Chapter 2 for the first-order model, and to relate this DLM to other popular applied forecasting methods and concepts. Constant DLMs, and in particular constant linear growth models, are central to this discussion and are considered in detail. Throughout the chapter, the observational variance sequence V_t is assumed known. This is purely to simplify discussion: the general variance learning procedure of Section 4.5 applies without affecting the primary features of the updating equations for the parameters and forecasts conditional upon known V_t.

Consider the canonical model $\{\mathbf{E}_n, \mathbf{J}_n(1), \cdot, \cdot\}$. For reasons that will be made clear below, denote the state vector of this model by $\boldsymbol{\lambda}_t$ rather than the usual $\boldsymbol{\theta}_t$, and the evolution error by $\partial\boldsymbol{\lambda}_t$ rather than $\boldsymbol{\omega}_t$. Then, with

$$\boldsymbol{\lambda}_t' = (\lambda_{t1}, \ldots, \lambda_{tn})$$

and

$$\partial\boldsymbol{\lambda}_t' = (\partial\lambda_{t1}, \ldots, \partial\lambda_{tn}),$$

the model equations can be written as

$$\begin{aligned}
&\text{Observation:} && Y_t = \lambda_{t1} + \nu_t,\\
&\text{System:} && \lambda_{tj} = \lambda_{t-1,j} + \lambda_{t-1,j+1} + \partial\lambda_{tj}, \quad (j = 1, \ldots, n-1),\\
& && \lambda_{tn} = \lambda_{t-1,n} + \partial\lambda_{tn}.
\end{aligned}$$

Here $\mu_t = \lambda_{t1}$ is the level at time t, that between times $t-1$ and t changes by the addition of $\lambda_{t-1,2}$ plus the noise $\partial\lambda_{t1}$. $\lambda_{t-1,2}$ represents a systematic change in level, that itself changes by the addition of $\lambda_{t-2,3}$ plus noise. Proceeding through the state parameters for $j = 1, \ldots, n-1$, each λ_{tj} changes systematically via the increment $\lambda_{t-1,j+1}$, and also by the addition of the noise term $\partial\lambda_{tj}$. The nth component λ_{tn} changes only stochastically. Although this is the canonical model form, interpretation of parameters leads us to prefer working with the similar model

$$\{\mathbf{E}_n, \mathbf{L}_n, \cdot\ ,\ \cdot\},$$

where $\mathbf{L}_n$ is the $n \times n$ upper triangular matrix of unit elements,

$$\mathbf{L}_n = \begin{pmatrix} 1 & 1 & 1 & \ldots & 1\\ 0 & 1 & 1 & \ldots & 1\\ 0 & 0 & 1 & \ldots & 1\\ \vdots & \vdots & \vdots & \ddots & \vdots\\ 0 & 0 & 0 & \ldots & 1 \end{pmatrix}.$$

The reader can easily verify that this model is similar to the canonical n^{th}-order polynomial DLM. As usual, denote this model's state parameter and evolution error by

$$\boldsymbol{\theta}_t = (\theta_{t1}, \ldots, \theta_{tn})'$$

and

$$\boldsymbol{\omega}_t = (\omega_{t1}, \dots, \omega_{tn})'.$$

Then the model equations are

$$\text{Observation:} \quad Y_t = \theta_{t1} + \nu_t,$$

$$\text{System:} \quad \theta_{tj} = \theta_{t-1,j} + \sum_{r=j+1}^{n} \theta_{t-1,r} + \omega_{tj}, \quad (j = 1, \dots, n-1),$$

$$\theta_{tn} = \theta_{t-1,n} + \omega_{tn}.$$

In this representation, ignoring the evolution noise terms, the state parameters can be thought of as "derivatives" of the mean response function $\mu_t = \theta_{t1}$. For each $j = 1, \dots, n$, θ_{tj} represents the jth derivative of the mean response. At time t, the expected future trajectory of θ_{tj} is a polynomial of degree $n - j$.

The difference between this and the canonical model lies simply in a time shift in the definition of the elements of the state vector. In this representation, the state parameters have the natural interpretation as derivatives at time t. In the canonical model, the higher-order terms in the state vector play the same role but are shifted back to time $t-1$. This is most easily seen by setting the evolution errors to zero in each model, when the parameters are related according to

$$\begin{aligned}\lambda_{t1} &= \theta_{t1},\\ \lambda_{tj} &= \theta_{tj} + \theta_{t,j+1}, \qquad (j = 2, \dots, n-1),\\ \lambda_{tn} &= \theta_{tn}.\end{aligned}$$

As previously pointed out in Section 5.3.4, for $n \geq 2$, unless carefully structured, the evolution error vector $\boldsymbol{\omega}_t$ introduces ambiguity, confusing parametric interpretation. This arises since in general, the parametric elements may be arbitrarily correlated through the variance matrix $\mathbf{W}_t$, although the basic notion of them as derivatives of the mean response remains sound.

A class of models with naturally interpretable and appropriate variance structure is defined by

$$\begin{aligned}\text{Observation:} \quad & Y_t = \theta_{t1} + \nu_t,\\ \text{System:} \quad & \theta_{tj} = \theta_{t-1,j} + \theta_{t,j+1} + \partial\theta_{tj}, \quad (j = 1, \dots, n-1),\\ & \theta_{tn} = \theta_{t-1,n} + \partial\theta_{tn},\\ & \partial\boldsymbol{\theta}_t \sim \mathrm{N}[\mathbf{0}, \mathrm{diag}(W_{t1}, \dots, W_{tn})].\end{aligned}$$

In such a model the "derivatives" θ_{tj} change stochastically between times $t-1$ and t via the increments $\theta_{t,j+1}$ and also by those affecting the higher-order derivatives $\partial\theta_{tk}$, $j < k \leq n$. With

$$\partial\boldsymbol{\theta}_t \sim \mathrm{N}[\mathbf{0}, \mathrm{diag}(W_{t1}, \dots, W_{tn})]$$

and

$$\omega_t = \mathbf{L}_n \partial\boldsymbol{\theta}_t$$

so that

$$\omega_t \sim \mathrm{N}[\mathbf{0}, \mathbf{W}_t] \quad \text{with} \quad \mathbf{W}_t = \mathbf{L}_n \mathrm{diag}(W_{t1}, \ldots, W_{tn})\mathbf{L}_n',$$

the DLM assumes its usual form, as follows.

Definition 7.2. An **nth-order polynomial growth DLM** is any model of the form

$$\{\mathbf{E}_n, \mathbf{L}_n, V_t\ , \mathbf{W}_t\},$$

where

$$\mathbf{W}_t = \mathbf{L}_n \mathrm{diag}(W_{t1}, \ldots, W_{tn})\mathbf{L}_n'.$$

The distinguishing features of *polynomial growth*, rather than simply *polynomial*, models are that

(a) the system matrix is not in canonical form as the alternative $\mathbf{L}_n$ matrix is used;
(b) the evolution variance matrix has the special form consistent with the definition of $\omega_t = \mathbf{L}_n \partial\boldsymbol{\theta}_t$, where the elements of $\partial\boldsymbol{\theta}_t$ are uncorrelated.

The special cases $n = 2$ and $n = 3$ are now explored in detail.

7.2 SECOND-ORDER POLYNOMIAL MODELS

7.2.1 The general model form

At any time t, a second-order polynomial DLM has a straight line forecast function of the form

$$f_t(k) = a_{t0} + a_{t1}k. \tag{7.1}$$

An alternative representation is

$$f_t(k) = f_t(0) + [f_t(1) - f_t(0)]k,$$

thus identifying the coefficients $\{a_{t0}, a_{t1}\}$ in terms of the first two values of the forecast function.

The canonical DLM takes the form

$$\left\{\begin{pmatrix}1\\0\end{pmatrix}, \begin{pmatrix}1 & 1\\0 & 1\end{pmatrix}, \cdot, \cdot\right\}. \tag{7.2}$$

Notice that in this special case of $n = 2$, $\mathbf{J}_2(1) = \mathbf{L}_2$, so that the similar model with system matrix $\mathbf{L}_2$ coincides with the canonical model. With parametrisation

$$\boldsymbol{\theta}_t = \begin{pmatrix} \theta_{t1} \\ \theta_{t2} \end{pmatrix} = \begin{pmatrix} \mu_t \\ \beta_t \end{pmatrix},$$

the usual DLM representation is

$$\begin{aligned} &\text{Observation:} && Y_t = \mu_t + \nu_t, && (7.3a)\\ &\text{System:} && \mu_t = \mu_{t-1} + \beta_{t-1} + \omega_{t1}, && (7.3b)\\ & && \beta_t = \beta_{t-1} + \omega_{t2}, && (7.3c)\\ & && (\boldsymbol{\theta}_{t-1} \mid D_{t-1}) \sim \mathrm{N}[\mathbf{m}_{t-1}, \mathbf{C}_{t-1}], && \end{aligned}$$

where

$$\boldsymbol{\omega}_t = (\omega_{t1}, \omega_{t2})' \sim \mathrm{N}[\mathbf{0}, \mathbf{W}_t], \qquad \nu_t \sim \mathrm{N}[0, V_t],$$

$$\mathbf{m}_{t-1} = \begin{pmatrix} m_{t-1} \\ b_{t-1} \end{pmatrix} \quad \text{and} \quad \mathbf{C}_{t-1} = \begin{pmatrix} C_{t-1,1} & C_{t-1,3} \\ C_{t-1,3} & C_{t-1,2} \end{pmatrix}.$$

Following (7.1), the forecast function is $f_t(k) = m_t + kb_t$. As usual, μ_t is the series level and now β_{t-1} represents incremental growth. As mentioned in the introduction, the observational variance is assumed known. Otherwise, the normal posterior distributions are simply replaced by Student T forms.

7.2.2 Updating equations

The general theory of Section 4.3 is used to provide the updating equations for this model in explicit terms rather than in vector and matrix form. Write the evolution variance matrix as

$$\mathbf{W}_t = \begin{pmatrix} W_{t1} & W_{t3} \\ W_{t3} & W_{t2} \end{pmatrix}.$$

The sequential analysis has the following components.

(1) Writing

$$\begin{aligned} R_{t1} &= C_{t-1,1} + 2C_{t-1,3} + C_{t-1,2} + W_{t1}, \\ R_{t2} &= C_{t-1,2} + W_{t2}, \end{aligned}$$

and

$$R_{t3} = C_{t-1,2} + C_{t-1,3} + W_{t3},$$

we have $(\boldsymbol{\theta}_t \mid D_{t-1}) \sim \mathrm{N}[\mathbf{a}_t, \mathbf{R}_t]$ where

$$\mathbf{a}_t = \begin{pmatrix} m_{t-1} + b_{t-1} \\ b_{t-1} \end{pmatrix} \quad \text{and} \quad \mathbf{R}_t = \begin{pmatrix} R_{t1} & R_{t3} \\ R_{t3} & R_{t2} \end{pmatrix}.$$

(2) The one-step forecast distribution is

$$(Y_t \mid D_{t-1}) \sim \mathrm{N}[f_t, Q_t]$$

with

$$f_t = f_{t-1}(1) = m_{t-1} + b_{t-1}$$

and

$$Q_t = R_{t1} + V_t.$$

(3) The adaptive vector is given by

$$\mathbf{A}_t = \begin{pmatrix} A_{t1} \\ A_{t2} \end{pmatrix} = \begin{pmatrix} R_{t1}/Q_t \\ R_{t3}/Q_t \end{pmatrix}.$$

(4) Writing $e_t = Y_t - f_t$, the posterior at time t is

$$(\boldsymbol{\theta}_t \mid D_t) \sim \mathrm{N}\left[\begin{pmatrix} m_t \\ b_t \end{pmatrix}, \begin{pmatrix} C_{t1} & C_{t3} \\ C_{t3} & C_{t2} \end{pmatrix}\right],$$

where

$$\begin{aligned} m_t &= m_{t-1} + b_{t-1} + A_{t1}e_t, \\ b_t &= b_{t-1} + A_{t2}e_t, \\ C_{t1} &= A_{t1}V_t, \\ C_{t2} &= R_{t2} - A_{t2}R_{t3}, \\ C_{t3} &= A_{t2}V_t. \end{aligned}$$

(5) It is of some interest to note the relationships

$$Q_t = V_t/(1 - A_{t1}) \quad \text{and} \quad \mathbf{C}_t = \begin{pmatrix} A_{t1} & A_{t2} \\ A_{t2} & c_t \end{pmatrix} V_t,$$

where $c_t = (r_t - A_{t2}^2)/(1 - A_{t1})$, with $r_t = (C_{t-1,2} + W_{t2})/Q_t$.

(6) The updating equations lead to an alternative representation of the observation series in terms of past observations and forecast errors. Use of this representation provides easy comparison of the Bayesian model with alternative forecasting techniques. The three identities

$$\begin{aligned} Y_t &= m_{t-1} + b_{t-1} + e_t, \\ m_t &= m_{t-1} + b_{t-1} + A_{t1}e_t, \\ b_t &= b_{t-1} + A_{t2}e_t \end{aligned}$$

lead directly to the second-order difference equation

$$Y_t - 2Y_{t-1} + Y_{t-2} = e_t - \psi_{t1}e_{t-1} + \psi_{t2}e_{t-2}$$

where

$$\psi_{t1} = 2 - A_{t-1,1} - A_{t-1,2}$$

and

$$\psi_{t2} = 1 - A_{t-2,1}.$$

This representation is a consequence of adopting the model, however inappropriate it may be for any particular series.

7.2.3 Constant models and limiting behaviour

Denote the standardised constant second-order polynomial DLM as

$$M = \left\{ \begin{pmatrix} 1 \\ 0 \end{pmatrix}, \begin{pmatrix} 1 & 1 \\ 0 & 1 \end{pmatrix}, 1, \mathbf{W} = \begin{pmatrix} W_1 & W_3 \\ W_3 & W_2 \end{pmatrix} \right\}.$$

Note that the observational variance is $V = 1$. This loses no generality since for a general value V, $\mathbf{W}$, and all variances of linear functions are simply scaled by the constant factor V. Further, if V is unknown, the standard variance learning procedure is applied. From Section 5.5, the updating equations have a stable limiting form, that is typically rapidly approached, as follows.

Theorem 7.1. *For the second-order polynomial constant DLM, M, writing $\mathbf{A} = (A_1, A_2)'$,*

$$\lim_{t\to\infty} \{\mathbf{A}_t, \mathbf{C}_t, \mathbf{R}_t, Q_t\} = \{\mathbf{A}, \mathbf{C}, \mathbf{R}, Q\}$$

exists such that

(1) $(1 - A_1)Q = 1$;
(2) $A_2^2 Q = W_2$;
(3) $(A_1^2 + A_1 A_2 - 2A_2)Q = W_1 - W_3$;
(4) *with* $r = A_1 A_2 Q - W_3 + W_2$ *and* $c = (r - A_2^2)/(1 - A_1)$,

$$\mathbf{C} = \begin{pmatrix} A_1 & A_2 \\ A_2 & c \end{pmatrix} \quad \text{and} \quad \mathbf{R} = \begin{pmatrix} A_1 & A_2 \\ A_2 & r \end{pmatrix} Q;$$

(5) *the feasible region for* $\mathbf{A}$ *is that satisfying*

$$0 < A_1 < 1, \qquad 0 < A_2 < 4 - 2A_1 - 4(1 - A_1)^{1/2} < 2.$$

Proof. Theorem 5.5 shows that the limit exists. Write

$$\mathbf{F} = \mathbf{E}_2, \qquad \mathbf{G} = \begin{pmatrix} 1 & 1 \\ 0 & 1 \end{pmatrix} \quad \text{and} \quad \mathbf{R} = \begin{pmatrix} R_1 & R_3 \\ R_3 & R_2 \end{pmatrix}.$$

Then $\mathbf{A} = \mathbf{R}\mathbf{E}_2 Q^{-1}$, so that $R_1 = A_1 Q$ and $R_3 = A_2 Q$ as in part (4). Next, $Q = \mathbf{E}_2'\mathbf{R}\mathbf{E}_2 + 1 = R_1 + 1 = A_1 Q + 1$ and (1) follows. Since $\mathbf{R} = \mathbf{G}\mathbf{C}\mathbf{G}' + \mathbf{W}$ and $\mathbf{C} = \mathbf{R} - \mathbf{A}\mathbf{A}'Q$, eliminating $\mathbf{C}$,

$$\mathbf{R} - \mathbf{G}^{-1}\mathbf{R}(\mathbf{G}')^{-1} = \mathbf{A}\mathbf{A}'Q - \mathbf{G}^{-1}\mathbf{W}(\mathbf{G}')^{-1},$$

leading to

$$\begin{pmatrix} 2R_3 - R_2 & R_2 \\ R_2 & 0 \end{pmatrix} =$$

$$\begin{pmatrix} A_1^2 & A_1A_2 \\ A_1A_2 & A_2^2 \end{pmatrix} Q - \begin{pmatrix} W_1 - 2W_3 + W_2 & W_3 - W_2 \\ W_3 - W_2 & W_2 \end{pmatrix}.$$

Equating components and rearranging leads to $R_2 = A_1A_2Q - W_3 + W_2$, $A_2^2Q = W_2$, and $(A_1^2 + A_1A_2 - 2A_2)Q = W_1 - W_3$, so completing the proofs of statements (1) to (4).

For the feasible region result (5): $0 < A_1 < 1$ since $A_1 = R_1/Q > 0$ and $1 - A_1 = Q^{-1} > 0$. Define $a = Q(A_1^2 + A_1A_2 - 2A_2)$. Then $|\mathbf{W}| \geq 0$ implies $W_1W_2 - W_3^2 \geq 0$. Substituting $W_2 = A_2^2Q$ from (2) and $W_3 = W_1 - a$ from (3) leads to $W_1^2 - W_1(A_2^2Q + 2a) + a^2 \leq 0$. At the boundary, the roots of this quadratic in W_1 must be real valued, so $(A_2^2Q + 2a)^2 - 4a^2 \geq 0$. This reduces to $A_2^2Q + 4a \geq 0$ so that on substituting for a, we obtain the quadratic $A_2^2 - 4A_2(2 - A_1) + 4A_1^2 \geq 0$. For this to be true, A_2 must lie below the lower root of this quadratic, and it follows that $A_2 \leq 4 - 2A_1 - 4(1 - A_1)^{1/2}$.

◇

Following note (6) of the previous section, the limiting representation of the observations in terms of forecast errors is given by

$$Y_t - 2Y_{t-1} + Y_{t-2} = e_t - \psi_1 e_{t-1} + \psi_2 e_{t-2},$$

where $\psi_1 = \lim_{t\to\infty} \psi_{t1} = 2 - A_1 - A_2$ and $\psi_2 = \lim_{t\to\infty} \psi_{t2} = 1 - A_1$.

7.2.4 Single discount models

Often in practice the discount concept will be applied to the trend DLM either when it stands alone or when it is a component of a larger model. The discount factor then defines the evolution variance matrices, as described for general block models in Section 6.3. Of course this leads to a non-constant model, although the practical differences are small. The following definition is simply a special case of the class of single discount DLMs of Section 6.3.

Definition 7.3. For any discount factor δ, $0 < \delta \leq 1$, a single discount, second-order polynomial growth DLM is any second-order model in which for all t,

$$\mathbf{W}_t = \frac{1-\delta}{\delta}\mathbf{L}_2\mathbf{C}_{t-1}\mathbf{L}_2'. \tag{7.4}$$

From a practical viewpoint, this is the most important class of second-order models. Note that a single discount factor is applied to the trend model as a whole, consistent with the ideas underlying component models.

However, this does not mean that different discount factors cannot be applied to the two model elements, μ_t and β_t, separately if desired, or indeed that some other form of evolution variance sequence be used. There may be isolated cases when such alternative approaches are suitable, but for the vast majority of applications, a single trend discount factor will be satisfactory. As shown in Harrison (1967), alternatives have very little to offer in terms of increased forecast accuracy while losing the simple discount interpretation. The single discount model generally converges rapidly to the constant DLM $\{\mathbf{E}_2, \mathbf{J}_2(1), 1, \mathbf{W}\}$, with $\mathbf{W} = \mathbf{L}_2\mathbf{C}\mathbf{L}_2'(1-\delta)/\delta$. The corresponding limiting value of $\mathbf{C}$ is easily derived according to

$$\mathbf{C}^{-1} = \delta \mathbf{L}_2'^{-1}\mathbf{C}^{-1}\mathbf{L}_2^{-1} + \mathbf{E}_2\mathbf{E}_2'.$$

The limiting values of the elements in the updating equations are

$$Q = \frac{1}{\delta^2}, \qquad \mathbf{C} = \begin{pmatrix} 1-\delta^2 & (1-\delta)^2 \\ (1-\delta)^2 & (1-\delta)^3/\delta \end{pmatrix},$$

$$\mathbf{A} = \begin{pmatrix} 1-\delta^2 \\ (1-\delta)^2 \end{pmatrix} \quad \text{and} \quad \mathbf{R} = \begin{pmatrix} 1-\delta^2 & (1-\delta)^2 \\ (1-\delta)^2 & (1-\delta)^3 \end{pmatrix} Q.$$

Generally, the limiting adaptation is much greater for the level μ_t than for the growth β_t. Typical values, corresponding to the case $\delta = 0.9$, are $A_1 = 0.19$ and $A_2 = 0.01$. Note also that the value of either A_1 or A_2 determines δ and all the limiting quantities.

7.2.5 Double discount models†

Usually, in practice, the whole polynomial trend is best viewed as a component to be discounted as a block using a single discount factor. Approaches using two discount factors, one for the trend and one for the growth, are, of course, possible, and although of very restricted practical interest, provide Bayesian analogues of standard double exponential smoothing techniques (Abraham and Ledolter 1983, Chapters 3 and 7; McKenzie 1976). Here comments are restricted to one of the approaches described in Section 6.4. Approach (b) of that section concerns the use of two discount factors δ_1 and δ_2, the former for the level and the latter for the growth.

Let $\boldsymbol{\Delta} = \text{diag}(\delta_1^{-1/2}, \delta_2^{-1/2})$ and apply the multiple discount strategy of Section 6.4. Then, with $\mathbf{G} = \mathbf{L}_2$,

$$\mathbf{C}_t^{-1} = \mathbf{R}_t^{-1} + \mathbf{F}\mathbf{F}' = \mathbf{G}'^{-1}\boldsymbol{\Delta}^{-1}\mathbf{C}_{t-1}^{-1}\boldsymbol{\Delta}^{-1}\mathbf{G}^{-1} + \mathbf{F}\mathbf{F}'$$

and so the limiting value of $\mathbf{C}$ satisfies

$$\mathbf{C}^{-1} = \mathbf{G}'^{-1}\boldsymbol{\Delta}^{-1}\mathbf{C}^{-1}\boldsymbol{\Delta}^{-1}\mathbf{G}^{-1} + \mathbf{F}\mathbf{F}'.$$

†This Section can be omitted without loss on a first reading.

Writing

$$\mathbf{C} = \begin{pmatrix} C_1 & C_3 \\ C_3 & C_2 \end{pmatrix},$$

we then have

$$\begin{pmatrix} C_2 & -C_3 \\ -C_3 & C_1 \end{pmatrix} = \begin{pmatrix} \delta_1 C_2 + |\mathbf{C}| & -(\delta_1 C_2 + dC_3) \\ -(\delta_1 C_2 + dC_3) & \delta_2 C_1 + \delta_1 C_2 + 2dC_3 \end{pmatrix},$$

where $d = (\delta_1 \delta_2)^{1/2}$. As a result,

$$\mathbf{C} = \begin{pmatrix} (1 - d^2) & (1-d)(1-\delta_2) \\ (1-d)(1-\delta_2) & (1-d)^2(1-\delta_2)/\delta_1 \end{pmatrix}$$

and

$$\mathbf{A} = \begin{pmatrix} 1 - d^2 \\ (1-d)(1-\delta_2) \end{pmatrix}.$$

Finally, $Q = 1 + \mathbf{F}'\mathbf{R}\mathbf{F}$, where the limiting value for $\mathbf{R}_t$ can be calculated from $\mathbf{R} = \mathbf{G}\boldsymbol{\Delta}\mathbf{C}\boldsymbol{\Delta}\mathbf{G}'$.

The limiting updating equations for the mean vector $\mathbf{m}_t = (m_t, b_t)'$ are

$$m_t = m_{t-1} + b_{t-1} + A_1 e_t,$$

$$b_t = b_{t-1} + A_2 e_t,$$

where with $0 < \delta_1,\ \delta_2 < 1$, it follows that $0 < A_1 < 1$ and

$$0 \leq A_2 \leq A_1[1 - (1 - A_1)^{1/2}].$$

7.3 LINEAR GROWTH MODELS

7.3.1 Introduction

In practice, the discount models of Section 7.2.4 are recommended for their simplicity, parsimony and performance. However, a class of second-order polynomial growth models, or *linear growth models*, that has different evolution variance structure is of great interest. Historically these models have been widely used by practitioners for modelling linear trends with easily interpretable parameters and stochastic components (Harrison 1965, 1967; Godolphin and Harrison 1975). As is shown below, constant linear growth models are such that the limiting adaptive coefficients in the vector **A** provide all the useful values, and so, in a very real practical sense, other second-order models are essentially redundant. A further reason for closely examining these models is for communication with users of other linear trend point forecasting methods and for comparison with such procedures.

Definition 7.4. A linear growth model is any second-order polynomial DLM equivalent to a DLM of the form

$$\left\{ \begin{pmatrix} 1 \\ 0 \end{pmatrix}, \begin{pmatrix} 1 & 1 \\ 0 & 1 \end{pmatrix}, V_t, \begin{pmatrix} W_{t1} + W_{t2} & W_{t2} \\ W_{t2} & W_{t2} \end{pmatrix} \right\}, \tag{7.5}$$

where W_{t1} and W_{t2} are scalar variances.

This definition is the special case of Definition 7.2 in which $n = 2$.[†] The standard equations for (7.5) are simply those in (7.3a, b and c). The evolution errors can now be expressed as

$$\boldsymbol{\omega}_t = \mathbf{L}_2 \partial\boldsymbol{\theta}_t,$$

where

$$\partial\boldsymbol{\theta}_t = \begin{pmatrix} \partial\mu_t \\ \partial\beta_t \end{pmatrix} \sim \mathrm{N}\left[\begin{pmatrix} 0 \\ 0 \end{pmatrix}, \begin{pmatrix} W_{t1} & 0 \\ 0 & W_{t2} \end{pmatrix}\right].$$

It follows that $\omega_{t1} = \partial\mu_t + \partial\beta_t$ and $\omega_{t2} = \partial\beta_t$. So the system equation $\boldsymbol{\theta}_t = \mathbf{L}_2\boldsymbol{\theta}_{t-1} + \boldsymbol{\omega}_t$ can be written as

$$\boldsymbol{\theta}_t = \mathbf{L}_2(\boldsymbol{\theta}_{t-1} + \partial\boldsymbol{\theta}_t).$$

In this form, β_t has the interpretation of *incremental growth* in the level of the series over the time interval from $t-1$ to t, evolving during that interval according to the addition of the stochastic element $\partial\beta_t$. The level μ_t at time t evolves systematically via the addition of the growth β_t and undergoes a further stochastic shift via the addition of $\partial\mu_t$. In terms of model equations, this implies the more familiar versions (e.g., Harrison and Stevens 1976)

$$\begin{aligned} &\text{Observation:} \quad && Y_t = \mu_t + \nu_t, \\ &\text{System:} && \mu_t = \mu_{t-1} + \beta_t + \partial\mu_t, \\ & && \beta_t = \beta_{t-1} + \partial\beta_t, \end{aligned}$$

with the zero-mean evolution errors $\partial\mu_t$ and $\partial\beta_t$ being uncorrelated.

7.3.2 Constant linear growth models

Suppose that the variances are constants with $V_t = V$, $W_{t1} = W_1$ and $W_{t2} = W_2$. Then using the results of Theorem 7.1, the limiting values for the linear growth model simplify as follows.

Theorem 7.2. *For the constant linear growth model, limiting variance values are*

$$\mathbf{C} = \begin{pmatrix} A_1 & A_2 \\ A_2 & A_2(A_1 - A_2)/(1 - A_1) \end{pmatrix} V \quad \text{and} \quad \mathbf{R} = \begin{pmatrix} A_1 & A_2 \\ A_2 & A_1 A_2 \end{pmatrix} Q.$$

The feasible region for the adaptive vector $\mathbf{A}$ *is given by*

$$0 < A_1 < 1 \quad \text{and} \quad 0 < A_2 < A_1^2/(2 - A_1) < A_1.$$

[†]Note that $\mathbf{W}$ in (7.5) is the simplest and preferred form for linear growth models although others are possible. For example, an equivalent model exists for which $\mathbf{W}_t = \mathrm{diag}(W_{t1}, W_{t2})$ although this is not pursued.

Proof. Limits **R**, **C**, and the bounds on A_1 follow directly from Theorem 7.1 by substituting the particular form of the **W** matrix. Specifically, replace W_1 with $W_1 + W_2$ and W_3 with W_2. To determine the bounds on A_2, note that since $R_2 = A_1 A_2 Q > 0$, then $A_2 > 0$. Also, it follows from Theorem 7.1 (3) that $A_1^2 - A_2(2 - A_1) > 0$, or $A_2 < A_1^2/(2 - A_1) < A_1$, as required.

◇

Figure 7.1a provides a graph of the boundary line $A_2 = A_1^2/(2 - A_1)$. The region below this dashed line and such that $0 < A_1 < 1$ is the feasible region for the adaptive coefficients given above.

The solid line in the figure, defined by $A_2 = 4 - 2A_1 - 4(1 - A_1)^{1/2}$, is the corresponding boundary for the general second-order polynomial model of Theorem 7.1. Clearly, the regions differ appreciably only for larger values of A_1. Such values are extremely unusual in practice; remember that these are limiting rates of adaptation in a closed model. For example, in stock control and monitoring applications, A_1 will rarely exceed 0.25, implying $A_2 < 0.036$. These values are widely applicable upper bounds on the adaptive coefficients, and so for most practical purposes, the linear growth model is an adequate subclass of all second-order models. Practitioners gain little by entertaining models outside this class. Figure 7.1b is a close-up of the graph over the range $0 < A_1 < 0.3$ when essentially the feasible regions coincide. Also graphed in Figures 7.1a and 7.1b are the possible values of adaptive coefficients in the single discount model of Definition 7.3. In this model, $\mathbf{W}_t$ is defined via a single quantity, and the limiting adaptive coefficients are related via $A_2 = 2 - A_1 - 2(1 - A_1)^{1/2}$. Finally, the double discount approach of Section 7.2.5 leads to the ranges $0 < A_1 < 1$ and $0 < A_2 < A_1[1 - (1 - A_1)^{1/2}]$, with the boundary line $A_2 = A_1[1 - (1 - A_1)^{1/2}]$ also appearing in Figures 7.1a and 7.1b. The ranges of limiting adaptive coefficients in this double discount model differ little from those in the linear growth class.

7.3.3 Limiting predictors in the constant model

Following Section 7.2.2, the limiting updating form of the closed, constant linear growth model is

$$m_t = m_{t-1} + b_{t-1} + A_1 e_t,$$
$$b_t = b_{t-1} + A_2 e_t,$$

with a limiting second difference representation

$$Y_t - 2Y_{t-1} + Y_{t-2} = e_t - \psi_1 e_{t-1} + \psi_2 e_{t-2}, \tag{7.6}$$

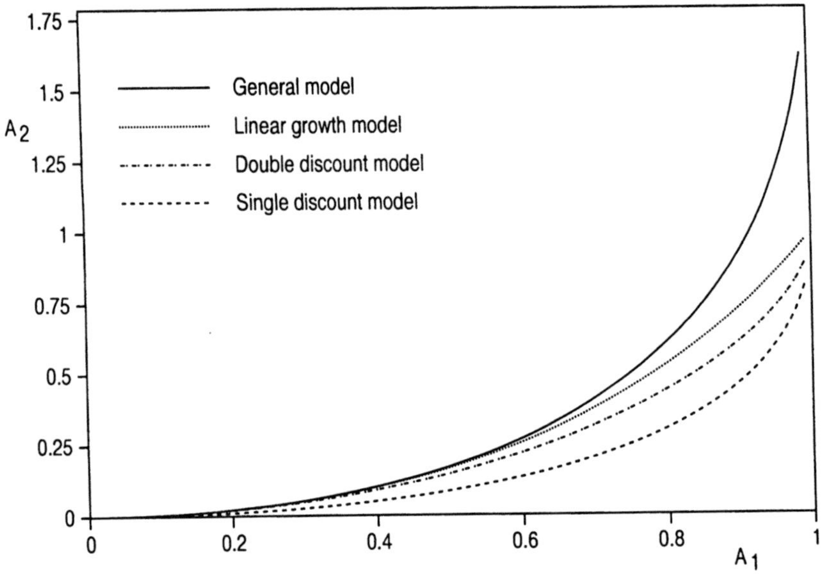

Figure 7.1a Ranges of adaptive coefficients for $0 < A_1 < 1$

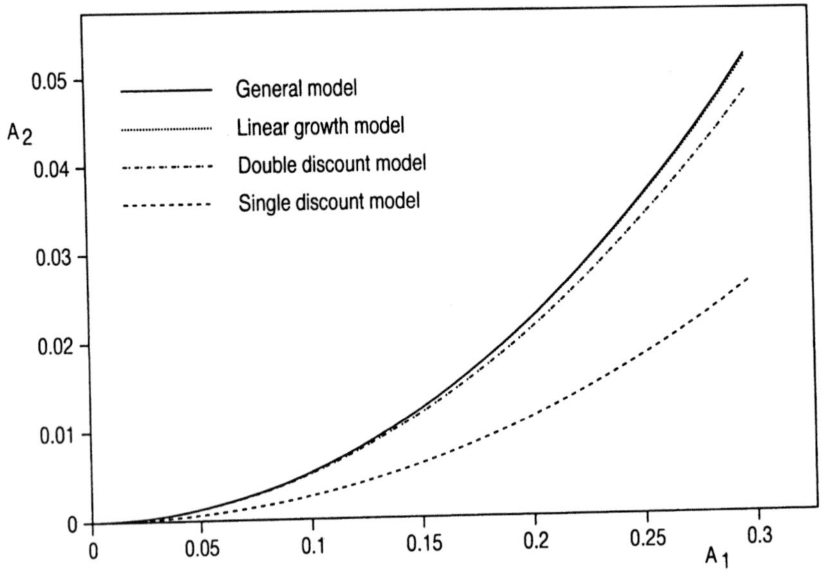

Figure 7.1b Ranges of adaptive coefficients for $0 < A_1 < 0.3$

where

$$\psi_1 = 2 - A_1 - A_2 \quad \text{and} \quad \psi_2 = 1 - A_1.$$

This can be written in terms of the backshift operator B as

$$(1-B)^2 Y_t = (1 - \psi_1 B + \psi_2 B^2) e_t.$$

A number of popular point prediction methods employ equations of the form (7.6), and the main ones are now discussed.

(1) **Holt's linear growth method** (Holt 1957).
For $k \geq 1$, the linear forecast function $f_t(k) = m_t + kb_t$ produces point forecasts that are sequentially updated according to

$$m_t = AY_t + (1-A)(m_{t-1} + b_{t-1}),$$
$$b_t = D(m_t - m_{t-1}) + (1-D)b_{t-1},$$

where $0 < A, D < 1$. Writing $e_t = Y_t - (m_{t-1} + b_{t-1})$ and rearranging,

$$m_t = m_{t-1} + b_{t-1} + Ae_t,$$
$$b_t = b_{t-1} + ADe_t.$$

So, if $A_1 = A$, $A_2 = AD$, and $D < A/(2-A)$, Holt's forecast function is just the *limiting* forecast function of the constant linear growth DLM.

(2) **Box and Jenkins' ARIMA(0,2,2) predictor.**
This Box and Jenkins (1976) predictor is based upon the model

$$Y_t - 2Y_{t-1} + Y_{t-2} = e_t - \psi_1 e_{t-1} + \psi_2 e_{t-2},$$

with uncorrelated errors $e_t \sim \mathrm{N}[0, Q]$.
The forecast function for this specific ARIMA model is defined by

$$f_t(1) = 2Y_t - Y_{t-1} - \psi_1 e_t + \psi_2 e_{t-1},$$
$$f_t(2) = 2f_t(1) - Y_t + \psi_2 e_t,$$
$$f_t(k) = 2f_t(k-1) - f_t(k-2), \qquad k > 2,$$

so

$$f_t(k) = f_t(1) + (k-1)[f_t(2) - f_t(1)], \qquad k \geq 1.$$

If $0 < \psi_2 < 1$ and $\psi_1 < 2\psi_2$, the forecast function is again the *limiting* forecast function of the constant linear growth DLM.

(3) **Exponentially weighted regression (EWR)** (Brown 1962).
Brown's EWR linear growth forecast function is $f_t(k) = m_t + kb_t$

for $k \geq 1$, where at time t given a discount factor $0 < \delta < 1$ and an infinite history of observations $Y_t, Y_{t-1}, \ldots$, the pair m_t and b_t are the values of μ and β, respectively, that minimise the *discounted* sum of squares

$$S_t(\mu, \beta) = \sum_{r=0}^{\infty} \delta^r (Y_{t-r} - \mu + r\beta)^2.$$

The appropriate values may be related via the recurrence equations

$$m_t = m_{t-1} + b_{t-1} + (1 - \delta^2)e_t,$$
$$b_t = b_{t-1} + (1 - \delta)^2 e_t,$$

where

$$e_t = Y_t - m_{t-1} - b_{t-1}.$$

So the EWR forecast function is the *limiting* forecast function of the single discount DLM of Section 7.2.4 and thus of any constant linear growth DLM for which $A_1 = 1 - \delta^2$ and $A_2 = (1 - \delta)^2$. In particular, the constant linear growth DLM, in which

$$W_1 = 2V(1 - \delta)^2/\delta \quad \text{and} \quad W_2 = V(1 - \delta)^4/\delta^2,$$

leads to such A values, with

$$Q = \frac{V}{\delta^2} \quad \text{and} \quad \mathbf{C} = \begin{pmatrix} 1 - \delta^2 & (1 - \delta)^2 \\ (1 - \delta)^2 & 2(1 - \delta)^3/\delta \end{pmatrix} V.$$

Quantitative relationships are given in the following table.

δ	A_1	A_2	V/W_1	W_1/W_2	Q/V
0.95	0.10	0.003	200	800	1.11
0.90	0.19	0.010	45	180	1.23
0.85	0.28	0.023	19	76	1.38
0.80	0.36	0.040	10	40	1.56
0.70	0.51	0.090	4	16	2.04
0.50	0.75	0.250	1	4	4.00

7.3.4 Discussion

Some further discussion of the above limiting predictors follows, although it is rather technical and may be omitted on first reading. However, for practitioners familiar with non-Bayesian methods, this section provides further insight into the nature of Bayesian models, sharply identifying points of difference.

The limiting updating equations for the constant DLM provide each of the limiting predictors, but with particular restrictions on the values of the limiting adaptive coefficients. Brown's values $A_1 = 1-\delta^2$ and $A_2 = (1-\delta)^2$ are constrained within the feasible region for the linear growth DLM (see Figures 7.1). However, Holt's region is defined by $0 < A_1, A_2 < 1$, and that of the ARIMA model by $0 < A_2 < 2$, $0 < A_1 + A_2 < 4$. These both allow values outside the feasible region determined in Theorem 7.1. Now it might be suggested that these regions should be contained within that for second-order polynomial models, since the latter class of models contains all those having a linear forecast function. The fact that this is not the case derives from a rather subtle and hidden point. The reason is that the predictors using $\mathbf{A}$ outside the feasible region from the DLM are unknowingly superimposing a moving average process, of order not more than 2, on the linear model. This leads to polynomial/moving average models that, in DLM terms, are obtained from the superposition of the polynomial model, as one component, with another component having one *zero* eigenvalue. The set of canonical models has the form

$$\left\{ \begin{pmatrix} \mathbf{E}_2 \\ 1 \end{pmatrix}, \begin{pmatrix} \mathbf{J}_2(1) & \mathbf{0} \\ \mathbf{0} & 0 \end{pmatrix}, V, \begin{pmatrix} \mathbf{W}_1 & \mathbf{W}_3 \\ \mathbf{W}_3' & W_2 \end{pmatrix}, \right\},$$

that will produce feasible regions $0 < A_1 < 2$, $0 < A_1 + A_2 < 4$ depending on the structure assigned to $\mathbf{W}$. In particular, if $\mathbf{W}_3 \neq \mathbf{0}$ then the value of $\mathbf{A}$ can lie outside the pure polynomial regions of Theorems 7.1 and 7.2. Then the forecast function is the sum of the forecast function of a second-order polynomial and the forecast function of the zero eigenvalue component, i.e., $f_t(0)$ is an arbitrary value, and $f_t(k) = m_t + kb_t$ for $k \geq 1$. This ties in with the foregoing popular point forecasting methods, that only define their linear forecast functions for $k \geq 1$. Without loss in generality, V may be set to zero, since it can be absorbed by W_2. Then $\mathbf{W}_3 \neq \mathbf{0}$ indicates that the observation and system noises are not independent.

Extending the zero eigenvalue block to the DLM

$$\left\{ \begin{pmatrix} \mathbf{E}_2 \\ \mathbf{E}_2 \end{pmatrix}, \begin{pmatrix} \mathbf{J}_2(1) & \mathbf{0} \\ \mathbf{0} & \mathbf{J}_2(0) \end{pmatrix}, V, \mathbf{W} \right\}$$

produces a forecast function that has arbitrary values for $f_t(0)$ and $f_t(1)$ and then becomes the linear forecast function $f_t(k) = m_t + kb_t$ for $k \geq 2$.

This subtlety applies to all DLMs. Referring, for example, to the first-order polynomial model of Chapter 2, the constant DLM has a limiting representation of the form

$$Y_t - Y_{t-1} = e_t - (1 - A)e_{t-1},$$

with $0 < A < 1$, whereas the ARIMA (0,1,1) predictor, having essentially the same form, allows $0 < A < 2$. Of course a value greater than 1 seems strange. For example, if $A = 1.9$, the representation is

$$Y_t - e_t = Y_{t-1} + 0.9e_{t-1},$$

and in informing about Y_t, Y_{t-1} is not being smoothed, but just the reverse. However, the region $0 < A < 2$ is precisely that valid for the constant DLM

$$\left\{\begin{pmatrix}1\\1\end{pmatrix}, \begin{pmatrix}1 & 0\\0 & 0\end{pmatrix}, 0, \begin{pmatrix}W_1 & W_3\\W_3 & V\end{pmatrix}\right\},$$

in which $\mathbf{A}$ may exceed unity if and only if $W_3 + V < 0$; this is evidently irrelevant from an applied perspective.

Generally, consider any TSDLM $\{\mathbf{F}, \mathbf{G}, V, \mathbf{W}\}$, with $\mathbf{G}$ of full rank n. Theorems 5.5 and 5.6 show that the updating equations have stable, limiting forms such that

$$\lim_{t\to\infty}\left\{\prod_{r=1}^{n}(1-\lambda_r B)Y_t - \prod_{r=1}^{n}(1-\rho_r B)e_t\right\} = 0,$$

where B is the backshift operator, $\lambda_1, \ldots, \lambda_n$ are the n eigenvalues of $\mathbf{G}$, and $\rho_1, \ldots, \rho_n$ are simply linear functions of the elements of the limiting adaptive vector $\mathbf{A}$. The feasible region for these ρ_r coefficients is a subset of the region

$$\{\rho_1, \ldots, \rho_n : |\rho_r| < 1, (r = 1, \ldots, n)\}.$$

However, this region can be enlarged by superimposing a zero eigenvalue component of dimension 1 on the model, leading to the $(n+1)$-dimensional DLM

$$\left\{\begin{pmatrix}\mathbf{F}\\1\end{pmatrix}, \begin{pmatrix}\mathbf{G} & \mathbf{0}\\\mathbf{0} & 0\end{pmatrix}, V, \mathbf{W}^*\right\}.$$

The forecast function may then have an arbitrary value for $k = 0$, but thereafter it follows the forecast function form of the sub-model $\{\mathbf{F}, \mathbf{G}, \cdot, \cdot\}$. The above limiting representation of the observation series holds for this extended model, and the ρ_r coefficients may take any values in the above region, depending on the structure of the evolution variance matrix $\mathbf{W}^*$.

More generally, by superimposing a zero eigenvalue component of dimension h on the model, leading to the $(n+h)$-dimensional DLM

$$\left\{\begin{pmatrix}\mathbf{F}\\\mathbf{E}_h\end{pmatrix}, \begin{pmatrix}\mathbf{G} & \mathbf{0}\\\mathbf{0} & \mathbf{J}_h(0)\end{pmatrix}, V, \mathbf{W}^*\right\},$$

the forecast function may have arbitrary values for $k \leq h$ before following the sub-model form. However, the above representation changes to include $h-1$ extra error terms, so that

$$\lim_{t\to\infty}\left\{\prod_{r=1}^{n}(1-\lambda_r B)Y_t - \prod_{r=1}^{n+h-1}(1-\rho_r B)e_t\right\} = 0.$$

Such model extensions are unnecessary in practice. However, here they serve to identify peculiarities of the classical techniques, warning practitioners to restrict the range of allowable values for adaptive coefficients well within the region previously suggested by advocates of such methods. For polynomial prediction, the zero eigenvalue component is superfluous, and its addition violates parsimony. Generally zero eigenvalue blocks are to be avoided in all applications unless they can be readily interpreted. Despite many years of experience in the chemical industry and in consultancy, neither of the authors has ever had cause to employ them.

7.4 THIRD-ORDER POLYNOMIAL MODELS

7.4.1 Introduction

By definition, a third-order polynomial DLM is any observable TSDLM that at time t, has a forecast function of the quadratic form

$$f_t(k) = a_{t0} + a_{t1}k + a_{t2}k^2, \qquad k \geq 0.$$

An alternative representation is given by

$$f_t(k) = f_t(0) + [f_t(1) - f_t(0)]k + [f_t(2) - 2f_t(1) + f_t(0)]k(k-1)/2,$$

where the coefficients of the quadratic forecast function are identified in terms of their first three values. These models are only occasionally used in practice, since for many short-term forecasting and micro-forecasting applications, local trends are adequately described using first- or second-order polynomials. However, when dealing with macro or aggregate data and when forecasting for longer lead times, the random variation measured by observational variances V_t is often small relative to movement in trend. In such circumstances, third-order polynomial descriptions may be needed. An example of such a case is given in the application to longer-term growth forecasting in Harrison, Gazard and Leonard (1977). That particular application concerned the preparation of forecasts of world mill consumption of fibres split into forty-eight categories according to fibre type and geographical region. These forecasts, that covered ten future years, were required and used for GATT negotiations by the well-known trouble-shooter Sir Harvey Jones, then at ICI Ltd. The main problem lay not so much in forecasting the total consumption as in predicting proportionate consumptions. Consequently, non-linear Gompertz-type functions, defined by three parameters, were locally approximated by a third-order Taylor expansion. A discount quadratic polynomial DLM was used for updating the corresponding three parameters. This form was then converted back and the full non-linear form used for the forecast function. The approach also employed a "top-down" hierarchical structure and used constrained and combined forecasting, as discussed later in Chapter 16. The success of this application led to the development of a computer programme used by the company for similar applications.

The canonical third-order polynomial DLM is $\{\mathbf{E}_3, \mathbf{J}_3(1), \cdot\,, \cdot\,\}$ but the similar DLM $\{\mathbf{E}_3, \mathbf{L}_3, \cdot\,, \cdot\,\}$ will be adopted in this chapter. Further, as with second-order models, a subclass of third-order models, namely those termed **quadratic growth** models, are essentially sufficient in terms of the possible values of limiting adaptive coefficients. Thus, attention is restricted to a brief description of the theoretical structure of models in this subclass. The evolution variance sequence in quadratic growth models has a particular, structured form, derived as the case $n = 3$ in Definition 7.2. In practice, the more parsimonious models based on single discount factors will typically be used without real loss of flexibility. Although apparently rather different in nature, these discount models are intimately related to the canonical quadratic growth models, and in constant DLMs, the rapidly approached limiting behaviour of the two are equivalent. This is directly analogous to the situation with first- or second-order models, and the analogy carries over to polynomials of higher order.

7.4.2 Quadratic growth models

Definition 7.5. A **quadratic growth DLM** is any model equivalent to a DLM of the form

$$\{\mathbf{E}_3, \mathbf{L}_3, V_t, \mathbf{W}_t\},$$

where for all t, $\mathbf{W}_t$ has the form

$$\mathbf{W}_t = \mathbf{L}_3 \text{diag}(W_{t1}, W_{t2}, W_{t3})\mathbf{L}_3'.$$

The model class is obtained as the defining variances $\mathbf{W}_t$ and V_t vary.

Writing $\nu_t \sim \mathrm{N}[0, V_t]$,

$$\boldsymbol{\theta}_t = \begin{pmatrix} \mu_t \\ \beta_t \\ \gamma_t \end{pmatrix} \quad \text{and} \quad \partial\boldsymbol{\theta}_t = \begin{pmatrix} \partial\mu_t \\ \partial\beta_t \\ \partial\gamma_t \end{pmatrix} \sim \mathrm{N}\left[\mathbf{0}, \begin{pmatrix} W_{t1} & 0 & 0 \\ 0 & W_{t2} & 0 \\ 0 & 0 & W_{t3} \end{pmatrix}\right],$$

a quadratic growth DLM can be written as

$$\begin{aligned} Y_t &= \mu_t + \nu_t, \\ \mu_t &= \mu_{t-1} + \beta_t + \partial\mu_t, \\ \beta_t &= \beta_{t-1} + \gamma_t + \partial\beta_t, \\ \gamma_t &= \gamma_{t-1} + \partial\gamma_t. \end{aligned}$$

The system equations may be compactly written $\boldsymbol{\theta}_t = \mathbf{L}_3(\boldsymbol{\theta}_{t-1} + \partial\boldsymbol{\theta}_t)$. The quantities μ_t, β_t and γ_t respectively represent level, growth and change in growth at time t. In a continuous-time analogue, β_t would represent the first derivative with respect to time, or gradient, of the expected level of the series at time t, and γ_t the second derivative of the expected level. The components of the evolution error $\partial\boldsymbol{\theta}_t$ represent the corresponding stochastic changes in the state vector components. Given the posterior mean $\mathrm{E}[\boldsymbol{\theta}_t | D_t] = \mathbf{m}_t = (m_t, b_t, g_t)'$, the forecast function is

$$f_t(k) = m_t + kb_t + k(k+1)g_t/2, \qquad (k \geq 0).$$

Finally, the updating equations lead to a limiting representation of the observation series, so that with adaptive vector $\mathbf{A}_t = (A_{t1}, A_{t2}, A_{t3})'$,

$$(1-B^3)Y_t = Y_t - 3Y_{t-1} + 3Y_{t-2} - Y_{t-3} = e_t - \psi_{t1}e_{t-1} + \psi_{t2}e_{t-2} - \psi_{t3}e_{t-3},$$

where the ψ_{ti} coefficients are given by

$$\begin{aligned} \psi_{t1} &= 3 - A_{t-1,1} - A_{t-1,2}, \\ \psi_{t2} &= 3 - 2A_{t-2,1} - A_{t-2,2} + A_{t-2,3}, \\ \psi_{t3} &= 1 - A_{t-3,1}. \end{aligned}$$

7.4.3 Constant quadratic growth model

Theorem 7.3. *In the constant quadratic growth model*

$$\{\mathbf{E}_3, \mathbf{L}_3, 1, \mathbf{L}_3\mathbf{W}\mathbf{L}_3'\},$$

with $\mathbf{W} = \mathrm{diag}(W_1, W_2, W_3)$, *the components* $\mathbf{A}_t$, $\mathbf{C}_t$, $\mathbf{R}_t$ *and* Q_t *have limiting values* $\mathbf{A}$, $\mathbf{C}$, $\mathbf{R}$ *and* Q *defined by the equations*

$$\begin{aligned} 1 &= (1-A_1)Q, \\ W_3 &= A_3^2 Q, \\ W_2 &= (A_1^2 - 2A_1A_3 - A_2A_3)Q, \\ W_1 &= (A_1^2 + A_1A_2 - 2A_2 + A_3)Q, \end{aligned}$$

and

$$\mathbf{R} = \begin{pmatrix} A_1 & A_2 & A_3 \\ A_2 & A_1A_2 - (1-2A_1-A_2)A_3 & (A_1+A_2)A_3 \\ A_3 & (A_1+A_2)A_3 & A_2A_3 \end{pmatrix} Q,$$

with

$$\mathbf{C} = \mathbf{R} - \mathbf{A}\mathbf{A}'Q.$$

In addition, the feasible region for $\mathbf{A}$ *is defined by*

$$\begin{aligned} \{\mathbf{A}: \quad 0 < A_3, A_1 < 1,\ 0 < A_2 < 2,\ 0 < A_2^2 - A_3(2A_1 + A_2), \\ 0 < A_1^2 + A_1A_2 - 2A_2 + A_3\}. \end{aligned}$$

Proof. The limits exist by Theorem 5.5. The proof of the limiting relationships follows that of Theorem 7.1. The technique is general and may be used to derive analogous limiting relationships for higher-order polynomial DLMs. The proof is only sketched, details being left to the reader.

From the defining equations, $\mathbf{R} = \mathbf{L}_3(\mathbf{C} + \mathbf{W})\mathbf{L}_3'$ and $\mathbf{C} = \mathbf{R} - \mathbf{A}\mathbf{A}'Q$, it follows that

$$\mathbf{L}_3^{-1}\mathbf{R}(\mathbf{L}_3')^{-1} = \mathbf{R} - \mathbf{A}\mathbf{A}'Q + \mathbf{W}.$$

The required representation of the components as functions of $\mathbf{A}$ and Q can now be deduced by matching elements in the above matrix identity, noting the special diagonal form of $\mathbf{W}$ and the identity

$$\mathbf{L}_3^{-1} = \begin{pmatrix} 1 & -1 & 1 \\ 0 & 1 & -1 \\ 0 & 0 & 1 \end{pmatrix}.$$

Finally, the feasible region for limiting adaptive coefficients can be deduced using the facts that $0 < W_1, W_2, W_3 < Q$, $1 < Q$ and the positive definiteness of $\mathbf{R}$.

◇

Note the following:

(1) Defining $\psi_1 = 3 - A_1 - A_2$, $\psi_2 = 3 - 2A_1 - A_2 + A_3$ and $\psi_3 = 1 - A_1$, the limiting representation of the observation series is

$$(1-B)^3 Y_t = (1 - \psi_1 B + \psi_2 B^2 - \psi_3 B^3)e_t = \prod_{v=1}^{3}(1 - \phi_v B)e_t.$$

The traditional ARIMA(0,3,3) predictor takes this form, subject to $|\phi_v| < 1$ for $v = 1, 2, 3$. The DLM limiting predictors thus correspond to a subset of the ARIMA predictors. As in Section 7.3.4, the full set of ARIMA predictors is obtained by extending the DLM to have a further zero eigenvalue component.

(2) Exponentially weighted regression (or discounted likelihood). Recurrences for the values m_t, b_t and g_t minimising the discounted sum of squares

$$S_t(\mu, \beta, \gamma) = \sum_{r=0}^{\infty} \delta^r [Y_{t-r} - \mu + \beta r - \gamma r(r+1)/2]^2,$$

with respect to μ, β and γ, correspond to the DLM updating equations with $A_{t1} = A_1 = 1 - \delta^3$, $A_{t2} = A_2 = 2 - 3\delta + \delta^3$ and $A_{t3} = A_3 = (1-\delta)^3$. Then the observation series has the form

$$(1-B)^3 Y_t = (1 - \delta B)^3 e_t.$$

This is also the limiting form given in a discount DLM with the evolution variance sequence structured using a single discount factor δ applied to the whole quadratic trend component.

(3) The above results generalise. In the constant n^{th}-order polynomial DLM $\{\mathbf{E}_n, \mathbf{L}_n, V, \mathbf{W}\}$, with $\mathbf{W} = \mathbf{L}_n \text{diag}(W_1, \ldots, W_n)\mathbf{L}_n'$, suppose

that $W_r = \binom{n}{r}c^r V$, $(r = 1, \ldots, n)$, where $c = (1-\delta)^2/\delta$ for some discount factor δ. Godolphin and Harrison (1975) show that the limiting representation of the observation series (a special case of equation 5.12) is

$$(1-B)^n Y_t = (1-\delta B)^n e_t.$$

This relationship obtains for EWR and in the limit, for the single discount n^{th}-order polynomial DLM, for which

$$\mathbf{R}_t = \mathbf{L}_n \mathbf{C}_t \mathbf{L}_n'/\delta.$$

The limiting value of the adaptive vector $\mathbf{A} = (A_1, \ldots, A_n)'$ is then given by

$$A_1 = 1 - \delta^n,$$

$$A_{r+1} = \binom{n}{r}(1-\delta)^r - A_r, \quad (r = 2, \ldots, n-1),$$

$$A_n = (1-\delta)^n,$$

with

$$Q = \frac{V}{\delta^n}.$$

7.5 EXERCISES

The following questions concern polynomial models with observational variances assumed known unless otherwise stated.

(1) Verify that the polynomial DLM $\{\mathbf{E}_n, \mathbf{L}_n, \cdot\,, \cdot\,\}$ is observable and similar to the canonical model $\{\mathbf{E}_n, \mathbf{J}_n(1), \cdot\,, \cdot\,\}$.

(2) The forecast function of the DLM $\{\mathbf{E}_n, \mathbf{L}_n, \cdot\,, \cdot\,\}$ is

$$f_t(k) = \mathbf{E}_n' \mathbf{L}_n^k \mathbf{m}_t, \quad k \geq 0.$$

Show directly that this is a polynomial of order n (degree $n-1$).

(3) Consider DLMs M and M_1 characterised by quadruples

$$M: \quad \{\mathbf{E}_n, \mathbf{J}_n(1), V_t, \mathbf{W}_t\},$$

$$M_1: \quad \{\mathbf{E}_n, \mathbf{L}_n, V_t, \mathbf{W}_{1t}\}.$$

(a) Calculate the respective observability matrices $\mathbf{T}$ and $\mathbf{T}_1$.

(b) Show that the similarity matrix $\mathbf{H} = \mathbf{T}^{-1}\mathbf{T}_1$ is given by

$$\mathbf{H} = \begin{pmatrix} 1 & 0 & 0 & 0 & \ldots & 0 & 0 \\ 0 & 1 & 1 & 0 & \ldots & 0 & 0 \\ 0 & 0 & 1 & 1 & \ldots & 0 & 0 \\ \vdots & \vdots & & & \ddots & \vdots & \vdots \\ 0 & 0 & 0 & 0 & \ldots & 1 & 1 \\ 0 & 0 & 0 & 0 & \ldots & 0 & 1 \end{pmatrix}.$$

(c) Interpret the meaning of the state parameters in each model.

(4) You apply the nth-order polynomial DLM $\{\mathbf{E}_n, \mathbf{L}_n, V, V\mathbf{W}\}$ to forecast a time series Y_t, the true generating process being M.

(a) Apply Theorem 5.6 to prove

$$\lim_{t\to\infty}\left\{(1-B)^n Y_t - e_t - \sum_{i=v}^{n}\beta_v e_{t-v}\right\} = 0,$$

for some real quantities β_v, $(v = 1, \ldots, n)$.

(b) Does the result of (a) mean that the observations are generated by an ARIMA$(0, n, n)$ process?

(c) Does the distribution of $(e_t|D_{t-1}, M)$ converge in probability to some specific form?

(d) Are the one-step errors e_t truly uncorrelated in the limit?

(e) If you answered "Yes" to (b), (c), or (d), you can take comfort from knowing that others suffer the same delusion! Adopting the DLM, the above relationship is true no matter what the real data-generating process. The distribution of e_t is dependent on the true model M, and each can have a completely different distribution from any other while their joint distributions are free to display marked dependencies.

(5) Suppose that the observation series Y_t really is generated by the nth-order polynomial normal DLM $M = \{\mathbf{E}_n, \mathbf{L}_n, V, \mathbf{W}\}$.

(a) Show directly that if $V > 0$, then for any $t > n$, the series can be represented by

$$(1-B)^n Y_t = \prod_{v=1}^{n}(1-\rho_v B)a_t,$$

where $a_t \sim \mathrm{N}[0, Q]$, independently, and for some real quantities $|\rho_v| < 1$, $(v = 1, \ldots, n)$.

(b) Writing $Z_t = (1-B)^n Y_{t+n}$, show that the joint distribution of the $\{Z_t\}$ series is such that $\mathrm{E}[Z_t] = 0$ and $\mathrm{C}[Z_{t+k}, Z_t] = 0$ for all $k > n$.

(c) Given data values $Y_1, \ldots Y_{t+n}$, how might you use them to support or refute the hypothesis that they are suitably modelled by the above model M?

(6) Suppose that the observation series Y_t really is generated by the nth-order polynomial normal DLM $M^* = \{\mathbf{E}_n, \mathbf{L}_n, V_t, \mathbf{W}_t\}$. Define $Z_t = (1-B)^n Y_{t+n}$.

(a) Prove that $\mathrm{E}[Z_t] = 0$ and $\mathrm{C}[Z_{t+k}, Z_t] = 0$ for all $k > n$.

(b) Given data values $Y_1, \ldots Y_{t+n}$, how might you use them to support or refute the hypothesis that they are suitably modelled by the above model M^*?

(c) How do you sort out whether the variances are constant, as in the previous question's model M, or whether they are time-dependent as in model M^*?

(7) Write a computer program to generate 102 observations according to the constant second-order polynomial DLM

$$\left\{ \begin{pmatrix} 1 \\ 0 \end{pmatrix}, \begin{pmatrix} 1 & 1 \\ 0 & 1 \end{pmatrix}, 100, 100 \begin{pmatrix} 1-\delta^2 & (1-\delta)^2 \\ (1-\delta)^2 & (1-\delta)^3 \end{pmatrix} \right\},$$

with $\mu_0 = 100$ and $\beta_0 = 0$. Classical model identification advocates looking at the autocorrelation structure of the derived series $X_t = Y_{t+1} - Y_t$ and of that of $Z_t = X_{t+1} - X_t = Y_{t+2} - 2Y_{t+1} + Y_t$. On the basis of no significant effect the autocorrelations are approximately $\mathrm{N}[0, 1/n]$, where n is the length of the series. Roughly, if $\mathrm{C}[X_t, X_{t+2}]$ is not significantly different from zero, then the data do not refute the first-order polynomial DLM. If it is significantly different and $\mathrm{C}[Z_t, Z_{t+3}]$ is not significantly different from zero, then the second-order polynomial DLM is not refuted. Produce series using discount factors $\delta = 0.95$, 0.9, and 0.8.

(a) Look at these series and draw an impression as to whether a first- or second-order model is suggested.

(b) Now examine the autocorrelation graphs to see what you conclude.

(c) Given the DLM, and for ease $\delta = 1$, what are the variances of X_t and Z_t? Draw your conclusions about the differencing approach to identification.

(8) Suppose that a series is generated according to the mechanism $Y_{t+1} - Y_t = a_t - 2a_{t-1}$ with $a_t \sim \mathrm{N}[0, 1]$ independently.

(a) If at $t = 0$, you precisely know a_1, show that your one-step prediction errors $e_t = Y_t - f_t$ satisfy $Y_{t+1} - Y_t = e_t - 2e_{t-1}$, where $e_t \sim \mathrm{N}[0, 1]$ independently.

(b) Suppose that you do not know a_1 precisely, but at time $t = 0$, it has distribution $(a_1 | D_0) \sim \mathrm{N}[0, Q_1]$, where the only restriction is $Q_1 > 0$. Prove that now your limiting one step ahead errors satisfy $Y_{t+1} - Y_t = e_t - 0.5e_{t-1}$, where $e_t \sim \mathrm{N}[0, 4]$ independently.

(c) Meditate on (a) and (b) particularly with respect to moving average processes and their supposedly unique representation.

(9) Consider the single discount second-order polynomial DLM of Definition 7.3, namely $\{\mathbf{E}_2, \mathbf{L}_2, V, \mathbf{W}_t\}$, where

$$\mathbf{W}_t = \frac{1-\delta}{\delta}\mathbf{L}_2\mathbf{C}_{t-1}\mathbf{L}', \quad \text{so that} \quad \mathbf{R}_t = \frac{1}{\delta}\mathbf{L}_2\mathbf{C}_{t-1}\mathbf{L}_2'.$$

(a) Use the identity $\mathbf{C}_t^{-1} = \mathbf{R}_t^{-1} + \mathbf{E}_2\mathbf{E}_2'V^{-1}$ to directly verify that

$$\lim_{t\to\infty} \mathbf{C}_t = \mathbf{C} = \begin{pmatrix} 1-\delta^2 & (1-\delta)^2 \\ (1-\delta)^2 & (1-\delta)^3/\delta \end{pmatrix} V,$$

$$\lim_{t\to\infty} Q_t = Q = \frac{V}{\delta^2}, \qquad \lim_{t\to\infty} \mathbf{A}_t = \mathbf{A} = \begin{pmatrix} 1-\delta^2 \\ (1-\delta)^2 \end{pmatrix},$$

and

$$\lim_{t\to\infty} \mathbf{R}_t = \mathbf{R} = \begin{pmatrix} A_1 & A_2 \\ A_2 & (1-\delta)A_2 \end{pmatrix} Q.$$

(b) Deduce that in this (rapidly approached) limiting form, the updating equations are just those of a constant DLM in which

$$\mathbf{W} = \begin{pmatrix} A_1 & A_2 \\ A_2 & W \end{pmatrix} \frac{(1-\delta)V}{\delta^3}$$

for *any* variance W (compare results in Section 5.3.4).

(10) In the constant linear growth DLM of Section 7.3.2

$$\left\{ \begin{pmatrix} 1 \\ 0 \end{pmatrix}, \begin{pmatrix} 1 & 1 \\ 0 & 1 \end{pmatrix}, V, \begin{pmatrix} W_\mu + W_\beta & W_\beta \\ W_\beta & W_\beta \end{pmatrix} \right\},$$

with known variances, calculate, for any $k \geq 1$,

(a) the k-step forecast distribution $p(Y_{t+k}|D_t)$;

(b) the k-step lead-time forecast distribution $p(X_{t,k}|D_t)$, where we define $X_{t,k} = \sum_{r=1}^{k} Y_{t+r}$.

(11) Using the limiting form of the updating equations in the constant linear growth model of the previous question, verify that the limiting forecast function can be written as

$$\begin{aligned} f_t(k) &= f_t(0) + [f_t(1) - f_t(0)]k, \qquad \text{for } k \geq 2, \\ f_t(0) &= Y_t - (2 - A_1 - A_2)e_t, \\ f_t(1) &= 2Y_t - Y_{t-1} - (2 - A_1 - A_2)e_t + (1 - A_1)e_{t-1}. \end{aligned}$$

(12) Verify the EWR recurrence equations of part (3) of Section 7.3.3, namely that given an infinite history $Y_t, Y_{t-1}, \ldots$ and defining $e_t = Y_t - m_{t-1} - b_{t-1}$, the unique values $\{m_t, b_t\}$ of $\{\mu, \beta\}$ that minimise the discounted sum of squares

$$S_t(\mu, \beta) = \sum_{v=0}^{\infty} \delta^v (Y_{t-v} - \mu + v\beta)^2$$

satisfy the recurrence relationships

$$\begin{aligned} m_t &= m_{t-1} + b_{t-1} + (1 - \delta^2)e_t, \\ b_t &= b_{t-1} + (1 - \delta)^2 e_t. \end{aligned}$$

(13) Given the infinite history of the previous question, a modeller decides to estimate the parameters of the forecast function $f_t(k) = \mu + k\beta$ using the method of discounted likelihood, that is equivalent to the model in which the data are independently distributed as

$$(Y_{t-v}|\mu, \beta) \sim \mathrm{N}[\mu_t - v\beta, V/\delta^v].$$

Show that the maximum likelihood estimates m_t and b_t are exactly the same as those for EWR.

(14) Show that with common initial information D_0, the first-order polynomial DLM $\{1, 1, V, W\}$ is equivalent to the DLM

$$\left\{ \begin{pmatrix} 1 \\ 1 \end{pmatrix}, \begin{pmatrix} 1 & 0 \\ 0 & 0 \end{pmatrix}, 0, \begin{pmatrix} W & 0 \\ 0 & V \end{pmatrix} \right\}.$$

(15) Following on from the last question, consider the extended, constant first-order polynomial DLM defined as

$$\left\{ \begin{pmatrix} 1 \\ 1 \end{pmatrix}, \begin{pmatrix} 1 & 0 \\ 0 & 0 \end{pmatrix}, 0, \begin{pmatrix} W & C \\ C & V \end{pmatrix} \right\}.$$

(a) Show that

$$\lim_{t \to \infty} (Y_t - Y_{t-1}) = e_t - (1 - A)e_{t-1}.$$

(b) Show that $\lim_{t\to\infty} Q_t = Q = (V + C)/(1 - A)$.

(c) Hence show that A may exceed 1 if and only if $C + V < 0$ and that if $W = (4 + \epsilon)V$ and $C = -2V$, then $\lim_{\epsilon \to 0} A = 2$.

CHAPTER 8

SEASONAL MODELS

8.1 INTRODUCTION

Cyclical or periodic behaviour is evident in many time series associated with economic, commercial, physical and biological systems. For example, annual seasonal cycles provide the basis of the agricultural calendar. Each year the earth revolves about the sun, the relative positions of the two bodies determining the earth's climatic conditions at any time. This natural cycle is uncontrollable and must be accepted.

It is important that the induced seasonality in product demand be recognised and included as a factor in forecasting models. Seasonal patterns can have enormous implications for stock control and production planning, especially in agribusiness, where, not uncommonly, demand patterns exhibit seasonal peak-to-trough ratios in excess of 10-to-1. Various other annual cycles, such as demand for fireworks, Valentines cards, Easter eggs and so forth, are even more pronounced.

Although many annual patterns arise from the natural solar cycle, not all do, nor is the effect always uncontrollable. One example we met concerned the demand for the most widely used anaesthetic of its time. Analysis revealed an annual cycle that initially surprised the marketing department, whose personnel were adamant in claiming that usage was steady throughout the year. Indeed they were right: usage is roughly constant. But because the ordering habit of the National Health Service was dominated by a three-monthly accounting period, delivery demands revealed a marked quarterly cycle that needed to be recognised in efficiently controlling stock, production and supplies. Similar seasonal variations in demand and sales arise partially as responses to advertising, with the promotional plan following the same broad lines each year.

In addition to identifying and anticipating cycles of this sort, decision-makers may wish to exert control in attempts to alter and reduce seasonal fluctuations. In a commercial environment, the aim is often to increase the utilisation of manufacturing plant and warehouse facilities. For example, in some countries milk-processing plants are almost 50% under-utilised due to the 20-to-1 peak-to-trough ratio in milk production, winter concentrate feeding being relatively costly compared to summer grazing. In order to counteract this, milk processors adopt a pricing policy that encourages dairy farmers to keep more cows over winter, thus dampening seasonal fluctuations in milk supplies. Another example that we encounterred in the pharmaceutical industry concerned a very well known skin cream. Initially this was marketed for winter chap, and seasonal demand had a pronounced peak at the end of autumn. However, it was noticed that people were also using the cream for sun protection and soothing. In response, marketing switched their major promotional effort to spring, with the effect that demand peaked massively in late spring. These examples illustrate the dynamic nature of seasonality and the need to model it accordingly.

The annual cycle, though of primary importance, is one amongst many of varying periods. The day/night, lunar and other planetary cycles are well known. So too are natural biorhythms in animals and oscillatory phenomena in the physical sciences.

Perhaps the least appreciated form of cyclic behaviour arises from system response mechanisms involving feedback and delay. Engineers are well aware of this in short-term process control systems, but politicians and management have a costly blind spot for such dynamics as they appear in social and economic systems, particularly when the response covers periods in excess of one year. Examples include the cobweb, pig and beef cycles, predator-prey interactions, boom and bust cycles such as the four-to-five year economic cycle and those christened Kitchin, Jugler and Kondratieff (van Duijn 1983). These are not inevitable cycles. They are simply system responses to imbalances in things such as supply and demand, that are often magnified by shortsighted reactive decision-makers. Such inadequate understanding, so rife amongst decision- and policy-makers, causes many economic, social and political crises. An illuminating example from agriculture, taken from Harrison and Quinn (1978), concerns beef markets. A real or supposed supply shortage, such as that predicted by the World Health Organisation in the early 1970s, prompts governmental agencies throughout the world to initiate incentive schemes. In response, beef breeders decide to increase their herd size. But this means bringing back an increased number of one-year-old heifers into the breeding herd. These heifers would normally have proceeded to the fattener. Consequently, despite the intention to increase the meat supply, the following year the supply drops well below the level it would have been if no action had been taken. The resulting price increase persuades breeders to expand their herds further. So again in the next year, there is a fall in meat supply and meat prices reach a record height. This is now drastically misinterpreted by breeders and decision-makers as indicating that the shortage was originally underestimated, leading to an even further expansion of breeding stock. The bubble bursts three years after the initial stimulus for expansion. Following the delay (due to impregnation, gestation, weaning and fattening), the increased cattle output, sometimes over 40% above the previous year, reaches the slaughterhouse. Disastrous consequences follow as prices collapse. The increased number of cows being fattened results in soaring feed and housing costs (up to 100% increase in concentrates and grass seed). Combined with low cattle prices, this destroys the breeder's profit margins (one cow in the west of Ireland went for 50 pence). The price of the newly produced calf may not even cover the vet's fees! The final response is now a massive culling of cows and calves at birth, sowing the seeds of the next shortage so that the cycle repeats. This example is not hypothetical.

In Britain, in one year in the mid-1970s, 23% of all Friesian calves were slaughtered at birth–not a widely published statistic. As a result, many breeders lost their farms with government misdirecting financial aid to the fatteners rather than to the breeding source. This was not just a European phenomenon: farmers in Australia were paid to shoot their cows, and the whole world was affected. Further, the effect mushroomed. Pig and poultry profit margins were shattered by the increased feed costs. The refrigerated lorry industry and the tanneries had a temporary bonanza, though they too were headed for deep trouble, and so on.

The key point of this example is that it well illustrates the common, but little understood, phenomenon of a delayed feedback response and cyclical behaviour that is often inflated by decision-makers' actions and that recurs in all economic sectors. The key to control lies in understanding the nature of the system, recognising the great time delays between action and effect, and the fact that once an imbalance occurs, the appropriate corrective action will typically cause an immediate worsening of the situation before balance is restored. It is also important to be aware of the consequences for related activities, such as milk, feed, pigs, leather goods, and so on, in the above example. The modern tendency to divide control responsibilities for separate areas as though they were independent only exaggerates crises such as beef shortages, butter mountains, and wine lakes, with the continued reactions of decision-makers causing problems in other areas and thus perpetuating imbalances and crises. The main message for the dynamic modeller is that cyclical patterns should not be automatically modelled as inevitable seasonality. If the modelling objectives are associated with major policy decisions, it is critical that the relevant system dynamics be studied, particularly with regard to the presence of feedback mechanisms, magnification and naive decision-makers.

Having identified these issues, this chapter now concentrates on descriptions of observed cyclical behaviour purely in terms of superficial seasonal factors. In practice, such simple representational models, either alone or combined with trend and/or regression components, often prove adequate for assessing current and historical seasonal patterns, for analysing changes over time, and for short-term forecasting. The term **seasonality** is used as a label for any cyclical or periodic behaviour, whether or not this corresponds to well-defined and accepted seasons. To begin, the basic structure of linear models for deterministic functions exhibiting pure seasonality is described, setting out the requisite notation and terminology. This leads to suitable DLMs via the consideration of cyclical forecast functions. Two important classes of models are detailed. The first uses seasonal factors and is termed the **form-free** approach since the form of the seasonal patterns is unrestricted. The second approach uses a functional representation of the seasonal factors in terms of trigonometric terms and is termed the **form** approach. Both approaches are useful in practice.

8.2 SEASONAL FACTOR REPRESENTATION OF CYCLICAL FUNCTIONS

Let $g(t)$ be any real-valued function defined on the non-negative integers $t = 0,\ 1, \ldots,$ where t is a time index. Note that the function is defined from time zero in order to conform with the usage of cyclical forecast functions in DLMs, as detailed below.

Definition 8.1.

(1) $g(t)$ is **cyclical** or **periodic** if for some integer $p > 1$ and for all integers $t, n \geq 0$, $g(t+np) = g(t)$.
(2) Unless otherwise stated, the smallest integer p such that this is true is called the **period** of $g(.)$.
(3) $g(\cdot)$ exhibits a single full **cycle** in any time interval containing p consecutive time points, such as $[t, t+p-1]$, for any $t > 0$.
(4) The **seasonal factors** of $g(\cdot)$ are the p values taken in any full cycle

$$\psi_j = g(j), \qquad (j = 0, \ldots, p-1).$$

Notice that for $t > 0$, $g(t) = g(j)$, where j is the remainder after division of t by p, denoted by $j = p|t$.
(5) The **seasonal factor vector** at time t is simply that permutation of the vector of seasonal factors that has its first element relating to time t, namely, when the current seasonal factor is ψ_j,

$$\boldsymbol{\psi}_t = (\psi_j, \psi_{j+1}, \ldots, \psi_{p-1}, \psi_0, \ldots, \psi_{j-1})'.$$

In particular, for any integers n and $k = np$, $\boldsymbol{\psi}_k = (\psi_0, \ldots, \psi_{p-1})'$.
(6) In any cycle, the time point corresponding to the relevant seasonal factor ψ_j is given a label M(j). This label then defines the timing within each cycle, as, for example, months within years, where M(0) may be January, M(1) February, and so forth. The labels are cyclic with period p: the label M(j) corresponding to time t if and only if $j = p|t$.
(7) When the p seasonal factors relating to a period p may take arbitrary real values, the seasonal pattern is termed **form-free.**

Definition 8.2. $\mathbf{E}_p$ and the $p \times p$ **permutation** matrix $\mathbf{P}$ are

$$\mathbf{E}_p = \begin{pmatrix} 1 \\ 0 \\ \vdots \\ 0 \\ 0 \end{pmatrix}, \qquad \mathbf{P} = \begin{pmatrix} 0 & 1 & 0 & \ldots & 0 \\ 0 & 0 & 1 & \ldots & 0 \\ \vdots & \vdots & \vdots & \ddots & \vdots \\ 0 & 0 & 0 & \ldots & 1 \\ 1 & 0 & 0 & \ldots & 0 \end{pmatrix} = \begin{pmatrix} \mathbf{0} & \mathbf{I}_{p-1} \\ 1 & \mathbf{0}' \end{pmatrix}.$$

Clearly $\mathbf{P}$ is p-cyclic, so that for any integer $n \geq 0$, $\mathbf{P}^{np} = \mathbf{I}_p$, and $\mathbf{P}^{k+np} = \mathbf{P}^k$ for $k = 1, \ldots, p$.

At any time t, the current value of $g(\cdot)$ is ψ_j, where $j = p|t$, given by

$$\psi_j = g(t) = \mathbf{E}_p'\boldsymbol{\psi}_t. \tag{8.1a}$$

Using the permutation matrix, it is clear that for all $t \geq 0$, the seasonal factors rotate according to

$$\boldsymbol{\psi}_t = \mathbf{P}\boldsymbol{\psi}_{t-1}. \tag{8.1b}$$

This relationship provides the initial step in constructing a purely seasonal DLM. Suppose the desired forecast function is cyclical in the sense that $f_t(k) = g(t+k)$. Equations (8.1) imply that the forecast function has the form of that in a time series DLM $\{\mathbf{E}_p, \mathbf{P}, \cdot, \cdot\}$. $\mathbf{P}$ has p distinct eigenvalues given by the p roots of unity, $\exp(2\pi ij/p)$ for $j = 1, \ldots, p$, so from Section 5.3, the model

$$\{\mathbf{E}_p, \mathbf{P}, ., .\}$$

and any observable, similar model, produces the desired forecast function.

8.3 FORM-FREE SEASONAL FACTOR DLMS

8.3.1 General models

Definition 8.3. The canonical form-free seasonal factor DLM of period $p > 1$ is defined, for any appropriate variances V_t and $\mathbf{W}_t$, as

$$\{\mathbf{E}_p, \mathbf{P}, V_t, \mathbf{W}_t\}.$$

With seasonal factor parameter vector $\boldsymbol{\psi}_t$ this DLM can be written

$$\begin{aligned} &\text{Observation equation:} \quad & Y_t &= \mathbf{E}_p'\boldsymbol{\psi}_t + \nu_t, \\ &\text{System equation:} \quad & \boldsymbol{\psi}_t &= \mathbf{P}\boldsymbol{\psi}_{t-1} + \boldsymbol{\omega}_t. \end{aligned} \tag{8.2}$$

(1) With $\mathrm{E}[\boldsymbol{\psi}_t \mid D_t] = \mathbf{m}_t = (m_{t0}, \ldots, m_{t,p-1})'$ the forecast function is

$$f_t(k) = \mathbf{E}_p'\mathbf{P}^k\mathbf{m}_t = m_{tj}, \qquad j = p|k.$$

(2) The model is observable with observability matrix $\mathbf{T} = \mathbf{I}_p$.
(3) Any similar model will be called a form-free seasonal factor DLM.

8.3.2 Closed, constant models

The constant form-free DLM $\{\mathbf{E}_p, \mathbf{P}, V, V\mathbf{W}\}$ may be written as

$$\begin{aligned} Y_t &= \psi_{t0} + \nu_t, \\ \psi_{tr} &= \psi_{t-1,r+1} + \omega_{tr}, \qquad (r = 0, \ldots, p-2), \\ \psi_{t,p-1} &= \psi_{t-1,0} + \omega_{t,p-1}, \end{aligned}$$

where $\nu_t \sim \mathrm{N}[0, V]$ and $\boldsymbol{\omega}_t = (\omega_{t0}, \ldots, \omega_{t,p-1})' \sim \mathrm{N}[0, V\mathbf{W}]$ with the usual independence assumptions. Consider the current seasonal level ψ_{t0}, supposing that the current time point is M(0). Having observed Y_t, no further observations are made directly on the seasonal factor for times labelled M(0) until time $t+p$, p observations later. Over that full period, the factors change stochastically via the addition of the p evolution errors. Any information gained about this particular seasonal factor is due entirely to the correlation structure in $\mathbf{W}$, and therefore the form of this matrix is of crucial importance. Generally it is specified according to the discount principle of Section 6.3, developed later. First, however, the theoretical structure of the model is further explored with a particularly simple diagonal evolution variance matrix. Throughout note that V is supposed known. If not, then the usual learning procedure applies without altering the essentials of the following discussion.

EXAMPLE 8.1. Consider the special DLM $\{\mathbf{E}_p, \mathbf{P}, V, W\mathbf{I}\}$ for which the individual errors ω_{tr} are uncorrelated for each time t. Then the model reduces to a collection of p first-order polynomial DLMs $\{1, 1, V, pW\}$. For clarity, suppose that $p = 12$ and the data is monthly over the year, M(0) being January, and so on. Then, each year in January, one observation is made on the current January level ψ_{t0}, the level then evolving over the next full year by the addition of 12 uncorrelated evolution error terms, each being distributed as $\mathrm{N}[0, W]$. The net result is that

$$\psi_{t+p,0} = \psi_{t0} + \omega_t,$$

with $\omega_t \sim \mathrm{N}[0, 12\,W]$. It is clear that the only link between the seasonal factors is that deriving from the initial prior covariance terms at time 0. The effect of this initial prior decays with time, and so for simplicity, assume that $(\boldsymbol{\psi}_0 \mid D_0) \sim \mathrm{N}[\mathbf{m}_0, \mathbf{C}_0]$, where $\mathbf{m}_0 = m_0\mathbf{1}$ and $\mathbf{C}_0 = C_0\mathbf{I}$ for some scalars m_0 and C_0. The following results may now be simply derived by applying the updating recurrences and limiting results from the closed, constant, first-order polynomial model of Section 2.3.

(1) For each t, $(\boldsymbol{\psi}_t \mid D_t) \sim \mathrm{N}[\mathbf{m}_t, \mathbf{C}_t]$, with $\mathbf{m}_t = (m_{t0}, \ldots, m_{t,p-1})'$ and variance matrix $\mathbf{C}_t = \mathrm{diag}(C_{t0}, \ldots, C_{t,p-1})$.
(2) Suppose that $t = np$ for some integer $n \geq 1$ so that n full periods have passed. Then the current time label is M(0) and the updating equations for the corresponding seasonal factor ψ_{t0} are

$$m_{t0} = m_{t-p,0} + A_t e_t,$$
$$C_{t0} = A_t V,$$

where $e_t = Y_t - m_{t-p,0}$, $R_t = C_{t-p,0} + pW$, and $A_t = R_t/\,(R_t + V)$. Similar comments apply to the next time intervals M(1), M(2), ..., with the subscript 0 updated to 1, 2, ..., respectively.

(3) Results of Section 2.3 apply to the p models $\{1, 1, V, prV\}$, where $r = W/V$, giving the following limiting results:

$$\lim_{t\to\infty} A_t = A = \left(\sqrt{1+4/pr}-1\right)pr/2,$$

$$\lim_{t\to\infty} C_{t0} = C = AV,$$

$$\lim_{t\to\infty} R_t = R = A/(1-A),$$

$$p(\psi_{tj} \mid D_t) \to \mathrm{N}[m_{t,j},\ AV + jW], \quad j = 0, \dots, p-1.$$

(4) The limiting analysis is equivalent to that obtained by applying the discount approach with a single discount factor $\delta = 1 - A$, leading to a model $\{\mathbf{E}_p, \mathbf{P}, V, \mathbf{W}_t\}$ with

$$\mathbf{W}_t = \frac{1-\delta}{\delta}\mathbf{P}\mathbf{C}_{t-1}\mathbf{P}'.$$

Following comment (4) above, it may also be shown that the limiting forecast function is equivalent to that derived using exponentially weighted regression techniques. This suggests a rephrasing of the form-free model and the use of more than one discount factor to structure the evolution variance matrix. Underlying this suggestion is the idea that a seasonal pattern is generally more stable than the underlying, deseasonalised level. Harrison (1965) discusses this idea. Thus the seasonal factors may be decomposed into an underlying level, plus seasonal deviations from this level. This provides the flexibility to model changes in two components separately using the ideas of component, or block, models described in Chapter 6. This decomposition is now described.

8.4 FORM-FREE SEASONAL EFFECTS DLMS

8.4.1 Introduction and definition

In decomposing a set of p seasonal factors into one *deseasonalised level* and p seasonal deviations from that level, the seasonal deviations are called *seasonal effects.* Although specifically concerned with seasonality, many of the points carry over to more general effects models. Practitioners familiar with standard statistical models will appreciate the idea of descriptions in terms of an overall mean for observations plus treatment, block and other effects. The effects for any treatment or block will be subject to an identifiability constraint that is imposed in one of several forms, either by aliasing one of the effects or by constraining an average of the effects. The commonest such constraint is the zero-sum constraint. The analogy for seasonality is that the seasonal deviations from the underlying level sum to zero in a full period, so that any $p - 1$ of them define the complete set. This zero sum constraint is used when discussing additive seasonality, although other constraints are possible and may typically be obtained by linear transformation of the seasonal factors. In generalised linear modelling using GLIM, for example, the effect at one chosen level, the first level by default, is constrained to be zero (Baker and Nelder 1978).

Initially, the underlying level of the series is set to zero for all t, so that the seasonal factors always sum to zero, producing a seasonal effects DLM. The superposition of this seasonal effect DLM and a first-order polynomial DLM then provides the constrained, seasonal effects component for a series with non-zero level.

Definition 8.4. A **form-free seasonal effects** DLM is any model

$$\{\mathbf{E}_p, \mathbf{P}, V_t, \mathbf{W}_t\} \tag{8.3}$$

with state vector $\boldsymbol{\phi}_t = (\phi_{t0}, \dots, \phi_{t,p-1})'$ satisfying $\mathbf{1}'\boldsymbol{\phi}_t = 0$ for all t.

The seasonal effects ϕ_{tj} represent seasonal deviations from their zero mean and are simply constrained seasonal factors. In terms of equations, such a model has the form given in (8.2), with the parameter notation changed from $\boldsymbol{\psi}$ to $\boldsymbol{\phi}$, and with the addition of the constraint $\mathbf{1}'\boldsymbol{\phi}_t = 0$.

8.4.2 Imposing constraints

The constraint (8.3) leads to the following model restrictions.

(1) **Initial prior.** Applying the constraint to $(\boldsymbol{\phi}_0 \mid D_0) \sim \mathrm{N}[\mathbf{m}_0, \mathbf{C}_0]$, since $(\mathbf{1}'\boldsymbol{\phi}_0 \mid D_0) \sim \mathrm{N}[\mathbf{1}'\mathbf{m}_0, \mathbf{1}'\mathbf{C}_0\mathbf{1}]$, necessarily

$$\begin{aligned} \mathbf{1}'\mathbf{m}_0 &= 0, \\ \mathbf{C}_0\mathbf{1} &= \mathbf{0}. \end{aligned} \tag{8.4}$$

So the initial prior means and the elements of each row (and each column) of the initial prior variance matrix must sum to zero.

(2) **Evolution variances.** Since $\boldsymbol{\phi}_t = \mathbf{P}\boldsymbol{\phi}_{t-1} + \boldsymbol{\omega}_t$, clearly $\boldsymbol{\omega}_t$ must also satisfy the zero-sum constraint

$$\mathbf{W}_t\mathbf{1} = \mathbf{0}. \tag{8.5}$$

Hence each row and column of $\mathbf{W}$ must sum to zero. Note that $\boldsymbol{\omega}_t$ already has zero mean, but if, more generally, it has a non-zero mean, then the constraint must also be applied to this mean.

Theorem 8.1. *In the form-free seasonal effects DLM, suppose the initial constraints $\mathbf{1}'\mathbf{m}_0 = 0$ and $\mathbf{C}_0\mathbf{1} = \mathbf{0}$, and that $\mathbf{W}_t\mathbf{1} = \mathbf{0}$ for all t. Then the posterior distributions $(\boldsymbol{\phi}_t \mid D_t) \sim \mathrm{N}[\mathbf{m}_t, \mathbf{C}_t]$ satisfy the constraints*

$$\mathbf{1}'\mathbf{m}_t = 0 \quad \text{and} \quad \mathbf{C}_t\mathbf{1} = \mathbf{0},$$

so that $\mathbf{1}'\boldsymbol{\phi}_t = 0$ with probability one.

Proof. The proof is by induction. If the constraints apply to the posterior at time $t-1$ then $\mathbf{1}'\mathbf{m}_{t-1} = 0$ and $\mathbf{C}_{t-1}\mathbf{1} = \mathbf{0}$. Proceeding to the prior, $(\phi_t \mid D_{t-1}) \sim \text{N}[\mathbf{a}_t, \mathbf{R}_t]$, with $\mathbf{a}_t = \mathbf{P}\mathbf{m}_{t-1}$ and $\mathbf{R}_t = \mathbf{P}\mathbf{C}_{t-1}\mathbf{P}' + \mathbf{W}_t$. Since $\mathbf{P1} = \mathbf{1}$, these prior moments also satisfy the constraints, with $\mathbf{1}'\mathbf{a}_t = 0$ and $\mathbf{R}_t\mathbf{1} = \mathbf{0}$. Updating to the posterior at t gives $\mathbf{m}_t = \mathbf{a}_t + \mathbf{R}_t\mathbf{E}_p Q_t^{-1} e_t$ and $\mathbf{C}_t = \mathbf{R}_t - \mathbf{R}_t\mathbf{E}_p\mathbf{E}_p'\mathbf{R}_t' Q_t^{-1}$, whence directly, $\mathbf{1}'\mathbf{m}_t = 0$ and $\mathbf{C}_t\mathbf{1} = \mathbf{0}$. The constraints therefore apply at $t = 1$, and so by induction to all t.

◇

Thus conditions (8.4) and (8.5) are consistent with condition (8.3). Two problems remain: to ensure that the initial prior satisfies (8.4) and to design a suitable sequence of evolution variance matrices satisfying (8.5).

8.4.3 Constrained initial priors

Practitioners are often only prepared to specify marginal priors, $(\phi_{0j} \mid D_0)$ for $j = 0, \ldots, p-1$, providing just the mean and variance of each effect. Rarely will they able or willing to produce covariance terms for the full, joint prior distribution. In such cases the usual practical procedure is to derive a coherent joint prior distribution satisfying (8.4) by applying constraint (8.3) to the incoherent seasonal effects joint distribution that corresponds to their specified moments together with zero covariances. This may be done formally within the following general framework:

Suppose the initial prior

$$(\phi_0 \mid D_0^*) \sim \text{N}[\mathbf{m}_0^*, \mathbf{C}_0^*] \tag{8.6}$$

may not satisfy constraint (8.4), as happens for a diagonal $\mathbf{C}_0^*$. Apply the following theorem.

Theorem 8.2. *Imposing the constraint* $\mathbf{1}'\phi_0 = 0$ *on the prior in* (8.6) *and writing* $U = \mathbf{1}'\mathbf{C}_0^*\mathbf{1}$ *and* $\mathbf{A} = \mathbf{C}_0^*\mathbf{1}/U$ *gives the revised joint prior*

$$\begin{aligned}(\phi_0 \mid D_0) &\sim \text{N}[\mathbf{m}_0, \mathbf{C}_0],\\ \mathbf{m}_0 &= \mathbf{m}_0^* - \mathbf{A}\mathbf{1}'\mathbf{m}_0^*,\\ \mathbf{C}_0 &= \mathbf{C}_0^* - \mathbf{A}\mathbf{A}'U.\end{aligned}$$

Proof. The joint distribution of ϕ_0 and their total $\theta = \mathbf{1}'\phi_0$ is

$$\left(\begin{matrix}\phi_0\\ \theta\end{matrix}\middle| D_0^*\right) \sim \text{N}\left[\begin{pmatrix}\mathbf{m}_0^*\\ \mathbf{1}'\mathbf{m}_0^*\end{pmatrix}, \begin{pmatrix}\mathbf{C}_0^* & \mathbf{A}U\\ \mathbf{A}'U & U\end{pmatrix}\right].$$

Then the conditional distribution of ϕ_0 is

$$(\phi_0 \mid \theta,\, D_0) \sim \text{N}[\mathbf{m}_0^* + \mathbf{A}(\theta - \mathbf{1}'\mathbf{m}_0^*),\, \mathbf{C}_0^* - \mathbf{A}\mathbf{A}'U].$$

Apply the zero-sum constraint by setting $\theta = 0$.

◇

Whatever initial prior is elicited, this theorem should be applied to ensure compliance with the constraint. Notice that if the specified prior satisfies the constraints, then no change occurs since $\mathbf{A} = \mathbf{0}$. Otherwise, in a general sense, the total variation in the constrained prior will always be less than that originally specified in the unconstrained prior, the difference being removed by imposing the deterministic constraint. Consequently, a practitioner might like to scale the revised prior variance so that its trace equals that of the elicited prior.

EXAMPLE 8.2. Suppose $p = 4$, with initial specification

$$(\phi_0 | D_0^*) \sim \mathrm{N}\left[\begin{pmatrix} 10 \\ 5 \\ 0 \\ -7 \end{pmatrix}, \mathbf{I}\right].$$

Imposing the constraints produces the revised prior

$$(\phi_0 | D_0) \sim \mathrm{N}\left[\begin{pmatrix} 8 \\ 3 \\ -2 \\ -9 \end{pmatrix}, \frac{1}{4}\begin{pmatrix} 3 & -1 & -1 & -1 \\ -1 & 3 & -1 & -1 \\ -1 & -1 & 3 & -1 \\ -1 & -1 & -1 & 3 \end{pmatrix}\right].$$

To rescale so that trace $\mathbf{C}_0 =$ trace $\mathbf{C}_0^*$, simply multiply $\mathbf{C}_0$ by $4/3$.

8.4.4 *Constrained evolution variances*

Many evolution variance matrix structures satisfy constraint (8.5). Harrison and Stevens (1976b) set $\boldsymbol{\omega}_t = \omega_t \mathbf{a}$, where $\mathbf{a}' = (p-1, -1, \ldots, -1)$ and $\omega_t \sim \mathrm{N}[0, W_t]$ for some $W_t > 0$. This structure imparts information from the observation to the current seasonal effect, and then only equally to the others via the renormalisation in applying the zero-sum constraint. In this case, $\mathbf{W}_t^* = W_t \mathbf{a}\mathbf{a}'$, and (8.5) is satisfied since $\mathbf{1}'\mathbf{a} = 0$. An alternative structure is derived by starting with $\mathbf{W}_t = W_t \mathbf{I}$ and applying the zero sum constraint as in Theorem 8.2 to obtain

$$\mathbf{W}_t = W_t(p\mathbf{I} - \mathbf{1}\mathbf{1}').$$

Historically, these two forms, and others, have been widely used in practice. However, they are not recommended for general application, being specifically designed to represent particular forms of change in the seasonal effects over time. Instead we recommend discount methods. For a single discount factor δ, possibly depending on t, the discount idea of Section 6.3 applied to the entire seasonal effects vector leads to

$$\mathbf{W}_t = \frac{1-\delta}{\delta}\mathbf{P}\mathbf{C}_{t-1}\mathbf{P}', \qquad (8.7)$$

that always satisfies (8.5) since $\mathbf{P}'\mathbf{1} = \mathbf{1}$ and $\mathbf{C}_{t-1}\mathbf{1} = \mathbf{0}$.

Definition 8.5. A single discount, form-free seasonal effects DLM is any form-free seasonal effects DLM $\{\mathbf{E}_p, \mathbf{P}, V_t, \mathbf{W}_t\}$ as defined by Definition 8.4 with evolution variance sequence $\mathbf{W}_t$ defined via (8.7).

8.5 TREND/FORM-FREE SEASONAL EFFECTS DLMS

The main reason for considering the form-free seasonal effects DLM is that it provides a widely applicable seasonal component that in a larger DLM, describes seasonal deviations from a *deseasonalised level*, or trend. The two most important such DLMs are those superimposing the seasonal effect DLM with either the first- or second-order polynomial DLM.

8.5.1 First-order polynomial/seasonal effects model

Definition 8.6. A **first-order polynomial trend/form-free seasonal effects DLM** is any DLM with parameter vector

$$\boldsymbol{\theta}_t = \begin{pmatrix} \mu_t \\ \boldsymbol{\phi}_t \end{pmatrix}$$

and quadruple

$$\left\{ \begin{pmatrix} 1 \\ \mathbf{E}_p \end{pmatrix}, \begin{pmatrix} 1 & \mathbf{0} \\ \mathbf{0} & \mathbf{P} \end{pmatrix}, V_t, \begin{pmatrix} W_{t,\mu} & \mathbf{0} \\ \mathbf{0} & \mathbf{W}_{t,\phi} \end{pmatrix} \right\}$$

satisfying the constraint $\mathbf{1}'\boldsymbol{\phi}_t = 0$, for all t.

Such models comprise the superposition of a first-order polynomial DLM (for the deseasonalised level) and a seasonal effects DLM. It is easily seen that in the absence of the constraint, this DLM is unobservable. However, the zero-sum constraint ensures that the DLM is constrained observable. The forecast function takes the form $f_t(k) = m_t + h_{tj}$, with $(j = p|k)$, where m_t is the expected value of the deseasonalised level at time $t+k$ and h_{tj} is the expected seasonal deviation from this level. The model may be written as

$$\begin{aligned} Y_t &= \mu_t + \phi_{t0} + \nu_t, \\ \mu_t &= \mu_{t-1} + \omega_t, \\ \phi_{tr} &= \phi_{t-1,r+1} + \omega_{tr}, \qquad (r = 0, \ldots, p-2), \\ \phi_{t,p-1} &= \phi_{t-1,0} + \omega_{t,p-1}. \end{aligned}$$

When subject to the zero-sum constraint, this model is similar to the seasonal factor model of Section 8.3. To see this, note that if $\mathbf{H}$ is the $p \times (p+1)$ matrix $\mathbf{H} = [\mathbf{1}, \mathbf{I}]$, then $\boldsymbol{\psi}_t = \mathbf{H}\boldsymbol{\theta}_t$ is the vector of seasonal factors

$$\psi_{tj} = \mu_t + \phi_{tj}, \qquad (j = 0, \ldots, p-1).$$

Given the constraint, it follows that for all t,

$$p^{-1}\sum_{j=1}^{p}\psi_{tj}=\mu_t,$$

so μ_t represents the average of the seasonal factors (the deseasonalised level). In the ψ_t parametrisation this is the factor model (8.2). By construction, this model represents the linear composition of the seasonal effects and first-order polynomial DLMs. Note that starting with the seasonal factor DLM and a $(p+1)\times p$ matrix $\mathbf{U}$, the relevant transformation is

$$\begin{pmatrix}\mu_t\\ \phi_t\end{pmatrix}=\mathbf{U}\psi_t=\begin{pmatrix}\frac{1}{p} & \frac{1}{p} & \cdots & \frac{1}{p}\\ 1-\frac{1}{p} & -\frac{1}{p} & \cdots & -\frac{1}{p}\\ -\frac{1}{p} & 1-\frac{1}{p} & \cdots & -\frac{1}{p}\\ \vdots & \vdots & \ddots & \vdots\\ -\frac{1}{p} & -\frac{1}{p} & \cdots & 1-\frac{1}{p}\end{pmatrix}\psi_t.$$

8.5.2 Second-order polynomial/seasonal effects model

Definition 8.7. A second-order polynomial trend/form-free seasonal effects DLM is any DLM with parameter vector

$$\boldsymbol{\theta}_t=\begin{pmatrix}\mu_t\\ \beta_t\\ \phi_t\end{pmatrix}$$

and quadruple

$$\left\{\begin{pmatrix}\mathbf{E}_2\\ \mathbf{E}_p\end{pmatrix},\begin{pmatrix}\mathbf{J}_2(1) & \mathbf{0}\\ \mathbf{0} & \mathbf{P}\end{pmatrix},V_t,\begin{pmatrix}\mathbf{W}_{t,\mu} & \mathbf{0}\\ \mathbf{0} & \mathbf{W}_{t,\phi}\end{pmatrix}\right\}$$

satisfying the constraint $\mathbf{1}'\phi_t=0$, for all t.

Again such a DLM is constrained observable. It is obtained from the superposition of a second-order polynomial and a seasonal effects DLM. The forecast function takes the form

$$f_t(k)=m_t+kb_t+h_{tj},\quad \text{with}\ \sum_{j=0}^{p-1}h_{tj}=0$$

where $j=p|k$. Further discussion of this useful class of TSDLMs is deferred until applications in later chapters.

8.6 *FOURIER FORM REPRESENTATION OF SEASONALITY*

8.6.1 *Introduction*

Alternative representations of cyclical patterns employ linear combinations of periodic functions. The particular approach we favour and develop uses the simplest and most natural class of periodic functions, namely trigonometric functions, leading to **Fourier form** representations of seasonality.

The main reasons for the use of form models, rather than the flexible and unrestricted seasonal effects models, are economy and interpretation. In some applications, observations result from sampling a simple waveform, so that sine/cosine waves provide a natural, economic characterisation. Simple phenomena exhibiting such behaviour abound in electrical and electronic systems, astronomy, marine depth soundings, and geophysical studies, including earthquake tremors. Many *pure* seasonal patterns also arise in response to the revolution of the earth about the sun. For example, in both the cases of British temperature and the Eire milk supply index, over 97% of the variation in the average monthly figures about their respective annual means may be characterised in terms of a single cosine wave of period 12. If such a representation is deemed acceptable, then it is defined in terms of only two quantities determining the phase and amplitude of the cosine waveform. By comparison, a monthly seasonal effects component requires 11 parameters, with a weekly component needing 51. The economy of form models is immediately apparent, and when appropriate, results in enhanced forecasting performance. Generally, compared to a full effects model, a Fourier form model can provide an acceptable representation of an apparently erratic seasonal pattern whilst economising on parameters.

8.6.2 *Fourier form representation of cyclical functions*

Consider the cyclical function $g(t)$ of Section 8.2 defined in terms of seasonal factors $\psi_0, \ldots, \psi_{p-1}$. The basic result is that any such p real numbers can be written as a linear combination of trigonometric terms. This representation depends on the parity of the period p. Throughout, let $\alpha = 2\pi/p$ and $h = p/2$ if p is even, but $h = (p-1)/2$ if p is odd.

Theorem 8.3. *Any p real numbers $\psi_0, \ldots, \psi_{p-1}$ can be represented as*

$$\psi_j = a_0 + \sum_{r=1}^{h} \left[a_r \cos(\alpha r j) + b_r \sin(\alpha r j)\right],$$

where h is the largest integer not exceeding $p/2$ and the real numbers $\{a_i, b_i\}$ are given by

$$a_r = \frac{2}{p}\sum_{j=0}^{p-1}\psi_j\cos(\alpha rj), \quad b_r = \frac{2}{p}\sum_{j=0}^{p-1}\psi_j\sin(\alpha rj), \qquad 1 \le r < \tfrac{p}{2}$$

$$a_0 = \frac{1}{p}\sum_{j=0}^{p-1}\psi_j, \quad a_{p/2} = \frac{1}{p}\sum_{j=0}^{p-1}(-1)^j\psi_j, \quad b_{p/2} = 0.$$

Proof. The results are trivially deduced by multiplying the ψ_j by appropriate sin/cosine terms and solving the resulting linear equations. The basic identities below are used in this solution. For example,

$$\sum_{j=0}^{p-1}\psi_j\cos(\alpha rj) = \sum_{j=0}^{p-1}a_r\cos^2(\alpha rj) = pa_r/2, \qquad r \neq p/2.$$

The full proof is left to the reader.

Basic trigonometric identities.

(1) For integer n and any x,

$$\cos(x + 2\pi n) = \cos(x) \quad \text{and} \quad \sin(x + 2\pi n) = \sin(x).$$

(2) For each $r = 1, \ldots, h$,

$$\sum_{j=0}^{p-1}\cos(\alpha rj) = \sum_{j=0}^{p-1}\sin(\alpha rj) = 0.$$

(3) For integers h and k,

$$\sum_{j=0}^{p-1}\cos(\alpha hj)\sin(\alpha kj) = 0,$$

$$\sum_{j=0}^{p-1}\cos(\alpha hj)\cos(\alpha kj) = \begin{cases} 0, & h \neq k, \\ p, & h = k = \frac{p}{2}, \\ \frac{p}{2}, & h = k \neq \frac{p}{2}, \end{cases}$$

$$\sum_{j=0}^{p-1}\sin(\alpha hj)\sin(\alpha hj) = \begin{cases} 0, & h \neq k, \\ 0, & h = k = \frac{p}{2}, \\ \frac{p}{2}, & h = k \neq \frac{p}{2}. \end{cases}$$

◇

Notation and terminology

(1) The quantities a_r and b_r are called the **Fourier coefficients.**

(2) For $r = 1, \ldots, h$, define the function $S_r(\cdot)$ by

$$\begin{aligned} S_r(\cdot) = S_{r\cdot} &= a_r \cos(\alpha r\cdot) + b_r \sin(\alpha r\cdot) \\ &= A_r \cos(\alpha r \cdot + \gamma_r). \end{aligned}$$

Then $S_r(\cdot)$ is called the $\mathbf{r}^{th}$ **harmonic** and takes values

$$S_{rj} = a_r \cos(\alpha rj) + b_r \sin(\alpha rj)$$

for $j = 0, \ldots, p-1$.

(3) The **amplitude** A_r and **phase** γ_r of the rth harmonic are

$$A_r = (a_r^2 + b_r^2)^{1/2} \quad \text{and} \quad \gamma_r = \arctan(-b_r/a_r).$$

The amplitude is the maximum value taken by the $S_{r\cdot}$, and the phase determines the position of that maximum.

(4) If $p = 2h$, the hth harmonic is called the **Nyquist** harmonic. Since $b_{p/2} = 0$, it follows that $A_{p/2} = |a_{p/2}|$ and $\gamma_{p/2} = 0$.

(5) The **frequency** of the rth harmonic is defined as $\alpha r = 2\pi r/p$, with π being the Nyquist frequency. The **cycle length** of the rth harmonic is p/r. In particular, the first harmonic is called the **fundamental** harmonic, having fundamental frequency α and fundamental cycle length p. Notice that the rth harmonic completes exactly r full cycles for each single, complete cycle of the fundamental harmonic.

Theorem 8.4. *Given any p seasonal factors $\psi_0, \ldots, \psi_{p-1}$, the total variation about their mean a_0 factorises into a linear sum of the squares of the h amplitudes, so that*

- *for p odd and with $h = (p-1)/2$,*

$$\sum_{j=0}^{p-1} (\psi_j - a_0)^2 = \frac{p}{2} \sum_{r=1}^{h} A_r^2;$$

- *for p even and with $h = p/2$,*

$$\sum_{j=0}^{p-1} (\psi_j - a_0)^2 = \frac{p}{2} \sum_{r=1}^{h-1} A_r^2 + pA_{p/2}^2.$$

The proof just uses the above identities and is left to the reader. The major consequence is that the importance of each harmonic, in terms of the percentage seasonal variation for which it accounts, is simply calculated; for the rth harmonic and odd p it is simply $100A_r^2/\sum_{v=1}^{h} A_v^2$.

EXAMPLE 8.3. As a simple illustration with $p = 12$, let the function $g(.)$ be defined by the following 12 seasonal factors, with mean $a_0 = 0$,

$$1.65, 0.83, 0.41, -0.70, -0.47, 0.40, -0.05, -1.51, -0.19, -1.02, -0.87, 1.52.$$

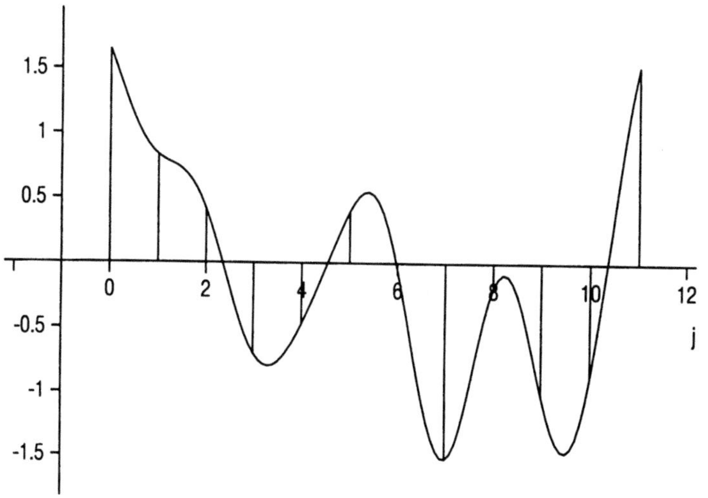

Figure 8.1a Seasonal factors: sum of harmonics

The Fourier coefficients, amplitudes and phases are as tabulated.

Harmonic r	1	2	3	4	5	6
a_r	0.80	0.75	0.25	−0.03	−0.20	0.08
b_r	0.30	−0.15	0.15	−0.60	0.01	0.00
Amplitude A_r	0.85	0.76	0.29	0.60	0.20	0.08
Phase γ_r	0.36	−0.20	0.55	1.53	−0.04	0.00

The factors are plotted as vertical lines against $j = 0, \ldots, 11$ in Figure 8.1(a). Superimposed is the full Fourier composition of the seasonal factors as a continuous function of t, clearly coinciding with the seasonal factors at the integer values. Figures 8.1(b), (c) and (d) display the corresponding 6 harmonic components S_{rj} as functions of j for $0 \leq j \leq 11$.

In the Fourier form representation, $g(t)$ is expressed as the sum of harmonics, so that for all integers $t > 0$, setting $j = p|t$, and utilising the fact that $S_{rt} = S_{rj}$, for $r = 1, \ldots, h$,

$$g(t) = \psi_j = a_0 + \sum_{r=1}^{h} S_{rj}. \tag{8.8}$$

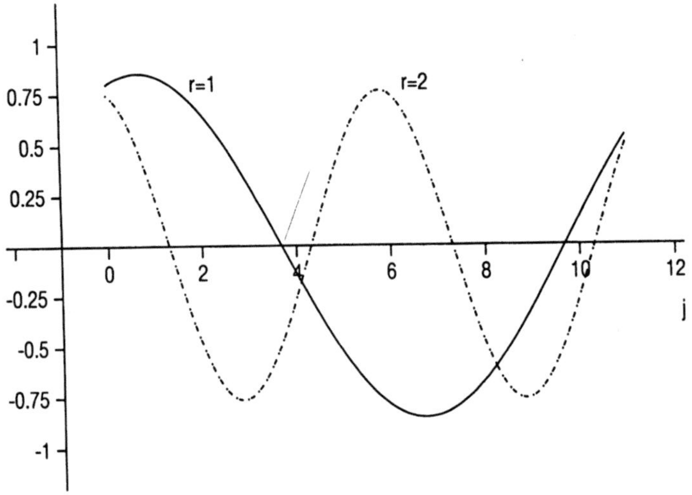

Figure 8.1b Harmonics 1 and 2

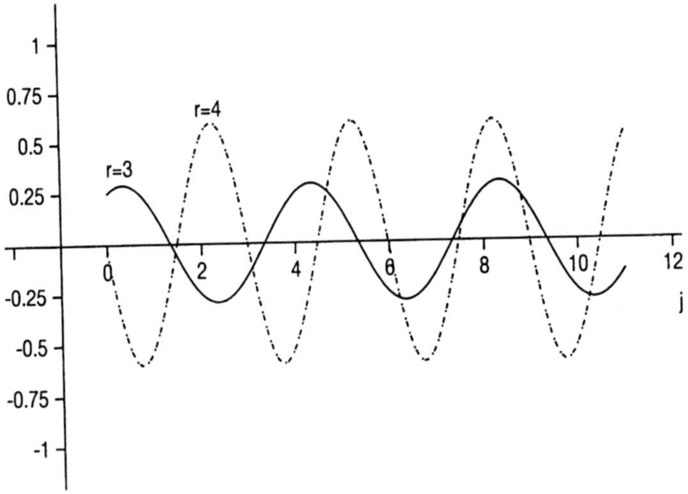

Figure 8.1c Harmonics 3 and 4

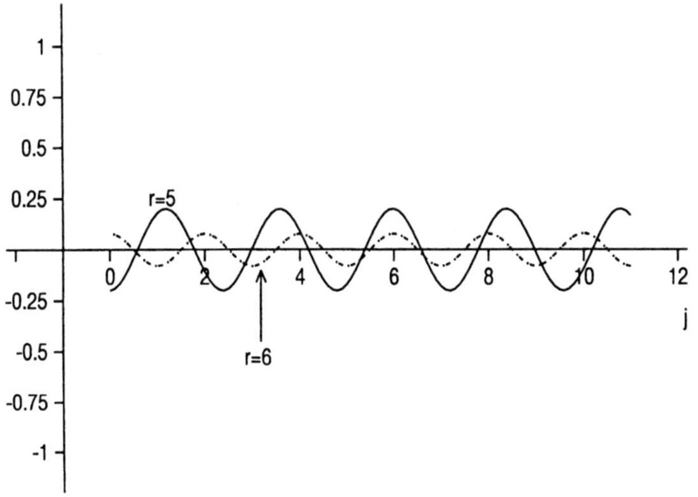

Figure 8.1d Harmonics 5 and 6

Remember that $\mathbf{J}_2(1,\omega) = \begin{pmatrix} \cos(\omega) & \sin(\omega) \\ -\sin(\omega) & \cos(\omega) \end{pmatrix}$, so that

$$\mathbf{J}_2^t(1,\omega) = \begin{pmatrix} \cos(\omega t) & \sin(\omega t) \\ -\sin(\omega t) & \cos(\omega t) \end{pmatrix} = \begin{pmatrix} \cos(\omega j) & \sin(\omega j) \\ -\sin(\omega j) & \cos(\omega j) \end{pmatrix}$$

and $\mathbf{J}_2^t(1,\omega) = \mathbf{J}_2^j(1,\omega)$. Hence S_{rt} may be written in deterministic DLM representation as

$$S_{rt} = \mathbf{E}_2'\mathbf{J}_2^t(1, r\omega)\begin{pmatrix} a_r \\ b_r \end{pmatrix}. \tag{8.9}$$

Consequently, for the stochastic case, the DLM representation is

$$\{\mathbf{E}_2, \mathbf{J}_2(1, r\omega), \cdot, \cdot\}.$$

In the case of $r = p/2$ this is not observable, and it simplifies to the observable DLM $\{1, -1, \cdot, \cdot\}$.

8.6.3 Harmonic component DLMs

Definition 8.8. An **harmonic component DLM** is defined
(a) for any frequency $\omega \in (0, \pi)$, as any DLM similar to

$$\{\mathbf{E}_2, \mathbf{J}_2(1,\omega), \cdot, \cdot\} = \left\{\begin{pmatrix} 1 \\ 0 \end{pmatrix}, \begin{pmatrix} \cos(\omega) & \sin(\omega) \\ -\sin(\omega) & \cos(\omega) \end{pmatrix}, \cdot, \cdot\right\},$$

(b) for the Nyquist frequency, $\omega = \pi$, as any DLM similar to

$$\{1, -1, \cdot, \cdot\}.$$

Note the following features, identified with items (a) and (b) of Definition 8.8 as appropriate:

(1) The component harmonic DLMs are observable, with respective observability matrices

$$\text{(a)}\ \ \mathbf{T} = \begin{pmatrix} 1 & 0 \\ \cos(\omega) & \sin(\omega) \end{pmatrix}, \qquad \text{(b)}\ \ \mathbf{T} = -1.$$

(2) With parameter $\boldsymbol{\theta}_t$, the respective forecast functions are

(a) $f_t(k) = a_t \cos(\omega k) + b_t \sin(\omega k)$, if $\mathrm{E}[\boldsymbol{\theta}_t \mid D_t] = (a_t, b_t)'$;

(b) $f_t(k) = (-1)^k a_t$, if $\mathrm{E}[\theta_t \mid D_t] = a_t$.

For any integer n, $f_t(k+np) = f_t(k)$, that in case (a) is a cosine wave of frequency ω, amplitude $(a_t^2 + b_t^2)^{1/2}$, and phase $\arctan(-b_t/a_t)$.

(3) As a theoretical aside, for (a) the forecast function can be written

$$\begin{aligned} f_t(k) &= \frac{1}{2}\left[(a_t - ib_t)\, e^{ik\omega} + (a_t + ib_t)\, e^{-ik\omega}\right] \\ &= d_t e^{ik\omega} + \bar{d}_t e^{-ik\omega}, \end{aligned}$$

and this is associated with canonical complex DLMs of the form

$$\left\{ \begin{pmatrix} 1 \\ 1 \end{pmatrix}, \begin{pmatrix} e^{ik\omega} & 0 \\ 0 & e^{-ik\omega} \end{pmatrix}, \cdot, \cdot \right\}.$$

So the component harmonic DLM comprises two sub-components. For real-valued time series, however, the complex model is of little interest since the associated parameter vector, $\boldsymbol{\alpha}_t$, comprises complex conjugate parameters. The two DLMs are similar models, being related by the invertible mapping

$$\boldsymbol{\theta}_t = \begin{pmatrix} 1 & 1 \\ i & -i \end{pmatrix} \boldsymbol{\alpha}_t.$$

(4) A point of interest is that for a constant DLM $\{\mathbf{E}_2, \mathbf{G}, V, V\mathbf{W}\}$, the derived series

$$Z_t = Y_t - 2Y_{t-1}\cos(\omega) + Y_{t-2}$$

is a moving average process of order 2.

8.6.4 Full seasonal effects DLMS

As described in Section 8.6.2, any p seasonal factors/effects can be expressed in terms of harmonic components. This section specifies DLMs having such cyclical forecast functions in terms of the equivalent collection of Fourier component DLMs, combined using the principle of superposition.

Definition 8.9. Define the $(p-1)$ vectors $\mathcal{F}$ and $(p-1)\times(p-1)$ matrices $\mathcal{G}$, for odd and even p, respectively, as

$$\{\mathcal{F}_o, \mathcal{G}_o\} = \left\{ \begin{pmatrix} \mathbf{E}_2 \\ \mathbf{E}_2 \\ \vdots \\ \mathbf{E}_2 \end{pmatrix}, \begin{pmatrix} \mathbf{J}_2(1,\omega) & \mathbf{0} & \dots & \mathbf{0} \\ \mathbf{0} & \mathbf{J}_2(1,2\omega) & \dots & \mathbf{0} \\ \vdots & \vdots & & \vdots \\ \mathbf{0} & \mathbf{0} & \dots & \mathbf{J}_2(1,h\omega) \end{pmatrix} \right\};$$

$$\{\mathcal{F}_e, \mathcal{G}_e\} = \left\{ \begin{pmatrix} \mathbf{E}_2 \\ \mathbf{E}_2 \\ \vdots \\ \mathbf{E}_2 \\ 1 \end{pmatrix}, \begin{pmatrix} \mathbf{J}_2(1,\omega) & \mathbf{0} & \dots & \mathbf{0} & \mathbf{0} \\ \mathbf{0} & \mathbf{J}_2(1,2\omega) & \dots & \mathbf{0} & \mathbf{0} \\ \vdots & \vdots & & \vdots & \vdots \\ \mathbf{0} & \mathbf{0} & \dots & \mathbf{J}_2(1,h\omega-\omega) & \mathbf{0} \\ \mathbf{0} & \mathbf{0} & \dots & \mathbf{0} & -1 \end{pmatrix} \right\}.$$

Definition 8.10. **A full effects Fourier form DLM** for a seasonal pattern of period p is any DLM of the form

$$\{\mathcal{F}_o, \mathcal{G}_o, \cdot, \cdot\}, \quad \text{if } p \text{ is odd,}$$
$$\{\mathcal{F}_e, \mathcal{G}_e, \cdot, \cdot\}, \quad \text{if } p \text{ is even.}$$

Such DLMs are usually formed from the superposition of the component harmonic DLMs of Definition 8.8, so that the system variance $\mathbf{W}$ has a block diagonal form corresponding to the relevant $\mathcal{G}$. Note the following:

(1) The forecast function is just the sum of the component forecast functions. With $\omega = 2\pi/p$ and $p-1$ parameter vector $\boldsymbol{\theta}_t$, write

$$\mathrm{E}[\boldsymbol{\theta}_t \mid D_t] = \mathbf{m}_t = (m_{t1}, m_{t2}; \dots; m_{t,p-1})',$$

the odd elements corresponding to the cosine coefficients a and the even to the sine coefficients b. Upon conveniently writing $m_{t,p} = 0$, we then have

$$f_t(k) = \sum_{r=1}^{h} S_{rk} = \sum_{r=1}^{h} [m_{t,2r-1}\cos(\omega r k) + m_{t,2r}\sin(\omega r k)].$$

(2) No constraints are required on any of the component distributions since by design, the Fourier DLM automatically ensures that the seasonal effects sum to zero.

(3) The DLM is observable, since each component is observable and no two components have a common eigenvalue.

(4) In most commercial applications a single discount factor is assigned to characterise the evolution variances, so that

$$\mathbf{R}_t = \frac{1}{\delta}\mathcal{G}\mathbf{C}_{t-1}\mathcal{G}' \qquad \text{and} \qquad \mathbf{W} = \frac{1-\delta}{\delta}\mathcal{G}\mathbf{C}_{t-1}\mathcal{G}'.$$

However, in some scientific cases it may be desirable to characterise the harmonics as evolving independently. Then the procedure of Section 6.3 can be employed, possibly using different discount factors for different component harmonics, consistent with a view that some are more durable than others.

8.6.5 *Deriving the seasonal effects from Fourier DLMs*

Given any DLM with a Fourier component DLM $\{\mathcal{F},\mathcal{G},\cdot,\cdot\}$, let the current state of information about the Fourier coefficients at time t be described by the marginal posterior

$$(\boldsymbol{\theta}_t \mid D_t) \sim \mathrm{N}[\mathbf{m}_t, \mathbf{C}_t].$$

Characterise the seasonal effect pattern at time t by $\boldsymbol{\phi}_t$, where the sum of its elements is zero. The p-vector $\boldsymbol{\phi}_t$ is defined by

$$\boldsymbol{\phi}_t = \mathbf{L}\boldsymbol{\theta}_t = \begin{pmatrix} \mathcal{F}' \\ \mathcal{F}'\mathcal{G} \\ \vdots \\ \mathcal{F}'\mathcal{G}^{p-1} \end{pmatrix} \boldsymbol{\theta}_t, \tag{8.10}$$

and it follows that

$$(\boldsymbol{\phi}_t|D_t) \sim \mathrm{N}[\mathbf{L}\mathbf{m}_t,\ \mathbf{L}\mathbf{C}_t\mathbf{L}'].$$

When estimating the observation variance V_t, the analogous result holds with a T distribution replacing the normal.

The reverse transformation is given by

$$\boldsymbol{\theta}_t = \mathbf{H}\boldsymbol{\phi}_t = (\mathbf{L}'\mathbf{L})^{-1}\mathbf{L}'\boldsymbol{\phi}_t,$$

where $\mathbf{H}$ is a $(p-1)\times p$ matrix. Upon using the trigonometric identities, it is clear that $\mathbf{L}'\mathbf{L}$ is always a diagonal matrix.

EXAMPLE 8.4. In the case $p=4$,

$$\mathbf{F} = \begin{pmatrix} 1 \\ 0 \\ 1 \end{pmatrix}, \qquad \mathbf{G} = \begin{pmatrix} 0 & 1 & 0 \\ -1 & 0 & 0 \\ 0 & 0 & -1 \end{pmatrix}, \qquad \mathbf{T} = \begin{pmatrix} 1 & 0 & 1 \\ 0 & 1 & -1 \\ -1 & 0 & 1 \end{pmatrix},$$

and we have

$$\mathbf{L} = \begin{pmatrix} 1 & 0 & 1 \\ 0 & 1 & -1 \\ -1 & 0 & 1 \\ 0 & -1 & -1 \end{pmatrix},$$

$$\mathbf{H} = \begin{pmatrix} 0.5 & 0 & -0.5 & 0 \\ 0 & 0.5 & 0 & -0.5 \\ 0.25 & -0.25 & 0.25 & -0.25 \end{pmatrix}.$$

8.6.6 Reduced Fourier form models

Sometimes a modeller requires a DLM that describes a seasonal pattern in simpler terms than that given by a full p seasonal effects DLM. The case of annual climatic cycles related to the earth's revolution about the sun are typical. In such cases, an economic representation in terms of only a few harmonics, or even a single component, may suffice. The construction of such reduced models is easy, just superimposing selected harmonic DLMs and omitting other insignificant harmonics. For example, for monthly data having an annual cycle, the DLM

$$\left\{ \begin{pmatrix} \mathbf{E}_2 \\ \mathbf{E}_2 \end{pmatrix}, \begin{pmatrix} \mathbf{J}_2(1,\omega) & \mathbf{0} \\ \mathbf{0} & \mathbf{J}_2(1,4\omega) \end{pmatrix}, \cdot, \cdot \right\}, \quad \text{with} \quad \omega = 2\pi/12,$$

confines the seasonal form to a composition of the first and fourth harmonics. Within this restriction the DLM can accommodate any composition of amplitudes and phases. Usually a single discount factor is applied to this DLM, but if required, different discount factors may be used for the two components to reflect the view that they develop independently and/or one is more durable than the other.

When appropriate, apart from being more economic and meaningful, a reduced form model produces a better forecasting performance than a full model. Clearly, included harmonic components that truly have very little or no effect degrade forecast performance since their assessment introduces extra variation and correlated forecast errors. For example, in a first-order polynomial trend/seasonal effects model, the full Fourier form representation leads to a simple updating for the current seasonal factor and no others. In contrast, a reduced form model updates the entire seasonal pattern, revising forecasts for the whole period rather than just individual time points. The message is that in addition to considering whether a reduced form model should be used for reasons related to the application area, the practical significance of components in the model should be assessed over time and possibly removed if deemed negligible. However this does not mean that such components will always be unnecessary, since future changes in seasonal patterns may require that they be reinstated.

As with full harmonic models, the Fourier coefficients in a reduced model are related to the constrained seasonal effects. In the (possibly reduced) n-dimensional Fourier DLM $\{\mathbf{F}, \mathbf{G}, \cdot, \cdot\}$, define the $p \times n$ matrix $\mathbf{L}$ and the $n \times p$ matrix $\mathbf{H}$ by

$$\mathbf{L} = \begin{pmatrix} \mathbf{F}' \\ \mathbf{F}'\mathbf{G} \\ \vdots \\ \mathbf{F}'\mathbf{G}^{p-1} \end{pmatrix} \qquad \text{and} \qquad \mathbf{H} = (\mathbf{L}'\mathbf{L})^{-1}\mathbf{L}'.$$

Then the orthogonality properties underlying the trigonometric identities of Section 8.6.2 apply just as in a full Fourier model. If this DLM has state vector $\boldsymbol{\theta}_t$ then the relationships between Fourier parameters and seasonal effects are simply $\boldsymbol{\phi}_t = \mathbf{L}\boldsymbol{\theta}_t$ and $\boldsymbol{\theta}_t = \mathbf{H}\boldsymbol{\phi}_t$, for all t.

EXAMPLE 8.5. For $p = 4$, using just the first harmonic, $n = 2$,

$$\mathbf{F} = \begin{pmatrix} 1 \\ 0 \end{pmatrix}, \qquad \mathbf{G} = \begin{pmatrix} 0 & 1 \\ -1 & 0 \end{pmatrix}, \qquad \mathbf{L} = \begin{pmatrix} 1 & 0 \\ 0 & 1 \\ -1 & 0 \\ 0 & -1 \end{pmatrix},$$

$$\mathbf{H} = \begin{pmatrix} 0.5 & 0 & -0.5 & 0 \\ 0 & 0.5 & 0 & -0.5 \end{pmatrix}.$$

8.6.7 *Assessing the importance of harmonic components*

Many readers will be familiar with spectral analysis and in particular, periodogram analysis, that provides a static assessment of the statistical significance of harmonics relative to a specified period p. The generalisation of this to dynamic analysis examines the statistical significance of harmonics based on the posterior distribution of the harmonic coefficients, thus allowing for changes in time and the presence of related variables and other dynamic components. Of course in practice, the modeller will finally judge the importance of harmonics by their practical significance: it is possible that a harmonic may be statistically significant without being practically significant and vice versa.

At time t, let the posterior distribution for the coefficient vector $\boldsymbol{\theta}_t$ be $(\boldsymbol{\theta}_t \mid D_t) \sim \mathrm{N}[\mathbf{m}_t, \mathbf{C}_t]$. As usual the observational variance sequence is assumed known, otherwise the normal distribution is simply replaced by a multivariate T, with modifications as noted below. Denote the marginal posteriors of the coefficients of the rth harmonic by

(a) $(\boldsymbol{\theta}_{tr} \mid D_t) \sim \mathrm{N}\left[(a_{tr}, b_{tr})', \mathbf{C}_{tr}\right]$, if $r \neq p/2$;
(b) $(\theta_{t,p/2} \mid D_t) \sim \mathrm{N}[a_{t,p/2}, C_{t,p/2}]$.

Notice that a_{tr} is the posterior estimate of the harmonic component of the seasonal pattern at time t, namely S_{tr}. A time plot of the estimates a_{tr}, (or the associated filtered values if filtering has been performed), with an indication of the associated uncertainty as measured by the posterior variance, provides a clear and useful visual indication of the contribution of the rth harmonic. A formal statistical assessment may be based on the calculation of HPD (highest posterior density) regions (Section 17.3.5) for the individual harmonics, as follows.

Writing $\mathbf{m}_{tr} = (a_{tr}, b_{tr})'$, we have the following distributions at time t and, of course, conditional on D_t :

(1) If the variances V_t are known, then with the usual χ^2_v as a random quantity having a standard chi-square distribution with v degrees of freedom,
  (a) $(\boldsymbol{\theta}_{tr} - \mathbf{m}_{tr})'\mathbf{C}_{tr}^{-1}(\boldsymbol{\theta}_{tr} - \mathbf{m}_{tr}) \sim \chi^2_2$, if $r \neq p/2$;
  (b) $(\theta_{t,p/2} - a_{t,p/2})^2/C_{t,p/2} \sim \chi^2_1$.
(2) If the variance V is unknown but is estimated in the usual way, with n_t degrees of freedom at time t, then with F_{v_1,v_2} representing the F distribution with v_1 and v_2 degrees of freedom,
  (a) $(\boldsymbol{\theta}_{tr} - \mathbf{m}_{tr})'\mathbf{C}_{tr}^{-1}(\boldsymbol{\theta}_{tr} - \mathbf{m}_{tr})/2 \sim F_{2,n_t}$, if $r \neq p/2$;
  (b) $(\theta_{t,p/2} - a_{t,p/2})^2/C_{t,p/2} \sim F_{1,n_t}$.

When considering the retention of a harmonic, the usual statistical tests simply calculate the following probabilities, high probabilities indicating that retention is statistically advisable.

(1) If the variances V_t are known,
  (a) $\Pr(\chi^2_2 \leq \mathbf{m}'_{tr}\mathbf{C}_{tr}^{-1}\mathbf{m}_{tr})$;
  (b) $\Pr(\chi^2_1 \leq a^2_{t,p/2}/C_{t,p/2})$.
(2) Or, when the variance is unknown,
  (a) $\Pr(F_{2,n_t} \leq \mathbf{m}'_{tr}\mathbf{C}_{tr}^{-1}\mathbf{m}_{tr}/2)$;
  (b) $\Pr(F_{1,n_t} \leq a^2_{t,p/2}/C_{t,p/2})$.

EXAMPLE 8.6. Gas consumption data. The data series in Table 8.1 is used to illustrate the above Fourier decomposition. The 65 observations are monthly totals of inland U.K. natural gas consumption over the period May 1979 to September 1984 inclusive, as derived from the Central Statistical Office Monthly Digest. Since gas usage follows the annual temperature cycle, the first harmonic is expected to dominate the seasonal pattern. However, as is evident from Figure 8.2, higher-frequency harmonics are necessary to account for industrial demand patterns and holiday effects.

The simplest model, a first-order polynomial, seasonal effects DLM with 12 parameters, is used, 6 harmonics representing the 11 seasonal effects. The final posterior distribution for the static model, $\{\mathbf{F}, \mathbf{G}, V, \mathbf{0}\}$ with a reference prior, is akin to a periodogram analysis, the F-values for the 6 harmonics at $t = 65$ being

Harmonic	1	2	3	4	5	6
F-value	837.8	1.18	6.86	68.14	12.13	0.32

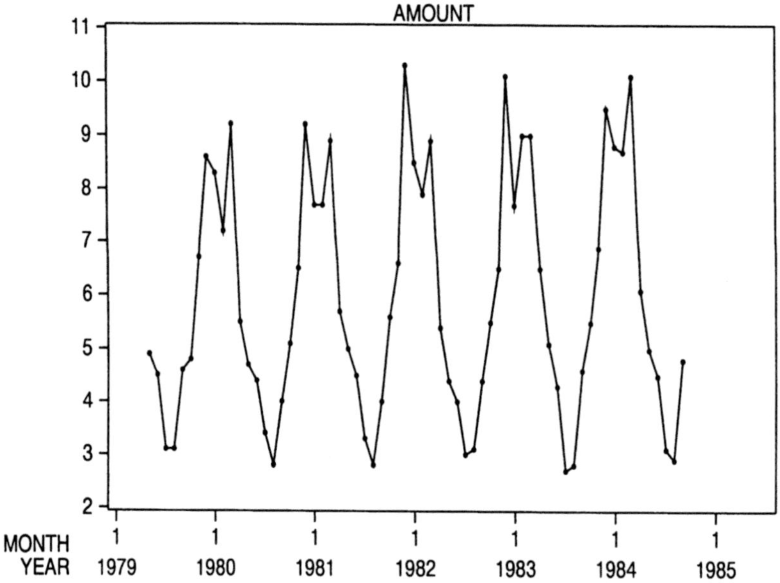

Figure 8.2 Gas consumption data

Table 8.1. UK gas consumption: amount in 10^6 tonnes coal equivalent

	Month											
Year	1	2	3	4	5	6	7	8	9	10	11	12
1979					4.9	4.5	3.1	3.1	4.6	4.8	6.7	8.6
1980	8.3	7.2	9.2	5.5	4.7	4.4	3.4	2.8	4.0	5.1	6.5	9.2
1981	7.7	7.7	8.9	5.7	5.0	4.5	3.3	2.8	4.0	5.6	6.6	10.3
1982	8.5	7.9	8.9	5.4	4.4	4.0	3.0	3.1	4.4	5.5	6.5	10.1
1983	7.7	9.0	9.0	6.5	5.1	4.3	2.7	2.8	4.6	5.5	6.9	9.5
1984	8.8	8.7	10.1	6.1	5.0	4.5	3.1	2.9	4.8			

In this static reference analysis the final degrees of freedom for the Student T posteriors is 53 (65 minus 12), thus the first five F-values may be compared to the $F_{2,53}$ distribution, and the final value for the Nyquist term to the $F_{1,53}$. On this basis, harmonics 1 and 4 are enormously significant, clearly indicating the dominance of the annual and quarterly cycles. Less significant, but still important, are harmonics 3 and 5. Harmonics 2 and 6 are insignificant. Whilst not definitive, this static "periodogram" reference analysis serves to identify the key harmonics.

For a more appropriate exploration of a dynamic model, consider a single discount model with discount factor of 0.95, so given the posterior

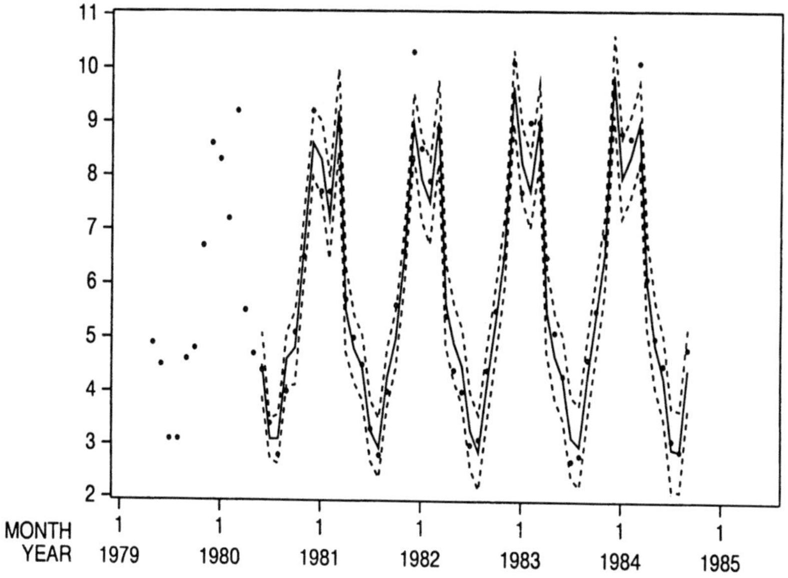

Figure 8.3 Gas consumption and the one–step point forecasts

variance matrix $\mathbf{C}_{t-1}$, $\mathbf{R}_t = 1.05\ \mathbf{G}\mathbf{C}_{t-1}\mathbf{G}'$. Whilst not optimised in terms of discount factors, this allows for changing parameter values. A reference prior-based analysis with this model is partially illustrated in Figures 8.3 to 8.10.

Figure 8.3 plots the data together with one-step ahead point forecasts and corresponding 90% Student T probability limits; the forecasts only appear in this reference analysis after 13 data points. The one-step forecasts look good, and the model adapts to the slight changes in pattern from year to year, such changes being particularly evident in the December levels of consumption. Figure 8.4 is produced after retrospective smoothing using the backwards filtering algorithms. This displays the retrospective estimates of the seasonal effects in each month, the vertical bar representing a 90% posterior probability interval symmetrically located about the posterior mode for the effect in that month. Figures 8.5 to 8.10 show the individual harmonic components of the seasonal pattern as retrospective posterior intervals for the harmonics $S_{tr} = \mathbf{E}_2'\boldsymbol{\theta}_{tr}$, $(r = 1, \ldots, 6)$. The dominance of harmonics 1 and 4 is clear from these plots, as is the nature of their contribution to the overall pattern.

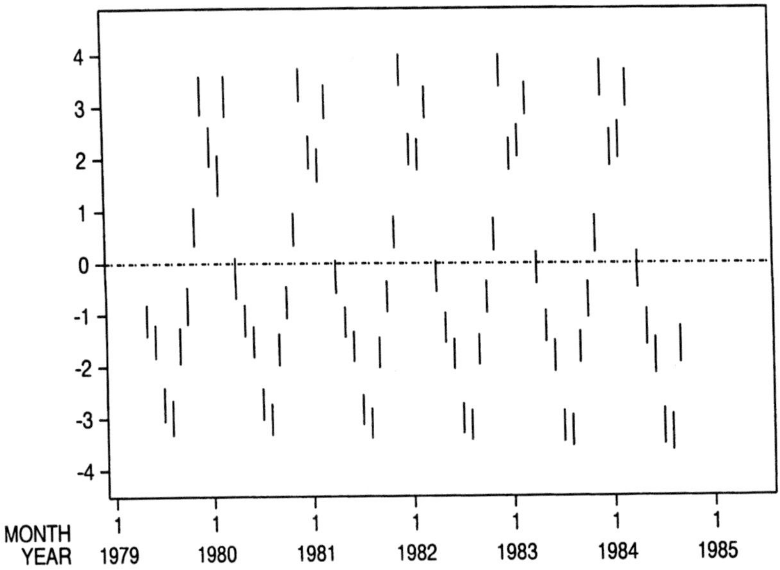

Figure 8.4 Estimated seasonal pattern in consumption series

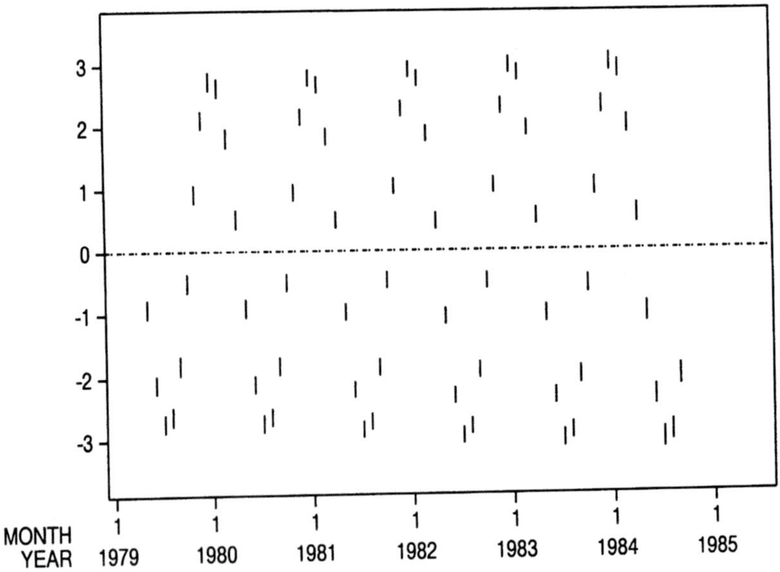

Figure 8.5 Harmonic 1

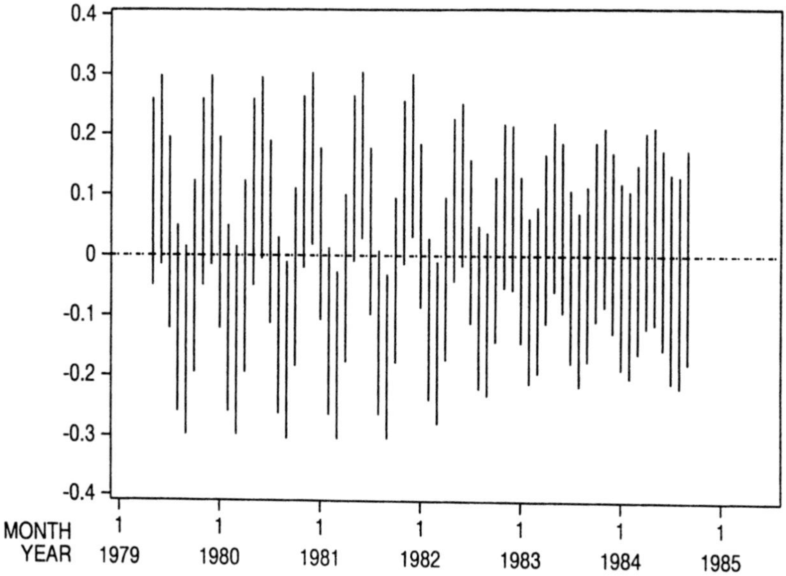

Figure 8.6 Harmonic 2

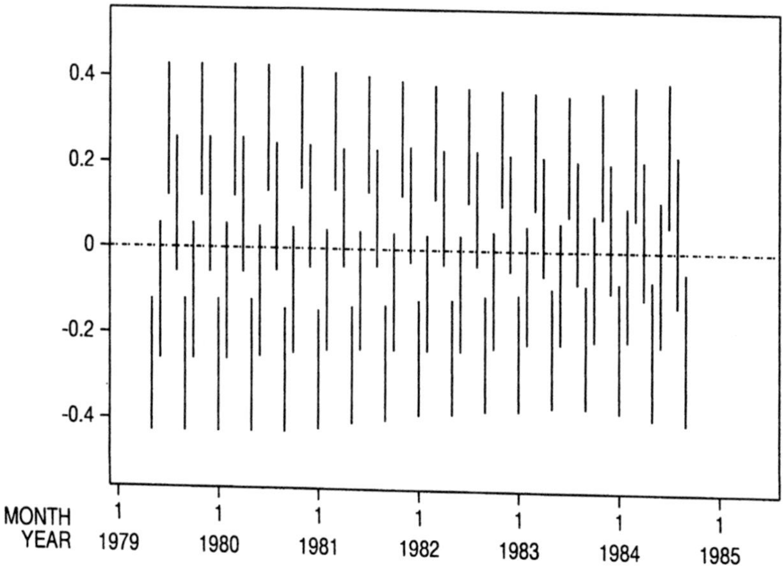

Figure 8.7 Harmonic 3

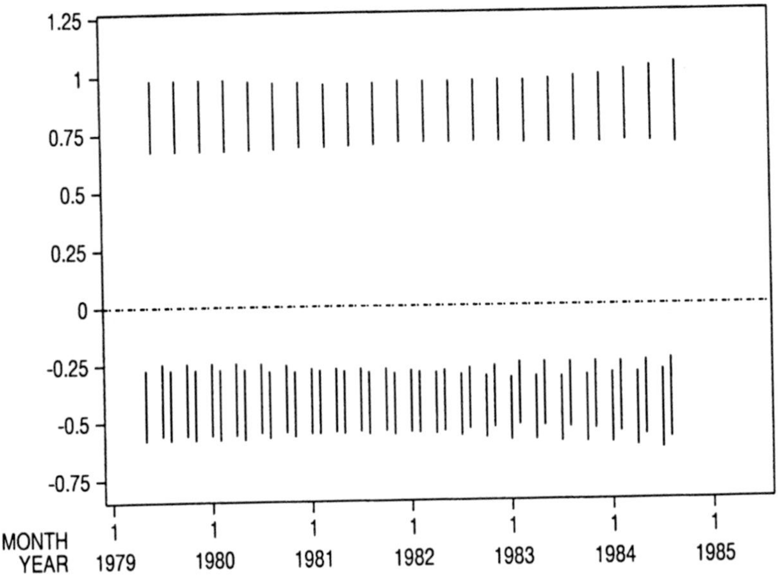

Figure 8.8 Harmonic 4

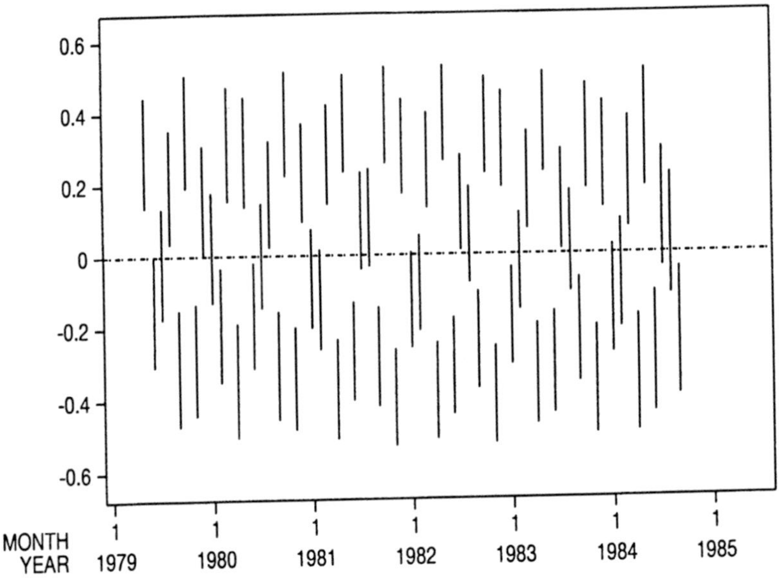

Figure 8.9 Harmonic 5

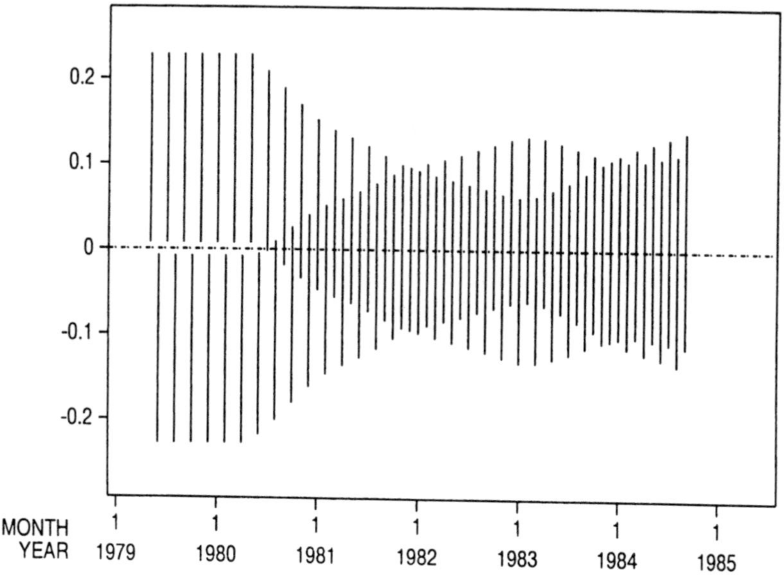

Figure 8.10 Harmonic 6

8.7 EXERCISES

These exercises concern seasonal DLMs with known variances. There is no problem in applying the usual procedures to extend the results to the unknown variance case. Some of the exercises are hard and the reader may simply wish to read the questions as part of the text.

(1) Consider the $p \times p$ permutation matrix

$$\mathbf{P} = \begin{pmatrix} \mathbf{0} & \mathbf{I}_{p-1} \\ 1 & \mathbf{0} \end{pmatrix}.$$

Verify that

(a) $\mathbf{P}$ is p-cyclic, so that $\mathbf{P}^{k+np} = \mathbf{P}^k$ for all integers k and n;

(b) the eigenvalues of $\mathbf{P}$ are the p roots of unity $\exp(2\pi ij/p)$ for $j = 0, \ldots, p-1$;

(c) the DLM $\{\mathbf{E}_p, \mathbf{P}, \cdot, \cdot\}$ is observable, with identity observability matrix.

(2) Using the basic trigonometric identities, or otherwise, verify that for integer $k \geq 1$, $p \geq 2$, and $\psi = 2k\pi/p$, the matrix

$$J_2(1, \psi) = \begin{pmatrix} \cos(\psi) & \sin(\psi) \\ -\sin(\psi) & \cos(\psi) \end{pmatrix}$$

is p-cyclic.

(3) A seasonal effects component DLM is used to model quarterly data, so that $p = 4$. A forecaster, unused to thinking about covariances, specifies initial prior moments for the seasonal factors as

$$(\boldsymbol{\theta}_0|D_0) \sim \mathrm{N}\left[\begin{pmatrix} -100 \\ 100 \\ -300 \\ 400 \end{pmatrix}, \begin{pmatrix} 200 & 0 & 0 & 0 \\ 0 & 200 & 0 & 0 \\ 0 & 0 & 200 & 0 \\ 0 & 0 & 0 & 400 \end{pmatrix}\right].$$

(a) Show that this is an invalid prior for seasonal effects.
(b) Derive a valid prior based on the provided information.

(4) In the first-order polynomial/seasonal effects Normal DLM,

$$\left\{\begin{pmatrix} 1 \\ \mathbf{E}_p \end{pmatrix}, \begin{pmatrix} 1 & \mathbf{0} \\ \mathbf{0} & \mathbf{P} \end{pmatrix}, V, \begin{pmatrix} W & \mathbf{W}_2 \\ \mathbf{W}_2' & \mathbf{W}_3 \end{pmatrix}\right\}, \quad \text{with} \quad \boldsymbol{\theta}_t = \begin{pmatrix} \mu_t \\ \boldsymbol{\phi}_t \end{pmatrix},$$

let $X_t = \sum_{r=1}^{p} Y_{t+r}$ be the annual demand. Suppose that

$$\left(\begin{matrix} \mu_t \\ \boldsymbol{\phi}_t \end{matrix}\middle| D_t\right) \sim \mathrm{N}\left[\begin{pmatrix} m_t \\ \mathbf{s}_t \end{pmatrix}, \begin{pmatrix} C & \mathbf{C}_2' \\ \mathbf{C}_2 & \mathbf{C}_3 \end{pmatrix}\right].$$

(a) If $\mathbf{W}_3 = \mathbf{0}$ and $\mathbf{W}_2 = \mathbf{0}$, prove that

$$(X_t \mid D_t) \sim \mathrm{N}\left[pm_t,\ pV + \frac{p(p+1)}{2}W + p^2C\right],$$

a result that does not depend on the information on the seasonal effects.

(b) If $\mathbf{W}_3 \neq \mathbf{0}$, in what way does the result in (a) change?

(5) The following table gives an Eire milk production figure for each month of 1975 in terms of millions of gallons:

Month	Jan	Feb	Mar	Apr	May	Jun
Milk	6	13	36	64	99	99
Month	Jul	Aug	Sep	Oct	Nov	Dec
Milk	89	82	64	43	19	9

(a) Express these figures in the Fourier form

$$a_0 + \sum_{r=1}^{5}(a_r \cos(2\pi rt/12) + b_r \sin(2\pi rt/12)) + a_6(-1)^t,$$

calculating the mean a_0 and the Fourier coefficients a_r and b_r.

(b) Derive the amplitudes A_k and phases γ_k associated with the kth harmonic, $k = 1, \ldots, 6$.

(c) Calculate the percentage of variation accounted for by each harmonic and draw a first impression as to how you might characterise the seasonality in a DLM.

(6) Consider the regression DLM $\{\mathbf{F}_t, \mathbf{I}_3, \cdot, \cdot\}$. $\mathbf{I}_3$ is the 3×3 identity matrix and $\mathbf{F}_{t+4} = \mathbf{F}_t$ for all t, so that $\mathbf{F}_t$ is 4-cyclic such that

$$\mathbf{F}_1 = \begin{pmatrix} 1 \\ 0 \\ 0 \end{pmatrix}, \quad \mathbf{F}_2 = \begin{pmatrix} 0 \\ 1 \\ 0 \end{pmatrix}, \quad \mathbf{F}_3 = \begin{pmatrix} 0 \\ 0 \\ 1 \end{pmatrix}, \quad \mathbf{F}_4 = \begin{pmatrix} -1 \\ -1 \\ -1 \end{pmatrix}.$$

(a) Show that the forecast function is that of a form-free seasonal effects DLM of period $p = 4$, so that this DLM offers an alternative way of modelling such a seasonal effects component.

(b) Generalise this regression DLM representation of form-free seasonal effects to all integers $p > 1$.

(7) With real $g \neq 0$, consider the DLM

$$\left\{ \begin{pmatrix} 1 \\ 1 \end{pmatrix}, \begin{pmatrix} 1 & g \\ -g & 1 \end{pmatrix} \cos(\omega), \cdot, \cdot \right\},$$

where $\omega = \arccos(1/\sqrt{1+g^2})$.

(a) Derive the forecast function $f_t(k)$.

(b) What is the period of the forecast function?

(c) Show that the DLM is similar to the DLM $\{\mathbf{E}_2, \mathbf{J}_2(1, \omega), \cdot, \cdot\}$.

(8) Consider the full seasonal effects Fourier form DLM $\{\mathcal{F}, \mathcal{G}, \cdot, \cdot\}$ with parameter $\boldsymbol{\theta}_t$ and observability matrix $\mathbf{T}$, as in Sections 8.6.4 and 8.6.5.

(a) Verify that the p seasonal effects $\boldsymbol{\phi}_t$ are given by the equation

$$\boldsymbol{\phi}_t = \mathbf{L}\boldsymbol{\theta}_t = \begin{pmatrix} \mathbf{T} \\ \mathcal{F}'\mathcal{G}^{p-1} \end{pmatrix},$$

where $\mathbf{L}$ is a $p \times (p-1)$ matrix.

(b) Verify that for integer p, $\mathbf{L}'\mathbf{L}$ is a non-singular diagonal matrix.

(c) Verify that the inverse relationship is $\boldsymbol{\theta}_t = \mathbf{H}\boldsymbol{\phi}_t$, where $\mathbf{H} = (\mathbf{L}'\mathbf{L})^{-1}\mathbf{L}'$ is a $(p-1) \times p$ matrix.

(d) Obtain the matrices $\mathbf{L}$ and $\mathbf{H}$ when $p = 5$.

(e) Obtain the matrices $\mathbf{L}$ and $\mathbf{H}$ when $p = 6$.

(9) In an analysis of a company's quarterly carpet sales, as calculated in the fourth quarter of 1994, the final marginal posterior for the first and second harmonic Fourier coefficients was

$$\left(\begin{matrix} a_1 \\ b_1 \\ a_2 \end{matrix} \middle| D \right) \sim \mathrm{N}\left[\begin{pmatrix} -6.2 \\ -59.0 \\ -0.3 \end{pmatrix}, \begin{pmatrix} 5.6 & -0.2 & -0.1 \\ -0.2 & 6.2 & -0.3 \\ -0.1 & -0.3 & 2.7 \end{pmatrix} \right].$$

(a) Assess the importance of the harmonics according to distribution of their amplitudes; in this case no formal analysis is really necessary.

(b) After discarding any negligible harmonics, derive the posterior $(\boldsymbol{\phi}|D)$ for the four seasonal quarters, making sure that these are not out of phase (Q3 should be the peak). You can use the relation $\boldsymbol{\phi} = \mathbf{L}\boldsymbol{\theta}$ of Example 8.5.

(10) The monthly carpet sales of the company in the previous question are modelled using a linear growth/Fourier seasonal effects DLM that includes both the first and third harmonic. In December 1994 write $\boldsymbol{\theta}_{dec} = (a_1, b_1, a_3, b_3)'$ for the current Fourier coefficients, and suppose that the marginal posterior is summarised in terms of the posterior mean and variance matrix of $\boldsymbol{\theta}_{dec}$, namely

$$\left[\begin{pmatrix} -12.3 \\ -16.4 \\ 3.4 \\ -2.5 \end{pmatrix}, \begin{pmatrix} 0.82 & 0.02 & -0.02 & 0.02 \\ 0.02 & 0.87 & -0.02 & 0.01 \\ -0.02 & -0.02 & 0.74 & -0.01 \\ 0.02 & 0.01 & -0.01 & 0.76 \end{pmatrix}\right].$$

(a) Write down the canonical form of the complete DLM in terms of $\{\mathbf{F}, \mathbf{G}\}$.

(b) Let the twelve seasonal effects be $\boldsymbol{\phi}_{jan} = (s_{jan}, \dots, s_{dec})$. Being careful about the phase, derive the posterior distribution of $\boldsymbol{\phi}_{jan} = \mathbf{M}\boldsymbol{\theta}_{dec}$, identifying the appropriate matrix $\mathbf{M}$. (The peak-to-trough difference should be about 47).

(11) Consider the single harmonic component normal DLM

$$\left\{\begin{pmatrix} 1 \\ 0 \end{pmatrix}, \begin{pmatrix} \cos(\omega) & \sin(\omega) \\ -\sin(\omega) & \cos(\omega) \end{pmatrix}, V_t, \begin{pmatrix} W_{t1} & W_{t2} \\ W_{t2} & W_{t3} \end{pmatrix}\right\}.$$

(a) Write down the observation and system equations of this DLM.

(b) If $Z_t = Y_t - 2Y_{t-1}\cos(\omega) + Y_{t-2}$, show that

$$Z_t = \nu_t - 2\nu_{t-1}\cos(\omega) + \nu_{t-2} + \rho_t,$$

where

$$\rho_t = w_{t1} - w_{t-1,1}\cos(\omega) + w_{t-1,2}\sin(\omega).$$

(c) Hence show that initially and based upon the *truth* of the DLM,

$$\mathrm{E}[Z_t] = 0 \quad \text{and} \quad \mathrm{C}[Z_t, Z_{t-k}] = 0, \text{ for all } k > 2.$$

(d) Additionally assuming a constant DLM, show that Z_t can be represented as a moving average process of order 2,

$$Y_t - 2Y_{t-1}\cos(\omega) + Y_{t-2} = a_t - \psi_1 a_{t-1} + \psi_2 a_{t-2},$$

where $a_t \sim \mathrm{N}[0, Q]$ independently.

(e) Using the limit theorems of Chapter 5 and remembering that

$$\begin{aligned} Y_{t+1} &= a_t\cos(\omega) + b_t\sin(\omega) + e_{t+1}, \\ a_t &= a_{t-1}\cos(\omega) + b_{t-1}\sin(\omega) + A_{1t}e_t, \\ b_t &= b_{t-1}\cos(\omega) - a_{t-1}\sin(\omega) + A_{2t}e_t, \end{aligned}$$

show that for this constant DLM,

$$\lim_{t\to\infty} \{Z_t - e_t + Be_{t-1} - (1 - A_1)e_{t-2}\} = 0,$$

where $\lim \mathbf{A}_t = (A_1, A_2)'$ and $B = (2 - A_1 - A_2 \tan(\omega))\cos(\omega)$. As a result, and using the earlier parts of this question, it should be clear that $\{Q, A_1, A_2\}$ may be derived in terms of V and $\mathbf{W}$.

(12) In full, the single harmonic discount DLM $\{\mathbf{E}_2, \mathbf{J}_2(1,\omega), 1, \mathbf{W}_t\}$ with discount factor $0 < \delta < 1$ is

$$\left\{ \begin{pmatrix} 1 \\ 0 \end{pmatrix}, \begin{pmatrix} \cos(\omega) & \sin(\omega) \\ -\sin(\omega) & \cos(\omega) \end{pmatrix}, 1, \begin{pmatrix} W_{1t} & W_{2t} \\ W_{2t} & W_{3t} \end{pmatrix} \right\},$$

with $\mathbf{R}_t = \mathbf{C}_{t-1}/\delta$ for all t.

(a) Verify that

$$\mathbf{J}_2^{-1}(1,\omega) = \mathbf{J}_2'(1,\omega).$$

(b) By considering the usual DLM relationship

$$\mathbf{C}_t^{-1} = \mathbf{R}_t^{-1} + \mathbf{F}V_t^{-1}\mathbf{F}',$$

without necessarily determining it, prove that $\lim_{t\to\infty} \mathbf{C}_t = \mathbf{C}$ exists and consequently so do limiting values $\{\mathbf{R}, \mathbf{A}, Q\}$ for $\{\mathbf{R}_t, \mathbf{A}_t, Q_t\}$.

(c) Show that the eigenvalues of the matrix $\mathbf{H} = \mathbf{C}\mathbf{R}^{-1}\mathbf{J}_2(1,\omega)$ are simply those of $\mathbf{J}_2^{-1}(1,\omega)$ multiplied by δ.

(d) Hence, using the result of Theorem 5.7 or otherwise show that

$$\lim_{t\to\infty} \left[Y_t - e_t - 2(Y_{t-1} - \delta e_{t-1})\cos(\omega) + Y_{t-2} - \delta^2 e_{t-2}\right] = 0.$$

(e) By comparison with the previous question deduce that

$$A_1 = 1 - \delta^2, \quad A_2 = (1-\delta)^2 \cot(\omega), \quad \text{and} \quad Q = 1/\delta^2.$$

(f) Show that

$$\mathbf{C} = \begin{pmatrix} A_1 & A_2 \\ A_2 & c \end{pmatrix} \quad \text{and} \quad \mathbf{R} = \frac{1}{\delta^2}\begin{pmatrix} A_1 & A_2 \\ A_2 & r \end{pmatrix},$$

where $c = (1-\delta)[3 - \delta + (1-\delta)^2/(\delta \sin^2(\omega))]$ and $r = \delta^2 c + A_2^2$.

(13) Write a computer program to implement a DLM with second-order polynomial and Fourier form seasonal components.† The program should provide the following facilities:

(a) Allow a user to select any number of harmonic components.

(b) Structure the evolution variance sequence in component discount form as in Definition 6.1, with one discount factor δ_T for the trend and one δ_S for the seasonal component. These discount factors are to be specified by the user.

---

†The BATS package (Pole, West, and Harrison 1994) implements this, as well as a wider class of models

(c) Allow for the choice of initialisation via either the reference analysis of Section 4.8, or through user input priors.

(d) Produce numerical and graphical summaries of analyses. Useful outputs are time graphs of on-line and filtered estimates, with uncertainties indicated by intervals about the estimates, of (i) the non-seasonal trend, (ii) the seasonal component, (iii) the individual harmonics, etc.

(14) Test the program by analysing the gas consumption data of Table 8.1. Gain experience in the use of the model by exploring different analyses of this series. In particular, with the full 6 harmonics, experiment with this data series with a range of different initial priors (including the reference analysis), and different values of the discount factors in the range $0.85 < \delta_T, \delta_S \leq 1$. Note that the analysis in Section 8.6.5 uses a model in which the trend is first-order, rather than second-order polynomial, and that this can be reproduced in the more general model by specifying zero mean and zero variance for the growth parameter.

(15) Reanalyse the gas consumption data using a reduced Fourier form model having only two harmonic components, $r = 1$ and $r = 4$, for the fundamental (annual) cycle and the fourth (quarterly) cycle.

(16) The data below are quarterly total sales (in thousands) of one-day-old turkey chicks from hatcheries in Eire over a period of years (taken from Ameen and Harrison 1985a).

Eire turkey chick data

	Quarter			
Year	1	2	3	4
1974	131.7	322.6	285.6	105.7
1975	80.4	285.1	347.8	68.9
1976	203.3	375.9	415.9	65.8
1977	177.0	438.3	463.2	136.0
1978	192.2	442.8	509.6	201.2
1979	196.0	478.6	688.6	259.8
1980	352.5	508.1	701.5	325.6
1981	305.9	422.2	771.0	329.3
1982	384.0	472.0	852.0	

Analyse these data using a reduced Fourier form model having only two harmonic components, $r = 1$ and $r = 4$, for the fundamental (annual) cycle and the fourth (quarterly) cycle. There is obvious, sustained growth in the series over the years and a marked annual seasonal pattern. Explore various models, discount factors, etc., comparing them through subjective exploration of graphical summaries together with numerical summaries such as the MSE, MAD and LLR measures. Verify that a static model ($\delta_T = \delta_S = 1$) performs poorly relative to dynamic models, and that in particular, there is a fair degree of change indicated in the seasonal pattern.

(17) Re-analyse the turkey data after transformation of the original Y_t to an alternative scale. Compare graphs of transformed series against time, considering transforms such as $\log(Y_t)$, $Y_t^{1/2}$ and $Y_t^{3/4}$, and try to identify a transformed scale on which the amplitude of the seasonality is most constant (although, as with most real data, there is a fair degree of change in seasonal pattern on all scales).

(18) **The Periodogram.** Consider np data points $Y_1, \ldots, Y_{np}$, where $n > 1$ is an integer and the seasonal period p is also an integer. You model this series according to a **static** first-order polynomial /full Fourier seasonal effects **constant** normal DLM with **reference prior**. Let there be h harmonics in the seasonal description, the period being p.

(a) Show that the final posterior estimate $\mathbf{m}_{np} = \mathrm{E}[\boldsymbol{\theta}_{np}|D_{np}]$ is the least squares/normal maximum likelihood estimate that minimizes

$$S(\boldsymbol{\theta}) = \sum_{v=0}^{np-1} (Y_{np-v} - \mathbf{F}'\mathbf{G}^{-v}\boldsymbol{\theta})^2.$$

(b) Let A_k be the estimate of the amplitude of the kth harmonic as calculated from the posterior estimate $\mathbf{m}_{np}$. Show that the total sum of squares $S = \mathbf{Y}'\mathbf{Y} - (np)\bar{Y}^2$ about the mean $\bar{Y}$ can be written as

$$S = \mathbf{Y}'\mathbf{Y} - (np)\bar{Y}^2 = \sum_{v=1}^{h} S_v^2 + R,$$

where $S_v = npA_v^2/2$ for $v \neq p/2$, $S_{p/2} = npA_{p/2}^2$, and R is the residual sum of squares.

(c) Plot the graph $\{A_v^2, v : v = 1, \ldots, h\}$. This is known as the Periodogram and is widely used in time series analysis. Like most simple techniques that exploit orthogonality, it was extremely useful before the advent of powerful computing facilities. Clearly it is still of limited use in initial data analysis but is superseded by facilities such as DLMs, that provide appropriate dynamic analyses incorporating stochastic seasonals, other components (such as stochastic trends, related variables), and full posterior distributions over all times using retrospective analysis, interventions, priors, etc.

CHAPTER 9

REGRESSION, AUTOREGRESSION, AND RELATED MODELS

9.1 INTRODUCTION

We now turn to models incorporating regression components, including regressions on independent variables and related classes of transfer function DLMs. We then consider, in some detail, classes of traditional, stationary time series components, namely the class of autoregressive, moving average models, or ARMA models. We provide some basic discussion of ARMA models in DLM contexts, and then more extensive development of AR component models, especially in connection with time series decompositions. Finally, an important class of time-varying autoregressive component models is discussed and illustrated.

9.2 THE MULTIPLE REGRESSION DLM

9.2.1 Definition

Consider modelling the series Y_t by regressing on a collection of n independent, or regressor, variables labelled $X_1, \ldots, X_n$. The value of the ith variable X_i at each time t is assumed known, denoted by X_{ti}, $(i = 1, \ldots, n;\ t = 1, \ldots)$. Usually a constant term is included in the model, in which case X_1 is taken as unity, $X_{t1} = 1$ for all t. For $t = 1, \ldots,$ let the regression vector $\mathbf{F}_t$ be given by $\mathbf{F}_t' = (X_{t1}, \ldots, X_{tn})$. Then the multiple regression DLM with regressors $X_1, \ldots, X_n$ is defined by the quadruple $\{\mathbf{F}_t, \mathbf{I}, V_t, \mathbf{W}_t\}$, for some observational variances V_t and evolution variance matrices $\mathbf{W}_t$. This is just as specified in Definition 3.1.

For each t then, the model equations are

$$\begin{aligned} &\text{Observation equation:} \quad & Y_t &= \mathbf{F}_t'\,\boldsymbol{\theta}_t + \nu_t\,, \quad & \nu_t &\sim \mathrm{N}[0, V_t], \\ &\text{System equation:} \quad & \boldsymbol{\theta}_t &= \boldsymbol{\theta}_{t-1} + \boldsymbol{\omega}_t\,, \quad & \boldsymbol{\omega}_t &\sim \mathrm{N}[\mathbf{0}, \mathbf{W}_t]. \end{aligned}$$

Write the elements of $\boldsymbol{\theta}_t$ as $\boldsymbol{\theta}_t' = (\theta_{t1}, \ldots, \theta_{tn})$. Then the observation equation can be written as

$$Y_t = \mu_t + \nu_t,$$

where the mean response μ_t is given by

$$\mu_t = \mathbf{F}_t'\boldsymbol{\theta}_t = \sum_{i=1}^{n} \theta_{ti} X_{ti}.$$

From this representation it is clear that the model can be viewed as being formed from the superposition of n straight-line regressions with zero origin, the simple models of Chapter 3.

In the regression DLM the state vector evolves only via the addition of a noise term ω_t. This is a dynamic generalisation of standard static regression models. Setting $\mathbf{W}_t = \mathbf{0}$ for all t provides the specialisation to static regression for then $\boldsymbol{\theta}_t = \boldsymbol{\theta}$ is constant over time, and the model equations reduce to

$$Y_t = \mathbf{F}_t'\boldsymbol{\theta} + \nu_t \quad \text{with} \quad \nu_t \sim \mathrm{N}[0, V_t].$$

Bayesian analyses of such standard models are well documented; see, for example, Broemling (1985), Box and Tiao, (1973), De Groot (1971), Press (1985) and Zellner (1971). Here $\boldsymbol{\theta}$ plays the role of a fixed regression vector and the time ordering of the observations is not so relevant. In the dynamic regression, stochastic variation over time is permitted through the noise terms ω_t, to model changes in regression relationships. The dynamic model, whilst retaining the basic linear structure, offers flexibility in adapting to observed data in which the relationships between the response series and the regressors cannot be adequately represented by a static linear model. See Chapter 3 for further discussion and illustrations, and examples in Ameen and Harrison (1984), Harrison and Johnston (1984), and Johnston et al (1986).

9.2.2 Common types of regressions

The DLM structure allows for a host of possible forms of relationships through appropriate choice of regression variables and combinations of them. The basic types are quantitative measurements; indicator variables to group the response data according to an underlying classificatory variable, or factor; and higher-order terms involving interactions between variables of these types. For examples of each of these, consider the sort of data series analysed in Section 3.3.4. Suppose that Y_t represents sales of a product, the data being quarterly figures over several years. Particular examples of the common model forms are now discussed.

(1) Straight-line regressions on quantitative variables
Suppose that $X = X_2$ is a related predictor variable, some form of economic indicator, for example. As with Y, X is a quantitative measurement, and often such measurements are viewed as effectively continuous, although this is by no means necessary. A basic straight-line regression model is formed by taking $X_1 = 1$ to define an intercept term, so that $\mu_t = \alpha_t + \beta_t X_t$. Here, of course, $\mathbf{F}_t' = (1, X_t)$ and $\boldsymbol{\theta}_t' = (\alpha_t, \beta_t)$.

(2) Multiple regressions on quantitative variables
Straight-line regressions may be extended to include other quantitative variables by adding in further terms such as $\gamma_t Z_t$ by superposition to give a canonical multiple regression on several variables, each contributing a linear term. Here Z_t is the value of a second independent variable, $\mathbf{F}_t' = (1, X_t, Z_t)$ and $\boldsymbol{\theta}_t' = (\alpha_t, \beta_t, \gamma_t)$.

(3) **Lagged variables**
Particular cases of multiple regression of importance in time series forecasting involve the use of regressor variables calculated as *lagged* values of basic independent variables. In the sales forecasting scenario above, it may be thought that the economic indicator X has predictive power for the sales series into the future as well as just for the current time. A similar example concerns forecasting of sales or demand with the independent variable X relating to advertising and other promotional activities. Here current promotional expenditure can be expected to impact not only on immediate sales, but also on sales further into the future. Hence, in modelling the mean response μ_t at time t, past, or lagged, values of the regressor variable X should be considered in addition to the current value X_t. Generally, suppose that it is felt that appreciable effects of the regressor variable may be sustained up to a maximum lag of k time points for some $k > 1$. The linear regression on lagged values of X then has the general form of

$$\begin{aligned}\mu_t =& \alpha_t + \sum_{i=0}^{k} \beta_{ti} X_{t-i}\\ =& \alpha_t + \beta_{t0} X_t + \beta_{t1} X_{t-1} + \cdots + \beta_{tk} X_{t-k}.\end{aligned}$$

Here $\mathbf{F}'_t = (1, X_t, X_{t-1}, \ldots, X_{t-k})$ and $\boldsymbol{\theta}'_t = (\alpha_t, \beta_{t0}, \beta_{t1}, \ldots, \beta_{tk})$. Much use of these models has been made, of course, in economic forecasting (e.g., Granger and Newbold 1977; Zellner 1971). We return to lagged relationships in detail in Section 9.3 below. Also note that autoregressions fall into this framework. Here the response series is directly regressed on lagged values of itself, with $X_t = Y_{t-1}$ for all t. Again, much further discussion of these models appears below.

(4) **Polynomial surfaces**
Higher-order terms can be used to refine the basic description by defining further variables. For example, the two-variable model $\mu_t = \alpha_t + \beta_t X_t + \gamma_t Z_t$ may be refined by including further regressor variables that are quadratic, and higher-order, functions of the two original variables X and Z. Examples include quadratic terms X^2 or Z^2, and cross-product terms such as XZ. These, and higher-order powers and cross-products, can be used to define polynomial regressions on possibly several variables, building up *response surface* descriptions of the regression function (Box and Draper 1987).

(5) **Classificatory variables**
Classificatory variables, or factors, can be included using dummy X variables to indicate the classifications for each response observation. Seasonality, for example, can be modelled this way as has already been seen in Chapter 8. Specifically in this context, consider the quarterly classification of the sales data, and suppose that a simple seasonal factor model is

desired, with different levels of sales in each quarter, those levels changing stochastically from year to year. This is just the model of Section 8.1, and it can be represented in regression DLM form as follows. Let $n = 4$ and define X_i as the indicator variable for the ith quarter of each year. Thus, for $i = 1, \dots, 4$,

$$X_{ti} = \begin{cases} 1, & \text{when } Y_t \text{ is in quarter } i \text{ of any year;} \\ 0, & \text{otherwise.} \end{cases}$$

The state vector is $\boldsymbol{\theta}_t' = (\phi_{t1}, \dots, \phi_{t4})$, where ϕ_{ti} is the seasonal factor for quarter i of the year at time t. The mean response $\mu_t = \sum_{i=1}^4 \phi_{ti} X_{ti}$ at time t is then given simply by $\mu_t = \phi_{ti}$, the relevant seasonal level, when time t corresponds to the ith quarter of the year.

The seasonal effects model, with effects representing seasonal deviations from an underlying, non-seasonal sales level, is also easily, and obviously, representable in regression form. This is necessary with classificatory variables in general if other regressors are to be included by superposition. Set

$$\alpha_t = \frac{1}{4} \sum_{i=1}^4 \phi_{ti},$$

and for $i = 1, \dots, 4$, $\theta_{ti} = \phi_{ti} - \alpha_t$. Clearly the θ_{ti} sum to zero at each time t, representing the seasonal effects. Redefine the state vector as $\boldsymbol{\theta}_t' = (\alpha_t, \theta_{t1}, \dots, \theta_{t4})$ and the regression vector as $\mathbf{F}_t' = (1, X_{t1}, \dots, X_{t4})$, where the X variables are the above indicators of the quarters. Then

$$\mu_t = \mathbf{F}_t' \boldsymbol{\theta}_t = \alpha_t + \sum_{i=1}^4 \theta_{ti} X_{ti},$$

subject to the zero-sum restriction on the final four elements of the state vector.

Grouping data according to an underlying factor in this way has wide uses in time series modelling, just as in other areas of statistics. Perhaps the most common use is in designed experiments where the factors relate to treatment groups, block effects, and so forth. One other important example is the use of dummy variables as intervention indicators, the corresponding parameter then representing a shift due to intervention.

(6) Several factors

As an example, suppose that the data are also classified according to sales area, there being just two areas for simplicity. The additive model formed by superposition has the form

$$\mu_t = \alpha_t + \sum_{i=1}^4 \theta_{ti} X_{ti} + \sum_{j=1}^2 \gamma_{tj} Z_{tj},$$

where, as with seasonality, γ_{t1} and γ_{t2} are the effects added to sales level in the two different sales regions (having zero-sum), with Z_{t1} and Z_{t2} being indicator variables for the two regions. Clearly this could be extended with the addition of further regressors of any types.

(7) **Factor by factor interactions**

Higher-order terms involve interactions of two basic types. Firstly, two classificatory variables may interact, producing what is often referred to as a factor by factor interaction. In the above example, this amounts to dealing with the two sales regions separately, having different seasonal factors within each region. One way of modelling this is to add in an *interaction* term of the form

$$\sum_{i=1}^{4}\sum_{j=1}^{2}\beta_{tij}U_{tij}.$$

Here the U_{tij} are dummy, indicator variables with $U_{tij} = 1$ if and only if observation Y_t corresponds to quarter i and sales area j. The β_{tij} are the interaction parameters, subject to zero-sum constraints $\sum_{i=1}^{4}\beta_{tij} = 0$ for $j = 1, 2$, and $\sum_{j=1}^{2}\beta_{tij} = 0$ for $i = 1, \ldots, 4$.

(8) **Other forms of interaction**

The second, and highly important, form of interaction is typified as follows. Consider a straight-line regression on the variable X combined by superposition with the seasonal effects model to give a mean response function of the form

$$\mu_t = \alpha_t + \beta_t X_t + \sum_{i=1}^{4}\theta_{ti}X_{ti}.$$

The effects of the variable X and the seasonality are additive, not interacting. Often it may be felt that the effect of X on the response is different in different quarters; more generally, that the regression coefficient of a variable takes different values according to the various levels of a classifying factor. Here the necessary refinement of the model is

$$\mu_t = \alpha_t + \sum_{i=1}^{4}(\theta_{ti} + \beta_{ti}X_t)X_{ti}.$$

In this case, we have $\mathbf{F}_t = (1; X_{t1}, \ldots, X_{t4};\ X_tX_{t1}, \ldots, X_tX_{t4})'$ and with state vector $\boldsymbol{\theta}_t = (\alpha_t; \theta_{t1}, \ldots, \theta_{t4};\ \beta_{t1}, \ldots, \beta_{t4})'$.

9.2.3 Summary of analysis

The analysis follows from the general theory of DLMs. The results are given in the case of constant, unknown observational variance, $V_t = V = 1/\phi$ for all t, with ϕ unknown, consistent with the summary in Section 4.6.

Thus, $\mathbf{W}_t$ is scaled by the current estimate S_{t-1} of $V = \phi^{-1}$ at each time. Then, as usual, the evolution/updating cycle is based on the following distributions:

$$\begin{aligned}
(\boldsymbol{\theta}_{t-1} \mid D_{t-1}) &\sim \mathrm{T}_{n_{t-1}}[\mathbf{m}_{t-1}, \mathbf{C}_{t-1}], \\
(\boldsymbol{\theta}_t \mid D_{t-1}) &\sim \mathrm{T}_{n_{t-1}}[\mathbf{a}_t, \mathbf{R}_t], \\
(\phi \mid D_{t-1}) &\sim \mathrm{G}[n_{t-1}/2, d_{t-1}/2] \text{ with } S_{t-1} = d_{t-1}/n_{t-1}, \\
(Y_t \mid D_{t-1}) &\sim \mathrm{T}[f_t, Q_t], \\
(\boldsymbol{\theta}_t \mid D_t) &\sim \mathrm{T}_{n_t}[\mathbf{m}_t, \mathbf{C}_t], \\
(\phi \mid D_t) &\sim \mathrm{G}[n_t/2, d_t/2] \text{ with } S_t = d_t/n_t,
\end{aligned}$$

where $\mathbf{a}_t = \mathbf{m}_{t-1}$, $\mathbf{R}_t = \mathbf{C}_{t-1} + \mathbf{W}_t$, $f_t = \mathbf{F}_t'\mathbf{m}_{t-1}$, $Q_t = \mathbf{F}_t'\mathbf{R}_{t-1}\mathbf{F}_t + S_{t-1}$, and the remaining elements are defined by the usual updating equations $n_t = n_{t-1} + 1$, $d_t = d_{t-1} + S_{t-1}e_t^2/Q_t$, $\mathbf{m}_t = \mathbf{m}_{t-1} + \mathbf{A}_t e_t$ and $\mathbf{C}_t = [\mathbf{R}_t - \mathbf{A}_t\mathbf{A}_t'Q_t](S_t/S_{t-1})$ where $e_t = Y_t - f_t$ and $\mathbf{A}_t = \mathbf{R}_t\mathbf{F}_t/Q_t$.

The usual filtering and forecasting equations apply similarly. The former are not reproduced here. For forecasting ahead to time $t+k$ from time t, it follows easily that

$$\begin{aligned}
(\boldsymbol{\theta}_{t+k} \mid D_t) &\sim \mathrm{T}_{n_t}[\mathbf{m}_t, \mathbf{R}_t(k)], \\
(Y_{t+k} \mid D_t) &\sim \mathrm{T}_{n_t}[f_t(k), Q_t(k)]
\end{aligned}$$

where $\mathbf{R}_t(k) = \mathbf{C}_t + \sum_{r=1}^{k} \mathbf{W}_{t+r}$, $f_t(k) = \mathbf{F}_{t+k}'\mathbf{m}_t$ and $Q_t(k) = S_t + \mathbf{F}_{t+k}'\mathbf{R}_t(k)\mathbf{F}_{t+k}$.

9.2.4 Comments

Various features of the regression model analysis require comment.

(1) **Stability**
Forecasting performance is achieved through the identification of stability in regression relationships. Thus models with small evolution noise terms are to be desired, static regressions in which $\boldsymbol{\theta}_t = \boldsymbol{\theta}$ is constant being ideal so long as they are appropriate. If the time-variation in $\boldsymbol{\theta}_t$ is significant, evidenced by large values on the diagonals of the evolution variance matrices $\mathbf{W}_t$, forecasting suffers in two ways. Firstly, when forecasting ahead the forecast distributions become very diffuse as the $\mathbf{R}_t(k)$ terms increase due to the addition of further evolution noise variance matrices. Secondly, in updating, the weight placed on new data is high, so that the posterior distributions adapt markedly from observation to observation. Thus, although very short-term forecasts may be accurate in terms of location and reasonably precise, medium- and longer-term forecasts may be poorly located and very diffuse.

As a consequence, effort is needed in practice to identify meaningful and appropriate independent variables to be used as regressors, often after transformation and combination, with a view to identifying regressions whose coefficients are as stable as possible over time. There will usually be a need for some minor stochastic variation to account, at the very least, for changing conditions in the environment of the series, and model misspecification. Thus, in practice, the $\mathbf{W}_t$ matrices determined by the forecaster will be relatively small.

(2) Structuring $\mathbf{W}_t$

Structuring the $\mathbf{W}_t$ sequence using discount factors can be done in various ways following the development of component models in Section 6.3. If the regressors are similar, related variables viewed as modelling an overall effect of an unobserved, underlying variable, then they should be considered together as one block or component for discounting. For example, the effects of a classifying factor variable should be viewed as a single component, as is the case in Chapter 8 and elsewhere, with seasonal models. Otherwise, considering regressors as contributing separately to the model implies a need for one discount factor for each regression parameter, the $\mathbf{W}_t$ matrix then being diagonal.

(3) Static regression

The static regression model obtains when $\mathbf{W}_t = \mathbf{0}$ for all t, implied by unit discount factors for all components. In this case, $\mathbf{R}_t = \mathbf{C}_{t-1}$ for all t, no information decaying on the state vector between observation stages. The update for the posterior variance matrix can be rewritten in terms of the precision, or information matrices $\mathbf{C}_t^{-1}$ and $\mathbf{C}_{t-1}^{-1}$ as,

$$\mathbf{C}_t^{-1} = [\mathbf{C}_{t-1}^{-1} + S_{t-1}^{-1}\mathbf{F}_t\mathbf{F}_t'](S_{t-1}/S_t).$$

It follows that in terms of the scale free matrices $\mathbf{C}_t/S_t$ for all t,

$$S_t\mathbf{C}_t^{-1} = S_{t-1}\mathbf{C}_{t-1}^{-1} + \mathbf{F}_t\mathbf{F}_t'.$$

Repeatedly applying this for times $t-1, t-2, \ldots, 1$ leads to

$$S_t\mathbf{C}_t^{-1} = S_0\mathbf{C}_0^{-1} + \sum_{r=1}^{t} \mathbf{F}_r\mathbf{F}_r'.$$

It also follows, in a similar fashion, that

$$\mathbf{m}_t = S_t^{-1}\mathbf{C}_t[S_0\mathbf{C}_0^{-1}\mathbf{m}_0 + \sum_{r=1}^{t} \mathbf{F}_r Y_r].$$

These results coincide, naturally, with the standard Bayesian linear regression results in static models (See, for example, DeGroot 1971, Chapter 11; Box and Tiao 1973, Chapter 2).

(4) Reference analysis

The reference analysis of Section 4.10 may be applied in cases of little initial prior information. This leads to the dynamic version of standard reference analyses of linear models (see above references). When sufficient data has been processed so that the posterior for $\boldsymbol{\theta}_t$ (and ϕ) is proper, the standard updating equations above apply. The main difference is that the degrees of freedom are not initially updated until sufficient data are available to make the posteriors proper. This results in reduced degrees of freedom n_t thereafter. Section 4.10 provides full details.

The meaning of "sufficient data" in this context brings in directly the notion of collinearity amongst the regressor variables. In the initial reference updating, the posterior distributions become proper at that time t such that the *precision* matrix $\sum_{r=1}^{t} \mathbf{F}_r\mathbf{F}_r'$ first becomes non-singular. At this stage, the standard updating may begin for $\boldsymbol{\theta}_t$, although one further observation is necessary to begin the updating for ϕ. The soonest this may occur is at $t = n$, the dimension of the parameter vector, when one observation has been observed for each parameter. If missing observations are encountered, then this increases by 1 for each. Otherwise, collinearity amongst the regressor variables can lead to the precision matrix being singular at time n. This is actually rather uncommon in practice, although much more often it is the case that the matrix is close to singularity, having a very small, positive determinant. This indicates strong relationships amongst the regressor variables, the well-known feature of multi-collinearity in regression. Numerical problems can be encountered in inverting the above precision matrix in such cases, so that care is needed. To avoid inverting the matrix, and also to allow for the case of precise singularity, a small number of further observations can be processed retaining the reference analysis updating equations. This means that further terms $\mathbf{F}_{t+1}\mathbf{F}_{t+1}'$, $\mathbf{F}_{t+2}\mathbf{F}_{t+2}'$, and so on are added to the existing precision matrix until it is better conditioned and does not suffer from numerical instabilities when inverted. Under certain circumstances, however, collinearity amongst regressors may persist over time. In such cases there is a need for action, usually to reduce the number of regressors and remove redundant variables, the model being over-parametrised. The problem of multi-collinearity here is precisely as encountered in standard static regression.

(5) Orthogonality

When dealing with regressors that are observed values of time series themselves, the static model concept of *orthogonality* of regressors also applies. The simplest, but important, use of this involves considering independent variables as deviations from some average value. In standard regression, it is common practice to standardise regressors; given a fixed sample of observations, the regressors are standardised primarily by subtracting the arithmetic mean, and secondarily, dividing by the standard deviation. The reason for subtracting the arithmetic mean is so that the regression effects are clearly separated from the intercept term in the model, being essentially orthogonal to it. Thus, for example, a static straight-line model on regressor X, $Y_t = \alpha + \beta X_t$ is rewritten as $Y_t = \alpha^* + \beta X_t^*$, where $X_t^* = X_t - \bar{X}$, with $\bar{X}$ the arithmetic mean of the X_t values in the fixed sample considered, and $\alpha^* = \alpha + \beta\bar{X}$ the new fixed-intercept term. The new regression vectors $\mathbf{F}_t' = (1, X_t^*)$ are such that $\sum \mathbf{F}_t\mathbf{F}_t'$ (the sum being over the fixed sample of observations) is now diagonal and so inverts easily. This *orthogonality*, and its more general versions with several regressors, allows simpler interpretation of the regression.

In dynamic regression the same principles apply, although now the time dependence clouds the issue. If a relatively short series of known number of observations is to be analysed, then the above form of standardisation to zero arithmetic mean for regressors may apply. It is usually to be expected that regression relationships do not change rapidly, and so the features of the static model may be approximately reproduced. Some problems do arise even under such circumstances, however. For a start, $\bar{X}$ may be unknown initially since X values later in the series have yet to be observed; this derives from the time series nature of the problem and applies even if the model is static. Some insight into what form of standardisation may be appropriate can be gained by considering the regression as possibly being derived from a more structured model, one in which Y_t and X_t are initially jointly normally distributed conditional on a collection of time-varying parameters that determine their mean vector and covariance matrix, forming a bivariate time series. Suppose specifically that Y_t has mean α_t and X_t has mean γ_t. It follows that

$$Y_t = \alpha_t + \beta_t(X_t - \gamma_t),$$

where β_t is the regression coefficient from the covariance matrix of Y_t and X_t. In the static model, $\gamma_t = \gamma$ is constant over time. The static model correction is now obvious; the population mean γ is simply estimated by the sample value $\bar{X}$ from the fixed sample of interest. More generally, if γ is assumed constant so that the X values are distributed about a common mean, then a sequentially updated estimate of γ is more appropriate. If, on the other hand, γ_t is possibly time-varying, a *local* mean for the X_t time series is appropriate. Such an estimate may be obtained from a separate time series model for the X_t series. Further development of this is left as an exercise to the reader.

(6) **Step ahead forecasting**

In forecasting ahead the future values of regressors are required. If some or all of the regressors are observed values of related time series, as is often the case in socio-economic modelling, for example, then the required future values may not be available at the time of forecasting. Various possible solutions exist for this problem. One general and theoretically exact method is to construct a joint model for forecasting the X variables as time series along with Y. This introduces the need for multivariate time series modelling and forecasting, a vast topic in its own right and beyond the scope of the present discussion (see Chapter 16 for multivariate DLMs).

Simpler alternative approaches involve the use of estimated values of the future regressors. These may be simply guessed at or provided by third parties, separate models, etc. Consider forecasting Y_{t+k} from time t, with $\mathbf{F}_{t+k}$ uncertain. Typically, information relevant to forecasting $\mathbf{F}_{t+k}$ separately will result in specification of some features of a forecast distribution, assumed to have a density $p(\mathbf{F}_{t+k} \mid D_t)$. Note that formally, the extra information relevant to forecasting $\mathbf{F}_{t+k}$ should be included in the conditioning here; without loss of generality, assume that this is already incorporated in D_t. Then the step ahead forecast distribution for Y_{t+k} can be deduced as

$$p(Y_{t+k} \mid D_t) = \int p(Y_{t+k} \mid \mathbf{F}_{t+k}, D_t) p(\mathbf{F}_{t+k} \mid D_t) d\mathbf{F}_{t+k}.$$

The first term in the integrand here is just the standard forecast T distribution from the regression, as specified in Section 9.2.3 above, with $\mathbf{F}_{t+k}$ assumed known and explicitly included in the conditioning. Features of the predictive density will depend on the particular forms of predictions for $\mathbf{F}_{t+k}$. Some generally useful features are available, as follows. Suppose that the forecast mean and variance matrix of $\mathbf{F}_{t+k}$ exist, denoted by $\mathbf{h}_t(k) = \mathrm{E}[\mathbf{F}_{t+k} \mid D_t]$ and $\mathbf{H}_t(k) = \mathrm{V}[\mathbf{F}_{t+k} \mid D_t]$ respectively. Then the forecast mean and variance of Y_{t+k} can be deduced. Simply note that when the degrees of freedom of the conditional T distribution for Y_{t+k} exceeds unity, $n_t > 1$, then

$$\begin{aligned} \mathrm{E}[Y_{t+k} \mid D_t] &= \mathrm{E}\{\mathrm{E}[Y_{t+k} \mid \mathbf{F}_{t+k}, D_t] \mid D_t\} \\ &= \mathrm{E}[\mathbf{F}'_{t+k}\mathbf{m}_t \mid D_t] = \mathbf{h}_t(k)'\mathbf{m}_t. \end{aligned}$$

Similarly, when $n_t > 2$,

$$\begin{aligned} \mathrm{V}[Y_{t+k} \mid D_t] &= \mathrm{E}\{\mathrm{V}[Y_{t+k} \mid \mathbf{F}_{t+k}, D_t] \mid D_t\} \\ &\qquad + \mathrm{V}\{\mathrm{E}[Y_{t+k} \mid \mathbf{F}_{t+k}, D_t] \mid D_t\} \\ &= \mathrm{E}[\frac{n_t}{n_t - 2} Q_t(k) \mid D_t] + \mathrm{V}[f_t(k) \mid D_t] \\ &= \frac{n_t}{n_t - 2}\{S_t + \mathrm{E}[\mathbf{F}'_{t+k}\mathbf{R}_t(k)\mathbf{F}_{t+k} \mid D_t]\} \\ &\qquad + \mathrm{V}[\mathbf{F}'_{t+k}\mathbf{m}_t \mid D_t] \\ &= \frac{n_t}{n_t - 2}[S_t + \mathbf{h}_t(k)'\mathbf{R}_t(k)\mathbf{h}_t(k) \\ &\qquad + \mathrm{trace}\{\mathbf{R}_t(k)\mathbf{H}_t(k)\}] + \mathbf{m}'_t\mathbf{H}_t(k)\mathbf{m}_t. \end{aligned}$$

In this way, uncertainty about the future regressor values are formally incorporated in forecast distributions for the Y series.

(7) **Posterior inferences**
Standard modes of inference about regression parameters in static models apply directly in the dynamic case. Consider any $q \leq n$ elements of the state vector $\boldsymbol{\theta}_t$, reordering the elements as necessary, so that the q of interest occupy the first q positions in the vector. Thus, $\boldsymbol{\theta}_t' = (\boldsymbol{\theta}_{t1}', \boldsymbol{\theta}_{t2}')$, where $\boldsymbol{\theta}_{t1}' = (\theta_{t1}, \ldots, \theta_{tq})$ is the subvector of interest. Then, with $\mathbf{m}_t$ and $\mathbf{C}_t$ conformably partitioned, $(\boldsymbol{\theta}_{t1} \mid D_t) \sim \mathrm{T}_{n_t}[\mathbf{m}_{t1}, \mathbf{C}_{t1}]$ in an obvious notation. Inferences about $\boldsymbol{\theta}_{t1}$ are based on this marginal posterior distribution. In particular, the contribution of the corresponding regressors to the model may be assessed by considering the support in the posterior for the values $\boldsymbol{\theta}_{t1} = \mathbf{0}$, consistent with no effect of the regressors. The posterior density $p(\boldsymbol{\theta}_{t1} \mid D_t)$ takes values greater than that at $\boldsymbol{\theta}_{t1} = \mathbf{0}$ whenever

$$(\boldsymbol{\theta}_{t1} - \mathbf{m}_{t1})'\mathbf{C}_{t1}^{-1}(\boldsymbol{\theta}_{t1} - \mathbf{m}_{t1}) < \mathbf{m}_{t1}'\mathbf{C}_{t1}^{-1}\mathbf{m}_{t1}.$$

The posterior probability that this occurs is given from the usual F distribution,

$$\begin{aligned}\Pr[(\boldsymbol{\theta}_{t1} - \mathbf{m}_{t1})'\mathbf{C}_{t1}^{-1}(\boldsymbol{\theta}_{t1} - \mathbf{m}_{t1}) < \mathbf{m}_{t1}'\mathbf{C}_{t1}^{-1}\mathbf{m}_{t1} \mid D_t] \\ = \Pr[\mathrm{F}_{q,n_t} < q^{-1}\mathbf{m}_{t1}'\mathbf{C}_{t1}^{-1}\mathbf{m}_{t1}],\end{aligned}$$

where F_{q,n_t} denotes a random quantity having the standard F distribution with q degrees of freedom in the numerator and n_t in the denominator. Thus the highest posterior density (HPD) based test of the hypothesis that $\boldsymbol{\theta}_{t1} = \mathbf{0}$ is based on the probability level

$$\alpha = \Pr[\mathrm{F}_{q,n_t} \geq q^{-1}\mathbf{m}_{t1}'\mathbf{C}_{t1}^{-1}\mathbf{m}_{t1}];$$

a small value of α indicates rejection of the hypothesised value $\boldsymbol{\theta}_{t1} = \mathbf{0}$ as unlikely.

(8) **Parameter constraints and relationships**
As in the general DLM, the variance matrices $\mathbf{C}_0$ and the sequence $\mathbf{W}_t$ may be structured in order to incorporate modeller's views about relationships amongst the parameters. At an extreme, the parameters may be subject to linear restrictions that relate components or condition some elements to taking known values. This then implies that $\mathbf{C}_0$ is singular with a specific structure determined by the linear constraints, the same form of constraints applying to the evolution variance matrices if the restrictions are to hold to $\boldsymbol{\theta}_t$ over time. More usually, initial views about relationships amongst the parameters will be modelled in terms of stochastic constraints of various kinds, usually leaving the variance matrices non-singular. One important example is the embodiment of beliefs about the likely decay of coefficients of lagged values of variables, related to the use of smoothness prior distributions in lagged regressions and autoregressions (Cleveland 1974; Leamer 1972; Young 1983; Zellner 1971). Similar structures may be applied to the coefficients of higher-order terms in polynomial regressions (Young 1977).

Other examples include the use of hierarchical models, often based on assumptions of exchangeability amongst subsets of elements of $\boldsymbol{\theta}_t$ (Lindley and Smith 1972). A typical example concerns symmetry assumptions about the effects of different levels of an underlying factor that groups the data; these may be initially viewed as exchangeable, with changes in the effects over time subject to the same assumption. Though interesting and important in application when appropriate, these topics are not developed further here in a general framework.

9.3 TRANSFER FUNCTIONS

9.3.1 Form-free transfer functions

Consider regression on current and past values of a single independent variable X, assuming initially that the regression parameters are constant over time. As in Section 9.2.2 above, regression on a fixed and finite number of lagged values falls within the standard regression DLM framework. Generally, if appreciable effects of the regressor variable are expected to be sustained up to a maximum lag of k time points, for some $k > 1$, the linear regression on lagged values determines the contribution to the mean response at time t as

$$\mu_t = \sum_{r=0}^{k} \beta_r X_{t-r} = \beta_0 X_t + \beta_1 X_{t-1} + \cdots + \beta_k X_{t-k}.$$

Here $\mathbf{F}_t' = (X_t, X_{t-1}, \ldots, X_{t-k})$ and $\boldsymbol{\theta}_t' = \boldsymbol{\theta}' = (\beta_0, \beta_1, \ldots, \beta_k)$. Projecting ahead from the current time t to times $t+r$, for $r \geq 0$, the *effect* of the current level X_t of the regressor variable is then simply the contribution to the mean response, namely $\beta_r X_t$ for $r = 0, 1, \ldots, k$, being zero for $r > k$. This defines the **transfer response function** of X,

$$\begin{cases} \beta_r X, & r = 0, 1, \ldots, k; \\ 0, & r > k. \end{cases}$$

In words this is just the effect of the current regressor value $X_t = X$ on the mean response at future times r, conditional on $X_{t+1} = \ldots = X_{t+r} = 0$. Obviously this model, for large enough k, provides a flexible method of modelling essentially any expected form of transfer response function, the coefficients β_r being arbitrary regression parameters to be specified or estimated. In the more practically suitable dynamic regression, the flexibility increases as stochastic variation in the parameters allows the model to adapt to changing responses, and also to cater for misspecification in the model. However, whilst the regression structure provides a very general model for lagged responses, there are often good reasons to consider functional relationships amongst the regression coefficients β_r that essentially alter the structure of the model, providing functional forms over time for the lagged effects of X. This leads to the considerations in the following sections.

9.3.2 *Functional form transfer functions*

One obvious feature of the regression model above is that the effect of X_t on Y_{t+r} is zero when $r > k$. The model is thus inappropriate for cases in which it is felt that lagged effects persist into the future, perhaps decaying smoothly towards zero as time progresses. One simple way of adapting the regression structure to incorporate such features is to consider regression, not on X directly, but on a constructed *effect* variable measuring the combined effect of current and past X values. Some examples provide insight.

EXAMPLE 9.1. Suppose that Y_t represents a monthly consumer series, such as sales or demand for a product, or consumer awareness of the product in the market. Currently, and prior to time $t = 1$, the Y series is supposed to follow a time series model with level parameter $\mu_t = \alpha_t$; this may, for example, be a simple steady model, or include other terms such as trend, seasonality and regressions. In month $t = 1$, the marketing company initiates a promotional campaign for the product involving expenditure on advertising and so forth, the expenditure being measured by a single independent variable X_t, $t = 1, 2, \ldots$. This may be a compound of various factors but is assumed, for simplicity, to measure investment in promoting the product. This is a simple instance of a very common event, and it is generally understood that with no other inputs, the effect of such promotional expenditure can be expected to be as follows: (a) in month t, the level of the Y series should increase, say in proportion to X_t; (b) without further expenditure at times $t+1$, $t+2$, ..., the effect of past expenditure will decay over time, often approximately exponentially; and (c) further expenditure in following months will have the same form of effect. In model terms, the anticipated mean response is given by the original level plus a second term, ξ_t,

$$\mu_t = \alpha_t + \xi_t,$$

where ξ_t is the effect on the current level of the series of current and past expenditure. This effect is modelled as

$$\xi_t = \lambda \xi_{t-1} + \psi X_t,$$

with $\xi_0 = 0$, there being no expenditure prior to $t = 1$. The parameter ψ determines the immediate, *penetration* effect of the monthly advertising, the level being raised initially on average by ψ per unit of expenditure; it is a positive quantity whose units depends on those of both the X and Y series. The parameter λ represents the *memory* of the market. Extrapolating ahead to time $t+k$, the model implies that

$$\xi_{t+k} = \lambda^k \xi_t + \psi \sum_{r=1}^{k} \lambda^{k-r} X_{t+r}.$$

Thus, if $X_{t+1} = X_{t+2} = \ldots = X_{t+k} = 0$, then

$$\xi_{t+k} = \lambda^k \xi_t.$$

This embodies point (b); with λ a dimensionless quantity in the unit interval, the effect of expenditure up to time t is reduced by a factor λ for each future time point, decaying asymptotically to zero at an exponential rate.

EXAMPLE 9.2. Example 9.1 concerns exponential decay of effects, where promotional advertising is not anticipated to sustain the sales/demand series at higher levels. Minor modification provides a closely related model for sustained growth or decay. An example concerns increases (or decreases) of sales to a new, higher (or lower) and sustained level following price reductions (or rises) for a product or products in a limited consumer market. Let the initial level α_t again represent previous information about the sales/demand series subject to an original pricing policy. Let X_t now represent the *reduction* in price in month t, either positive, implying a decrease in price, or negative implying an increase. It is to be expected that in a finite market, sales will tend to increase as the price is reduced, eventually levelling off at some *saturation* level. Similarly, sales tend to decay towards zero as prices increases. This can be modelled via

$$\xi_t = \xi_{t-1} + \theta_t,$$

where

$$\theta_t = \lambda \theta_{t-1} + \psi X_t,$$

with $\xi_0 = \theta_0 = 0$. Suppose, for example, that a single price change is made at time t, with $X_{t+1} = X_{t+2} = \ldots = 0$ and $\xi_{t-1} = \theta_{t-1} = 0$. It follows that the immediate effect on the mean response is simply $\xi_t = \theta_t = \psi X_t$, the positive quantity ψ again measuring the immediate unit response. Projecting to time $t+r$, $(r = 1, \ldots, k)$, under these conditions, we have $\theta_{t+r} = \lambda^r \theta_t$ and thus

$$\xi_{t+k} = \sum_{r=0}^{k} \lambda^r \theta_t = \theta_t(1 - \lambda^{k+1})/(1 - \lambda)$$

if $0 < \lambda < 1$. Thus, if X_t is positive, so that the price decreases, then ξ_{t+k} is an increasing function of k, tending to the limit $\psi X_t(1-\lambda)$ as k increases. Given the initial penetration factor ψ, λ determines both the rate of increase and the eventual saturation level. Similarly, of course, a negative value of X_t consistent with price increase implies a decay in level.

These two examples typify a class of structures for lagged effects of a single independent variable. The general form of such models detailed here is apparently an extension of the usual DLM representation, but can easily be rewritten as a standard DLM, as will be seen below. The initial definition is given in terms of the extended representation for interpretability.

Definition 9.1. Let X_t be the value of an independent, scalar variable X at time t. A general **transfer function model** for the effect of X on the response series Y is defined by

$$\begin{aligned} Y_t =& \mathbf{F}'\boldsymbol{\theta}_t + \nu_t, & (9.1a)\\ \boldsymbol{\theta}_t =& \mathbf{G}\boldsymbol{\theta}_{t-1} + \boldsymbol{\psi}_t X_t + \partial\boldsymbol{\theta}_t, & (9.1b)\\ \boldsymbol{\psi}_t =& \boldsymbol{\psi}_{t-1} + \partial\boldsymbol{\psi}_t, & (9.1c) \end{aligned}$$

with terms defined as follows: $\boldsymbol{\theta}_t$ is an n-dimensional state vector, $\mathbf{F}$ a constant and known n-vector, $\mathbf{G}$ a constant and known evolution matrix, and ν_t and $\partial\boldsymbol{\theta}_t$ are observation and evolution noise terms. (Note the use of the ∂ notation for the latter rather than the usual $\boldsymbol{\omega}_t$ notation). All these terms are precisely as in the standard DLM, with the usual independence assumptions for the noise terms holding here. The term $\boldsymbol{\psi}_t$ is an n-vector of parameters, evolving via the addition of a noise term $\partial\boldsymbol{\psi}_t$, assumed to be zero-mean normally distributed independently of ν_t (though not necessarily of $\partial\boldsymbol{\theta}_t$).

The state vector $\boldsymbol{\theta}_t$ carries the effect of current and past values of the X series through to Y_t in equation (9.1a); this is formed in (9.1b) as the sum of a linear function of past effects, $\boldsymbol{\theta}_{t-1}$, and the current effect $\boldsymbol{\psi}_t X_t$, plus a noise term.

Suppose that conditional on past information D_t, the posterior point estimates of the two vectors $\boldsymbol{\theta}_t$ and $\boldsymbol{\psi}_t$ are denoted by

$$\mathbf{m}_t = \mathrm{E}[\boldsymbol{\theta}_t|D_t] \qquad \text{and} \qquad \mathbf{h}_t = \mathrm{E}[\boldsymbol{\psi}_t|D_t].$$

Extrapolating expectations into the future in (9.1b and c), it follows that

$$\mathrm{E}[\boldsymbol{\theta}_{t+k} \mid D_t] = \mathbf{G}^k\mathbf{m}_t + \sum_{r=1}^{k}\mathbf{G}^{k-r}\mathbf{h}_t X_{t+r}. \tag{9.2}$$

Then, from (9.1a), the forecast function is

$$f_t(k) = \mathrm{E}[Y_{t+k} \mid D_t] = \mathbf{F}'\mathbf{G}^k\mathbf{m}_t + \mathbf{F}'\sum_{r=1}^{k}\mathbf{G}^{k-r}\mathbf{h}_t X_{t+r}. \tag{9.3}$$

Let a_t denote the first term here, $a_t = \mathbf{F}'\mathbf{G}^k\mathbf{m}_t$, summarising the effects of past values of the X series; a_t is known at time t. Also, consider the special case in which after time $t+1$, there are no input values of the regressor variable, so that

$$X_{t+r} = 0, \qquad (r = 2, \ldots, k). \tag{9.4}$$

Then (9.3) implies that

$$f_t(k) = a_t + \mathbf{F}'\mathbf{G}^{k-1}\mathbf{h}_t X_{t+1}. \tag{9.5}$$

This can be seen as determining the way in which any particular value of the X series in the next time period is expected to influence the response into the future, the dependence on the step ahead index k coming, as in TSDLMS, through powers of the system matrix $\mathbf{G}$.

In the special case that $\boldsymbol{\psi}_t = \boldsymbol{\psi}$ is constant over time and known, $\boldsymbol{\psi} = \mathbf{h}_t$, the transfer response function of X is given by $f_{t-1}(k+1)$ subject to (9.4) with past effects $a_{t-1} = 0$. Under these circumstances, (9.5) then leads to

$$\mathbf{F}'\mathbf{G}^k\boldsymbol{\psi}X$$

being the expected effect on the response due to $X_t = X$.

EXAMPLE 9.1 (continued). In the exponential decay model as described earlier, we have dimension $n = 1$, $\boldsymbol{\theta}_t = \xi_t$, the effect variable, $\psi_t = \psi$ for all t, $\mathbf{F} = 1$ and $\mathbf{G} = \lambda$, all noise terms assumed zero. Note that $\mathbf{G}$ is just the memory decay term λ, assumed known. The transfer response function of X is simply $\lambda^k\psi X$.

EXAMPLE 9.2 (continued). In the second example of growth or decay to a new level, $n = 2$,

$$\boldsymbol{\theta}_t = \begin{pmatrix} \xi_t \\ \theta_t \end{pmatrix}, \quad \boldsymbol{\psi}_t = \boldsymbol{\psi} = \begin{pmatrix} 0 \\ \psi \end{pmatrix}, \quad \mathbf{F} = \begin{pmatrix} 1 \\ 1 \end{pmatrix}, \quad \mathbf{G} = \begin{pmatrix} 1 & 1 \\ 0 & \lambda \end{pmatrix},$$

with zero noise terms. The transfer response function is simply $\psi(1 - \lambda^{k+1})/(1-\lambda)$.

The general model (9.1) can be rewritten in the standard DLM form as follows. Define the new, $2n$-dimensional state parameters vector $\tilde{\boldsymbol{\theta}}_t$ by catenating $\boldsymbol{\theta}_t$ and $\boldsymbol{\psi}_t$, giving

$$\tilde{\boldsymbol{\theta}}_t' = (\boldsymbol{\theta}_t', \boldsymbol{\psi}_t').$$

Similarly, extend the $\mathbf{F}$ vector by catenating an n-vector of zeros, giving a new vector $\tilde{\mathbf{F}}$ such that

$$\tilde{\mathbf{F}}' = (\mathbf{F}', 0, \ldots, 0).$$

For the evolution matrix, define

$$\tilde{\mathbf{G}}_t = \begin{pmatrix} \mathbf{G} & X_t\mathbf{I}_n \\ \mathbf{0} & \mathbf{I}_n \end{pmatrix},$$

where $\mathbf{I}_n$ is the $n \times n$ identity matrix. Finally, let $\boldsymbol{\omega}_t$ be the noise vector defined by

$$\boldsymbol{\omega}_t' = (\partial\boldsymbol{\theta}_t' + X_t\partial\boldsymbol{\psi}_t', \partial\boldsymbol{\psi}_t').$$

Then the model (9.1) can be written as

$$\begin{aligned} Y_t &= \tilde{\mathbf{F}}'\tilde{\boldsymbol{\theta}}_t + \nu_t, \\ \tilde{\boldsymbol{\theta}}_t &= \tilde{\mathbf{G}}_t\tilde{\boldsymbol{\theta}}_{t-1} + \boldsymbol{\omega}_t. \end{aligned} \tag{9.6}$$

Thus the transfer function model (9.1) has standard DLM form (9.6) and the usual analysis applies. Some particular features of the model in this setting require comment.

(a) The model as discussed provides just the transfer function for the variable X. In practice, this will usually be combined by superposition with other components (as in Example 9.1), such as trend, seasonality, regression and maybe even transfer functions of other independent variables.
(b) The unknown parameters in $\boldsymbol{\psi}_t$ play a role similar to the regression parameters in the dynamic regression model of Section 9.2 and are likely to be subject to some variation over time in particular applications. Thus, in some cases, more appropriate versions of the models in the two preceding examples would have ψ time dependent.
(c) Some of the elements of $\boldsymbol{\psi}_t$ may be fixed and known. This can be modelled by setting to zero the corresponding values of the initial prior variance matrix $\mathbf{C}_0$ for $\tilde{\boldsymbol{\theta}}_0$ and those of the evolution variance matrices $\mathbf{W}_t = \mathrm{V}[\boldsymbol{\omega}_t]$. However, it will often be the case that from the structure of the model, there are zero elements in $\boldsymbol{\psi}_t$, as is the case with the integrated transfer response function in Example 9.2. Then for practical application, the general model (9.6) may be reduced in dimension to include just the non-zero elements of $\boldsymbol{\psi}_t$. This is obviously desirable from a computational viewpoint if n is at all large. In general, suppose that just $p < n$ of the n elements of $\boldsymbol{\psi}_t$ are non-zero. The reader may verify that the model (9.6) may be reduced to one of dimension $n+p$, rather than $2n$, in which $\tilde{\mathbf{F}}' = (\mathbf{F}', 0, \ldots, 0)$, having p trailing zero elements, and

$$\tilde{\mathbf{G}}_t = \begin{pmatrix} \mathbf{G} & X_t\mathbf{H} \\ \mathbf{0} & \mathbf{I}_p \end{pmatrix},$$

where $\mathbf{H}$ is an $n \times p$ matrix with just one unit element in each column, all other elements being zero. For instance, the model of Example 9.2 may be written as a 3-dimensional DLM with $\mathbf{F}' = (1, 1, 0)$ and

$$\tilde{\mathbf{G}}_t = \begin{pmatrix} 1 & 1 & 0 \\ 0 & \lambda & X_t \\ 0 & 0 & 1 \end{pmatrix}.$$

Thus, although the general model (9.6) always applies, it is often the case that a reduced form will be utilised.
(d) As specified, the model is developed from (9.1) and this results in a particular, structured form for the evolution noise vector, $\boldsymbol{\omega}_t' = (\partial\boldsymbol{\theta}_t' + X_t\partial\boldsymbol{\psi}_t', \partial\boldsymbol{\psi}_t')$. There are various ways of specifying the evolution variance matrices. The most direct and appropriate is to assume the evolution noise terms $\partial\boldsymbol{\theta}_t$ and $\partial\boldsymbol{\psi}_t$ uncorrelated with variance matrices $\mathrm{V}[\partial\boldsymbol{\theta}_t] = \mathbf{U}_t$ and $\mathrm{V}[\partial\boldsymbol{\psi}_t] = \mathbf{Z}_t$ respectively. It

then follows that

$$\mathbf{W}_t = \begin{pmatrix} \mathbf{U}_t + X_t^2 \mathbf{Z}_t & X_t \mathbf{Z}_t \\ X_t \mathbf{Z}_t & \mathbf{Z}_t \end{pmatrix}.$$

Choice of $\mathbf{U}_t$ and $\mathbf{Z}_t$ is most simply guided by discount factors. The two subvectors $\boldsymbol{\theta}_t$ and $\boldsymbol{\psi}_t$ are naturally separated as distinct components of the model and so the component discounting concept of Section 6.3 applies.

(e) The specific form of the stochastic structure in the evolution equation is not vital to the model, deriving as it does from the assumptions underlying (9.1). We can simply model the series directly using (9.2) and then impose *any* form on $\mathbf{W}_t$, simplifying the modelling process by choosing simple, discount based forms, for example.

(f) The discussion in Section 9.2.2 about the problems arising in forecasting ahead when future X values are unknown at the time is pertinent here. The only technical differences between the models arise through the appearance of the regressors in the evolution rather than the observation equation.

(g) More general models involve the concept of *stochastic* transfer responses, as the following example illustrates. Suppose a company uses various alternative styles of advertising films or campaigns in promoting its products. Suppose that advertising effort is characterised by a simple measure X_t, such as advertising expenditure, and a basic transfer response model relates X_t to an output variable Y_t such as consumer demand or estimated awareness of the products. Then, although the model may adequately describe the relationship between X and Y over time during any given advertising campaign, it does not capture wider qualitative aspects of differences between films and their effects may differ widely. An appropriate extension of the model to allow for this sort of additional variation is to assume that the transfer response parameters are sampled from a population of such parameters. With reference to the simple decay of effects in Example 9.1, the transfer response function of film i may be taken as $\lambda^k \psi^{(i)} X$. The stochastic (at any given time) nature of the response is modelled by assuming, for example, that $\psi^{(i)} \sim \mathrm{N}[\psi, U]$, independently over i. This can obviously be incorporated within the DLM form, as can be verified by the reader, with extension to time-variation in the overall expected response parameter ψ about which the individual, film specific parameters $\psi^{(i)}$ are distributed. The variance U describes a second level of variation over and above any variation over time in ψ.

(h) $\mathbf{G}$, hence $\tilde{\mathbf{G}}_t$, typically depends on parameters that must be specified for the usual analysis to apply. In the two above examples, the decay parameter λ enters into $\mathbf{G}$. In many applications, it will be possible to specify values in advance; otherwise it may be desired to widen the analysis to allow for uncertainty in some of the elements of $\mathbf{G}$ and to incorporate learning about them. Some preliminary discussion of these sorts of problems appears in the next section, further development being left to later chapters.

9.3.3 Learning about parameters in $\mathbf{G}$*: introductory comments*

The final point in the above discussion raises, for the first time, issues of estimation in *non-linear models*, models in which the mean response function of the model has non-linear terms in some parameters. The estimation problems raised typify those of much more general, non-linear models. General concepts and techniques of non-linear estimation appear in later chapters. Here we restrict discussion to some basic, introductory comments in the context of simple transfer function models.

Consider the dynamic version of the model in Example 9.2, a DLM in which the evolution equation is

$$\theta_t = \lambda\theta_{t-1} + \omega_t.$$

For known λ, the usual analysis applies. Otherwise, if it is desired to learn about λ from the data, a much more complicated analysis is implied. With λ constant over time, though unknown, the formal analysis proceeds as follows:

(a) For each value of λ, in this example $0 < \lambda < 1$, specify an initial prior for θ_0 (and the observational variance V if unknown) of standard form. This may depend on λ and so is denoted by $p(\theta_0 \mid \lambda, D_0)$. Also, specify the initial prior distribution, of any desired form, for λ, denoted by $p(\lambda \mid D_0)$.

(b) For each value of λ, process the data according to the usual DLM with $\mathbf{G} = \lambda$. The distributions $p(Y_t \mid \lambda, D_{t-1})$, $p(\theta_t \mid \lambda, D_t)$ etc., become available at time t, defined by the usual normal or T forms, but with moments depending on the particular value of λ.

(c) Learning about λ proceeds via the sequential updating of the posterior distributions

$$p(\lambda \mid D_t) \propto p(\lambda \mid D_{t-1}) p(Y_t \mid \lambda, D_{t-1}),$$

starting from the initial distribution provided in (a); here the observed one-step forecast density from (b) provides the likelihood function for λ. At each t, this posterior for λ is easily calculated as above up to a constant of normalisation, this constant being determined by integrating over the parameter space for λ, in this case the unit interval.

(d) Posterior inferences at time t for θ_t, Y_{t+1}, and other quantities of inference are based on their posterior distributions *marginal* with respect to λ. For example, the posterior for θ_t is defined by

$$p(\theta_t \mid D_t) = \int_0^1 p(\theta_t \mid \lambda, D_t)p(\lambda \mid D_t)d\lambda.$$

The first term in the integrand is the conditional normal or T density from (b), the second the posterior for λ from (c). Point estimates of θ_t and probabilities etc. may be calculated from this posterior via further integrations.

This formally defines the extended analysis. In practice, of course, it is impossible to perform this analysis exactly since it requires an infinite number of DLM analyses to be performed corresponding to the infinite number of values for λ. Thus approximations are used in which the parameter space is discretised, a finite, and often fairly small number of values of λ being considered, resulting in a discrete posterior distribution for each t in (c). Thus the integrals appearing in (d) become summations. This defines what may be called *multi-process models*, comprising a collection of DLMs analysed in parallel and mixed for inferences with respect to the posterior probabilities over values of λ. Chapter 12 is devoted to multi-process modelling. The particular analysis here can be viewed as the use of numerical integration in approximating the various integrals appearing in the formal theory. Uses of these and other, related and more sophisticated techniques of numerical integration appear in Chapters 13 and 15.

9.3.4 Non-linear learning: further comments

Alternative approaches to the analysis of the DLM, extended to include parameters such as λ above, are based on analytic approximation such as a *linearisation*. Again in the above simple model, group λ with θ_t in the vector $\boldsymbol{\theta}_t' = (\theta_t, \lambda)'$ as a new model state vector, evolving according to the *non-linear* evolution equation

$$\boldsymbol{\theta}_t = \mathbf{g}(\boldsymbol{\theta}_{t-1}) + \boldsymbol{\omega}_t, \tag{9.7}$$

where $\mathbf{g}(\boldsymbol{\theta}_{t-1})$ is the non-linear vector function $\lambda(\theta_{t-1}, 1)'$ and $\boldsymbol{\omega}_t = (\omega_t, 0)'$. Major computational problems now arise in the model analysis due to the non-linearity. The linearity basic to the DLM combines with the assumed normality to provide neat, tractable and efficiently updated sufficient summaries for all prior/posterior and predictive distributions of interest. Once this is lost, such distributions, although theoretically easily defined, can only be calculated through the use of numerical integration as described in the previous section. Approaches using analytic approximations are typically quite easy to develop and apply, and are very commonly used since often more refined approximations are unnecessary. One of the important features of such approaches is that they naturally extend to cover models in which parameters such as λ here are themselves dynamic, varying over time. Various approximations based on linearisation of non-linear functions exist, the most obvious, and widely applied, being based on Taylor series approximations. Assume that at time $t-1$, the posterior distribution for $(\boldsymbol{\theta}_{t-1} \mid D_{t-1})$ is adequately approximated by the usual distribution $(\boldsymbol{\theta}_{t-1} \mid D_{t-1}) \sim \mathrm{T}_{n_{t-1}}[\mathbf{m}_{t-1}, \mathbf{C}_{t-1}]$. A linear approximation to the non-linear function $\mathbf{g}(.)$ in (9.7) based on a Taylor series expansion about the estimate $\boldsymbol{\theta}_{t-1} = \mathbf{m}_{t-1}$ leads to the *linearised* evolution equation

$$\boldsymbol{\theta}_t \approx \mathbf{g}(\mathbf{m}_{t-1}) + \mathbf{G}_t(\boldsymbol{\theta}_{t-1} - \mathbf{m}_{t-1}) + \boldsymbol{\omega}_t = \mathbf{h}_t + \mathbf{G}_t\boldsymbol{\theta}_{t-1} + \boldsymbol{\omega}_t, \qquad (9.8)$$

where $\mathbf{h}_t = \mathbf{g}(\mathbf{m}_{t-1}) - \mathbf{G}_t\mathbf{m}_{t-1}$ and $\mathbf{G}_t$ is the 2×2 matrix derivative of $\mathbf{g}(\boldsymbol{\theta}_{t-1})$ evaluated at $\boldsymbol{\theta}_{t-1} = \mathbf{m}_{t-1}$; in this particular model,

$$\frac{\partial \mathbf{g}(\boldsymbol{\theta}_{t-1})}{\partial \boldsymbol{\theta}'_{t-1}} = \begin{pmatrix} \lambda & \theta_{t-1} \\ 0 & 1 \end{pmatrix}.$$

When evaluated at $\boldsymbol{\theta}_{t-1} = \mathbf{m}_{t-1}$, this provides a known matrix $\mathbf{G}_t$ and results in a DLM, albeit with a constant term $\mathbf{h}_t$ added to the evolution equation (a simple extension is discussed at the end of Section 4.3). This approximate model can be analysed as usual to give approximate prior, posterior and forecast distributions in standard forms. In particular, it follows from the linearised evolution equation that $(\boldsymbol{\theta}_t \mid D_{t-1}) \sim \mathrm{T}_{n_{t-1}}[\mathbf{a}_t, \mathbf{R}_t]$, where

$$\mathbf{a}_t = \mathbf{h}_t + \mathbf{G}_t\mathbf{m}_{t-1} = \mathbf{g}(\mathbf{m}_{t-1})$$

and

$$\mathbf{R}_t = \mathbf{G}_t\mathbf{C}_{t-1}\mathbf{G}'_t + \mathbf{W}_t.$$

Note that $\mathbf{R}_t$ has the usual form, and the mean $\mathbf{a}_t$ is precisely the non-linear function $\mathbf{g}(.)$ evaluated at the estimated value of $\boldsymbol{\theta}_{t-1}$. Thus the linearisation technique, whilst leading to a standard analysis, retains the non-linearity in propagating the state vector through time. Note that as mentioned earlier, this approach clearly extends easily to cover cases in which λ is dynamic, incorporating the relevant terms in the evolution noise. Note finally that the linearised model, though derived as an approximation, may be interpreted as a perfectly valid DLM in its own right without reference to approximation. See Chapter 13 for further details of this, and other, approaches to non-linear models, and Chapter 15 for Markov chain Monte Carlo simulation methods of analysis for dealing with a much broader range of non-linear learning problems.

9.3.5 Further comments on transfer functions

The regression structure for form-free transfer functions has several attractions that make it the most widely used approach to modelling lagged effects. One such is the simplicity and familiarity of the regression structure. A second, and probably the most important, attraction is the flexibility of regression models. Whatever the nature of a transfer function, the regression on lagged X values will allow the model to adapt adequately to the observed relationship so long as there is sufficient data and information available to appropriately estimate the parameters. In addition, and in particular with dynamic regression, as time evolves, the estimates will adapt to changes in the series allowing for changes in the response function. A related point is that the coefficients can also rapidly adapt to changes and inaccuracies in the timing of observations that can distort the observed relationship. Johnston and Harrison (1980) provide an interesting application in which form-free transfer functions are used as components of larger forecasting models in consumer sales forecasting.

By contrast, the approach using functional representations of transfer effects is rather less flexible since such models impose a particular form on the transfer function. There is some flexibility to adapt in parameter estimation, particularly with time-varying parameters, but the imposed form still must be approximately appropriate for the model to be useful. If the form is basically adequate, then the advantages of the functional form model relative to the form-free regression model are apparent. Primarily, there will usually be many more parameters in a regression model in order to adequately represent a particular form of response. As mentioned earlier, it will usually be desirable to model anticipated relationships amongst regression coefficients using structured initial priors and evolution variance matrices, thus implicitly recognising the form nature of the response. A form model, having fewer parameters, provides a more efficient and parsimonious approach. Finally, as earlier mentioned, regression models directly truncate the effects of lagged X whereas form models allow for smooth decay over time without truncation.

9.4 ARMA MODELS

9.4.1 Introduction

Consider writing the observation equation of the DLM as $Y_t = \mu_t + X_t + \nu_t$, where $\mu_t = \mathbf{F}_t'\boldsymbol{\theta}_t$ is the usual mean response function; ν_t is the observational error, with the usual independence structure; and X_t is a new, unobservable term introduced to describe any additional time series variation not explained through μ_t and the evolution model for $\boldsymbol{\theta}_t$. This variation, whilst expected to be zero, may be partially predictable through a model for X_t as a stochastic process over time, whose values are expected to display some form of dependence. Here we discuss standard, *stationary* linear process models, namely the autoregressive, moving-average models (or ARMA models) that form the basis of much of classical time series analysis. In earlier chapters, we have already linked simple DLMs to point prediction methods based on autoregressive models; we now move to more explicit representation of component time series structure using such models. In this section, we briefly review the basics of stationary linear processes, and comment on issues of modelling and inference with ARMA component DLMs. We discuss basic reference material in AR models, referring to well-known and accessible sources, including Abraham and Ledolter (1983), Box and Jenkins (1976), Broemling (1985), Harvey (1981), Young (1984), and Zellner (1971), for further reading. The following section takes AR models much further, introducing some new results on time series decomposition that are illustrated there and form a prelude to further material on AR models in Chapter 15.

9.4.2 Stationarity

Stationarity of the time series X_t involves assumptions of symmetry over time, as discussed following the definition in Section 5.6. The definition is conditional on any specified model structure for X_t, as given, for example, by a specified observation and evolution equation of a DLM, and conditional on all required model parameters. This state of information is denoted by M (for model). Recall that a time series X_t, $(t = 1, 2, \dots)$, is stationary if the joint distribution (conditional on M) of any collection of k values is invariant with respect to arbitrary shifts of the time axis. In terms of joint densities,

$$p(X_{t_1}, \dots, X_{t_k} \mid M) = p(X_{s+t_1}, \dots, X_{s+t_k} \mid M)$$

for any integers $k \geq 1$ and $s \geq 0$, and any k time points $t_1, \dots, t_k$. The series is weakly stationary, or second-order stationary, if for all integers $t > 0$ and $s < t$,

$$\begin{aligned} \mathrm{E}[X_t \mid M] &= \mu, \qquad \text{constant over time,} \\ \mathrm{V}[X_t \mid M] &= W, \qquad \text{constant over time,} \end{aligned}$$

and

$$\mathrm{C}[X_t, X_{t-s} \mid M] = \gamma_s = \rho_s W, \qquad \text{independent of } t,$$

whenever these moments are finite.

Stationarity clearly implies weak stationarity for series whose second-order moments exist, stationarity being a much more far-reaching symmetry assumption. A weakly stationary series has a fixed mean μ, variance W and autocovariances γ_s depending only on the lag between values, not on the actual timings. The autocorrelations ρ_s, $(s = 0, 1, \dots)$, thus determine the second-order structure of the series; note that $\rho_0 = 1$ and $\gamma_0 = W$. Also, if any collection of any number of values are assumed jointly normally distributed given M, then weak stationarity together with normality implies stationarity. In the normal framework of DLMs, therefore, the two definitions coincide.

Suppose that X_t is a weakly stationary series. A basic result in the theory of stationary processes is that X_t can be *decomposed* into the sum of the mean μ and a linear combination of values in a (possibly infinite) sequence of zero-mean, uncorrelated random quantities. In representing this result mathematically, it is usual to extend the time index t backwards to zero and negative values. This is purely for convenience in mathematical notation and is adopted here for this reason and for consistency with general usage. Introduce a sequence of random quantities ϵ_t, $(t = \dots, -1, 0, 1, \dots)$, such that $\mathrm{E}[\epsilon_t \mid M] = 0$, $\mathrm{V}[\epsilon_t \mid M] = U$ and $\mathrm{C}[\epsilon_t, \epsilon_s \mid M] = 0$ for all $t \neq s$ and some variance U. Then the representation of X_t is given by

$$X_t = \mu + \sum_{r=0}^{\infty} \psi_r \epsilon_{t-r}, \tag{9.9}$$

for some (possibly infinite) sequence of coefficients $\psi_0, \psi_1, \dots$, with $\psi_0 = 1$. Note that this formally extends the time series X_t to $t \leq 0$. The representation (9.9) implies some obvious restrictions on the coefficients. Firstly, $\mathrm{V}[X_t \mid M] = W$ implies that $W = U \sum_{r=0}^{\infty} \psi_r^2$, so that the sum of squared coefficients must converge. Secondly, the autocovariances γ_s may be written in terms of sums of products of the ψ_r, that must also converge.

There is a huge literature on mathematical and statistical theory of stationary stochastic processes and time series analysis for stationary series. From the time series viewpoint, Box and Jenkins (1976) provide comprehensive coverage of the subject, with many references. Classical time series analysis is dominated by the use of models that can be written in the form (9.9), special cases of which are now introduced. The above reference provides much further and fuller development. Firstly, however, recall the *backshift operator* B such that for any series X_t and any t, $B^r X_t = X_{t-r}$ for all $r \geq 0$. Write (9.9) as

$$X_t = \mu + \sum_{r=0}^{\infty} \psi_r B^r \epsilon_t = \mu + \psi(B)\epsilon_t,$$

where $\psi(.)$ is a polynomial function (of possibly infinite degree), given by $\psi(x) = 1 + \psi_1 x + \psi_2 x^2 + \dots$, for any x with $|x| < 1$. This equation may often be written as $\phi(B)(X_t - \mu) = \epsilon_t$, where $\phi(.)$ is another polynomial of the form $\phi(x) = 1 - \phi_1 x - \phi_2 x^2 - \dots$ satisfying the identity $\psi(x)\phi(x) = 1$ for all x with $|x| < 1$. If this representation exists, the function $\psi(\cdot)$ is said to be *invertible*. In such cases, the coefficients ϕ_r are determined as functions of the ψ_r, and (9.9) has the equivalent representation

$$X_t - \mu = \epsilon_t + \sum_{r=1}^{\infty} \phi_r B^r (X_t - \mu) = \epsilon_t + \phi_1(X_{t-1} - \mu) + \phi_2(X_{t-2} - \mu) + \dots \tag{9.10}$$

9.4.3 Autoregressive models

Suppose (9.10) to hold with $\phi_r = 0$ when $r > p$ for some integer $p \geq 1$. Then X_t is said to be an *autoregressive process of order* p, denoted AR(p) for short and written as $X_t \sim \mathrm{AR}(p)$. Here

$$X_t = \mu + \sum_{r=1}^{p} \phi_r (X_{t-r} - \mu) + \epsilon_t,$$

the value at time t depending linearly on a finite number of past values. It is an easy consequence (see, for example, Abraham and Ledolter 1983, Chapter 5; Box and Jenkins 1976) that the autocorrelations ρ_s may be non-zero for large values of s, though they decay exponentially eventually. The AR(p) process is Markovian in nature, with

$$p(X_t \mid X_{t-1}, \ldots, X_{t-p}, X_{t-p-1}, \ldots, M) = p(X_t \mid X_{t-1}, \ldots, X_{t-p}, M).$$

If normality is assumed, for $t > p$ and any specified initial information D_0,

$$(X_t \mid X_{t-1}, \ldots, X_{t-p}, D_0) \sim \mathrm{N}[\mu + \sum_{r=1}^{p} \phi_r (X_{t-r} - \mu), U].$$

EXAMPLE 9.3: AR(1) model. If $p = 1$ then $X_t = \mu + \phi_1 (X_{t-1} - \mu) + \epsilon_t$. Here it easily follows, on calculating variances, that $W = \phi_1^2 W + U$. Hence, if $|\phi_1| < 1$, $W = U/(1 - \phi_1^2)$. The condition that the first-order autoregression coefficient be less than unity in modulus is called the *stationarity* condition. Further, autocorrelations are given by $\rho_s = \phi_1^s$, so that stationarity is necessary in order that the autocorrelations be valid. Note then that ρ_s decays exponentially to zero in s, although it is non-zero for all s. The polynomial $\phi(.)$ here is simply $\phi(x) = 1 - \phi_1 x$, with

$$\psi(x) = 1/\phi(x) = 1 + \phi_1 x + \phi_1^2 x^2 + \ldots \qquad (|x| < 1)$$

Thus, in terms of (9.9), $\psi_r = \phi_1^r = \rho_r$ for $r \geq 0$.

Higher order models, AR(p) for $p > 1$, can provide representations of stationary series with essentially any form of correlation structure that is sustained over time but decays eventually (and exponentially) towards zero. In higher-order models too, the ϕ coefficients are subject to stationarity restrictions that lead to valid autocorrelations.

9.4.4 Moving-average models

Suppose (9.9) to hold with $\psi_r = 0$ for $r > q \geq 1$. Then X_t is said to be a *moving-average process of order* q, denoted MA(q) for short and written as $X_t \sim \mathrm{MA}(q)$. Thus X_t depends on only a finite number $q + 1$ of the ϵ_t,

$$X_t = \mu + \epsilon_t + \psi_1 \epsilon_{t-1} + \ldots + \psi_q \epsilon_{t-q}.$$

It is an easy consequence that X_t and X_{t-s} are unrelated for $s > q$, depending as they do entirely on distinct ϵ terms that are uncorrelated. Thus

$\rho_s = 0$ for $s > q$. For q relatively small, therefore, the MA process is useful for modelling *local* dependencies amongst the X_t.

EXAMPLE 9.4: MA(1) model. If $q = 1$ then $X_t = \mu + \epsilon_t + \psi_1 \epsilon_{t-1}$, values more than one step apart being uncorrelated. It easily follows that $W = (1 + \psi_1^2)U$ and $\rho_1 = \psi_1/(1 + \psi_1^2)$. Note that $|\rho_1| \leq 0.5$. There is an identifiability problem inherent in this, and all other, MA representations of a stationary process; note that replacing ψ_1 by ψ_1^{-1} leads to the same value for ρ_1, so that there are two essentially equivalent representations for the MA(1) process with a given value of ρ_1. However, here $\psi(x) = 1 + \psi_1 x$, so that the representation in terms of (9.10) leads to $X_t = \mu + \epsilon_t - \psi_1(X_{t-1} - \mu) + \psi_1^2(X_{t-2} - \mu) - \ldots$ For $|\psi_1| < 1$, the powers ψ_1^r decrease exponentially to zero, naturally reflecting a rapidly diminishing effect of values in the past on the current value. If, however, $|\psi_1| > 1$, then this implies an increasing dependence of X_t on values into the distant past, which is obviously embarrassing. For this (and other, similarly ad-hoc) reasons, the MA model is typically subjected to invertibility conditions that restrict the values of the ψ coefficients, just as the stationarity conditions that apply in AR models. Invertibility gives an illusion of uniqueness and serves to identify a single model whose limiting errors converge in probability to the ϵ_t. In this case the restriction is to $|\psi_1| < 1$ (see the earlier references for full discussion of such conditions).

9.4.5 ARMA models

Given the required stationarity conditions, any AR(p) model $\phi(B)(X - \mu) = \epsilon_t$ implies the representation (9.9) with coefficients ψ_r that decay towards zero as r increases. Thus, as an approximation, taking q sufficiently large implies that any AR model can be well described by an MA(q) model, possibly with q large. Similarly, an invertible MA(q) model can be written in autoregressive form with coefficients ϕ_r decaying with r, and similar thinking indicates that for large enough p, this can be well described by an AR(p) model. Thus, with enough coefficients, any stationary process can be well approximated by using either AR or MA models. Combining AR and MA models can, however, lead to adequate representations with many fewer parameters, and much classical time series modelling is based on autoregressive moving-average models, ARMA for short. An ARMA(p, q) model for X_t is given by

$$X_t = \mu + \sum_{r=1}^{p} \phi_r (X_{t-r} - \mu) + \sum_{r=1}^{q} \psi_r \epsilon_{t-r} + \epsilon_t,$$

where as usual, the ϵ_t are zero-mean, uncorrelated random quantities with constant variance U. The above references fully develop the mathematical and statistical theory of ARMA models. They describe a wide range of possible stationary autocorrelation structures, subject to various restrictions on the ϕ and ψ coefficients.

9.4.6 ARMA models in DLM form

All ARMA noise models can be written in DLM form in a variety of ways and for a variety of purposes. Suppose here that the series X_t is stationary with zero-mean, $\mu = 0$, and is actually observed. In the context of modelling Y_t in Section 9.4.1, this is equivalent to supposing that $Y_t = X_t$ for all t.

EXAMPLE 9.5: Autoregressive DLMs: regression model form. In the AR(p) case, the model may be written as a simple, static regression sequentially defined over time as mentioned in Section 9.1. Simply note that

$$X_t = \mathbf{F}_t'\boldsymbol{\theta} + \nu_t,$$

where $\nu_t = \epsilon_t$, $\mathbf{F}_t' = (X_{t-1}, \dots, X_{t-p})$ and $\boldsymbol{\theta}' = (\phi_1, \dots, \phi_p)$. This is useful in leading to sequential learning of the AR parameters by simply applying standard DLM results. The standard updating equations lead to posterior distributions that relate directly to those in traditional reference analyses from a non-sequential viewpoint. Some of the basic structure is detailed here as it is relevant to further methodological development in Chapter 15.

Note first that the standard normal theory analysis requires that at time t, the regression vector $\mathbf{F}_t'$ be known. In particular, this implies the need for known or assumed initial values $\mathbf{F}_1' = (X_0, \dots, X_{-p+1})$ at the (arbitrary) origin $t = 1$. Assuming such values, the standard theory applied to sequentially update posteriors $(\boldsymbol{\theta}|D_t) \sim \mathrm{N}[\mathbf{m}_t, \mathbf{C}_t]$ via the static regression updating equations summarised in item (3) of Section 9.2.4. Assuming known observation variance U, and writing in terms of the precision, or information matrices $\mathbf{C}_t^{-1}$, we thus have, at time $t = n$,

$$\begin{aligned}\mathbf{C}_n^{-1} &= \mathbf{C}_0^{-1} + U^{-1}\sum_{t=1}^{n}\mathbf{F}_t\mathbf{F}_t',\\ \mathbf{m}_n &= \mathbf{C}_t[\mathbf{C}_0^{-1}\mathbf{m}_0 + U^{-1}\sum_{t=1}^{n}\mathbf{F}_tX_t].\end{aligned} \qquad (9.11)$$

These results coincide with standard Bayesian linear regression results, as in DeGroot (1971, Chapter 11) and Box and Tiao (1973, Chapter 2), and, in the linear model formulation of AR processes, specifically Zellner (1971). These results are most useful in a later development, in Chapter 15, of computational algorithms for analysis of DLMs with AR components.

The case of a reference prior for $\boldsymbol{\theta}$, in which $\mathbf{C}_0^{-1} \to \mathbf{0}$, results in the standard reference normal posterior distribution for $(\boldsymbol{\theta}|D_n)$ (conditional also on the assumed initial values for the X_t process), in which the above equations reduce to

$$\mathbf{C}_n = U(\sum_{t=1}^{n} \mathbf{F}_t\mathbf{F}_t')^{-1},$$
$$\mathbf{m}_n = (\sum_{t=1}^{n} \mathbf{F}_t\mathbf{F}_t')^{-1}\sum_{t=1}^{n} \mathbf{F}_t X_t. \tag{9.12}$$

The value $\mathbf{m}_n$ in (9.12) coincides with the usual maximum likelihood estimate of $\boldsymbol{\theta}$, conditional on the assumed initial values for the series.

Forecasting ahead leads to computational problems in calculating forecast distributions, however, since future values are needed as regressors. Zellner (1971) discusses such problems. See also Broemling (1985) and Schnatter (1988).

EXAMPLE 9.6: An alternative AR model in DLM form. There are various alternative representations of AR models in DLM form that have uses in contexts when the AR parameters ϕ_j are specified. One such form that we use in Chapter 15 is as follows. Given the standard AR(p) model with zero mean, as in Example 9.5 above, write $\mathbf{E} = \mathbf{E}_p = (1, 0, \dots, 0)'$ and

$$\mathbf{G} = \begin{pmatrix} \phi_1 & \phi_2 & \phi_3 & \cdots & \phi_p \\ 1 & 0 & 0 & \cdots & 0 \\ 0 & 1 & 0 & \cdots & 0 \\ \vdots & & \ddots & \cdots & \vdots \\ 0 & 0 & \cdots & 1 & 0 \end{pmatrix}.$$

Also, let $\boldsymbol{\omega}_t = \epsilon_t\mathbf{E} = (\epsilon_t, 0, \dots, 0)'$, a p-variate normal evolution error term with (singular) variance matrix

$$\mathbf{W} = \begin{pmatrix} U & 0 & \cdots & 0 \\ 0 & 0 & \cdots & 0 \\ \vdots & \vdots & \cdots & \vdots \\ 0 & 0 & \cdots & 0 \end{pmatrix}.$$

It easily follows that

$$X_t = \mathbf{E}'\mathbf{F}_t \quad \text{and} \quad \mathbf{F}_t = \mathbf{G}\mathbf{F}_{t-1} + \boldsymbol{\omega}_t,$$

representing the AR model in DLM form with no independent observational error terms.

EXAMPLE 9.7: ARMA models in DLM form. There are similarly various equivalent ways of representing a specified ARMA(p, q) model in DLM form. We detail one here, noting that the MA(q) is an obvious special case when $p = 0$. This also provides an alternative to the above representation for pure autoregressive models when $q = 0$.

For the usual ARMA model with $\mu = 0$,

$$X_t = \sum_{r=1}^{p} \phi_r X_{t-r} + \sum_{r=1}^{q} \psi_r \epsilon_{t-r} + \epsilon_t,$$

define $n = \max(p, q+1)$, and extend the ARMA coefficients to $\phi_r = 0$ for $r > p$, and $\psi_r = 0$ for $r > q$. Introduce the n-vector $\boldsymbol{\theta}_t$ whose first element is the current observation X_t. Further, define $\mathbf{E} = \mathbf{E}_n = (1, 0, \dots, 0)'$ as above, with

$$\mathbf{G} = \begin{pmatrix} \phi_1 & 1 & 0 & \dots & 0 \\ \phi_2 & 0 & 1 & \dots & 0 \\ \vdots & \vdots & \vdots & \ddots & \vdots \\ \phi_{n-1} & 0 & 0 & \cdots & 1 \\ \phi_n & 0 & 0 & \dots & 0 \end{pmatrix}$$

and

$$\boldsymbol{\omega}_t = (1, \psi_1, \dots, \psi_{n-1})' \epsilon_t.$$

With these definitions, it can be verified that the ARMA model may be rewritten as

$$\begin{aligned} X_t &= \mathbf{E}' \boldsymbol{\theta}_t, \\ \boldsymbol{\theta}_t &= \mathbf{G} \boldsymbol{\theta}_{t-1} + \boldsymbol{\omega}_t. \end{aligned} \tag{9.13}$$

Thus, conditional on the defining parameters and also on D_{t-1}, the ARMA process has the form of an observational noise-free DLM of dimension n. The evolution noise has a specific structure, giving evolution variance matrix

$$\mathbf{U} = U(1, \psi_1, \dots, \psi_{n-1})'(1, \psi_1, \dots, \psi_{n-1}). \tag{9.14}$$

Given an initial prior for $\boldsymbol{\theta}_0$, the standard analysis applies, *conditional* on values of the defining parameters ϕ_r and ψ_r being specified. Note that the model is a TSDLM and so the forecast function is given simply by

$$f_t(k) = \mathrm{E}[X_{t+k} \mid D_t] = \mathbf{E}' \mathbf{G}^k \mathbf{m}_t, \qquad (k = 1, 2, \dots),$$

where $\mathbf{m}_t = \mathrm{E}[\boldsymbol{\theta}_t \mid D_t]$ as usual. It follows, as with all TSDLMs, that the eigenvalues of $\mathbf{G}$ determine the form of the forecast function and that the model has a canonical Jordan block representation (see Exercise 12). In addition, Theorem 5.1 implies that there is a stable limiting form for the updating equations determined by a limiting value for the sequence of posterior variance matrices $\mathbf{C}_t$ of $\boldsymbol{\theta}_t$.

EXAMPLE 9.3 (continued). In the AR(1) case $p = 1$ and $q = 0$, so that $n = 1$, $\mathbf{G} = \phi_1$, and directly, $X_t = \phi_1 X_{t-1} + \epsilon_t$. Forecasting ahead gives $f_t(k) = \phi_1^k m_t$, where $m_t = X_t$. The stationarity condition $|\phi_1| < 1$ implies an exponential decay to zero, the marginal mean of X_{t+k}, as k increases.

EXAMPLE 9.4 (continued). In the MA(1) case $p = 0$ and $q = 1$, so that $n = 2$. Here

$$\mathbf{G} = \begin{pmatrix} 0 & 1 \\ 0 & 0 \end{pmatrix} = \mathbf{J}_2(0),$$

giving the component model $\{\mathbf{E}_2, \mathbf{J}_2(0), ., .\}$. Setting $\boldsymbol{\theta}_t' = (X_t, \theta_t)$, the evolution equation gives $X_t = \theta_{t-1} + \epsilon_t$ and $\theta_t = \psi_1\epsilon_t$. The standard MA(1) representation follows by substitution, namely $X_t = \epsilon_t + \psi_1\epsilon_{t-1}$. Here $\mathbf{G}^k = \mathbf{0}$ for $k > 1$, so that, with $\mathbf{m}_t' = (X_t, m_t)$, the forecast function is given by $f_t(1) = f_{t+1} = m_t$ and $f_t(k) = 0$ for $k > 1$.

EXAMPLE 9.8. In an MA(q) model for any q, it similarly follows that the forecast function takes irregular values up to $k = q$, being zero, the marginal mean of X_{t+k}, thereafter. The canonical component model is $\{\mathbf{E}_{q+1}, \mathbf{J}_{q+1}(0), ., .\}$. (Incidentally, it is possible to incorporate the observational noise component ν_t directly into this noise model, if desired.)

Related discussion of similar representations of ARMA models (usually referred to as state-space representations) can be found in Abraham and Ledolter (1983), Harvey (1981), Priestley (1980), and Young (1984), for example. Harvey in particular discusses initial conditions and limiting forms of the updating equations.

9.4.7 ARMA components in DLMs

In classical ARMA modelling, practical application is approached by first transforming the Y_t series in an attempt to achieve a transformed series X_t that is, at least approximately, zero mean and stationary. In addition to the usual possibilities of instantaneous, non-linear data transformations to correct for variance inhomogeneities, data series are usually subject to *differencing* transforms. The first-order difference of Y_t is simply defined as $(1 - B)Y_t = Y_t - Y_{t-1}$, the second-order $(1 - B)^2Y_t = Y_t - 2Y_{t-1} + Y_{t-2}$, and higher-order differences similarly defined by $(1 - B)^kY_t$. In attempting to derive stationarity this way, it is important that the original Y_t series be non-seasonal. If seasonality of period p is evident, then Y_t may be first subjected to seasonal differencing, producing the transformed series $(1 - B^p)Y_t$. Thus, the classical strategy is to derive a series of the form $X_t = (1 - B)^k(1 - B^p)Y_t$, for some k and p, that may be assumed to be stationary. Once this is approximately obtained, residual autocorrelation structure in the X_t series is modelled within the above ARMA framework. Full discussion of differencing, and the relationship with the assumptions of stationarity of differenced series, can be found in the references in Section 9.4.3.

By comparison, DLM approaches tend to model the original time series Y_t directly, representing the series through trend, regression, seasonal and residual components without such direct data transformations. In contexts of often highly non-stationary development over time, application of differencing transformations can confuse the interpretation of the model, confound the components, and highlight noise at the expense of meaningful interpretations. In particular, with regression effects the differencing carries over to transform independent variables in the model, obscuring the nature of the regression relationships. This confusion is exacerbated by the appearance of time-varying parameters that are so fundamental to short-term forecasting performance. Furthermore, the occurrence of abrupt changes in time series structure that may be directly modelled through a DLM representation evidence highly non-stationary behaviour that cannot usually be removed by differencing or other data transformations. Consistent with this view is the use of stationary time series models as *components* of an overall DLM, and there is much interest in such component models, especially in time series analysis in physical science and biomedical applications. Autoregressive components, in particular, have real utility in modelling actual physical structure in a time series not captured in trend, seasonal or regression terms, and also as a catch-all noise model used as a purely empirical representation of residual variation. To take a specific example, consider a series Y_t following an underlying polynomial trend μ_t with a correlated process X_t superimposed, and in the context of additive observational errors. Then we have

$$Y_t = \mu_t + X_t + \nu_t,$$

where both μ_t and X_t have representations in terms of specific component DLMs. For example, suppose μ_t is a first-order polynomial trend. At time t, denote by ω_t the evolution error in the trend, supposing $\omega_t \sim \mathrm{N}[0, W]$, and assume the above AR(p) representation for the process X_t. Then we have

$$Y_t = \mathbf{F}'\boldsymbol{\theta}_t + \nu_t \quad \text{and} \quad \boldsymbol{\theta}_t = \mathbf{G}\boldsymbol{\theta}_{t-1} + \boldsymbol{\omega}_t,$$

where $\boldsymbol{\theta}_t = (\mu_t, X_t, X_{t-1}, \dots, X_{t-p+1})'$, a $(p+1) \times 1$ state vector at time t, $\mathbf{F} = (1, 1, 0, \dots, 0)'$,

$$\mathbf{G} = \begin{pmatrix} 1 & 0 & 0 & 0 & \cdots & 0 \\ 0 & \phi_1 & \phi_2 & \phi_3 & \cdots & \phi_p \\ 0 & 1 & 0 & 0 & \cdots & 0 \\ 0 & 0 & 1 & 0 & \cdots & 0 \\ \vdots & \vdots & & \ddots & \cdots & \vdots \\ 0 & 0 & 0 & \cdots & 1 & 0 \end{pmatrix},$$

and the evolution error $\boldsymbol{\omega}_t$ has two lead elements corresponding to the trend increment and AR process innovation at time t, all other elements being zero, and thus has the singular variance matrix

$$\mathbf{W} = \begin{pmatrix} W & 0 & 0 & \cdots & 0 \\ 0 & U & 0 & \cdots & 0 \\ 0 & 0 & 0 & \cdots & 0 \\ \vdots & \vdots & \vdots & \cdots & \vdots \\ 0 & 0 & 0 & \cdots & 0 \end{pmatrix}.$$

Note now that the AR process X_t is unobserved, or latent. DLM analysis leads to sequential learning on X_t in addition to the trend component, and standard normal theory applies conditional on specified values of the AR parameters ϕ_j. If these parameters are uncertain, the problems of parameter estimation and inference become essentially non-linear, so that analytic approximations or numerical methods are needed. We defer further discussion of this important class of problems until Chapter 15, where advanced numerical methods based on Markov chain Monte Carlo simulation techniques are developed, with this specific component model as a key example.

Two final comments are relevant. First, in many applications purely observational noise, including measurement errors, sampling errors, and so forth, are significant and induce variation in the observed series that is comparable with the levels of variation of model components. Thus, even if underlying trends are negligible, it is important to be able to embed AR and other ARMA models in the above DLM form in order to distinguish, and ultimately infer, the process structure from contaminating noise. Again, this is revisited in examples in Chapter 15. Second, we note extensions to time-varying parameter AR models. In some applications, AR component forms are very useful locally in time, but their global applicability is questionable unless the defining AR parameters change through time. This leads to extended dynamic models with the ϕ_j coefficients themselves evolving; these are the focus of Section 9.6 below, following some important developments in time series decomposition.

9.5 TIME SERIES DECOMPOSITIONS

The use of autoregressive component DLMs in modelling and identifying latent, quasi-cyclical structure in time series is of wide interest in time series analysis in various fields, as discussed and exemplified in West (1996a-c, 1997). In such applications, one is often interested in decomposition of time series into hidden, or latent, components that have physical interpretation. In this connection, some new results on time series decomposition in the DLM framework are relevant. These results are quite general, and so are given here in a general context. The special cases of autoregressive structure are, perhaps, of most practical interest, and so are given special attention here. The next section develops these results in time-varying autoregressions, and the results are applied further in Section 15.3. The key result on decomposition is derived essentially directly from the theory of superposition and model structuring of Chapters 5 and 6, though the specific and constructive use of this theory given here is relatively new, first discussed in West (1997).

The focus is on decomposition and structure of components DLMs, i.e., of a process X_t in a model $Y_t = \mu_t + X_t + \nu_t$, where μ_t represents additional components and ν_t observational noise. We consider explicitly components X_t that arise from sub-models of the form

$$X_t = \mathbf{E}'\boldsymbol{\theta}_t \quad \text{and} \quad \boldsymbol{\theta}_t = \mathbf{G}\boldsymbol{\theta}_{t-1} + \boldsymbol{\omega}_t, \tag{9.15}$$

with the usual zero-mean and independence assumptions for the evolution error sequence, and in which $\mathbf{E} = \mathbf{E}_p = (1, 0, \ldots, 0)'$, so that X_t is the first element of the state vector $\boldsymbol{\theta}_t$. Autoregressive components are a particular example of interest. Note that for clarity the development is given in terms of a constant state matrix $\mathbf{G}$; we mention extensions to time-varying $\mathbf{G}$ below.

Suppose that the state matrix $\mathbf{G}$ has distinct eigenvalues arranged in the diagonal matrix $\mathbf{A} = \text{diag}(\alpha_1, \ldots, \alpha_p)$; we note that similar development is possible in cases of repeat eigenvalues, though that is not pursued here. Then, following the development of similar models in Section 5.3, $\mathbf{G}$ is diagonalisable as $\mathbf{G} = \mathbf{BAB}^{-1}$, and with $\mathbf{G}^k = \mathbf{BA}^k\mathbf{B}^{-1}$ for all $k \geq 0$. It follows that as with forecast function calculations in Section 5.3.2,

$$\text{E}[X_{t+k}|\boldsymbol{\theta}_t] = \mathbf{E}'\mathbf{BA}^k\mathbf{B}^{-1}\boldsymbol{\theta}_t$$

as a function of the lead-time $k \geq 0$. This reduces to

$$\text{E}[X_{t+k}|\boldsymbol{\theta}_t] = \sum_{j=1}^{p} x_{t,j}\alpha_j^k, \tag{9.16}$$

where the $x_{t,j}$ are obtained as element-wise products of the two p-vectors $\mathbf{B}'\mathbf{E}$ and $\mathbf{B}^{-1}\boldsymbol{\theta}_t$, i.e., $x_{t,j} = (\mathbf{B}'\mathbf{E})_j(\mathbf{B}^{-1}\boldsymbol{\theta}_t)_j$, for $j = 1, \ldots, p$. Now, at $k = 0$ we obtain $\text{E}[X_t|\boldsymbol{\theta}_t] = X_t$, as X_t is the first element of $\boldsymbol{\theta}_t$. Hence, from (9.16) at $k = 0$, it follows that

$$X_t = \sum_{j=1}^{p} x_{t,j}. \tag{9.17}$$

This is the basic decomposition of X_t into p component, latent sub-series, one corresponding to each of the eigenvalues of $\mathbf{G}$. Generally, the eigen-structure will contain complex values and vectors, so that some or all of the $x_{t,j}$ processes will be complex-valued. In such cases, they will be paired with conjugate processes in order that the sum be real-valued, as required. The specific form of decomposition of an AR(p) model provides an important special case for illustration.

EXAMPLE 9.6 (continued). In the AR(p) model of Example 9.6,

$$\mathbf{G} = \begin{pmatrix} \phi_1 & \phi_2 & \phi_3 & \cdots & \phi_p \\ 1 & 0 & 0 & \cdots & 0 \\ 0 & 1 & 0 & \cdots & 0 \\ \vdots & & \ddots & \cdots & \vdots \\ 0 & 0 & \cdots & 1 & 0 \end{pmatrix}$$

has eigenvalues that are the p reciprocal roots of the AR characteristic equation $\phi(u) = 0$, where $\phi(u) = 1 - \phi_1 u - \ldots - \phi_p u^p$; thus the eigenvalues satisfy $\prod_{j=1}^{p}(1 - \alpha_j u) = 0$. It is usual that the roots are distinct. Suppose that the p eigenvalues occur as c pairs of complex conjugates and $r = p - 2c$ distinct real values. Write the complex eigenvalues as $a_j \exp(\pm 2\pi i/\lambda_j)$ for $j = 1, \ldots, c$, noting that the real-valued, non-zero elements λ_j correspond to the periods of quasi-cyclical component behaviour in the series. Correspondingly, we write the real eigenvalues as a_j, for $j = 2c+1, \ldots, p$. In a stationary process, $|a_j| < 1$ for each j for real and complex roots; the development permits non-stationary processes, as one or more of the eigenvalues, real or complex, may have modulus outside this range.

We can now rewrite (9.17) as

$$X_t = \sum_{j=1}^{c} z_{t,j} + \sum_{j=1}^{r} y_{t,j}, \tag{9.18}$$

with summands $z_{t,j}$ corresponding to the complex roots, and $y_{t,j}$ corresponding to the real roots. This easily is done by directly computing the factors $x_{t,j}$ based on the specified values of the ϕ_j and observed values of X_t. Using the theory of similar models from Section 5.3, it easily follows that the real-valued processes $y_{t,j}$ follow individual AR(1) models, namely

$$y_{t,j} = a_j y_{t-1,j} + \delta_{t,j},$$

for some zero-mean evolution error $\delta_{t,j}$, related across component indices j. The complex components occur in conjugate pairs. Thus, any complex component process, $x_{t,d}$, say, is paired with a conjugate, $x_{t,h}$, say, such that $z_{t,j} = x_{t,d} + x_{t,h}$ is real for all t. In this case, the model similarity transform implies the quasi-periodic structure of the canonical similar model (as in Section 5.4.4 of chapter 5) of the form

$$z_{t,j} = 2a_j \cos(2\pi/\lambda_j) z_{t-1,j} - a_j^2 z_{t-2,j} + \epsilon_{t,j}$$

for additional zero-mean terms $\epsilon_{t,j}$. For each j, the errors $\epsilon_{t,j}$ themselves follow individual AR(1) models, so that each $z_{t,j}$ series is a quasi-cyclical ARMA(2, 1) process. The $\epsilon_{t,j}$ are related across component series j and are also correlated with the innovations $\delta_{t,j}$ of the AR(1) components.

Hence the entire time series is decomposed into the sum of time-varying components corresponding to the autoregressive roots. This results in a decomposition (9.18) of X_t as the sum of r real-valued latent component series, each representing short-term correlation consistent with AR(1) behaviour, and a collection of c components $z_{t,j}$, each having the form of a stochastically time-varying, damped harmonic component of period λ_j. We stress that this decomposition result is constructive; with specified ϕ_j values and observed data X_t over a time interval, we can easily compute the implied latent processes that combine additively via (9.18). Some of the practical relevance of these results is illustrated in Section 15.3.4, and, following extension to time-varying system matrices, in the following section.

These basic results can be extended to cover models with time-varying evolution matrices. Assume a time-dependent evolution matrix $\mathbf{G}_t$ evolving according to a model in which $\mathrm{E}[\mathbf{G}_t|\mathbf{G}_{t-1}, \boldsymbol{\theta}_{t-1}, D_{t-1}] = \mathbf{G}_{t-1}$, i.e., the scalar entries follow random walks. An example is time-varying autoregression, in the next section. Then the time-dependent, "instantaneous" decomposition is as follows. Suppose that for each t, $\mathbf{G}_t$ has distinct eigenvalues arranged in the diagonal matrix $\mathbf{A}_t = \mathrm{diag}(\alpha_{t,1}, \dots, \alpha_{t,p})$, so that $\mathbf{G}_t = \mathbf{B}_t\mathbf{A}_t\mathbf{B}_t^{-1}$, where $\mathbf{B}_t$ is the eigenvector matrix at time t. Then equations (9.17) and (9.18) are essentially unchanged; now the latent processes $x_{t,j}$ are obtained as element-wise products of the two time-varying vectors $\mathbf{B}_t'\mathbf{E}$ and $\mathbf{B}_t^{-1}\boldsymbol{\theta}_t$, i.e., $x_{t,j} = (\mathbf{B}_t\mathbf{E})_j(\mathbf{B}_t^{-1}\boldsymbol{\theta}_t)_j$, for $j = 1, \dots, p$. Hence the decomposition results hold even though we are now dealing with a system matrix $\mathbf{G}_t$ whose eigenstructure may be varying through time. There are some intricacies in dealing with questions of time-variation in the numbers of real and complex eigenvalues, and hence in the relative numbers of components $z_{t,j}$ and $y_{t,j}$ in (9.18), but the decomposition has real practical utility in the time-varying context too, as is illustrated in the following section.

9.6 TIME-VARYING AUTOREGRESSIVE DLMS

9.6.1 Model structure and analysis

An important extension of autoregressive component models is that of time-varying autoregressions. Modelling time-variation in AR parameters adds flexibility to adapt models to time-varying patterns of dependence and non-stationarities; these models embody the notion of "local stationarity" but "global" non-stationarity of time series structure. We develop and illustrate these models here, in the context of a time-varying AR model for an observed series, free from additional observational error and other model components. Early work in this area appears in Kitegawa and Gersch (1985). The development here follows recent work, including that of Prado and West (1997) where further details and more extensions of analysis appear.

Extend the basic AR model of Section 9.4.6 to incorporate AR coefficients that follow a random walk through time. We thus have

$$X_t = \mathbf{F}_t'\boldsymbol{\phi}_t + \epsilon_t \quad \text{and} \quad \boldsymbol{\phi}_t = \boldsymbol{\phi}_{t-1} + \boldsymbol{\omega}_t,$$

where $\mathbf{F}_t' = (X_{t-1}, \dots, X_{t-p})$ and $\boldsymbol{\phi}_t = \boldsymbol{\theta}_t = (\phi_{t,1}, \dots, \phi_{t,p})'$. With $\boldsymbol{\omega}_t \sim \mathrm{N}[\mathbf{0}, \mathbf{W}_t]$ and $\mathbf{W}_t$ specified, we have a standard DLM framework, assuming the initial values $\mathbf{F}_1$ are known elements of D_0. Then standard updating and filtering equations apply, resulting in the usual collections of posterior distributions for the $\boldsymbol{\phi}_t$ sequence through time.

We note that the model can be equivalently expressed in the alternative AR DLM form of Example 9.6, but now with time-varying evolution matrix. Thus, with $\mathbf{E} = (1, 0, \dots, 0)'$ and

$$\mathbf{G}_t = \begin{pmatrix} \phi_{t,1} & \phi_{t,2} & \phi_{t,3} & \cdots & \phi_{t,p} \\ 1 & 0 & 0 & \cdots & 0 \\ 0 & 1 & 0 & \cdots & 0 \\ \vdots & & \ddots & \cdots & \vdots \\ 0 & 0 & \cdots & 1 & 0 \end{pmatrix},$$

we have $X_t = \mathbf{E}'\mathbf{F}_t$ and $\mathbf{F}_t = \mathbf{G}_t\mathbf{F}_{t-1} + (\epsilon_t, 0, \dots, 0)'$. This form is just that required for the "instantaneous" time series decomposition result of the previous section to apply. At each time t, therefore, we can directly compute the decompositions (9.18) based on any specified $\boldsymbol{\phi}_t$ vector, such as the posterior mean $\mathrm{E}[\boldsymbol{\phi}_t|D_n]$ as used in the example below. This relies on computing the eigenvalues and eigenvectors of the above $\mathbf{G}_t$ matrix at each time point at the specified estimate of $\boldsymbol{\phi}_t$. We note that though not developed here, it is possible to evaluate more formal inferences about the eigenstructure by simulating posterior distributions of $\boldsymbol{\phi}_t$ and hence deducing simulation-based posterior inferences about the eigenstructure and resulting time series decomposition.

Note that as the AR vector is time-varying, then so too are the moduli and frequencies/periods of the latent subseries. In particular, the real-valued roots generate latent AR(1) components $y_{t,j} = a_{t,j}y_{t-1,j} + \delta_{t,j}$ with time-varying moduli $a_{t,j}$, and the pairs of complex roots generate latent ARMA(2, 1) components, $z_{t,j} = 2a_{t,j}\cos(2\pi/\lambda_{t,j})z_{t-1,j} - a_{t,j}^2 z_{t-2,j} + \epsilon_{t,j}$, with time-varying moduli and wavelengths/periods $\lambda_{t,j}$ and with correlated innovations $\epsilon_{t,j}$. Patterns of change over time are usually best illustrated by exploring estimates of the $a_{t,j}$ and $\lambda_{t,j}$ through time, as in the example in the following section.

9.6.2 An illustration

The data in Figure 9.1 are electroencephalogram (EEG) voltage levels recorded during a brain seizure of an individual undergoing electroconvulsive therapy (ECT). ECT is a major tool in brain seizure treatment, and EEG monitoring is the primary method of observation of brain ac-

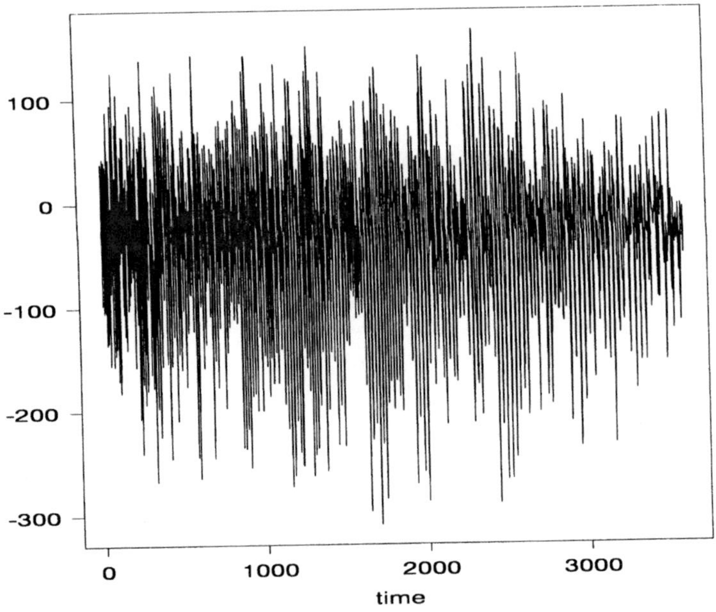

Figure 9.1 EEG series (units in millivolts)

tivity during ECT (as in other contexts). Activity is monitored by scalp electrodes measuring resulting EEG waveforms at various scalp locations; this collection of data is a series subsampled from one of several EEG channels recorded in a specific seizure. The original sampling rate was 256 observations per second, and we have subsampled every sixth observation for this analysis, taken from Prado and West (1997). The experiment was performed by Dr Andrew Krystal of Duke University, who kindly provided the data, background information, and his expert opinion. The data exhibit quasi-cyclical behaviour over time, with evident time-varying frequency characteristics and high-frequency distortions; Figure 9.2 shows shorter sections of the data at several separated intervals, clearly showing up changes in the basic waveform at different stages of the seizure. Our analysis is based on a time-varying AR(20) model capable of capturing the basic waveform structure and its evolution over time. Models of various orders were explored and compared to identify the chosen order $p = 20$. The analysis structures $\mathbf{W}_t$ in terms of a single discount factor for the evolution of ϕ_t, chosen similarly by exploring several values. The reported analysis is based on a model discount factor of 0.994. More details on model selection appear in Prado and West (1997).

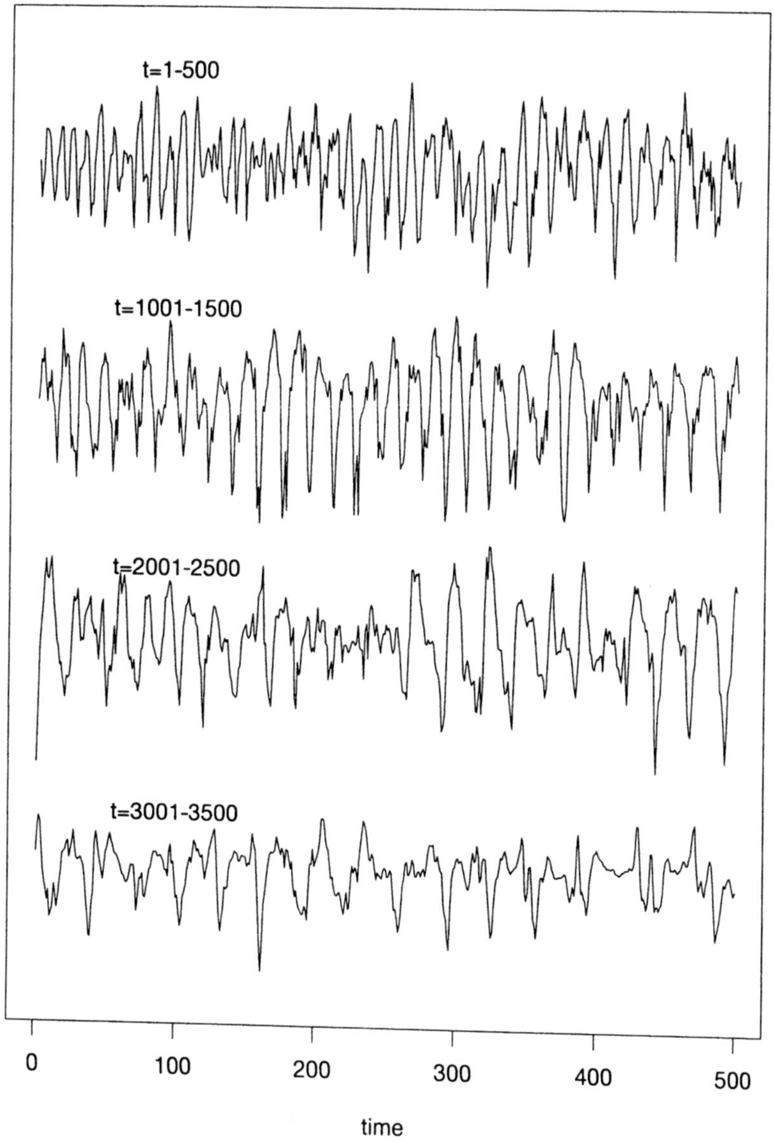

Figure 9.2 Four sections of 500 consecutive values of the EEG series, taken from near the beginning, central portion and end of the full data series. This clearly illustrates the changing nature of the seizure waveform throughout the seizure episode

Some summary inferences appear in Figures 9.3, 9.4 and 9.5, all based on the posterior means $E[\phi_t|D_n]$ over $t = 1, \ldots, n = 3600$. First, the time series decomposition result is applied. At each t, the eigenvalues of $\mathbf{G}_t$ (i.e., the reciprocal characteristic roots of the AR polynomial at the current estimate $E[\phi_t|D_n]$) are computed. It is found that each set of roots contains a complex pair with modulus generally close to unity and a wavelength, $\lambda_{t,1}$ in the $10-30$ range, together with several more pairs of complex roots of lower moduli. Applying the decomposition results in the identification of a dominant component with this higher wavelength, $\lambda_{t,1}$; the estimated amplitude of this component is much greater than those of the remaining components, indicating this as the dominant feature of the EEG waveform. Two further quasi-cyclical components of much smaller, but meaningful, amplitudes, have longer wavelengths; the second is in the $5 - 10$ range, the others lower. The estimated latent components are graphed in Figure 9.3, ordered according to estimated amplitudes, as is evident in the figures. A part of this decomposition graph is redrawn in Figure 9.4, restricting attention for clarity to a section of just 500 consecutive observations in the central part of the seizure episode. The figures indicate the dominance of the first component, and suggest that the first two components together carry the EEG signal, the additional components representing higher frequency signal distortions due to physiological and experimental noise.

Figure 9.5 displays estimated time trajectories of the wavelengths $\lambda_{t,j}$ of the first three components. We note that the components have been ordered in terms of estimated amplitudes; there is no inherent ordering of components in the time series decomposition result, so one must be imposed for identification. It happens that the component of longest wavelength dominates in amplitude and is clearly distinguished from the higher frequency components. The amplitudes of the remaining components are more similar, and problems arise in identifying components at the start of the series as a result. Note the minor erratic appearance of the trajectories of $\lambda_{t,2}$ and $\lambda_{t,3}$ at the very start of the time interval. This does not imply such behaviour in the wavelength processes; rather, it is simply a result of the inherent identification problem. These components have similar estimated amplitudes initially, so the ordering by amplitudes may alternate between the two, and result in the "switching" effects appearing in the graph. With this in mind, it is evident that the wavelength of the second component varies smoothly around a level of $8 - 10$ time units, that of the third component being smaller, around 4 units. The wavelength ranges of the two main components correspond to approximate frequency ranges of $1.5 - 4.5$ cycles per second for the dominant component, and $4 - 9$ cycles per second for the second component. These correspond nicely with expected ranges of frequencies for the dominant "slow wave" typical of seizure activity, and the higher frequency waveforms of normal brain activity that are relatively suppressed during seizures (Dyro 1989). The third and higher-frequency components evidently represent corrupting noise in the data.

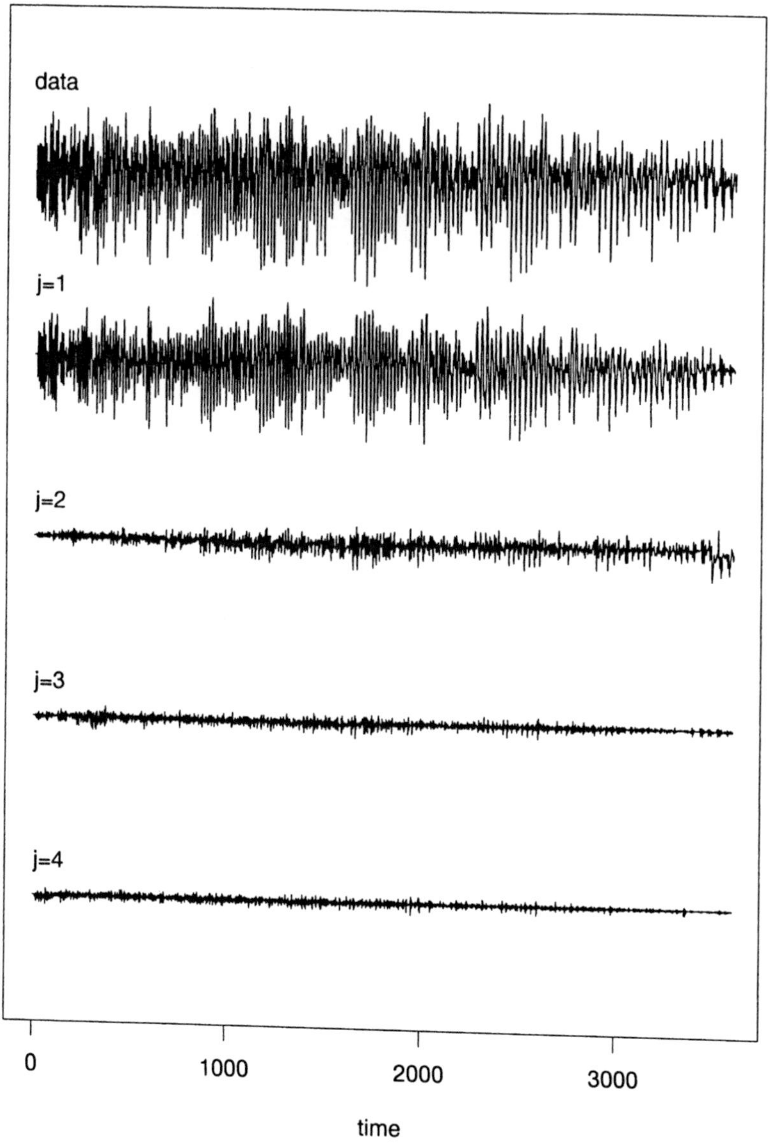

Figure 9.3 Estimated trajectories of the dominant latent components ($j = 1, 2, 3, 4$) of the EEG series based on time series decomposition

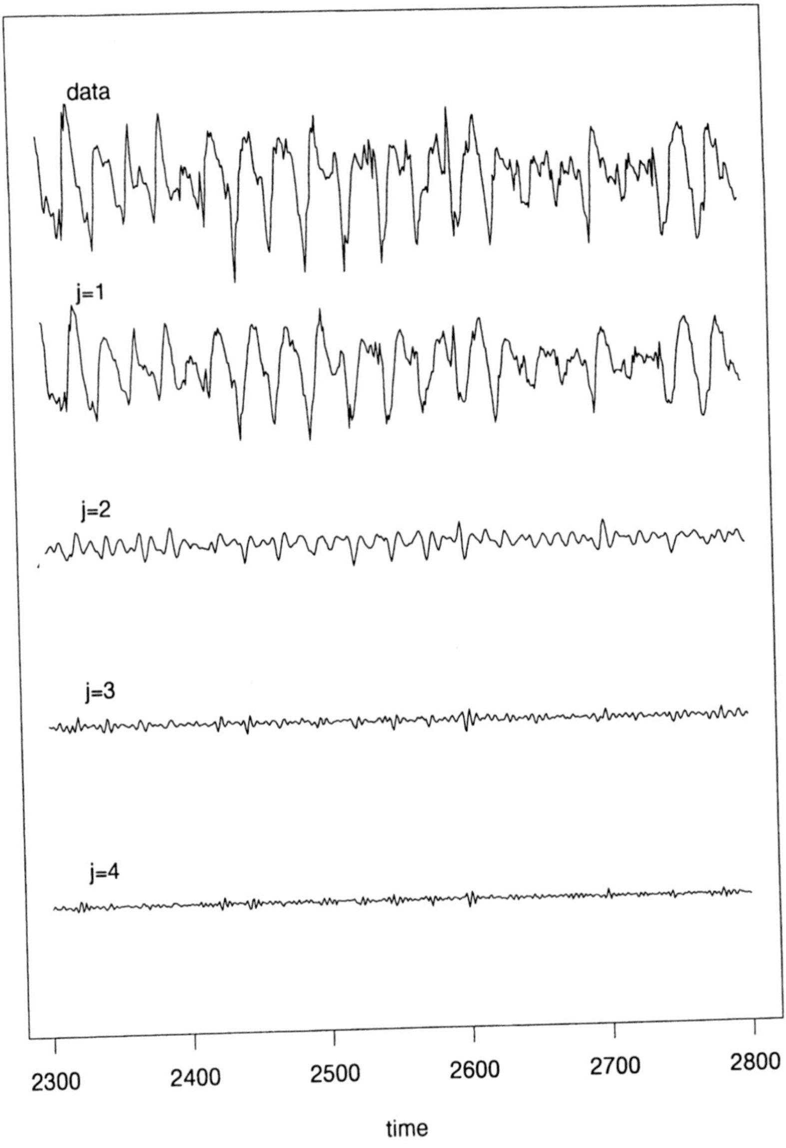

Figure 9.4 Estimated trajectories of the dominant latent components ($j = 1, 2, 3, 4$) of the EEG series, as in Figure 9.3, now restricted to a central section of 500 time points

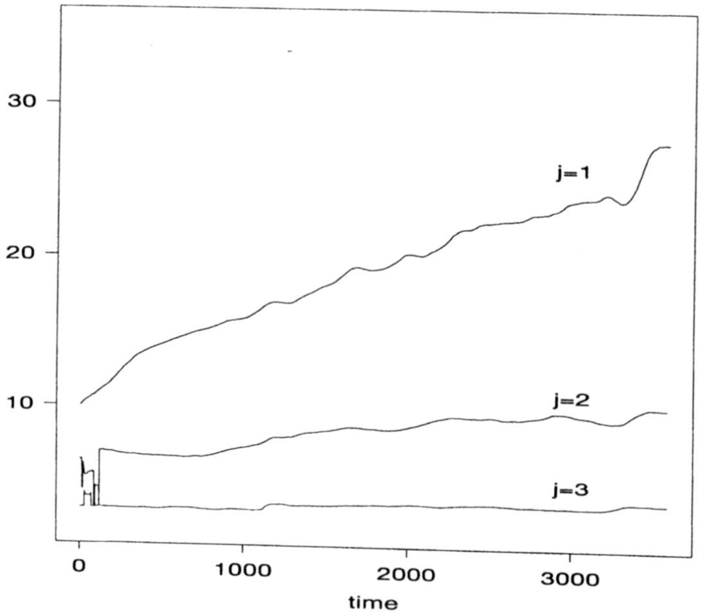

Figure 9.5 Estimated time trajectories of the wavelength parameters $\lambda_{t,j}$ of the three dominant latent components ($j = 1, 2, 3$) in the EEG analysis

There is evident time-variation in the wavelengths, captured by the time-variation modelled in $\boldsymbol{\phi}_t$. The general increase in the wavelength of the dominant component corresponds to a gradual lengthening of the waveform as the seizure matures and gradually tails off. The wavelength of the second component is relatively stable, indicative of its underlying constancy in nature. The higher-frequency components appear essentially constant too, as is to be expected under the interpretation that they represent background experimental noise.

9.7 EXERCISES

(1) Consider the discount regression DLM in which V_t is known for all t and $\mathbf{R}_t = \mathbf{C}_{t-1}/\delta$. Show that the updating equations can be written as

$$\mathbf{m}_t = \mathbf{C}_t(\delta^t \mathbf{C}_0^{-1}\mathbf{m}_0 + \sum_{r=0}^{t-1} \delta^r \mathbf{F}_{t-r} V_{t-r}^{-1} Y_{t-r})$$

and

$$\mathbf{C}_t^{-1} = \delta^t \mathbf{C}_0^{-1} + \sum_{r=0}^{t-1} \delta^r \mathbf{F}_{t-r}\mathbf{F}'_{t-r}V_{t-r}^{-1}.$$

Deduce that if $\mathbf{C}_0^{-1} \approx \mathbf{0}$, $\mathbf{m}_t$ is approximately given by the exponentially weighted regression (EWR) form

$$\mathbf{m}_t \approx (\sum_{r=0}^{t-1} \delta^r \mathbf{F}_{t-r}\mathbf{F}'_{t-r}V_{t-r}^{-1})^{-1}(\sum_{r=0}^{t-1} \delta^r \mathbf{F}_{t-r}V_{t-r}^{-1}Y_{t-r}).$$

(2) Consider the reference prior analysis of the static model $\{\mathbf{F}_t, \mathbf{I}, V, \mathbf{0}\}$, referring to the general theory of reference analysis in Section 4.10. As in that section, let

$$\mathbf{K}_t = \sum_{r=1}^{t} \mathbf{F}_r\mathbf{F}'_r, \qquad \mathbf{k}_t = \sum_{r=1}^{t} \mathbf{F}_r Y_r,$$

and $t = [n]$ be the first time $t \geq n$ such that $\mathbf{K}_t$ is non-singular. Using the results of Theorem 4.6 and Corollary 4.5, verify that the reference analysis leads to

$$\mathbf{C}_{[n]} = \mathbf{K}_{[n]}^{-1}$$

and

$$\mathbf{m}_{[n]} = \mathbf{C}_{[n]}\mathbf{k}_{[n]}.$$

(3) Table 9.1 provides data on weekly sales of a product over a number of standardised, four-weekly months, together with the corresponding values of a compound index of market buoyancy and product competitiveness. Take Sales as response, Index as independent variable, and consider fitting dynamic straight line regression models to the series to explain and predict Sales based on Index. Note that several observations are missing, being denoted by asterisks in the table.

Consider single discount models with $\mathbf{W}_t = \mathbf{C}_{t-1}(\delta^{-1} - 1)$ for all t for updating, and assume that the observational variance is constant. Assume that the initial information is summarised by $\mathbf{a}_1 = (100, 0)'$, $\mathbf{R}_1 = \text{diag}(25, 10)$, $n_0 = 10$ and $S_0 = 1.5$.

(a) Explore sensitivity to the value of δ by fitting the model to the data for values of δ representing the range $0.75 \leq \delta \leq 1$. Examine sensitivity by plotting the on-line and filtered estimates of the time trajectories of the intercept parameter α_t and the regression parameter β_t.

(b) Assess support from the data for the values of δ by calculating the aggregate LLR measure for each value considered.

(4) Analyse the Sales/Index series above with $\delta = 0.95$, recording the values of the posterior summary quantities $\mathbf{m}_{80}$, $\mathbf{C}_{80}$, S_{80} and n_{80}

Table 9.1. Sales and advertising Index series
(Missing data indicated by $*$)

	SALES Y_t				INDEX X_t			
	Week				Week			
Mth	1	2	3	4	1	2	3	4
1	102.29	101.18	100.49	100.31	0.00	0.00	0.17	0.26
2	99.39	101.69	99.96	105.25	0.21	0.22	0.23	0.29
3	103.71	99.21	99.82	100.95	0.28	0.05	0.00	0.00
4	101.91	100.46	102.81	101.59	0.13	0.18	0.24	0.05
5	104.93	101.95	101.01	99.03	0.21	0.05	0.00	0.14
6	100.80	101.90	98.35	99.21	0.15	0.03	0.00	0.00
7	99.87	101.30	98.42	98.66	0.15	0.23	0.05	0.14
8	103.16	102.52	103.08	101.37	0.16	0.27	0.24	0.18
9	101.75	100.76	97.21	100.47	0.25	0.21	0.25	0.05
10	99.10	***	***	***	0.00	***	***	***
11	***	***	***	***	***	***	***	***
12	105.72	106.12	103.41	97.75	0.17	0.27	0.06	0.00
13	101.18	101.61	105.78	98.35	0.00	0.22	0.21	0.04
14	104.08	102.02	100.33	101.34	0.17	0.04	0.00	0.13
15	100.37	99.49	103.39	101.32	0.03	0.15	0.24	0.20
16	100.26	102.70	100.40	101.85	0.28	0.20	0.03	0.24
17	103.50	95.41	99.19	103.05	0.06	0.00	0.14	0.03
18	99.79	103.49	102.93	102.63	0.13	0.15	0.27	0.06
19	102.72	101.52	98.55	101.07	0.00	0.00	0.16	0.04
20	99.65	104.59	104.24	100.74	0.16	0.20	0.23	0.05

at the final observation stage $t = 80$, corresponding to week 4 of month 20. Looking ahead, suppose that the step ahead values of Index over the coming four weeks are calculated as

$$X_{81} = 0.15, \quad X_{82} = 0.22, \quad X_{83} = 0.14 \quad \text{and} \quad X_{84} = 0.03.$$

Taking $\mathbf{W}_{80+k} = \mathbf{W}_{81} = \mathbf{C}_{80}(\delta^{-1} - 1)$ constant for forecasting ahead from time 80 (as usual in discount models), calculate the following forecast distributions:

(a) The marginal forecast distributions $p(Y_{80+k}|D_{80})$ for each of $k = 1, 2, 3$ and 4.

(b) The full, joint forecast distribution for the 4 quantities

$$p(Y_{81}, \ldots, Y_{84}|D_{80}).$$

(c) $p(Z|D_{80})$, where $Z = \sum_{r=81}^{84} Y_r$, the cumulative total Sales over the coming month.

(d) The conditional forecast distribution

$$p(Y_{81}, \ldots, Y_{84}|Z, D_{80})$$

for any given value of the total Z.

(5) A forecast system uses the first-order polynomial model $\{1, 1, 100, 5\}$ as a base model for forecasting sales of a group of products. Initially, $(\mu_0|D_0) \sim \mathrm{N}[400, 20]$. The effect of a price increase at $t = 1$ is modelled via the addition of a transfer response effect E_t, giving

$$Y_t = \mu_t + E_t + \nu_t, \qquad (t = 1, 2, \dots),$$

where

$$E_t = 0.9E_{t-1} + \partial E_t, \qquad (t = 2, 3, \dots),$$

and

$$(\partial E_t|D_0) \sim \mathrm{N}[0, 25],$$

independently of $(\mu_1|D_0)$ and $(\omega_t|D_0)$ for all t. It is also assumed that initially,

$$(E_1|D_0) \sim \mathrm{N}[-40, 200],$$

independently of $(\mu_1|D_0)$ and $(\omega_t|D_0)$.

(a) Calculate the forecast distributions $p(Y_t|D_0)$ for each of $t = 1, 2, 3$ and 4.

(b) Given $Y_1 = 377$, calculate $p(\mu_1, E_1|D_1)$ and revise the forecast distributions for the future, calculating $p(Y_t|D_1)$ for each of $t = 2, 3$ and 4.

(c) Perform similar calculations at $t = 2$ given $Y_2 = 404$. Calculate $p(\mu_2, E_2|D_2)$ and $p(Y_t|D_2)$ for $t = 3$ and 4.

(6) It is desired to produce a component DLM that models the form response of stock-market indicators to stock market news. The general form of response anticipated is that of an increase (or decrease) to a new level after damped, cyclic variation due to under/over reaction of market makers. Discuss the forms of the forecast functions of the two models

$$\left\{ \begin{pmatrix} 1 \\ 0 \end{pmatrix}, \begin{pmatrix} 1 & 1 \\ 0 & \phi \end{pmatrix}, \cdot, \cdot \right\}$$

and

$$\left\{ \begin{pmatrix} 1 \\ 0 \end{pmatrix}, \lambda \begin{pmatrix} \cos(\omega) & \sin(\omega) \\ -\sin(\omega) & \cos(\omega) \end{pmatrix}, \cdot, \cdot \right\},$$

where $0 < \phi,\ \lambda < 1$ and $0 < \omega < 2\pi$ with $\omega \neq \pi$. Verify that a model formed by the superposition of these or any similar models has qualitatively the desired form of the forecast function.

(7) Identify the form of the forecast function of the model of Definition 9.1 when

$$\mathbf{F} = \begin{pmatrix} 1 \\ 1 \\ 1 \end{pmatrix} \quad \text{and} \quad \mathbf{G} = \begin{pmatrix} 1 & 0 & 0 \\ 0 & \lambda & 0 \\ 0 & 0 & \phi \end{pmatrix},$$

for some λ and ϕ such that $0 < \lambda, \phi < 1$. Describe the form of the implied transfer response function, and explain under what circumstances this form would be appropriate as a description of the response of a sales series to a price increase X_t.

(8) Consider the form response model of Definition 9.1 in which $\mathbf{F} = 1$, $\mathbf{G} = \lambda$, $(0 < \lambda < 1)$, $\boldsymbol{\theta}_t = \theta_t$ is scalar, $\psi_t = \psi$, a scalar constant, and $\boldsymbol{\delta\theta}_t = \boldsymbol{\delta\psi}_t = 0$ for all t. Suppose also that $\nu_t \sim \mathrm{N}[0, V]$ with V known, $\theta_0 = 0$ and $(\psi|D_0) \sim \mathrm{N}[h_0, H_0]$ with known moments. Finally, suppose that $X_1 = 1$ and $X_t = 0$ for $t > 1$. With $(\psi|D_t) \sim \mathrm{N}[h_t, H_t]$, verify that

$$H_t^{-1} = H_0^{-1} + V^{-1}(1 - \lambda^{2t})/(1 - \lambda^2)$$

and

$$h_t = H_t(H_0^{-1}h_0 + V^{-1}\sum_{r=1}^{t} \lambda^{r-1}Y_r).$$

Find the limiting posterior distribution of $(\psi|D_t)$ as $t \to \infty$.

(9) Consider the first-order polynomial plus AR(1) noise model defined by equations

$$\begin{aligned} Y_t &= \mu_t + X_t, \\ \mu_t &= \mu_{t-1} + \omega_t, \\ X_t &= \phi X_{t-1} + \epsilon_t, \end{aligned}$$

where $\epsilon_t \sim \mathrm{N}[0, U]$ independently of $\omega_t \sim \mathrm{N}[0, W]$. Show how the model may be written in DLM form and derive the updating equations for the elements of $\mathbf{m}_t$ and $\mathbf{C}_t$. Identify the form of the forecast function $f_t(k)$.

(10) Consider the stationary AR(1) plus noise model defined by

$$\begin{aligned} Y_t &= X_t + \nu_t, \\ X_t &= \phi X_{t-1} + \epsilon_t, \end{aligned}$$

where $\epsilon_t \sim \mathrm{N}[0, U]$ independently of $\nu_t \sim \mathrm{N}[0, V]$.

(a) Show that the autocorrelations of X_t are given by $\rho_j = \phi^j$.

(b) Show also that Y_t is a stationary process and find the autocorrelations of Y_t; confirm that the autocorrelations of Y_t are damped towards zero relative to those of X_t, and that this damping increases with increasing U. Interpret this result.

(c) Show that Y_t has the autocorrelation structure of a stationary ARMA(1, 1) process. Generalise this result in the case that X_t is AR(p).

(11) Consider the first-order polynomial plus MA(1) noise model defined by equations

$$\begin{aligned} Y_t &= \mu_t + X_t, \\ \mu_t &= \mu_{t-1} + \omega_t, \\ X_t &= \psi\epsilon_{t-1} + \epsilon_t, \end{aligned}$$

where $\epsilon_t \sim \mathrm{N}[0, U]$ independently of $\omega_t \sim \mathrm{N}[0, W]$. Show that the model can be written in DLM form. Derive the updating equations for the elements of $\mathbf{m}_t$ and $\mathbf{C}_t$, and identify the form of the forecast function $f_t(k)$.

(12) Suppose that Y_t follows an AR(1) model $\{Y_{t-1}, 1, V, 0\}$ with autoregressive coefficient ϕ.

(a) Suppose that $(\phi|D_1) \sim \mathrm{N}[m_1, C_1]$ for some known m_1 and C_1. Deduce that for all $t > 1$, $(\phi|D_t) \sim \mathrm{N}[m_t, C_t]$ and identify the moments as functions of the past data $Y_t, Y_{t-1}, \ldots, Y_1$.

(b) Letting $C_1^{-1} \to 0$, verify that

$$C_t^{-1} \to V^{-1} \sum_{r=2}^{t} Y_{r-1}^2.$$

(c) On the basis of the model assumptions, the series Y_t is stationary with variance $\mathrm{V}[Y_t|\phi] = V(1-\phi^2)^{-1}$ for all t. Deduce that as $t \to \infty$, $tC_t \to (1-\phi^2)$. Comment on the implications for the precision with which ϕ can be estimated in the cases (i) $\phi = 0.9$ and (ii) $\phi = 0.1$.

(13) Consider the ARMA model of (9.11) with $n > p$ and

$$\phi(B) = \prod_{i=1}^{s} (1 - \lambda_i B)^{n_i},$$

where $\sum_{i=1}^{s} n_i = p$, and the λ_i are distinct and non-zero. Obtain $\mathbf{F}$ and $\mathbf{G}$ of the similar canonical DLM, showing that one of the blocks of the system matrix is $\mathbf{J}_{n-p}(0)$.

CHAPTER 10

ILLUSTRATIONS AND EXTENSIONS OF STANDARD DLMS

10.1 INTRODUCTION

In the preceding chapters the focus has been on the theoretical structure and analysis of DLMs, with little reference to practical aspects of modelling. Here we switch the focus to the latter to consolidate what has been developed in theory, illustrating many basic concepts via analyses of typical data sets. We consider both retrospective analysis of a time series as well as forecasting with an existing model, and describe a variety of modelling activities using the class of models built up from the trend, seasonal and regression components of Chapters 7, 8 and 9. Together these three components provide for the majority of forms of behaviour encountered in commercial and economic areas, and thus this class, of what may be referred to as *standard* models, forms a central core of structures for the time series analyst. In approaching the problem of modelling a new series, the basic trend and seasonal components are a useful first attack. If retrospective analysis is the primary goal, then these simple and purely descriptive models may be adequate in themselves, providing estimates of the trend (or deseasonalised series), seasonal pattern (detrended series) and irregular or random component over time. In addition to retrospective time series decomposition, these models can prove adequate for forecasting in the short term. The inclusion of regression terms is the next step, representing an attempt to move away from simple descriptions via explanatory relationships with other variables. Linking in such variables is also the route to firmer, more credible and reliable short/medium-term forecasting, the key idea being that future changes in a series modelled using a simpler trend/seasonal description may be adequately predicted (at least qualitatively) by one or a small number of regression variables. Identifying the important variables is, of course, and as usual in statistics generally, the major problem.

The next two sections consider several models corresponding to different levels of complexity, inputs based on different states of prior information, and different objectives in modelling. Initially we concentrate on a reference analysis, as developed in Section 4.10, in order to present the main features of model analyses free from the effects of chosen initial priors. In addition, we assume closed models throughout. That is, once the model is specified, fitting proceeds without further interventions. The final sections of this chapter concern practical issues of error analysis and model assessment, data irregularities, transformations and other modifications and extensions of the basic normal DLM.

10.2 *BASIC ANALYSIS: A TREND/SEASONAL DLM*

10.2.1 *Data and basic model form*

Consider the data set given in Table 10.1 and plotted over time in Figure 10.1. The single time series is of quarterly observations providing a measure, on a standard scale, of the total quarterly sales of a company in agricultural markets over a period of 12 years. Quarters are labelled Qtr., running from 1 to 4 within each year, and years are labelled Year, running from 73 (1973) to 84 (1984). The series, referred to simply as Sales, is evidently highly seasonal, driven by the annual cycle of demand for agricultural supplies following the natural annual pattern of farming activities. This can be clearly seen from the graph or by examining the figures in the table, making comparisons between rows.

Table 10.1. Agricultural sales data

Qtr.	Year											
	73	74	75	76	77	78	79	80	81	82	83	84
1	8.48	8.94	9.20	9.13	9.23	9.49	9.37	9.56	9.71	9.72	9.82	10.11
2	8.70	8.86	9.11	9.23	9.21	9.54	9.66	9.98	9.60	9.88	9.90	9.90
3	8.09	8.45	8.69	8.65	8.68	9.06	9.03	9.19	9.18	9.11	8.87	9.47
4	8.58	9.00	8.87	8.84	9.20	9.35	9.44	9.50	9.53	9.49	9.38	9.47

While the form of seasonality is stable, the actual quantified pattern does change somewhat from year to year. Underlying the seasonal series is a changing trend that grows at differing rates during different periods of the 12 years recorded.

The most basic, possibly appropriate descriptive model is that comprising a polynomial trend plus seasonal effects as in Section 8.5. Given that any trend term will be quantified only locally, a second-order trend term is chosen. Quarterly changes in level and growth components of such a trend are assumed to model the apparent changes in direction of the trend in the series. This model is very widely used in practice in this sort of context as a first step in retrospective time series analysis. Any movement in trend is ascribed to changes in level and growth parameters of the model rather than being explained by possible regression variables.

In usual DLM notation, the first component of the model is a linear trend

$$\{\mathbf{E}_2,\ \mathbf{J}_2(1),\ \cdot\ ,\ \cdot\}.$$

Added to this is the seasonal component. Since a full seasonal pattern is specified, this may be in terms of either 4 seasonal factors, constrained to zero sum, or the corresponding 3 Fourier coefficients. We work, as usual, in terms of the latter. The Fourier representation has just 2 harmonics: the first providing the fundamental annual cycle, the second adding in the Nyquist frequency. In DLM notation, the Fourier model is, following

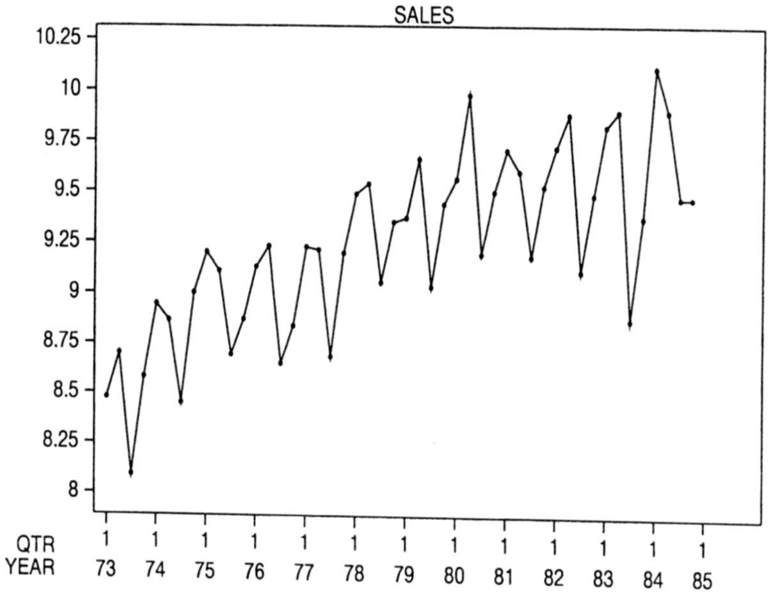

Figure 10.1 Agricultural sales series

Section 8.6,

$$\left\{ \begin{pmatrix} 1 \\ 0 \\ 1 \end{pmatrix}, \begin{pmatrix} c & s & 0 \\ -s & c & 0 \\ 0 & 0 & -1 \end{pmatrix}, \ldots, \cdot \right\},$$

where $\omega = \pi/2$, $c = \cos(\omega) = 0$ and $s = \sin(\omega) = 1$. The full 5-dimensional TSDLM thus has

$$\mathbf{F} = (1, 0;\ 1, 0; 1)'$$

and

$$\mathbf{G} = \begin{pmatrix} 1 & 1 & 0 & 0 & 0 \\ 0 & 1 & 0 & 0 & 0 \\ 0 & 0 & 0 & 1 & 0 \\ 0 & 0 & -1 & 0 & 0 \\ 0 & 0 & 0 & 0 & -1 \end{pmatrix}.$$

The corresponding parameter vector has 5 elements, $\boldsymbol{\theta}_t = (\theta_{t1}, \ldots, \theta_{t5})'$, with the following meaning:

- θ_{t1} is the underlying, deseasonalised level of the series at time t;
- θ_{t2} is the deseasonalised growth between quarters $t-1$ and t;
- θ_{t3} and θ_{t4} are the Fourier coefficients of the first harmonic, the full annual sine/cosine wave, with the contribution of this harmonic to

the seasonal factor at time t being, in the notation of Section 8.6, simply $S_{1t} = \theta_{t3}$;
- similarly, $S_{2t} = \theta_{t5}$ is the Nyquist contribution to the seasonal factor.

The seasonal factors are obtained from the Fourier coefficients as in Section 8.6. As before, the factors (or effects since they are constrained to have zero sum) at time t are $\boldsymbol{\phi}_t = (\phi_{t0}, \cdots, \phi_{t3})'$. Thus ϕ_{t0} is the seasonal factor for quarter t. Then, following Example 8.4,

$$\boldsymbol{\phi}_t = \mathbf{L}(\theta_{t3}, \theta_{t4}, \theta_{t5})',$$

where

$$\mathbf{L} = \begin{pmatrix} 1 & 0 & 1 \\ 0 & 1 & -1 \\ -1 & 0 & 1 \\ 0 & -1 & -1 \end{pmatrix}.$$

Finally, we assume that the observational variance V of the model is constant in time and unknown, requiring, as is typical, the modified analysis for on-line variance learning. The model, with 6 parameters (including V), is now essentially specified. In line with the development of reference analyses in Section 4.10, we assume that there is no change in model parameters during the first 6 quarters, there being no information available to inform on any such changes. The reference updating equations apply up to $t = 6$ when the posterior distribution for $\boldsymbol{\theta}_6$ and V given D_6 is fully specified in the usual conjugate normal/inverse gamma form. To proceed from this time point on, the sequence of evolution variance matrices $\mathbf{W}_t$ is required to allow for time-varying parameters.

10.2.2 Discount factors for components

The evolution variance matrices are specified, as usual, following the concept of block discounting consistent with the use of component models, as described in Section 6.3. Two discount factors provide the flexibility to allow for a full range of rates of change over time in the trend and seasonal components of the model; these are denoted by δ_T (for the trend component) and δ_S (for the seasonal component). The former determines the rate at which the trend parameters θ_{t1} and θ_{t2} are expected to vary between quarters, with $100(\delta_T^{-1} - 1)\%$ of information (as measured by reciprocal variance or precision) about these parameters decaying each quarter, between observations. The factor δ_S plays the same role for the seasonal (Fourier) parameters, or equivalently, the seasonal effects.

Thus, at times $t > 6$, the model evolution equation is as follows. With

$$(\boldsymbol{\theta}_{t-1}|D_{t-1}) \sim \mathrm{T}_{n_{t-1}}[\mathbf{m}_{t-1}, \mathbf{C}_{t-1}],$$

then

$$(\boldsymbol{\theta}_t|D_{t-1}) \sim \mathrm{T}_{n_{t-1}}[\mathbf{a}_t, \mathbf{R}_t],$$

where

$$\mathbf{a}_t = \mathbf{G}\mathbf{m}_{t-1} \qquad \text{and} \qquad \mathbf{R}_t = \mathbf{P}_t + \mathbf{W}_t,$$

with

$$\mathbf{P}_t = \mathbf{G}\mathbf{C}_{t-1}\mathbf{G}',$$

and the evolution variance matrix is defined as

$$\mathbf{W}_t = \text{block diag}\{\ \mathbf{P}_{tT}(\delta_T^{-1} - 1),\ \ \mathbf{P}_{tS}(\delta_S^{-1} - 1)\ \},$$

where $\mathbf{P}_{tT}$ is the upper-left 2×2 block of $\mathbf{P}_t$ and $\mathbf{P}_{tS}$ is the lower-right 3×3 block.

For the trend discount factor, values in the range 0.8 to 1.0 are typically appropriate, with smaller values anticipating greater change. δ_S is typically larger than δ_T to reflect

(1) relatively less information being obtained about the seasonal parameters than the trend in each quarter, and
(2) the expectation that seasonal patterns will tend to have more sustained, stable forms relative to the potentially more rapidly changing deseasonalised trend.

This second point is consistent with the view that in using this simple, descriptive TSDLM, we are not attempting to anticipate sustained movement and changes in trend and seasonality, but just to detect and estimate them. Omitted related explanatory variables are more likely to have a marked impact on changes in the underlying trend in the series than on the seasonal pattern, the latter viewed as more durable over time.

Forecasting accuracy comes through the identification of stable structure and relationships, and in the best of all worlds, discount factors of unity would be used giving constant parameters over time. We may start at this ideal, comparing the predictive performance of such a model with that based on more realistic values, possibly comparing several such models. This sort of activity is described further in Section 10.2.5 below. Here we consider the particular case in which $\delta_T = 0.85$ and $\delta_S = 0.97$.

10.2.3 Model fitting and on-line inferences

From time $t = 6$ onwards then, the usual updating equations apply, one-step ahead forecast and posterior distributions for model parameters being sequentially calculated for times $t = 7, \dots, 48$. Recall that in the reference analysis, the (posterior) degrees of freedom at time t is t minus the dimension of the model, so here it is $t - 5$. Thus, for $t > 6$, the one-step ahead forecast distribution at $t - 1$ has $n_{t-1} = t - 6$ degrees of freedom,

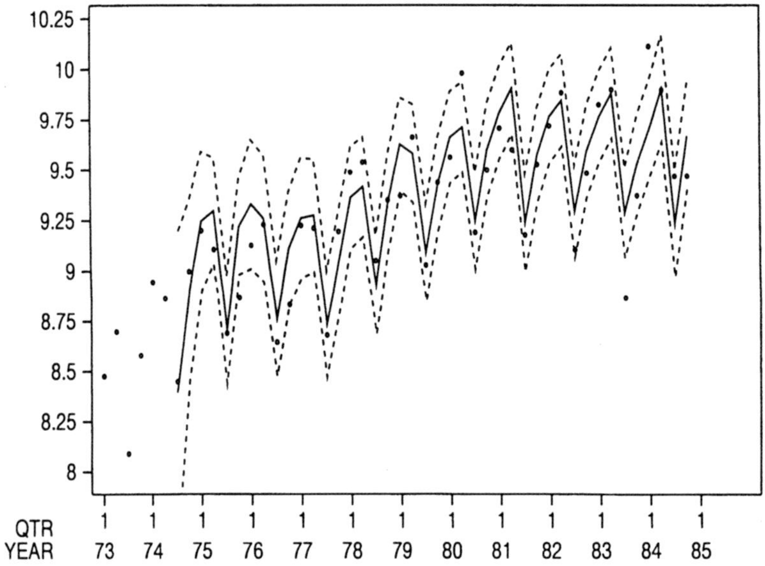

Figure 10.2 Agricultural sales with one-step ahead forecasts

$(Y_t|D_{t-1}) \sim \mathrm{T}_{t-6}[f_t, Q_t]$. Figure 10.2 displays information about these sequentially calculated forecasts. Here, superimposed on a scatter plot of the data over time, we have the point forecasts f_t joined up as the solid line, and 90% probability limits about the point forecasts as dotted lines. The latter provide the 90% HPD region for Y_t given D_{t-1} from the relevant T predictive distribution, an interval symmetrically located (hence with equal 5% tails) about f_t. Due to the reference prior used, no forecasts are available initially. Clearly, most of the points lie within the 90% limits; five or six exceptions are clear. Of course, roughly 5 out of 48 are expected to lie outside the 90% limits. Further insight into the one-step ahead forecasting performance is given by examining Figure 10.3. Here the standardised forecast errors $e_t/Q_t^{1/2}$ are plotted over time with 90% probability limits from the one-step ahead forecast distributions. The several points noted above show up clearly here as leading to large (in absolute value) residuals. Having reached time $t = 48$ and the end of the data, we can proceed to examine model inferences by considering the estimated model components over time. Central to this is the following general result: Let $\mathbf{x}_t$ be any 5-vector, and set $x_t = \mathbf{x}_t'\boldsymbol{\theta}_t$. Then from the posterior for $\boldsymbol{\theta}_t$ at time t, it follows that

$$(x_t \,|D_t) \sim \mathrm{T}_{t-5}[\mathbf{x}_t'\mathbf{m}_t, \mathbf{x}_t'\mathbf{C}_t\mathbf{x}_t].$$

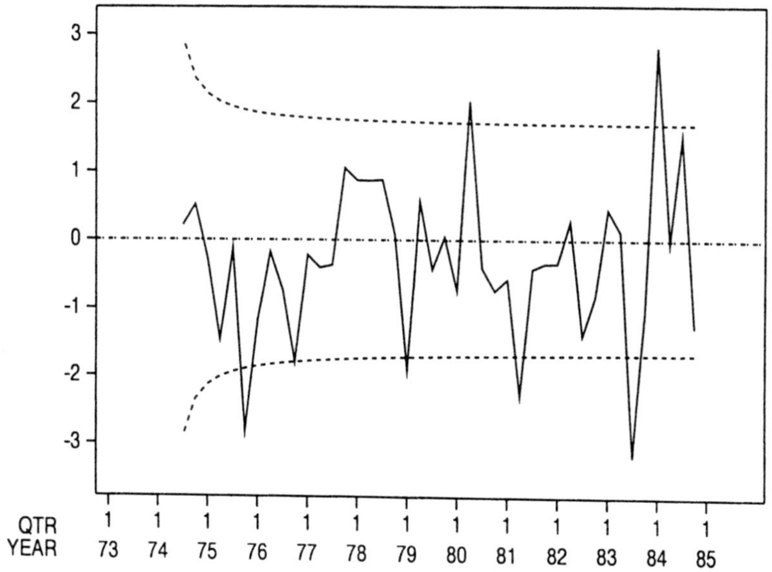

Figure 10.3 Standardised one-step ahead forecast errors

The mode of this T distribution, $\mathbf{x}_t'\mathbf{m}_t$, is referred to as the *on-line* estimated value of the random quantity x_t. Inferences about model components are made by appropriately choosing $\mathbf{x}_t$ as follows.

(1) DESEASONALISATION
If $\mathbf{x}_t' = (1, 0, 0, 0, 0)$, then x_t is the underlying, deseasonalised level of the series $x_t = \theta_{t1}$. From the sequence of posterior T distributions, 90% HPD intervals are calculated and plotted in Figure 10.4 in a similar fashion to the forecast intervals in Figure 10.2. The local linearity of the trend is evident, as are changes, both small and more abrupt, in the growth pattern over time. In particular, there is a marked change in trend from positive growth during the first two years or so, to little, even negative growth during the third and fourth years. This abrupt change is consistent with the general economic conditions at the time, and could, perhaps, have been accounted for if not forecast. With respect to the simple, closed model illustrated here, however, adaptation to the change is apparent over the next couple of observations, although responding satisfactorily in general to changes that are greater than anticipated requires some form of intervention, as discussed in later chapters. Taking $\mathbf{x}_t' = (0, 1, 0, 0, 0,)$ similarly provides on-line inference about the annual changes in trend.

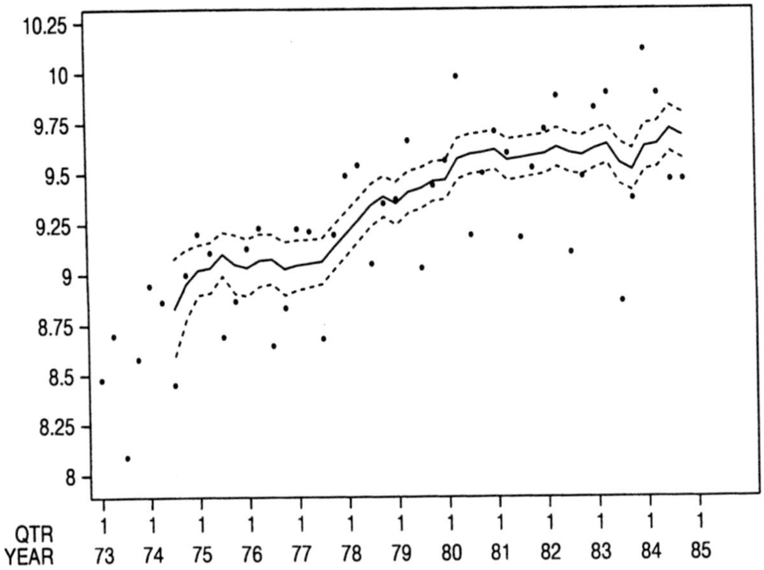

Figure 10.4 On-line estimated trend in sales

(2) SEASONAL FACTORS/EFFECTS

The seasonal effect at time t is given by $\phi_{t0} = x_t = \mathbf{x}_t'\boldsymbol{\theta}_t$, with $\mathbf{x}_t' = (0,0,1,0,1)$, to select the two harmonic contributions to the seasonal pattern from the state vector. Thus, as a function over time, this x_t provides the detrended seasonal pattern of the Sales series. Figure 10.5 displays information from the posterior distributions for x_t at each time, analogous to that for the trend although displayed in a different form. At each t on the graph the vertical bar represents the 90% HPD interval for the seasonal effect, the mid-point of the interval being the on-line estimated value. This is a plot similar to those displayed in Figure 8.4. There is evidently some minor variation in the seasonal pattern from year to year, although the basic form of a peak in spring and summer and deep trough in autumn is sustained, consistent with expected variation in demand for agricultural supplies.

Further investigation of the seasonal pattern is possible using the ideas of Section 8.6 to explore harmonics. Taking $\mathbf{x}_t = (0,0,1,0,0)'$ implies x_t is the contribution of the first harmonic, and taking $\mathbf{x}_t = (0,0,0,0,1)'$ leads to the Nyquist harmonic.

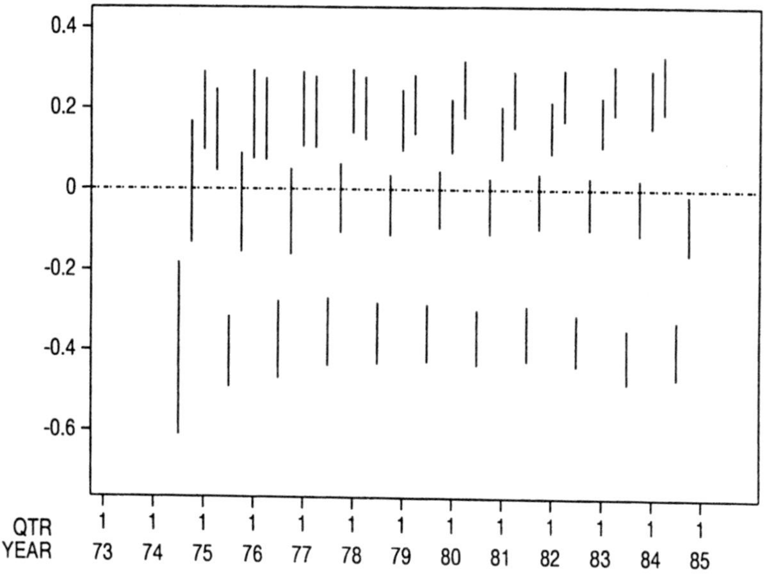

Figure 10.5 On-line estimated seasonal pattern in sales

10.2.4 Step ahead forecasting

Consider finally step ahead forecasting of the Sales series from time 4/84, the end of the data series. Without intervention, we simply forecast by projecting based on the existing model and fit to the past 12 years of data, as summarised by the posterior distributions at $t = 48$, using the theory summarised in Section 4.6. The posteriors at time 48 have defining quantities

$$\mathbf{m}_{48} = (9.685,\ 0.015,\ -0.017,\ 0.304,\ 0.084)'$$

and

$$\mathbf{C}_{48} = 0.0001 \begin{pmatrix} 45.97 & 3.77 & -2.30 & 1.86 & -1.03 \\ & 0.66 & -0.20 & 0.15 & -0.09 \\ & & 12.87 & -0.10 & -0.13 \\ & & & 13.23 & -0.24 \\ & & & & 6.38 \end{pmatrix}$$

(only the upper part of the symmetric matrix $\mathbf{C}_{48}$ being displayed);

$$n_{48} = 43 \qquad \text{and} \qquad S_{48} = 0.016.$$

Forecasting ahead requires that we specify, at time $t = 48$, the evolution variance matrices for the future, $\mathbf{W}_{t+r}$ for $r = 1, \ldots, k$, where k is the maximum step ahead of interest. We use the discount model strategy described in Section 6.3. using the existing one-step ahead matrix $\mathbf{W}_{t+1}$

in forecasting ahead. Thus, at any time t, forecasting ahead is based on $\mathbf{W}_{t+r} = \mathbf{W}_{t+1}$ for $r = 2, \ldots, k$. Although in updating to future times we use evolution variance matrices defined via discounts based on the information available at those times, in forecasting ahead we use slightly different values given by the currently implied $\mathbf{W}_{t+1}$. As discussed in Section 6.3, this approach is simple to implement and the sequence $\mathbf{W}_{t+r} = \mathbf{W}_{t+1}$ into the future appropriate for forecasting ahead from our current position at time t. Proceeding to time $t+1$, and observing Y_{t+1}, our view of the future evolution variances for times $t+2$ onwards is changed slightly, being based on the new, current (at $t+1$) value. This may be interpreted most satisfactorily as an intervention to change the evolution variance sequence.

Thus, in the example, the evolution variance matrix for forecasting from time 48 is given by $\mathbf{W}_{t+1} = \text{block diag}[\mathbf{W}_T,\ \mathbf{W}_S\ ]$ where, by applying the discount concept,

$$\mathbf{W}_T = 0.0001 \begin{pmatrix} 9.56 & 0.78 \\ & 0.12 \end{pmatrix}$$

and

$$\mathbf{W}_S = 0.0001 \begin{pmatrix} 0.41 & 0.003 & 0.008 \\ & 0.40 & -0.004 \\ & & 0.20 \end{pmatrix},$$

(again, of course, each of these is a symmetric matrix), the latter being roughly equal to 0.0001 diag(0.4, 0.4, 0.2).

It is now routine to apply the results of Section 4.6 to project forecast distributions into the future, and this is done up to $k = 24$, giving 6 full years of forecast up to 4/90. The forecast distributions are all Student T with 43 degrees of freedom, therefore roughly normal, with increasing scale parameters as we project further ahead. Figure 10.6 displays the point forecasts $f_{48}(k) = \text{E}[Y_{48+k}|D_{48}]$, $(k = 1, \ldots, 24)$, joined up as a solid line, and 90% prediction intervals symmetrically located about the point forecasts.

10.2.5 Numerical summaries of predictive performance

Finally, for easy and informal comparison of alternative models, we mention three basic measures of model predictive performance that are easily calculated from the one-step forecast errors, or *residuals*, e_t. Each of these is useful only in a relative sense, providing comparison between different models.

Total absolute deviation is defined as $\sum |e_t|$. With proper initial priors, the first few errors will reflect the appropriateness of the initial priors as well as that of the model, and their contribution to the sum may be important. In reference analyses, however, there are no defined errors until $t = 6$ (more generally, $n+1$ in an n-dimensional model), and so the sum starts at $t = 6$.

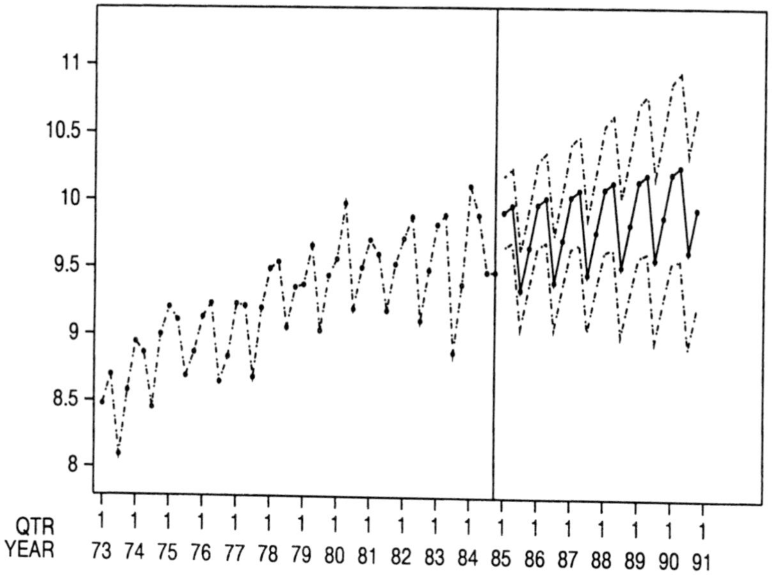

Figure 10.6 Step-ahead forecasts of agricultural sales

The *mean absolute deviation*, or MAD, measure is simply this total divided by the number of errors in the sum. In this example, the MAD based on the final 43 observations is calculated as 0.128.

Total square error is $\sum e_t^2$, the *mean square error*, or MSE, being this divided by the number of terms in the sum. In this example, the MSE based on the final 43 observations is calculated as 0.028. Both MAD and MSE are simple measures of actual forecasting accuracy.

Model likelihood, a rather more formal measure of goodness of predictive performance, was introduced in examples in Chapters 2 and 3. West (1986a) gives very general discussion and development of the ideas in this section, and the notation used there is adopted here. For each observation Y_t, the observed value of the one-step predictive density $p(Y_t|D_{t-1})$ is larger or smaller according to whether the observation accords or disagrees with the forecast distribution. This accounts for raw accuracy of the point forecast f_t, and also for forecast precision, a function of the spread of the distribution. The aggregate product of these densities for observations $Y_t, Y_{t-1}, \ldots, Y_s$, $(s = 1, \ldots, t)$, provides the overall measure

$$p(Y_t|D_{t-1})p(Y_{t-1}|D_{t-2})\ldots p(Y_s|D_{s-1}) = p(Y_t, Y_{t-1}, \ldots, Y_s|D_{s-1}),$$

just the joint predictive density of those observations from time $s-1$. If proper priors are used, then this can extend back to time $s = 1$ if required. In the reference analysis, the densities are, of course, only defined for $s > n+1$ in a n-dimensional model.

This measure, based on however many observations are used, provides a likelihood (in the usual statistical sense of the term) for the parameters, such as discount factors, determining the model. Just as with MAD and MSE, these measures may be used as informal guides to model choice when comparing two or more models differing only in values of defining parameters. Suppose we consider two models with the same mathematical structure, model 0 and model 1, differing only, for example, in the values of discount factors. To make clear the dependence of the model on these values, suffix the predictive densities by 0 and 1 respectively. Then the *relative likelihood* of model 0 versus model 1 based on the observation Y_t at time t is just the ratio

$$H_t = p_0(Y_t|D_{t-1})/p_1(Y_t|D_{t-1}).$$

More generally, the observations $Y_t, Y_{t-1}, \dots, Y_s$, $(s = 1, \dots, t)$, provide the the overall likelihood ratio

$$\begin{aligned} H_t(t-s+1) &= \prod_{r=s}^{t} H_r \\ &= p_0(Y_t, Y_{t-1}, \dots, Y_s|D_{s-1})/p_1(Y_t, Y_{t-1}, \dots, Y_s|D_{s-1}). \end{aligned}$$

These likelihood ratios, alternatively called *Bayes' Factors*, or *weights of evidence* (Jeffreys 1961; Good 1985 and references therein; West 1986a), provide a basic measure of predictive performance of model 0 relative to model 1. Note that the evidence for or against the models accumulates multiplicatively as observations are processed via

$$H_t(t-s+1) = H_t \; H_{t-1}(t-s),$$

for $t > s$, with boundary values $H_t(1) = H_t$ when $s = t$. Many authors, including those referenced above, consider the log-likelihood ratio, or log Bayes' factor/weight of evidence, as a natural and interpretable quantity. On the log scale, evidence is additive, with

$$\log[H_t(t-s+1)] = \log(H_t) + \log[H_{t-1}(t-s)].$$

Following Jeffreys (1961), a log Bayes' factor of $1(-1)$ indicates evidence in favour of model $0(1)$, a value of 2 or more (-2 or less) indicating the evidence to be strong. Clearly the value 0 indicates no evidence either way.

EXAMPLE 10.1. Let model 0 be the model used throughout this section, having discount factors $\delta_T = 0.85$ and $\delta_S = 0.97$. For comparison, let model 1 be a *static* model, having constant parameters, defined by unit discount factors $\delta_t = \delta_S = 1$. Figure 10.7 provides a plot versus time t of the individual log weights $\log(H_t)$ and a similar plot of the cumulative log Bayes' factors $\log[H_t(t-6)]$. The latter drops below 0 for only a few quarters in year 75, being positive most of the time, essentially increasing up to a final value between 7 and 8, indicating extremely strong evidence in favour of the dynamic model relative to the static. For additional comparison, the

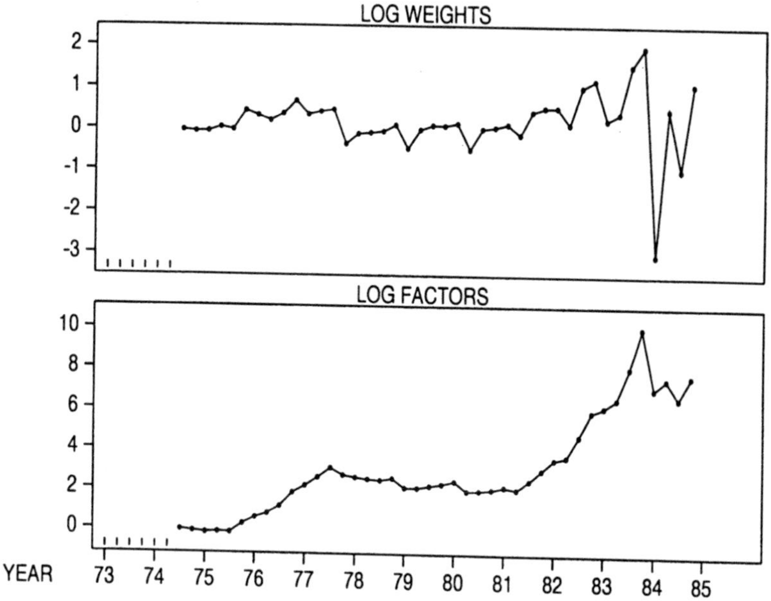

Figure 10.7 Log weights and Bayes' factors plotted over time

static model has MAD=0.158 and MSE=0.039, both well in excess of those in the dynamic model. Similar comparisons with models having different values of discounts indicate support for that used here as acceptable, having MAD and MSE values near the minimum and largest likelihood.

Visual inspection of residual plots and retrospective residual analyses are related and important assessment activities. However, whilst providing useful guides to performance, we **do not** suggest that the choice of a model for forecasting be automated by choosing one (in the example, the values of discount factors) that minimises MAD/MSE and/or maximises the likelihood measure. All such activities pay essentially no regard to the longer-term predictive ability of models, focusing as they do on how well a model has performed in one-step ahead forecasting in the past. In considering forecasting ahead, a forecaster must be prepared to accept that often, a model found to be adequate in the past may not perform as well in the future, and that continual model monitoring and assessment of predictive performance is a must. We consider such issues in Chapter 11.

10.3 A TREND/ SEASONAL/ REGRESSION DLM

10.3.1 Data and basic model form

The second illustration concerns commercial sales data from the food industry. The response series, Sales, is the monetary value of monthly total sales, on a standardised, deflated scale, of a widely consumed and established food product, covering a variety of brands and manufacturers, in UK markets. The data runs over the full 6 years 1976 to 1981 inclusive; thus there are 72 observations, given in Table 10.2. The second series, Index, is used to partially explain the movement in trend of sales. This is a compound measure constructed by the company concerned, based on market prices, production and distribution costs, and related variables, and is to be used as an independent regressor variable in the model for Sales. The Index variable is standardised by subtracting the arithmetic mean and dividing by the standard deviation, both mean and deviation being calculated from the 72 observations presented. The two data series are graphed over time in Figure 10.8. The response series clearly has a varying trend over the 6 years, is apparently inversely related to Index, and is evidently seasonal with a rather variable seasonal pattern from year to year.

The Sales figures refer, as mentioned above, to a mature, established consumer product. In consequence for short-term forecasting, any sustained growth (positive or negative!) in Sales over time should be attributable to regressor variables relating to pricing policies, costs, advertising and competitor activity, and general market/economic conditions; hence the introduction of the Index series. The model for Sales does not, therefore, include a descriptive growth term, the initial component being a simple steady model, the first-order polynomial component of Chapter 2, $\{1, 1, \cdot, \cdot\}$. The second component is the regression DLM for the effect of the Index series. Note that we include only contemporaneous values of Index, lagged effects being ignored for this illustration, it being assumed that the Index has been constructed to account for these. Thus the explanatory component is the regression with zero origin of Chapter 3, $\{X_t, 1, \cdot, \cdot\}$, where X_t is the value of the Index series at t. The single parameter of the first-order polynomial component now provides the (time-varying) intercept for the dynamic regression of Sales on Index. In models such as these, the possible effects of related, omitted variables are partially accounted for by the purely random movement allowed in the intercept term and regression parameter.

The third model component is a seasonal term describing monthly variation within a year free from the effects of regression variables. The most general model is, of course, a full seasonal effects model with 12 monthly seasonal effects (with zero sum). As usual, we adopt the Fourier representation of seasonal effects. To proceed, we consider hypothetical prior information and its implications for model specification, forecasting ahead from time 0, and subsequent data analysis.

Table 10.2. Sales and Index series

Year	Month 1	2	3	4	5	6
	SALES					
1976	9.53	9.25	9.36	9.80	8.82	8.32
1977	9.93	9.75	10.57	10.84	10.77	9.61
1978	10.62	9.84	9.42	10.01	10.46	10.66
1979	10.33	10.62	10.27	10.96	10.93	10.66
1980	9.67	10.40	11.07	11.08	10.27	10.64
1981	10.58	10.52	11.76	11.58	11.31	10.16
	INDEX					
1976	0.54	0.54	1.12	1.32	1.12	1.12
1977	−0.03	0.15	−0.03	−0.22	−0.03	0.15
1978	−0.22	0.93	1.32	0.93	0.15	0.15
1979	0.54	0.54	0.34	−0.03	−0.03	0.15
1980	−1.14	−0.96	−0.96	−0.59	−0.41	−0.03
1981	−1.84	−1.84	−2.70	−2.19	−2.19	−1.84

Table 10.2 (continued)

Year	Month 7	8	9	10	11	12
	SALES					
1976	7.12	7.10	6.59	6.31	6.56	7.73
1977	8.95	8.89	7.89	7.72	8.37	9.11
1978	11.03	10.54	10.02	9.85	9.24	10.76
1979	10.58	9.66	9.67	10.20	10.53	10.54
1980	11.03	9.63	9.08	8.87	8.65	9.27
1981	10.14	9.81	10.27	8.80	8.62	9.46
	INDEX					
1976	1.12	1.32	1.93	2.13	1.52	1.32
1977	0.15	0.15	0.73	0.54	0.34	−0.03
1978	0.34	0.34	−0.41	1.12	0.54	−0.22
1979	−0.03	−0.22	−0.03	−0.59	−0.59	−0.78
1980	0.15	0.34	0.54	0.54	0.73	0.54
1981	−0.96	−1.67	−1.67	−0.96	−1.14	−0.96

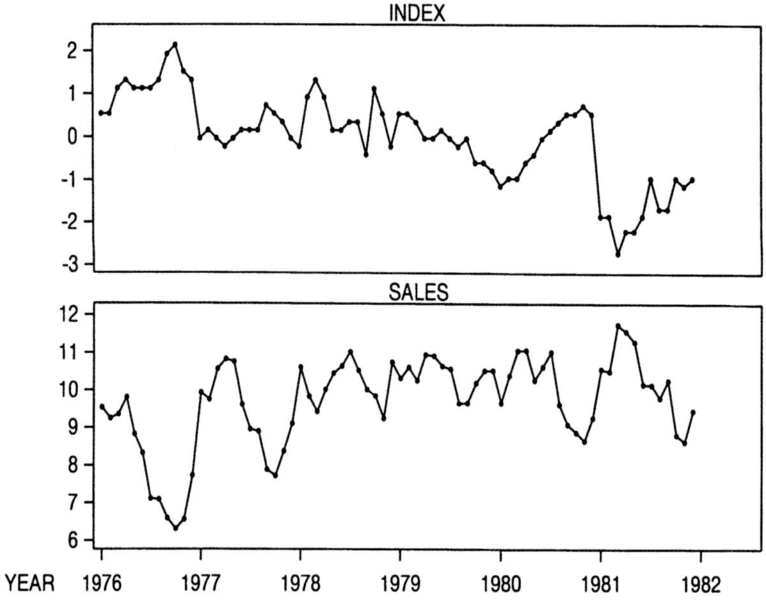

Figure 10.8 Monthly Sales and Index series

10.3.2. Hypothetical initial prior information

Suppose that at December 1975, where $t = 0$, we are in the position of having to provide step ahead forecasts of Sales based on existing information D_0. Such information includes past data (prior to 1976) on the Sales and Index series, past experience with modelling their relationship and forecasting, and past experience with similar products. In practice, if such a model had been used previously, then there would exist a full posterior distribution for the model parameters at $t = 0$, providing the prior for the new data beginning in January 1976, at $t = 1$. Suppose, however, that past modelling activities have only been partially communicated, and that we adopt initial priors based on the following initial information at $t = 0$. Note that rather than communicating the posterior for $(\boldsymbol{\theta}_0 \mid D_0)$ as usual, we are specifying information directly for the first time point and will determine the prior for $(\boldsymbol{\theta}_1 \mid D_0)$ directly.

- The parameter $\theta_{1,1}$, the first element of the DLM state vector at $t = 1$, represents the non-seasonal intercept term, the non-seasonal level if the Index value is $X_1 = 0$. Given D_0, the value $m_{1,1} = 9.5$ is taken as the prior mean. Information on the intercept is viewed as rather vague, particularly since the parameter can be expected to change to account for effects of omitted variables, and values as low as 9 or as high as 10 are plausible.

- At $t = 1$, the prior view of the regression effect of Index is that the regression coefficient is negative, and that to achieve an expected change in Sales of 1 unit, the Index would have to change by about 1.5 units. This suggests a value near $-2/3$ for the prior estimate of the regression parameter at $t = 1$, namely $\theta_{1,2}$, the second element of the DLM parameter vector. The value adopted is $m_{1,2} = -0.7$. Uncertainty is such that values as high as -0.5 or as low as -0.9 are plausible.
- Prior information on the seasonal effects $\boldsymbol{\phi}_1 = (\phi_{1,0}, \dots, \phi_{1,11})'$ is provided initially in terms of the prior estimates

  $$0.8,\ 1.3,\ 1,\ 1.1,\ 1.2,\ 0,\ -1.2,\ -1,\ -1,\ -1.5,\ -0.5,\ 0,$$

  although the uncertainties are fairly large, with values possibly ranging up to plus or minus 0.4 or so from each of these estimates. Note that these estimates do not sum to zero, and will therefore have to be constrained to do so later.
- The observational variance is viewed as constant over time, initially estimated at $S_0 = 0.0225 = 0.15^2$, having fairly low associated degrees of freedom $n_0 = 6$. The 90% equal-tails interval for the standard deviation $V^{1/2}$ is easily calculated as $(0.1, 0.29)$.

This information suffices to determine marginal priors for the intercept parameter, the regression parameter and the seasonal effects separately. There is no information provided to determine prior correlation between the components, so they are taken as uncorrelated. In the usual notation then, marginal priors are as follows:

INTERCEPT:
$(\theta_{1,1} \mid D_0) \sim \mathrm{T}_6[9.5,\ 0.09]$, with 90% HPD interval $(8.92,\ 10.08)$;

REGRESSION:
$(\theta_{1,2} \mid D_0) \sim \mathrm{T}_6[-0.7,\ 0.01]$, with 90% interval $(-0.89,\ -0.51)$;

SEASONAL:
$(\boldsymbol{\phi}_1 \mid D_0) \sim \mathrm{T}_6$, with mean vector given by the estimates above, and scale matrix $0.04\,\mathbf{I}$, although since neither conform to the zero-sum constraint, they must be adjusted according to the theory of Section 8.4. Applying Theorem 8.2 leads easily to the constrained mean vector

$$\begin{aligned}\mathbf{m}_{\phi_1} = (&0.783,\ 1.283,\ 0.983,\ 1.083,\ 1.183,\ -0.017,\\ &-1.217,\ -1.017,\ -1.017,\ -1.517,\ -0.517,\ -0.017)'\end{aligned}$$

(to 3 decimal places). The constrained scale matrix, denoted by $\mathbf{C}_{\phi_1}$, has diagonal elements equal to 0.0367 and off-diagonal elements -0.0034 so that the pairwise correlations between the effects are all -0.092. On this basis, 90% prior HPD intervals for the effects are given by the means ± 0.356.

10.3.3. Seasonal component prior

The prior for the seasonal effects suggests a seasonal pattern that grows from 0 in December to a plateau in February to May, decays rapidly thereafter to a trough in October, and returns to 0 by the end of the year. If used in the model, this is the pattern that will apply for step ahead forecasts. With such large uncertainty about the effects, as indicated by the 90% interval, it is sensible to proceed with a full seasonal model, having all Fourier harmonics, in order that forthcoming data provide further information about an unrestricted pattern. The main problem with restricting the seasonal pattern lies in the possibility that the prior may be doubtful and that any restrictions imposed on the basis of the prior may be found to be unsatisfactory once further data are observed. For illustration here, however, the seasonal pattern is restricted by a reduced Fourier form representation that is consistent with this initial prior. This is obtained by first considering a full harmonic representation and examining the implications of the prior for the individual harmonic components.

Using the theory of Section 8.6, we can deduce by linear transformation the initial prior for the eleven Fourier coefficients in the full harmonic model. If $\mathbf{H}$ is the transformation matrix of Section 8.6, then the coefficients are given by $\mathbf{H}\boldsymbol{\phi}_1$, having a T_6 prior distribution with mean vector $\mathbf{Hm}_{\phi_1}$ and scale matrix $\mathbf{HC}_{\phi_1}\mathbf{H}'$. It is left as an exercise for the reader to verify that this is essentially a diagonal matrix. This follows since harmonics are always orthogonal, so we know in advance that the Fourier coefficients from one harmonic will be uncorrelated with those of any other. Although the two coefficients of any harmonic may be correlated in general, these correlations are negligible in this case. Thus the prior can be summarised in terms of prior means and scale factors for the marginal T_6 distributions of the individual Fourier coefficients. These are given in Table 10.3. Also provided are the F values associated with the hypotheses of zero contribution from the harmonics, although since the coefficients are largely uncorrelated, this is redundant information; the assessment may be made from inferences on the T-distributed, uncorrelated coefficients directly.

Recall that the first 5 F values refer to the $F_{2,6}$ distribution, the final one for the Nyquist harmonic to the $F_{1,6}$ distribution. Clearly, the first harmonic dominates in the prior specification of the seasonal pattern, the third and fourth harmonics being also of interest, but the rest are negligible. Accordingly, our earlier comments about the possible benefits of proceeding with the full, flexible seasonal model notwithstanding, the second, fifth and sixth (Nyquist) harmonics are dropped from the model. We assume the corresponding coefficients are zero at time $t = 1$, consistent with the initial prior, and also for all times $t = 1, \ldots, 72$ for the full span of the data to be processed. This restriction can be imposed properly by taking the prior T distribution for the 11 coefficients, and calculating the T conditional distribution of the 6 coefficients for the 3 harmonics retained, given that

Table 10.3. Initial prior summary for Fourier coefficients

Harmonic	Coefficient	Prior		F value
		Mean	Scale	
1	cos	0.691	0.0067	136.5
	sin	1.159	0.0067	
2	cos	−0.033	0.0067	0.3
	sin	−0.058	0.0067	
3	cos	0.283	0.0067	6.2
	sin	−0.050	0.0067	
4	cos	−0.217	0.0067	5.1
	sin	0.144	0.0067	
5	cos	0.026	0.0067	0.7
	sin	0.091	0.0067	
6	cos	0.033	0.0029	0.3

the other 5 coefficients are zero. However, in view of the fact that the coefficients are essentially uncorrelated and the specified prior rather vague, the marginal distributions already specified are used directly.

10.3.4. Model and initial prior summary

The model is 8 dimensional, given at time t by $\{\mathbf{F}_t,\ \mathbf{G},\ .\ ,\ ,\ \}$, with

$$\mathbf{F}_t' = (1;\quad X_t;\quad \mathbf{E}_2';\ \mathbf{E}_2';\ \mathbf{E}_2'\ )$$

and

$$\mathbf{G} = \text{block diag}[1;\quad 1;\quad \mathbf{G}_1;\quad \mathbf{G}_3;\quad \mathbf{G}_4\ ],$$

where the harmonic terms are defined, as usual, by

$$\mathbf{E}_2 = \begin{pmatrix} 1 \\ 0 \end{pmatrix} \qquad \text{and} \qquad \mathbf{G}_r = \begin{pmatrix} \cos(\pi r/6) & \sin(\pi r/6) \\ -\sin(\pi r/6) & \cos(\pi r/6) \end{pmatrix},$$

for $r = 1, 3$ and 4.

The initial prior is given by $n_1 = 6$ and $S_1 = 0.15^2$, determining the gamma prior for $(V^{-1} \mid D_0)$ and, for $(\boldsymbol{\theta}_1 \mid D_0) \sim \mathrm{T}_6[\mathbf{a}_1,\ \mathbf{R}_1]$, we have

$$\mathbf{a}_1' = (9.5;\quad -0.7;\quad 0.691,\ 1.159;\quad 0.283,\ -0.050;\quad -0.217,\ 0.144\ )$$

and

$$\mathbf{R}_1 = \text{block diag}\ [0.09;\quad 0.01;\quad 0.0067\ \mathbf{I}_6].$$

Finally, note that having converted to a restricted seasonal pattern, the implied seasonal effects are constrained at each time to accord with the combination of just 3 harmonics, and so the initial prior for the effects will be rather different to that for the effects originally unrestricted (apart from zero sum) in Section 10.3.2. The theory of Section 8.6 can be used to transform the marginal prior above for the 6 Fourier coefficients to that for the full 12 seasonal effects. It is routine to apply this theory, which leads to the following estimates (the new prior mean for $(\phi_1 \mid D_0)$):

$$0.76, 1.36, 1.05, 0.99, 1.18, -0.09, -1.19, -0.89, -1.08, -1.43, -0.71, 0.05$$

(to 2 decimal places). These restricted values differ only slightly from the unrestricted values, as is to be expected since the differences lie in the omission of what are insignificant harmonics.

It remains to specify discount factors determining the extent of anticipated time variation in the parameters. This illustration uses, as usual, the block discounting technique of Section 6.3, requiring one discount factor for each of the intercept, regression and seasonal components. The values used are 0.9, 0.98 and 0.95 respectively. The 0.98 for regression reflects the anticipation that the quantified regression relationship will tend to be rather durable, and 0.9 for the intercept anticipates greater change in the constant term to allow for effects of independent variables not explicitly recognised and model misspecification generally. These three discount factors now complete the model definition, and we can proceed to forecast and process the new data.

10.3.5. 'No-data' step ahead forecasting

Initially, step ahead forecast distributions are computed for the three years 1976 to 1978 from time $t = 0$, December 1975. In discount models the evolution variances for forecasting step ahead are taken to be constant, fixed at the currently implied value for forecasting one-step ahead, as described in Section 6.3. Thus, at $t = 0$, we require the current evolution variance $\mathbf{W}_1$, using this as the value in forecasting ahead to times $k > 0$. We derive this here after commenting on a feature of general interest and utility.

For general time t, recall the notation $\mathbf{P}_t = \mathbf{G}\mathbf{C}_{t-1}\mathbf{G}'$ and that $\mathbf{R}_t = \mathbf{P}_t + \mathbf{W}_t$. Recall also that in a component model with r components having discount factors $\delta_1, \dots, \delta_r$, if the diagonal block variance matrices from $\mathbf{P}_t$ are denoted by $\mathbf{P}_{t1}, \dots, \mathbf{P}_{tr}$, then the evolution variance matrix is defined as $\mathbf{W}_t = \text{block diag}\ [\mathbf{P}_{t1}(\delta_1^{-1} - 1),\ \dots, \mathbf{P}_{tr}(\delta_r^{-1} - 1)]$. This defines $\mathbf{W}_t$ in terms of $\mathbf{P}_t$. In the special case that $\mathbf{P}_t$ (equivalently $\mathbf{R}_t$) is block diagonal, the r components being uncorrelated, then it follows immediately that $\mathbf{W}_t$ can be written in terms of $\mathbf{R}_t$ also, viz.,

$$\mathbf{W}_t = \text{block diag}\ [\mathbf{R}_{t1}(1 - \delta_1),\ \dots, \mathbf{R}_{tr}(1 - \delta_r)].$$

In our example, we have $r = 3$ components and have specified $\mathbf{R}_1$ above in block diagonal (actually diagonal) form. Thus, given the discount factors

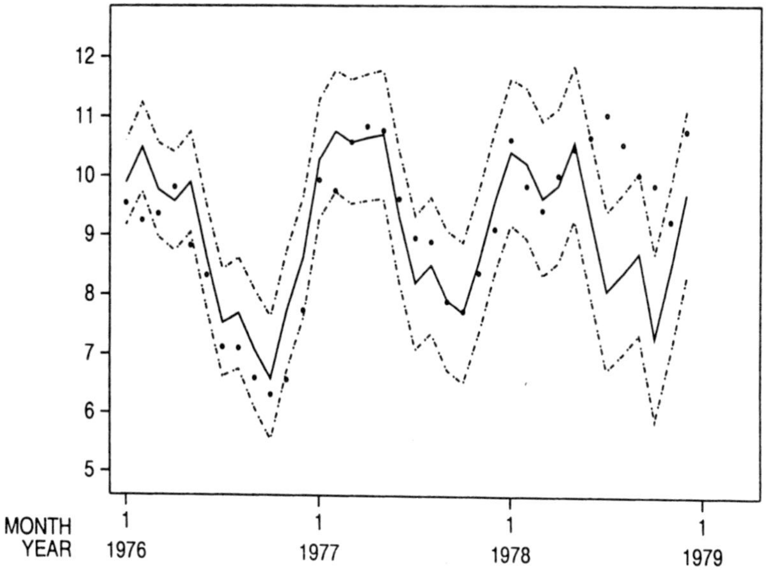

Figure 10.9 Forecasts of Sales made in December 1975

for the 3 components at 0.9, 0.98 and 0.95, we have

$$\begin{aligned}\mathbf{W} = \mathbf{W}_1 &= \text{block diag}\,[0.09(1-0.9);\quad 0.01(1-0.98);\quad 0.0067(1-0.95)\mathbf{I}]\\ &= \text{block diag}\,[\,0.009;\quad 0.0002;\quad 0.0003\ \mathbf{I}].\end{aligned}$$

Thus, viewed step ahead from $t = 0$, the state parameter vector evolves according to the system equation with evolution noise term having scale matrix $\mathbf{W}$.

Forecast distributions are now routinely calculated from the theory summarised in Section 4.6, although, since a regression component is included in the model, values are required for the regression variable, Index, in order that $\mathbf{F}_t$ be available. In this illustration, we are interested in forecasting over the first 36 months. As an hypothetical study, suppose that the first 36 values of the Index series (the values that actually arise in the future) are provided as values to use in forecasting Sales. These X_t values can be viewed either as point estimates or forecasts of the Index series made prior to 1976, or as values provided for a *What if?* study to assess the consequences for Sales of these particular future possible values of Index. For the purposes of our example, of course, this is ideal since forecast errors will be totally derived from the unpredictable, observational noise, and from model misspecification, rather than errors in forecasting the Index series. Finally, note that these three years of forecasts are *no-data* forecasts, based purely on the model and the initial priors.

Figure 10.9 provides graphical display of the forecast means and 90% symmetric intervals for the outcomes based on the forecast T distributions. Also plotted are the actual observations that later become available. If these forecasts had actually been available at December 1975, then the resulting accuracy would have been fairly impressive up to the early part of the third year, 1978. However, although most of the first two and a half years of data lie within the 90% intervals, there is clear evidence of systematic variation in the data not captured by the model. Almost all of the point forecast errors in the first year or so are negative, the model overforecasting to a small degree. The second year is rather better, but things appear to break down in late 1978 with radical under-forecasting in the last 6 or 7 months. Proceeding to analyse the data in the next section, we should therefore expect the model to adapt to the apparent change in data structure seen here as the observations are processed.

10.3.6. Data analysis and one-step ahead forecasting of Sales

Figure 10.10 provides a graph of the data with one-step point forecasts and associated 90% intervals as sequentially calculated during the data analysis.

During 1976, the model clearly learns and adapts to the data, the forecasts being rather accurate and not suffering the systematic deficiencies of those produced at December 1975. Figure 10.11 displays actual one-step forecast errors, Figure 10.12 gives a similar plot but now of standardised errors with 90% intervals, symmetrically located about 0, from the forecast T distributions. Clearly the model adapts rapidly in the first year, the forecast errors being acceptably small and appearing unrelated. Having adapted to the lower level of the data in the first year (lower, that is, than predicted at time 0), however, there are several consecutive, positive errors in 1977, indicating a higher level than forecast, until the model adapts again towards the end of that year.

Note that the model has 8 parameters, so that there is an inevitable delay of several months before it can completely adapt to change. Subsequent forecasts for late 1977 and early 1978 are acceptable.

In mid-1978, forecast performance breaks down dramatically, with one very large error and several subsequent positive errors. The changing pattern of Sales at this time is not predicted by the model, and again, several observations are required for learning about the change. The nature of the change is made apparent in Section 10.3.7 below. For the moment, note that this sort of event, so commonly encountered with real series, goes outside the model form and so will rarely be adequately handled by a closed model. What is required is some form of intervention, either to allow for more radical change in model parameters than currently modelled through the existing discount factors, or to incorporate some explanation

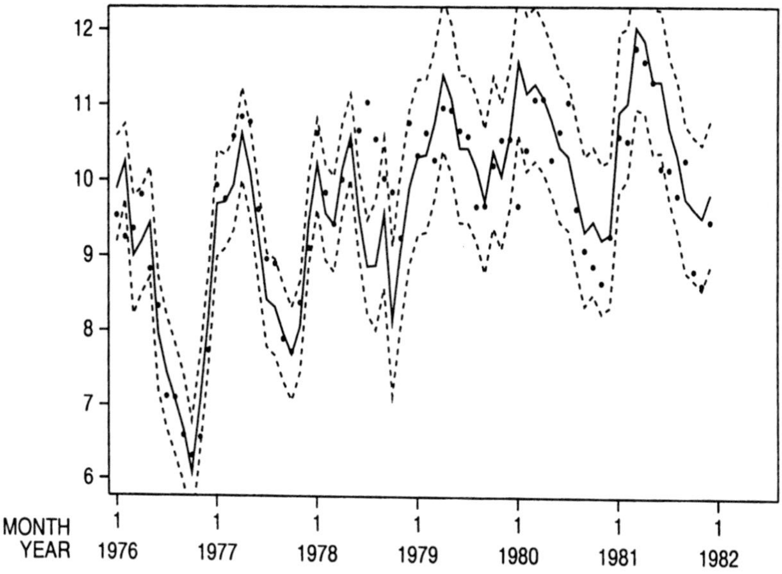

Figure 10.10 One-step ahead forecasts of Sales

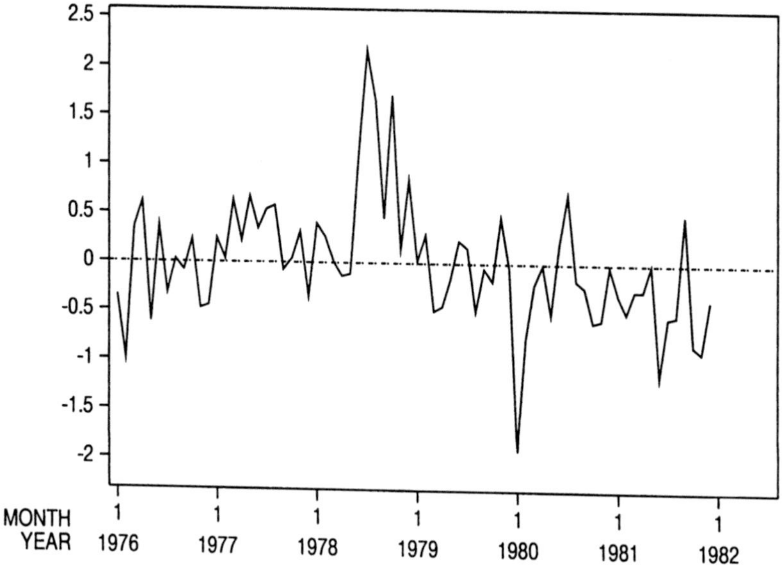

Figure 10.11 One-step ahead forecast errors

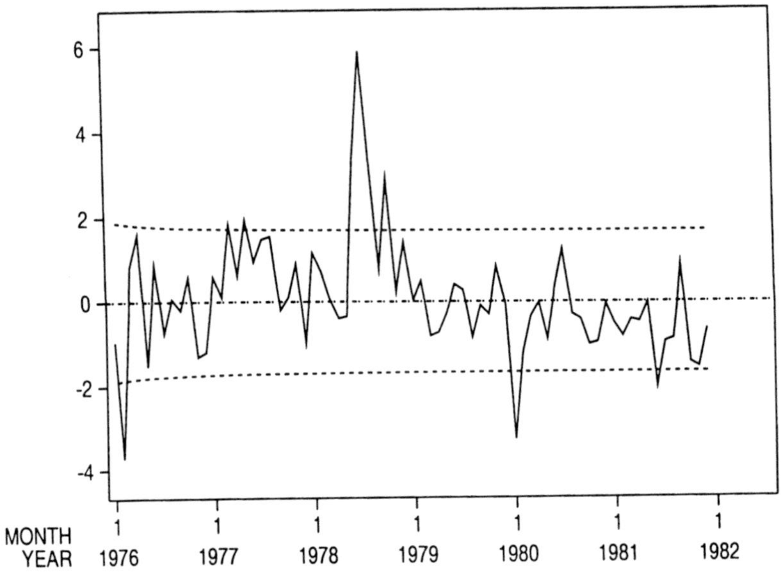

Figure 10.12 Standardised one-step ahead forecast errors

of the change via intervention effects or other independent variables. We proceed, however, with the model as it stands, closed to intervention.

After 1978, the analysis continues with reasonable success during 1979 and 1980, a single, large negative error in January 1980 being essentially an outlier. During late 1980 and 1981, negative errors (although acceptably small in absolute value), predominate, and in particular, the model has difficulty in forecasting the seasonal troughs, reflecting, perhaps, a change in seasonal pattern greater than anticipated.

One important feature clearly apparent in Figure 10.10 is that the widths of the one-step forecast 90% intervals increase markedly towards the end of 1978, being much wider from then on than in early years. This effect is explained by the presence of the large forecast errors in late 1978. In the updating for the observational variance, recall that the estimate at time t is updated via

$$S_t = S_{t-1}(n_{t-1} + e_t^2/Q_t)/(n_{t-1} + 1).$$

Thus large standardised errors $e_t/Q_t^{1/2}$ lead to an inflation of the estimate, the posterior distribution for V then favouring larger values, and consequently, the forecast distribution for the next observation will be more diffuse. This is an interesting feature; the model/data mismatch that leads to large errors shows up in the variance estimate, which can therefore be used as a simple diagnostic, and the model responds in the way of increased uncertainty about the future. Note, however, that this is due entirely to the fact that V is unknown, the effect being diminished for larger degrees of freedom n_{t-1}.

10.3.7. Retrospective time series analysis after filtering

Having processed all the data and examined the one-step forecasting activity through Figures 10.10 to 10.12, we now move to retrospective time series analysis more formally by performing the backwards filtering and smoothing with the model, obtaining posterior distributions for the model parameters at all times $t = 1, \ldots, 72$, now conditional on all the data information D_{72}. Of course these are T distributions with $n_{72} = 78$ degrees of freedom (the initial prior 6 plus 1 for each observation), and model components may be extracted as in Section 10.2.3 (although there we used the on-line posteriors rather than the filtered posteriors). In the notation of Section 4.7, we have posteriors

$$(\boldsymbol{\theta}_t \mid D_{72}) \sim \mathrm{T}_{72}[\mathbf{a}_{72}(t-72),\ \mathbf{R}_{72}(t-72)]$$

with moments determined by the filtering algorithms of Theorem 4.4 and Corollary 4.3. Thus, for any vector $\mathbf{x}_t$ if $x_t = \mathbf{x}_t'\boldsymbol{\theta}_t$ then

$$(x_t \mid D_t) \sim \mathrm{T}_{72}[\mathbf{x}_t'\mathbf{a}_{72}(t-72),\ \mathbf{x}_t'\mathbf{R}_t(t-72)\mathbf{x}_t],$$

the mean here providing a *retrospective* estimate of the quantity x_t. Various choices of $\mathbf{x}_t$ are considered.

(1) SEASONAL PATTERN
$\mathbf{x}_t' = (0; 0; 1, 0; 1, 0; 1, 0)$ implies that $x_t = \theta_{t3} + \theta_{t5} + \theta_{t7} = \phi_{t0}$, the seasonal factor at time t. Figure 10.13 displays 90% intervals for the effects over time, symmetrically located, as usual, about the filtered, posterior mean. Changes in seasonal pattern are evident from this graph. The amplitude of seasonality drops noticeably in early 1978, is stable over the next 3 years, but increases again in 1981. The phase also shifts markedly in early 1978 (recall the deterioration in forecasting performance during this period), and continues to shift slightly from year to year from then on. Note also the increase in uncertainty about the effects in later years due to the inflation in estimation of V earlier described.

(2) NON-SEASONAL TREND
$\mathbf{x}_t' = (1; X_t; 0, 0; 0, 0; 0, 0)$ implies that $x_t = \theta_{t1} + \theta_{t2}X_t$ is the deseasonalised trend at time t, including the regression effect of Index. The corresponding posterior mean plus 90% limits appear in Figure 10.14.

(3) REGRESSION EFFECT
A related case is given by $\mathbf{x}_t' = (0; X_t; 0, 0; 0, 0; 0, 0)$, so that $x_t = \theta_{t2}X_t$ is the regression effect of Index free from the time-varying intercept term θ_{t1}. The corresponding plot is given in Figure 10.15. This figure is useful in retrospectively assessing the effects of the historic changes in values of

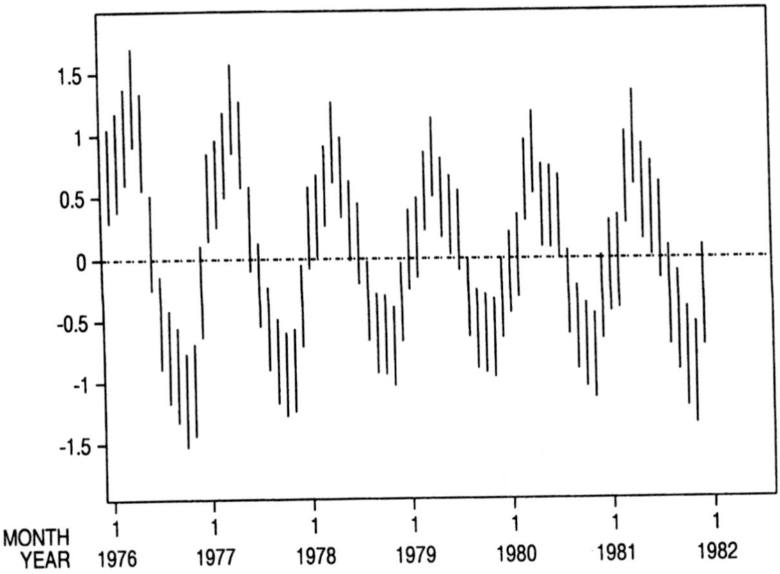

Figure 10.13 Estimated monthly pattern in Sales

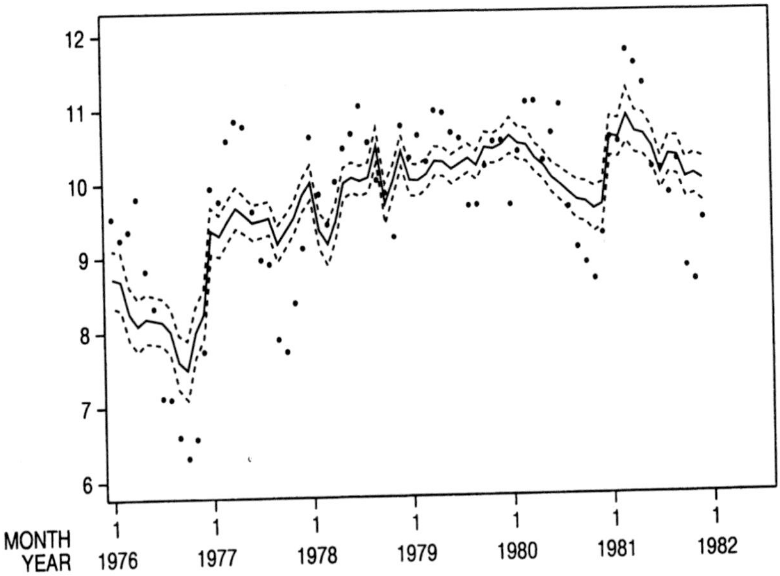

Figure 10.14 Estimated non-seasonal variation in Sales

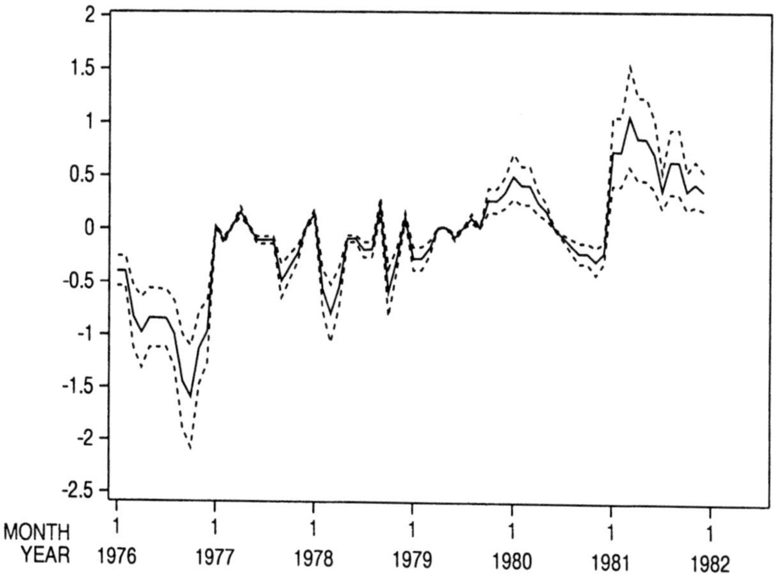

Figure 10.15 Estimated effect of Index on Sales

the independent variable series. If Index is subject to some form of control by the company (such as via pricing policies), then this graph provides information about the effects of past attempts at controlling the Sales series via changes to Index.

(4) REGRESSION COEFFICIENT
Taking $\mathbf{x}_t' = (0;1;0,0;0,0;0,0)$ leads to $x_t = \theta_{t2}$, the coefficient in the regression on Index at time t. The corresponding plot is given in Figure 10.16. There is obvious movement in the coefficient, the Sales series being apparently less strongly related to Index in the later years, the coefficient drifting upwards to around 0.4, although the uncertainty about the values is fairly large.

(5) INTERCEPT COEFFICIENT
Figure 10.17 is the display for the intercept term θ_{t1} obtained from (2) above with $X_t = 0$, so that $\mathbf{x}_t' = (1;0;0,0;0,0;0,0)$. Again there is obvious change here, consistent with the features noted in discussing the model adaptation to data in Section 10.3.6. It is of interest to the company to attempt, retrospectively, to explain this movement in the parameter; if independent variable information exists and can be related to this smooth movement, then future changes may be anticipated and forecasting accuracy improved.

Note that the estimated time evolution of the parameters as described above are a consequence of the model analysis. Here the data series are,

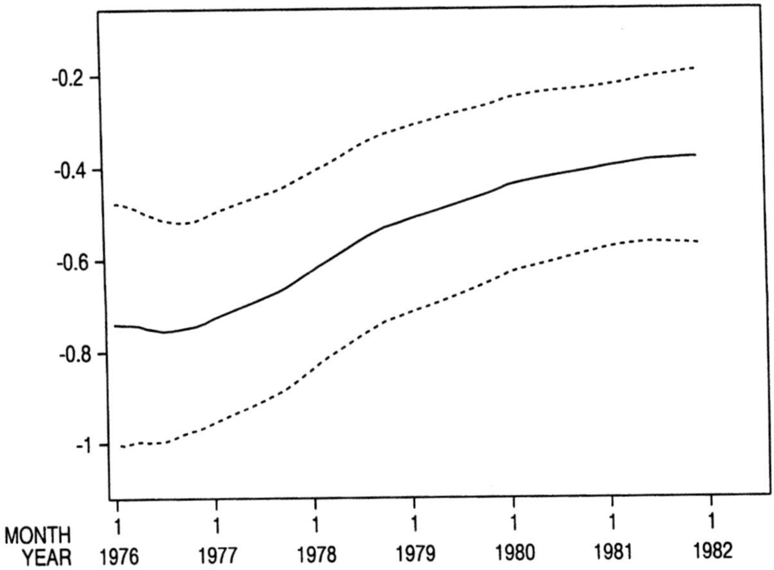

Figure 10.16 Estimated coefficient of Index

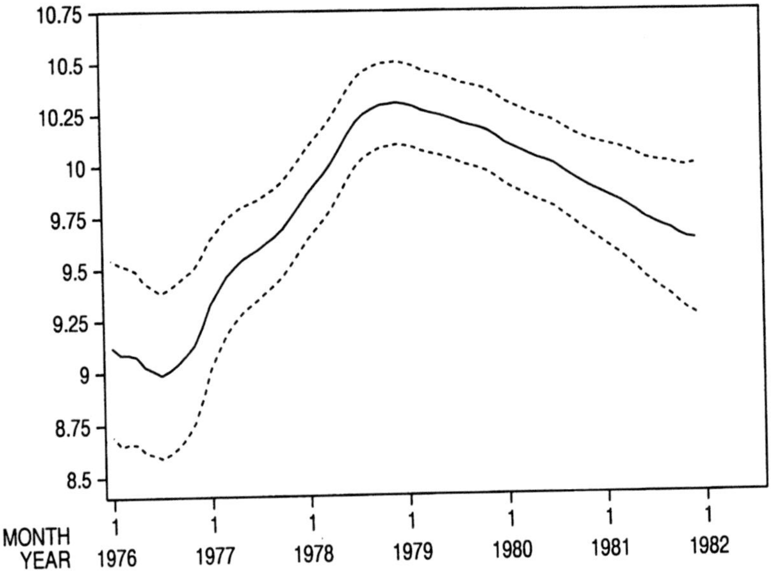

Figure 10.17 Estimated intercept parameter

at times, rather ill-behaved relative to the model form specified, and the changes estimated may be partially explained as the model's response to the model/data mismatch. It is inevitable that in allowing for change in model components, some change may be derived from the effects of variables omitted that could otherwise have explained some of the observed deviations away from model form. If such explanations could be found, then the estimation of existing components would be protected from possible biases induced by their omission. Hence care is needed in interpreting the changes identified by any given model.

10.3.8. Step ahead forecasting: What if? analysis

Finally, at time $t = 72$, December 1982, consider forecasting step ahead based solely on the model and current posteriors. To forecast Sales during 1982 to December 1984 requires values for the Index series as inputs. For illustration of a *What if?* analysis, the values are taken as $X_t = 2$ ($t = 73, \dots, 84$), $X_t = 0$ ($t = 85, \dots, 96$) and $X_t = -2$ ($t = 97, \dots, 108$). The values of Index over 1976 to the end of 1981 range from a low of -2.70 to a high of 2.13. Thus, the step ahead values used here allow exploration of the implications of fairly extreme values of Index; during 1982, Index is unfavourable at 2, and during 1984 favourable at -2. During 1983, the value of 0 leads to forecasts based only on the seasonal pattern. Figure 10.18 displays the historical data and forecasts over these three years based on these *What if?* inputs.

10.4 ERROR ANALYSIS

10.4.1 General comments

There is a large literature on statistical techniques for error, or residual, analysis in linear models generally and classical time series models in particular. Almost all of the formal testing techniques are non-Bayesian, though most of the useful approaches are informal and do not require adherence to any particular inferential framework. Fair coverage and reference to error analysis in time series can be found in Box and Jenkins (1976) and Abraham and Ledolter (1985).

General Bayesian theory for residual diagnostics appears in Box (1980), and Smith and Pettit (1985). In our opinion, the most useful error analyses consist of informal examination of forecast errors, looking at graphical displays and simple numerical summaries to obtain insight into possible model inadequacies with a view to refinement. The use of formal tests of goodness of fit, though possibly of value in certain contexts, does not address the key issues underlying error analyses. These can be summarised as

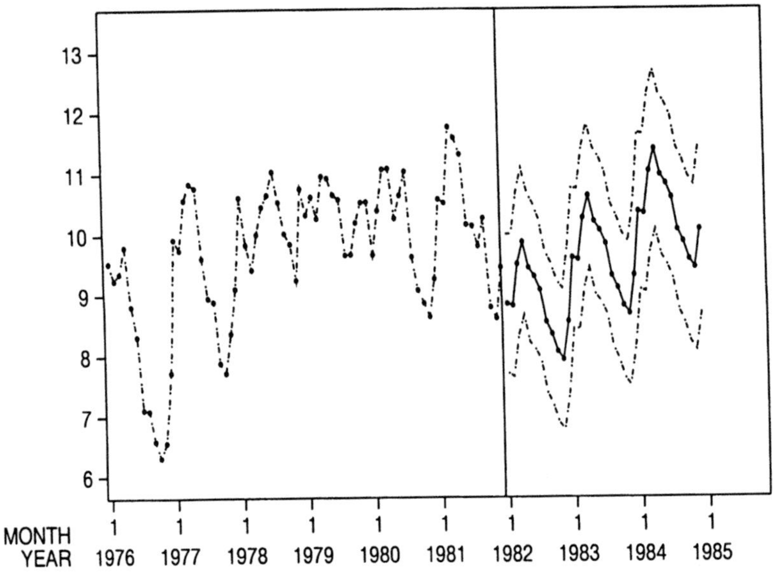

Figure 10.18 'What-if?' step-ahead forecasts

(1) identifying periods of time in the history of the series analysed where model performance deteriorated;
(2) suggesting explanations for this deterioration, or breakdown, in model performance; and
(3) modifying the model for the future in the light of such explanations.

The focus in classical error analysis is usually on *fitted* errors, or residuals, measuring the retrospective departure of the data from the model. How closely the historical data fit the model in retrospect is not really the point in time series forecasting; more incisive are investigations of the *predictive fit* of the model, embodied in the sequence of one-step ahead forecast distributions. The one-step forecast errors e_t are thus the raw material for model assessment. Thus, we have presented plots of the e_t, and their unstandardised analogues, in the illustrations of Sections 10.2 and 10.3. Under the assumptions of the model, the forecast distributions are, as usual, $(Y_t \mid D_{t-1}) \sim \mathrm{T}_{n_{t-1}}[f_t, Q_t]$, so that as a random quantity prior to observing Y_t, the error e_t has the predictive distribution

$$(e_t \mid D_{t-1}) \sim \mathrm{T}_{n_{t-1}}[0, Q_t].$$

Of course, when n_{t-1} is large, or the observational variance sequence is known, these distributions are essentially normal. Also,

$$p(e_t, \dots, e_1 \mid D_0) = \prod_{r=1}^{t} p(e_r \mid D_{r-1}),$$

and so $(e_t \mid D_{t-1})$ is conditionally independent of $(e_s \mid D_{s-1})$ for $s < t$. The observed errors are thus the realisation of what is, under the model assumptions, a sequence of independent random quantities with the above distributions. Observed deviations in the error sequence away from this predicted behaviour are indicative of either irregularities in the data series, or model inadequacies, or both. The raw errors e_t should be standardised with respect to their scale parameters for examination. So consider the sequence of standardised errors

$$u_t = e_t/\sqrt{Q_t},$$

hypothetically (conditionally) independent and distributed as

$$(u_t \mid D_{t-1}) \sim \mathrm{T}_{n_{t-1}}[0, 1], \qquad (t = 1, 2, \ldots).$$

The plot over time of the u_t sequence is the most useful starting point in error analysis. Deviations of the error sequence from model predictions show up rather easily in such plots. Common deviations of interest and importance from the model assumptions are as follows:

(a) Individual, extreme errors, possibly due to outlying observations;
(b) Groups, or patches, of a few errors of the same sign suggesting either a local drift away from zero location or the development of positive correlation between the errors;
(c) Patches of negatively related errors, the signs tending to alternate;
(d) Patches of errors large in absolute value, without any clear pattern of relationship.

Though not exclusive, these features are common and account for many types of deviation of the data from the model, and are now examined.

10.4.2 Outliers

A single large error u_t indicates that Y_t is a single observation that deviates markedly from forecast under the model. The notion of extremity with respect to the forecast distribution of u_t involves considering values in the tails of the distribution. The fact that the degrees of freedom may differ for each error is important; Figures 10.3 and 10.12 provide, as a graphical guide, 90% forecast intervals symmetrically located about the mode of zero. Note that these are both equal-tailed and HPD intervals since the T densities are symmetric about zero. Values of u_t outside these intervals occur only 10% of the time, on average. Supplementing these plots with the corresponding 95% intervals would be a useful addition. Such observations occur in every area of application and are commonly referred to as outliers. With socio-economic data series, it is not uncommon to encounter such wild values one observation in twenty, and often they are rather more frequent. In the context under study, there may be many possible explanations for outliers. A true outlier provides no information about the observation process, such as the case if the observation has been misrecorded in the data collection process. The same may be true even though the observation is correctly recorded, when there exists auxiliary information as to why the value is wild. For example, in a consumer sales series the announcement of a future price rise can lead to consumers stocking up in advance at the old price, leading to a single, very large observation that is totally correct and acceptable, but that is quite irrelevant to the underlying pattern of sales generally. In ideal circumstances, all outliers should be explained in this way, being of potentially major importance to decision makers. From the point of view of retrospective error analysis, outliers can be identified by their single, large standardised errors. From the point of view of modifying the model for the future in view of these outliers, there is often little to be done other than inform decision makers that outliers can be expected, but not predicted unless suitable auxiliary information arises in advance. Without such information, if outliers are to be anticipated, the main way of reflecting their existence is to somehow increase forecast uncertainty; very large values that are rare under the existing model will be less rare under a model generating forecast distributions with heavier tails. Thus, a common model modification is to extend the basic, normal DLM to include an observational error distribution that has heavy tails. Some possibilities are developed in Harrison and Stevens (1976), West (1981), and West, Harrison and Migon (1985). Usually the extensions are to heavier-tailed, symmetric error distributions. If large errors of a particular sign predominate, then the suggestion may be that the errors have a skew distribution, possibly indicating the need for a data transformation (see Section 10.6).

A typical example outlier shows up in the January 1980 error in the example of Section 10.3. Figure 10.12 shows the error in this month to be way outside the 90% interval, and in fact it is outside the 99% limit. The error corresponds to a much lower sales figure in that month than forecast. Another apparent outlier occurs in February 1976, but this one is explained as due to the inappropriateness of the initial prior estimate of the seasonal factor for that month.

10.4.3 Correlation structure

Continuing with an examination of the errors in Figure 10.12, the very wild observation in the sales series in mid-1978 may be initially thought to be an outlier, but the sequence of positive errors immediately following this one point indicate something more sustained has occurred over a period of a few months. This is consistent with item (b) of Section 10.4.1, the occurrence of a patch of consecutive, positive errors. In this example, there was a marked change in the seasonal pattern of the sales series for the year of 1978 that led to these errors. This is a very common occurrence. If changes occur that are much more marked than allowed for through the existing dynamics of the model, then there will be a subsequent run of errors that are positively correlated. Gradually, the model adapts to the change, and after sufficient observations, the errors are back under control. Note that this feature of autocorrelated errors involves the development, from time to time, of *local* correlation structure in the errors that eventually decays as the model adapts. In a model in which the basic dynamic is too constrained, due to discount factors that are too large, the natural movement in the series will tend to be under-predicted throughout and therefore the positive correlation between consecutive errors will tend to be sustained. This feature appears again during the last few months of the data set, where there is a run of several, negative errors as the model fails to adapt rapidly enough to the lower sales levels in the recession of late 1981.

There are various ways of modifying the model for the future to deal with this. Notably

(1) Forecasting and updating in the future subject to some form of monitoring of the model to detect the development of correlation between forecast errors. This may simply involve the forecaster in subjective monitoring, calling for intervention when correlation develops. The most useful and usual form of intervention involves corrective action to decrease discount factors, often just temporarily, on some or all of the existing model components. This has the effect of increasing adaptivity to change and the model then responds more rapidly in the future.
(2) Explanation of consecutive, positively related deviations from forecast via the inclusion in the model of additional terms. For example, a temporary jump in level of the series can be modelled by including an additional, possibly slowly varying level parameter for a few time periods. Regression-type terms are also useful if independent variables can be identified. Seasonal patterns not modelled will also show up in correlated errors.
(3) The extension of the model to include a noise process term to model the local correlation structure and use past errors to forecast those to come in the short term; see the development of noise models in Section 9.4. It should be stressed that such terms be included only to describe local error structure that cannot be otherwise, more constructively, explained via parameter changes or regression type terms. It is sometimes tempting to explain more global movement in the series by such noise models when in fact they should be attributed to changes in trend or other components of the basic DLM. These comments notwithstanding, short-term forecasting accuracy can be improved by incorporating a simple noise model when local correlations develop. Most useful is the basic AR(1) type model of

Section 9.3, with, typically, a coefficient expected to be fairly low, between 0 and 0.5, say.

Consecutive observations exhibiting negative correlation structure also occur, though perhaps less commonly than positive correlation. One obvious explanation for this is the use of a dynamic model that is too adaptive, having discount factors that are too small. The result is that the posterior distributions for model parameters adapt markedly to each observation, and thus short-term predictive accuracy is lost. Negatively correlated errors result as the model swings from one observation to the next. Again, corrections to the model properly require adjustments to discount factors, although simple noise models will improve short-term forecasting accuracy.

10.4.4 Observational variance structure

Patches of errors that are large in absolute value but that do not show any clear structure can arise, and often do, if the assumptions about the observational variances in the normal DLM are inappropriate. Particular reasons for such features are

(1) The possibility that each observation Y_t is a function of several or many basic quantities, such as a total or average, leading to variances that may differ widely depending on the number of such components of each of the Y_t;
(2) Observational variances that depend on the underlying level of the series, due to original non-normality of the Y_t; and
(3) Possible random fluctuations in the observational variance about the level, representing volatility explained only by the absence of relevant independent variables from the model, i.e., model misspecification that is evidenced, not by systematic, local correlations in the errors, but rather by random changes in variation.

These and other features can be modelled adequately by extending and modifying the basic, normal DLM as in the next section.

10.5 DATA MODIFICATIONS AND IRREGULARITIES

10.5.1 Outliers and missing values

Following the discussion of outliers in Section 10.4.2, consider the treatment of wild observations. For routine updating in the model, the fact that explainable or otherwise, an observation is quite unrelated to the underlying process means that for forecasting the future, it provides little or no information and should be discarded. Formal techniques for doing this by extending the model to include outlier generating distributions can be used (Harrison and Stevens 1976; West 1981; and West, Harrison and Migon 1985). Some such models are developed in later chapters. Often, however, simply discarding the observation from the model updating is the best approach. That outlying observations can seriously adversely affect the analysis of a model and degrade predictions is well known and not specific to time series. There is a large literature on outlier handling and modelling techniques. See, for example, Barnett and Lewis (1978), Box (1980), Smith and Pettit (1985), West (1984a, 1985a), and references therein. Work specifically concerning outlier problems in time series can be found in Box and Tiao (1968), Fox (1972), Harrison and Stevens (1976), Kleiner, Martin and Thompson (1981), Martin, R.D.Thompson, D.J. Masreliez and Martin (1977), West (1981, 1986a), and West and Harrison (1986). In the standard updating equations for the DLM, recall that the error e_t enters in the equations for the estimates of both $\boldsymbol{\theta}_t$ and V, namely

$$\mathbf{m}_t = \mathbf{a}_t + \mathbf{A}_t e_t$$

and

$$S_t = S_{t-1}(n_{t-1} + e_t^2/Q_t)/(n_{t-1} + 1).$$

Thus a large, outlying value of $|e_t|$ leads to a major correction to both $\mathbf{a}_t$ and S_{t-1} to obtain the posterior quantities $\mathbf{m}_t$ and S_t. The variance estimate, a quadratic function of the error, can be inflated enormously when n_{t-1} is small. To guard against this, the most appropriate action is to discard the observation from the updating of the model if it has been identified as an outlier. Thus, given this identification (which can sometimes be rather difficult to achieve, but is assumed here), Y_t provides no information about the model; the information set at time t is just $D_t = D_{t-1}$, and the posterior distributions are equal to the priors, viz., $\mathbf{m}_t = \mathbf{a}_t$, $\mathbf{C}_t = \mathbf{R}_t$, $S_t = S_{t-1}$ and $n_t = n_{t-1}$. We stress again that this is appropriate if Y_t has been identified as uninformative about the model. In Chapters 11 and 12 we return to problems of outlier detection and modelling.

An observation that is truly missing, being unrecorded or lost, is treated in just the same way, of course. Other approaches to time series, in particular using ARIMA type models, have enormous difficulties in handling missing values. The sequential Bayesian approach, in marked contrast, trivially accounts for them: no information leads to no changes in distributions for model parameters. In some applications observations associated with discrete, external events may be viewed as missing in order simply to separate the events from the routine model. This can be done, for example, to model the effects of individual, irregular events such as strikes in production plants, responses to budget and tax announcements, irregular holidays, and so forth. This is a simple alternative to modelling the effects of such events with extra parameters to be estimated, appropriate if all that is required is protection of the standard model from the observation.

10.5.2 Irregular timing intervals

Another problem of practical importance is the irregular timing of observations. So far we have assumed, as with all classical models, that the observations are equally spaced in time, and that the timing is precise. There will often be timing errors, some random, some systematic, that will be impossible to identify and quantify. Such minor discrepancies are assumed to be accounted for by the error sequences ν_t and ω_t in the model. Of greater significance is the problem of observations arriving at intervals of irregular length. For example, monthly data may be recorded on a particular day towards the end of each month, but the precise day can vary between months. Daily observations may be recorded only to the nearest hour. In these cases, where the data are reported at an aggregate level but may be based on different numbers of time intervals at a lower level of aggregation, it is clear that the timing should, if possible, be accounted for in the model. If the timing is unknown then little can be done. Otherwise, the approach is simple. The answer lies in the fact that there is a basic, original time unit that provides a precise time scale. In the first example above the unit is the day, in the second it is the hour. If we define our model so that the observations are indexed by this basic unit, then the observations are equally spaced, but some, often many of them, are missing. Thus, with monthly data in the first example, if t indexes days and $t = 1$ represents January 1st, then we may receive data, $Y_{29}, Y_{57}, Y_{86}, \ldots$, one in each month, the others, $Y_1, Y_2, \ldots, Y_{28}, Y_{30}, \ldots$, being viewed as missing.

Note that if the data are aggregates, the Y_t above, for example, purporting to be monthly totals or averages of daily data, then further, obvious modifications are necessary to provide an appropriate model at the daily level.

10.5.3 Discount models with missing data

Some comments are in order concerning the use of discount models with missing data. The inherent, one-step ahead nature of the discount technique and the overly rapid decay of information if it is used to project step ahead, have been discussed in Sections 6.3 and 10.2.4. This feature is apparent when missing values occur. When $\boldsymbol{\theta}_t$ evolves with no incoming data, whether truly missing or due to irregular timing, the evolution error variance matrix should be based on that most recently defined using the standard discount factors. For example, suppose that we evolve at $t = 4$ as usual with $\mathbf{W}_4$ determined by discount factors. If Y_4 and Y_5, say, are missing, then the evolution over the corresponding time intervals is based on $\mathbf{W}_5 = \mathbf{W}_6 = \mathbf{W}_4$, that value most recently used in a time interval when an observation was made. Only when the next observation Y_7 arrives is the discount technique applied again, determining $\mathbf{W}_8$ for evolution to $t = 8$.

10.6 DATA TRANSFORMATIONS

10.6.1 Non-normality of observations

It is rather commonly found that observed data do not adequately conform to the assumptions of normality and constancy of variance of the observational errors. We noted in Section 10.4.4 that the forecast errors can signal apparent changes in variance, consistent with original non-normality of the sampling structure of the series. A typical feature of commercial and socio-economic series of positive data (by far the majority of data in these areas are positive) is an increasing variance with level. Possible, closely related reasons are that the data are aggregated or that a non-normal model is appropriate. As an example, consider sales or demand data for batches of items, the batch sizes being uncorrelated with constant mean f and variance V. For a batch of size n, the number sold has mean nf and variance nV. As a function of varying n (with f and V fixed), the variance of the batch size is then proportional to the mean batch size. With Poisson data, the variance equals the mean. Other non-normal observational distributions for positive data, such as lognormal, gamma or inverse gamma, various compound Poisson forms, and others, have the feature that the variance increases as a function of the mean of the distribution. Empirical or theoretical evidence that such distributions may underlie a series therefore will suggest such a mean/variance relationship. Often a transformation, such as logs, will approximately remove this dependence of variation on level, leading to a transformed series with roughly constant variance and a distribution closer to normality, or at least symmetry. Such data transformations are widely used in statistics, and can provide a satisfactory solution, the data being modelled on the transformed scale. Uncritical use of transformations for these reasons of convenience should, however, be guarded against. Often a suitable transformation will be difficult to identify and use of a simple substitute may complicate or obscure patterns and relationships evident on the original data scale. The benefits of the original scale are obvious when we consider the interpretation of a given model and the communication of its implications to others. In particular, the construction of linear models in terms of components for distinct features neatly separates the random error term from the structural component of a series, and operating on a transformed scale will usually mean that the benefits arising from component modelling are diminished.

Generally then, it is preferable to model the series on the original data scale. Various techniques are available when we want to retain the original scale but the data are essentially non-normal. The obvious approach, in principle, is to use a more appropriate sampling distribution for the raw data, and we do this in Chapter 14. Here we stay with normal models, considering modifications of the basic DLM framework to allow for certain, commonly encountered features of non-normality.

10.6.2 Transformations and variance laws

The above cautionary notes about transformations notwithstanding, consider operating with a DLM for a transformed series. Let $g(.)$ be a known continuous and monotone function and define the transformed series Z_t by

$$Y_t = g(Z_t), \qquad \text{or} \qquad Z_t = g^{-1}(Y_t), \qquad\qquad t = 1, 2, \dots ,$$

the inverse of g existing by definition. Often, though not always, it is the case that Z_t is real-valued, the transformation g mapping the real line to the (possibly bounded) original range for Y_t. Most commonly the Y_t are positive, and this is assumed throughout this section. The associated observational error distributions are typically skewed to the right, with variances increasing as the location increases, indicating greater random, unpredictable variation in the data at higher levels. Various underlying error distributions can be supported on theoretical and empirical grounds. Some examples and general theory are given in Stevens (1974) and Morris (1983). The primary aims of transformation are to achieve a roughly normal, or more realistically, roughly symmetric distribution for the Z_t with a variance not depending on the location. We concentrate here on the latter feature, the normality being very much of secondary importance. In fact, the transformations usually used to achieve constancy of variance also tend to lead to rough symmetry.

For convenience of notation throughout this section, we drop the dependence of the distributions on the time index t, the discussion being, in any case, rather general. Thus $Y = Y_t$ and $Z = Z_t$ are related via $Y = g(Z)$. Approximate, guiding relationships between the moments of Y and those of Z can be used if we assume, as is usually appropriate, that the function g is at least twice differentiable, having first and second derivatives g' and g'' respectively. If this is so, then simple approximations from Lindley (1965, Part 1, Section 3.4) apply. Suppose that the transformed quantity Z has mean μ and variance V, whether the distribution be approximately normal or otherwise. Neither need, in fact, be known; we simply assume that these are the moments of Z conditional on their values being assumed known, writing $\mu = \mathrm{E}[Z]$ and $V = \mathrm{V}[Z]$. Then, following Lindley as referenced above, we have

(a) $\mathrm{E}[Y] \approx g(\mu) + 0.5g''(\mu)V$;
(b) $\mathrm{V}[Y] \approx \{g'(\mu)\}^2 V$.

Thus, approximately, the variance will be constant (as a function of μ) on the transformed scale, if and only if the original Y has variance $\mathrm{V}[Y] \propto \{g'(\mu)\}^2$. If we ignore the second term in (a), assuming that the first term dominates, then $\mu \approx g^{-1}(\mathrm{E}[Y])$, and the requirement becomes

$$\mathrm{V}[Y] \propto \{g'[g^{-1}(\mathrm{E}[Y])]\}^2. \qquad (10.1)$$

This is a particular case of a *variance law* for Y, the variance being functionally related to the mean.

Particular cases of (10.1) are examined in connection with the most widely used class of *power* transformations (Box and Cox 1964). Here

$$Z = \begin{cases} (Y^\lambda - 1)/\lambda, & \lambda \neq 0; \\ \log(Y), & \lambda = 0, \end{cases}$$

for some *power index* λ. Interesting values of this index, in our context of positive data, lie in the range $0 \leq \lambda < 1$. These provide decreasing transformations that shrink larger values of Y towards zero, reducing the skewness of the sampling distribution. Note that the inverse transformation is just

$$g(Z) = \begin{cases} (\lambda Z + 1)^{1/\lambda}, & \lambda \neq 0; \\ \exp(Z), & \lambda = 0. \end{cases}$$

Using this in (10.1) leads to

$$\mathrm{V}[Y] \propto \mathrm{E}[Y]^{2(1-\lambda)},$$

and the following special cases:

(a) $\lambda = 0$ leads to $\mathrm{V}[Y] \propto \mathrm{E}[Y]^2$, with the log transform appropriate. If in fact Y is lognormally distributed, then the variance is exactly a quadratic function of the mean and the log transform exactly appropriate.

(b) $\lambda = 0.5$ gives $\mathrm{V}[Y] \propto \mathrm{E}[Y]$ and the square root transformation is appropriate, consistent with a Poisson distribution, for example.

(c) $\lambda = 0.25$ gives $\mathrm{V}[Y] \propto \mathrm{E}[Y]^{1.5}$, consistent with compound Poisson-like distributions often empirically supported in commercial and economic applications (Stevens 1974).

10.6.3 Forecasting transformed series

Given a DLM for the transformed Z series, forecasting the original Y series is, in principle, straightforward within the Bayesian framework. We suppose that from a DLM for the Z series, a forecast distribution (one-step or otherwise) is available in the usual T form, namely

$$Z \sim \mathrm{T}_n[f, Q].$$

Note again that we have simplified the notation and dependence on the time series context for clarity and generality. Here n, f and Q are all known at the time of forecasting. Let P_Z denote the cumulative distribution function, or cdf, of this T distribution, so that $P_Z(z) = \Pr[Z \leq z]$ for all real z. Since $g(.)$ is continuous and monotonic, then the corresponding forecast cdf for Y is defined as

$$P_Y(y) = \Pr[Y \leq y] = \Pr[Z \leq g^{-1}(y)] = P_Z(g^{-1}(y)).$$

Point forecasts, uncertainty measures and probabilities can, in principle, now be calculated for Y. In practice, these calculations may be difficult to perform without numerical integrations. Practically useful transformations, such as the power transformations, lead to distributions that are not of standard form, so that unlike the usual T distributions, a little more effort is required to summarise them adequately. Some generally useful features are as follows:

(1) The forecast distribution will usually (though beware, not always) be unimodal, so that point estimates such as means, modes and medians provide reliable guides as to the location of the distribution.
(2) The moments of Y are often obtainable only using numerical integration. Simple quadrature, for example, is relatively straightforward since Y is a scalar. The approximate results used in the last section provide rough values.
(3) The median of Y can always be found easily, being the solution in y to the equation

$$0.5 = P_Y(y) = P_Z(g^{-1}(y)).$$

Thus, since the mode f of P_Z is also the median, then the median of Y is simply $g(f)$.
(4) The result (3) is a special case of the result that percentage points of P_Y are obtained as the transformed values of those of P_Z. Simply replace 0.5 in the equation in (3) by any required probability to see this. Thus, for example, intervals for Z with stated probabilities under the T distribution for Z provide intervals for Y with the same probabilities, simply by transforming the endpoints.

10.6.4 Log transforms and multiplicative models

The most important transformation is the natural log transform. Consider, for illustration, a simple dynamic regression for the transformed series Z_t given by

$$Z_t = \alpha_t + \beta_t X_t^* + \nu_t, \qquad \nu_t \sim \mathrm{N}[0, V],$$

where X_t^* is the observed value of a regressor variable. The level at time t is $\mu_t = \alpha_t + \beta_t X_t^*$, and the transformed series follows a standard DLM with constant, possibly unknown variance V. The original series is given by $Y_t = \exp(Z_t)$, corresponding to the power transformation with index $\lambda = 0$. Some points of relevance to the model for the original series are as follows:

(1) Conditional on the model parameters $\boldsymbol{\theta}_t = (\alpha_t, \beta_t)'$ and V,

$$Y_t = e^{\mu_t + \nu_t} = e^{\mu_t} e^{\nu_t}.$$

Now, $\exp(\nu_t)$ is a lognormal random quantity, so that Y_t has a lognormal distribution. Properties of lognormal distributions are fully discussed in Aitchison and Brown (1957). In particular, $\mathrm{E}[Y_t \mid \boldsymbol{\theta}_t, V] = \exp(\mu_t + V/2)$, the mode of Y_t is $\exp(\mu_t - V/2)$, and the median is $\exp(\mu_t)$. Also, the variance is proportional to the square of the mean, the constant of proportionality being given by $1 - \exp(-V)$.

(2) The model for Y_t is multiplicative, given by

$$Y_t = \gamma_t X_t^{\beta_t} e^{\nu_t},$$

where $\gamma_t = \exp(\alpha_t)$ and $X_t = \exp(X_t^*)$. Often X_t is an original positive quantity transformed to the log scale along with Y_t.

(3) This form of model, widely used to represent socio-economic relationships, is particularly of use in studying regressions in terms of *rates* of change of the response with the independent variable. An increase of $100\epsilon\%$ in X_t, $(0 \leq \epsilon \leq 1)$, leads to an expected increase of $100(1+\epsilon)^{\beta_t}\%$ in Y_t. Thus, inferences about β_t from the DLM for Z_t lead directly to inferences about expected *percentage* changes on the original scale.

Similar features arise, of course, in more general DLMs for the log values.

10.7 MODELLING VARIANCE LAWS

10.7.1 Weighted observations

Before considering modelling of variance laws generally, the simple case of weighted observations, involving observational variances known up to a constant scale parameter, is discussed. As a motivating example, consider monthly data Y_t that are arithmetic means, or averages, over daily data in each month t. If the raw daily data are assumed uncorrelated, with common, constant variance V, then the variance of Y_t is $V_t = V/n_t$, where n_t is the known number of days in month t, obviously varying from month to month. The observation is said to be *weighted*, having a known, positive *weight*, or *variance divisor*, $k_t = n_t$. An observation with little or no weight has a large or infinite variance; one with high weight has a small variance. Unit weight $n_t = 1$ corresponds to the original unweighted model. This sort of weighting applies generally when the data are aggregates, of some form, of more basic quantities. Note also that the treatment of outliers, missing values and irregularly spaced data of Section 10.5 is a special case, with missing values essentially receiving no weight.

The basic DLM requires only a minor modification to allow for weighted observations. The usual theory applies directly to the general model

$$\{\mathbf{F}_t, \mathbf{G}_t, k_t V, \mathbf{W}_t V\},$$

where k_t is, for each t, a known, positive constant, giving Y_t a weight k_t^{-1}. The modification to the updating and forecasting equations is that the scale parameter Q_t now includes the weight. Thus,

$$(Y_t \mid D_{t-1}) \sim \mathrm{T}_{n_{t-1}}[f_t, Q_t],$$

where $f_t = \mathbf{F}_t'\mathbf{a}_t$ as usual, but

$$Q_t = k_t S_{t-1} + \mathbf{F}_t'\mathbf{R}_t\mathbf{F}_t.$$

Similarly, future values k_{t+r} appear in the expressions $Q_t(r)$ for the scale parameters of step ahead forecast distributions made at time t for times $t+r$.

10.7.2 Observational variance laws

Suppose we wish to use a DLM for the Y_t series on the original, positive scale, but recognise that the data are non-normal with a variance law identified, at least approximately. Forecasting and updating can usually satisfactorily proceed with a weighted model, the weights provided as rough estimates of the variance law, varying as the level of the data varies. Precise values for the variance multipliers are not essential. What is important is that the variance multipliers change markedly as the level of the series changes markedly, providing an approximate indication of the relative degrees of observational variation at different levels. Denote the variance law by $k(.)$, so that the variance at time t is expected to be approximately given by

$$\mathrm{V}_t = k(\mu_t)V,$$

where $\mu_t = \mathbf{F}_t'\boldsymbol{\theta}_t$ is the level of the series. Examples are

(a) $k(\mu_t) = 1 + b\mu_t^p$, for constants $b > 0$ and $p > 0$;
(b) $k(\mu_t) = \mu_t^p$ for $p > 0$.

The latter has been much used in practice, as in Stevens (1974), Harrison and Stevens (1971, 1976b), Smith and West (1983), West, Harrison and Pole (1987), for example. Recall from Section 10.6.2 that a power-law variance function of the form in (b) corresponds roughly to a power transformation of Y_t to constant variance with power index $\lambda = 1 - p/2$. Note that generally, $k(.)$ may depend on t, although this is not explicitly considered here. Obviously, the model analysis would be lost if variances were allowed to depend on $\boldsymbol{\theta}_t$ in this way, the prior, posterior and forecast distributions no longer being analytically tractable. We stay within the ambit of standard theory, however, if the value of the variance law $k(.)$ is replaced by a known quantity, a variance multiplier. The obvious multiplier is obtained by replacing the conditional mean μ_t in the law by its prior mean, the forecast value f_t, thus using the model $\{\mathbf{F}_t,\ \mathbf{G}_t,\ k_tV,\ \mathbf{W}_tV\}$, where $k_t = k(f_t)$ is known. This simply, and appropriately, accounts for changing variance with *expected* level of the series. Note that in forecasting ahead from time t, the variance multipliers used for future observations will now depend upon the time t, being functions of the current (at time t) expected values of the future observations.

10.8 STOCHASTIC CHANGES IN VARIANCE

10.8.1 General considerations

The unknown variance scale parameter V appearing in the observational variance, whether weighted or not, has been assumed throughout to be constant over time. The updating for V based on gamma prior and posterior distributions for the precision parameter $\phi = 1/V$ provides a coherent, effective learning algorithm that eventually leads to convergence. To see this, recall that the posterior at time t is given by $(\phi \mid D_t) \sim \mathrm{G}[n_t/2, d_t/2]$, where the degrees of freedom parameter n_t updates by 1 for each observation. Thus, as $t \to \infty$, $n_t \to \infty$ and the posterior converges about the mode. With the usual point estimate $S_t = d_t/n_t$ of V, the posterior asymptotically degenerates with $|V - S_t| \to 0$ with probability one. Now, although this is an asymptotic result, the posterior can become quite precise rapidly as n_t increases, leading to under-adaptation to new data so far as learning on the variance is concerned. The problem lies with the assumption of constancy of V, conflicting somewhat with the underlying belief in change over time applied to the parameters in $\boldsymbol{\theta}_t$. Having explored above the possibility of changes in observational variance due to non-normality, we now consider the possibility that whether using a variance weight or not, the scale parameter V may vary stochastically and unpredictably over time. Some supporting arguments for such variation are as follows.

(1) There may actually be additional, stochastic elements affecting the observational error sequence that have not been modelled. Some possible sources of extra randomness are inaccuracies in the quoted timing of observations, truncation of readings, changes in data recording and handling procedures, etc. Some of these are present in the application discussed by Smith and West (1983), for example. Some such unexplained errors may be systematic, but if not identified, a general method of allowing for possible extra observational variation that may change in time is to simply suppose that V may change, albeit slowly and steadily.

(2) The variance V_t may change deterministically in a way not modelled, or modelled inappropriately. A variance function may be improperly specified, or omitted, the changes in variance thus not being adequately predicted. If such changes are not too dramatic, then they may be adequately estimated by allowing for stochastic drift in the scale parameter V.

(3) More generally, all features of model misspecification that will become apparent in the forecast error sequence can be attributed to changes in V. If V is allowed to vary stochastically, then the estimated trajectory over time of V will provide indications of times of improvement or deterioration in forecast performance; the estimation of V will tend to favour larger values in the latter case, for example. Thus a simple model for slow, purely random variance changes can be a useful diagnostic tool.

10.8.2 A model for change in V: discounted variance learning

Consider the general DLM $\{\mathbf{F}_t,\ \mathbf{G}_t,\ k_tV,\ \mathbf{W}_tV\}$, with known weights k_t^{-1}. Assuming the unknown scale V to have been constant up to time t, the usual analysis leads to the posterior

$$(\phi \mid D_{t-1}) \sim \mathrm{G}[n_{t-1}/2, d_{t-1}/2]$$

for precision $\phi = 1/V$. Suppose, however, that we now believe V to be subject to some random disturbance over the time interval $t-1$ to t. The simplest way of modelling steady, stochastic variation is via some form of random walk for V or some function of V. Make the time dependence explicit with time subscripts, so that the variance and precision at time t are V_t and $\phi_t = 1/V_t$, respectively. Then at $t-1$, suppose the precision ϕ_{t-1} has the usual posterior

$$(\phi_{t-1} \mid D_{t-1}) \sim \mathrm{G}[n_{t-1}/2, d_{t-1}/2]. \tag{10.2}$$

Proceeding to time t, it is desirable to retain the gamma form of distribution for the resulting distribution $p(\phi_t|D_{t-1})$, as it is conjugate to the likelihood function for updating based on the next observation Y_t. This desired constraint led to the development of a method of "variance discounting" to model a decay of information about the precision, and hence the variance, between time points, while retaining the gamma form of posterior and prior distributions. Explicitly, introduce a variance discount factor δ at time t, with $0 < \delta < 1$. Based on the time $t-1$ posterior (10.2), suppose that ϕ_t is derived from ϕ_{t-1} by some "random walk" model resulting in the time t prior distribution

$$(\phi_t \mid D_{t-1}) \sim \mathrm{G}[\delta n_{t-1}/2, \delta d_{t-1}/2]. \tag{10.3}$$

This has the same location as (10.2), i.e., $\mathrm{E}[\phi_t|D_{t-1}] = \mathrm{E}[\phi_{t-1}|D_{t-1}] = 1/S_{t-1}$, but increased dispersion through the discounting of the degrees of freedom parameter, i.e., $\delta n_{t-1} < n_{t-1}$. Hence the mapping from (10.2) to (10.3) neatly represents stochastic evolution in terms of loss of information, or increased dispersion, in a way completely analogous to the standard use of discount factors in DLMs. This idea was originally developed in Ameen and Harrison (1985) and applied in Harrison and West (1986, 1987).

Discount techniques are implemented in the BATS software in Pole, West and Harrison (1994).

A formal model underlies this use of variance discounting, deriving from special cases of results in Uhlig (1994). In particular, consider the following multiplicative model for generating ϕ_t from ϕ_{t-1}. Suppose γ_t to be a beta distributed random variable, independent of ϕ_{t-1}, with density

$$p(\gamma_t|D_{t-1}) \propto \gamma_t^{\delta n_{t-1}/2-1}(1-\gamma_t)^{(1-\delta)n_{t-1}/2-1},$$

for $0 < \gamma_t < 1$. This beta density, namely

$$(\gamma_t|D_{t-1}) \sim \text{Beta}[\delta n_{t-1}/2, (1-\delta)n_{t-1}/2],$$

is such that $\text{E}[\gamma_t|D_{t-1}] = \delta$. Given ϕ_{t-1}, set

$$\phi_t = \gamma_t \phi_{t-1}/\delta. \tag{10.4}$$

It is easily deduced that under the time $t-1$ prior (10.2), the resulting distribution of ϕ_t from (10.4) is the gamma distribution (10.3). This is true for any n_{t-1}, d_{t-1} and δ. Hence, the evolution model (10.4) may be introduced, applying over all time t, to formally model stochastic variation in the observation precision sequence. This shows that the variance discounting result arises from a stochastic evolution in which the ϕ_t sequence changes as a result of independent random "shocks" γ_t/δ. The maintenance of gamma prior and posterior distributions at each time enables continued, closed-form sequential updating, with the minor modification that the degrees of freedom quantity n_{t-1} is discounted between successive updates.

Proceeding to observe Y_t, the prior (10.3) updates to the posterior gamma via the usual updating equations, now including the variance discount factor. Thus

$$n_t = \delta n_{t-1} + 1 \quad \text{and} \quad d_t = \delta d_{t-1} + S_{t-1}e_t^2/Q_t, \tag{10.5}$$

in the usual notation. This analysis is summarised in Table 10.4 in a form analogous to the summary in Section 4.6, providing the simple generalisation of the updating and one-step ahead forecasting equations to include discounting of the variance learning procedure. We note that practically suitable variance discount factors δ take values near unity, typically between 0.95 and 0.99. Clearly, $\delta = 1$ leads to the original, constant variance model, $V_t = V$ for all t. Extensions to time-varying discount factors, applying a factor δ_t at time t, are immediate.

10.8.3 Limiting behaviour of constant, discounted variance model

The asymptotic behaviour of the discounted variance learning procedure provides insight into the nature of the effect of the discounting. Note from the summary that

$$n_t = 1 + \delta n_{t-1} = \ldots = 1 + \delta + \delta^2 + \ldots + \delta^{t-1}n_1 + \delta^t n_0$$

Table 10.4. Summary of updating with variance discounting

Univariate DLM with variance discounting	
Observation:	$Y_t = \mathbf{F}_t'\boldsymbol{\theta}_t + \nu_t, \quad \nu_t \sim \mathrm{N}[0, k_t/\phi_t]$
System:	$\boldsymbol{\theta}_t = \mathbf{G}_t\boldsymbol{\theta}_{t-1} + \boldsymbol{\omega}_t, \quad \boldsymbol{\omega}_t \sim \mathrm{T}_{n_{t-1}}[\mathbf{0}, \mathbf{W}_t]$
Precision:	$\phi_t = \gamma_t\phi_{t-1}/\delta$ with $\gamma_t \sim \mathrm{Beta}[\delta n_{t-1}/2, (1-\delta)n_{t-1}/2]$
Information:	$(\boldsymbol{\theta}_{t-1} \mid D_{t-1}) \sim \mathrm{T}_{n_{t-1}}[\mathbf{m}_{t-1}, \mathbf{C}_{t-1}]$ $(\boldsymbol{\theta}_t \mid D_{t-1}) \sim \mathrm{T}_{n_{t-1}}[\mathbf{a}_t, \mathbf{R}_t]$ with $\mathbf{a}_t = \mathbf{G}_t\mathbf{m}_{t-1}, \ \mathbf{R}_t = \mathbf{G}_t\mathbf{C}_{t-1}\mathbf{G}_t' + \mathbf{W}_t,$ $(\phi_{t-1} \mid D_{t-1}) \sim \mathrm{G}[n_{t-1}/2, d_{t-1}/2]$ $(\phi_t \mid D_{t-1}) \sim \mathrm{G}[\delta n_{t-1}/2, \delta d_{t-1}/2]$ with $S_{t-1} = d_{t-1}/n_{t-1}$
Forecast:	$(Y_t \mid D_{t-1}) \sim \mathrm{T}_{\delta n_{t-1}}[f_t, Q_t]$ with $f_t = \mathbf{F}_t'\mathbf{a}_t, Q_t = \mathbf{F}_t'\mathbf{R}_t\mathbf{F}_t + S_{t-1}$
Updating Recurrence Relationships	
	$(\boldsymbol{\theta}_t \mid D_t) \sim \mathrm{T}_{n_t}[\mathbf{m}_t, \mathbf{C}_t]$ $(\phi_t \mid D_t) \sim \mathrm{G}[n_t/2, d_t/2]$ with $\mathbf{m}_t = \mathbf{a}_t + \mathbf{A}_t e_t,$ $\mathbf{C}_t = (S_t/S_{t-1})[\mathbf{R}_t - \mathbf{A}_t\mathbf{A}_t'Q_t],$ $n_t = \delta n_{t-1} + 1, \ d_t = \delta d_{t-1} + S_{t-1}e_t^2/Q_t$ $S_t = d_t/n_t,$ where $e_t = Y_t - f_t$ and $\mathbf{A}_t = \mathbf{R}_t\mathbf{F}_t/Q_t$.

and

$$\begin{aligned} d_t =&(S_{t-1}e_t^2/Q_t) + \delta d_{t-1} = \dots \\ =&(S_{t-1}e_t^2/Q_t) + \delta(S_{t-2}e_{t-1}^2/Q_{t-1}) + \dots + \delta^{t-1}(S_0 e_1^2/Q_1) + \delta^t d_0. \end{aligned}$$

As $t \to \infty$, with $0 < \delta < 1$, then

$$n_t \to (1-\delta)^{-1}.$$

Also $S_t = d_t/n_t$, so that for large t,

$$S_t \approx (1-\delta)\sum_{r=0}^{t-1} \delta^r (e_{t-r}^2/Q_{t-r}^*),$$

with $Q_t^* = Q_t/S_{t-1}$ for all t, the scale-free one-step forecast variance at time t. More formally,

$$\lim_{t\to\infty}\{S_t - (1-\delta)\sum_{r=0}^{t-1}\delta^r(e_{t-r}^2/Q_{t-r}^*)\} = 0$$

with probability one. Thus n_t converges to the constant, limiting degrees of freedom $(1-\delta)^{-1}$. The value $\delta = 0.95$ implies a limit of 20, $\delta = 0.98$ a limit of 50. The usual, static variance model has $\delta = 1$, so that of course, $n_t \to \infty$. Otherwise, the fact that the changes in variance are to be expected implies a limit to the accuracy with which the variance at any time is estimated, this being defined by the limiting degrees of freedom, directly via the discount factor. The point estimate S_t has the limiting form of an exponentially weighted moving average of the standardised forecast errors; e_{t-r}^2/Q_{t-r}^* asymptotically receives weight $(1-\delta)\delta^r$, the weights decaying with r and summing to unity. Thus, the estimate continues to adapt to new data, whilst further discounting old data, as time progresses.

Note finally that if it is desired to allow for greater variation in ϕ_t at a given time in response to external information about possible marked changes, a smaller intervention value of the discount factor can replace δ for just that time point, so implying a resulting increase in uncertainty about γ_t.

10.8.4 Filtering with discounted variance

We now turn to extension of the filtering analyses of Sections 4.7 and 4.8, to include the modification to time-varying observational precision parameters. In the model with (10.4), the problem is that of updating the on-line posteriors defined by (10.2) in the light of data received afterwards. This is of considerable practical interest since it is the revised, filtered distributions that provide the retrospective analysis, identifying periods of stability and points of marked change in the observational variance sequence.

At time t with information set D_t, consider the precision ϕ_{t-k}, for integers $(1 \le k \le t-1)$. We are interested in evaluating features of the distribution $p(\phi_{t-k} \mid D_t)$, and do so recursively as follows. Under the Markov evolution (10.4), note that

$$p(\phi_{t-k} \mid \phi_{t-k+1}, D_t) \propto p(\phi_{t-k} \mid D_{t-k})p(\phi_{t-k+1} \mid \phi_{t-k}, D_{t-k}).$$

The first term on the left-hand side here is simply the gamma posterior (10.2) with $t-1$ replaced by $t-k$. The second term is derived from the beta evolution (10.4). It follows, after some calculus left to the reader, that this has the form

$$p(\phi_{t-k} \mid \phi_{t-k+1}, D_t) \propto (\phi_{t-k} - \delta\phi_{t-k+1})^{(1-\delta)n_{t-k}/2-1}\exp(-d_{t-k}\phi_{t-k}/2),$$

for $\phi_{t-k} > \delta\phi_{t-k+1}$. We can represent this density via

$$\phi_{t-k} = \eta_{t-k} + \delta\phi_{t-k+1},$$

where $\eta_{t-k} \sim \mathrm{G}[(1-\delta)n_{t-k}/2, d_{t-k}/2]$. One implication is that

$$\mathrm{E}[\phi_{t-k}|\phi_{t-k+1}, D_t] = (1-\delta)S_{t-k}^{-1} + \delta\phi_{t-k+1}.$$

This implies, in particular, that

$$\mathrm{E}[\phi_{t-k}|D_t] = (1-\delta)S_{t-k}^{-1} + \delta\mathrm{E}[\phi_{t-k+1}|D_t],$$

or

$$S_t(-k)^{-1} = (1-\delta)S_{t-k}^{-1} + \delta S_t(-k+1)^{-1},$$

where $S_t(-k)$ is the filtered posterior estimate of V_t, just the harmonic mean of $(V_{t-k}|D_t)$ for each $k < t$, initialised at $k = 0$ with $S_t(0) = S_t$.

This very neat and elegant result shows that the filtered estimate of precision at time $t-k$ is obtained by averaging the posterior estimate from that time, S_{t-k}^{-1}, with the filtered estimate at time $t-k+1$; the relative weights, defined in terms of δ, are of a form familiar in exponential smoothing. The standard results for constant precision are recovered by setting $\delta = 1$. The obvious extension to time-dependent discount factors is left to the reader.

It should be clear, however, that the filtered distributions are no longer gamma; the closed form sequential updating of gamma distributions does not carry over to filtering, unfortunately. Arguing by analogy, however, appropriate gamma approximations to the filtered distributions have the form

$$p(\phi_{t-k}|D_t) \approx \mathrm{G}[n_t(-k)/2, d_t(-k)/2],$$

where $d_t(-k) = n_t(-k)S_t(-k)$, with filtered degrees of freedom defined by

$$n_t(-k) = (1-\delta)n_{t-k} + \delta n_{t-k+1},$$

initialised at $n_t(0) = n_t$. With this approximation assumed, the filtering algorithms for the ϕ_t sequence are completed. Note that the filtering algorithms for the state vector $\boldsymbol{\theta}_t$, in Corollaries 4.3 and 4.4, must be modified to incorporate this; specifically, the values n_t and S_t appearing there for the original case of static observational variance are replaced by $n_t(-k)$ and $S_t(-k)$ throughout. Note finally that extensions to time-dependent variance discount factors are immediate; details are left to the reader.

10.9 EXERCISES

(1) Reanalyse the agricultural sales series of Section 10.2 using the same linear trend/seasonal model. Extend the discussion of Example 10.1 to a fuller exploration of sensitivity of the analysis to variation in the dynamic controlled through the discount factors. Do this as follows.

(a) For any values $\boldsymbol{\delta} = (\delta_T, \delta_S)'$ of the trend and seasonal discount factors, explicitly recognise the dependence of the analysis on $\boldsymbol{\delta}$ by including it in the conditioning of all distributions. Then the aggregate predictive density at time t is $p(Y_t, \ldots, Y_1|\boldsymbol{\delta}, D_0)$, calculated sequentially as in Section 10.2.5. As a function of $\boldsymbol{\delta}$, this defines the likelihood function for $\boldsymbol{\delta}$ from the data. Calculate this quantity at $t = 48$ for each pair of values of discount factors with $\delta_T = 0.8, 0.85, \ldots, 1$ and $\delta_S = 0.9, 0.925, \ldots, 1$. What values of $\boldsymbol{\delta}$ are supported by the data? In particular, assess the support for the static model defined by $\boldsymbol{\delta} = (1, 1)'$ relative to other dynamic models fitted.

(b) Explore sensitivity of inferences to variation in $\boldsymbol{\delta}$ amongst values supported by the data from (a). In particular, do the estimated time trajectories of trend and seasonal components change significantly as $\boldsymbol{\delta}$ varies? What about step ahead forecasts over one or two years from the end of the data series?

(c) Can you suggest how you might combine forecasts (and other inferences) made at time t from two (or more) models with different values of $\boldsymbol{\delta}$?

(d) Compare models with different values of $\boldsymbol{\delta}$ using MSE and MAD measures of predictive performance as alternatives to the above model likelihood measures. Do the measures agree as to the relative support for different values of $\boldsymbol{\delta}$? If not, describe how and why they differ.

(2) Consider the prior specification for the seasonal factors ϕ_1 of the model for the Sales/Index series in Section 10.3.2, given initially by

$$(\phi_1|D_0) \sim \mathrm{T}_6[\mathbf{m}^*_{\phi_1}, \mathbf{C}^*_{\phi_1}],$$

with $\mathbf{m}^*_{\phi_1}$ given by

$$(0.8,\ 1.3,\ 1,\ 1.1,\ 1.2,\ 0,\ -1.2,\ -1,\ -1,\ -1.5,\ -0.5,\ 0)'$$

and $\mathbf{C}^*_{\phi_1} = (0.2)^2\mathbf{I}$.

(a) Verify that this initial prior does not satisfy the zero-sum constraint on seasonal factors. Verify also that Theorem 8.2 leads to the constrained prior as given in Section 10.3.2, namely

$$(\phi_1|D_0) \sim \mathrm{T}_6[\mathbf{m}_{\phi_1}, \mathbf{C}_{\phi_1}],$$

with $\mathbf{m}_{\phi_1}$ given by

$$(0.783,\ 1.283,\ 0.983,\ 1.083,\ 1.183,\ -0.017,$$
$$-1.217,\ -1.017,\ -1.017,\ -1.517,\ -0.517,\ -0.017)'$$

and

$$\mathbf{C}_{\phi_1} = (0.19)^2 \begin{pmatrix} 1 & -0.09 & -0.09 & \dots & -0.09 \\ -0.09 & 1 & -0.09 & \dots & -0.09 \\ -0.09 & -0.09 & 1 & \dots & -0.09 \\ \vdots & \vdots & \vdots & \ddots & \vdots \\ -0.09 & -0.09 & -0.09 & \dots & 1 \end{pmatrix}.$$

(b) Let $\mathbf{H}$ be the transformation matrix of Section 8.6.5 used to transform from seasonal effects to Fourier coefficients. Calculate $\mathbf{H}$ and verify the stated results of Section 10.3.3, Table 10.3, by calculating the moments $\mathbf{H}\mathbf{m}_{\phi_1}$ and $\mathbf{H}\mathbf{C}_{\phi_1}\mathbf{H}'$.

(3) Reanalyse the Sales/Index data of Section 10.3 using the same model. Instead of initialising the analysis with the prior used in that section, perform a reference analysis as described in Section 4.8 (beginning with $\mathbf{W}_t = \mathbf{0}$, corresponding to unit discount factors, over the reference period of the analysis). Describe the main differences between this analysis and that discussed in Section 10.3.

(4) Consider the transformation $Y = g(Z)$ of Section 10.6.2 with $\mathrm{E}[Z] = \mu$ and $\mathrm{V}[Z] = V$. By expanding $g(Z)$ in a Taylor series about $Z = \mu$ and ignoring all terms after the quadratic in $Z - \mu$, verify the approximate values for the mean and variance of Y given by $\mathrm{E}[Y] \approx g(\mu) + 0.5g''(\mu)V$ and $\mathrm{V}[Y] \approx \{g'(\mu)\}^2 V$.

(5) Suppose that Y_t is the sum of an uncertain number k_t of quantities Y_{tj}, $(j = 1, \dots, k_t)$, where the basic quantities are uncorrelated, with common mean β_t and common, known variance V_t. This arises when items are produced and sold in batches of varying sizes.

(a) Show that $\mathrm{E}[Y_t|k_t, \beta_t] = \mu_t$, where $\mu_t = k_t\beta_t$ and $\mathrm{V}[Y_t|k_t, \beta_t] = k_t V_t$.

(b) Suppose that a forecaster has a prior distribution for the batch size k_t with finite mean and variance, and that k_t is viewed as independent of β_t. Show that unconditional on k_t, $\mathrm{E}[Y_t|\beta_t] = \beta_t \mathrm{E}[k_t]$ and $\mathrm{V}[Y_t|\beta_t] = V_t\mathrm{E}[k_t] + \beta_t^2\mathrm{V}[k_t]$.

(c) In the special case of Poisson batch or lot sizes, $\mathrm{E}[k_t] = \mathrm{V}[k_t] = a$, for some $a > 0$. Letting $\mu_t = \mathrm{E}[Y_t|\beta_t]$ as usual, show that Y_t has a quadratic variance function $\mathrm{V}[Y_t|\beta_t] = V_t^*[1 + b_t\mu_t^2]$, where $V_t^* = aV_t$ and $b_t^{-1} = a^2 V_t$.

(6) Reanalyse the agricultural sales data of Section 10.2. Use the same model but now allow for changes in observational variation about the level through a variance discount factor δ as in Section 10.8. (Note that the reference analysis can be performed as usual assum-

ing that all discount factors, including that for the variance, are unity initially over the reference part of the analysis).

(a) Perform an analysis with $\delta = 0.95$. Verify that relative to the static variance model, variance discounting affects only the uncertainties in forecasting and inferences about time trajectories of model components.

(b) Analyse the data for several values of δ over the range $\delta = 0.90, 0.92, \ldots, 1$, and assess relative support from the data using the likelihood function over δ provided by the aggregate predictive densities from each model.

(7) Consider the two series in the table below. The first series, Sales, is a record of the quarterly retail sales of a confectionary product over a period of eleven years. The product cost has a major influence on Sales, and the second series, Cost, is a compound, quarterly index of cost to the consumer constructed in an attempt to explain nonseasonal changes in Sales.

	SALES				COST			
	Quarter				Quarter			
Year	1	2	3	4	1	2	3	4
1975	157	227	240	191	10.6	8.5	6.7	4.1
1976	157	232	254	198	1.9	0.4	1.1	1.9
1977	169	234	241	167	0.2	2.9	4.1	−1.4
1978	163	227	252	185	−4.0	−4.5	−5.3	−8.4
1979	179	261	264	196	−12.8	−13.2	−10.1	−4.6
1980	179	248	256	193	−1.1	−0.1	0.0	−2.5
1981	186	260	270	210	−5.1	−6.4	−8.0	−6.5
1982	171	227	241	170	−3.7	−1.3	6.1	16.5
1983	140	218	208	193	22.9	23.9	18.0	8.3
1984	184	235	245	209	2.9	0.7	−2.4	−7.0
1985	206	260	264	227	−9.8	−10.6	−12.3	−13.2

Consider fitting DLMs with first-order polynomial, regression on Cost and full seasonal components. These data are analysed using a similar model in West, Harrison and Pole (1987b), with extensive summary information from the analysis provided there. Based on previous years of data, the forecaster assesses initial prior information as follows.

- The initial, underlying level of the series when Cost is zero is expected to be about 220, with a nominal standard error of 15.
- The regression coefficient of Cost is estimated as −1.5 with a standard error of about 0.7.

- The seasonal factors for the four quarters of the first year are expected to be −50(25), 25(15), 50(25) and −25(15), the nominal standard errors being given in parentheses.
- The trend, regression and seasonal components are initially assumed to be uncorrelated.
- The observational variance is estimated as 100, with initial degrees of freedom of 12.

Analyse the series along the following lines.

(a) Using the information provided for the seasonal factors above, apply Theorem 8.2 to derive the appropriate intial prior that satisfies the zero-sum constraint.

(b) Identify the full initial prior quantities for the 6-dimensional state vector $\boldsymbol{\theta}_1$ and the observational variance V. Write down the defining quantities $\mathbf{a}_1$, $\mathbf{R}_1$, n_0 and S_0 based on the above initial information.

(c) Use three discount factors to structure the evolution variance matrices of the model: δ_T for the constant intercept term, δ_R for the regression coefficient, and δ_S for the seasonal factors. Following West, Harrison and Pole (1987b), consider initially the values $\delta_T = \delta_S = 0.9$ and $\delta_R = 0.98$. Fit the model and perform the retrospective, filtering calculations to obtain filtered estimates of the state vector and all model components over time.

(d) Based on this analysis, verify the findings in the above reference to the effect that the regression parameter on Cost is, in retrospect, rather stable over time.

(e) Produce step ahead forecasts from the end of the data in the fourth quarter of 1985 for the next three years. The estimated values of Cost to be used in forecasting ahead are given by

Year	Quarter			
	1	2	3	4
1986	8.4	10.6	7.2	13.0
1987	−2.9	−0.7	−6.4	−7.0
1988	−14.9	−15.9	−18.0	−22.3

(8) Refer to the models and concepts underlying the variance discounting method in Section 10.8.

(a) Derive the distribution (10.3) from the gamma posterior (10.2) combined with the multiplicative evolution model (10.4).

(b) Verify the result stated for $p(\phi_{t-k} \mid \phi_{t-k+1}, D_t)$ in deriving filtering recurrences in Section 10.8.4, and deduce the formula for $\mathrm{E}[\phi_{t-k} \mid D_t]$.

CHAPTER 11

INTERVENTION AND MONITORING

11.1 INTRODUCTION

In Section 2.3.2 we introduced simple intervention ideas and considered in detail intervention into a first-order polynomial model. That intervention provided feed-forward information anticipating a major change in the level of a time series, modelled by altering the prior distribution for the level parameter to accommodate the change. Had the model used in that example been closed to intervention, then the subsequent huge change in the level of the series would have been neither forecast nor adequately estimated afterwards, the match between model and data breaking down entirely. In practice, all models are only components of forecasting systems that include the forecasters as integral components. Interactions between forecasters and models are necessary to adequately allow for events and changes that go beyond the existing model form. This is evident also in the illustrations of standard, closed models in Chapter 10, where deterioration in forecasting performance, though small, is apparent. In this chapter, we move closer to illustrating forecasting systems rather than simply models, considering ways in which routine interventions can be incorporated into existing DLMs, and examples of why and when such interventions may be necessary to sustain predictive performance. The mode of intervention used in the example of Section 2.3.2 was simply to represent departures from an existing model in terms of major changes in the parameters of the model. This is the most widely used and appropriate method, although others, such as extending the model to include new parameters, are also important. An example data set highlights the need for intervention.

EXAMPLE 11.1. The data set in Table 11.1 perfectly illustrates many of the points to be raised in connection with intervention. This real data series, referred to as CP6, provides monthly total sales, in monetary terms on a standard scale, of tobacco and related products marketed by a major company in the UK. The time of the data runs from January 1955 to December 1959 inclusive. The series is graphed over time in Figure 11.1. Initially, during 1955, the market clearly grows at a fast but steady rate, jumps markedly in December, then falls back to pre-December levels and flattens off for 1956. There is a major jump in the sales level in early 1957, and another in early 1958. Throughout the final two years, 1958 and 1959, there is a steady decline back to late 1957 levels. An immediate reaction to this series, that is not atypical of real series in consumer markets, is that it is impossible to forecast with a simple time series model. This is correct. However, the sketch in Figure 11.2 suggests that in fact a simple model may be appropriate for short-term forecasting (up to twelve months

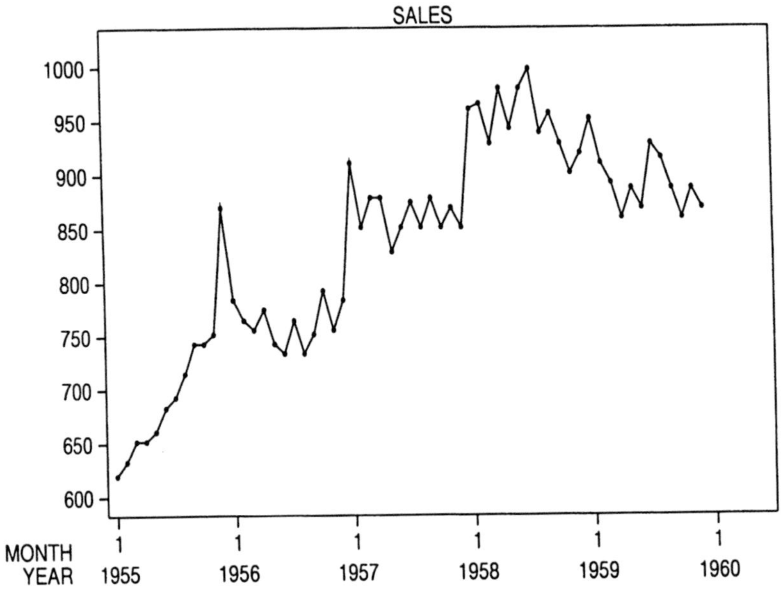

Figure 11.1 CP6 sales series

Table 11.1. CP6 Sales data

Year	Month 1	2	3	4	5	6	7	8	9	10	11	12
1955	620	633	652	652	661	683	693	715	743	743	752	870
1956	784	765	756	775	743	734	765	734	752	793	756	784
1957	911	852	879	879	829	852	875	852	879	852	870	852
1958	961	966	929	980	943	980	998	939	957	929	902	920
1959	952	911	893	861	888	870	929	916	888	861	888	870

ahead, say), *if* that model is open to interventions to explain (and possibly anticipate) some of the discontinuities the series exhibits. From this figure, it is apparent that a simple possible description of the trend in the series is as a sequence of roughly linear segments (i.e., *piecewise linear*), with major changes in the quantified linear form at three places. Thus, a second-order polynomial model, providing a linear predictor in the short term, may be used so long as the abrupt changes in parameter values are allowed for. Other features to note are the possible *outliers* at two points in the series. These observations deviate markedly from the general pattern and should be considered individually, perhaps being omitted from the analysis.

This is not to say that such extreme observations should be ignored, of course, since they may be critical commercially. For updating the model

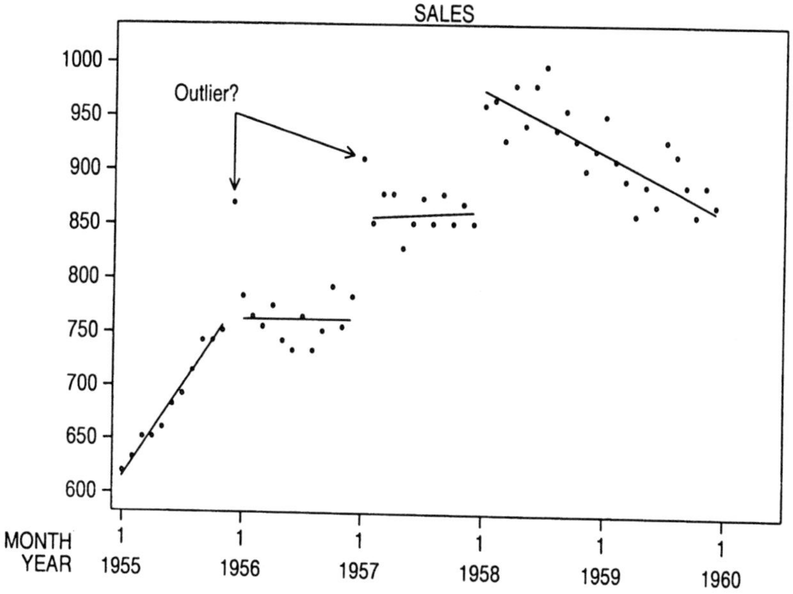

Figure 11.2 Heuristic model for CP6 sales

however, they are ignored since they convey little or no information relevant to forecasting the underlying trend within the model structure. Finally, note that the random scatter in Sales about the sketched trend is apparently higher at higher levels of the series, suggesting that intervention, or some other appropriate device such as a variance law as in Section 10.7, be required to adjust the observational variance to model greater randomness at such levels.

The sources of information available to a forecaster are not always restricted to historical data and other, related information, but also include information about forthcoming events affecting the environment of the series. In cases when it is perceived that these events may materially affect the development of the series, then action is taken to intervene into the existing model to feed-forward information, allowing the model to anticipate and predict change. For example, a marketing department of a company may know that an export licence for their products has just been granted, that a new type of "special offer" campaign is to be launched, that a patent expires or that unexpectedly a competitive product has been banned. In each case, the likely effects on demand, sales and inventory must be assessed in order to anticipate changes and plan accordingly. The decisions resting on forecasts in such cases may be of major importance to the company. Some of the events in the CP6 series may have been anticipated at the time in this way. In forming views of the likely outcomes, the forecasters may have prior experience of related events to base their judgements upon, but clearly, such views are usually largely subjective. This does, of course, raise questions about possible personal biases and errors in intervention, and thus such facilities require caution in use and should always be subject to calibration and retrospective assessment.

Interventions can be roughly classified as either *feed-forward* or *feed-back*. The former is anticipatory in nature, as in the above examples. The latter is corrective, responding to events that had not been foreseen or adequately allowed for. Corrective actions often arise when it is seen that forecasting performance has deteriorated, thus prompting the forecaster to examine the environment of the series for explanation. In such instances, any such information, that should ideally have been available beforehand, must be used retrospectively in an attempt to adjust the model appropriately to the current, local conditions. It is evident that a complete forecasting system should be open to both feed-forward and feed-back interventions. In ideal circumstances, forecasting systems should operate according to a principle of *Management by Exception*. That is, based upon experience and analysis, a formal statistical model is adopted and is used routinely to process data and information, providing forecasts and inferences that are used unless exceptional circumstances arise. The exceptions occur in the two ways introduced above; the first relating to the receipt of information providing the basis for feed-forward interventions, the second provided by the detection of deterioration in forecasting performance, usually detected by some form of forecast monitoring activity. Such monitoring may be largely subjective and informal, involving the forecaster in considering the forecast performance from a subjective standpoint, or may take the form of an automatic, statistical error analysis or control scheme that continually monitors the match between model and data and issues signals of breakdown as necessary. Often it is the case that exceptions matter most in forecasting. When circumstances change markedly there are major opportunities for both losses and gains. Thus, the more vigilant and successful the forecaster is in anticipating major events, the more effective the decisions. Also, concerning feed-back interventions, the more perceptive and immediate the diagnosis is, the better. Hence it is to be recommended that there should always be a user response to exceptions identified by a monitoring system. However, in some circumstances there may be no identifiable reason for the exceptions, but the very fact that something has occurred raises uncertainty about the future. Then the forecaster's response will not necessarily identify possible sources of model breakdown, but it must communicate the problem to the model by increasing total uncertainty in the model, making it more adaptive to new data, so that changes that may have taken place are rapidly identified and estimated.

Many of these ideas are made more concrete in this chapter, where an existing model is assumed to be operating subject to both forms of intervention. Much of the development is based on West and Harrison (1989).

We consider throughout a single intervention at time t into the existing model; clearly the concepts and theory apply generally to analyses subject to possibly many interventions. In addition, although we consider interventions at time t that represent changes to the model at time t, it should be clear that feed-forward interventions can also be made for times after t, although this is not specifically described.

For clarity, the development is in terms of a model with observational variance sequence known, so that all prior, posterior and predictive distributions are normal. The results apply, of course and with no essential difference, to the case of variance learning when all such distributions are T rather than normal. We comment on minor differences between the cases as necessary.

Thus, for reference, we have the model at the current time t given, as usual, by

$$Y_t = \mathbf{F}_t'\boldsymbol{\theta}_t + \nu_t, \qquad \nu_t \sim \mathrm{N}[0, V_t], \tag{11.1}$$

$$\boldsymbol{\theta}_t = \mathbf{G}_t\boldsymbol{\theta}_{t-1} + \boldsymbol{\omega}_t, \qquad \boldsymbol{\omega}_t \sim \mathrm{N}[\,\mathbf{0}, \mathbf{W}_t], \tag{11.2}$$

with the usual assumptions of Sections 4.2 and 4.3. The historical information D_{t-1} (including past data and any previous interventions) is summarised in terms of the posterior for $\boldsymbol{\theta}_{t-1}$, namely

$$(\boldsymbol{\theta}_{t-1} \mid D_{t-1}) \sim \mathrm{N}[\mathbf{m}_{t-1},\ \mathbf{C}_{t-1}], \tag{11.3}$$

where the mean vector and variance matrix are known. Thus, the prior for the state vector at the current time is, via the evolution equation (11.2) as usual,

$$(\boldsymbol{\theta}_t \mid D_{t-1}) \sim \mathrm{N}[\mathbf{a}_t,\ \mathbf{R}_t], \tag{11.4}$$

with

$$\mathbf{a}_t = \mathbf{G}_t\mathbf{m}_{t-1} \qquad \text{and} \qquad \mathbf{R}_t = \mathbf{G}_t\mathbf{C}_{t-1}\mathbf{G}_t' + \mathbf{W}_t.$$

Suppose now that feed-forward intervention is to be made at the current time into the model (11.1-11.4). This may be a response to additional information that can be viewed as extra observational data; a response to new information about related variables, such as competitor activity in a consumer market, legislation changes, and so forth; forecasts from other individuals or models; control actions, such as changes in advertising campaigns, pricing policy, etc.; or it may simply reflect a dissatisfaction with the current prior and the forecasts of the future that it implies, thus being a purely subjective intervention by the forecaster. Whatever the case, we represent the intervention at time t by I_t. This is an information set that identifies time t as a point of intervention and includes all the information used to effect the intervention, such information being made specific below in the various modes of intervention we consider. Note now that following the intervention, the available information set prior to observing Y_t is $\{I_t, D_{t-1}\}$, rather than just D_{t-1}.

The following section details the basic modes of intervention into the existing model at time t and the theory necessary to combine interventions of essentially any desired form into the existing model, allowing the intervention effects to be used both in forecasting from the current time onwards and in retrospective time series analysis.

11.2 MODES OF FEED-FORWARD INTERVENTION

11.2.1 Ignoring observation Y_t

The first mode of intervention involves simply treating Y_t as an outlier. Examples include the effects of strikes on sales and inventory levels, pricing changes that lead to forward purchasing of goods, and other interventions in the environment of the time series that may lead to a single observation being quite discrepant and essentially unrelated to the rest of the series. In such cases, although the observation is of critical importance to the company, perhaps it should not be used in updating the model for forecasting the future since it provides no relevant information.

EXAMPLE 11.1 (continued). In the CP6 series, such an event is (retrospectively) apparent at December 1955, where Sales leaps upwards by about 15% for just that month. This is the market response, in terms of immediate purchasing of stocks, to a company announcement of a forthcoming price rise. A second possible outlier is the very high value in January 1957 that presages a change in the overall level of Sales, but to a lower level than the single point in January.

When such information exists, so that a discrepant observation is anticipated, one possible, fail-safe reaction is to just omit the observation from the analysis, treating it as if it were a true missing or unrecorded value. Y_t is uninformative about the future and so should be given no weight in updating the model distributions. Thus,

$$I_t = \{ \ Y_t \text{ is missing } \},$$

so that $D_t = \{I_t, \ D_{t-1}\}$ is effectively equal to D_{t-1} alone. The posterior for the state vector at time t is just the prior, with $(\boldsymbol{\theta}_t \mid D_t) \sim \mathrm{N}[\mathbf{m}_t, \ \mathbf{C}_t]$, where

$$\mathbf{m}_t = \mathbf{a}_t \qquad \text{and} \qquad \mathbf{C}_t = \mathbf{R}_t.$$

Formally, this can be modelled in the DLM format in a variety of ways, the simplest, and most appropriate, is just to view the observation as having a very large variance V_t; formally, let V_t tend to infinity, or V_t^{-1} tend to 0, in the model equations, so that the observation provides no information for $\boldsymbol{\theta}_t$ (nor for the scale parameter in the case of variance learning). In the updating equations, the one-step ahead forecast variance Q_t tends to infinity with V_t, and it follows easily that the posterior for $\boldsymbol{\theta}_t$ is just the prior, as required. Formally, then,

$$I_t = \{ V_t^{-1} = 0 \}$$

in this case.

A modification of this mode of intervention is often desirable, stemming from the uncertainties associated with the sorts of events that suggest Y_t be ignored. Immediately following such events, it may be that the series will develop in a rather different way than currently forecast, exhibiting delayed effects of the events. Thus, for example, a huge increase in sales at time t that represents forward buying before a previously announced increase in price at $t+1$ can be expected to be followed by a drop-off in sales at time $t+1$, and possibly later times, after the price increase takes place. In order to adapt to the changing pattern after the omitted observation, an additional intervention may be desired to increase uncertainty about components of $\boldsymbol{\theta}_t$. In CP6, for example, the possible outlier in January 1957 is followed by Sales at a higher level than during 1956. This calls for intervention of the second type, considered in the following section.

11.2.2 Additional evolution noise

A common response to changes in conditions potentially affecting the development of the series is increased uncertainty about the future, reflected by increased uncertainties about some or all of the existing model parameters. The withdrawal of a major competing product in a consumer market may be anticipated by feeding forward an estimated increase in the level of sales, but it is surely the case that the uncertainty about the new level will be greater than about the current level, possibly much greater. In other instances, the response to an exception identified by a forecast monitoring system may be simply to increase prior variances of some or all components of the model. This is an appropriate reflection of the view that although *something* has changed, it is difficult to attribute the change to a particular component. As a catch-all measure, the entire variance matrix $\mathbf{R}_t$ may be altered to reflect increased uncertainty about all parameters without a change in the prior mean $\mathbf{a}_t$ that would anticipate the direction of change. This does, of course, lead to a loss of information on the entire state vector and is therefore an omnibus technique to be used in cases of complete neutrality as to sources of change. Otherwise it is well to be selective, increasing uncertainties only on those components that are viewed as potentially subject to major change. In the above example, the change in the sales market as it expands to meet the increased demand will evidence itself in a marked increase in sales level, and also possibly in the observational variance about the level, but is unlikely, for example, to seriously impact upon the phases of components of seasonal patterns in sales.

Generally, the model is open to interventions on components of $\boldsymbol{\theta}_t$ that increase uncertainties in this way, simply by adding in further evolution noise terms paralleling those in equation (11.2). As in the sales example, it is common that such interventions include a shift in mean in addition to an inflation in uncertainty. All such interventions can be formally represented in DLM form by extending the model to include a second evolution of the state vector in addition to that in the routine model in (11.2).

Generally, suppose the intervention information to be given by

$$I_t = \{\mathbf{h}_t,\ \mathbf{H}_t\},$$

where $\mathbf{h}_t$ is the mean vector and $\mathbf{H}_t$ the covariance matrix of a random quantity $\boldsymbol{\xi}_t$, with

$$\boldsymbol{\xi}_t \sim \mathrm{N}[\mathbf{h}_t,\ \mathbf{H}_t].$$

Suppose also that $\boldsymbol{\xi}_t$ is uncorrelated with $(\boldsymbol{\theta}_{t-1} \mid D_{t-1})$ and with $\boldsymbol{\omega}_t$, so that it is also uncorrelated with $(\boldsymbol{\theta}_t \mid D_{t-1})$. The intervention is effected by adding the additional noise term $\boldsymbol{\xi}_t$ to $\boldsymbol{\theta}_t$ after (11.2); equivalently, the post-intervention prior distribution is defined via the extended evolution equation

$$\boldsymbol{\theta}_t = \mathbf{G}_t\boldsymbol{\theta}_{t-1} + \boldsymbol{\omega}_t + \boldsymbol{\xi}_t, \tag{11.5}$$

replacing (11.2). Thus,

$$(\boldsymbol{\theta}_t \mid I_t,\ D_{t-1}) \sim \mathrm{N}[\mathbf{a}_t^*,\ \mathbf{R}_t^*],$$

where

$$\mathbf{a}_t^* = \mathbf{a}_t + \mathbf{h}_t \qquad \text{and} \qquad \mathbf{R}_t^* = \mathbf{R}_t + \mathbf{H}_t.$$

This allows for arbitrary shifts in the prior mean vector to the revised value $\mathbf{a}_t^*$. Some elements of $\mathbf{h}_t$ may be zero, not anticipating the direction of changes in the corresponding parameters. In practice, these mean changes may be assigned by directly choosing the $\mathbf{h}_t$ vector, the expected increment in $\boldsymbol{\theta}_t$ due to intervention. Alternatively, the adjusted mean $\mathbf{a}_t^*$ can be specified directly and then $\mathbf{h}_t$ deduced as $\mathbf{h}_t = \mathbf{a}_t^* - \mathbf{a}_t$.

This mode of intervention allows for many possible and practically useful increases in variance through $\mathbf{H}_t$. Note that some of the variances on the diagonal of $\mathbf{H}_t$ (and the corresponding covariance terms off-diagonal) may be zero, with the result that the corresponding elements of $\boldsymbol{\theta}_t$ are not subject to change due to intervention. This allows the forecaster the flexibility to protect some components of the model from intervention, when they are viewed as durable and unlikely to be subject to change. There will, of course, be some correlation due to the off-diagonal terms in $\mathbf{R}_t$. Again, it is sometimes the case that the intervention variance matrix $\mathbf{H}_t$ will be specified directly as the variance of the change in $\boldsymbol{\theta}_t$ due to intervention, the elements representing the uncertainty as to the nature and extent of the change forecast as $\mathbf{h}_t$. Although obviously very flexible, it is difficult in general to assign appropriate values here, although as a general rule it is desirable to err on the side of caution, with any change forecast via $\mathbf{h}_t$ being hedged with sufficient uncertainty that the model will be adaptive to future data, rapidly identifying and estimating changes. In line with the use of discount factors in structuring evolution variance matrices in standard models, it is appropriate to extend the discount concept to intervention. Thus, if $\mathbf{W}_t$ is structured using a standard set of discount factors, a matrix $\mathbf{H}_t$, with the same structure, models an extra change in $\boldsymbol{\theta}_t$ with the same correlation pattern in the evolution noise. The appropriateness of this is apparent when the intervention is incorporated into standard DLM form, as follows.

THEOREM 11.1. *Conditional on information $\{I_t,\ D_{t-1}\}$, the DLM (11.1-11.4) holds with the evolution equation (11.2) amended according to (11.5), written now as*

$$\boldsymbol{\theta}_t = \mathbf{G}_t\boldsymbol{\theta}_{t-1} + \boldsymbol{\omega}_t^*,$$

where $\boldsymbol{\omega}_t^* = \boldsymbol{\omega}_t + \boldsymbol{\xi}_t$ is distributed as

$$\boldsymbol{\omega}_t^* \sim \mathrm{N}[\mathbf{h}_t,\ \mathbf{W}_t^*],$$

with $\mathbf{W}_t^* = \mathbf{W}_t + \mathbf{H}_t$. In addition, $\boldsymbol{\omega}_t$ is independent of $\boldsymbol{\theta}_{t-1}$.

Proof. Obvious from (11.5) and left to the reader.

◇

This theorem confirms that the intervention can be written in the usual DLM form, with a generalisation to a possibly non-zero mean for the evolution noise vector.

This simple sort of structuring is particularly appropriate when $\mathbf{h}_t = \mathbf{0}$, when the addition of $\boldsymbol{\xi}_t$ simply increases uncertainty about $\boldsymbol{\theta}_t$. An automatic intervention technique described and illustrated in Section 11.5 below uses this approach, and more discussion appears there. Similar techniques are used in West and Harrison (1986a, 1989).

EXAMPLE 11.1 (continued). Consider the marked jump in level of CP6 Sales in January 1958. Suppose that this change was anticipated prior to occurrence, being the result of planned market expansion. In line with the previous sections, this change could be modelled through intervention, feeding forward prior information about the change and allowing the level parameter of any model to change. For concreteness, suppose a second-order polynomial model with level and growth parameters μ_t and β_t at time t corresponding to January 1958. Thus, $n = 2$ and $\boldsymbol{\theta}_t' = (\mu_t,\ \beta_t)$, $\mathbf{F}_t' = \mathbf{E}_2' = (1, 0)$ and

$$\mathbf{G}_t = \mathbf{G} = \mathbf{J}_2(1) = \begin{pmatrix} 1 & 1 \\ 0 & 1 \end{pmatrix}.$$

Suppose also that $V_t = 15$, $\mathbf{a}_t' = (865, 0)$, and

$$\mathbf{R}_t = \begin{pmatrix} 100 & 10 \\ 10 & 2 \end{pmatrix}.$$

Available information is such that a jump of roughly 100 is anticipated in the level of Sales. Possible interventions, in the mode of this section, include

(1) a simple shift in level, setting $\mathbf{h}_t' = (100, 0)$ and $\mathbf{H}_t = \mathbf{0}$;
(2) more realistically, a shift in level as above but hedged with uncertainty about the size of the shift via $\mathbf{H}_t = \text{diag}(100, 0)$, for example;
(3) as in (2) but with increased uncertainty about the new growth as well as the new level, via $\mathbf{H}_t = \text{diag}(100, 25)$, say;
(4) as in (3) but including correlation between the changes in level and growth via, for example,

$$\mathbf{H}_t = \begin{pmatrix} 100 & 25 \\ 25 & 25 \end{pmatrix},$$

thus suggesting that larger changes in growth will be associated with larger changes in level.

11.2.3 Arbitrary subjective intervention

The most general mode of intervention into the existing model is simply to change the prior moments of $\boldsymbol{\theta}_t$ to new values anticipating changes in the series. Thus, suppose that the intervention information is given by

$$I_t = \{\mathbf{a}_t^*,\ \mathbf{R}_t^*\},$$

where the *post-intervention* values $\mathbf{a}_t^*$ and $\mathbf{R}_t^*$ are given by the forecaster. Note that this covers the case of the previous section, where in addition to a possible mean shift, the post-intervention uncertainty always exceeds that pre-intervention. It goes well beyond that special case, however. As an extreme example, taking $\mathbf{R}_t^* = \mathbf{0}$ implies that $\boldsymbol{\theta}_t = \mathbf{a}_t^*$ with probability one; thus intervention informs precisely on the values of the parameters at time t. More practically, it allows for cases in which uncertainty about some of the parameters may decrease.

For forecasting Y_t and further into the future, the post-intervention moments replace those in (11.4), and they are then updated as usual when data are observed. A problem arises, however, when considering filtering and smoothing the series for retrospective analysis. The problem is that the post-intervention prior

$$(\boldsymbol{\theta}_t \mid I_t,\ D_{t-1}) \sim \text{N}[\mathbf{a}_t^*,\ \mathbf{R}_t^*] \tag{11.6}$$

is no longer consistent with the model (11.1-11.4). Filtering requires a joint distribution for $\boldsymbol{\theta}_t$ and $\boldsymbol{\theta}_{t-1}$ conditional on D_{t-1} *and* the intervention information I_t, and the arbitrary changes made on the moments of $\boldsymbol{\theta}_t$ to incorporate intervention do not provide a coherent joint distribution. To do so, we need to be able to express the changes due to intervention in a form consistent with the DLM, in a way similar to that used in the previous section in Theorem 11.1. This can be done in Theorem 11.2 below, based on the following, general result.

Lemma 11.1. *Let $\mathbf{K}_t$ be an n-square, upper triangular, non-singular matrix, and $\mathbf{h}_t$ any n-vector, and define*

$$\boldsymbol{\theta}_t^* = \mathbf{K}_t\boldsymbol{\theta}_t + \mathbf{h}_t,$$

where $\mathrm{E}[\boldsymbol{\theta}_t] = \mathbf{a}_t$ and $\mathrm{V}[\boldsymbol{\theta}_t] = \mathbf{R}_t$. Then $\boldsymbol{\theta}_t^*$ has moments $\mathbf{a}_t^*$ and $\mathbf{R}_t^*$ if $\mathbf{K}_t$ and $\mathbf{h}_t$ are chosen as follows:

$$\mathbf{K}_t = \mathbf{U}_t\mathbf{Z}_t^{-1},$$
$$\mathbf{h}_t = \mathbf{a}_t^* - \mathbf{K}_t\mathbf{a}_t,$$

where $\mathbf{U}_t$ and $\mathbf{Z}_t$ are the unique, upper-triangular, non-singular square root matrices of $\mathbf{R}_t^*$ and $\mathbf{R}_t$ respectively; thus $\mathbf{R}_t^* = \mathbf{U}_t\mathbf{U}_t'$ and $\mathbf{R}_t = \mathbf{Z}_t\mathbf{Z}_t'$.

Proof. The matrices $\mathbf{U}_t$ and $\mathbf{Z}_t$ exist and are unique since $\mathbf{R}_t^*$ and $\mathbf{R}_t$ are symmetric, positive definite matrices (see, for example, Graybill 1969). They define the Cholesky decomposition of these variance matrices and are easily computed. From the definition of $\boldsymbol{\theta}_t^*$ it follows that

$$\mathbf{a}_t^* = \mathbf{K}_t\mathbf{a}_t + \mathbf{h}_t,$$

and so the expression for $\mathbf{h}_t$ is immediate for any given $\mathbf{K}_t$. Secondly,

$$\mathbf{R}_t^* = \mathbf{K}_t\mathbf{R}_t\mathbf{K}_t';$$

thus

$$\mathbf{U}_t\mathbf{U}_t' = (\mathbf{K}_t\mathbf{Z}_t)(\mathbf{K}_t\mathbf{Z}_t)'.$$

Now $\mathbf{K}_t\mathbf{Z}_t$ is a square, non-singular, upper-triangular matrix, and since the matrix $\mathbf{U}_t$ is unique, it follows that $\mathbf{U}_t = \mathbf{K}_t\mathbf{Z}_t$. The expression for $\mathbf{K}_t$ follows since $\mathbf{Z}_t$ is non-singular.

◇

The Lemma shows how a second evolution of $\boldsymbol{\theta}_t$ to $\boldsymbol{\theta}_t^*$ can be defined to achieve any desired moments in (11.6). It is useful to think of the intervention in these terms, but for calculations it is often desirable to incorporate this second intervention into the original DLM, as follows.

THEOREM 11.2. *Suppose that the moments $\mathbf{a}_t^*$ and $\mathbf{R}_t^*$ in (11.6) are specified to incorporate intervention, and define $\mathbf{K}_t$ and $\mathbf{h}_t$ as in Lemma*

11.1. Then (11.6) is the prior obtained in the DLM (11.1-11.4) with evolution equation (11.2) amended to

$$\boldsymbol{\theta}_t = \mathbf{G}_t^* \boldsymbol{\theta}_{t-1} + \boldsymbol{\omega}_t^*, \qquad \boldsymbol{\omega}_t^* \sim \mathrm{N}[\mathbf{h}_t, \mathbf{W}_t^*], \tag{11.7}$$

where, given D_{t-1} and I_t, $\boldsymbol{\omega}_t^*$ is uncorrelated with $\boldsymbol{\theta}_{t-1}$ and

$$\begin{aligned}
\mathbf{G}_t^* &= \mathbf{K}_t \mathbf{G}_t, \\
\boldsymbol{\omega}_t^* &= \mathbf{K}_t \boldsymbol{\omega}_t + \mathbf{h}_t, \\
\mathbf{W}_t^* &= \mathbf{K}_t \mathbf{W}_t \mathbf{K}_t'.
\end{aligned}$$

Proof. An easy deduction from the lemma, and left to the reader.

◇

Thus, any interventions modelled by (11.6) can be formally, and routinely, incorporated into the model by appropriately amending the evolution equation at time t, reverting to the usual equations for future times not subject to intervention. As with the intervention modes in Sections 11.2.1 and 11.2.3, this is important since it means that the usual updating, forecasting, filtering and smoothing algorithms apply directly with the post-intervention model. Note that in forecasting more than one-step ahead, interventions for future times can be simply incorporated in the same fashion, by appropriately changing the model based on the forecast moments for the $\boldsymbol{\theta}$ vector at those times, pre- and post-intervention.

EXAMPLE 11.1 (continued). Consider again intervention into the CP6 model in January 1958. The four example interventions earlier considered can all be phrased in terms of (11.7). The details are left as exercises for the reader. Additionally, of course, other, arbitrary changes can be accommodated via this mode of intervention. As an example, suppose that $\mathbf{a}_t^{*\prime} = (970, 0)$ and $\mathbf{R}_t^* = \mathrm{diag}(50, 5)$. This represents direct intervention to anticipate a new level of 970, with variance 50 *decreased* from the pre-intervention value of 100, new growth estimated at 0 with increased variance of 5. Additionally, the post-intervention level and growth are uncorrelated.

11.2.4 Inclusion of intervention effects

There is one further mode of intervention to be explored. The preceding modes allow for the information I_t by appropriately amending the model at the time (or in advance if the information is available); in each case the dimension n of the model remains fixed, the intervention providing changes to the model parameters. Sometimes it is of interest to isolate the effects of an intervention, providing extra parameters that define the model changes.

EXAMPLE 11.1 (continued). Consider once more the CP6 intervention. To isolate and estimate the jump in level let γ_t represent the change, so that post-intervention, the new level is $\mu_t^* = \mu_t + \gamma_t$. The quantity γ_t is an additional model parameter that can be included in the state vector from now on. Thus, at the intervention time t, extend the model to three parameters,

$$\boldsymbol{\theta}_t^* = (\mu_t^*, \beta_t, \gamma_t)'.$$

The DLM at time t is now subject to an additional evolution after (11.2), namely

$$\begin{pmatrix} \mu_t^* \\ \beta_t \\ \gamma_t \end{pmatrix} = \begin{pmatrix} 1 & 0 \\ 0 & 1 \\ 0 & 0 \end{pmatrix} \begin{pmatrix} \mu_t \\ \beta_t \end{pmatrix} + \begin{pmatrix} 1 \\ 0 \\ 1 \end{pmatrix} \gamma_t,$$

with the distribution assigned to γ_t determining the expected change in level. In vector form

$$\boldsymbol{\theta}_t^* = \mathbf{K}_t \boldsymbol{\theta}_t + \boldsymbol{\xi}_t,$$

where

$$\mathbf{K}_t = \begin{pmatrix} 1 & 0 \\ 0 & 1 \\ 0 & 0 \end{pmatrix} \quad \text{and} \quad \boldsymbol{\xi}_t = \begin{pmatrix} 1 \\ 0 \\ 1 \end{pmatrix} \gamma_t.$$

For example, an anticipated change of 100 with a variance of 50 implies $\gamma_t \sim \mathrm{N}[100, 50]$. As a consequence,

$$(\boldsymbol{\theta}_t^* \mid I_t,\ D_{t-1}) \sim \mathrm{N}[\mathbf{a}_t^*,\ \mathbf{R}_t^*],$$

where

$$\begin{aligned} \mathbf{a}_t^* &= \mathbf{K}_t \mathbf{a}_t + \mathbf{h}_t, \\ \mathbf{R}_t^* &= \mathbf{K}_t \mathbf{R}_t \mathbf{K}_t' + \mathbf{H}_t, \end{aligned}$$

with $\mathbf{h}_t' = E[\boldsymbol{\xi}_t' \mid I_t,\ D_{t-1}] = 100(1,0,1) = (100,0,100)$ and

$$\mathbf{H}_t = V[\boldsymbol{\xi}_t \mid I_t,\ D_{t-1}] = 50 \begin{pmatrix} 1 & 0 & 1 \\ 0 & 0 & 0 \\ 1 & 0 & 1 \end{pmatrix}.$$

In addition, the observation equation is altered so that $\mathbf{F}_t' = (1,0,0)$ at time t. From t onwards, the model remains 3-dimensional, the $\mathbf{F}_t$ vectors all extended to have a third element of 0, the $\mathbf{G}_t$ matrices extended from $\mathbf{J}_2(1)$ to

$$\begin{pmatrix} 1 & 1 & 0 \\ 0 & 1 & 0 \\ 0 & 0 & 0 \end{pmatrix},$$

and similarly, the 3×3 evolution variance matrices for times after t having third rows and columns full of zeros. Thus, as new data are processed, the posterior for γ_t is revised, learning about the change that took place at time t. An important variation on this technique is to allow for such changes in level to occur over two (or more) time periods, with incremental changes γ_t, γ_{t+1}, for example, leading to a gradual step up to a new level at time $t+1$. This requires a model extension to add more parameters. As an aside, note that it is possible and sometimes appropriate to relate the parameters in order to restrict the increase in dimension to just one, as is done in modelling advertising campaign effects in Harrison (1988), for example. Adding parameters is particularly useful following intervention *action* in the environment of the series to attempt control and produce changes. Changes in pricing policy or advertising strategy, for example, are effected in an attempt to change sales levels. Intervention effects such as γ_t (and γ_{t+1}, etc.) then provide measures of just how much change occurred.

This sort of model extension intervention mode can be phrased in DLM terms as in Theorem 11.2, although in this case the matrix $\mathbf{G}_t^* = \mathbf{K}_t\mathbf{G}_t$ will not be square. Extending the model to include an additional k parameters at a particular time t means that $\mathbf{G}_t^*$ will be $(n+k)\times n$ if the parameters are unrelated. The details of such extensions, similar to those in the example, are essentially as in Theorem 11.2 with the additional feature of an increase in model dimension. See also Harrison (1988) for approaches in which the increase in dimensionality is restricted.

11.2.5 Model analysis with intervention

The main reason for expressing all modes of intervention in DLM form is that the existing theory can be applied to formally incorporate interventions of essentially arbitrary forms, with the model $\{\mathbf{F}_t, \mathbf{G}_t, V_t, \mathbf{W}_t\}$ amended according to the mode of intervention. It then follows that the theory applies, with, for example, the elements $\mathbf{a}_t$ and $\mathbf{R}_t$ replaced throughout by their post-intervention (starred) values. In particular, in filtering and smoothing for retrospective time series analysis and estimation of the time trajectories of model components and parameters over time, it is vital that these subjective changes at intervention times be formally incorporated in model form so that their implications for times past, as well as for forecasting the future, are properly understood.

There are certainly other modes of intervention and subjective interference with a routine statistical model that may be used in practice. In particular, it is worth pointing out that if, in retrospect, it is seen that an intervention in the past was inappropriate, then in filtering back beyond that point of intervention it may be well to ignore the intervention. Thus, for example, the filtered distribution for time $t-1$ just prior to intervention at t can be taken as it was at the time, simply the posterior given D_{t-1}, and filtering continuing on back in time from this new starting point.

Finally note that the concepts apply directly to models with unknown observational variances, the actual techniques being directly appropriate (although T distributions replace normal distributions, as usual). Also, it should be apparent that the specific normal, linear structure used throughout is quite secondary, the intervention models being obviously appropriate and useful more generally. If we simply drop the normality assumptions and use distributions only partially specified in terms of means and variance matrices, for example, the techniques apply directly. However, no further development is given here of this, or other, intervention ideas. We proceed in the next section with illustrations of the main modes of intervention detailed above.

11.3 ILLUSTRATIONS

11.3.1 CP6 Sales

An illustrative example is given using the CP6 Sales data. The objective of this example is to demonstrate just how effective simple interventions can be if relevant feed-forward information exists and is appropriate. This example is purely hypothetical, the interventions being used are easily seen to be adequate descriptions of the discontinuities in the series. In practice, of course, the interventions are usually made speculatively, and though often they may be dramatically effective, will usually not be ideal.

The second-order polynomial model introduced in Example 11.1 for the CP6 Sales data are used here. The basic time variation in monthly level and growth parameters is determined by a single discount factor of 0.9; thus, each month, the uncertainty about level and growth increases by roughly 10%. The observational variance is constant, set initially at $V_t = V = 64$, with standard deviation of 8 determining purely random variation about the underlying level.

The features of interest are as follows.

(1) At time $t = 1$, January 1955, the analysis begins with an initial prior for the level and growth $\boldsymbol{\theta}_1 = (\mu_1, \beta_1)'$ specified by prior mean $\mathbf{a}_1 = (600, 10)'$, and variance matrix $\mathbf{R}_1 = \text{diag}(10000, 25)$. The prior standard deviation for the level, set at 100, is extremely large, representing a very vague initial prior. The one-step forecast distribution for Y_1 thus has mean 600, and variance 10064; the 90% forecast interval symmetrically located about the mean appearing in Figure 11.3 reflects this uncertainty. As data are sequentially processed as usual, the one month ahead forecasts are calculated, and for each month, the forecast means with 90% forecast intervals appear in the figure. During the first year of data, Sales closely follows a steep, linear growth and one-step forecasts are accurate.

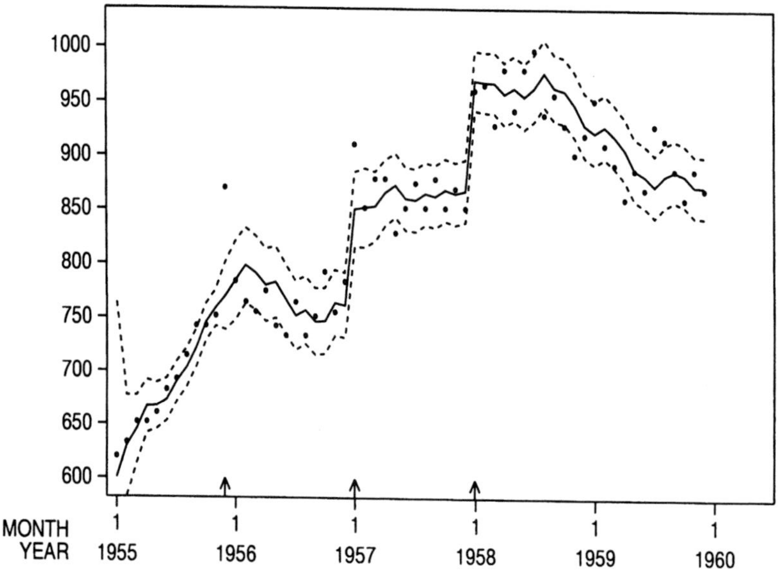

Figure 11.3 CP6 forecasts with ideal interventions

(2) Intervention is made at December 1955 to allow for an anticipated outlier; this is predicted as a response to forward-buying due to a future price rise announcement. Thus, for $t = 12$, the observational variance is infinite, or $V_{12}^{-1} = 0$ as in Section 11.2.1.

(3) In addition, following the price rise, a change in growth is anticipated in the new year and so, rather than waiting until January to anticipate the change, an additional intervention is made to the model in December. The intervention is neutral, specifying an increase in uncertainty about level and growth with no specific direction of change in mind. Specifically, as in Section 11.2.2, the prior variance matrix

$$\mathbf{R}_{12} = \begin{pmatrix} 41.87 & 6.11 \\ 6.11 & 1.23 \end{pmatrix}$$

is increased to $\mathbf{R}_{12}^*$ by the addition of

$$\mathbf{H}_{12} = \begin{pmatrix} 100 & 25 \\ 25 & 25 \end{pmatrix}.$$

Thus, additional zero-mean normal changes to level and growth have standard deviations of 10 and 5 respectively, and correlation of 0.5, so that positive growth changes are associated with positive level changes.

(4) A third change is made at $t = 12$, this being to alter the observational variance for the future from $V = 64$ to $V = 225$, the standard deviation increasing from 8 to 15. This reflects the views that firstly, variation is likely to be more erratic after this first year of fast market growth, and secondly, that higher variation in Sales is expected at higher levels.

The effects of these interventions are clear in Figure 11.3, an arrow on the time axis indicating the intervention. After ignoring the December 1955 observation, the forecast interval width increases, reflecting greater uncertainty about $\boldsymbol{\theta}_{12}$ following intervention and the increased value of V. Observing Sales sequentially throughout 1956, the model adapts and proceeds adequately to the end of the year. Note also the wider forecast intervals due to the larger value of V.

(5) The second set of interventions takes place for $t = 25$, January 1957, where the current prior for level and growth is given by

$$\mathbf{a}_{25} = \begin{pmatrix} 770.8 \\ 1.47 \end{pmatrix} \quad \text{and} \quad \mathbf{R}_{25} = \begin{pmatrix} 114.4 & 13.2 \\ 13.2 & 2.2 \end{pmatrix}.$$

Anticipating a marked increase in Sales level following a takeover, an estimated change in level of 80 units is adopted with a variance of 100. No effects on growth are expected, or allowed, by using an additional evolution term $\boldsymbol{\xi}_{25}$ as in Section 11.2.2, with mean and variance matrix

$$\mathbf{h}_{25} = \begin{pmatrix} 80 \\ 0 \end{pmatrix} \quad \text{and} \quad \mathbf{H}_{25} = \begin{pmatrix} 100 & 0 \\ 0 & 0 \end{pmatrix},$$

respectively.

(6) In addition to this estimated change in level, the January 1957 observation is discarded as an outlier, reflecting a view that the marked change anticipated in the new year will begin with a maverick value, as the products that are to be discontinued are sold cheaply.

The interventions have a clear effect on short-term forecast accuracy, seen in the figure. Again, for the remainder of 1957 things are stable.

(7) The third intervention comes in January 1958. Another jump in level is anticipated, this time of about 100 units. Unlike the previous change, however, there is a feeling of increased certainty about the new level. Also, it is anticipated that growth may change somewhat more markedly than already modelled through the routine discount factor, and so the prior variance of the growth is to be increased. This intervention, then, is of the mode in Section 11.2.3. The prior moments, namely

$$\mathbf{a}_{37} = \begin{pmatrix} 864.5 \\ 0.86 \end{pmatrix} \quad \text{and} \quad \mathbf{R}_{37} = \begin{pmatrix} 91.7 & 9.2 \\ 9.2 & 1.56 \end{pmatrix},$$

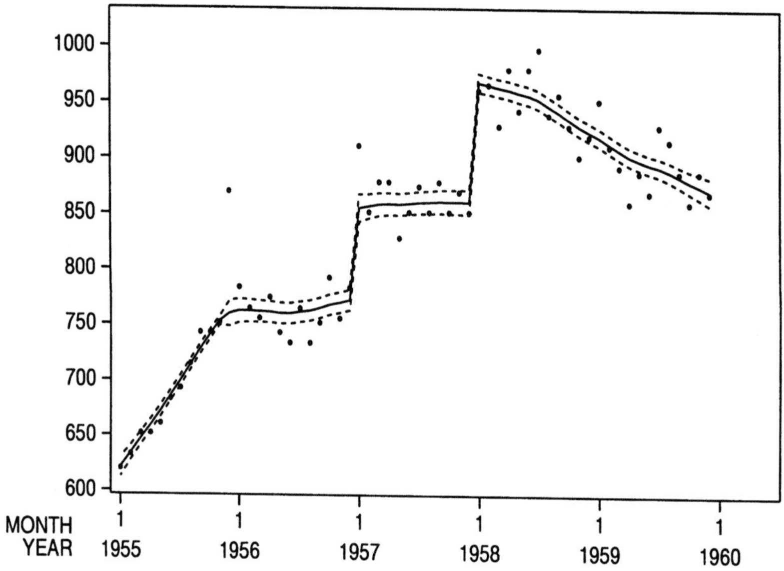

Figure 11.4 Filtered trend in CP6 with ideal interventions

are simply altered to

$$\mathbf{a}_{37}^* = \begin{pmatrix} 970 \\ 0 \end{pmatrix} \quad \text{and} \quad \mathbf{R}_{37}^* = \begin{pmatrix} 50 & 0 \\ 0 & 5 \end{pmatrix}.$$

The forecasts in Figure 11.3 adapt accordingly, and analysis proceeds as normal. As a whole, the one-step forecasting performance has been good due to the appropriate, though entirely subjective, interventions.

(8) At the end of the analysis, $t = 60$ in December 1959, backward filtering provides a retrospective look at the development over time. Figure 11.4 provides a plot of the retrospectively estimated level of the series, with mean and 90% interval taken from the filtered posterior distributions $p(\mu_t|D_{60})$ $(t = 1, \ldots, 60)$. Note, of course, that D_{60} incorporates the interventions. The local linearity of the trend is evident, and startlingly so, as is the fairly high precision with which the trend is estimated. Clearly, the random noise in the data about the trend is a dominant feature of the series, the underlying trend being smooth and sustained apart from the abrupt changes allowed for by intervention.

11.3.2 UK marriages

The second illustration concerns the data in Table 11.2, providing quarterly total numbers of marriages registered in the UK during the years 1965 to 1970 inclusive. The data, taken from the UK Monthly Digest of Statistics, are given in thousands of registrations.

Table 11.2. Numbers of marriages in UK (in thousands)

Year	Quarter			
	1	2	3	4
1965	111.2	83.5	129.5	97.8
1966	114.7	84.6	131.1	106.3
1967	117.5	80.6	143.3	97.6
1968	131.3	77.5	145.9	108.0
1969	88.1	112.6	152.0	98.5
1970	91.2	117.5	160.0	102.2

Before examining the data more closely, consider the pattern that the marriage figures could be expected to take over a year on general grounds. It might be strongly believed that the majority of marriages take place in late spring and summer in the UK, many people preferring to get married when there is a reasonable chance of good weather. Certainly there are short exceptional times, such as Christmas and Easter, that are particularly popular, but generally, a strong annual cycle is anticipated, with a high peak in the third quarter, that contains the favoured holiday months of July and August. In addition, the winter months of November, December, January and February should see the trough in the seasonal pattern.

Consider now the data in the table, restricting attention to the first four years, 1965 to 1968 inclusive. The data for these four years are plotted as the first part of Figure 11.5. Over these years there is certainly a strong seasonal pattern evident, though not of the form anticipated in the previous paragraph. The summer boom in marriages is there in the third quarter, but there is a secondary peak in the first quarter, the months January, February and March. In addition, the major trough in numbers comes in the second, spring and early summer, quarter rather than in the fourth, early winter quarter. These features are rather surprising though clearly apparent in each of these four years as they were prior to 1965. It can be seen that the amplitude of the seasonality increases somewhat in mid-1967 and that this increase is sustained during 1968, but the form of the pattern is still the same: secondary peak in quarter 1; deep trough in quarter 2; high peak in the summer, quarter 3; and secondary trough in quarter 4.

Before proceeding to explain the rather surprising seasonality here, consider a simple model for forecasting the series as it stands. Suppose that we use a second-order polynomial for the trend in the non-seasonal level

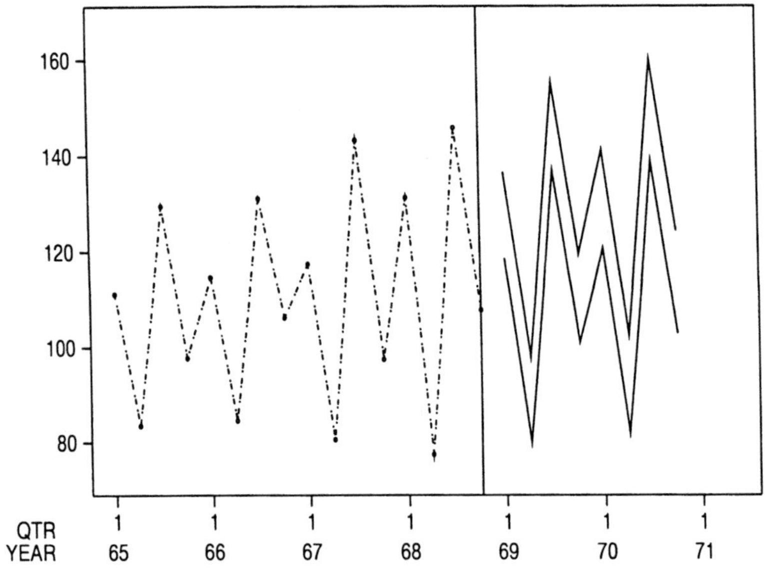

Figure 11.5 Marriage data and forecasts: no interventions

of marriage, with a full four seasonal effects provided, as usual, by the Fourier representation with three Fourier coefficients. Thus, the model has five parameters in $\boldsymbol{\theta}_t$: the level, growth and three Fourier coefficients. In addition, assume now that the observational variance is to be estimated, being assumed constant, $V_t = V$, for all time. The initial prior information assumed at $t = 1$, the first quarter of 1965, is as follows.

- The observational variance has an initial inverse gamma distribution with 12 degrees of freedom and estimate of 9. Thus, the prior estimate of standard deviation is 3. Equivalently, V has a scaled, inverse χ^2_{12} distribution.
- The initial posterior for $\boldsymbol{\theta}_0$, the parameters at the end of 1964 given historical data and experience, is multivariate T_{12}, the mean and scale matrix taken as

$$\mathbf{m}_0 = (100,\ 1;\ -7.5,\ -7.5,\ 17.5)'$$

and

$$\mathbf{C}_0 = \text{diag}(16,\ 1;\ 4.5,\ 4.5,\ 2.5).$$

Thus, the level in late 1964 is estimated as 100, with scale 16, thus having variance (from the T_{12} distribution) of $16 \times 12/10 = 19.2$. Similarly, the quarterly growth in level at the time is initially estimated as 1 with scale of 1. Transforming from Fourier coefficients to seasonal factors (Section 8.6.4), the prior for the 3

Fourier coefficients, with mean $(-7.5,\ -7.5,\ 17.5)'$ and scale matrix diag(4.5, 4.5, 2.5), is seen to be consistent with initial estimates of seasonal effects given by $(10, -25, 25, -10)'$, these being based on pre-1965 data and experience and clearly anticipating the seasonal pattern over 1965-1968.

- Two discount factors, one for the linear trend and one for the seasonal pattern, are required to complete the model definition. Both components are fairly stable over time, and so the discount factors are chosen to be fairly high, both set at 0.95. Concerning the seasonal component, this should allow for adequate adaptation to the observed increases in amplitude of the seasonal pattern at higher levels of the data.

With this model, the sequential, one-step ahead forecasting and updating analysis proceeds and is illustrated in the first part of Figure 11.7. Up to the end of 1968, the graph gives one-step ahead forecast means and symmetric 90% intervals about the means, with the data superimposed. The strong seasonal pattern comes through in the forecasts, and it is clear that forecast accuracy is reasonably good. There is a deterioration in accuracy during late 1967 and early 1968, when the amplitude of the seasonal swings in the data increases, but the model adapts to this in late 1968, the final 2 observations in that year being well inside the forecast intervals.

Up to this point, everything is essentially routine, the data being forecast in the short-term with a simple, standard model closed to interventions. At the end of 1968, time $t = 16$, the prior distribution for the next quarter, $t = 17$, is summarised by

$$(\boldsymbol{\theta}_{17} \mid D_{16}) \sim \mathrm{T}_{28}[\mathbf{a}_{17},\ \mathbf{R}_{17}],$$

where

$$\mathbf{a}_{17} = (117.1,\ 0.84;\ -8.35,\ -9.81,\ 19.03)' \tag{11.8}$$

and

$$\mathbf{R}_{17} \approx \text{block diag}\left\{\begin{pmatrix} 5.71 & 0.56 \\ 0.56 & 0.07 \end{pmatrix};\ 2.79,\ 2.66,\ 1.35\right\}. \tag{11.9}$$

The first 2×2 matrix here refers to the level and growth elements, the final three to the (essentially uncorrelated) Fourier coefficients. For the observational variance, V has a scaled, inverse χ^2_{28} distribution with point estimate $S_{16} = 16.16$. The latter part of Figure 11.5 provides a graphical display of the implications for the future. Forecasting from the final quarter of 1968 with no changes to the model, the intervals displayed are, as usual, 90% forecast intervals symmetrically located about the step ahead forecast means.

Consider an explanation for the seasonal pattern and why, in late 1968, this would have changed our view of the future of the marriage series, prompting an intervention at the time. The explanation lies in the UK income tax laws. Income is only taxed above certain threshold levels, the non-taxable portion of income being referred to as a tax-free allowance, and every employed person has a basic allowance. On getting married, one of the marriage partners is eligible for a higher tax-free allowance, referred to as the married person's allowance (although during the time span of this data it was called a married *man's* allowance). Now, in the good old days of the 1960s, the law was such that a couple could claim this extra allowance for the entire tax year during which they were married, the tax years running from April to March inclusive. Thus, for example, a wedding in late March of 1967 led to an entitlement to reclaim some portion of tax paid during the previous twelve months when both partners were single. Delaying this wedding for a week or two to early April would mean that this extra income would be lost since a new tax year has begun. The seasonal pattern is now explained; many marriages were obviously held in the first quarter of each year to maximise financial benefit, and the second quarter saw a huge slump in numbers since the resulting tax benefits were minimal.

In forecasting ahead from late 1968 over the next two years, the latter part of Figure 11.5 is perfectly acceptable on the basis of the historical information as it stands. However, in early 1968 there was an announcement that the tax laws were being revised. The change was simply that beginning in 1969, this entitlement to reclaim tax paid during the current tax year was abolished; from then on, there would be no tax incentive for couples to avoid marrying in the second quarter. Knowing this and understanding the effect that the old law had on marriages, a change in the forecasting model is called for. On the grounds of a simple cycle following the UK seasons as earlier discussed, it is assumed in late 1968 that the secondary peak in the first quarter will disappear, with marriages transferring to the more attractive second and third quarters. The levels in the second quarter will increase markedly from the totally artificial trough, those in the third quarter rather less so.

As an illustration of intervention to incorporate this view, suppose that in predicting from the final quarter of 1968, the expected seasonal effects are taken as $(-16,\ 0,\ 40,\ -24)'$. Thus, the anticipation is that numbers of marriages grow consistently throughout the calendar year to a high peak during the summer, then crash down to a low trough during the final quarter of the year. Transforming to Fourier coefficients (Section 8.6.4), it is seen that these expected seasonal effects are consistent with expected Fourier coefficients of about $(-28,\ 12,\ 12)'$, although there is clearly a fair degree of uncertainty about just how the pattern will change. What is fairly acceptable, however, is the view that the tax changes are unlikely to affect the non-seasonal level of marriages, nor the rate of growth of this level. In the mode and notation of Section 11.2.3, an arbitrary change is made to the current prior moments $\mathbf{a}_{17}$ and $\mathbf{R}_{17}$ to accommodate this view.

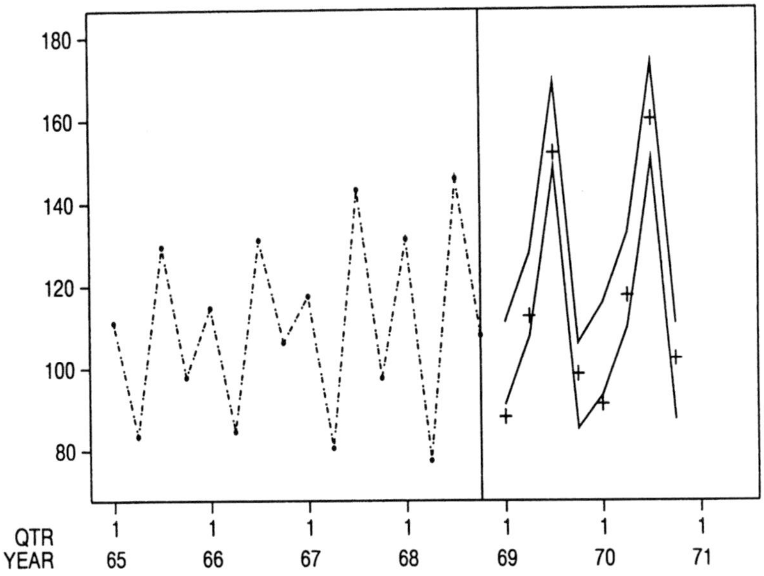

Figure 11.6 Marriage data and step ahead forecasts

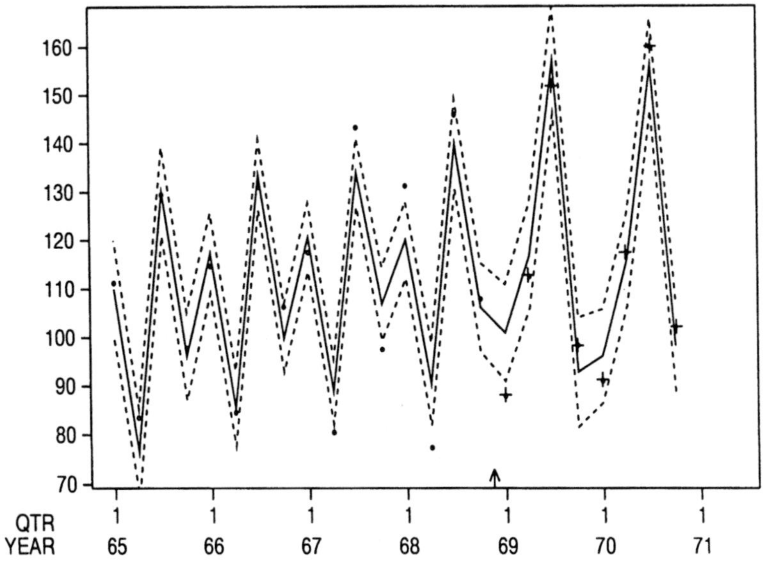

Figure 11.7 One-step ahead Marriage forecasts

The revised values in (11.6) are taken as

$$\mathbf{a}_{17}^* = (117.1,\ 0.84;\ -28,\ 12,\ 12)'$$

and

$$\mathbf{R}_{17}^* = \text{block diag}\left\{\begin{pmatrix} 5.71 & 0.56 \\ 0.56 & 0.07 \end{pmatrix};\ 8,\ 8,\ 4\right\}.$$

These replace the values in (11.8) and (11.9). Note that the level and growth components are unchanged, and they are taken as uncorrelated with the seasonal parameters. Figure 11.6 displays the post-intervention step ahead forecasts made in late 1968 for the next 2 years, the display being analogous to that in Figure 11.5 made pre-intervention. The forecast intervals are slightly wider, reflecting the increase in uncertainty about the seasonal pattern. The crosses on the graph are the actual data for those 2 years. Note that the pattern did in fact change markedly in line with the intervention-based forecasts, although several of the actual values are rather low relative to forecast. This suggests, in retrospect, that the non-seasonal level was very slightly over-estimated at the end of 1968. Moving through years 1969 and 1970, the data are sequentially processed and the model adapts as usual. The latter part of Figure 11.7 now completes the display of the one-step forecasting activity, and the adequacy of the model is apparent.

The main point here to be restressed is that an understanding of the mechanisms influencing, and possibly driving, systems is the key to effective interventions necessary to adapt to changing conditions.

11.4 MODEL MONITORING

11.4.1 Bayes' factors for model assessment

The focus is now switched to problems of feed-back intervention, considering models operating subject to continual monitoring to detect deteriorations in predictive performance that are consistent with some form of model breakdown (e.g., changes in parameters, etc.). This section discusses automatic methods of sequentially monitoring the forecasting activity to detect breakdowns, the assessment of model performance being based on purely statistical measures of accuracy. At this most basic level, the problem of model assessment is simply one of examining the extent to which the observed values of the time series are consistent with forecasts based on the model. In the DLM framework, the focus is on consistency of each observation with the corresponding one-step ahead forecast distribution. Equivalently, the assessment can be made on the basis of standardised, one-step ahead forecast errors, measuring the extent to which they deviate from model hypothesis of standard normal, or T distributed, uncorrelated quantities. There are many statistical techniques with which to examine such questions, some specifically designed with time series data in mind, others of use more widely. Central to all such techniques is the notion of assessing model performance relative to that obtained through using one or more alternative models. With the objectives of detecting changes in parameters, outlying observations, and so forth in mind, these alternatives should be designed to allow for the forms of behaviour in the series that are consistent with such changes, wild observation etc. There are clearly many possible forms that these alternatives can take, the specific forms chosen to be based on the specific forms of departure from the existing, routine model that are anticipated. We consider some possible, generally useful forms below and describe how they may be used in automatic model monitoring and assessment. Initially, however, the basic concepts of model assessment and sequential monitoring are detailed in a general setting. Much of the basic material here is developed from West (1986a). The basic mathematical ingredients of sequential model monitoring have already been introduced in informal settings in previous chapters, notably Section 10.2.5, the keystone being provided by Bayes' factors, defined formally below.

Consider any two models with the same mathematical structure, differing only through the values of defining parameters, for example, in the values of discount factors. Denote any model by the symbol M, and let these two models be denoted by M_0 and M_1. M_0 will have a special status, being the routine or standard model that is used subject to continual assessment. M_1, (and possibly further models M_2, etc., at a later stage), is an alternative that is introduced to provide assessment of M_0 by comparison. At time t, each model provides a predictive distribution for Y_t given D_{t-1}, as usual. Formally including the specification of the models in the conditioning, we write these densities as

$$p(Y_t|D_{t-1}, M_i), \qquad (i = 0, 1)$$

with D_{t-1} being the historical information that is common to the two models at time t. The inclusion of M_i differentiates between the two models. For simplicity of notation in this section, however, we temporarily discard this general notation, making clear the dependence of the distributions on the model using the subscript 0 or 1. Thus, the predictive densities at time t are here denoted by

$$p_i(Y_t|D_{t-1}) = p(Y_t|D_{t-1}, M_i), \qquad (i = 0, 1)$$

respectively.

Definitions 11.1.

(i) The **Bayes' factor** for M_0 versus M_1 based on the observed value of Y_t is defined as

$$H_t = p_0(Y_t|D_{t-1})/p_1(Y_t|D_{t-1}).$$

(ii) For integers $k = 1, \dots, t$, the Bayes' factor for M_0 versus M_1 based on the sequence of k consecutive observations $Y_t, Y_{t-1}, \dots, Y_{t-k+1}$

is defined as

$$H_t(k) = \prod_{r=t-k+1}^{t} H_r = \frac{p_0(Y_t, Y_{t-1}, \ldots, Y_{t-k+1}|D_{t-k})}{p_1(Y_t, Y_{t-1}, \ldots, Y_{t-k+1}|D_{t-k})}.$$

These Bayes' factors, or *weights of evidence* (Jeffreys 1961; Good 1985 and references therein; West 1986a), provide the basic measures of predictive performance of M_0 relative to M_1. For each k, $H_t(k)$ measures the evidence provided by the most recent (up to and including time t) k consecutive observations. Some basic features of Bayes' factors are noted:

(1) Setting $k = 1$ in (ii) leads to the special case (i): $H_t(1) = H_t$, for all t.
(2) Taking $k = t$, the Bayes' factor based on all the data are $H_t(t)$.
(3) The Bayes' factors for M_1 versus M_0 are the reciprocals of those for M_0 versus M_1, $H_t(k)^{-1}$.
(4) Evidence for or against the model M_0 accumulates multiplicatively as data are processed. Specifically, for each $t > 1$,

$$H_t(k) = H_t \, H_{t-1}(k-1), \qquad (k = 2, \ldots, t.)$$

(5) On the log scale, evidence is additive, with

$$\log[H_t(k)] = \log(H_t) + \log[H_{t-1}(k-1)], \qquad (k = 2, \ldots, t.)$$

(6) Following Jeffreys (1961), a log Bayes' factor of 1 (−1) indicates evidence in favour of model 0 (1), a value of 2 or more (−2 or less) indicating the evidence to be strong. Clearly, the value 0 indicates no evidence either way.

The definition of Bayes' factors is general, obviously applying outside the confines of normal DLMs. In considering specifically the possibilities that in model M_0, Y_t may be a wild, outlying observation or that the defining parameters in M_0 may have changed at (or before) time t, the alternative M_1 should provide for the associated forms of departure of the observations from prediction under M_0. Some possible forms for $p_1(Y_t \mid D_{t-1})$ are discussed in Section 11.4.3. Notice, however, that it is not actually necessary to construct a fully specified alternative model for the data; all that is needed is a suitable sequence of alternative, one-step forecast densities that provide the denominators of the Bayes' factors. Once appropriate densities are defined, they may be used without further consideration of particular forms of departure from M_0 that may lead to such a density. This is important in practice since the derived Bayes' factor is easily computed and forms the basis for sequential monitoring even though a formal model M_1 has not been constructed.

11.4.2 Cumulative Bayes' factors

In considering how the Bayes' factors may be applied in monitoring M_0, assume for the moment that appropriate densities have been defined; thus the *sequence* $p_1(Y_t|D_{t-1})$, $t = 1, \ldots$, is available to base the Bayes' factors upon. Assume here that the alternatives are appropriate for the deviations from M_0 and forms of departure of the data from M_0 that are of interest.

The overall Bayes' factor $H_t(t)$ is a basic tool in overall model assessment. In the monitoring context, however, the focus is on *local* model performance, and here the individual measures H_t and the cumulative measures $H_t(k)$ for $k < t$ are key. As an illustration of this, suppose that $t = 6$ and that $H_r = 2$ for $r < 6$. Thus, each of the first five observations are well in accord with the standard model M_0, their individual Bayes' factors all being equal to 2. Consequently, $H_5(5) = 32$, representing the cumulative evidence for M_0 relative to M_1 from the first five observations. Suppose that Y_6 is very unlikely under M_0, out in the tails of the forecast distribution. Then even though it may be extremely discrepant, the cumulative Bayes' factor $H_6(6)$ may still exceed 1, and even be much larger than 1, thus indicating no evidence against M_0. Clearly, we would require $H_6 \leq 1/32$, an extremely small Bayes' factor, to even begin to doubt M_0. The problem is that the evidence in favour of the standard model from earlier observations *masks* that against it at time 6 in the overall measure. Hence the need to consider the Bayes' factors from individual observations as they arise. In addition, it is important to look back over groups of recent observations when considering the possibility of small or gradual change in a time series, consistent with a shift in the Bayes' factors from favouring M_0 to favouring M_1. In such cases, where observations gradually drift away from forecasts, the individual Bayes' factors, although small, may not be small enough that they signal the changes individually, needing to be cumulated to build up evidence against M_0. For each k, $H_t(k)$ assesses the fit of the most recent k observations. A single small value of $H_t(1) = H_t$ provides a warning of a possible outlier or the onset of change in the series at time t. A small $H_t(k)$ for $k > 1$ is indicative of possible changes having taken place (at least) k steps back in the past. To focus on the most likely point of change, we can identify the most discrepant group or recent, consecutive observations by minimising the Bayes' factors $H_t(k)$ with respect to k. This may be done simply, sequentially as the standard model M_0 is updated, using the following result (West 1986a).

THEOREM 11.3. *With H_t and $H_t(k)$ as in Definition 11.1, let*

$$L_t = \min_{1 \leq k \leq t} H_t(k),$$

with $L_1 = H_1$. *Then the quantities* L_t *are updated sequentially over time by*

$$L_t = H_t \min\{1,\ L_{t-1}\}$$

for $t > 1$. *The minimum at time* t *is taken at* $k = l_t$, *with* $L_t = H_t(l_t)$, *where the integers* l_t *are sequentially updated via*

$$l_t = \begin{cases} 1 + l_{t-1}, & \textit{if } L_{t-1} < 1, \\ 1, & \textit{if } L_{t-1} \geq 1. \end{cases}$$

Proof. Since $H_t(1) = H_t$ and for $2 \leq k \leq t$, $H_t(k) = H_t H_{t-1}(k-1)$, then

$$\begin{aligned} L_t &= \min\{H_t,\ \min_{2 \leq k \leq t} H_t H_{t-1}(k-1)\} \\ &= H_t\ \min\{1,\ \min_{2 \leq k \leq t} H_{t-1}(k-1)\} \\ &= H_t\ \min\{1,\ \min_{1 \leq j \leq t-1} H_{t-1}(j)\} \\ &= H_t\ \min\{1,\ L_{t-1}\}, \end{aligned}$$

as stated. Note that $L_t = H_t$ if and only if $l_t = 1$, otherwise $L_t = H_t L_{t-1}$ and $l_t = 1 + l_{t-1}$, providing the stated results.

◇

The focus on the local behaviour of the series is evident in L_t. If at time $t-1$, the evidence favours the standard model M_0 so that $L_{t-1} \geq 1$, then $L_t = H_t$, and decisions about possible inadequacies of the model are based on Y_t alone. If H_t is very small, then Y_t is a possible outlier or may indicate the onset of change. If the evidence is against M_0 before time t with $L_{t-1} < 1$, then evidence is cumulated via $L_t = H_t L_{t-1}$, and l_t increases by 1. l_t is termed the *run-length* at time t, counting the number of recent, consecutive observations that contribute to the minimum Bayes' factor. The local focus is geared to detecting slow, gradual change. To see this, note that a relatively slow or gradual change beginning at time t leads to a sequence of consecutive values $H_{t+1}, H_{t+2}, \ldots$ that are small, though not exceedingly so. Hence, starting from $L_t = 1$, subsequent values $L_{t+1}, \ldots$ will drop rapidly as the evidence against M_0 is built up. The fact that M_0 was adequate prior to time t does not now mask this local breakdown since that evidence, being of historical relevance only, has been discarded in moving at time t to $L_t = 1$.

The sequence $\{L_t\}$ provides a sequential monitor, or tracking, of the predictive performance of M_0 relative to M_1. The simplest mode of operation involves monitoring the sequence until evidence of inadequacy of M_0, as measured by a small value of L_t at the time, is sufficiently great to warrant intervention. This simple detection of model breakdown is the

primary goal here; once detected, the forecaster may intervene in feed-back mode, attempting to correct the model retrospectively. This use of the Bayes' factors has much in common with standard tracking signals based on sequential probability ratio tests (SPRTs), that has a long history, going back to the early work of Wald in the 1940s (Page 1954; Barnard 1959; Berger 1985, Chapter 7). Specifically, let τ be a prespecified threshold for Bayes' factors, defining the lower limit on acceptability of L_t; τ lies between 0 and 1, values between 0.1 and 0.2 being most appropriate. If L_t exceeds τ, then M_0 operates as usual. Even though the evidence may be against M_0 in the sense that $L_t < 1$, it is not viewed as strong evidence unless $L_t < \tau$. From Theorem 11.3, if in fact $\tau < L_t < 1$, then $L_{t+1} < L_t$ if and only if $H_{t+1} < 1$. If, however, L_t falls below the threshold, breakdown in predictive performance of M_0 is indicated, and we have the following considerations:
If $L_t < \tau$, then

- If $l_t = 1$, then $L_t = H_t$ and the single observation Y_t has led to the monitor signal. There are several possible reasons for such an extreme or discrepant observation. Y_t may be an outlier under M_0, in which case it may be rejected and M_0 used as usual for the next observation stage. Alternatively, Y_t may represent a major departure from M_0 at time t, such as a level change, for example, that M_1 is designed to allow for. This inability to distinguish between an outlying observation and true model changes based on a single, discrepant observation dogs any automatic monitoring scheme. The need for some form of intervention is paramount.
- If $l_t > 1$ then there are several observations contributing individual Bayes' factors to L_t, suggesting departure of the series from M_0 at some time past. The suggested time of onset of departure is l_t steps back at time $t - l_t + 1$.

It is obviously possible to extend the approach to consider two or more alternative forecast densities, leading to a collection of monitoring signals. For example, with two alternatives M_1 and M_2, the application of Theorem 11.3 leads to sequentially updated quantities $L_{i,t}$ for $i = 1$ and 2 respectively. Then the standard analysis will proceed as usual unless either of the minimum Bayes' factors $L_{1,t}$ and $L_{2,t}$ falls below prespecified thresholds τ_1 and τ_2.

11.4.3 Specific alternatives for the DLM

Within the DLM framework, the predictive distributions are, of course, normal or T depending on whether the observational variances are known or unknown. We consider the normal case for illustration, the following fundamental framework being used throughout this section. Suppose that the basic, routine model M_0 to be assessed is a standard normal DLM producing one-step ahead forecast distributions $(Y_t \mid D_{t-1}) \sim \mathrm{N}[f_t, Q_t]$ as usual. Assessing consistency of the observed values Y_t with these distributions is essentially equivalent to assessing consistency of the standardised forecast errors $e_t/Q_t^{1/2} = (Y_t - f_t)/Q_t^{1/2}$ with their forecast distributions, standard normal, and the hypothesis that they are uncorrelated. Thus, for discussion in this section, Bayes' factors are based on the predictive densities of the forecast errors, simply linear functions of the original observations. In the context of sequential monitoring, the focus lies not on the historical performance of the model, but on the local performance; it is the extent to which the current and most recent observations accord with the model that determines whether or not some form of intervention is desirable. To start then, consider only the single observation, equivalently the single forecast error, and without loss of generality, take $f_t = 0$ and $Q_t = 1$, so that under M_0, the forecast distribution for $e_t = Y_t$ is simply

$$(e_t \mid D_{t-1}) \sim \mathrm{N}[0, 1],$$

and so

$$p_0(e_t \mid D_{t-1}) = (2\pi)^{-1/2}\exp\{-0.5e_t^2\}.$$

Within the normal model, there are various possible alternatives M_1 that provide for the types of departure form M_0 encountered in practice. Key examples are as follows.

EXAMPLE 11.2. An obvious alternative is the level change model M_1 in which e_t has a non-zero mean h with

$$p_1(e_t \mid D_{t-1}) = (2\pi)^{-1/2}\exp\{-0.5(e_t - h)^2\}.$$

For any fixed shift h, the Bayes' factor at time t is

$$H_t = p_0(e_t \mid D_{t-1})/p_1(e_t \mid D_{t-1}) = \exp\{0.5(h^2 - 2he_t)\}.$$

Ranges of appropriate values of h may be considered by reference to values of the Bayes' factor at various, interesting values of the error e_t. As an illustration, suppose that the error is positive and consider the point at which $H_t = 1$, so that there is no evidence from e_t alone to discriminate between M_0 and M_1. At this point, $\log(H_t) = 0$ and so, since h is non-zero, $h = 2e_t$. To be indifferent between the models on the basis of an error $e_t = 1.5$, for example, (at roughly the upper 90% point of the forecast distribution) suggests that $h = 3$. Similarly, suppose that a threshold $H_t = \tau$, $(0 < \tau << 1)$, is specified, below which the evidence is accepted as a strong indication that e_t is inconsistent with M_0. A threshold of -2 for the log-Bayes' factor implies $\tau = e^{-2} \approx 0.135$, for example. Fixing e_t at an acceptably extreme value when $H_t = \tau$ leads to a quadratic equation for h, namely $h^2 - 2he_t - 2\log(\tau) = 0$. Thus, $e_t = 2.5$ (roughly the upper 99% point of the forecast distribution) and $\log(\tau) = -2$ implies $h = 1$ or $h = 4$. Thus, values of h between 3 and 4 lead to indifference between M_0

and M_1 when e_t is near 1.5, and fairly strong evidence ($\tau = e^{-2}$) against M_0 for values as high as $e_t = 2.5$. Note that if e_t is negative, the sign of h must be reversed.

EXAMPLE 11.3. Consider the previous example, and suppose that (as will typically be the case in practice) it is desired to allow for the possibilities of change in either direction. Then two alternatives are needed. Treating the cases symmetrically, suppose that the first, denoted by M_1, has the form in Example 11.2 with mean shift $h > 0$, and the second, M_2, has the same form though with mean shift $-h$. The mode of operation of the sequential monitor now involves applying Theorem 11.3 to each alternative and leads to minimum Bayes' factors $L_{i,t}$, $(i = 1, 2)$. The standard analysis will now proceed as usual unless either of the minimum Bayes' factors $L_{1,t}$ and $L_{2,t}$ falls below prespecified thresholds. Consistent with the symmetric treatment of the alternatives, suppose that these thresholds are both equal to some value τ. It follows from Example 11.2 that with a common run-length l_t,

$$\log(L_{1,t}) = 0.5h^2 l_t - hE_t,$$

where

$$E_t = \sum_{r=0}^{l_t - 1} e_{t-r}$$

is the sum of the most recent l_t errors. Similarly,

$$\log(L_{2,t}) = 0.5h^2 l_t + hE_t.$$

Hence

$$\begin{aligned} L_{1,t} \geq 1 \quad &\text{if and only if} \quad E_t \leq \ \ 0.5hl_t, \\ L_{2,t} \geq 1 \quad &\text{if and only if} \quad E_t \geq -0.5hl_t. \end{aligned}$$

In either case, $|E_t| \leq 0.5hl_t$, consistent with the standard model and monitoring is reinitialised with $l_t = 0$ and $L_{1,t} = L_{2,t} = 1$ before proceeding to time $t + 1$. The monitor will signal in favour of one of the alternatives if either Bayes' factor drops below the prespecified threshold τ,

$$\begin{aligned} L_{1,t} \leq \tau \quad &\text{if and only if} \quad E_t \geq \ \ 0.5hl_t - \frac{\log(\tau)}{h}, \\ L_{2,t} \leq \tau \quad &\text{if and only if} \quad E_t \leq -0.5hl_t + \frac{\log(\tau)}{h}. \end{aligned}$$

Thus, the double monitoring is based on E_t alone, the operation here being equivalent to standard *backward cusum* techniques (Harrison and Davies 1964, and later in Section 11.6.2). E_t is a cusum (cumulative sum) of the most recent l_t errors. The standard model is accepted as satisfactory at time t so long as E_t satisfies

$$0.5hl_t - \frac{\log(\tau)}{h} \geq E_t \geq -0.5hl_t + \frac{\log(\tau)}{h}.$$

If E_t is so large or so small as to lie outside one or other of these bounds, then the monitor signals in favour of the corresponding alternative model, and intervention is needed. If, on the other hand, E_t lies within the bounds and, in fact, $|E_t| < 0.5hl_t$, then monitoring is reinitialised with $l_t = 0$ and $L_{1,t} = L_{2,t} = 1$ before moving to time $t+1$. Otherwise, monitoring proceeds as usual. The monitoring bounds are sometimes referred to as *moving V-masks* for the backward cusum (Harrison and Davies 1964).

EXAMPLE 11.4. A useful single alternative is the scale shift model M_1 in which e_t has standard deviation k rather than unity, with

$$p_1(e_t \mid D_{t-1}) = (2\pi k^2)^{-1/2}\exp\{-0.5(e_t/k)^2\}.$$

The Bayes' factor at time t is then

$$H_t = k\ \exp\{-0.5e_t^2(1-k^{-2})\}.$$

Modelling a scale inflation with $k > 1$ provides a widely useful alternative. It is immediately clear that it is appropriate for changes in observational variance, or *volatility*, in the series. It is also a rather robust alternative for more general changes, being designed as it is to allow for observations that are extreme relative to M_0. The use of such models has a long history in Bayesian statistics, particularly in modelling outliers (Box and Tiao 1968; and Smith and Pettit 1985). The effects of different values of k can be assessed as in the previous example by considering indifference and extreme points, this being left to exercises for the reader. With a view to allowing for large errors, we draw on experience with outlier models, as just referenced, anticipating that above a certain level, the particular value of k chosen is largely irrelevant. The key point is that the alternative provides a larger variance than M_0; thus large errors will tend to be more consistent with M_1 no matter how large the variance inflation is. Some indication of the variation with respect to k appears in Figure 11.8. The curves plotted here provide Bayes' factor as a function of $|e_t|$ over the range 2-4, for $k =$ 2, 3, 4 and 5. This range of values of $|e_t|$ focuses attention on the region in which we would doubt the model M_0. Obviously, evidence in favour of M_0 decreases as $|e_t|$ increases. Of major interest is the fact that when the error becomes extreme relative to M_0, the choice of k matters little, the Bayes' factors all being close to 0.2 when the error is 2.5, and below 0.1 when the error is at 3. A scale inflation of $k = 3$ or $k = 4$ with a threshold of $\tau \approx 0.15$ provides a useful, general alternative.

In considering specifically the possibilities that in model M_0, e_t may be a wild, outlying observation or that the defining parameters in M_0 may have changed at (or before) time t, then either of these alternatives may be suitable. The level shift model allows for a jump in the series in a particular

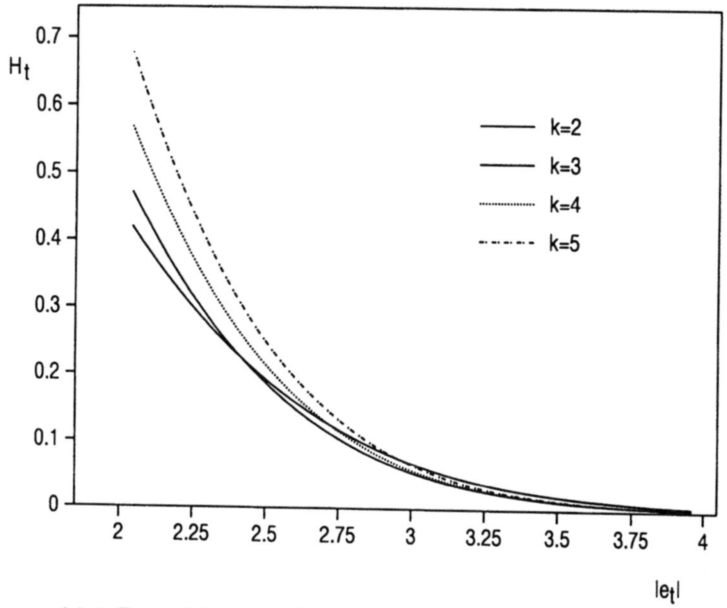

Figure 11.8 Bayes' factors H_t in normal scale inflation model

direction determined by the sign of h. The scale shift model allows for changes in either direction, and also for changes in variation. In this sense, the latter alternative may be viewed as a general alternative, allowing for a variety of types of changes in addition to outlying observations.

11.5 FEED-BACK INTERVENTION

11.5.1 Automatic exception detection and diagnosis

Identifying model breakdown using simple monitoring techniques is the first step in automatic exception handling. Most appropriately, such signals should prompt a user response, suggesting subjective, feed-back interventions to correct M_0 and adapt to new conditions. In automatic mode, the monitoring of model performance to *detect* deterioration in predictions needs to be supplemented with techniques for *diagnosis* of the problem, and subsequently with *adaptation* to control and correct for the problem. When changes in parameter values are the primary causes of breakdown of M_0, adaptation is required rapidly to improve future predictions. The possibility of outlying observations confuses the diagnosis issue; on the basis of a single, discrepant observation and no further information, parametric change is indistinguishable from a wild data point. Generally, if an outlier can be identified as such, the policy of omitting the observation from the analysis, treating it as a missing value, is a simple and sound one. The updating of distributions in M_0 will then be unaffected by the wild value, and possible causes and consequences can be explored separately. The following logical scheme provides a guide to the use of the Bayes' factors in detecting and diagnosing model breakdown in such circumstances.

At time t, whatever has occurred previously, proceed with the monitor as follows:

(A) Calculate the single Bayes' factor H_t. If $H_t \geq \tau$, then Y_t is viewed as consistent with M_0; proceed to (B) to assess the possibilities of model failure (i.e., marked changes in parameters) prior to time t. If, on the other hand, $H_t < \tau$, then Y_t is a *potential* outlier and should be omitted from the analysis, being treated as a missing value (from the point of view of updating and revising M_0). However, the possibility that Y_t presages change in model parameters must be allowed for after rejecting the observation; thus the need for intervention is identified and we proceed to (C).

(B) Calculate the cumulative Bayes' factor L_t and the corresponding run-length l_t to assess the possibility of changes prior to time t. If $L_t \geq \tau$, then M_0 is satisfactory, and so proceed to (D) to perform standard updates, etc. Otherwise, $L_t < \tau$ indicates change that should be indicated, requiring intervention; proceed to (C). Note that the sensitivity of this technique to slow changes can be increased by inferring a possible breakdown of M_0 if either $L_t < \tau$ or $l_t > 3$ or 4, say. The rationale here is that several (3 or 4) recent observations may provide evidence very marginally favouring M_1 over M_0, but this may be so small that L_t, whilst being less than 1, still exceeds τ.

(C) Issue signal of possible changes consistent with deterioration of predictions from M_0 and call for feed-back interventions to adapt the model for the future. Following such interventions, update the time index to $t+1$ for the next observation stage, and proceed to (A), reinitialising monitoring by setting $l_t = 0$ and $L_t = 1$.

(D) Perform usual analysis and updating with M_0, proceeding to (A) at time $t+1$.

With this scheme in mind, it remains to specify the forms of intervention at points of possible changes that are detected. The full range of user interventions of Section 11.2 are, of course, available. Automatic alternatives for routine use are described in the next section.

11.5.2 Automatic adaptation in cases of parametric change

It is a general principle that the onset of change brings with it increased uncertainties that if appropriately incorporated in the model, lead naturally to more rapid adaptation in the future. Thus, models can be made self-correcting if uncertainties about parameters can be markedly increased at points of suspected change. This is quite consistent with the use of increased variances on model parameters in the various forms of subjective intervention in Section 11.2, and the mode of intervention described there suggests a general, automatic procedure for adapting to changes once detected. The basic idea is simple: on detecting change, the prior variances of model parameters can be increased by the addition of further evolution noise terms to allow further data to influence more heavily the updating to posterior distributions. It is possible to use this automatic mode of intervention retrospectively, feeding in greater uncertainty about parameters at the most likely point of change l_t observations past, although for forecasting, it currently matters little that the change was not allowed for at the time. What is important is a response now to adapt for the future.

With this in mind, consider again the scheme (A) to (D) above, and suppose that we are at (C), having identified that changes beyond those allowed for in M_0 may have occurred. An intervention of the form (11.5), additional evolution noise at the current time, permits greater changes in $\boldsymbol{\theta}_t$, and if the extra evolution variance $\mathbf{H}_t$ is appropriately large, leads to automatic adaptation of M_0 to the changes. Additionally, unless particular parameters are identified by the user as subject to changes in preferred directions, then the additional evolution noise should not anticipate directions of change, therefore having zero mean. Thus, the feed-back interventions called for at (C) may be effected simply, automatically and routinely via the representation from Theorem 11.1: $\boldsymbol{\theta}_t = \mathbf{G}_t\boldsymbol{\theta}_{t-1} + \boldsymbol{\omega}_t + \boldsymbol{\xi}_t$, where $\boldsymbol{\omega}_t \sim \mathrm{N}[\mathbf{0}, \mathbf{W}_t]$, as usual, being independent of the automatic intervention noise term $\boldsymbol{\xi}_t \sim \mathrm{N}[\mathbf{0}, \mathbf{H}_t]$. The scheme (A) to (D) with this form of automatic intervention is represented in the flowchart of Figure 11.9.

The additional variance $\mathbf{H}_t$ must be specified. It is important that $\mathbf{H}_t$ provide increased uncertainty for those parameters most subject to abrupt change, and obviously the user has much scope here to model differing degrees of durability of parameters, just as in designing the standard evolution matrix $\mathbf{W}_t$. It is possible, as an extreme example, to take some of the elements of $\mathbf{H}_t$ to be zero, indicating that no additional changes are allowed in the corresponding parameters of $\boldsymbol{\theta}_t$. As a generally useful approach, the magnitude and structure of $\mathbf{H}_t$ may be modelled on that of the standard evolution variance matrix $\mathbf{W}_t$. In particular, if all parameters are subject to possible changes, an appropriate setting, neutral as to source and magnitude of changes, is simply to take

$$\mathbf{H}_t = (c-1)\mathbf{W}_t,$$

where $c > 1$. This has the effect of an additional noise term with the same covariance structure as the usual $\boldsymbol{\omega}_t$, but with an inflation of the overall variation in $\boldsymbol{\theta}_t$ by a factor of c.

A generalisation of this simple technique has been used extensively in connection with the use of discount factors to routinely structure $\mathbf{W}_t$ (West

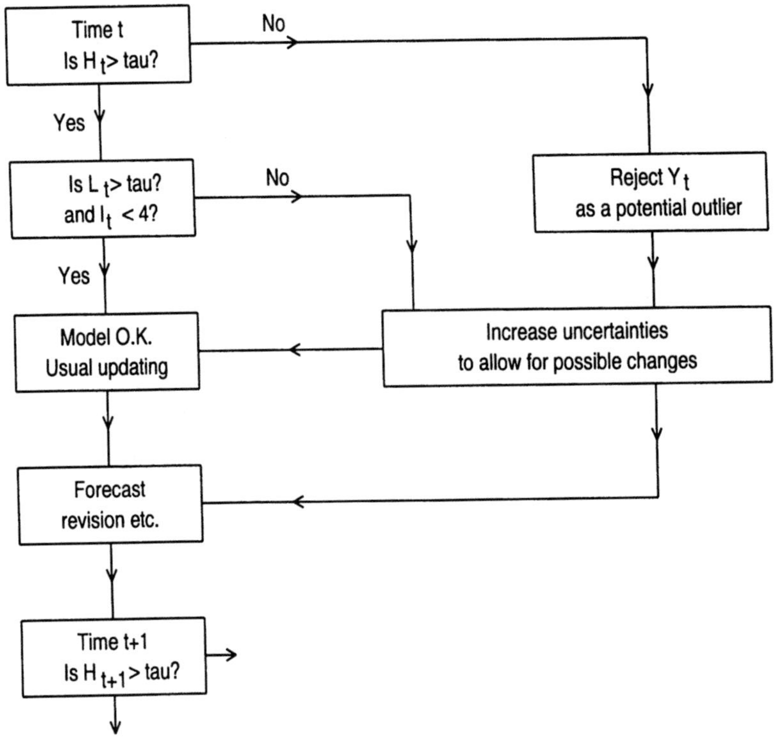

Figure 11.9 Exception detection and handling routine

and Harrison 1986a; Harrison and West 1986, 1987; West, Harrison and Pole 1987; Pole, West and Harrison 1994). Discount factors for components of $\boldsymbol{\theta}_t$ provide suitable structure and magnitudes for the standard evolution noise variance matrix $\mathbf{W}_t$, controlling the nature and extent of the basic dynamic. At points of abrupt change that go beyond this basic dynamic, a more heavily discounted version of the matrix $\mathbf{W}_t$ is an obvious way of simply and automatically extending the model. Thus, $\mathbf{H}_t$ may be specified so that $\mathbf{W}_t + \mathbf{H}_t$ is just of that form; this is based on the discount concept that applies to $\mathbf{W}_t$ but with smaller discount factors than standard. The following section illustrates the use of this in automatic adaptation to abrupt changes.

11.5.3 Industrial sales illustration

Illustration is provided in an analysis combining the use of the above scheme with simple feed-forward intervention based on externally available information. The series of interest is displayed in Figure 11.10, the data given in Table 11.3.

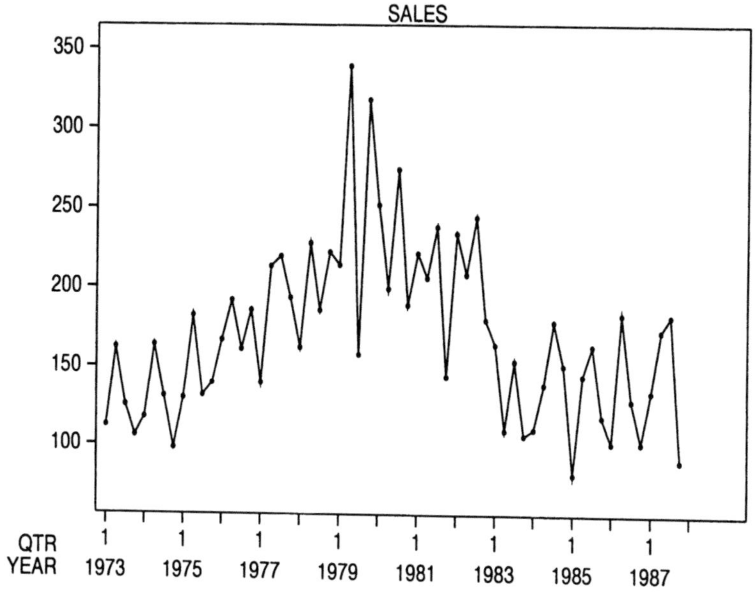

Figure 11.10 Quarterly industrial sales series

The analysis reported was performed using the BATS package (West, Harrison and Pole 1987), as were most others throughout the book. (Updated software for just these analyses, with a variety of additional illustrations, is given by Pole, West and Harrison (1994), as mentioned in the Preface.) The data are quarterly sales figures over a period of years, representing total sales of a chemical product in international industrial markets. Seasonality is evident, though the extent and durability over time of the seasonal component is unclear from the graph, the non-seasonal trend also changing markedly from time to time.

Consider the position of a forecaster at $t = 1$, corresponding to the first quarter of 1973, using a basic first-order polynomial component DLM for the trend in sales, plus a full seasonal effects component. For all t, $\boldsymbol{\theta}_t$ has 6 elements comprising the level growth, and 4 seasonal factors constrained to have zero sum. Suppose the initial prior for $\boldsymbol{\theta}_1$ is specified as

$$(\boldsymbol{\theta}_1|D_0) \sim \mathrm{T}_{20}[\mathbf{a}_1, \mathbf{R}_1],$$

and for the observational precision,

$$(\phi|D_0) \sim \mathrm{G}[20, 4500],$$

where

$$\mathbf{a}_1 = (130, 0;\ 0, 0, 0, 0)'$$

Table 11.3. Quarterly industrial sales data

Year	Quarter			
	1	2	3	4
1973	112.08	162.08	125.42	105.83
1974	117.50	163.75	131.25	97.92
1975	130.00	182.08	131.67	139.58
1976	166.67	191.67	160.83	185.42
1977	139.58	212.88	219.21	193.33
1978	162.00	227.50	185.42	221.92
1979	213.79	339.08	157.58	317.79
1980	251.83	199.08	274.13	188.79
1981	221.17	205.92	238.17	143.79
1982	233.75	207.92	244.17	179.58
1983	164.04	109.58	153.67	106.25
1984	110.42	138.75	178.33	150.83
1985	81.67	144.17	163.33	118.33
1986	101.67	182.92	128.75	101.67
1987	134.17	172.50	182.08	90.42

and

$$\mathbf{R}_1 = \begin{pmatrix} 225 & 0 & 0 & 0 & 0 & 0 \\ 0 & 100 & 0 & 0 & 0 & 0 \\ 0 & 0 & 300 & -100 & -100 & -100 \\ 0 & 0 & -100 & 300 & -100 & -100 \\ 0 & 0 & -100 & -100 & 300 & -100 \\ 0 & 0 & -100 & -100 & -100 & 300 \end{pmatrix}.$$

Thus, the initial estimate of $V = 1/\phi$ is $S_0 = 4500/20 = 225$, an estimated observational standard deviation of 15, with 20 degrees of freedom. The expected level of the series is 130; thus the prior on the variance represents a reasonably strong belief that the random variation will have standard deviation of about 10 or 11% of the expected level. $\mathbf{R}_1$ indicates a rather uncertain view about the components initially, particularly in the seasonal component where no pattern is anticipated, with large variances so that the model will rapidly adapt to the data over the first couple of years. The model specification is completed through the values of discount factors. The component discount approach provides $\mathbf{W}_t$ in block diagonal form (like $\mathbf{R}_1$) for all t through the use of two discount factors, one for trend and one for seasonality. Trend and seasonality are viewed to be of similar durability, and this is reflected in the choice of discount factors of 0.95 for each component. In addition, some minor variation over time in V is modelled through the use of an observational variance discount factor of 0.99 (Section 10.8).

The model is subject to monitoring and automatic adaptation at points of change identified. At such points, the above standard values of the discount factors are dropped to lower values consistent with more marked change in the parameters. Note that this automatic shift to smaller, exceptional discount factors applies only at points identified as change points through the routine monitoring scheme of the previous section. Otherwise the standard values apply. Thus, at points of possible change, the use of smaller discounts simulates a user intervention to increase uncertainty about $\boldsymbol{\theta}_t$ and V above and beyond that already modelled through the standard discounts. In the analysis reported here, these exceptional values are 0.1 for the trend and seasonal components and 0.9 for the observational variance. Concerning the trend and seasonal components, the standard discount of 0.95 implies an increase of roughly 2.5% in the standard deviations of the parameters between observations in the standard analysis. At exceptions, the drop to 0.1 implies roughly a three-fold increase, allowing much more marked adaptation to forthcoming data. Monitoring is based on the use of the single, robust scale-inflation alternative in Example 11.4. Thus, predictive performance of the model under standard conditions is compared through the Bayes' factors with an alternative whose one-step ahead forecast distributions have the same location but increased spread, the forecast standard deviations under the alternative inflated by a factor $k > 1$. The value used for k here is 2.5, the threshold level τ taken as 0.2. Now the one-step forecast distributions are Student T, but the discussion of the case of normal distributions in the examples of the previous section provides a guide to the implications of these chosen values (and the comments about insensitivity to them apply). With a normal forecast distribution, the Bayes' factor H_t, based on a single observation, will be unity for a standardised forecast error of roughly ± 1.5. For errors larger in absolute value, the evidence weighs against the standard model, reaching the threshold $\tau = 0.2$ at standardised errors of roughly ± 2.5. Following the discussion about monitor sensitivity in cases of less marked change, the decision is taken to signal for automatic intervention based on the size of the minimum Bayes' factor L_t and the run-length l_t combined, whenever $L_t < \tau$ *or* $l_t \geq 3$.

The data are sequentially observed and processed subject to the monitoring and adaptation scheme thus defined. The background to the data series puts the data in context and provides information relevant to modelling the development over time. In the early years of the data, representing the early life of the product in newly created markets, the product sales grow noticeably, the company manufacturing the product essentially dominating the market. In early 1979, however, information becomes available to the effect that a second major company is to launch a directly competitive product later that year. In addition to the direct effect on sales that this is expected to have over coming years, it is anticipated that it will also encourage other competing companies into the market. Thus, change in trend, and possibly seasonality, is expected. For the analysis here, however, such information is not fed-forward, all interventions to adapt to observed changes being left to the automatic scheme. One feed-forward intervention is made, however, on the basis of company actions taken immediately on receiving the information about the competitor. In an attempt to promote the existing product to consumers and to combat the initial marketing of the competitor, the company launches a marketing drive in early 1979 that includes price reductions throughout that year. The market response to this is anticipated to lead to a major boost in sales, but there is great uncertainty about the response, the effects of the promotional activities possibly continuing to the end of the year. In order that the underlying market trend, including competitor effects, be estimated as accurately as possible free from the effects of this promotion, some form of intervention is necessary. The simplest such intervention is to assume that the effects of the promotions lead to observations that are essentially uninformative about the underlying market trend and seasonal pattern, and the decision is taken simply to omit the observations in the last three quarters of 1979 when the promotions are expected to significantly impact on sales. Thus, these three observations are effectively ignored by the analysis, treated as missing values.

As the analysis proceeds from $t = 0$, updating as usual, the values of L_t and l_t are calculated from the simple recursions in Theorem 11.3. The logged values of the former are plotted against t in Figure 11.11. There are six points at which the monitor signals:

(1) The first point at which evidence weighs against the standard model is at $t = 16$, where $L_{16} = H_{16}$ is less than unity for the first time. The monitor does not signal, however, since it remains above the threshold and $l_{16} = 1$. At $t = 17$, corresponding to the first quarter of 1977, $H_{17} < 1$ too, and $L_{17} = H_{16}H_{17} < \tau$, indicating breakdown based on two observations, $l_{17} = 2$.

(2) After automatic intervention using the exceptional discount factors at $t = 17$, $L_t > 1$, so that $l_t = 1$ until $t = 23$. At $t = 25$, the first quarter of 1979, L_{25} exceeds τ, but $l_{25} = 3$ and the monitor signals. Three consecutive observations provide evidence against the model that though not so extreme as to cross the chosen threshold, is viewed as requiring intervention. Note that this use of the run-length in addition to the Bayes' factor may be alternatively modelled by directly linking the threshold to the run-length.

(3) A similar exception occurs at $t = 37$, the first quarter of 1982, again based on the run-length.

(4) At $t = 41$, the single observation Y_{41} is discrepant enough so that $H_{41} < \tau$, and the monitor signals with $l_{41} = 1$.

(5) At times $t = 46, 47$ and 48 L_t is low, close to the threshold though not dropping below it. There is clear evidence here accumulating

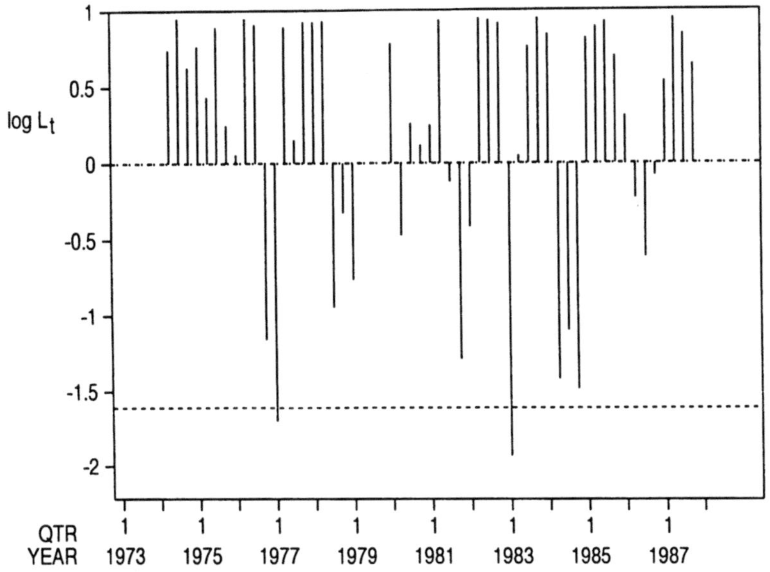

Figure 11.11 Tracking signal for industrial sales data analysis

against the model, and the monitor signals when $l_{48} = 3$, in the fourth quarter of 1984.

(6) Finally, the run-length reaches 3 again at $t = 56$, the fourth quarter of 1986, and an exception is identified. In contrast to the previous exception, note that the evidence here is only marginally against the standard model, certainly compared to that at the preceding exception.

These events are interpreted and discussed with reference to the graphs in Figures 11.12 to 11.17. Figure 11.12 displays the usual on-line estimate, with two standard deviation intervals, of the non-seasonal trend component at each time t. The adequacy of the locally linear trend description is apparent, with marked changes in the trend at three points of intervention. The trend apparently grows roughly linearly up to the start of the 1980s, plateaus off for a few years, and drops markedly and rapidly during 1983 from levels near 200 to between 100 and 150, remaining at these, lower levels thereafter. Uncertainty about this on-line estimated trend is fairly high, particularly after the crash to lower levels in 1983. At the intervention times, the interval about the estimated trend increases in width

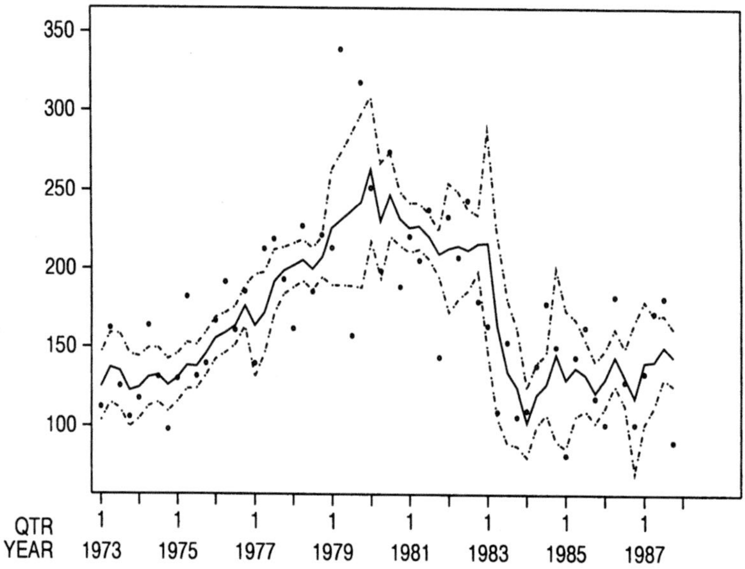

Figure 11.12 On-line estimated trend in industrial sales

due to the intervention, allowing the model to adapt to any changes. In 1979, the intervention in the first quarter has this effect and is followed by three quarters in which, due to the feed-forward intervention to isolate the effects of the company promotions, the observations are ignored. As a consequence, the uncertainty about the trend, and the other model components, increases throughout 1979, no information being obtained.

As always, the on-line estimates of model components can appear rather erratic, responding as they do to data as they are processed. The interventions increase this response, and so on-line trajectories provide only a first, tentative indication of the pattern of behaviour over time. More appropriate for retrospective analysis are the smoothed and filtered distributions for model components calculated using the filtering algorithms backwards over time from the end of the series. Figure 11.13 provides a plot of the data (with the three observations omitted from analysis removed and indicated on the time axis), together with a plot of the filtered estimates of the mean response function. The latter, appearing as a line through the data over time, provide smoothed, or retrospectively fitted, values of the sales figures. The points of intervention are indicated by vertical lines. Note that uncertainty about the smoothed values is not indicated on the graph, so as not to obscure the main features noted here. Uncertainties are indicated, in terms of 2 standard deviation intervals, in the corresponding

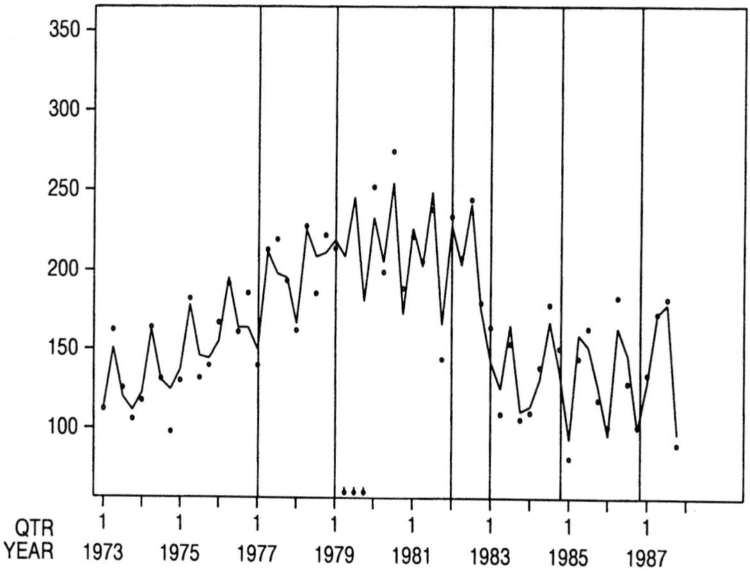

Figure 11.13 Retrospective fitted values of industrial sales

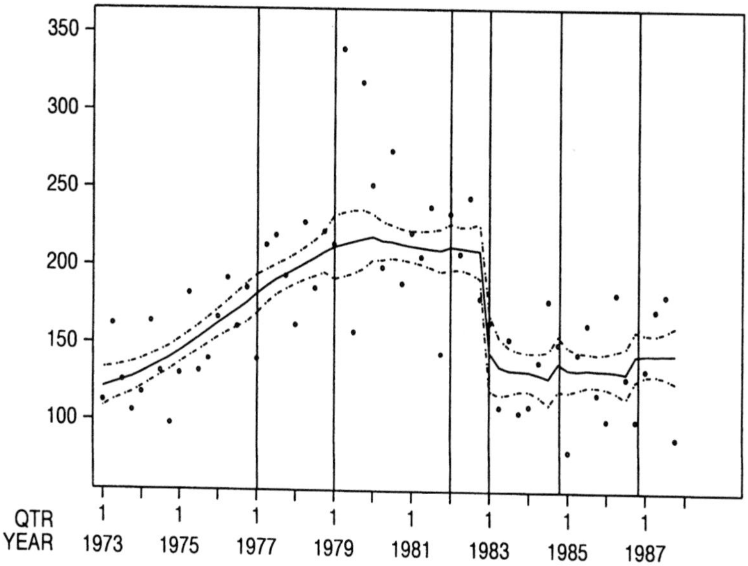

Figure 11.14 Filtered trend in industrial sales

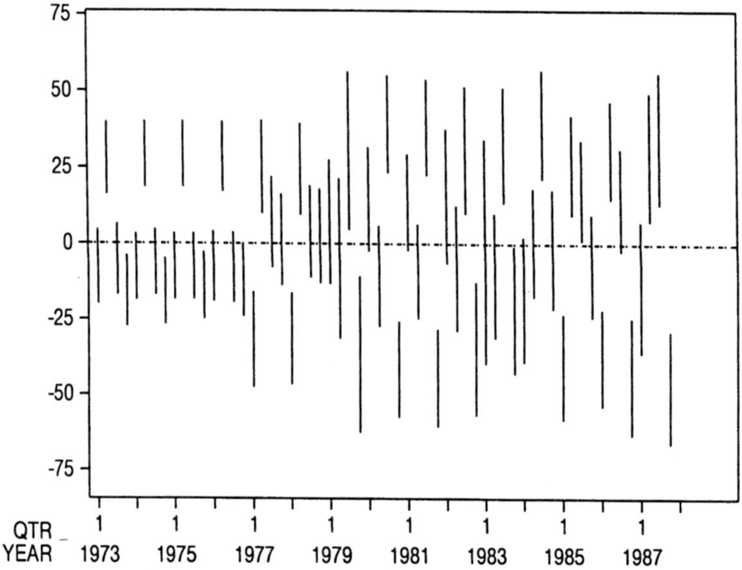

Figure 11.15 Filtered seasonal component

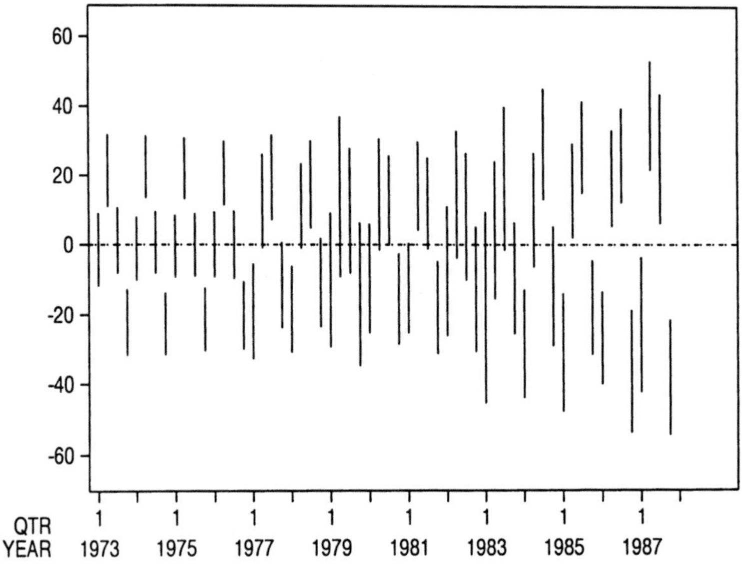

Figure 11.16 Filtered first harmonic

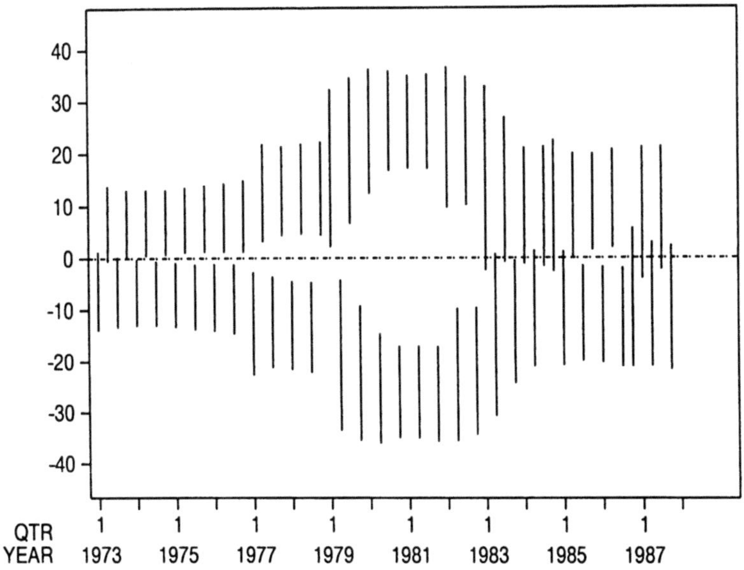

Figure 11.17 Filtered second harmonic

plots for the trend and seasonal components appearing in Figures 11.14 and 11.15. Additionally, the seasonal form is decomposed into the two harmonic components; filtered estimates of each of these, with intervals, appear in Figures 11.16 and 11.17.

Figure 11.14, the smoothed version of Figure 11.12, indicates that the non-seasonal trend in sales grows steadily until 1979, then flattens off at around 220, decaying very slightly until the end of 1982. At this point there is a huge drop in level to around 130, growth thereafter being small though increasing towards the end of the series in late 1986. Consider now the effects of the automatic interventions that resulted from the monitor signals. At each of these six times, the exceptional discount factors used allow for abrupt change in all model components. The trend, however, changes markedly only at the beginning of 1983. Clearly, the monitor signal here was vital in terms of rapidly adapting to this change. The earlier signals do not appear to have led to the trend adapting, so that they are attributable to some other feature of mismatch between the model and the data. The intervention in early 1979, followed by further increased uncertainty in the model due to the three omitted observations in that year, inflates uncertainty sufficiently that the levelling off in trend is identified and appropriately estimated without further interventions. This behaviour may, in retrospect, be attributed to the competitor incursion into the market. The huge drop in sales level in early 1983 bears study, possibly relating

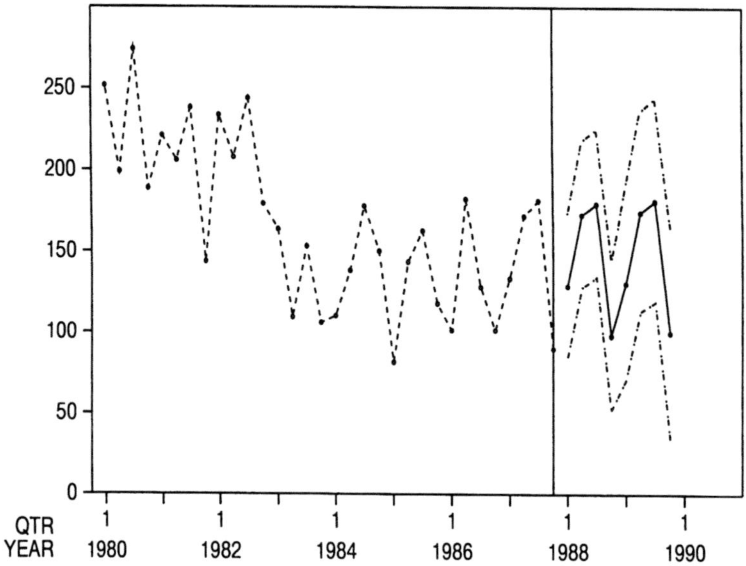

Figure 11.18 Forecast industrial sales for 1988/89

to several factors. Company decisions to withdraw from part of the market is one possibility that if true, could and should have led to further feed-forward intervention to anticipate the change. Major and sustained incursion by competitors is another. The international industrial recession is a third, important consideration, that could easily explain the crash in sales level through the withdrawal of demand from one or a small number of consumer companies who had hitherto comprised a major part of the market. Whatever reasons are considered retrospectively, the model, operating without feed-forward information, identifies and appropriately estimates the crash. The final two interventions, in 1984 and 1986, appear to be attributable, at least in part, to minor, though important, increases in non-seasonal trend.

The form of the seasonal pattern and its changes over time are evident from the fitted sales values in Figure 11.13 and the seasonal component in Figure 11.15, together with the harmonics in Figures 11.16 and 11.17. From Figure 11.13, the seasonal form appears stable up to the first intervention point in early 1977. Here, and again at the second intervention in early 1979, the form of seasonality apparently changes. In particular, the contribution of the second harmonic is more marked after each intervention, very noticeably so after that in 1979. After the drop in level in 1983, seasonality again changes noticeably, the second harmonic becoming again less important whilst the amplitude of the first is much increased.

In retrospect, the three observations omitted from the analysis can now be clearly explained as the response to promotion. The first of these, in the second quarter of 1979, is a huge peak in sales resulting from the price-cutting promotional activity. In the third quarter, sales drop off, partially due to seasonal demand but probably also due to stocks having been depleted in the previous quarter. Sales bounce back to high levels in the final quarter as supplies come through again.

Figure 11.18 displays step ahead forecast sales, with 2 standard deviation intervals, made for each quarter in 1988 and 1989 from the end of the series in the fourth quarter of 1987. Here the effects of the later interventions show up in the forecasts, particularly through the final shift to a higher non-seasonal level in sales.

11.6 A BAYES' DECISION APPROACH TO MODEL MONITORING

11.6.1 Introduction

The Bayes' factors model monitoring approach of Section 11.4, like that using the associated sequences of sequential probability ratio tests (SPRTs), is concerned with statistical rather than practical significance and ignores explicit loss functions. Readers familiar with Bayes' decision theory will question the wisdom of this, not least because of the disturbing features associated with distributions with inverse polynomial tails, where huge jumps and outliers will not be signalled. For example, in monitoring the mean of a univariate Student T distribution, if $Y \sim \mathrm{T}_v[\mu, Q]$, then for any given means μ_1 and μ_2, as y increases, the Bayes' factor tends to 1 since

$$\lim_{Y \to \infty} \frac{p(Y|\mu_1)}{p(Y|\mu_2)} = \lim_{Y \to \infty} \left[\frac{vQ + (Y - \mu_2)^2}{vQ + (Y - \mu_1)^2} \right]^{\frac{v+1}{2}} = 1.$$

In order to overcome the drawbacks associated with such statistical tests, this section develops a simple, powerful decision theoretic monitoring scheme. In the applications of interest this results in decisions based upon cumulative sums (cusums) of a function of the observations, that in the case of the exponential family of distributions, includes the Bayes' factor schemes as particular cases. So, whilst retaining good features of the Bayes' factor schemes, this approach provides acceptable schemes for those cases in which the Bayes' factor schemes are unsatisfactory.

Cusum forecast monitoring schemes have been employed since 1959, following the procedures of Harrison and Davies (1964). The justification has largely been empirical success together with an appeal to the fact that in many cases, they are equivalent to sequences of Wald sequential probability ratio tests and thus statistically efficient.

After defining a single sided cusum decision scheme, the equivalence of exponential family Bayes' factor schemes and cusums is given. The Bayes' decision approach based upon Harrison and Veerapen (1994), and as extended by Harrison and Lai (1999), is then detailed. The general idea is that at each time t, the decision problem concerns a set of three point decision problems $\{\delta_s, \delta_a, \delta_c\}$. At any time the loss function of each decision, relative to that of continuing the current run, δ_c, is assumed to be linear. Accepting a proposed set of requirements, little more than two indifference points need be specified in order to establish the decision procedure. In particular, the exponential family Bayes' factor schemes are shown to correspond to a formal decision theoretic approach using a simple linear loss function.

11.6.2 Cusums

After each observation a cusum monitoring scheme makes one of three decisions:

δ_s : questionable model performance; issue monitor signal;
δ_a : acceptable model performance; reinitialize monitor;
δ_c : continue monitor with a further observation.

In quality control applications these decisions often relate to two simple hypotheses about a parameter μ, namely the Acceptable Quality Level μ_0, and the Rejectable Quality Level μ_1. In forecasting, μ_0 relates to the satisfactory performance of the routine DLM and μ_1 to some specific inadequacy. In operation, any number of such inadequacies may be specified and monitored using multiple cusums, that are simply compositions of single-sided schemes.

Definition 11.2. Let $X_1, \dots, X_r$, be observations since the last decision δ_a or δ_s was made, where r is called the run length. An upper cusum decision scheme $CD(k, g, h)$, with $g < h$, operates so that the decision based upon

$$C_r = \sum_{i=1}^{r} (X_i - k) = C_{r-1} + X_r - k$$

is

$$\begin{aligned} &\delta_s \ : \ \text{if } C_r \geq h, \\ &\delta_a \ : \ \text{if } C_r \leq g, \\ &\delta_c \ \ \text{otherwise.} \end{aligned}$$

If either of the decisions $\{\delta_s, \delta_a\}$ is made, the scheme is reinitialised, setting $r = C_0 = 0$.

In quality control an upper cusum relates to $\mu_1 > \mu_0$. A lower cusum scheme is then appropriate when $\mu_1 < \mu_0$ and simply reverses the test inequality signs. The symmetric graphical V-mask cusum scheme, as given in Woodward and Goldsmith (1964) and Harrison and Veerapen (1994), is simply a composition of two independent schemes, an upper CD$(k, 0, h)$ and a lower CD$(k^*, 0, h)$, where $k + k^* = \mu_0$.

Definition 11.3. An upper cusum decision scheme $CD(k, \mathbf{g}, \mathbf{h})$, with $\mathbf{g} = (g_1, \dots)$, $\mathbf{h} = (h_1, \dots)$ and $g_r < h_r$ for all r, is such that the decision based upon C_r is

$$\begin{aligned} \delta_s &: \text{ if } C_r \geq h_r, \\ \delta_a &: \text{ if } C_r \leq g_r, \\ \delta_c &\ \text{ otherwise.} \end{aligned}$$

If either of the decisions $\{\delta_s, \delta_a\}$ is made, then after appropriate intervention in the case of δ_s, the scheme is reinitialised, setting $r = C_0 = 0$.

11.6.3 Exponential family/cusum equivalences

The exponential family of distributions includes many frequently used distributions (normal, gamma, binomial, beta, etc.). Consequently, results for these various distributions can be expressed in a general form, as in Chapter 14 that deals with exponential family dynamic models. Consider independent identically distributed exponential family random variables Y_i with continuous natural scalar parameter η, known scale factor ϕ, $X_i = X(Y_i)$ as a single sufficient statistic for η given Y_i, and generalised conditional pdf for Y_i given η

$$p(Y_i|\eta) = b(Y_i, \phi) \exp\{\phi(X_i \eta - a(\eta))\}.$$

Here $a(\eta)$ is convex and twice differentiable so that $\mu = E(X_i|\eta) = \dot{a}(\eta)$ and μ increases with η.

Phrase the monitoring issue in terms of two possible values of the natural parameter $\eta_1 > \eta_0$, and let the initial prior odds ratio on η_1 versus η_0 be $h(0)$. After r observations the posterior odds ratio $h(r)$ in favour of η_1 versus η_0 is given by

$$\log h(r) = \log h(0) + \log H_r$$

where H_r is the Bayes' factor for η_1 versus η_0 of Definition 11.1, i.e.,

$$\log H_r = \phi(\eta_1 - \eta_0) \sum_{i=1}^{r} \left[X_i - \frac{a(\eta_1) - a(\eta_0)}{\eta_1 - \eta_0} \right].$$

With $h(0) = 1$, the quantity $h(r)$ reduces to the Bayes' factor. Now, for prespecified thresholds $h > g > 0$, the decision is

$$\begin{aligned} \delta_s &: \text{ if } h(r) \geq h, \\ \delta_a &: \text{ if } h(r) \leq g, \\ \delta_c &: \text{ otherwise.} \end{aligned}$$

So this monitor is equivalent to the cusum scheme

$$\mathrm{CD}\left(\frac{a(\eta_1)-a(\eta_0)}{\eta_1-\eta_0},\ \frac{\log g-\log h(0)}{\phi(\eta_1-\eta_0)},\ \frac{\log h-\log h(0)}{\phi(\eta_1-\eta_0)}\right).$$

If $g = h(0)$ this reduces to the cusum scheme $\mathrm{CD}(k, 0, h)$; when $g = h(0) = 1$ it reduces to the Bayes' factor test of Section 11.4.

EXAMPLE 11.5. Monitoring a normal mean μ.
Let $(X_i|\mu) \sim \mathrm{N}[\mu, 1]$ be independent, identically distributed normal random variables with mean μ, and without loss in generality, unit variance. Let $\mu_1 > \mu_0$ and the prior odds on μ_1 versus μ_0 be $h(0) = g$. Then

$$\log h(r) = \log h(0) + (\mu_1 - \mu_0)\sum_{i=1}^{r}\left(X_i - \frac{\mu_0+\mu_1}{2}\right),$$

and the Bayes' factor test is equivalent to the cusum decision scheme

$$\mathrm{CD}\left(\frac{\mu_1+\mu_0}{2},\ 0,\ \frac{\log h-\log h(0)}{\mu_1-\mu_0}\right).$$

11.6.4 A Bayesian decision approach

Consider a single-sided monitoring scheme concerned with signalling when the scalar parameter μ, relating to a general probability distribution, is unusually large. Generally, for any given run length of r observations, the loss functions $l(\delta, \mu)$ associated with the three possible decisions are such that

$l(\delta_s, \mu)$ decreases monotonically with μ;
$l(\delta_a, \mu)$ increases monotonically with μ;
$l(\delta_c, \mu)$ is relatively flat, intersecting each of the other loss functions at just one value of μ.

For each run length r, define $a_r < s_r$ as the two unique values of μ for which the relative loss functions corresponding to information relating to that run length satisfy

$$l(\delta_s, s_r) - l(\delta_c, s_r) = 0$$

and

$$l(\delta_a, a_r) - l(\delta_c, a_r) = 0.$$

These relative loss functions are usually smooth and well-represented as linear functions of μ over their respective critical ranges, i.e., locally in the neighbourhoods of a_r and s_r, and if necessary, with appropriate choice of parametrisation. Based upon information D_r relating to a run length r, and with v_r and w_r as positive proportionality constants that usually depend upon r, consider the loss functions as

$$l(\delta_s, \mu) - l(\delta_c, \mu) = w_r(s_r - \mu) \tag{11.10a}$$

and

$$l(\delta_a, \mu) - l(\delta_c, \mu) = v_r(\mu - a_r). \tag{11.10b}$$

Each time a decision δ_a or δ_s is taken, then after appropriate intervention, the sequential decision process is reinitialised with information $(\mu|D_0)$, mean $E[\mu|D_0] = m_0$, and precision equivalent to n_0 observations. After r subsequent observations, given $D_r = \{D_0, Y_1, \ldots, Y_r\}$, the precision associated with $(\mu|D_r)$ is then $n_r = n_0 + r$ equivalent observations.

The following requirements will be demanded of the loss functions:

(i) As the run length r increases, so the utility of deciding δ_c decreases. Otherwise, a long run may hide a recent process change and delay detection. Consequently, a desirable property is that s_r and a_r be monotonic in r and that $s_r - a_r$ decreases to 0. Define k such that

$$\lim_{r\to\infty} a_r = \lim_{r\to\infty} s_r = k.$$

(ii) Within a run, the information provided by the rth observation Y_r relative to D_r is inversely proportional to n_r. It is then arguable that the utility of taking decision δ_c relative to taking one of the other two decisions is adequately characterised as decreasing in inverse proportion to some function of "the precision" n_r. So, for some constants $b, c \geq 0$ and some positive monotonic increasing function of n_r, $\psi(\cdot)$, define

$$s_r = k + \frac{c}{\psi(n_r)} = k + \frac{\psi(n_0)}{\psi(n_r)}(s_0 - k), \tag{11.11a}$$

$$a_r = k - \frac{b}{\psi(n_r)} = k - \frac{\psi(n_0)}{\psi(n_r)}(k - a_0). \tag{11.11b}$$

This decision scheme simply requires the specification of two unique initial indifference values $\{a_0, s_0\}$, an ultimate indifference value k, and the function ψ. There is no need to specify the proportionality constants $\{v_r, w_r\}$, the only requirement being that they are positive.

Theorem 11.4. *In monitoring a scalar parameter μ, let the prior mean be* $\mathrm{E}[\mu|D_0] = m_0$ *with an associated precision n_0, and given $D_r = \{y_r, D_{r-1}\}$, the posterior mean, $m_r = \mathrm{E}[\mu|D_r]$, be such that if X_r is some known function of Y_r, then*

$$n_r m_r = n_0 m_0 + \sum_{i=1}^{r} X_i.$$

Then, given D_r, $a_0 < k < s_0$, $\psi(\cdot)$ and the loss functions defined by equations 11.10 and 11.11, the Bayesian decision scheme is a cusum decision *scheme CD*$(k, \mathbf{g}, \mathbf{h})$, *where for* $r = 1, \ldots,$

$$g_r = n_0(k - m_0) + \frac{n_r \psi(n_0)}{\psi(n_r)}(a_0 - k)$$

and

$$h_r = n_0(k - m_0) + \frac{n_r \psi(n_0)}{\psi(n_r)}(s_0 - k).$$

So, writing $C_r = \sum_{i=1}^{r}(X_i - k)$, *the Bayes' decision is*

$$\begin{aligned} \delta_s &: \ \textit{if} \ \ C_r \geq h_r, \\ \delta_a &: \ \textit{if} \ \ C_r \leq g_r, \end{aligned}$$

and

$$\delta_c \ : \ \textit{otherwise.}$$

Proof. Given D_r, and with $n_r = n_0 + r$, the expected losses $l(\delta)$ satisfy

$$\begin{aligned} l_r(\delta_a) - l_r(\delta_c) &\propto \frac{1}{n_r}\left(n_0 m_0 + \sum_{i=1}^{r} X_i\right) - k + \frac{\psi(n_0)}{\psi(n_r)}(k - a_0) \\ &\propto C_r + n_0(m_0 - k) + \frac{n_r \psi(n_0)}{\psi(n_r)}(k - a_0). \end{aligned}$$

Similarly,

$$l_r(\delta_s) - l_r(\delta_c) \propto -\left[C_r + n_0(m_0 - k) + \frac{n_r \psi(n_0)}{\psi(n_r)}(k - s_0)\right].$$

Notice that at most one of these two differences of expected losses can be negative. The unique Bayes' decision is δ_a if the first is negative, δ_s if the second is negative, and δ_c otherwise.

◇

11.6.5 A typical Bayes' decision setting

The case of $\psi(n) \propto n$ is of particular interest, providing many equivalent Bayes', SPRT, and traditional linear cusum schemes, since then

$$s_r = s_0 + \frac{r}{n_0 + r}(k - s_0) \qquad \text{and} \qquad a_r = a_0 + \frac{r}{n_0 + r}(k - a_0),$$

producing the cusum scheme

$$\text{CD}(k,\ n_0[a_0 - m_0],\ n_0[s_0 - m_0]).$$

This gives a decision theoretic basis for the Bayes' factor schemes relating to the natural parameter of the exponential family, but also generates acceptable schemes for parameters from other distributions. Further, in sequences of sequential tests, practitioners typically set the prior mean such that $a_0 = m_0$, resulting in an equivalent traditional cusum decision scheme

$$CD(k,\ 0,\ n_0[s_0 - m_0]).$$

We now look at some important examples, each of which satisfies the theorem requirements that

$$n_r = n_0 + r \qquad \text{and} \qquad n_r m_r = n_0 m_0 + \sum_{i=1}^{r} X_i.$$

EXAMPLE 11.6. The normal mean.
In DLM forecasting with known variances, X_i is often the standardised one-step ahead error and $m_0 = 0$. With the usual DLM notation, we have

$$X_i = (Y_i - f_i)/\sqrt{Q_i}$$

and adopt

$$(X_i|\mu, D_{i-1}) \sim \mathrm{N}[\mu, 1], \quad \text{and} \quad (\mu|D_0) \sim \mathrm{N}\left[m_0,\ 1/n_0\right],$$

so that for $r \geq 0$,

$$(\mu|D_r) \sim \mathrm{N}\left[m_r,\ 1/n_r\right].$$

EXAMPLE 11.7. Conjugate analysis for the exponential family. .
In many applications concerning members of the exponential family, the sampling and prior distributions of the natural parameter μ may be written as

$$p(Y_r|\mu, D_{r-1}) \propto \exp\{\phi[X_r\mu - a(\mu)]\}$$

and

$$p(\mu|D_{r-1}) \propto \exp\{n_{r-1}\phi[m_{r-1}\mu - a(\mu)]\},$$

when the posterior, given $D_r = \{y_r, D_{r-1}\}$, is easily verified as

$$p(\mu|D_r) \propto \exp\{n_r\phi[m_r\mu - a(\mu)]\},$$

where

$$n_r = n_{r-1} + 1$$

and

$$n_r m_r = n_{r-1}m_{r-1} + X_r.$$

When $\mathrm{E}[\mu|D_r] = m_r$ it is evident that the Bayes' decision procedure results in a cusum decision scheme that is also a Bayes' factor (SPRT) scheme when $\psi(n_r) \propto n_r$. Since this applies to many frequently met cases, such as normal means, binomial means, Poisson means, etc., the reader may feel that this gives strong support to the proposition that in many applications, it is sufficient to characterise $\psi(n_r)$ as proportional to n_r.

EXAMPLE 11.8. Variance monitoring.
In monitoring the DLM observation variance V, the key quantity is

$$X_i = \frac{(Y_i - f_i)^2}{Q_i} = \frac{e_i^2}{Q_i},$$

where, writing $\mu = V$ in the foregoing,

$$(X_i|V, D_{i-1}) \sim \mathrm{G}\left[1/2,\ 1/2V\right].$$

With the prior

$$(V^{-1}|D_0) \sim \mathrm{G}\left[1 + n_0/2,\ n_0 m_0/2\right],$$

such that $\mathrm{E}[V|D_0] = m_0$, we see that

$$(V^{-1}|D_r) \sim \mathrm{G}\left[1 + n_r/2\ n_r m_r/2\right],$$

with $\mathrm{E}[V|D_r] = m_r$, where $n_r = n_0 + r$ and $n_r m_r = n_0 m_0 + \sum_{i=1}^{r} X_i$.

EXAMPLE 11.9. Control of a normal mean with unknown variance.
Without loss in generality, consider the DLM case in which, within the run commenced at time $t = 1$,

$$X_i = \frac{e_i}{\sqrt{S_0 Q_i}},$$

$$(X_i|\mu, D_0) \sim \mathrm{T}_{v_0}\left[\mu, 1\right],$$

and

$$(\mu|D_0) \sim \mathrm{T}_{v_0}\left[m_0,\ 1/n_0\right].$$

For monitoring the mean, it is a good idea to keep the run prior $(V|D_0)$ as the constant working distribution of V within each run, updating it at the start of each new run (although the schemes can easily be modified to incorporate variance learning within runs). Then the marginal posterior distribution for μ is

$$(\mu|D_r) \sim \mathrm{T}_{v_0}\ \left[m_r,\ 1/n_r\right].$$

Control of the mean now follows an acceptable cusum scheme that has no trouble in signalling very large changes in μ and outliers. Since the T distribution is not a member of the exponential family, this scheme differs markedly from the Bayes' factor scheme that fails to deal adequately with very large mean shifts and outliers.

11.7 EXERCISES

(1) Verify the general results in Theorem 11.1 and Theorem 11.2.

(2) In the first-order polynomial model $\{1, 1, V_t, W_t\}$ with state parameter $\mu_t = \boldsymbol{\theta}_t$, suppose that $(\mu_{t-1}|D_{t-1}) \sim \mathrm{N}[m_{t-1}, C_{t-1}]$ as usual. Intervention is to be performed to achieve the prior $(\mu_t|D_{t-1}) \sim \mathrm{N}[m_{t-1}, R_t^*]$, where $R_t^* = C_{t-1} + W_t + H_t = R_t + H_t$ for some $H_t > 0$. This can obviously be done directly through Theorem 11.1 by adding a further evolution noise term $\xi_t \sim \mathrm{N}[0, H_t]$ to the existing evolution equation. Alternatively, the more general approach through Theorem 11.2 can be used. Show that this approach leads to the revised evolution equation at time t given by

$$\mu_t = G_t^* \mu_{t-1} + \omega_t^*, \qquad \omega_t^* \sim \mathrm{N}[h_t, W_t^*],$$

where

$$G_t^* = (1 + H_t/R_t)^{1/2},$$
$$h_t = (1 - G_t^*)m_{t-1},$$

and

$$W_t^* = (1 + H_t/R_t)W_t.$$

(3) Suppose that $\mathbf{R}$ is a 2×2 variance matrix,

$$\mathbf{R} = \begin{pmatrix} R_1 & R_3 \\ R_3 & R_2 \end{pmatrix}.$$

Let $\mathbf{Z}$ be the unique, upper-triangular matrix

$$\mathbf{Z} = \begin{pmatrix} Z_1 & Z_3 \\ 0 & Z_2 \end{pmatrix}$$

satisfying $\mathbf{R} = \mathbf{ZZ}'$. Prove that $\mathbf{Z}$ is given by

$$Z_1 = (R_1 - R_3^2/R_2)^{1/2},$$
$$Z_2 = R_2^{1/2}$$

and

$$Z_3 = R_3/R_2^{1/2}.$$

(4) Consider Example 11.1 supposing that at $t = 37$ (January 1958), we have

$$\mathbf{a}_{37} = \begin{pmatrix} 865 \\ 0 \end{pmatrix} \quad \text{and} \quad \mathbf{R}_{37} = \begin{pmatrix} 100 & 10 \\ 10 & 2 \end{pmatrix}.$$

A forecaster considers five possible interventions, leading to the five possible post-intervention moment pairs $\mathbf{a}_{37}^*$ and $\mathbf{R}_{37}^*$ in (a)-(e) below. In each case, use the result of the previous exercise to calculate the quantities $\mathbf{G}_t^*$, $\mathbf{h}_t$ and $\mathbf{W}_t^*$ in the amended evolution equation as in Theorem 11.2.

(a)
$$\begin{pmatrix} 970 \\ 0 \end{pmatrix} \quad \text{and} \quad \begin{pmatrix} 100 & 10 \\ 10 & 2 \end{pmatrix}.$$

(b)
$$\begin{pmatrix} 970 \\ 0 \end{pmatrix} \quad \text{and} \quad \begin{pmatrix} 100 & 0 \\ 0 & 0 \end{pmatrix}.$$

(c)
$$\begin{pmatrix} 970 \\ 0 \end{pmatrix} \quad \text{and} \quad \begin{pmatrix} 100 & 0 \\ 0 & 25 \end{pmatrix}.$$

(d)
$$\begin{pmatrix} 970 \\ 0 \end{pmatrix} \quad \text{and} \quad \begin{pmatrix} 100 & 25 \\ 25 & 25 \end{pmatrix}.$$

(e)
$$\begin{pmatrix} 970 \\ 0 \end{pmatrix} \quad \text{and} \quad \begin{pmatrix} 50 & 5 \\ 5 & 2 \end{pmatrix}.$$

(5) A quarterly seasonal effects component of a DLM at time t has the usual normal prior distribution $(\boldsymbol{\theta}_t|D_{t-1}) \sim \mathrm{N}[\mathbf{a}_t, \mathbf{R}_t]$ with

$$\mathbf{a}_t = \begin{pmatrix} 25 \\ 131 \\ -90 \\ -66 \end{pmatrix} \quad \text{and} \quad \mathbf{R}_t = \begin{pmatrix} 12 & -4 & -4 & -4 \\ -4 & 12 & -4 & -4 \\ -4 & -4 & 12 & -4 \\ -4 & -4 & -4 & 12 \end{pmatrix}.$$

(a) Verify that this prior is consistent with the zero-sum constraint $\mathbf{1}'\boldsymbol{\theta}_t = 0$.

(b) A forecaster considers it likely that the peak level in the second quarter can be expected to increase from 131 to about 165 before the next observation, but that there is additional uncertainty about this change. This is modelled by an additional evolution noise term, and the moments are directly revised to

$$\mathbf{a}_t^* = \begin{pmatrix} 25 \\ 165 \\ -90 \\ -66 \end{pmatrix} \quad \text{and} \quad \mathbf{R}_t = \begin{pmatrix} 28 & -4 & -4 & -4 \\ -4 & 12 & -4 & -4 \\ -4 & -4 & 12 & -4 \\ -4 & -4 & -4 & 12 \end{pmatrix}.$$

Show that this revised prior does not now satisfy the zero-sum constraint, and apply Theorem 8.2 to appropriately correct it.

(6) In monitoring based on the scale inflation alternative of Example 11.4, define E_t to be the sum of squares of the most recent l_t forecast

errors at time t,

$$E_t = \sum_{r=0}^{l_t-1} e_{t-r}^2.$$

(a) Let x be any real quantity such that $0 < x \leq 1$. Prove that $\log(L_t/x) \leq 0$ if and only if

$$E_t \geq al_t + b(x),$$

where

$$a = 2k^2\log(k)/(k^2 - 1)$$

and

$$b(x) = -2k^2\log(x)/(k^2 - 1).$$

Verify that $a > 0$ and $b(x) \geq 0$ for all x such that $0 < x \leq 1$.

(b) Using (a), prove that $L_t \leq 1$, so that evidence weighs against the standard model when $l_t^{-1}E_t \geq a$. Comment on this result.

(c) Show also that the monitor signals model failure when $E_t \geq al_t + b(\tau)$.

(7) In Exercise 6, suppose that the data actually follow the standard model being monitored, so that the errors e_t are independent, standard normally distributed random quantities.

(a) Deduce that $E_t \sim \chi^2_{l_t}$, a chi-square random quantity with l_t degrees of freedom.

(b) With $k = 2.5$ and $\tau = 0.2$, calculate the probability that $L_t \leq \tau$ for each of $l_t = 1, 2, 3$ and 4. Comment on the meaning of these probabilities.

(c) Recalculate the above probabilities for each combination of pairs of values of k and τ given by $k = 2$ or 3 and $\tau = 0.1, 0.2$ and 0.3. Comment on the sensitivity of the derived probabilities as functions of k and τ and the implications for the monitor in practice.

(8) In Exercises 6 and 7, suppose now that $e_t \sim \mathrm{N}[0, k^2]$ for all t.

(a) Deduce the distribution of E_t.

(b) Perform the probability calculations in (b) above and comment on the meaning of these probabilities.

(9) Following Example 11.3, describe how the scale inflation alternative can be extended to two alternatives, with scale factors $k > 1$ and k^{-1} respectively, to monitor for the possibilities of either increased or decreased variation.

(10) Write a computer program to implement the monitoring and automatic adaptation routine described in Section 11.5, based on the scale inflation alternative model of Example 11.4 for the monitor. This should apply to any component DLM using component discounting, the automatic adaptation being modelled through the use

of exceptional discount factors as in Sections 11.5.2 and 11.5.3. Test the program by performing the analysis of the industrial sales data in 11.5.3, reproducing the results there.

(11) Analyse the CP6 sales data in Table 11.1 and Figure 11.1 using a first-order polynomial model with a single discount factor $\delta = 0.9$ for the trend component and a variance discount factor $\beta = 0.95$. Apply the automatic monitoring as programmed in the previous example with exceptional trend and variance discount factors of 0.05 and 0.5 respectively.

(12) Reanalyse the CP6 sales data as in the previous example. Instead of discounting the observational variance to account for changes, explore models that rather more appropriately, use variance laws as in Section 10.7.2. In particular, consider models in which the observational variance at time t is estimated by the weighted variance $V_t = k(f_t)V$, where V is unknown as usual and $k(f) = f^p$, for some $p > 1$. Explore the relative predictive fit of models with values of p in the range $1 \leq p \leq 2$.

(13) Reanalyse the confectionary sales and cost series from Exercise 10.7 subject to monitoring and adaptation using the program from Exercise 7 above. Note that these data are analysed using a related model in West, Harrison and Pole (1987), with extensive summary information from the analysis provided there.

CHAPTER 12

MULTI-PROCESS MODELS

12.1 INTRODUCTION

Discussion of interventionist ideas and monitoring techniques is an implicit acknowledgement of the principle that although an assumed DLM form may be accepted as appropriate for a series, the global behaviour of the series may be adequately mirrored only by allowing for changes from time to time in model parameters and defining structural features. Intervention allows for parametric changes, and goes further by permitting structural changes to be made to the defining quadruple $\{\mathbf{F},\ \mathbf{G},\ V,\ \mathbf{W}\}_t$. In using automatic monitoring techniques as in Section 11.4, the construction of specific alternative models rather explicitly recognises the global inadequacy of any single DLM, introducing possible explanations of the inadequacies. The simple use of such alternatives to provide comparison with a standard, chosen model is taken much further in this chapter. We formalise the notion of explicit alternatives by considering classes of DLMs, the combination of models across a class providing an overall super-model for the series. This idea was originally developed in Harrison and Stevens (1971), and taken further in Harrison and Stevens (1976a). We refer to such combinations of basic DLMs as *multi-process* models; any single DLM defines a process model; the combination of several defines a multi-process model. Loosely speaking, the combining is effected using discrete probability mixtures of DLMs, and so multi-process models may be alternatively referred to simply as mixture models. Following Harrison and Stevens (1976a), we distinguish two classes of multi-process models, Class I and Class II, that are fundamentally different in structure and serve rather different purposes in practice. The two classes are formally defined and developed throughout the chapter after first providing a general introduction.

Generically, we have a DLM defined by the usual quadruple at each time t, here denoted by

$$M_t \quad : \quad \{\mathbf{F},\ \mathbf{G},\ V,\ \mathbf{W}\}_t,$$

conditional on initial information D_0. Suppose that any defining parameters of the model that are possibly subject to uncertainty are denoted by $\boldsymbol{\alpha}$. Examples include discount factors, transformation parameters for independent variables, eigenvalues of $\mathbf{G}$ in the case when $\mathbf{G}_t = \mathbf{G}$ constant for all time, and so forth. Previously, all such quantities, assumed known, had been incorporated in the initial information set D_0 and therefore not made explicit. Now we are considering the possibility that some of these quantities are uncertain, and so we explicitly include them in the conditioning of all distributions in the model analysis, reserving the symbol D_0 for all other known and certain quantities. Represent the dependence of the model on these uncertain quantities by writing

$$M_t = M_t(\boldsymbol{\alpha}), \qquad (t = 1, 2, \ldots).$$

The initial prior in the model may also depend on $\boldsymbol{\alpha}$ although this possibility is not specifically considered here. For any given value of $\boldsymbol{\alpha}$, $M_t(\boldsymbol{\alpha})$ is a standard DLM for each time t. It is the possibility that we do not precisely know, or are not prepared to assume that we know, the value of $\boldsymbol{\alpha}$ that leads us to consider multi-process models. Let $\mathcal{A}$ denote the set of possible values for $\boldsymbol{\alpha}$, whether it be uncountably infinite, with $\boldsymbol{\alpha}$ taking continuous values; discrete and finite; or even degenerate at a single value. The class of DLMs at time t is given by

$$\{\ M_t(\boldsymbol{\alpha}) \ : \ \boldsymbol{\alpha} \in \mathcal{A}\ \}. \tag{12.1}$$

The two distinct possibilities for consideration are as follows:

(I) For some $\boldsymbol{\alpha}_0 \in \mathcal{A}$, $M_t(\boldsymbol{\alpha}_0)$ holds for all t. (12.2a)

(II) For some sequence of values $\boldsymbol{\alpha}_t \in \mathcal{A}$, $(t = 1, 2, \ldots)$,
$M_t(\boldsymbol{\alpha}_t)$ holds at time t. (12.2b)

In (I), a single DLM is viewed as appropriate for all time, but there is uncertainty as to the "true" value of the defining parameter vector $\boldsymbol{\alpha} = \boldsymbol{\alpha}_0$. In (II), by contrast, and usually more realistically, there is no single DLM accepted as adequate for all time. The possibility that different models are appropriate at different times is explicitly recognised and modelled through different defining parameters $\boldsymbol{\alpha}_t$.

EXAMPLE 12.1. Suppose that $\boldsymbol{\alpha}$ is a set of discount factors used to determine the evolution variance matrices $\mathbf{W}_t$. Under (I), a fixed set of discount factors is assumed appropriate, although just which set of values is uncertain. Sometimes this assumption, leading to a stable and sustained degree of variation in model parameters $\boldsymbol{\theta}_t$ over time, is tenable. This might be the case, for example, with time series observed in physical or electronic systems with stable driving mechanisms. More often in socio-economic areas, this assumption will be plausible only over rather short ranges of time, these ranges being interspersed by marked discontinuities in the series. To cater adequately for such discontinuities within an existing DLM, the parameters $\boldsymbol{\theta}_t$ must change appropriately, and this may often be effected by altering discount factors temporarily. The interventionist ideas of Chapter 11 are geared to this sort of change, and (II) above formalises the notion in terms of alternative models.

Other examples include unknown power transformation parameters, and analogously, variance power law indices (see Section 10.6). Whichever form of model uncertainty, (12.2a) or (12.2b), is assumed to be appropriate, the basic analysis of multi-process models rests upon manipulation of collections of models within a discrete mixture framework. The basics are developed in the next section in connection with the first approach, (I) of (12.2a).

12.2 MULTI-PROCESS MODELS: CLASS I

12.2.1 General framework

The basic theory underlying mixture models is developed in the context of the introduction under assumption (I) of (12.2a). Much of this follows Harrison and Stevens (1976a), and general background material can be found in Titterington, Smith and Makov (1985).

Thus, Y_t follows a DLM $M_t(\boldsymbol{\alpha})$, the precise value $\boldsymbol{\alpha} = \boldsymbol{\alpha}_0$ being uncertain. The following components of analysis are basic:

(1) Given any particular value $\boldsymbol{\alpha} \in \mathcal{A}$, the DLM $M_t(\boldsymbol{\alpha})$ may be analysed as usual, producing sequences of prior, posterior and forecast distributions that are sequentially updated over time as data are processed. The means, variances and other features of these distributions all depend, usually in complicated ways, on the specific value $\boldsymbol{\alpha}$ under consideration. We make this dependence explicit in the conditioning of distributions and densities. At time t, let $\mathbf{X}_t$ be any vector of random quantities of interest; for example, the full $\boldsymbol{\theta}_t$ state vector, a future observation Y_{t+k} $(k > 0)$, etc. Inference about $\mathbf{X}_t$ in $M_t(\boldsymbol{\alpha})$ is based on the density

$$p(\mathbf{X}_t \mid \boldsymbol{\alpha}, D_t). \tag{12.3}$$

Many such densities may exist, one for each $\boldsymbol{\alpha} \in \mathcal{A}$.

(2) Starting with an initial prior density $p(\boldsymbol{\alpha} \mid D_0)$ for $\boldsymbol{\alpha}$, information is sequentially processed to provide inferences about $\boldsymbol{\alpha}$ via the posterior $p(\boldsymbol{\alpha} \mid D_t)$ at time t. This is sequentially updated, as usual, using Bayes' theorem,

$$p(\boldsymbol{\alpha} \mid D_t) \;\propto\; p(\boldsymbol{\alpha} \mid D_{t-1})p(Y_t \mid \boldsymbol{\alpha}, D_{t-1}). \tag{12.4}$$

Here $p(Y_t \mid \boldsymbol{\alpha}, D_{t-1})$ is the usual, one-step ahead predictive density from $M_t(\boldsymbol{\alpha})$, simply (12.3) at time $t-1$ and with $\mathbf{X}_{t-1} = Y_t$. The posterior $p(\boldsymbol{\alpha} \mid D_t)$ informs about $\boldsymbol{\alpha}$, identifying interesting values and indicating the relative support, from the data and initial information, for the individual DLMs $M_t(\boldsymbol{\alpha})$ $(\boldsymbol{\alpha} \in \mathcal{A})$.

(3) To make inferences about $\mathbf{X}_t$ without reference to any particular value of $\boldsymbol{\alpha}$, the required unconditional density is

$$p(\mathbf{X}_t \mid D_t) = \int_{\mathcal{A}} p(\mathbf{X}_t \mid \boldsymbol{\alpha}, D_t)p(\boldsymbol{\alpha} \mid D_t)d\boldsymbol{\alpha}, \tag{12.5}$$

or, simply, the expectation of (12.3) with respect to $\boldsymbol{\alpha}$ having density (12.4). If, for example, $\mathbf{X}_t = Y_{t+1}$, then (12.5) is the one-step forecast density for Y_{t+1} at time t.

This is, in principle, how we handle uncertain parameters $\boldsymbol{\alpha}$. In practice, all is not so straightforward; it is rare that (12.4) is a tractable, easily calculated and manageable density. The calculations implicitly required are as follows:

(i) For each $\mathbf{X}_t$ of interest, calculate (12.3) for all possible values of $\boldsymbol{\alpha}$.
(ii) For each $\boldsymbol{\alpha}$, calculate (12.4), and then integrate over $\mathcal{A}$ to normalise the posterior density $p(\boldsymbol{\alpha} \mid D_t)$.
(iii) For each $\mathbf{X}_t$ of interest, perform the integration in (12.5) and the subsequent integrations for posterior moments, probabilities, etc., for $\boldsymbol{\alpha}$.

Unfortunately, we will typically not now have a neat set of sequential updating equations for these calculations. The parameters in $\boldsymbol{\alpha}$ will enter into the likelihood $p(Y_t \mid \boldsymbol{\alpha}, D_{t-1})$ in (12.4) in such a complicated way that there will be no conjugate form for the prior for $\boldsymbol{\alpha}$, no simple summary of D_{t-1} for $\boldsymbol{\alpha}$ in terms of a fixed and small number of sufficient statistics. Thus, all the required operations in (i) to (iii) must be done numerically. Generally this will be a daunting task if the parameter space $\mathcal{A}$ is at all large, and impossible to carry out if, in particular and as is often the case, $\boldsymbol{\alpha}$ is continuous. The computations are really only feasible when the parameter space is discrete and relatively small. Two cases arise, namely when either (a) $\boldsymbol{\alpha}$ truly takes only a small number of discrete values, and (b) when a discrete set of values is chosen as representative of a large and possibly continuous "true" space $\mathcal{A}$. Under (b) large spaces $\mathcal{A}$ can often be adequately approximated for some purposes by a fairly small discrete set that somehow spans the larger space, leading to the consideration of a small number of distinct DLMs. We now consider the use of such collections of models, whether they be derived under (a) or (b).

12.2.2 Definitions and basic probability results

Definition 12.1. Suppose that for all t, $M_t(\boldsymbol{\alpha})$ holds for some $\boldsymbol{\alpha} \in \mathcal{A}$, the parameter space being the finite, discrete set $\mathcal{A} = \{\boldsymbol{\alpha}_1, \dots, \boldsymbol{\alpha}_k\}$ for some integer $k \geq 1$. Then the series Y_t is said to follow a **multi-process, class I** model.

The general theory outlined in the previous section applies here with appropriate specialisation the discretisation of $\mathcal{A}$. Thus, the densities for $\boldsymbol{\alpha}$ in (12.4) and (12.5) are mass functions. For convenience, the notation is simplified in the discrete case as follows:

Definition 12.2.

(i) $p_t(j)$ is the posterior probability at $\boldsymbol{\alpha} = \boldsymbol{\alpha}_j$ in (12.4), namely

$$p_t(j) = p(\boldsymbol{\alpha}_j \mid D_t) = \Pr[\boldsymbol{\alpha} = \boldsymbol{\alpha}_j \mid D_t], \qquad (j = 1, \dots, k)$$

for all t, with specified initial prior probabilities $p_0(j)$.

(ii) $l_t(j)$ is the value of the density $p(Y_t \mid \boldsymbol{\alpha}_j, D_{t-1})$, providing the likelihood function for $\boldsymbol{\alpha}$ in (12.4) as j varies,

$$l_t(j) = p(Y_t \mid \boldsymbol{\alpha}_j, D_{t-1}), \qquad (j = 1, \ldots, k)$$

for each t.

The discrete versions of (12.4) and (12.5) are as follows:

- The posterior masses are updated via $p_t(j) \propto p_{t-1}(j)l_t(j)$, or

$$p_t(j) = c_t p_{t-1}(j)l_t(j), \tag{12.6}$$

where $c_t^{-1} = \sum_{j=1}^{k} p_{t-1}(j)l_t(j)$. Note that c_t^{-1} is just the observed value of the unconditional predictive density for Y_t, namely

$$c_t^{-1} = \sum_{j=1}^{k} p(Y_t \mid \boldsymbol{\alpha}_j, D_{t-1})p_{t-1}(j).$$

- Marginal posterior densities are now simply

$$p(\mathbf{X}_t \mid D_t) = \sum_{j=1}^{k} p(\mathbf{X}_t \mid \boldsymbol{\alpha}_j, D_t)p_t(j). \tag{12.7}$$

From (12.7) it follows that all posterior distributions for linear functions of $\boldsymbol{\theta}_t$, and predictive distributions for future observations, are *discrete probability mixtures* of the standard T or normal distributions. Some basic features of such mixtures, fundamental to their use in inference, are as follows:

(a) Probabilities are calculated as discrete mixtures. Generally, for any set of values of interest $\mathbf{X}_t \in \mathcal{X}$,

$$\Pr[\mathbf{X}_t \in \mathcal{X} \mid D_t] = \sum_{j=1}^{k} \Pr[\mathbf{X}_t \in \mathcal{X} \mid \boldsymbol{\alpha}_j, D_t]p_t(j),$$

the conditional probabilities $\Pr[\mathbf{X}_t \in \mathcal{X} \mid \boldsymbol{\alpha}_j, D_t]$ being calculated from the relevant T or normal distribution as usual.

(b) Moments are similar mixtures, thus,

$$\mathrm{E}[\mathbf{X}_t \mid D_t] = \sum_{j=1}^{k} \mathrm{E}[\mathbf{X}_t \mid \boldsymbol{\alpha}_j, D_t]p_t(j),$$

with the conditional moments calculated from the relevant T or normal distribution.

(c) Particular results of use derived from (b) are as follows: Let $\mathbf{h}_t = \mathrm{E}[\mathbf{X}_t \mid D_t]$ and $\mathbf{H}_t = \mathrm{V}[\mathbf{X}_t \mid D_t]$. Similarly, for $j = 1, \ldots, k$, let $\mathbf{h}_t(j) = \mathrm{E}[\mathbf{X}_t \mid \boldsymbol{\alpha}_j, D_t]$, and $\mathbf{H}_t(j) = \mathrm{V}[\mathbf{X}_t \mid \boldsymbol{\alpha}_j, D_t]$. Then we have

- Posterior means:

$$\mathbf{h}_t = \sum_{j=1}^{k} \mathbf{h}_t(j) p_t(j),$$

- Posterior variances:

$$\mathbf{H}_t = \sum_{j=1}^{k} \{\mathbf{H}_t(j) + [\mathbf{h}_t(j) - \mathbf{h}_t][\mathbf{h}_t(j) - \mathbf{h}_t]'\} p_t(j).$$

In previous chapters we have used Bayes' factors in model comparison. The formalism of alternatives within the multi-process framework here brings Bayes' factors into the picture. Consider any two elements of $\mathcal{A}$, namely $\boldsymbol{\alpha}_i$ and $\boldsymbol{\alpha}_j$ for some i and j. From (12.6) it follows that

$$\frac{p_t(i)}{p_t(j)} = \frac{p_{t-1}(i)}{p_{t-1}(j)} \frac{l_t(i)}{l_t(j)} = \frac{p_{t-1}(i)}{p_{t-1}(j)} H_t(\boldsymbol{\alpha}_i, \boldsymbol{\alpha}_j),$$

where, in an extension of the earlier notation for Bayes' factors,

$$H_t(\boldsymbol{\alpha}_i, \boldsymbol{\alpha}_j) = \frac{l_t(i)}{l_t(j)} = \frac{p(Y_t \mid \boldsymbol{\alpha}_i, D_{t-1})}{p(Y_t \mid \boldsymbol{\alpha}_j, D_{t-1})},$$

just the Bayes' factor for $\boldsymbol{\alpha} = \boldsymbol{\alpha}_i$ relative to $\boldsymbol{\alpha} = \boldsymbol{\alpha}_j$ based on the single observation Y_t. Thus, at each observation stage, the ratio of posterior probabilities for any two values of $\boldsymbol{\alpha}$, hence any two of the DLMs in the multi-process mixture, is modified through multiplication by the corresponding Bayes' factor. As information builds up in favour of one value, the Bayes' factors for that value relative to each of the others will increase, resulting in an increased posterior probability on the corresponding DLM.

12.2.3 Discussion and illustration

In application, forecasts and decisions will usually be based on information relating to the entire multi-process model, inferences from individual DLMs $M_t(\boldsymbol{\alpha}_j)$ in the mixture being combined, as detailed above, in proportion to their current posterior model probabilities $p_t(j)$. It is worth reiterating that a single DLM representation of a series is often a rather hopeful and idealistic assumption, although it can be useful if subject to careful user management and monitoring. These comments notwithstanding, the mixture modelling procedure provided by a multi-process, class I model may be used in a manner similar to classical model identification and parameter estimation techniques to identify a single DLM, or a restricted class of them, for future use. The posterior probabilities across $\mathcal{A}$ identify supported values of $\boldsymbol{\alpha}$, those with low weight that contribute little to the multi-process. It is well to be aware that these probabilities can change markedly over time, different DLMs best describing the data series at different times reflecting the global inadequacy of any single DLM. Under

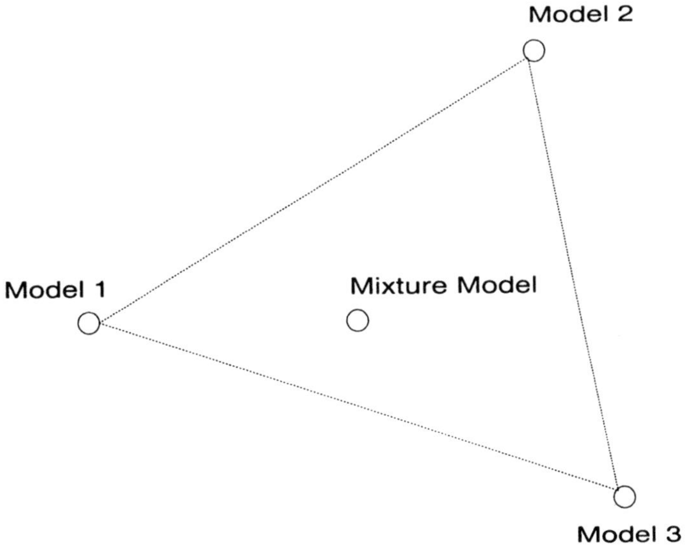

Figure 12.1 Mixtures of three distinct models

very general conditions, however, the posterior probabilities will converge to zero on all but one value in $\mathcal{A}$, the probability on that value tending to unity. This convergence may take a very large number of observations, but if suitable in the context of the application, can serve to identify a single value for $\boldsymbol{\alpha}$ and hence a single DLM. The use of a representative set of values $\mathcal{A}$ hopefully suitably spanning a larger, continuous space is the usual setup, and here convergence to a particular value identifies that model closest to the data, even though no single model actually generates the series. This basic principle of mixture modelling is schematically illustrated in Figure 12.1, where four DLMs are symbolically displayed. The generic model $M_t(\boldsymbol{\alpha})$ is seen to lie in the convex hull of the mixture components $M_t(\boldsymbol{\alpha}_j)$ $(j = 1, 2, 3)$ and is identified as one possible DLM represented as a mixture of the three components with respect to particular values of the posterior model probabilities.

If, as is usually the case in practice, no single DLM actually generates the series, the mixture approach allows the probabilities to vary as the data suggest thus adapting to process change.

A rather different context concerns cases when $\boldsymbol{\alpha}$ is simply an index for a class of distinct, and possibly structurally different, DLMs. An example concerns the launch of a new product in a consumer market subject to seasonal fluctuations in demand, the product being one of several or many such similar products. Initially, there are no sales data on which to base forecasts and decisions; the only relevant information is subjective and derived from market experience with similar products. In the market of interest, suppose that there are a small number of rather distinct patterns of demand across items, reflected in differences in seasonal patterns. Here $\boldsymbol{\alpha}$ simply indicates which of these structurally different patterns of demand the sales of the new product will follow, it being believed that one and only one such pattern will hold for each product line. The particular DLMs may thus differ generally in structure and dimensions of $\mathbf{F}_t$, $\mathbf{G}_t$. After processing a relatively small amount of data, say several months, major decisions are to be made concerning purchasing of raw materials and stock levels for the new product. At this stage, it is desirable to choose a particular, single DLM to base such decisions upon, the mixture only being used initially due to the uncertainty about the likely seasonal form.

EXAMPLE 12.2. Reconsider the exchange rate (\$ USA/£ UK) index data analysed in Section 2.6. The analyses there concerned a first-order polynomial model for the series, various models differing in the value of the defining discount factor being explored. Here that study is extended, the model and initial priors are as used in Section 2.6, to which the reader may refer. The single discount factor defining the evolution variance sequence is δ, viewed here as uncertain so that $\boldsymbol{\alpha} = \delta$. Now, δ is a continuous quantity, taking values in the range 0-1, typically between 0.7 and 1. With $k = 4$, the four distinct values $\mathcal{A} = \{1.0,\ 0.9,\ 0.8,\ 0.7\}$ are taken as representing the a priori plausible range of values, hopefully spanning the range of values appropriate for the series. For example, a mixture of models 1 and 2, with discount factors of 1.0 and 0.9 respectively, can be viewed as providing the flexibility to approximate the forms of behaviour of models with discount factors between 0.9 and 1. Initial probabilities are assigned as $p_0(j) = 0.25$ $(j = 1, \ldots, 4)$ not favouring any particular value in $\mathcal{A}$. The level parameter of the series is $\mu_t = \boldsymbol{\theta}_t$. Conditional on any particular value $\boldsymbol{\alpha} = \boldsymbol{\alpha}_j \in \mathcal{A}$, we have

$$(Y_t \mid \boldsymbol{\alpha}_j, D_{t-1}) \sim \mathrm{T}_{n_{t-1}}[f_t(j), Q_t(j)] \tag{12.8}$$

and

$$(\mu_t \mid \boldsymbol{\alpha}_j, D_t) \sim \mathrm{T}_{n_t}[m_t(j), C_t(j)], \tag{12.9}$$

for each $t = 1, 2, \ldots$, where the quantities $f_t(j)$, $Q_t(j)$, etc., are sequentially calculated as usual. Now the dependence on δ is made explicit through the arguments j. Recall that Figure 2.7 displays a plot of the data with point forecasts from the models with discount factors of 1.0 and 0.8, models corresponding to $j = 1$ and $j = 3$. In this example, the initial degrees of freedom parameter is $n_0 = 1$ as in Section 2.6, so that $n_t = t + 1$ for each value of j, not differing between the DLMs. Thus, in (12.6) we have

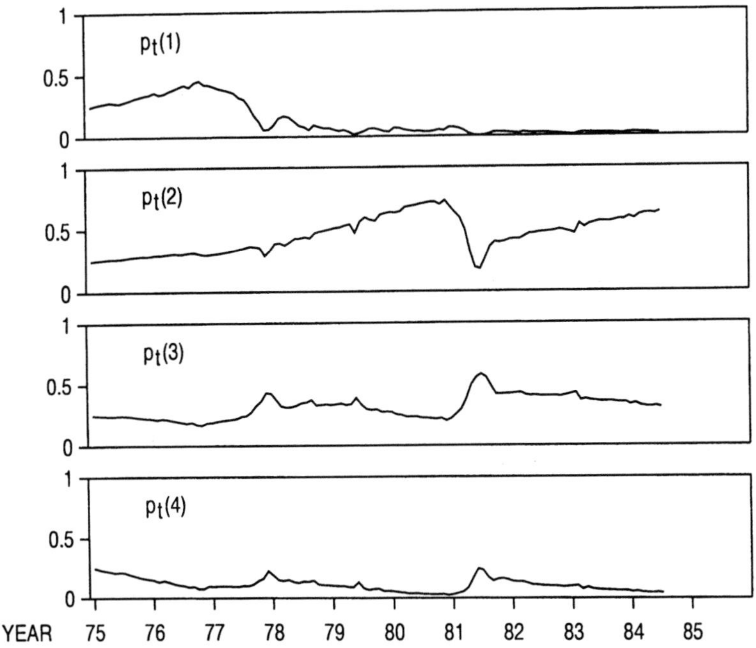

Figure 12.2 USA/UK index: Posterior model probabilities

$p_t(j) \propto p_{t-1}(j)l_t(j)$, with

$$l_t(j) = \frac{\Gamma[(n_{t-1}+1)/2]}{\Gamma[n_{t-1}/2]\sqrt{n_{t-1}\pi Q_t(j)}}\left\{1 + \frac{(Y_t - f_t(j))^2}{n_{t-1}Q_t(j)}\right\}^{-(n_{t-1}+1)/2}.$$

Figure 12.2 displays a plot over time of the four model probabilities $p_t(j)$, $(j = 1, \ldots, 4;\ t = 1, 2, \ldots)$, from this multi-process model. In examining the time variation in these probabilities the reader should refer also to the plot of the data in Figure 2.7. Overall, $p_t(2)$ increases up to high values around 0.6, $p_t(3)$ increases and levels off near 0.4, the other two probabilities becoming essentially negligible. This indicates the suitability of discount values between 0.8 and 0.9, the larger probabilities $p_t(2)$ indicating more support for values nearer 0.9. Certainly model 1, with discount factor of 1, has essentially no support from the data. The extremely adaptive model 4, with discount factor 0.7, has slightly higher probabilities, but they are still essentially negligible. In addition to these general conclusions, note the occurrence of one or two rather marked, though transient, changes in the general trend in the probabilities. In early 1981, in particular, the posterior probabilities on the more adaptive models 3 and to a lesser extent 4, increase at the expense of those on the generally favoured, though less adaptive, model 2. This continues for a few months until late 1981

Table 12.1. Summary of multi-process model at $t = 34$.

			μ_{34}		Y_{35}	
j	$\boldsymbol{\alpha}_j = \delta$	$p_{34}(j)$	$m_{34}(j)$	$\sqrt{C_{34}(j)}$	$f_{35}(j)$	$\sqrt{Q_{35}(j)}$
1	1.0	0.18	−0.008	0.005	−0.008	0.027
2	0.9	0.36	0.002	0.008	0.002	0.026
3	0.8	0.32	0.009	0.011	0.009	0.027
4	0.7	0.14	0.013	0.013	0.013	0.028

when $p_t(2)$ begins to grow again. The reason for this can be clearly seen in Figure 2.7. During early 1981, the level of the series drops markedly over a period of months, clearly deviating from the steady behaviour anticipated under a first-order polynomial model with a relatively high discount factor. Models 3 and 4, having lower discount factors than model 2, respond more rapidly to this drift in level with a consequent improvement in forecasting performance, as judged by Bayes' factors, relative to models 1 and 2. Later in the year, model 2 has had time to adapt to the data and the subsequent, steady behaviour of the series is again more consistent with this model.

This example also serves to illustrate some features of inference with discrete mixtures of distributions that users of multi-process models need to be aware of. In each of the DLMs $M_t(\boldsymbol{\alpha}_j)$, the prior, posterior and forecast distributions are all standard, unimodal distributions, of well-known and well-understood forms, whether they be normal or T (or inverse χ^2 for the observational variance). Inferences use point estimates and forecasts that are posteriors means or modes, uncertainty measures such as standard deviations, and probabilities that are easily calculated from the standardised forms of the normal or T distributions. If one of these models has a very high posterior probability at the time of interest, then it may be used alone for inference. Otherwise, the formal procedure is to use the full, unconditional mixture provided by (12.7), and the features of such mixtures can be far from standard (Titterington, Smith and Makov 1985). In particular, simple point forecasts such as means or modes of mixtures can mislead if they are used without further investigation of the shape of the distribution and, possibly, calculation of supporting probabilities, hpd regions, etc. Consider, for example, inferences made at $t = 34$, corresponding to October 1977, about the level of the series $(\mu_{34} \mid D_{34})$ and the next observation $(Y_{35} \mid D_{34})$. The relevant conditional posterior and forecast distributions, equations (12.8) and (12.9) with $t = 34$ and $n_t = 35$, are summarised in Table 12.1.

Consider first inference about μ_{34}. The individual posterior densities (12.9) summarised in the table are graphed in Figure 12.3. They are clearly rather different, model 1, in particular, favouring negative values of μ_{34} whilst the others favour positive values. More adaptive models

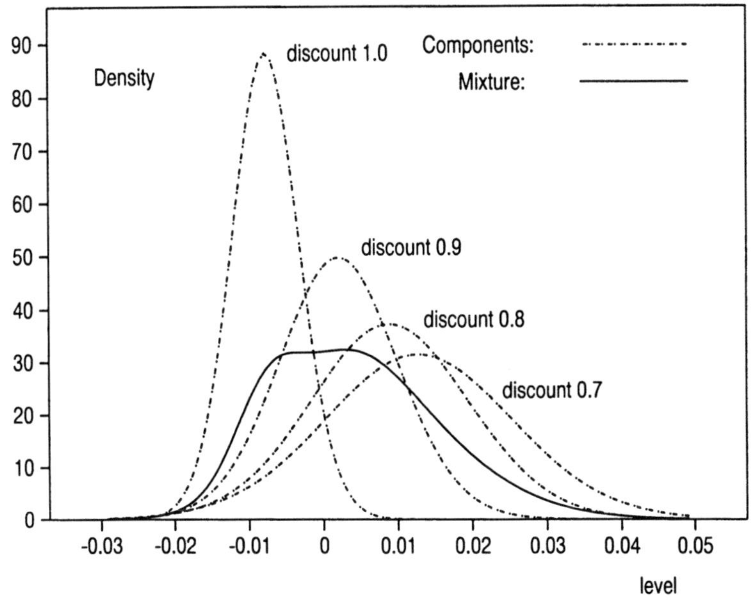

Figure 12.3 Posteriors for μ_{34} at $t = 34$

with smaller discount factors have more diffuse posteriors. The two central densities, with discounts of 0.9 and 0.8, have probabilities 0.36 and 0.32 respectively, dominating the mixture, as is clear from the overall mixture density graphed in the figure. The mixture is skewed to the right and very flat in the central region between roughly -0.007 and 0.006. The mean, calculated from the formula in Section 12.2.2, is 0.004. In fact, the density is unimodal with the mode near 0.005. If the probability on model 1, actually 0.18, had been slightly higher, then the mixture would have more density on negative values and could have become bimodal. Generally, with a mixture of k symmetric and unimodal distributions (as is the case here with four T distributions), there is the possibility of multimodality with up to k modes. Note finally that the mixture is rather spread, reflecting appreciable posterior probability on each of the rather distinct components.

Consider the problem of forecasting $(Y_{35} \mid D_{34})$ based on the mixture of components (12.8) at $t = 35$. The component one-step ahead T densities are graphed in Figure 12.4, together with the mixture. By contrast with Figure 12.3, the components here are very similar. Although they have the same means as the posteriors for μ_{34}, the scale parameters are very much larger; note the differences between the figures in the scales on the axes. The reason for this is that in each of the DLMs, the uncertainty in the Y_t series is dominated by observational variation about the level, the differences between models evident in the posteriors in Figure 12.3 being totally masked, and quite unimportant, for forecasting Y_{35}. As a

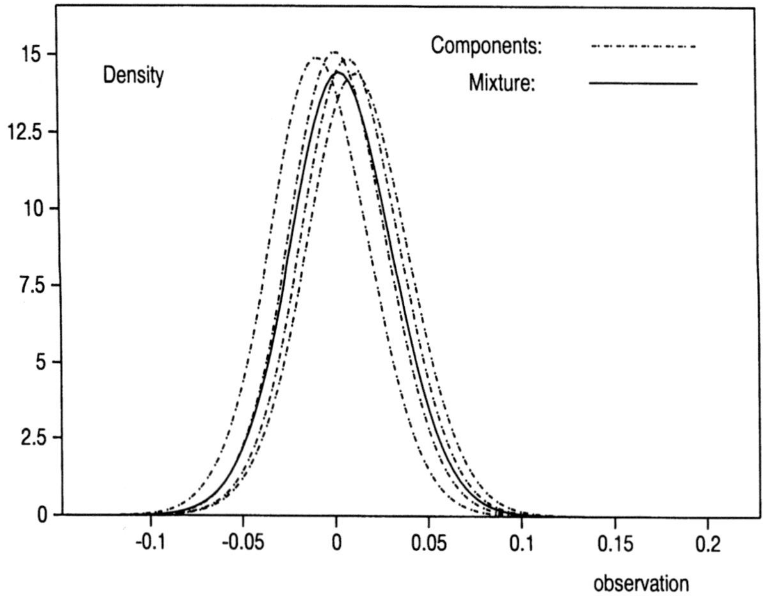

Figure 12.4 Forecast densities for Y_{35} at $t = 34$

consequence, the overall mixture density is very similar to each of the components.

Further illustration of inference using mixtures, and much more discussion, appears in the sections below concerning Multi-process, Class II models.

12.2.4 Model identification via mixtures

In the previous section we referred to the possible use of multi-process class I models as an automatic model identification technique. Some further comments about this aspect of mixture modelling are given here, specifically concerning the feature of convergence of the posterior model probabilities as t increases. The basic asymptotic theory for posterior distributions is rather general, and so we revert to the framework of Section 12.2, embedding the discrete set of values of $\boldsymbol{\alpha}$ in a general, continuous parameter space $\mathcal{A}$ for the purposes of this section. We do not provide formal, mathematical theory relating to convergence, preferring only to discuss concepts and motivate the key results. Theoretical details of convergence posterior distributions, albeit in a much less general framework, can be found in Berk (1970). Related material on asymptotic normality of posterior distributions, in a rather general framework, appears in Heyde and Johnstone (1979), and Sweeting and Adekola (1987). Related results specific to multi-process, normal DLMs, though less general, appear in Anderson and Moore (1979, Section 10.1).

In the general setting with $\boldsymbol{\alpha}$ continuous, suppose that $p(\boldsymbol{\alpha} \mid D_0) > 0$ for all $\boldsymbol{\alpha} \in \mathcal{A}$. For any two, distinct values $\boldsymbol{\alpha}_1$ and $\boldsymbol{\alpha}_2$, the Bayes' factor measuring the evidence from Y_t in favour of $\boldsymbol{\alpha} = \boldsymbol{\alpha}_1$ relative to $\boldsymbol{\alpha} = \boldsymbol{\alpha}_2$ is

$$H_t(\boldsymbol{\alpha}_1, \boldsymbol{\alpha}_2) = p(Y_t \mid \boldsymbol{\alpha}_1, D_{t-1})/p(Y_t \mid \boldsymbol{\alpha}_2, D_{t-1}).$$

The focus below is on the log-Bayes' factors $\log[H_t(\boldsymbol{\alpha}_1, \boldsymbol{\alpha}_2)]$, assumed to be finite for all t and all $\boldsymbol{\alpha}_1,\ \boldsymbol{\alpha}_2 \in \mathcal{A}$. The corresponding overall log-Bayes' factor, or log-likelihood ratio, from the first t observations $Y_t, \dots, Y_1$, is simply

$$\log[\prod_{r=1}^{t} H_r(\boldsymbol{\alpha}_1, \boldsymbol{\alpha}_2)] = \sum_{r=1}^{t} \log[H_r(\boldsymbol{\alpha}_1, \boldsymbol{\alpha}_2)].$$

Denote the average value of the individual log-factors by

$$L_t(\boldsymbol{\alpha}_1, \boldsymbol{\alpha}_2) = t^{-1} \sum_{r=1}^{t} \log[H_r(\boldsymbol{\alpha}_1, \boldsymbol{\alpha}_2)].$$

Now, for $j = 1, 2$, we have

$$p(\boldsymbol{\alpha}_j | D_t) = p_t(j) \propto p_0(j) \prod_{r=1}^{t} p(Y_r | \boldsymbol{\alpha}_j, D_{r-1}),$$

and so

$$p(\boldsymbol{\alpha}_1 | D_t)/p(\boldsymbol{\alpha}_2 | D_t) = [p_0(1)/p_0(2)] \prod_{r=1}^{t} H_r(\boldsymbol{\alpha}_1, \boldsymbol{\alpha}_2)$$

for all t. It follows that

$$\log[p(\boldsymbol{\alpha}_1 | D_t)/p(\boldsymbol{\alpha}_2 | D_t)] = \log[p_0(1)/p_0(2)] + t\ L_t(\boldsymbol{\alpha}_1, \boldsymbol{\alpha}_2).$$

As t increases, the effect of the initial prior on the posterior for $\boldsymbol{\alpha}$ becomes negligible in the sense that as $t \to \infty$,

$$t^{-1}\log[p(\boldsymbol{\alpha}_1 \mid D_t)/p(\boldsymbol{\alpha}_2 \mid D_t)] - L_t(\boldsymbol{\alpha}_1, \boldsymbol{\alpha}_2) \to 0.$$

Thus, the limiting behaviour of the posterior is defined by that of the functions L_t. Two assumptions provide the required structure for convergence.

(A) As t increases, $L_t(\boldsymbol{\alpha}_1, \boldsymbol{\alpha}_2)$ converges, in an appropriate probabilistic sense, to a finite limiting value for all $\boldsymbol{\alpha}_1$ and $\boldsymbol{\alpha}_2$. Denote this value by $L(\boldsymbol{\alpha}_1, \boldsymbol{\alpha}_2)$, so that

$$L(\boldsymbol{\alpha}_1, \boldsymbol{\alpha}_2) = \lim_{t \to \infty} L_t(\boldsymbol{\alpha}_1, \boldsymbol{\alpha}_2).$$

(B) There exists a unique value $\boldsymbol{\alpha}_0 \in \mathcal{A}$ such that for all $\boldsymbol{\alpha} \neq \boldsymbol{\alpha}_0$,

$$L(\boldsymbol{\alpha}_0, \boldsymbol{\alpha}) > 0.$$

These assumptions require some comment. Concerning (A), the mode of convergence, whether it be with probability one, or in probability, etc., is of little importance. When the underlying mechanism generating the data can be assumed to give rise to a joint distribution for the data $Y_t, \ldots, Y_1$, for all t, then (A) can be made more concrete. Note that this does not assume that this true underlying distribution is known, nor that it bears any relation to the class of models used. An interesting example is, however, the hypothetical case when the data truly follow a DLM within the class, identified by a particular, unknown value of $\boldsymbol{\alpha}$, although this is not necessary. Whatever this true distribution may be, (B) applies (with probability one) if for all t, $\mathrm{E}\{\log[H_t(\boldsymbol{\alpha}_1, \boldsymbol{\alpha}_2)] \mid D_0\}$ is finite, where the expectation is with respect to the full t random quantities $Y_t, \ldots, Y_1$ having this true, joint distribution. If this is the case, then

$$L(\boldsymbol{\alpha}_1, \boldsymbol{\alpha}_2) = \lim_{t\to\infty} t^{-1} \sum_{r=1}^{t} \mathrm{E}\{\log[H_r(\boldsymbol{\alpha}_1, \boldsymbol{\alpha}_2)] \mid D_0\}.$$

Assumptions (A) and (B) are both valid in cases when the posterior for $\boldsymbol{\alpha}$ is asymptotically normal. This will usually be the case. Formal regularity conditions required, and proofs of the asymptotic normality, can be found in Heyde and Johnstone (1979), and Sweeting and Adekole (1987). In such cases, for large values of t the posterior density $p(\boldsymbol{\alpha} \mid D_t)$ concentrates increasingly about a unique posterior mode $\hat{\boldsymbol{\alpha}}_t$. Assuming the posterior to be twice differentiable, then for $\boldsymbol{\alpha}$ near $\hat{\boldsymbol{\alpha}}_t$,

$$\begin{aligned}\log[p(\boldsymbol{\alpha} \mid D_t)] \approx &- 0.5t(\boldsymbol{\alpha} - \hat{\boldsymbol{\alpha}}_t)'\mathbf{I}_t(\hat{\boldsymbol{\alpha}}_t)(\boldsymbol{\alpha} - \hat{\boldsymbol{\alpha}}_t)\\ &+ \text{constants not involving } \boldsymbol{\alpha}\text{, and other, negligible terms,}\end{aligned}$$

where

$$\mathbf{I}_t(\boldsymbol{\alpha}) = -t^{-1}\frac{\partial^2}{\partial\boldsymbol{\alpha}\,\partial\boldsymbol{\alpha}'}\log[p(\boldsymbol{\alpha} \mid D_t)],$$

so that $\mathbf{I}_t(\hat{\boldsymbol{\alpha}}_t)$ is a symmetric, positive definite matrix. Thus, for large t,

$$L_t(\boldsymbol{\alpha}_1, \boldsymbol{\alpha}_2) \approx t^{-1}\log[p(\boldsymbol{\alpha}_1 \mid D_t)/p(\boldsymbol{\alpha}_2 \mid D_t)] \approx (\boldsymbol{\alpha}_2 - \boldsymbol{\alpha}_1)'\mathbf{I}_t(\hat{\boldsymbol{\alpha}}_t)(\bar{\boldsymbol{\alpha}} - \hat{\boldsymbol{\alpha}}_t),$$

where $\bar{\boldsymbol{\alpha}} = (\boldsymbol{\alpha}_1 + \boldsymbol{\alpha}_2)/2$. Under general conditions (as given in the above references), the posterior eventually becomes degenerate at a particular value $\boldsymbol{\alpha} = \boldsymbol{\alpha}_0$, with $\lim_{t\to\infty} \hat{\boldsymbol{\alpha}}_t = \boldsymbol{\alpha}_0$ and a positive definite limit $\mathbf{I}_0 = \lim_{t\to\infty} \mathbf{I}_t(\hat{\boldsymbol{\alpha}}_t)$. Thus,

$$L(\boldsymbol{\alpha}_1, \boldsymbol{\alpha}_2) = \lim_{t\to\infty} L_t(\boldsymbol{\alpha}_1, \boldsymbol{\alpha}_2) = (\boldsymbol{\alpha}_2 - \boldsymbol{\alpha}_1)'\mathbf{I}_0(\bar{\boldsymbol{\alpha}} - \boldsymbol{\alpha}_0).$$

In this case, assumptions (A) and (B) clearly hold, the limiting posterior distribution identifying the value $\boldsymbol{\alpha}_0$ in the formal sense that

$$\lim_{t\to\infty} \Pr[||\boldsymbol{\alpha} - \boldsymbol{\alpha}_0|| > \epsilon \mid D_t] = 0$$

for all $\epsilon > 0$.

Under assumptions (A) and (B) it now follows that for all $\boldsymbol{\alpha}_1, \boldsymbol{\alpha}_2 \in \mathcal{A}$,

$$\lim_{t\to\infty} t^{-1}\log[p(\boldsymbol{\alpha}_1 \mid D_t)/p(\boldsymbol{\alpha}_2 \mid D_t)] = L(\boldsymbol{\alpha}_1, \boldsymbol{\alpha}_2).$$

Thus,

$$\lim_{t\to\infty} [p(\boldsymbol{\alpha}_1 \mid D_t)/p(\boldsymbol{\alpha}_2 \mid D_t)] = \begin{cases} \infty, & \text{if } L(\boldsymbol{\alpha}_1, \boldsymbol{\alpha}_2) > 0; \\ 0, & \text{if } L(\boldsymbol{\alpha}_1, \boldsymbol{\alpha}_2) < 0. \end{cases}$$

This applies in the general case of continuous $\mathcal{A}$, and also as a corollary when working with finite, discrete set $\{\boldsymbol{\alpha}_1, \ldots, \boldsymbol{\alpha}_k\}$ in a multi-process model. Note that we can *always* introduce a larger, continuous space in which to embed a discrete set $\mathcal{A}$, and so the following results apply to multi-process class I models.

THEOREM 12.1. *In the multi-process class I model of Definitions 12.1 and 12.2, suppose that $\mathcal{A} = \{\boldsymbol{\alpha}_1, \ldots, \boldsymbol{\alpha}_k\}$ is embedded in a continuous space such that assumptions (A) and (B) above hold, with $\boldsymbol{\alpha}_0$ defined as in (B) (although unknown and not necessarily in $\mathcal{A}$). Suppose also that for a single element $\alpha_i \in \mathcal{A}$,*

$$L(\boldsymbol{\alpha}_0, \boldsymbol{\alpha}_i) = \min_{\boldsymbol{\alpha}_j \in \mathcal{A}} L(\boldsymbol{\alpha}_0, \boldsymbol{\alpha}_j).$$

Then the posterior model probabilities converge with limits

$$\lim_{t\to\infty} p_t(i) = 1,$$

and

$$\lim_{t\to\infty} p_t(j) = 0, \quad (j \neq i).$$

The numbers $L(\boldsymbol{\alpha}_0, \boldsymbol{\alpha}_i)$ are all non-negative, being positive if, as is usual, $\boldsymbol{\alpha}_0 \notin \mathcal{A}$. These numbers are the limiting versions of the average Bayes' factors, or log-likelihood ratios, comparing the values $\boldsymbol{\alpha}_j$ with the "optimal" value $\boldsymbol{\alpha}_0$. As a corollary to the theorem, suppose that there exists an unknown, underlying joint distribution for the data $Y_t, \ldots, Y_1$, for all t, as previously discussed. Now, for all j,

$$\begin{aligned} L(\boldsymbol{\alpha}_0, \boldsymbol{\alpha}_j) &= \lim_{t\to\infty} t^{-1} \sum_{r=1}^{t} \mathrm{E}\{\log[H_r(\boldsymbol{\alpha}_0, \boldsymbol{\alpha}_j)] \mid D_0\} \\ &= \lim_{t\to\infty} t^{-1} \sum_{r=1}^{t} \mathrm{E}\{\log[p(Y_r \mid \boldsymbol{\alpha}_0, D_{r-1})] \mid D_0\} \\ &\qquad - \lim_{t\to\infty} t^{-1} \sum_{r=1}^{t} \mathrm{E}\{\log[p(Y_r \mid \boldsymbol{\alpha}_j, D_{r-1})] \mid D_0\}. \end{aligned}$$

The first term here does not involve j and so can be ignored if the second term is finite. We thus have the following result.

Corollary 12.1. *In the framework of Theorem 12.1, suppose that for all* $\boldsymbol{\alpha}_j \in \mathcal{A}$, *the quantity*

$$L^*(\boldsymbol{\alpha}_j) = \lim_{t\to\infty} t^{-1} \sum_{r=1}^{t} \mathrm{E}\{\log[p(Y_r \mid \boldsymbol{\alpha}_j, D_{r-1})] \mid D_0\}$$

exists, the expectation being with respect to the true distribution of the random quantities $(Y_t, \ldots, Y_1 \mid D_0)$. *Then the posterior probability on model* i *converges to unity, where*

$$L^*(\boldsymbol{\alpha}_i) = \max_{\boldsymbol{\alpha}_j \in \mathcal{A}} L^*(\boldsymbol{\alpha}_j).$$

The quantity $L^*(\boldsymbol{\alpha})$, for any $\boldsymbol{\alpha}$ of interest, is the limiting average log-likelihood of the model $M_t(j)$ with respect to the true distribution of the series. This is related to the limiting average **Kullback-Leibler divergence**, otherwise directed divergence, measuring the discrepancy between the model and the data. A basic reference is Kullback and Leibler (1951), although such *entropy*-based measures have been used extremely widely in statistics and other areas for many years and under many names. Kullback (1983) gives further details. The measure $L^*(\boldsymbol{\alpha})$ is the limiting average of the individual quantities $\mathrm{E}\{\log[p(Y_t \mid \boldsymbol{\alpha}, D_{t-1})] \mid D_0\}$.

Some further comments are in order.

(1) These results apply when there is a single value i, $(1 \le i \le k)$ such that α_i *uniquely* maximises $L^*(\alpha)$ over $\alpha \in \mathcal{A}$, i.e., minimises $L(\boldsymbol{\alpha}_0, \boldsymbol{\alpha})$ as in the statement of Theorem 12.1. In some (rare) cases $\mathcal{A}$ may contain two or more values of α for which $L^*(\alpha)$ attains the maximum over $\mathcal{A}$. The above discussion then applies with minor modification to show that the posterior model probabilities converge to zero on all but these maximising values of α, being asymptotically uniform across these values.

(2) Assume that a model of the form $M_t(\boldsymbol{\alpha}_0)$ actually generates the series. If $\boldsymbol{\alpha}_0 \in \{\boldsymbol{\alpha}_1, \ldots, \boldsymbol{\alpha}_k\}$ then the limiting, preferred value is $\boldsymbol{\alpha}_i = \boldsymbol{\alpha}_0$, the true value. If, on the other hand, $\boldsymbol{\alpha}_0 \notin \{\boldsymbol{\alpha}_1, \ldots, \boldsymbol{\alpha}_k\}$, then $\boldsymbol{\alpha}_i$ is closest to $\boldsymbol{\alpha}_0$ in the sense of divergence, that usually corresponds to closeness in terms of Euclidean distance.

(3) If $M_j(t)$ is a time series DLM (TSDLM) for each j, Theorem 5.1 implies that $p(Y_t \mid \boldsymbol{\alpha}, D_{t-1})$ has a stable, limiting form. In particular, as $t \to \infty$,

$$(Y_t \mid \boldsymbol{\alpha}_j, D_{t-1}) \sim \mathrm{N}[f_t(j), Q(j)],$$

for some constant variances $Q(j)$. Thus, for large t,

$$\begin{aligned}\mathrm{E}\{\log[p(Y_t \mid \boldsymbol{\alpha}_j, D_{t-1}] \mid D_0\} \\ &= -0.5\log(2\pi Q(j)) - 0.5\mathrm{E}[(Y_t - f_t(j))^2 \mid D_0]/Q(j) \\ &= -0.5\log(2\pi Q(j)) - 0.5V_0(j)/Q(j),\end{aligned}$$

say, where $V_0(j) = \mathrm{E}[(Y_t - f_t(j))^2 \mid D_0]$ is assumed to exist. Then for all j,

$$L^*(\boldsymbol{\alpha}_j) = -0.5\log(2\pi Q(j)) - 0.5V_0(j)/Q(j).$$

EXAMPLE 12.3. In the first-order polynomial model of Example 12.2 where $\boldsymbol{\alpha} = \delta$, the discount factor, suppose the observational variance to be known and equal to V. Suppose that the data are generated by such a model with true discount factor of δ_0. For any $\delta_j \in \{\delta_1, \ldots, \delta_k\}$, recall from Section 2.4 that asymptotically, $(Y_t \mid \delta_j, D_{t-1}) \sim \mathrm{N}[f_t(j), V/\delta_j]$, and the errors $Y_t - f_t(j)$ are zero mean even though the model is incorrect, having true variances

$$V_0(j) = \mathrm{V}[Y_t - f_t(j) \mid D_0] = [1 + (\delta_j - \delta_0)^2/(1 - \delta_j)^2]V/\delta_0.$$

Thus, from note (2) above, the limiting, preferred discount factor δ_i is that value maximising

$$\log(\delta_j/\delta_0) - (\delta_j/\delta_0)[1 + (\delta_j - \delta_0)^2/(1 - \delta_j)^2].$$

It is left as an exercise for the reader to explore this as a function of δ_j.

12.3 MULTI-PROCESS MODELS: CLASS II

12.3.1 Introduction and definition

The reader will by now be conversant with the basic underlying principle of Bayesian forecasting, that at any time, historical information relevant to forecasting the future is sufficiently summarised in terms of posterior distributions for model parameters. Given such summaries, a model used in the past may be altered or discarded at will in modelling for the future, whether such modifications be based on formal interventions or otherwise. Thus, the notion that there may be uncertainty as to which of a possible class of DLMs is most appropriate at any time, irrespective of what has happened in the past, is natural and acceptable as a norm. This notion is formalised via the multi-process, class II models (Harrison and Stevens 1976a) that embody the assumption (12.2b).

Definition 12.3. Suppose that at each time t, $\boldsymbol{\alpha}$ takes a value in the discrete set $\mathcal{A} = \{\boldsymbol{\alpha}_1, \ldots, \boldsymbol{\alpha}_k\}$, the values possibly differing over time. Then irrespective of the mechanism by which the values of $\boldsymbol{\alpha}$ are chosen, the Y_t series is said to follow a **multi-process, class II model.**

This situation, in which no single DLM is assumed to adequately describe the series but any one of a discrete collection may obtain at each observation stage, rather often describes the real situation in practice. The variety of possible multi-process, class II models is clearly enormous. We motivate the development of an important special class, which has been used rather widely with success, in the following example.

EXAMPLE 12.3. As in the multi-process, class I case in Example 12.2, let $\boldsymbol{\alpha} = \delta$, the discount factor of a first-order polynomial model. Take the parameter space to contain just two values, $\mathcal{A} = \{0.9,\ 0.1\}$. Under multi-process class II assumptions, δ takes either the value 0.9 or the value 0.1 at each time t. If $\delta = 0.9$ the dynamic movement in the level of the series is steady, information decaying at a typical rate of about 10% between observations. It is to be expected that this will be the case most of the time. If $\delta = 0.1$ the movement allowed in the level is much greater, the change in level having a standard deviation that is 9 times that obtained with the discount of 0.9. This value is appropriate, usually rather infrequently, as an alternative to permit more marked, abrupt changes in level. Compare the use of similar, alternative discounts in automatic interventions in Section 11.5.

In order to utilise the multi-process idea, it remains to specify the mechanisms by which a particular value of $\boldsymbol{\alpha}$ is chosen at each time. There are many possibilities, including, for example, subjective intervention by the forecaster to determine the values. We restrict attention, however, to probabilistic mechanisms that provide multi-processes based on discrete probability mixtures of DLMs.

Definition 12.4. Suppose that in the multi-process, class II framework of Definition 12.3, the value $\boldsymbol{\alpha} = \boldsymbol{\alpha}_j$ at time t, defining the model $M_t(j)$, is selected with known probability

$$\pi_t(j) = \Pr[M_t(j) \mid D_{t-1}].$$

Then the series Y_t follows a **multi-process, class II mixture model.**

Here $\pi_t(j) = \Pr[\boldsymbol{\alpha} = \boldsymbol{\alpha}_j \text{ at time } t \mid D_{t-1}]$ may, in general, depend upon the history of the process. Some practically important possibilities follow, in order of increasing complexity of the resulting analyses.

(1) Fixed model selection probabilities

$$\pi_t(j) = \pi(j) = \Pr[M_t(j) \mid D_0]$$

for all t. Thus, prior to observing Y_t, $M_t(j)$ has prior probability $\pi(j)$ of obtaining independently of what has previously occurred.

(2) First-order Markov probabilities in which the model obtaining at time t depends on which of the models obtained at time $t-1$, but not on what happened prior to $t-1$. Here we have fixed, and known,

transition probabilities

$$\pi(j \mid i) = \Pr[M_t(j) \mid M_{t-1}(i), D_0],$$

for all i and j, $(i = 1, \dots, k;\ j = 1, \dots, k)$, and all t. Given these probabilities, this model supposes that

$$\Pr[M_t(j) \mid M_{t-1}(i), D_{t-1}] = \pi(j \mid i).$$

Thus, the chance of any model obtaining at any time depends on which model obtained at the immediately preceding time, but not on any other features of the history of the series. The *marginal* probabilities of models at time t are calculated via

$$\begin{aligned}\pi_t(j) &= \Pr[M_t(j) \mid D_{t-1}] \\ &= \sum_{i=1}^{k} \Pr[M_t(j) \mid M_{t-1}(i), D_{t-1}]\Pr[M_{t-1}(i) \mid D_{t-1}] \\ &= \sum_{i=1}^{k} \pi(j \mid i)p_{t-1}(i),\end{aligned}$$

where $p_{t-1}(i)$ is the posterior probability, at time $t-1$, of model $M_{t-1}(i)$.

(3) Higher-order Markov probabilities that extend the dependence to models at times $t-2, t-3, \dots$, etc.

EXAMPLE 12.3 (continued). In the discount factor example, fixed selection probabilities imply the same chance of the lower discount factor $\delta = 0.1$ applying at each time. Viewing $\delta = 0.1$ as consistent with the possibility of marked changes, or jumps, in the level of the series, this models a series subject to the possibility of jumps at random times. If $\pi(2) = \Pr[\delta = 0.1] = 0.05$, for example, then jumps may occur roughly 5% of the time. A first-order Markov model as in (2) refines this random jump mechanism according to expected behaviour of the series. For example, if $\delta = 0.1$ at time $t-1$, it may be felt that $\delta = 0.1$ at time t is rather less likely than otherwise, consistent with the view that jumps are unlikely to occur in runs. This can be modelled by taking $\pi(1 \mid 2) > \pi(2 \mid 2)$. The reverse would be the case if jumps were expected to occur in groups or runs.

Theoretical developments of the basic probability results for Markov models are, in principle, straightforward, though obviously rather more complicated than those with fixed selection probabilities. We concentrate exclusively on the latter for the remainder of the chapter. Details of analyses for Markov models are therefore left to the interested reader. Although working with mixture models as in Definition 12.4, the mixture terminology is dropped for simplicity; we refer to the models simply as multi-process, class II models, the mixture feature being apparent and understood.

12.3.2 Fixed selection probability models: Structure of analysis

The analysis of multi-process, class I models in Section 12.2 introduced the use of discrete mixtures of standard DLMs. These are also central to the analysis of class II models. We simplify the model notation here and throughout the rest of the chapter. The possible DLMs, hitherto indexed by parameters α, are now distinguished only by integer indices, setting $\alpha_j = j$, $(j = 1, \ldots, k)$. Thus, the model index set is now simply $\mathcal{A} = \{1, \ldots, k\}$, and we refer to $M_t(j)$ as model j at time t. Some further notation serves to simplify the presentation.

Definition 12.5. For each t and integer h, $(0 \le h < t)$, define the probabilities

$$p_t(j_t, j_{t-1}, \ldots, j_{t-h}) = \Pr[M_t(j_t), M_{t-1}(j_{t-1}), \ldots, M_{t-h}(j_{t-h}) \mid D_t].$$

Thus,

$$p_t(j_t) = \Pr[M_t(j_t) \mid D_t],$$

consistent with previous usage,

$$p_t(j_t, j_{t-1}) = \Pr[M_t(j_t), M_{t-1}(j_{t-1}) \mid D_t],$$

and so on.

The analysis of class II models is far more complicated than that of class I. The nature of the complications can be appreciated by considering the position at time $t = 1$, assuming that $(\boldsymbol{\theta}_0 \mid D_0)$ has a normal or T distribution.

At $t = 1$, there are k possible DLMs $M_1(j_1)$ with prior probabilities $\pi(j_1)$, $(j_1 = 1, \ldots, k)$. Within each DLM, analysis proceeds in the usual way, providing posterior distributions for the state vector $\boldsymbol{\theta}_1$ that generally differ across models. In DLM j_1 the state vector has posterior $p(\boldsymbol{\theta}_1 \mid M_1(j_1), D_1)$ and the DLM has posterior probability

$$p_1(j_1) = \Pr[M_1(j_1) \mid D_1] \propto p(Y_1 \mid M_1(j_1), D_0)\pi(j_1).$$

Unconditionally, inferences about the state vector are based on the discrete mixture

$$p(\boldsymbol{\theta}_1 \mid D_1) = \sum_{j_1=1}^{k} p(\boldsymbol{\theta}_1 \mid M_1(j_1), D_1) p_1(j_1).$$

Proceeding to time $t = 2$, any of the k possible DLMs $M_2(j_2)$, $(j_2 = 1, \ldots, k)$, may be selected, again with probabilities $\pi(j_2)$. Now it is only possible to retain the components of standard DLM analyses if in addition, the particular models possible at $t = 1$ are considered. Thus, conditional on both $M_2(j_2)$ and $M_1(j_1)$ applying, for some j_2 and j_1, the posterior for $\boldsymbol{\theta}_2$ given D_2 follows from the usual DLM analysis, depending, of course, on j_2 and j_1, denoted by $p(\boldsymbol{\theta}_2 \mid M_2(j_2), M_1(j_1), D_2)$. Unconditionally, the posterior for inference is

$$p(\boldsymbol{\theta}_2 \mid D_2) = \sum_{j_2=1}^{k} \sum_{j_1=1}^{k} p(\boldsymbol{\theta}_2 \mid M_2(j_2), M_1(j_1), D_2) p_2(j_2, j_1).$$

Again this is a discrete mixture of standard posteriors. However, whereas at $t = 1$ the mixture contained k components, one for each possible model at time $t = 1$, this contains k^2 components, one for each combination of models possible at time $t = 1$ and $t = 2$. Another way of looking at the mixture is to write

$$p(\boldsymbol{\theta}_2 \mid M_2(j_2), D_2) = \sum_{j_1=1}^{k} p(\boldsymbol{\theta}_2 \mid M_2(j_2), M_1(j_1), D_2) \Pr[M_1(j_1) \mid D_2].$$

Thus, conditional on model j_2 at time $t = 2$, the posterior is a mixture of k standard forms, depending on which of the model obtained at time $t = 1$. Then unconditionally,

$$p(\boldsymbol{\theta}_2 \mid D_2) = \sum_{j_2=1}^{k} p(\boldsymbol{\theta}_2 \mid M_2(j_2), D_2) p_2(j_2).$$

This development continues as time progresses. Then at time t, the posterior density can be written hierarchically as

$$p(\boldsymbol{\theta}_t \mid D_t) = \sum_{j_t=1}^{k} p(\boldsymbol{\theta}_t \mid M_t(j_t), D_t) p_t(j_t), \tag{12.10a}$$

or equivalently, as

$$\sum_{j_t=1}^{k} \sum_{j_{t-1}=1}^{k} p(\boldsymbol{\theta}_t \mid M_t(j_t), M_{t-1}(j_{t-1}), D_t) p_t(j_t, j_{t-1}), \tag{12.10b}$$

and so on down to the final stage

$$\sum_{j_t=1}^{k} \sum_{j_{t-1}=1}^{k} \cdots \sum_{j_1=1}^{k} p(\boldsymbol{\theta}_t \mid M_t(j_t), M_{t-1}(j_{t-1}), \ldots, M_1(j_1), D_t) \times p_t(j_t, j_{t-1}, \ldots, j_1). \tag{12.10c}$$

Only at the final stage of elaboration in (12.10c), where the sequence of models obtaining at each of the times $1, \ldots, t$ are assumed, are the posteriors

$$p(\boldsymbol{\theta}_t \mid M_t(j_t), M_{t-1}(j_{t-1}), \ldots, M_1(j_1), D_t)$$

of standard DLM form. Thus, to obtain the marginal posterior as a mixture, it is necessary to consider all k^t possible combinations that may apply. Within any particular combination $M_t(j_t), M_{t-1}(j_{t-1}), \ldots, M_1(j_1)$, the usual DLM analysis applies. The posterior probabilities on each combination, namely $p_t(j_t, j_{t-1}, \ldots, j_1)$, weight the components in the overall mixture.

At each level of elaboration of the mixtures in equations (12.10), the number of components of the mixture corresponds to the number of steps back in time that are being explicitly considered. In (12.10a) there are k components, corresponding to k possible models at time t. In (12.10b) there are k^2 possible combinations of models at times t and $t-1$, and so on down to the full k^t possibilities in (12.10c). At the intermediate levels (12.10a,b), etc., the conditional posteriors in the mixture are themselves discrete mixtures of more basic components.

Obvious problems arise with such an analysis, largely relating to the computational demands made by this explosion of the number of mixture components as time progresses. In each possible combination of models, the posterior for the state vector, predictive distributions for future observations, etc., are summarised in terms of a fixed number of means, variances and so forth. After t observations there are then k^t collections of such quantities requiring calculation and storage. If k is relatively small and the analysis is to be performed only on a very short series, then the computations may be feasible with sufficient computing capacity. See Schervish and Tsay (1988) for illustrations of analysis of series of up to t around 200 or so observations in models with $k = 4$. In practice, however, the possibly immense computational demands of the analysis will often be prohibitive. Fortunately, it is usually possible to avoid the explosion of the size of mixture models by exploiting further features of the particular class II structure used, and a full analysis with all the computational problems will often not be necessary. On general grounds, mixtures of posterior distributions with many components tend to suffer major redundancies in the sense that components can be grouped together with others that are similar in location and spread. Thus, for all practical purposes, it may be possible to reduce the number of components by approximations, leading to a smaller, manageable mixture. This reduction can be quite dramatic in terms of numbers of discarded components. Some general discussion of this follows.

12.3.3 Approximation of mixtures: General discussion

It is basic to dynamic modelling that as time progresses, what occurred in the past becomes less and less relevant to inference made for the future. This applies to mixtures, the possible models obtaining in the past losing relevance to inferences made at the current time t as t increases. It is therefore to be expected that the full, conditional posterior $p(\boldsymbol{\theta}_t \mid M_t(j_t), M_{t-1}(j_{t-1}), \ldots, M_1(j_1), D_t)$ will depend negligibly on early models $M_1(j_1)$, $M_2(j_2)$, etc., when t is large. Depending on the dimension n of the state vector and the complexity of the multi-process model, it is thus to be expected that for some fixed integer $h \geq 1$, the full posterior will depend essentially upon those models applying only up to h-steps back in time, viz.,

$$\begin{aligned} p(\boldsymbol{\theta}_t \mid M_t(j_t), M_{t-1}(j_{t-1}), \dots, M_1(j_1), D_t) \\ \approx p(\boldsymbol{\theta}_t \mid M_t(j_t), M_{t-1}(j_{t-1}), \dots, M_{t-h}(j_{t-h}), D_t). \end{aligned} \tag{12.11}$$

If this is assumed, the number of components of the mixture posterior at any time will not exceed k^{h+1}. As a consequence, it is only necessary to consider models up to h-steps back in time when performing the analysis. Thus, the full mixture (12.10c) will be approximately

$$\begin{aligned} \sum_{j_t=1}^{k} \sum_{j_{t-1}=1}^{k} \cdots \sum_{j_{t-h}=1}^{k} p(\boldsymbol{\theta}_t \mid M_t(j_t), M_{t-1}(j_{t-1}), \dots, M_{t-h}(j_{t-h}), D_t) \\ \times\, p_t(j_t, j_{t-1}, \dots, j_{t-h}). \end{aligned} \tag{12.12}$$

This is a mixture whose components are all still of standard DLM form, but now containing a fixed number k^{h+1} components rather than the original, increasing number k^t. The posterior model probabilities weighting the posteriors in this mixture are now calculated. Firstly, and as usual, by Bayes' Theorem,

$$\begin{aligned} p_t(j_t, j_{t-1}, \dots, j_{t-h}) \propto & \Pr[M_t(j_t), \dots, M_{t-h}(j_{t-h}) \mid D_{t-1}] \\ & \times p(Y_t \mid M_t(j_t), \dots, M_{t-h}(j_{t-h}), D_{t-1}). \end{aligned} \tag{12.13}$$

The second term in (12.13) is given by

$$\begin{aligned} & p(Y_t \mid M_t(j_t), \dots, M_{t-h}(j_{t-h}), D_{t-1}) = \\ & \quad \sum_{j_{t-h-1}=1}^{k} p(Y_t \mid M_t(j_t), \dots, M_{t-h}(j_{t-h}), M_{t-h-1}(j_{t-h-1}), D_{t-1}) \\ & \qquad \times \Pr[M_{t-h-1}(j_{t-h-1}) \mid D_{t-1}], \end{aligned}$$

an average of the normal or T one-step predictive densities for Y_t under each of the models in the conditionings. The averaging is with respect to models $(h+1)$-steps back, those at time $t-h-1$. The probabilities weighting these terms are available from the identity

$$\Pr[M_{t-h-1}(j_{t-h-1}) \mid D_{t-1}] = \sum_{j_{t-1}=1}^{k} \cdots \sum_{j_{t-h}=1}^{k} p_{t-1}(j_{t-1}, \dots, j_{t-h}, j_{t-h-1}).$$

The first term in (12.13) is similarly calculated via

$$\begin{aligned}
&\Pr[M_t(j_t), \dots, M_{t-h}(j_{t-h}) \mid D_{t-1}] \\
&=\Pr[M_t(j_t) \mid M_{t-1}(j_{t-1}), \dots, M_{t-h}(j_{t-h}), D_{t-1}]p_{t-1}(j_{t-1}, \dots, j_{t-h}) \\
&=\pi(j_t)p_{t-1}(j_{t-1}, \dots, j_{t-h}) \\
&=\pi(j_t) \sum_{j_{t-h-1}=1}^{k} p_{t-1}(j_{t-1}, \dots, j_{t-h-1}). \qquad (12.14)
\end{aligned}$$

This is directly available since the summands here are just the k^{h+1} posterior model probabilities at time $t-1$.

Further approximations to the mixture (12.12) can often be made to reduce the number of components. Three key considerations are as follows:

(A) Ignore components that have very small posterior probabilities.
(B) Combine components that are roughly equal into a single component, also combining the probabilities.
(C) Replace the contribution of a collection of components by a component that somehow represents their contribution.

These points apply generally to the use of mixtures, not only to the time series context. To provide insight, consider the following example where we drop the notation specific to the time series context for simplification.

EXAMPLE 12.4. Suppose $\boldsymbol{\theta}_t = \theta$ has density

$$p(\theta) = \sum_{j=1}^{4} p_j(\theta)p(j),$$

where for $j = 1, \dots, 4$, $p_j(.)$ is a T density with mode, scale and degrees of freedom possibly depending on the index j. Thus, in model j,

$$\theta \sim \mathrm{T}_{n(j)}[m(j), C(j)].$$

Note that this can be viewed as a very special case of (12.12). The approach (A) to approximating the mixture would apply if, for example, $p(4) = 0.005$, the fourth component receiving only 0.5% of the probability. In such a case,

$$p(\theta) \approx \sum_{j=1}^{3} p_j(\theta)p^*(j),$$

where $p^*(j) = p(j)/(1-p(4))$, $(j = 1, 2, 3)$, in order that $\sum_{j=1}^{3} p^*(j) = 1$. Case (B) would apply in the ideal, and extreme, case $p_3(\theta) = p_4(\theta)$ for all θ. Then

$$p(\theta) = p_1(\theta)p(1) + p_2(\theta)p(2) + p_3(\theta)[p(3) + p(4)],$$

the final two, equal, components being combined to leave a mixture of only three components. More realistically, if the two densities are very similar rather than exactly equal, then the same form of approximation is appropriate. Suppose, for example, $n(3) = n(4)$, $C(3) = C(4)$ and $m(3) = m(4) + \epsilon$, where ϵ is small relative to the scale of the distributions, so that the densities look similar (as do some of the T densities in Figure 12.4, for example). Then the contribution $p_3(\theta)p(3) + p_4(\theta)p(4)$ to the overall density may be approximated by a single component $p^*(\theta)[p(3) + p(4)]$ where $p^*(.)$ is a T density with the same degrees of freedom and scale as the two component densities, and mode $am(3) + (1 - a)m(4)$ where $a = p(3)/[p(3) + p(4)]$. This is in the spirit of approximations developed in Harrison and Stevens (1971 and 1976a), and those termed *quasi-Bayes* in Titterington, Smith and Makov (1985). See West (1992c) for more recent related developments. One interpretation is that $p^*(.)$ is a form of average of the two densities it replaces. Using the third approach (C), we can often reduce the size of a mixture by approximation even though the components removed are not, apparently, very similar. The basic technique is to replace a mixture of components of a given functional form with a single density of the same form. With T distributions here, and generally in DLM models, the combination in a mixture may be unimodal and roughly symmetric. Suppose $p(\theta)$ is the density in Figure 12.4, with components also graphed there. The mixture is unimodal and apparently very close to symmetry, suggesting that it can be well approximated by a single T density. This is the case here, and quite often in practice. More generally, some subset of the components of a mixture may be approximated in this way, and substituted with some combined probability. It remains, of course, to specify how the approximating density is chosen, this being the subject of the next section. Sometimes, however, it will not be appropriate. Suppose, for example, that the mixture is as displayed in Figure 12.3. The mixture is far from symmetric and cannot be well approximated by a single T density. Recall that mixtures can become multi-modal, clearly pin-pointing the dangers of uncritical use of unimodal approximations.

12.3.4 Approximation of mixtures: Specific results

The mixture approximating, or collapsing, techniques derived in this section are fundamental to the application of multi-process, class II models (Harrison and Stevens 1971, 1976a and b; Smith and West 1983). What is needed is a method by which an approximating density can be chosen to represent a mixture in cases when such an approximation is desirable and sensible. We begin rather generally, again ignoring the time series context and notation since the results here are not specific to that context, assuming that the density of the random vector $\boldsymbol{\theta}$ is the mixture

$$p(\boldsymbol{\theta}) = \sum_{j=1}^{k} p_j(\boldsymbol{\theta})p(j). \tag{12.15}$$

Here the component densities may generally take any forms, although often they will have the same functional form, such as normal or T, differing only through defining parameters such as means, variances, etc. The probabilities $p(j)$ are known. This density is to be approximated by a density $p^*(\boldsymbol{\theta})$ of specified functional form, the parameters defining the approximation to be chosen. Thus, a mixture of normal densities may be approximated by a normal with mean and variance matrix to be chosen in some optimal way. The notion that $p^*(.)$ should be *close* to $p(.)$ brings in the concept of a measure of how close, and the need for a *distance* measure between densities, or distributions. There are many such measures, some leading to similar or equivalent results, and we focus on just one.

Viewing $p(.)$ as the true density of $\boldsymbol{\theta}$ to be approximated by $p^*(.)$, consider the quantity

$$-\mathrm{E}\{\log[p^*(\boldsymbol{\theta})]\} = -\int \log[p^*(\boldsymbol{\theta})]p(\boldsymbol{\theta})d\boldsymbol{\theta}. \tag{12.16}$$

For any approximating distribution with density $p^*(.)$, this *entropy* related quantity is a natural measure of the closeness of approximation to the true distribution. Similar measures abound in Bayesian, and non-Bayesian, statistics. Recall, for example, the appearance of this sort of measure in the convergence results in multi-process, class I models of Section 12.2.4. Choosing the approximating density to achieve a small value of (12.16) is clearly equivalent to attempting to minimise the quantity $K(p^*)$ defined by

$$K(p^*) = \mathrm{E}\{\log\left[\frac{p(\boldsymbol{\theta})}{p^*(\boldsymbol{\theta})}\right]\} = \int \log\left[\frac{p(\boldsymbol{\theta})}{p^*(\boldsymbol{\theta})}\right] p(\boldsymbol{\theta})d\boldsymbol{\theta}. \tag{12.17}$$

This is the Kullback-Leibler directed divergence between the approximating distribution whose density is $p^*(.)$ and the true distribution with density $p(.)$. The divergence is defined for continuous and discrete distributions alike, although we restrict attention here to continuous distributions and assume that $p(.)$ and $p^*(.)$ have the same support. Thus, $p(\boldsymbol{\theta}) > 0$ if and only if $p^*(\boldsymbol{\theta}) > 0$. Although not a true distance measure, the properties of the divergence are appropriate to the problem of density approximation. Two key properties are that

- $K(p^*) \geq 0$ for all densities $p^*(.)$, and
- $K(p^*) = 0$ if and only if $p^* = p$, the true density, almost everywhere.

From now on, the divergence is used to measure closeness of approximations to mixtures, its use being illustrated in examples pertinent to the multi-process context. Related discussion can be found in Titterington, Makov and Smith (1985). Some of the results quoted, and used in exercises, are to be found in Quintana (1987), and related material appears in Amaral and Dunsmore (1980).

EXAMPLE 12.5. Suppose $\boldsymbol{\theta} = \theta$ is scalar whose true distribution has finite mean and variance. Suppose also that the approximation $p^*(.)$ is a normal density with mean m and variance C to be determined. It can be easily shown that

$$2K(p^*) = c + \log(C) + \mathrm{V}[\theta]/C + (m - \mathrm{E}[\theta])^2/C$$

for some constant c. It follows directly that the optimal normal approximation to any distribution with finite mean and variance just matches the moments, setting $m = \mathrm{E}[\theta]$ and $C = \mathrm{V}[\theta]$.

EXAMPLE 12.6. The multivariate generalisation of Example 12.5 assumes that the vector $\boldsymbol{\theta}$ has true distribution with finite mean vector and variance matrix, and that the approximating distribution is multivariate normal, $\mathrm{N}[\mathbf{m}, \mathbf{C}]$. Then as a function of $\mathbf{m}$ and $\mathbf{C}$,

$$\begin{aligned} 2K(p^*) &= c + \log(|\mathbf{C}|) + \mathrm{E}\{(\boldsymbol{\theta} - \mathbf{m})'\mathbf{C}^{-1}(\boldsymbol{\theta} - \mathbf{m})\} \\ &= c + \log(|\mathbf{C}|) + \mathrm{trace}(\mathbf{C}^{-1}\mathrm{V}[\boldsymbol{\theta}]) + (\mathrm{E}[\boldsymbol{\theta}] - \mathbf{m})'\mathbf{C}^{-1}(\mathrm{E}[\boldsymbol{\theta}] - \mathbf{m}), \end{aligned}$$

for some constant c. It is left as an exercise to the reader to verify that this is minimised by taking $\mathbf{m} = \mathrm{E}[\boldsymbol{\theta}]$ and $\mathbf{C} = \mathrm{V}[\boldsymbol{\theta}]$. In the case when $p(.)$ is a mixture as in (12.15), suppose that the mixture components $p_j(.)$ have means $\mathbf{m}(j)$ and variance matrices $\mathbf{C}(j)$. Then the optimal values for the approximating moments are the true moments, namely

$$\mathbf{m} = \sum_{j=1}^{k} \mathbf{m}(j)p(j)$$

and

$$\mathbf{C} = \sum_{j=1}^{k} \{\mathbf{C}(j) + (\mathbf{m} - \mathbf{m}(j))(\mathbf{m} - \mathbf{m}(j))'\}p(j).$$

In the multi-process context, various mixture collapsing procedures may be considered. The basic approach for the posteriors for the state vector $\boldsymbol{\theta}_t$ uses the normal mixture technique of Example 12.6. However, this needs a little refinement to cover the cases when the observational variance sequence is unknown and estimated, leading to mixtures of T posteriors for the state vector and gamma mixtures for the reciprocal variance, or precision, parameters. The following examples provide the relevant theory for these cases.

EXAMPLE 12.7. Suppose $\boldsymbol{\theta} = \phi$, a scalar, having a true distribution with finite mean. Suppose also that the approximating density $p^*(\phi)$ is gamma, $\mathrm{G}[n/2, d/2]$ for some degrees of freedom parameter $n > 0$ and $d > 0$ to be identified.

We consider first a special case of relevance to multi-process, in which the degrees of freedom parameter n is fixed in advance, the optimisation problem therefore being restricted to the choice of d. It is easily shown that as a function of d alone, $2K(p^*) = d\mathrm{E}[\phi] - n\,\log(d)$, which is minimised by setting $n/d = \mathrm{E}[\phi]$. This simply equates the mean of the gamma distribution to the true mean and provides $d = n/\mathrm{E}[\phi]$.

If n is to be considered too, the problem is rather more complex. Here, as a function of n and d,

$$2K(p^*) = constant + d\mathrm{E}[\phi] - n\,\log(d/2) - n\mathrm{E}[\log(\phi)] + 2\,\log[\Gamma(n/2)].$$

It follows by differentiation that the minimising values of n and d satisfy

$$\text{(a)}\quad \mathrm{E}[\phi] = n/d,$$

$$\text{(b)}\quad \mathrm{E}[\log(\phi)] = \gamma(n/2) + \log(2\mathrm{E}[\phi]/n).$$

Here $\gamma(.)$ is the digamma function defined, for all $x > 0$, by $\gamma(x) = \Gamma'(x)/\Gamma(x)$. Let $S = d/n$. Then from (a), $S^{-1} = \mathrm{E}[\phi]$, the approximating Gamma distribution having the true mean. The value of n can be found from (b), although the unique solution can be found generally only via numerical methods. Given S from (a), (b) leads to $\mathrm{E}[\log(\phi)] = \gamma(n/2) - \log(nS/2)$, an implicit equation in n that may be solved numerically. If it is clear that n is relatively large, then the approximation $\gamma(x) \approx \log(x) - 1/(2x)$, for x large, may be applied, leading from (b) to $n^{-1} = \log(\mathrm{E}[\phi]) - \mathrm{E}[\log(\phi)]$. This approximation is certainly adequate for $n > 20$.

Consider the special case that $p(.)$ is a discrete mixture of gamma distributions, the component $p_j(.)$ being $\mathrm{G}[n(j)/2, d(j)/2]$ with means $S(j)^{-1} = n(j)/d(j)$. Now the required quantities $\mathrm{E}[\phi]$ and $\mathrm{E}[\log(\phi)]$ in (a) and (b) are given by

$$\mathrm{E}[\phi] = \sum_{j=1}^{k} S(j)^{-1} p(j),$$

$$\mathrm{E}[\log(\phi)] = \sum_{j=1}^{k} \{\gamma(n(j)/2) - \log(d(j)/2)\} p(j).$$

EXAMPLE 12.8. Specifically in connection with the multi-process DLM, suppose that the q-vector $\boldsymbol{\theta}$ and the scalar ϕ have a joint distribution that is a mixture of normal/gamma forms. The mixture has k components, the jth component being defined as follows:

(1) Given ϕ, $(\boldsymbol{\theta} \mid \phi) \sim \mathrm{N}[\mathbf{m}(j), \mathbf{C}(j)/\{S(j)\phi\}]$ for some mean vector $\mathbf{m}(j)$, variance matrix $\mathbf{C}(j)$, and estimate $S(j) > 0$ of ϕ^{-1}.
(2) Marginally, $\phi \sim \mathrm{G}[n(j)/2, d(j)/2]$ for positive $n(j)$ and $d(j)$, having mean $\mathrm{E}[\phi] = S(j)^{-1} = n(j)/d(j)$.
(3) Putting (1) and (2) together and integrating out ϕ gives the marginal, multivariate T distribution in model j, $\boldsymbol{\theta} \sim \mathrm{T}_{n(j)}[\mathbf{m}(j), \mathbf{C}(j)]$.

Now, suppose that the mixture is to be approximated by $p^*(\boldsymbol{\theta}, \phi)$, a single, normal/gamma distribution defined by parameters $\mathbf{m}$, $\mathbf{C}$, n and d, giving $(\boldsymbol{\theta} \mid \phi) \sim \mathrm{N}[\mathbf{m}, \mathbf{C}/\{S\phi\}]$, where $S = d/n$, $\phi \sim \mathrm{G}[n/2, d/2]$ and $\boldsymbol{\theta} \sim \mathrm{T}_n[\mathbf{m}, \mathbf{C}]$. Calculation of the Kullback-Leibler divergence requires integration of $\log[p^*(\boldsymbol{\theta}, \phi)]$ with respect to the joint mixture distribution of $\boldsymbol{\theta}$ and ϕ. We can write

$$\begin{aligned} K(p^*) &= constant - \mathrm{E}\{\log[p^*(\boldsymbol{\theta}, \phi)]\} \\ &= constant - \mathrm{E}\{\log[p^*(\boldsymbol{\theta} \mid \phi)] + \log[p^*(\phi)]\}, \end{aligned}$$

specifying the joint density in terms of the conditional/marginal pair in (1) and (2). It is left as an exercise to the reader to verify that on substituting the densities from (1) and (2), this results in

$$\begin{aligned} 2K(p^*) = &\ constant \\ &- n\log(n/2) + 2\log[\Gamma(n/2)] - (n+q-2)\mathrm{E}[\log(\phi)] \\ &- n\log(S) + nS\mathrm{E}[\phi] + \log(|S^{-1}\mathbf{C}|) \\ &+ S\sum_{j=1}^{k} S(j)^{-1}\{\mathrm{trace}(\mathbf{C}^{-1}\mathbf{C}(j)) \\ &\qquad + (\mathbf{m} - \mathbf{m}(j))'\mathbf{C}^{-1}(\mathbf{m} - \mathbf{m}(j))\}p(j). \end{aligned}$$

It follows that minimisation with respect to $\mathbf{m}$, $\mathbf{C}$, S and n is achieved as follows:

(a) Minimisation with respect to $\mathbf{m}$ leads, as in Example 12.6, to

$$\mathbf{m} = \left\{\sum_{j=1}^{k} S(j)^{-1}p(j)\right\}^{-1} \sum_{j=1}^{k} S(j)^{-1}\mathbf{m}(j)p(j).$$

(b) Minimisation with respect to $\mathbf{C}$ leads, again as in Example 12.6, to

$$\mathbf{C} = S\sum_{j=1}^{k} S(j)^{-1}\{\mathbf{C}(j) + (\mathbf{m} - \mathbf{m}(j))(\mathbf{m} - \mathbf{m}(j))'\}p(j).$$

(c) Minimisation with respect to S leads, as in Example 12.7(a), to

$$S^{-1} = \mathrm{E}[\phi] = \sum_{j=1}^{k} S(j)^{-1}p(j).$$

This also implies that in (a) and (b),

$$\mathbf{m} = \sum_{j=1}^{k} \mathbf{m}(j)p^*(j)$$

and

$$\mathbf{C} = \sum_{j=1}^{k} \{\mathbf{C}(j) + (\mathbf{m} - \mathbf{m}(j))(\mathbf{m} - \mathbf{m}(j))'\} p^*(j),$$

where the revised weights $p^*(j) = p(j)S/S(j)$ sum to unity.

(d) Minimisation with respect to n leads, as in Example 12.7(b), to

$$\mathrm{E}[\log(\phi)] = \gamma(n/2) - \log(nS/2).$$

With S obtained from (c), this again must generally be solved numerically for n. Since $p(\phi)$ is a mixture of gamma densities, then $\mathrm{E}[\log(\phi)]$ may be calculated as in Example 12.7.

Two special cases for consideration are as follows:

(i) Results (a) and (b) clearly specialise to the corresponding versions in Example 12.6 if it is assumed that $\phi = S^{-1}$ is known. Formally, let each $n(j)$ tend to infinity and $S(j) = S$.

(ii) As in Example 12.7, consider the case that the $n(j)$ are equal. The above results apply, of course, to define the optimal, approximating gamma distribution, the resulting value of n being typically different to that common to the mixture components. It may be viewed as appropriate to fix the degrees of freedom n at the common value, in which case the optimisation problem is restricted to the choice of S, defined in (c), part (d) being irrelevant.

12.4 CLASS II MIXTURE MODELS ILLUSTRATED

12.4.1 Second-order polynomial models with exceptions

As an illustration, we consider a widely used class II model that was introduced in Harrison and Stevens (1971, 1976a and b). See also Green and Harrison (1973), Smith and West (1983), and Ameen and Harrison (1985b) for extensions and applications. The model is designed for series that behave generally according to a second-order polynomial, or *locally linear* model, but that are subject to exceptional events including outliers and changes in level or growth. These are just the sorts of commonly occurring exceptions discussed in Chapter 11 and that were handled using subjective intervention and simple, automatic exception detection and adaptation methods. The multi-process models developed here can be seen as a formal, model based technique for monitoring and adaptation. Clearly, the multi-process technique may also be applied to other, more complex models.

To begin, consider the usual, second-order polynomial model for the Y_t series, assuming the observational variance sequence known. We use the linear growth form of Section 7.3 with level and growth parameters μ_t and

β_t comprising the state vector $\boldsymbol{\theta}_t = (\mu_t, \beta_t)'$ at time t. Stochastic changes are given in 'delta' notation so that the model is

$$\begin{aligned} Y_t &= \mu_t + \nu_t, \qquad\qquad \nu_t \sim \mathrm{N}[0, V_t], \\ \mu_t &= \mu_{t-1} + \beta_t + \delta\mu_t, \\ \beta_t &= \beta_{t-1} + \delta\beta_t, \end{aligned}$$

where $\delta\mu_t$ and $\delta\beta_t$ are the changes in level and growth, having a joint normal distribution with zero-mean vector and variance matrix to be specified. In matrix notation, the evolution equation is

$$\boldsymbol{\theta}_t = \mathbf{G}\boldsymbol{\theta}_{t-1} + \boldsymbol{\omega}_t,$$

where

$$\mathbf{G} = \begin{pmatrix} 1 & 1 \\ 0 & 1 \end{pmatrix}$$

and

$$\boldsymbol{\omega}_t = \begin{pmatrix} \delta\mu_t + \delta\beta_t \\ \delta\beta_t \end{pmatrix} = \mathbf{G}\begin{pmatrix} \delta\mu_t \\ \delta\beta_t \end{pmatrix}.$$

Thus, the evolution variance matrix $\mathbf{W}_t$ is given by

$$\mathbf{W}_t = \mathbf{G}\mathrm{V}[(\delta\mu_t, \delta\beta_t)']\mathbf{G}'. \tag{12.18}$$

In addition, the usual independence assumptions hold. As a specific example, suppose that $V_t = V = 1$, $\mathrm{V}[\delta\mu_t] = 0.1$, $\mathrm{V}[\delta\beta_t] = 0.01$ and $\mathrm{C}[\delta\mu_t, \delta\beta_t] = 0$ for all t. Then

$$\mathbf{W}_t = \mathbf{W} = \begin{pmatrix} 0.11 & 0.01 \\ 0.01 & 0.01 \end{pmatrix}.$$

The model is constant, having an evolution variance matrix $\mathbf{W}$ of a linear growth form, and the constant defining quadruple is

$$\left\{ \begin{pmatrix} 1 \\ 0 \end{pmatrix}, \begin{pmatrix} 1 & 1 \\ 0 & 1 \end{pmatrix}, 1, \begin{pmatrix} 0.11 & 0.01 \\ 0.01 & 0.01 \end{pmatrix} \right\}. \tag{12.19}$$

Note that the model is standardised to have unit observational variance for convenience here, so that $\mathbf{W}_t$ is scale-free.

Consider modelling outliers and changes in trend in a series thought to behave generally according to this model. An outlying observation Y_t may be modelled via a large (positive or negative) observational error ν_t. Under existing model assumptions, $\Pr[|v_t| > 3] < 0.01$, so that values larger than 3 in absolute value are exceptional. Larger values can be generated by replacing $V = 1$ in the model with a larger variance. A single, extreme value $\nu_t = 5$, for example, is an exception in the model as it stands but perfectly consistent with a model having a larger observational variance, say $V_t = 100$, for example, at time t alone. If outliers of this sort of magnitude are expected to occur some (small) percentage of the time, then the alternative DLM

$$\left\{ \begin{pmatrix} 1 \\ 0 \end{pmatrix}, \begin{pmatrix} 1 & 1 \\ 0 & 1 \end{pmatrix}, 100, \begin{pmatrix} 0.11 & 0.01 \\ 0.01 & 0.01 \end{pmatrix} \right\} \tag{12.20}$$

will adequately model them, whilst (12.19) remains appropriate for uncontaminated observations. This sort of outlier model, commonly referred to as a scale shift or scale inflation model, has been widely used in modelling outliers in Bayesian analyses, and its use is not restricted to time series (Box and Tiao 1968; Smith and Pettit 1985). The larger observational variance leads to larger observational errors than in the standard model (12.19) but remains neutral as to the sign, the errors still having zero mean.

Similarly, changes in level μ_t much greater than predicted by (12.19) can be accommodated by replacing $\mathbf{W}$ with an alternative in which the variance of $\delta\mu_t$ is inflated. An inflation factor of 100, as used in (12.20) for the outlier model, for example, leads to

$$\mathbf{W} = \begin{pmatrix} 10.01 & 0.01 \\ 0.01 & 0.01 \end{pmatrix} \tag{12.21}$$

for just the times of change. With this particular matrix, the increment $\delta\mu_t$ to the level parameter at time t can be far greater in absolute value than under (12.19), leading to the possibility of marked, abrupt changes or jumps in level. Note again that the direction of change is not anticipated in this model.

Finally, the same idea applies to modelling jumps in growth of the series. Taking $\mathrm{V}[\delta\beta_t] = 1$, for example, to replace the original variance of 0.01, leads to

$$\mathbf{W} = \begin{pmatrix} 1.1 & 1 \\ 1 & 1 \end{pmatrix} \tag{12.22}$$

for the times of possible abrupt changes in growth.

In line with the discussion of Sections 12.3.1 and 12.3.2 consider collections of DLMs each of which may apply at any given time, the selection being according to fixed model probabilities. Suppose, for example, that

- the standard DLM (12.19) applies with probability 0.85;
- the outlier generating DLM (12.20) applies with probability 0.07;
- the level change DLM (12.21) applies with probability 0.05;
- the growth change DLM (12.22) applies with probability 0.03.

With these probabilities, the series is expected to accord to the standard DLM (12.19) about 85% of the time. The chance of an outlier at any time is 7%; thus one outlying observation in 14 is to be expected. Abrupt changes in the trend of the series are viewed as likely to occur about 8% of the time, with level changes more likely at 5% then growth changes at 3%. This sort of setup is appropriate for many real series; the precise values of the probabilities chosen here are also representative of the forms of behaviour evident in many macro commercial and economic series.

This is an example of a mixture of DLMs with $k = 4$ possible models applying at any time. Some important variations in this particular context are as follows:

(i) Reduction to 3 DLMs by combining level and growth changes into a single model. For example, taking (12.19) with $\mathbf{W}$ replaced by

$$\mathbf{W} = \begin{pmatrix} 11 & 1 \\ 1 & 1 \end{pmatrix}$$

would give an appropriate trend change model, allowing for changes in either or both components μ_t and β_t of the trend. This reduces the number of models and is sensible if it is not of interest to distinguish between changes in trend and changes in growth. Also, it can often be difficult to distinguish between the two, particularly when growth changes are small, in which case the combination of the two into a single, overall model loses little in practice.

(ii) It may be desired to use discount factors to assign evolution variances $\mathbf{W}_t$ rather than using a constant matrix. If this is so, then exceptionally small discount factors apply to model abrupt changes as in Section 11.5. However, the specific structure of the linear growth model here cannot be obtained by using the standard discounting technique. Approaches using discount factors can take the following forms:

Firstly, and generally often appropriately, the quantities $\delta\mu_t$ and $\delta\beta_t$ may be taken as uncorrelated, with variances determined as fractions of the posterior variances of μ_{t-1} and β_{t-1}. Thus, given the usual posterior

$$(\boldsymbol{\theta}_{t-1} \mid D_{t-1}) \sim \mathrm{N}[\mathbf{m}_{t-1}, \mathbf{C}_{t-1}]$$

at $t-1$, write

$$\mathbf{C}_{t-1} = \begin{pmatrix} C_{t-1,\mu} & C_{t-1,\mu\beta} \\ C_{t-1,\mu\beta} & C_{t-1,\beta} \end{pmatrix}.$$

Separate discount factors for level and growth, denoted by δ_μ and δ_β respectively, now define $W_{t,\mu} = C_{t-1,\mu}(\delta_\mu^{-1} - 1)$ and $W_{t,\beta} = C_{t-1,\beta}(\delta_\beta^{-1} - 1)$. The linear growth structure is retained with evolution variance matrix

$$\mathbf{W}_t = \begin{pmatrix} W_{t,\mu} + W_{t,\beta} & W_{t,\beta} \\ W_{t,\beta} & W_{t,\beta} \end{pmatrix}. \tag{12.23}$$

The possibilities of abrupt changes are modelled using smaller, exceptional discount factors δ_μ^* and δ_β^*, simply inflating the elements of $\mathbf{W}_t$ appropriately.

(iii) All alternative discounting methods lead to evolution variance matrices that do not have the linear growth structure (Section 7.3). However, they are still perfectly valid second-order polynomial models and the differences in practical application are typically small. In line with the usual component modelling ideas applied to discounting, the key alternative approach is appropriate if the level and growth are viewed as a single trend component, the two parameters changing together at the same rate. Here a single discount factor δ controls the time variation. With

$$\mathrm{V}[\boldsymbol{\theta}_t \mid D_{t-1}] = \mathbf{P}_t = \mathbf{G}\mathbf{C}_{t-1}\mathbf{G}',$$

we obtain

$$\mathbf{W}_t = \mathbf{P}_t(\delta^{-1} - 1).$$

Again, abrupt changes in trend are modelled by altering δ to a smaller, exceptional value δ^* to define a trend change DLM.

We proceed to analyse and apply the multi-process with 4 possible DLMs. By way of general notation,

$$\mathbf{F} = \mathbf{E}_2 = \begin{pmatrix} 1 \\ 0 \end{pmatrix} \quad \text{and} \quad \mathbf{G} = \mathbf{J}_2(1) = \begin{pmatrix} 1 & 1 \\ 0 & 1 \end{pmatrix}.$$

Suppose initially, for notational simplicity, that the observational variance sequence V_t is known (the extension to the case of an unknown, constant variance will be summarised below). The basic DLM quadruple is then

$$\{\mathbf{F},\ \mathbf{G},\ V_t,\ \mathbf{W}_t\}.$$

The special cases for the 4 possible models in the multi-process are as follows:

(1) Standard DLM: $\{\mathbf{F},\ \mathbf{G},\ V_tV(1),\ \mathbf{W}_t(1)\}$, where $V(1) = 1$ and $\mathbf{W}_t(1) = \mathbf{W}_t$, a standard evolution variance matrix.
(2) Outlier DLM: $\{\mathbf{F},\ \mathbf{G},\ V_tV(2),\ \mathbf{W}_t(2)\}$, where $V(2) > 1$ is an inflated variance consistent with the occurrence of observations that would be extreme in the standard DLM, and $\mathbf{W}_t(2) = \mathbf{W}_t(1) = \mathbf{W}_t$.
(3) Level change DLM: $\{\mathbf{F},\ \mathbf{G},\ V_tV(3),\ \mathbf{W}_t(3)\}$, where $V(3) = 1$ and $\mathbf{W}_t(3)$ is an evolution variance matrix consistent with level changes.
(4) Growth change DLM: $\{\mathbf{F},\ \mathbf{G},\ V_tV(4),\ \mathbf{W}_t(4)\}$, where $V(4) = 1$ and $\mathbf{W}_t(4)$ is an evolution variance matrix consistent with growth changes.

Formally, we have models $M_t(j)$ indexed by $\alpha_j = j$, $(j = 1, \ldots, 4)$, the generic parameter α taking values in the index set $\mathcal{A}$.

12.4.2 Model analysis (V_t known)

In providing for the possibility of various exceptions, it is clear that after the occurrence of any such event, further observations will be required before the nature of the event can be identified. Thus, an outlier is indistinguishable from the onset of change in either level or growth, or both, until the next observation is available. Hence, in approximating mixtures by collapsing over possible models in the past (Section 12.3.3), it will be usual that $h = 1$ at least in (12.11). The original material in Harrison and Stevens (1971, 1976a) and Smith and West (1983) use $h = 1$, and indeed this will often be adequate. In some applications $h = 2$ may be necessary, retaining information relevant to all possible models up to two steps back in time. This may be desirable, for example, if exceptions are rather frequent, possibly occurring consecutively over time. It will be very rare, however, that $h > 2$ is necessary. For illustration here we follow the above-referenced authors in using $h = 1$.

We have models

$$M_t(j_t): \qquad \{\mathbf{F},\ \mathbf{G},\ V_tV(j_t),\ \mathbf{W}_t(j_t)\}, \qquad (j_t = 1, \ldots, 4).$$

For each j_t, it is assumed that $M_t(j_t)$ applies at time t with fixed and prespecified probability $\pi(j_t) = \Pr[M_t(j_t) \mid D_{t-1}] = \Pr[M_t(j_t) \mid D_0]$. Thus, at time t, the model is defined by observation and evolution equations

$$(Y_t \mid \boldsymbol{\theta}_t, M_t(j_t)) \sim \mathrm{N}[\mathbf{F}'\boldsymbol{\theta}_t, V_tV(j_t)] \tag{12.24}$$

and

$$(\boldsymbol{\theta}_t \mid \boldsymbol{\theta}_{t-1}, M_t(j_t)) \sim \mathrm{N}[\mathbf{G}\boldsymbol{\theta}_{t-1}, \mathbf{W}_t(j_t)], \tag{12.25}$$

with probability $\pi(j_t)$, conditionally independently of the history of the series D_{t-1}.

Assume also that at $t = 0$, the initial prior for the state vector is the usual normal form

$$(\boldsymbol{\theta}_0 \mid D_0) \sim \mathrm{N}[\mathbf{m}_0, \mathbf{C}_0],$$

irrespective of possible models obtaining at any time, where $\mathbf{m}_0$ and $\mathbf{C}_0$ are known and fixed at $t = 0$. Given this setup, the development of Section 12.3.3 applies in this special case of $h = 1$; the position at any time $t - 1$ is now summarised.

Historical information D_{t-1} is summarised in terms of a 4-component mixture posterior distribution for $\boldsymbol{\theta}_{t-1}$, the mixture being with respect to the four possible models obtaining at time $t - 1$. Within each component, the posterior distributions have the usual conjugate normal forms. Thus,

(a) For $j_{t-1} = 1, \ldots, 4$, model $M_{t-1}(j_{t-1})$ applied at time $t - 1$ with posterior probability $p_{t-1}(j_{t-1})$. These probabilities are now known and fixed.

(b) Given $M_{t-1}(j_{t-1})$ and D_{t-1}, $\boldsymbol{\theta}_{t-1}$ is

$$\mathrm{N}[\mathbf{m}_{t-1}(j_{t-1}), \mathbf{C}_{t-1}(j_{t-1})]. \tag{12.26}$$

Note that generally, the quantities defining these distributions depend on the model applying at $t-1$, hence the index j_{t-1}.

Evolving to time t, statements about $\boldsymbol{\theta}_t$ and Y_t depend on the combinations of possible models applying at both $t-1$ and t.

(c) Thus, from (12.25) and (12.26), for each j_{t-1} and j_t we have

$$(\boldsymbol{\theta}_t \mid M_t(j_t), M_{t-1}(j_{t-1}), D_{t-1}) \sim \mathrm{N}[\mathbf{a}_t(j_{t-1}), \mathbf{R}_t(j_t, j_{t-1})],$$

where $\mathbf{a}_t(j_{t-1}) = \mathbf{G}\mathbf{m}_{t-1}(j_{t-1})$ and

$$\mathbf{R}_t(j_t, j_{t-1}) = \mathbf{G}\mathbf{C}_{t-1}(j_{t-1})\mathbf{G}' + \mathbf{W}_t(j_t).$$

Note that $\mathbf{a}_t(j_{t-1})$ does not differ across the $M_t(j_t)$, depending only upon possible models applying at $t-1$ since $\mathbf{G}$ is common to these models.

(d) Similarly, the one-step ahead forecast distribution is given, for each possible combination of models, by the usual form

$$(Y_t \mid M_t(j_t), M_{t-1}(j_{t-1}), D_{t-1}) \sim \mathrm{N}[f_t(j_{t-1}), Q_t(j_t, j_{t-1})], \tag{12.27}$$

where

$$f_t(j_{t-1}) = \mathbf{F}'\mathbf{a}_t(j_{t-1}),$$

also common across the models at time t, and

$$Q_t(j_t, j_{t-1}) = \mathbf{F}'\mathbf{R}_t(j_t, j_{t-1})\mathbf{F} + V_t V(j_t).$$

Note again that due to the particular structure of the multi-process, the means of these sixteen possible forecast distributions take only four distinct values,

$$\mathrm{E}[Y_t \mid M_t(j_t), M_{t-1}(j_{t-1}), D_{t-1}] = f_t(j_{t-1})$$

for each of the four values j_t. This follows since the $M_t(j_t)$ differ only through scale parameters and not in location. Note, however, that this feature is specific to the particular model here and may be present or absent in other models. Calculation of the forecast distribution unconditional on possible models then simply involves the mixing of these standard normal components with respect to the relevant probabilities, calculated as follows:

(e) For each j_t and j_{t-1},

$$\begin{aligned}&\Pr[M_t(j_t), M_{t-1}(j_{t-1}) \mid D_{t-1}]\\ &\quad = \Pr[M_t(j_t) \mid M_{t-1}(j_{t-1}), D_{t-1}]\Pr[M_{t-1}(j_{t-1}) \mid D_{t-1}].\end{aligned}$$

Now, by assumption, models apply at t with constant probabilities $\pi(j_t)$ irrespective of what happened previously, so that

$$\Pr[M_t(j_t) \mid M_{t-1}(j_{t-1}), D_{t-1}] = \pi(j_t);$$

additionally,

$$\Pr[M_{t-1}(j_{t-1}) \mid D_{t-1}] = p_t(j_{t-1}),$$

and then

$$\Pr[M_t(j_t), M_{t-1}(j_{t-1}) \mid D_{t-1}] = \pi(j_t)p_{t-1}(j_{t-1}).$$

This is just a special case of the general formula (12.14) with $h = 1$.

(f) The marginal predictive density for Y_t is a mixture of the $4^2 = 16$ components (12.27) with respect to these probabilities,

$$p(Y_t \mid D_{t-1}) = \sum_{j_t=1}^{4} \sum_{j_{t-1}=1}^{4} \{\pi(j_t)p_{t-1}(j_{t-1}) \times p(Y_t \mid M_t(j_t), M_{t-1}(j_{t-1}), D_{t-1})\}. \tag{12.28}$$

Now consider updating the prior distributions in (c) to posteriors when Y_t is observed. Given j_{t-1} and j_t, the standard updating equations apply within each of the sixteen combinations, the posterior means, variances, etc. obviously varying across combinations.

(g) Thus,

$$(\boldsymbol{\theta}_t \mid M_t(j_t), M_{t-1}(j_{t-1}), D_t) \sim \mathrm{N}[\mathbf{m}_t(j_t, j_{t-1}), \mathbf{C}_t(j_t, j_{t-1})], \tag{12.29}$$

where

$$\begin{aligned}
\mathbf{m}_t(j_t, j_{t-1}) &= \mathbf{a}_t(j_{t-1}) + \mathbf{A}_t(j_t, j_{t-1})e_t(j_{t-1}),\\
\mathbf{C}_t(j_t, j_{t-1}) &= \mathbf{R}_t(j_t, j_{t-1}) \\
&\quad - \mathbf{A}_t(j_t, j_{t-1})\mathbf{A}_t(j_t, j_{t-1})'Q_t(j_t, j_{t-1}),\\
e_t(j_{t-1}) &= Y_t - f_t(j_{t-1})
\end{aligned}$$

and

$$\mathbf{A}_t(j_t, j_{t-1}) = \mathbf{R}_t(j_t, j_{t-1})\mathbf{F}/Q_t(j_t, j_{t-1}).$$

Posterior probabilities across the sixteen possible models derive directly from the general formula (12.13) in the case $h = 1$.

(h) Thus,

$$\begin{aligned}
p_t(j_t, j_{t-1}) &= \Pr[M_t(j_t), M_{t-1}(j_{t-1}) \mid D_t]\\
&\propto \pi(j_t)p_{t-1}(j_{t-1})p(Y_t \mid M_t(j_t), M_{t-1}(j_{t-1}), D_{t-1}).
\end{aligned}$$

The second term here is the observed value of the predictive density (12.27), providing the model likelihood, and so these probabilities are easily calculated. These are simply given by

$$p_t(j_t, j_{t-1}) = \frac{c_t \pi(j_t) p_{t-1}(j_{t-1})}{Q_t(j_t, j_{t-1})^{1/2}} \exp\{-0.5 e_t(j_t, j_{t-1})^2 / Q_t(j_t, j_{t-1})\}, \tag{12.30}$$

where c_t is the constant of normalisation such that

$$\sum_{j_t=1}^{4} \sum_{j_{t-1}=1}^{4} p_t(j_t, j_{t-1}) = 1.$$

Inferences about $\boldsymbol{\theta}_t$ are based on the unconditional, sixteen component mixtures that average (12.29) with respect to the posterior model probabilities (12.30).

(i) Thus,

$$p(\boldsymbol{\theta}_t \mid D_t) = \sum_{j_t=1}^{4} \sum_{j_{t-1}}^{4} p(\boldsymbol{\theta}_t \mid M_t(j_t), M_{t-1}(j_{t-1}), D_t) p_t(j_t, j_{t-1}). \tag{12.31}$$

with components given by the normal distributions in (12.29).

These calculations essentially complete the evolution and updating steps at time t. To proceed to time $t+1$, however, we need to remove the dependence of the joint posterior (12.31) on possible models obtaining at time $t-1$. If we evolve (12.31) to time $t+1$ directly, the mixture will expand to $4^3 = 64$ components for $\boldsymbol{\theta}_{t+1}$, depending on all possible combinations of $M_{t+1}(j_{t+1})$, $M_t(j_t)$ and $M_{t-1}(j_{t-1})$. However, the principle of approximating such mixtures by assuming that the effects of different models at $t-1$ are negligible for time $t+1$ applies. Thus, in moving to $t+1$, the sixteen-component mixture (12.31) will be reduced, or collapsed, over possible models at $t-1$. The method of approximation of Example 12.6 applies here. For each $j_t = 1, \ldots, 4$, it follows that

$$\begin{aligned} &\bullet \quad p_t(j_t) = \Pr[M_t(j_t) \mid D_t] = \sum_{j_{t-1}=1}^{4} p_t(j_t, j_{t-1}), \\ &\bullet \quad \Pr[M_{t-1}(j_{t-1}) \mid D_t] = \sum_{j_t=1}^{4} p_t(j_t, j_{t-1}), \qquad (12.32) \\ &\bullet \quad \Pr[M_{t-1}(j_{t-1}) \mid M_t(j_t), D_t] = p_t(j_t, j_{t-1}) / p_t(j_t). \end{aligned}$$

The first equation here gives current model probabilities at time t. The second gives posterior probabilities over the possible models one-step back in time, at $t-1$. These one-step back, or *smoothed,* model probabilities are of great use in retrospective assessment of which models were likely at the previous time point. The third equation is of direct interest here in collapsing the mixture (12.31) with respect to time $t-1$; it gives the posterior (given D_t) probabilities of the various models at time $t-1$ *conditional* on possible models at time t. To see how these probabilities feature in the posterior (12.31), note that this distribution can be rewritten as

$$p(\boldsymbol{\theta}_t \mid D_t) = \sum_{j_t=1}^{4} p(\boldsymbol{\theta}_t \mid M_t(j_t), D_t) p_t(j_t), \tag{12.33}$$

the first terms of the summands being given by

$$p(\boldsymbol{\theta}_t \mid M_t(j_t), D_t) = \\ \sum_{j_{t-1}=1}^{4} p(\boldsymbol{\theta}_t \mid M_t(j_t), M_{t-1}(j_{t-1}), D_t) p_t(j_t, j_{t-1})/p_t(j_t). \tag{12.34}$$

In (12.33), the posterior is represented as a four-component mixture, the components being conditional only on models at time t and being calculated as four-component mixtures themselves in (12.34). Only in the latter mixture are the component densities of standard normal form. Now (12.33) is the exact posterior, the dependence on possible models at time t being explicit, whilst that on models at $t-1$ is implicit through (12.34). Thus, in moving to time $t+1$, the posterior will have the required form (12.26) if each of the components (12.34) is replaced by a normal distribution. For each j_t, the mixture in (12.34) has precisely the form of that in Example 12.6 and may be collapsed to a single approximating normal (optimal in the sense of minimising the Kullback-Leibler divergence) using the results of that example. For each j_t, define the appropriate mean vectors $\mathbf{m}_t(j_t)$ by

$$\mathbf{m}_t(j_t) = \sum_{j_{t-1}=1}^{4} \mathbf{m}_t(j_t, j_{t-1}) p_t(j_t, j_{t-1})/p_t(j_t);$$

the corresponding variance matrices $\mathbf{C}_t(j_t)$ are given by

$$\sum_{j_{t-1}=1}^{4} \{\mathbf{C}_t(j_t, j_{t-1}) + (\mathbf{m}_t(j_t) - \mathbf{m}_t(j_t, j_{t-1}))(\mathbf{m}_t(j_t) - \mathbf{m}_t(j_t, j_{t-1}))'\} \\ \times p_t(j_t, j_{t-1})/p_t(j_t).$$

Then (12.34) is approximated by the single normal posterior having the same mean and variance matrix, namely

$$(\boldsymbol{\theta}_t \mid M_t(j_t), D_t) \sim \mathrm{N}[\mathbf{m}_t(j_t), \mathbf{C}_t(j_t)]. \tag{12.35}$$

The distributions (12.35) replace the components (12.34) of the mixture (12.33), collapsing from sixteen to four standard normal components. In doing so, we complete the cycle of evolution, updating and collapsing; the resulting four-component mixture is analogous to the starting four-component mixture defined by components (12.26) with the time index updated from $t-1$ to t.

12.4.3 Summary of full model analysis

The above analysis extends to include learning about a known and constant observational variance scale parameter $V_t = V$ for all t. The conditionally conjugate normal/gamma analysis applies as usual within any collection of DLMs applying at all times. Differences arise only through the requirement that mixtures be approximated by collapsing with respect to models h-steps back before evolving to the next time point. Since we now have an extended, conditional normal/gamma posterior at each time, this collapsing is based on that in Example 12.8. Full details of the sequential analysis are summarised here.

At $t = 0$, the initial prior for the state vector and observational scale has the usual conjugate form irrespective of possible models obtaining at any time. With precision parameter $\phi = V^{-1}$, we have

$$\begin{aligned}(\boldsymbol{\theta}_0 \mid V, D_0) &\sim \mathrm{N}[\mathbf{m}_0, \mathbf{C}_0 V/S_0],\\ (\phi \mid D_0) &\sim \mathrm{G}[n_0/2, d_0/2],\end{aligned}$$

where $\mathbf{m}_0$, $\mathbf{C}_0$, n_0 and d_0 are known and fixed at $t = 0$. The initial point estimate S_0 of V is given by $S_0 = d_0/n_0$, and the prior for $\boldsymbol{\theta}_0$ marginally with respect to V is just $(\boldsymbol{\theta}_0 \mid D_0) \sim \mathrm{T}_{n_0}[\mathbf{m}_0, \mathbf{C}_0]$.

At times $t-1$ and t, the components of analysis are now described. Historical information D_{t-1} is summarised in terms of a four-component mixture posterior distribution for the state vector $\boldsymbol{\theta}_{t-1}$ and the variance scale parameter V, the mixture being with respect to the four possible models obtaining at time $t-1$. Within each component, the posterior distributions have the usual conjugate normal/gamma forms and the corresponding model probabilities $p_{t-1}(j_{t-1})$ are currently known. Given $M_{t-1}(j_{t-1})$, $\boldsymbol{\theta}_{t-1}$ and ϕ have a joint normal/gamma posterior with marginals

$$\begin{aligned}(\boldsymbol{\theta}_{t-1} \mid M_{t-1}(j_{t-1}), D_{t-1}) &\sim \mathrm{T}_{n_{t-1}}[\mathbf{m}_{t-1}(j_{t-1}), \mathbf{C}_{t-1}(j_{t-1})],\\ (\phi \mid M_{t-1}(j_{t-1}), D_{t-1}) &\sim \mathrm{G}[n_{t-1}/2, d_{t-1}(j_{t-1})/2],\end{aligned} \tag{12.36}$$

where $S_{t-1}(j_{t-1}) = d_{t-1}(j_{t-1})/n_{t-1}$ is the estimate of $V = \phi^{-1}$ in model $M_{t-1}(j_{t-1})$. The quantities defining these distributions generally depend on the model applying at $t-1$, hence on the index j_{t-1}. The exception here is the degrees of freedom parameter n_{t-1} common to each of the four possible models.

Evolving to time t, statements about $\boldsymbol{\theta}_t$ and Y_t depend on the combinations of possible models applying at both $t-1$ and t. The prior for $\boldsymbol{\theta}_t$ and

ϕ is normal/gamma with marginals, for each j_{t-1} and j_t, given by

$$(\boldsymbol{\theta}_t \mid M_t(j_t), M_{t-1}(j_{t-1}), D_{t-1}) \sim \mathrm{T}_{n_{t-1}}[\mathbf{a}_t(j_{t-1}), \mathbf{R}_t(j_t, j_{t-1})],$$
$$(\phi \mid M_t(j_t), M_{t-1}(j_{t-1}), D_{t-1}) \sim \mathrm{G}[n_{t-1}/2, d_{t-1}(j_{t-1})/2],$$

where $\mathbf{a}_t(j_{t-1}) = \mathbf{G}\mathbf{m}_{t-1}(j_{t-1})$, and $\mathbf{R}_t(j_t, j_{t-1}) = \mathbf{G}\mathbf{C}_{t-1}(j_{t-1})\mathbf{G}' + \mathbf{W}_t(j_t)$. Note that $\mathrm{E}[\boldsymbol{\theta}_t \mid M_t(j_t), M_{t-1}(j_{t-1}), D_{t-1}] = \mathbf{a}_t(j_{t-1})$ does not differ across the $M_t(j_t)$, depending only upon possible models applying at $t-1$ since $\mathbf{G}$ is common to these models. Also,

$$p(\phi \mid M_t(j_t), M_{t-1}(j_{t-1}), D_{t-1}) = p(\phi \mid M_{t-1}(j_{t-1}), D_{t-1}).$$

Forecasting one-step ahead,

$$(Y_t \mid M_t(j_t), M_{t-1}(j_{t-1}), D_{t-1}) \sim \mathrm{T}_{n_{t-1}}[f_t(j_{t-1}), Q_t(j_t, j_{t-1})], \quad (12.37)$$

where

$$f_t(j_{t-1}) = \mathbf{F}'\mathbf{a}_t(j_{t-1}),$$

also common across the models at time t, and

$$Q_t(j_t, j_{t-1}) = \mathbf{F}'\mathbf{R}_t(j_t, j_{t-1})\mathbf{F} + S_{t-1}(j_{t-1})V(j_t).$$

As in the previous section with V known, the particular structure of the multi-process leads to the modes (and means when $n_{t-1} > 1$) of these sixteen possible forecast distributions taking only four distinct values,

$$\mathrm{E}[Y_t \mid M_t(j_t), M_{t-1}(j_{t-1}), D_{t-1}] = f_t(j_{t-1})$$

for each of the four values j_t. Calculation of the forecast distribution unconditional on possible models then simply involves the mixing of these standard T components with respect to the relevant probabilities calculated as in part (e) of the previous section. Forecasting Y_t is based on the related mixture

$$p(Y_t \mid D_{t-1}) = \sum_{j_t=1}^{4} \sum_{j_{t-1}=1}^{4} p(Y_t \mid M_t(j_t), M_{t-1}(j_{t-1}), D_{t-1})\pi(j_t)p_{t-1}(j_{t-1}). \quad (12.38)$$

Updating proceeds as usual given j_{t-1} and j_t. The implied normal/gamma posterior has margins

$$\begin{aligned}(\boldsymbol{\theta}_t \mid M_t(j_t), M_{t-1}(j_{t-1}), D_t) &\sim \mathrm{T}_{n_t}[\mathbf{m}_t(j_t, j_{t-1}), \mathbf{C}_t(j_t, j_{t-1})],\\ (\phi \mid M_t(j_t), M_{t-1}(j_{t-1}), D_t) &\sim \mathrm{G}[n_t/2, d_t(j_t, j_{t-1})/2],\end{aligned} \quad (12.39)$$

where

$$
\begin{aligned}
\mathbf{m}_t(j_t, j_{t-1}) &= \mathbf{a}_t(j_{t-1}) + \mathbf{A}_t(j_t, j_{t-1})e_t(j_{t-1}),\\
\mathbf{C}_t(j_t, j_{t-1}) &= [S_t(j_t, j_{t-1})/S_{t-1}(j_{t-1})]\\
&\quad \times [\mathbf{R}_t(j_t, j_{t-1}) - \mathbf{A}_t(j_t, j_{t-1})\mathbf{A}_t(j_t, j_{t-1})'Q_t(j_t, j_{t-1})],\\
e_t(j_{t-1}) &= Y_t - f_t(j_{t-1}),\\
\mathbf{A}_t(j_t, j_{t-1}) &= \mathbf{R}_t(j_t, j_{t-1})\mathbf{F}/Q_t(j_t, j_{t-1}),
\end{aligned}
$$

with $n_t = n_{t-1} + 1$ and

$$d_t = d_{t-1}(j_{t-1}) + S_{t-1}(j_{t-1})e_t(j_t, j_{t-1})^2/Q_t(j_t, j_{t-1}),$$

and resulting variance estimate $S_t(j_t, j_{t-1}) = d_t(j_t, j_{t-1})/n_t$. Note that n_t is common across models. Posterior model probabilities are given, following part (h) of the previous section, from equation (12.30) with the normal one-step forecast density replaced by the corresponding T form here. Note that since the degrees of freedom of the T distributions (12.39) are all n_{t-1}, then the probabilities are simply given by

$$p_t(j_t, j_{t-1}) = \frac{c_t\pi(j_t)p_{t-1}(j_{t-1})}{Q_t(j_t, j_{t-1})^{1/2}\{n_{t-1} + e_t(j_t, j_{t-1})^2/Q_t(j_t, j_{t-1})\}^{n_t/2}}, \tag{12.40}$$

where c_t is a constant of normalisation.

Inferences about $\boldsymbol{\theta}_t$ and V are based on the sixteen-component mixtures that average (12.39) with respect to the posterior model probabilities (12.40). The marginal for $\boldsymbol{\theta}_t$ is

$$p(\boldsymbol{\theta}_t \mid D_t) = \sum_{j_t=1}^{4}\sum_{j_{t-1}}^{4} p(\boldsymbol{\theta}_t \mid M_t(j_t), M_{t-1}(j_{t-1}), D_t)p_t(j_t, j_{t-1}), \tag{12.41}$$

with components given by the T distributions in (12.39). Similarly, the posterior for $\phi = 1/V$ is a mixture of sixteen gamma distributions.

In evolving to $t+1$, the sixteen-component mixture (12.41) is collapsed over possible models at $t-1$. The method of approximation of Example 12.8 applies here. For each $j_t = 1, \ldots, 4$, it follows, as in the previous section, that the collapsed posterior for $\boldsymbol{\theta}_t$ and V is defined by

$$p(\boldsymbol{\theta}_t, \phi \mid D_t) = \sum_{j_t=1}^{4} p(\boldsymbol{\theta}_t, \phi \mid M_t(j_t), M_{t-1}(j_{t-1}), D_t)p_t(j_t), \tag{12.42}$$

the component densities in this sum being approximated as follows (cf. Example 12.8).

For each j_t, define the variance estimates $S_t(j_t)$ by

$$S_t(j_t)^{-1} = \sum_{j_{t-1}=1}^{4} S_t(j_t, j_{t-1})^{-1}p_t(j_t, j_{t-1})/p_t(j_t),$$

with

$$d_t(j_t) = n_t S_t(j_t).$$

Define the weights

$$p_t^*(j_t) = S_t(j_t)S_t(j_t, j_{t-1})^{-1}p_t(j_t, j_{t-1})/p_t(j_t),$$

noting that they sum to unity, viz., $\sum_{j_t=1}^4 p_t^*(j_t) = 1$. Further, define the mean vectors $\mathbf{m}_t(j_t)$ by

$$\mathbf{m}_t(j_t) = \sum_{j_{t-1}=1}^{4} \mathbf{m}_t(j_t, j_{t-1})p_t^*(j_{t-1}),$$

and the variance matrices $\mathbf{C}_t(j_t)$ by the formulae

$$\sum_{j_{t-1}=1}^{4} \{\mathbf{C}_t(j_t, j_{t-1}) + (\mathbf{m}_t(j_t) - \mathbf{m}_t(j_t, j_{t-1}))(\mathbf{m}_t(j_t) - \mathbf{m}_t(j_t, j_{t-1}))'\}p_t^*(j_{t-1}).$$

For each j_t, the mixture posterior $p(\boldsymbol{\theta}_t, \phi | M_t(j_t), M_{t-1}(j_{t-1}), D_t)$ is then approximated by single normal/gamma posteriors having marginals

$$(\boldsymbol{\theta}_t \mid M_t(j_t), D_t) \sim \mathrm{T}_{n_t}[\mathbf{m}_t(j_t), \mathbf{C}_t(j_t)],$$
$$(\phi \mid M_t(j_t), D_t) \sim \mathrm{G}[n_t/2, d_t(j_t)/2].$$

These approximate the components in the mixture (12.42), thus collapsing from sixteen to four standard normal/gamma components. In doing so, we complete the cycle of evolution, updating and collapsing; the resulting four-component mixture is analogous to the starting four-component mixture defined by components (12.36) with $t-1$ updated to t.

12.4.4 Illustration: CP6 series revisited

The model analysis is illustrated using the CP6 series. The basic linear growth form described in Section 12.4.1 is used, the linear growth evolution variance sequence being defined by separate level and growth discount factors δ_μ and δ_β as in (12.23). The four component models at time t have defining quantities as follows:

(1) Standard model: $V(1) = 1$ and $\mathbf{W}_t(1) = \mathbf{W}_t$ given by (12.23) with $\delta_\mu = \delta_\beta = 0.9$, having model probability $\pi(1) = 0.85$;
(2) Outlier model: $V(2) = 100$ and $\mathbf{W}_t(2) = \mathbf{W}_t$, having model probability $\pi(2) = 0.07$;
(3) Level change model: $V(3) = 1$ and $\mathbf{W}_t(3)$ given by (12.23) with $\delta_\mu = 0.01$ and $\delta_\beta = 0.9$, having model probability $\pi(3) = 0.05$; and
(4) Growth change model: $V(4) = 1$ and $\mathbf{W}_t(4)$ given by (12.23) with $\delta_\mu = 0.9$ and $\delta_\beta = 0.01$, having model probability $\pi(4) = 0.03$.

Initial priors are defined by $\mathbf{m}_0 = (600, 10)'$, $\mathbf{C}_0 = \mathrm{diag}(10000, 25)$, $d_0 = 1440$ and $n_0 = 10$. It should be remarked that these values are chosen to

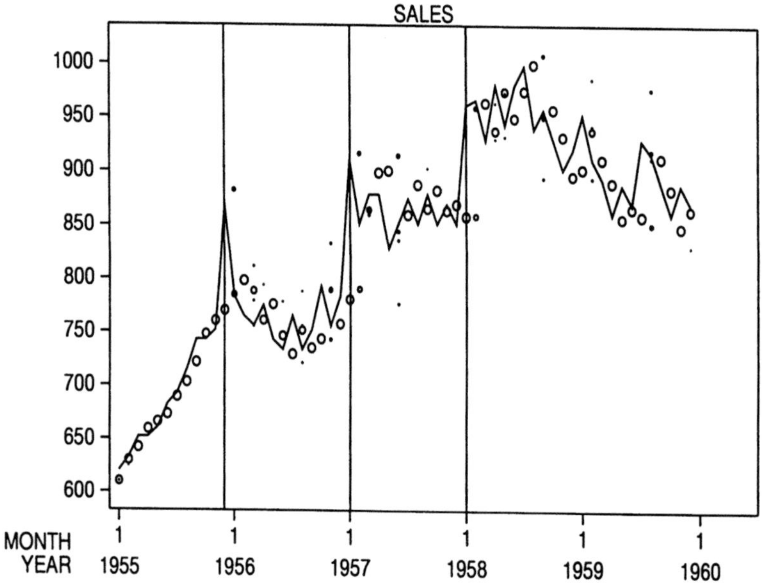

Figure 12.5 Point forecasts for CP6 sales series

be in line with previous analyses of the series (Section 11.3.1). The values of discount factors are not optimised in any sense. Also, it is clear that the variance scale factor V apparently changes at later stages of the data, inflating with the level of the series. Although this was catered for and modelled in earlier analyses, V is assumed constant, though uncertain, here for clarity. Figures 12.5-12.9 illustrate selected features of the multi-process analysis, these features being common to the use of the models generally and chosen in an attempt to identify key points. Before discussing the analysis, recall that from earlier analyses, the notable features of the CP6 series are (i) an outlier at $t = 12$, followed by a switch from positive to negative growth; (ii) a jump to higher levels between $t = 24$ and $t = 26$, with a possible outlier at $t = 25$; (iii) a further jump at $t = 37$; and (iv) higher random variation at higher levels of the data in later stages, with one or two events possibly classifiable as level/growth changes, though obscured by the greater random variation.

An overall impression of the analysis can be obtained from Figure 12.5. Here the data are plotted and joined over time by the solid line (unlike most previous plots where the forecasts etc. are joined, the data plotted as separate symbols; this is a temporary change of convention for clarity in this example). Superimposed on the data are one-step ahead point forecasts from the distributions (12.38) for all t. Each forecast distribution has sixteen component T distributions whose modes are taken as individual point forecasts, although, as noted following equation (12.37), the particular structure of the model means that there are only four distinct such point forecasts at each time. This follows from (12.37), where it is clear that the mode in the component conditional on $(M_t(j_t), M_{t-1}(j_{t-1}), D_{t-1})$ is simply $f_t(j_{t-1})$, for all 4 values of j_t. Thus, $f_t(1)$ is the point forecast for Y_t in each of the four models that include $M_{t-1}(1)$, and so applies with probability $p_{t-1}(1)$. Generally, the forecast $f_t(j_{t-1})$ applies with probability $p_{t-1}(j_{t-1})$. Now, rather than plotting all four modes at each t, only those modes that apply with reasonably large probabilities are drawn. Specifically, $f_t(j_{t-1})$ is plotted only if $p_{t-1}(j_{t-1}) > 0.05$. In addition, all forecasts plotted appear as circles whose radii are proportional to the corresponding probabilities; thus more likely point forecasts are graphed as larger circles. This serves to give a relatively simple visual summary of overall forecasting performance (although without indications of uncertainty). In stable periods where the basic, linear growth form is adequate, the standard model (whose prior probability at any time is 0.85) clearly dominates. This results in a preponderance of large circles denoting the point forecasts $f_t(1)$ at each time, the mode of Y_t conditional on the standard model applying at the previous time point. At times of instability, the radii of these larger circles decrease, reflecting lower probability, and up to three further circles appear denoting those remaining forecasts with probabilities in excess of 0.05. These tend to be graphed as circles with very small radii, reflecting low probability, sometimes just appearing as points. This occurs at those times t when the previous observation Y_{t-1} (and sometimes the previous one or two observations) are poorly forecast under the standard model. In such cases, the probabilities $p_{t-1}(j_{t-1})$ spread out over the four models, reflecting uncertainty as to the behaviour of the series at time $t-1$. This uncertainty feeds directly through to the forecast distribution for Y_t as illustrated on the graph. Some of the occurrences are now described.

Consider the position at November 1955, time 11. Up until this time, the series is stable and is well modelled by the standard DLM; hence the single point forecasts in Figure 12.5. This is clearly evident in Figure 12.9a. Here the four posterior model probabilities $p_t(j_t)$ at each time t are plotted as vertical bars. Up to $t = 11$, the standard model has posterior probability near 1 at each time. Figures 12.6a and 12.7a reflect existing information given D_{11}. Figure 12.7a displays the posterior density of the current level parameter, $p(\mu_{11} \mid D_{11})$, the mixture of components (12.36) with $t-1 = 11$. The four components (12.36) are also plotted on the graph. The components are all located between roughly 750 and 760, and are similarly spread over 730 to 780. That based on $M_{11}(1)$ is the most peaked and has corresponding posterior probability $p_{11}(1)$ very close to 1. This reflects the previous stability of the series and consistency with the standard DLM. As a result, the mixture of these four components is essentially equal to the first, standard component. This is a typical picture in stable periods. Similar comments apply to the one-step ahead forecast

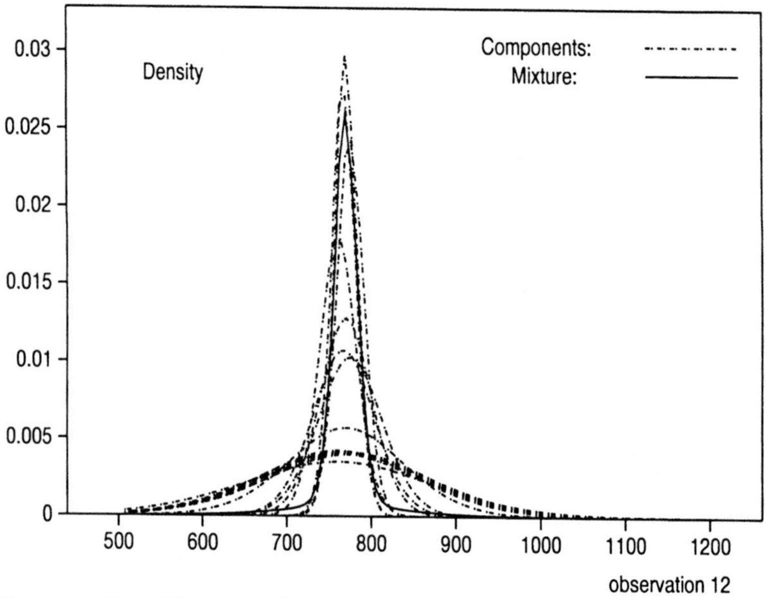

Figure 12.6a Forecasts for Y_{12} at $t = 11$

distribution $p(Y_{12} \mid D_{11})$ (from (12.28) with t=12), graphed in Figure 12.6a. This is a sixteen-component mixture, the components also appearing on the graph. It is difficult to distinguish the overall mixture since it corresponds closely to the highly peaked components in the centre. The only additional feature of note is that several of the components are much more diffuse than the majority here, being those that condition on the outlier model at time 12. The observational variance inflation factor of 100 in the definition of $M_t(2)$ produces this spread. In combining the components, however, these outlier components (along with those for level and growth changes) have small probability and so contribute little to the mixture. Again, this is a typical picture in stable periods.

Proceed now to December 1955, time 12. $Y_{12} = 870$ is a very wild observation relative to the standard forecast distribution, and most of the probability under the mixture density in Figure 12.6a is concentrated between 700 and 820. The outlier components, however, give appreciable probability to values larger than 870. Hence, in updating to posterior model probabilities given D_{12}, those four components that include $M_{12}(2)$ will receive much increased weights. This is clear from Figure 12.9a; the outlier and level change models share most of the posterior probability at $t = 12$, the former being the more heavily weighted due to the initial balance of prior probabilities that slightly favour the outlier model. Figure 12.7b plots the posterior density $p(\mu_{12} \mid D_{12})$ together with the four components; this

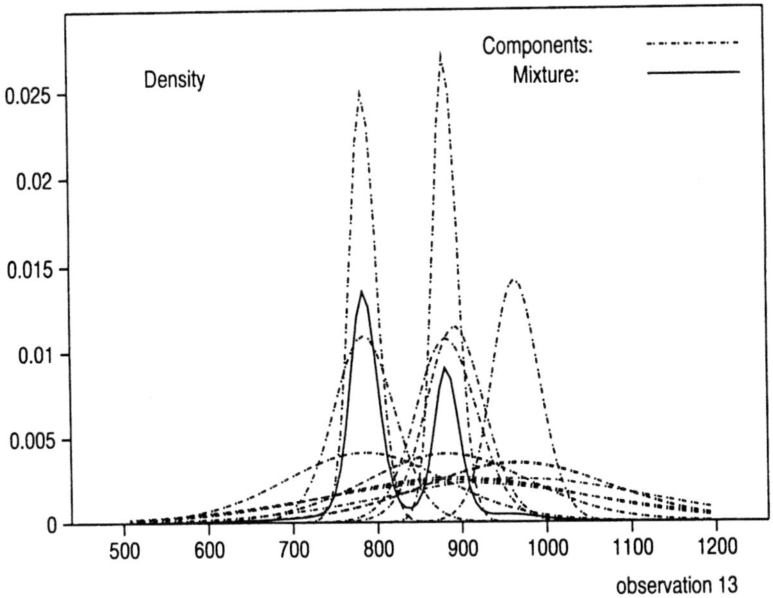

Figure 12.6b Forecasts for Y_{13} at $t = 12$

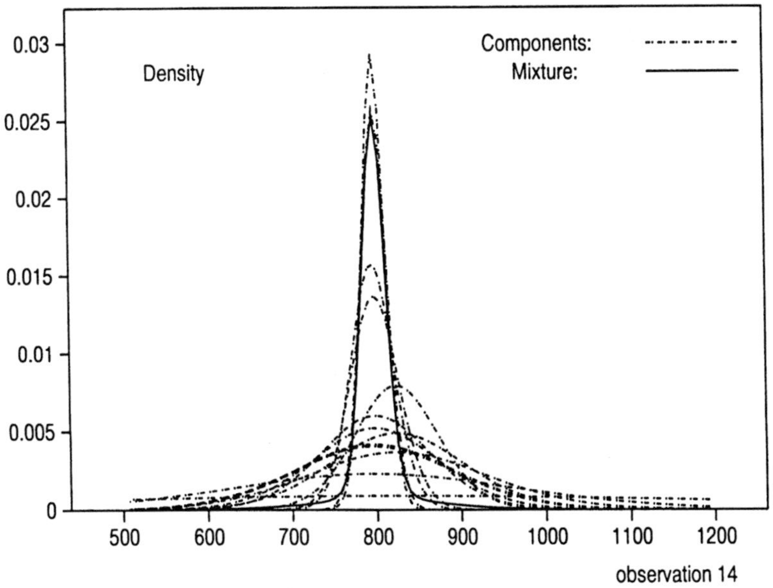

Figure 12.6c Forecasts for Y_{14} at $t = 13$

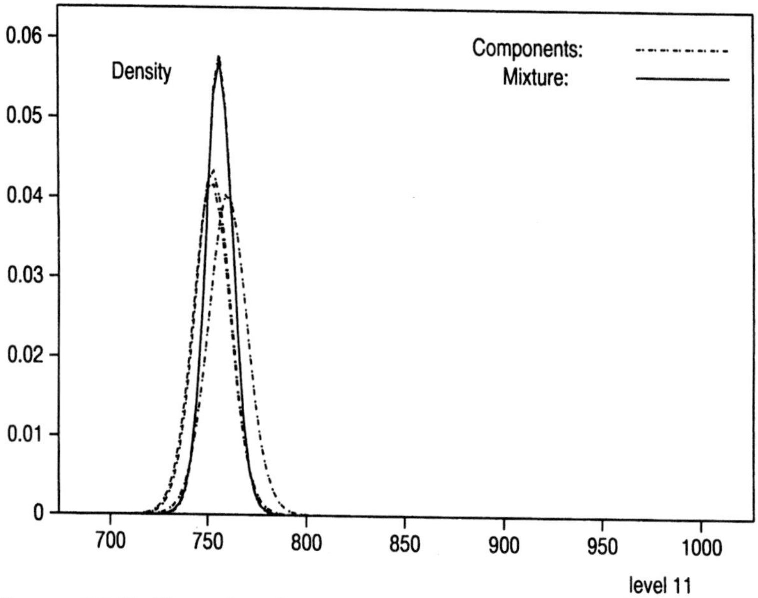

Figure 12.7a Posteriors for μ_{11} at $t = 11$

is the analogue at time 12 of Figure 12.7a at time 11. Here, in contrast to Figure 12.7a, the components are rather disparate. The peaked component located near 770 corresponds to the outlier model for Y_{12}, being the density $p(\mu_{12} \mid M_{12}(2), D_{12})$. In this case, the observation has effectively been ignored as an outlier, the inference being that the level remains between about 740 and 800.

The two peaked components located in the region of $Y_{12} = 870$ come from the level and growth change models, the former providing the more peaked posterior. If the observation is due to marked change in level and/or growth parameters, then the inference is that the current level actually lies in the region of 840 to 900. The fourth, more diffuse component located near 840 is $p(\mu_{12} \mid M_{12}(1), D_{12})$, the posterior from the standard model at time 12. If Y_{12} is a reliable observation and no level or growth change has occurred, then clearly the posterior for the level is a compromise between the prior, located near 770, and the likelihood from Y_{12}, located at 870. The extra spread in this posterior is now explained. Conditional on $M_{12}(1)$, the extreme observation leads to very large forecast errors resulting in inflated estimates of V in the corresponding components; thus, for each $j_{11} = 1, \ldots, 4$ the variance estimates $S_{12}(1, j_{11})$ are all inflated, and consequently, $S_{12}(1)$ is much larger than $S_{12}(j_{12})$ for $j_{12} > 1$. As a result, the four posterior T distributions $p(\mu_{12} \mid M_{12}(1), M_{11}(j_{11}), D_{12})$ are all rather diffuse and hence so is $p(\mu_{12} \mid M_{12}(1), D_{12})$, the plotted den-

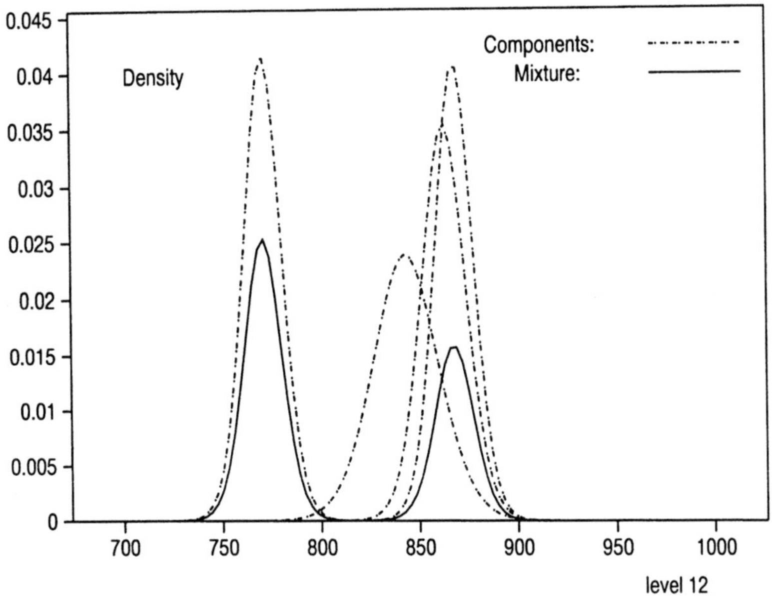

Figure 12.7b Posteriors for μ_{12} at $t = 12$

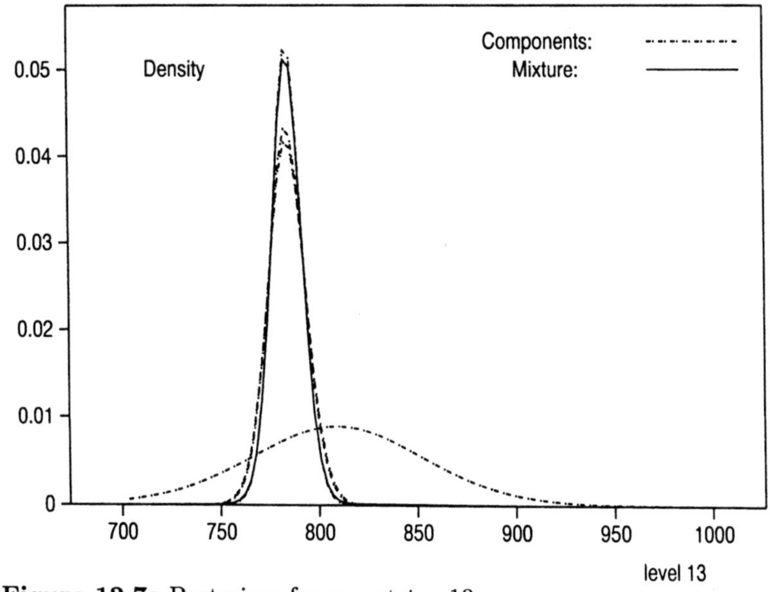

Figure 12.7c Posteriors for μ_{13} at $t = 13$

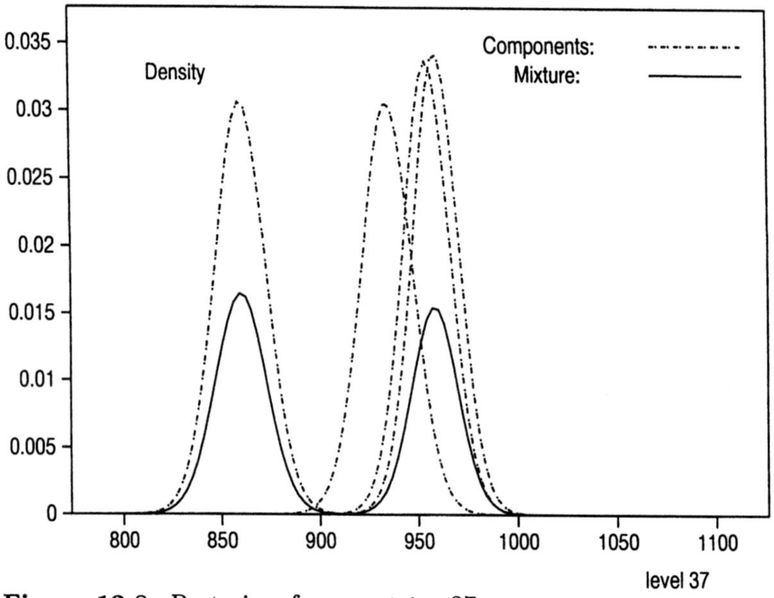

Figure 12.8a Posteriors for μ_{37} at $t = 37$

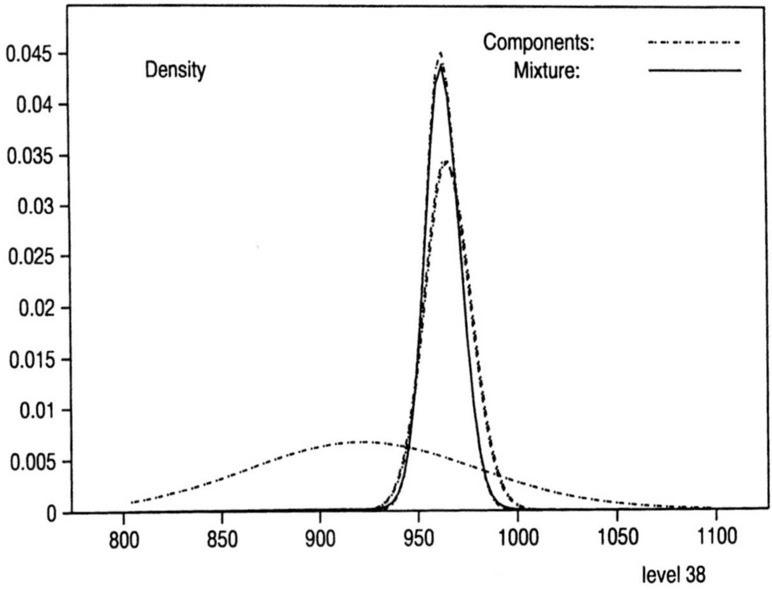

Figure 12.8b Posteriors for μ_{38} at $t = 38$

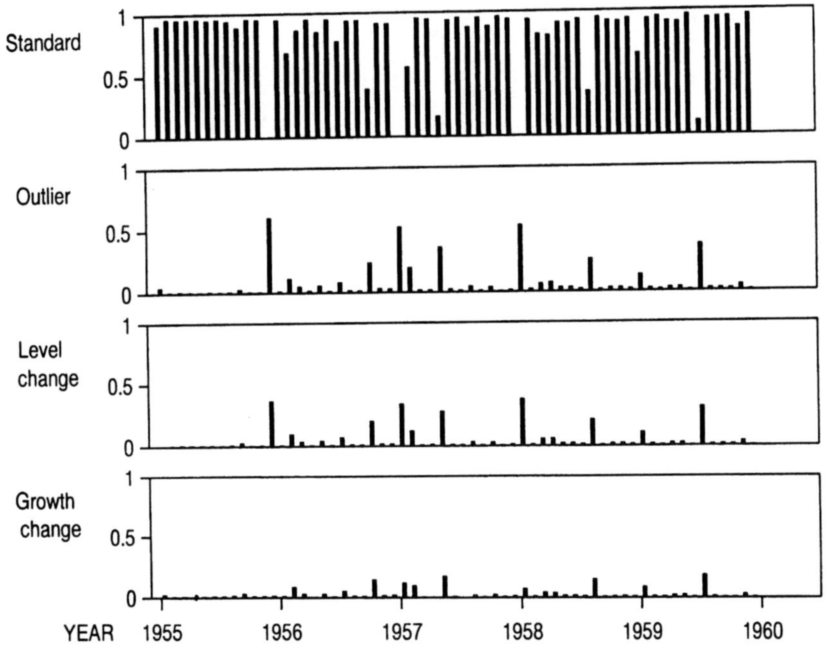

Figure 12.9a Posterior model probabilities

sity. As it happens, this obviously inappropriate (in retrospect) density receives essentially no posterior probability, as can be seen from Figure 12.9a, contributing negligibly to the overall mixture posterior plotted in Figure 12.7b. The posterior is clearly bimodal. This represents the ambiguity as to whether Y_{12} is an outlier or whether it indicates the onset of a change in level or growth. If the former is true, then the more peaked mode near 770 is the correct location, otherwise the second mode near 870 is correct. Until further information is processed, there is a complete split between the two inferences. Forecasting ahead to $t = 13$, January 1956, the bimodality carries over to $p(Y_{13} \mid D_{12})$, graphed, along with the sixteen components, in Figure 12.6b. The appearance of two distinct point forecasts with appreciable probability in Figure 12.5 also evidences the bimodality. An additional feature to note here concerns the components located around 960 to 980. These correspond to the four model combinations that include $M_{12}(4)$, the growth change model at time 12, although they receive little weight in the overall mixture.

Moving on now to observe $Y_{13} = 784$ it becomes clear that Y_{12} was in fact an outlier. Figure 12.9b presents posterior model probabilities analogous to those in Figure 12.9a, although these are the *one-step back*, or

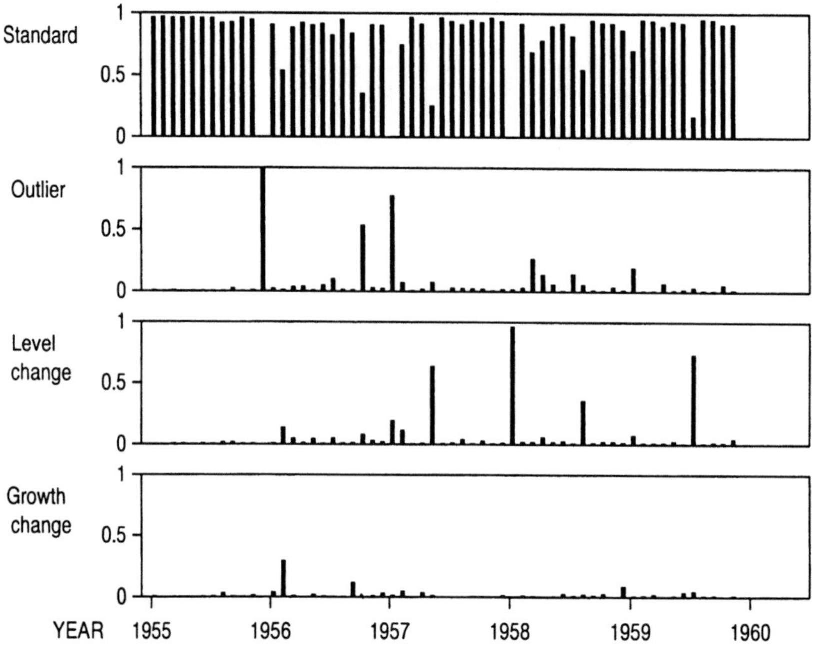

Figure 12.9b One-step back model probabilities

smoothed, probabilities referring to models at the previous time point, calculated as in (12.32). For each t, the vertical bars in the figure are the retrospective probabilities $\Pr[M_{t-1}(j_{t-1}) \mid D_t]$ for each $j_{t-1} = 1, \ldots, 4$. These probabilities, calculated in (12.34), are very useful for retrospective assessment and diagnosis of model occurrence at any time given one further, confirmatory observation. Thus, for example, observing Y_{13} clarifies the position at time 12, with $\Pr[M_{12}(2) \mid D_{13}] \approx 1$. The outlier is clearly identified and in the updating has therefore been essentially ignored. After updating, $p(\mu_{13} \mid D_{13})$ appears in Figure 12.7c, along with its four components. The corresponding one-step ahead forecast densities appear in Figure 12.6c. From these graphs it is clear that things have reverted to stability with unimodal distributions once the outlier has been identified and accommodated.

In summary, the posteriors for model parameters tend to comprise components with similar locations in stable periods. These components separate out at the onset of an event, with posterior probabilities reflecting ambiguity as to whether the event relates to an outlier or a change-point. Further observations usually identify the event and posteriors reflect this as the series reverts to stability and consistency with the standard DLM.

To further illustrate the analysis, consider the level change at $t = 37$, January 1958. Figures 12.8a and 12.8b present $p(\mu_t \mid D_t)$ for $t = 37$ and 38, along with the four components of each $p(\mu_t \mid M_t(j_t), D_t)$. Here the extreme observation Y_{37} initially leads to bimodality in the posterior for μ_{37} at the time; see Figure 12.8a. The observation is either an outlier or indicates a change-point, but the model cannot as yet distinguish the possibilities. Observing Y_{38} apparently confirms a level change, from around 860 to near 950, and the posterior for the level at time 38 reflects this. Finally, current and one-step back probabilities indicate the switching between models over time and the diagnosis of events.

12.4.5 Other applications

Several notable applications of this and related multi-process models exist in commercial and economic fields, where models have been developed and implemented primarily for short term forecasting (Harrison and Stevens 1971, 1976a, b), Johnston and Harrison (1980). The adaptability of the multi-process approach to abrupt changes in trend can be of crucial benefit to decision-makers in such areas.

An interesting application in medicine is documented in West (1982), Smith and West (1983), Smith et al (1983), and Trimble et al (1983). That application concerns a problem in clinical monitoring typical of many situations (not restricted to medicine) in which time series observations relate to the state of a monitored subject, and the detection and interpretation of abrupt changes in the pattern of the data are of paramount importance. Often series are noisy and difficult to interpret using simple methods, and the changes of interest are obscured by noise inherent in the measurement process and outliers in the data. In addition, series may be subject to abrupt changes of several types, with only a subset corresponding to the changes of interest. It is therefore important to be able to distinguish, or diagnose, types of change as well as simply detect change of any kind.

The medical application concerns the monitoring of the progress of kidney function in individual patients who had recently received a transplant. The level of renal functioning is indicated by the rate at which chemical substances are cleared from the blood, and this can be inferred indirectly from measurements of blood and/or urine levels of such substances. This process involves the use of well-determined physiological relationships, but is subject to various sources of error, thus producing noise-corrupted measurements of filtration rates. Following transformations and corrections made for both physiological and statistical reasons, this process results in a response series that is inversely related to kidney well-being, and that can be expected to behave essentially according to a linear growth model in periods of consistent kidney function. In particular, (a) if the kidney functions at a stable, roughly constant rate, the response series should exhibit no growth; (b) if kidney function is improving with roughly constant growth between equally spaced observations, the response series should decay roughly linearly; and (c) the reverse is the case if kidney function deteriorates at a roughly constant rate. Each of these cases is encountered in practice. Stable function (a) may occur at different times during post-operative patient care, with a successful transplant eventually leading to stable function at a high level, corresponding to a low level in the response series. Case (b) is anticipated immediately post-transplant as the transplanted organ is accepted by the patient's body and begins to function, eventually reaching normal levels of chemical clearance. Case (c) is expected if at any time after the transplant, kidney function deteriorates. This may be attributable to rejection of the transplant and is obviously of paramount importance. One of the main objectives in monitoring the data series is thus to detect an abrupt change from either (a) or (b) to (c), consistent, in model terms, with an abrupt change in growth in the series from non-positive to positive. The appropriateness of the multi-process model is apparent, as is the emphasis on the use of the posterior model probabilities to detect and diagnose change when it occurs. A high posterior probability for the growth change model is used to signal the possibility of such a change. However, since only changes from non-positive to positive growth are consistent with rejection, then only a subset of all possible growth changes give cause for concern. For example, an abrupt change from positive to negative growth is consistent with a marked and sustained improvement in renal function. These particular features of this application raise questions that are easily answered using the posterior distributions for growth parameters; a change at time t of the sort consistent with rejection is indicated if $\beta_{t-1} \leq 0$ and $\beta_t > 0$, that can be readily assessed using the relevant posterior $p(\beta_t \mid D_t)$ and $p(\beta_{t-1} \mid D_t)$ (Note that the latter is a filtered distribution whose calculation has not hitherto been discussed in the multi-process context; see Section 12.4.6 below).

The renal monitoring application has various other interesting features from a modelling viewpoint. Concerning the use of the four-component multi-process mixture, the response series are certainly subject to outlying observations a small fraction (about 5%) of the time due to mistakes in data transcription, equipment/operator malfunction and blood/urine sample contamination. The level change model is also very necessary, being consistent with marked increases in level of kidney function when the patient is subject to dialysis treatment to assist renal function. This is common and frequent in early stages of post-operative care. Further, minor features of the application include timing errors and irregularly spaced observations and the use of quadratic power law functions for the observational variance. Fuller details of these, and other, features of the application are found in the above references.

12.4.6 Step ahead forecasting and filtering

Further important components of the analysis of multi-process mixture models are step ahead forecast distributions and retrospective filtered distributions for model parameters in the past. Some features of these distributions are detailed here, fuller development being left to the interested reader. The development here is in terms of the four-component linear growth model analysed above, providing a concrete setting for the discussion of features that apply more generally.

All step ahead forecast distributions are, of course, mixtures. Consider the position at time t with the posterior distributions summarised in (12.42). The one-step ahead forecast distribution for Y_{t+1} is then simply the analogue of that in (12.38) with time index increased from t to $t+1$, namely the 4^2-component mixture

$$p(Y_{t+1} \mid D_t) = \sum_{j_{t+1}=1}^{4} \sum_{j_t=1}^{4} p(Y_{t+1} \mid M_{t+1}(j_{t+1}), M_t(j_t))\pi(j_{t+1})p_t(j_t).$$

The component densities of the mixture are the standard normal or T forecast densities derived under the standard DLM determined by $M_{t+1}(j_{t+1})$ and $M_t(j_t)$ jointly holding. Consider now forecasting two steps ahead for Y_{t+2}. Given any combination of models at times t, $t+1$ and $t+2$, it follows that the two step ahead forecast distribution is a standard form with density

$$p(Y_{t+2} \mid M_{t+2}(j_{t+2}), M_{t+1}(j_{t+1}), M_t(j_t), D_t)$$

for each j_t, j_{t+1} and j_{t+2}, each running from 1 to 4. The overall distribution is a mixture of these 4^3 components with mixing probabilities

$$\begin{aligned}
&\Pr[M_{t+2}(j_{t+2}), M_{t+1}(j_{t+1}), M_t(j_t) \mid D_t] \\
&\quad = \Pr[M_{t+2}(j_{t+2}) \mid M_{t+1}(j_{t+1}), M_t(j_t), D_t] \\
&\qquad \times \Pr[M_{t+1}(j_{t+1}) \mid M_t(j_t), D_t]\Pr[M_t(j_t) \mid D_t] \\
&\quad = \pi(j_{t+2})\pi(j_{t+1})p_t(j_t).
\end{aligned}$$

Specifically,

$$\begin{aligned}
&p(Y_{t+2} \mid D_t) \\
&\quad = \sum_{j_{t+2}=1}^{4} \sum_{j_{t+1}=1}^{4} \sum_{j_t=1}^{4} p(Y_{t+2} \mid M_{t+2}(j_{t+2}), M_{t+1}(j_{t+1}), M_t(j_t)D_t) \\
&\qquad\qquad \times \pi(j_{t+2})\pi(j_{t+1})p_t(j_t).
\end{aligned}$$

Forecasting further ahead obviously increases the number of components in the mixture to account for all possible models obtaining between time t and the forecast time point. The general details are left to the reader.

In specific applications there may be features of the model that can be exploited to simplify the problem of summarising these forecast distributions. In the model here, for example, the structure of the multi-process at each time is such that the $M_t(j)$ differ only through evolution and observational variances, for all t and all j. This implies that whilst the components of all mixture forecast distributions may have widely differing variances, the differences in point forecasts, the modes or means of the components, derive only from the differences in the means of $\boldsymbol{\theta}_t$ across the models $M_t(j_t)$. Specifically, $\mathrm{E}[Y_{t+1} \mid M_{t+1}(j_{t+1}), M_t(j_t), D_t]$ does not depend on j_{t+1}, nor does $\mathrm{E}[Y_{t+2} \mid M_{t+2}, (j_{t+2}), M_{t+1}(j_{t+1}), M_t(j_t), D_t]$ depend on either of j_{t+2} or j_{t+1}, and so forth. Hence there are only distinct point forecasts for each step ahead, clearly reducing the numbers of calculations required. Further simplifications usually derive from consideration of the comments on approximation of mixtures in Section 12.3.3. For example, if one of the $p_t(j_t)$ is close to 1, then the corresponding components of forecast distributions dominate the mixtures and the others may be ignored.

Consider now the calculation of filtered distributions for retrospective analysis. As with forecast distributions, it should be immediately clear that filtered distributions are mixtures of standard components. The calculations for one-step back from time t can now be performed. Conditional on $M_t(j_t)$ and $M_{t-1}(j_{t-1})$, the one-step back filtered distribution for the state vector $\boldsymbol{\theta}_{t-1}$ has the standard form derived in Section 4.7, with density

$$p(\boldsymbol{\theta}_{t-1} \mid M_t(j_t), M_{t-1}(j_{t-1}), D_t).$$

There are 4^2 such component densities in the distribution

$$p(\boldsymbol{\theta}_{t-1} \mid D_t) = \sum_{j_t=1}^{4} \sum_{j_{t-1}=1}^{4} p(\boldsymbol{\theta}_{t-1} \mid M_t(j_t), M_{t-1}(j_{t-1}), D_t) p_t(j_t, j_{t-1}),$$

where the mixing probabilities

$$p_t(j_t, j_{t-1}) = \Pr[M_t(j_t), M_{t-1}(j_{t-1}) \mid D_t]$$

are as previously calculated in (12.40). Often, filtering one-step back suffices, recent data providing little additional information about state vectors further back in time. Two, and more, step back distributions are calculated as multi-component mixtures in a similar fashion, the details again being left to the interested reader.

12.4.7 Further comments and extensions

There are other features of the analysis of even this relatively simple multi-process mixture model that may assume importance in application. Some particular features and extensions of the basic model are noted here. Again, these points apply to general models, although they are highlighted with reference to the particular model used here for the purposes of exposition.

Consider first the structuring of the multi-process. Here the four components of the model at each time differ only through the evolution and observational variances. These are designed to model the possibilities of evolution and observational errors that are significantly larger in absolute value than usual, whilst the retention of zero-means implies that the signs of the errors remain unanticipated. The sizes of the variances in the outlier and parametric-change states are determined in advance. The particular values are not critical, all that is necessary is that they provide significantly larger values than the standard model. Robustness to these values is similar to that in more usual mixture models for outliers, and the experience of others working with such models may be drawn upon in assigning values (Box and Tiao 1968; Box 1980; Smith and Pettit 1985). See also the related discussion in Section 11.4.3. The use of discount factors to specify the evolution variances in the level and growth change states is a point of difference, although the basic ideas are similar. It is, of course, possible to use alternative methods, such as choosing the evolution variance matrices in the change models as multiples of those in the standard, that would be more in line with earlier uses of multi-processes (Harrison and Stevens 1971, 1976a, b; Smith and West 1983). However, the self-consistency and simplicity of the discount approach makes it the more attractive, and the end results would be similar. Note finally that the design of the multi-process structure is similar in principle whatever approach to discounting is taken; recall the various alternatives described in Section 12.4.1.

Other multi-processes might have component DLMs differing in one or more of the other components $\mathbf{F}_t$ and $\mathbf{G}_t$. This also allows for components having different numbers of parameters in the state vector, a possibility not hitherto considered, although related ideas were discussed in connection with subjective intervention in Section 11.2.4.

The remaining inputs to the model required from the user are the prior model probabilities $\pi(j)$. These must be assessed and specified in advance, reflecting the forecaster's opinion as to the likely rate of occurrence of outliers, level changes and growth changes. Again, experiences with outlier modelling more widely, as referenced above, for example, guide the choice of values of the outlier model probability. This will typically be near 0.05, rarely very much smaller but usually less than 0.1. Very precise values for this, and the change model probabilities, are unnecessary since the occurrence of a forecast error deemed extreme under the standard model will lead to posterior model probabilities that heavily favour the outlier and change models so long as their prior probabilities are not negligible. The discussion in Section 11.4.3 of crossover points between the predictive density under the standard model and those in the outlier/change models is again relevant here. It is also usual, though not a panacea, that outliers occur more frequently than changes, so that the prior probabilities will often marginally favour the former. This is certainly true of data arising in the majority of commercial applications, though it may not be the case in other areas of application. Thus, when *something* happens at a given time t, the posterior probabilities $p_t(j_t)$ will tend to favour the outlier model. A further observation will identify what has happened, as illustrated in Figures 12.9a and b in the CP6 example.

Often, previous data on the series, or on similar, related series, will be available to guide the choice of prior probabilities. Also, although here assumed constant over time, there are obvious ways in which the probabilities may be updated as data are processed to reflect a changing view as to the frequency of occurrence of events. This may be based solely on the data, may use externally available information, or be based purely on subjective opinion. Going further with this notion, the model probabilities may even be linked to independent variables to allow for change with external circumstances, time, and so forth. The extension to incorporate Markov transitions between models discussed in Section 12.3.1 is a simple case of this; others might incorporate independent variable information that attempts to predict change through a separate model for the transition probabilities. In this way it is possible, for example, to produce models related to catastrophe theoretic ideas (Harrison and Smith 1979), and the threshold switching models of classical time series (Tong and Lim 1980; Tong 1983).

The modelling of change points in dynamic models using multi-processes complements more conventional Bayesian approaches to change-point analysis in regression and related models. Key works in change-point modelling have been done by Smith (1975, 1980), Smith and Cook (1980). See also Pole and Smith (1985). Various related approaches and many references are found in Broemling and Tsurumi (1988).

One important extensions of the basic multi-process model used here is to include a component model that is designed to allow for marked change in the observational variance V, hitherto assumed constant. In many areas of application, time series are subject to changes in volatility that can be represented by abrupt increases or decreases in V. Share indices, exchange rates and other financial time series are prime examples. In addition, inappropriate data transformation or variance functions lead to observational variances that change with level so that if the level shifts abruptly, then so will the variance. In Section 10.8 variance discounting was introduced to cater for minor, sustained, stochastic changes in V. The effect produced was simply to discount the degrees of freedom parameter n_t between observations, reducing it by multiplying by a discount factor less than, but close to, unity. This extends directly to model abrupt changes by including a component in the multi-process that is the same as the standard model except for the inclusion of a small variance discount factor. Details are straightforward and left to the reader; the main, technical point of difference is that in this variance change state, the degrees of freedom parameter will be less than that in other states so that in collapsing mixture posterior distributions as in Section 12.4.2, the general results of Example 12.8 apply.

Concerning the collapsing of mixtures of posteriors more widely, the approximation here based on ignoring dependence on models more than one step back in time is adequate in the context of this particular model but may need relaxing in others. In particular, when working with higher-dimensional models with possibly more components, it may take up to three or more observations after an exceptional event to determine the nature of the event and appropriately estimate the state parameters. In such cases, $h > 1$ is needed, and the number of components in all mixture distributions may increase dramatically. Schervish and Tsay (1988) provide illustrations and discussions of several analyses in which h is effectively allowed to increase with the number of observations analysed. Quite often, of course, there will be little to gain in doing this, as is the case when the series is essentially well-behaved, not deviating markedly from the standard DLM description. In such circumstances, the technical and computational complexity of the multi-process model is highly redundant.

This potential redundancy of the multi-process approach can be addressed by combining simpler monitoring methods with multi-processes, as in Ameen and Harrison (1985b) , for example. Typically, a series will be stable over periods of observations, conforming to the standard DLM a good deal of the time. Thus, over long stretches of the series, the multi-process is redundant, posterior model probabilities being close to unity on the standard model. Only at times of exceptional events, and over one or a few time intervals following exceptions, is the full complexity of the multi-process really necessary. This is clearly illustrated with the CP6 analysis where the posterior model probabilities in Figures 12.9a and b clearly indicate that the mixture is only really necessary a small proportion of the time. The approach developed in this reference recognises this, and develops along the following lines:

(a) Model the series with the standard DLM subject to a simple monitoring scheme such as the sequential Bayesian tests or cusums described in Section 11.4. So long as the monitoring indicates the data to be consistent with the standard model, proceed with the usual DLM analysis. In parallel, use alternative DLMs that provide for the anticipated forms of departure from the standard. Although this is similar to a multi-process, class I approach, the parallel DLMs only come into play at stage (b) and are simply updated to provide initialisation at that stage.

(b) When the monitor signals deterioration in forecast performance of the standard DLM, begin a period of operation with a multi-process model, class II model, introducing the parallel DLMs from (a).

(c) Continue this until the the exceptional event that triggered the monitor signal has been identified, and the multi-process analysis settles down to conformity with the standard DLM. This occurs when the posterior model probability on the standard model is close to unity.

(d) At this stage, revert to operation with only the standard DLM taken from the relevant component of the multi-process. Return to (a) and continue.

Under such a scheme, the use of mixtures is restricted to periods of uncertainty about the series where they are needed to identify exceptional events. In stable periods, the standard DLM analysis is used, the computational economy being evident. Additionally, the problems of inference and prediction are simplified since for much of the time, a single DLM is used rather than a mixture of possibly many.

12.5 EXERCISES

(1) Refine the multi-process analysis of Example 12.2 by considering a finer grid of values for the discount factor δ. Specifically, take

$$\mathcal{A} = \{1.0, 0.975, 0.95, \ldots, 0.725, 0.7\},$$

with initial probabilities $p_0(j) = 1/13$, $(j = 1, \ldots, 13)$. For $t = 34$ and $t = 116$, plot the posterior probabilities $p_t(j)$ versus j. Comment on the support provided by the data for discount factors over the chosen range of possible values.

(2) In the framework of Example 12.3, verify that the posterior distribution for δ concentrates at that value δ_i such that $|\delta_i - \delta_0| = \min_j |\delta_j - \delta_0|$.

(3) Consider the multi-process, class I model defined for $t = 1$ by

$$M(j): \qquad \{1, 1, 100, W(j)\}, \qquad (j = 1, 2),$$

where $W(1) = 0$ and $W(2) = 50$. Initially, $p_0(j) = \Pr[M(j)|D_0] = 0.5$ and

$$(\mu_0|D_0, M(j)) \sim \mathrm{N}[100, 100]$$

for each model, $j = 1, 2$.

(a) With $p_1(j) = \Pr[M(j)|D_1]$, show that $p_1(1) < p_1(2)$ if and only if $|e_1| > 14.94$, where $e_1 = Y_1 - 100$, the forecast error common to the two models.

(b) Deduce that

$$\Pr[p_1(1) \geq p_1(2)|M(1), D_0] = 0.74$$

and

$$\Pr[p_1(1) < p_1(2)|M(2), D_0] = 0.34.$$

(c) Comment on the results in (b).

(4) Consider a multi-process model for Y_1 given by

$$M(j): \qquad \{1, 1, V(j), W\}, \qquad (j = 1, 2),$$

where $V(1) = 1$ and $V(2) = 33$, and $p_0(1) = \Pr[M(1)|D_0] = 0.9$. Suppose also that $(\mu_1|D_0) \sim N[0, 3]$.

(a) Show that $p(Y_1|D_0)$ is a mixture of two normals,

$$(Y_1|D_0) \sim \begin{cases} N[0,4], & \text{with probability } 0.9; \\ N[0,36], & \text{with probability } 0.1. \end{cases}$$

(b) Show that $p(\mu_1|D_1)$ is also a mixture of two normals,

$$(\mu_1|D_1) \sim \begin{cases} N[3Y_1/4, 3/4], & \text{with probability } p_1(1) = q(Y_1); \\ N[Y_1/12, 11/4], & \text{with probability } 1 - q(Y_1), \end{cases}$$

where

$$q(Y_1)^{-1} = 1 + 0.037e^{Y_1^2/9}.$$

(c) Deduce that

$$E[\mu_1|D_1] = [1 + 8q(Y_1)]Y_1/12.$$

(d) Write a computer program to plot $q(Y_1)$, $E[\mu_1|D_1]$ and $V[\mu_1|D_1]$ as functions of Y_1 over the interval $-5 \le Y_1 \le 5$. Comment on the form of this functions as Y_1 varies.

(e) Write a computer program to graph $p(\mu_1|D_1)$ as a function of μ_1 for any given value of Y_1. Comment on the form of the posterior for the three possible values $Y_1 = 0$, $Y_1 = 4$ and $Y_1 = 8$.

(5) A forecast system uses the model $\{1, 1, V_t, W_t\}$ as a base model for forecasting sales of a group of products. Intervention information is available to incorporate into the model analysis. Initially, $(\mu_0|D_0) \sim N[400, 20]$. Produce forecast distributions $p(Y_t|D_0, I)$ for each of $t = 1, 2, 3$ and 4 given the following intervention information I available at $t = 0$:

(a) $V_1 = 100$ and $\omega_1 \sim N[100, W_1]$ with $W_1 = 400$. This non-zero mean evolution error describes the likely impact on sales level of the receipt of a new export licence in the first time period. Verify that $(Y_1|D_0, I) \sim N[500, 520]$.

(b) $\nu_2 \sim N[500, V_2]$ with $V_2 = 200$ describing the anticipated effect of an additional spot order at $t = 2$, and $W_2 = 5$. Verify that $(Y_2|D_0, I) \sim N[1000, 625]$.

(c) The expected effect of a price increase at $t = 3$ is modelled via a transfer response function, with

$$Y_t = \mu_t + E_t + \nu_t, \qquad (t = 3, 4, \dots),$$

where $V_t = 100$, $W_t = 5$ and

$$E_t = 0.8E_{t-1}, \qquad (t = 4, 5, \dots),$$

with

$$(E_3|D_0, I) \sim N[-40, 200],$$

independently of $(\mu_0|D_0)$.
Verify that $(Y_3|D_0, I) \sim \mathrm{N}[460, 730]$.

(d) At $t = 4$, there is a prior probability of 0.4 that a further export licence will be granted. If so, then $\omega_4 \sim \mathrm{N}[200, 600]$, otherwise $\omega_4 \sim \mathrm{N}[0, 5]$ as usual. Show that $p(Y_4|D_0, I)$ is a mixture of two normals,

$$(Y_4|D_0, I) \sim \begin{cases} \mathrm{N}[668, 1258], & \text{with probability } 0.4; \\ \mathrm{N}[468, 663], & \text{with probability } 0.6. \end{cases}$$

Plot the corresponding forecast density over the range $400 < Y_4 < 750$. Calculate the forecast mean $\mathrm{E}[Y_4|D_0, I]$ and variance $\mathrm{V}[Y_4|D_0, I]$. From (c), calculate the joint distribution of $(Y_3, E_3|D_0, I)$ and deduce the posterior $(E_3|Y_3, D_0, I)$. Verify that $(E_3|Y_3 = 387, D_0, I) \sim \mathrm{N}[-60, 145]$.

(e) Calculate the joint posterior $p(\mu_4, E_4|Y_4, D_0, I)$ as a function of Y_4, and deduce the marginal posterior for μ_4.

(6) Write a computer program to implement the multi-process, class II model used in Section 12.4. Verify the program by reproducing the analysis of the CP6 Sales series in 12.4.3.

CHAPTER 13

NON-LINEAR DYNAMIC MODELS: ANALYTIC AND NUMERICAL APPROXIMATIONS

13.1 INTRODUCTION

In previous chapters we have encountered several models that depend on parameters that introduce parameter non-linearities into otherwise standard DLMs. Although the full class of DLMs provides an enormous variety of useful models, it is the case that sometimes, elaborations to include models with *unknown* parameters result in such non-linearities, thus requiring extensions of the usual linear model analysis. Some typical, and important, examples are as follows.

EXAMPLE 13.1. In many commercial series in which seasonality is a major factor, an apparent feature of the form of seasonal patterns is that amplitudes of seasonal components appear to increase markedly at higher levels of the series. This occurs rather commonly in practice with positive data series. With a dynamic linear model for seasonality, it is certainly possible to adapt to changes in seasonal parameters by allowing for major variation through the evolution errors. However, when the changes are very marked and have a systematic form related to the level of the series, they are more appropriately modelled in an alternative way. Let α_t represent the underlying, non-seasonal level of the series at time t, that may include trend and regression terms. The usual, linear model for seasonality of period p is defined via the mean response $\mu_t = \alpha_t + \phi_t$, where ϕ_t is the seasonal effect at time t. Now the observed inflation in seasonal deviations from level as α_t increases can be directly modelled via a mean response that is non-linear in α_t and ϕ_t, the simplest such model being multiplicative of the form

$$\mu_t = \alpha_t(1 + \phi_t).$$

More elaborate non-linearities are also possible, but this simple multiplicative model alone is of major use in practice, and is referred to as a *multiplicative seasonal effects model.* Two possible models for Y_t use this multiplicative form.

(a) A DLM obtained via log-transformation. With $Y_t > 0$ (as is essentially always the case when multiplicative seasonality is apparent), assume that

$$Y_t = \alpha_t(1 + \phi_t)e^{\nu_t},$$

where ν_t is normally distributed observational noise. Then Y_t follows a lognormal model (Section 10.6), and

$$\log(Y_t) = \alpha_t^* + \phi_t^* + \nu_t,$$

where $\alpha_t^* = \log(\alpha_t)$ and $\phi_t^* = \log(1+\phi_t)$. Thus, the series may be modelled using a standard seasonal DLM after log transformation, α_t^* and ϕ_t^* both being linear functions of the state vector. Note also that this is consistent with an observational variance increasing with level on the original scale.

(b) For reasons of interpretation of model parameters (and others, see Section 10.6) it is often desirable to avoid transformation, suggesting a model of the form

$$Y_t = \alpha_t(1+\phi_t) + \nu_t.$$

Note that this representation may also require the use of a variance power law to model increased observational variation with level.

Model (b) is considered here. The state vector $\boldsymbol{\theta}_t' = (\boldsymbol{\theta}_{t1}', \boldsymbol{\theta}_{t2}')$ comprises trend and/or regression parameters $\boldsymbol{\theta}_{t1}$ such that $\alpha_t = \mathbf{F}_{t1}'\boldsymbol{\theta}_{t1}$, and seasonal parameters $\boldsymbol{\theta}_{t2}$ such that $\phi_t = \mathbf{F}_{t2}'\boldsymbol{\theta}_{t2}$, for some known vectors $\mathbf{F}_{t1}$ and $\mathbf{F}_{t2}$. Thus,

$$Y_t = \mathbf{F}_{t1}'\boldsymbol{\theta}_{t1}[1 + \mathbf{F}_{t2}'\boldsymbol{\theta}_{t2}] + \nu_t,$$

a bilinear function of the state vector. In summary, the model here involves parameter non-linearities in the observational equation simply through the mean response function, and the parameters involved are standard time-varying elements of the state vector.

See also Abraham and Ledolter (1983, Chapter 4), Harrison (1965), Harrison and Stevens (1971) and Gilchrist (1976, Chapter 8) for further discussion of multiplicative seasonality. In West, Harrison and Pole (1987), multiplicative seasonal models are implemented as components of trend, seasonal and regression DLMs, the analysis being based on the use of linearisation as described in Section 13.2.

EXAMPLE 13.2. Recall the simple transfer response model in Example 9.1 in which a DLM is defined by the equations $Y_t = \mu_t + \nu_t$ and $\mu_t = \lambda\mu_{t-1} + \psi_t X_t$, X_t being the observed value of a regressor variable at t. With λ assumed known, this is a DLM. Otherwise, as introduced in Section 9.3.3, the evolution of the state parameter μ_t involves a non-linear term, multiplicative in λ and μ_{t-1}, and the DLM analysis is lost. The same applies to the general transfer function models of Definition 9.1, or the alternative representation in equation (9.6). In these models, the evolution equation for the state vector at time t involves bilinear terms in state vector parameters from time $t-1$ and other, constant but unknown, parameters from $\mathbf{G}$.

EXAMPLE 13.3. Long-term growth towards an asymptote may be modelled with a variety of parametric forms, often called growth or trend curves (Harrison and Pearce 1972; Gilchrist 1976, Chapter 9). Most such curves are basically non-linear in their defining parameters. A model similar to those in Example 13.2 is obtained when modelling long term growth using Gompertz growth curves, as follows. Suppose that $\mu_t = \alpha - \beta\exp(-\gamma t)$ for some parameters α, β and γ, with $\gamma > 0$. Then $\mu_0 = \alpha - \beta$, $\lim_{t\to\infty} \mu_t = \alpha$ and μ_t grows exponentially over time from $\alpha - \beta$ to β. The form

$$e^{\mu_t} = \exp[\alpha - \beta e^{-\gamma t}], \qquad (t > 0),$$

is known as a Gompertz curve, and often provides an appropriate qualitative description of growth to an asymptote. If such growth is assumed for the series Y_t, then transforming to logs suggests the model $\log(Y_t) = \mu_t + \nu_t$. Now, let $\alpha_t = \alpha$ and $\beta_t = -\beta\exp(-\gamma t)$, so that $\beta_t = \lambda\beta_{t-1}$, where $\lambda = \exp(-\gamma)$ and $0 < \lambda < 1$. Then $\log(Y_t)$ follows a DLM $\{\mathbf{F}, \mathbf{G}, V_t, \mathbf{0}\}$, where

$$\mathbf{F} = \mathbf{E}_2 = \begin{pmatrix} 1 \\ 0 \end{pmatrix} \qquad \text{and} \qquad \mathbf{G} = \begin{pmatrix} 1 & \lambda \\ 0 & \lambda \end{pmatrix}.$$

For fixed λ, this is an evolution noise-free DLM (with a second-order polynomial as the special case when $\lambda = 1$). Such a model with a non-zero evolution variance matrix $\mathbf{W}_t$ provides the basic Gompertz form with parameters that vary stochastically in time, allowing for random deviation in the growth away from an exact Gompertz curve. In practice, it is usually desirable to learn about λ, so that the evolution equation becomes bilinear in unknown parameters.

EXAMPLE 13.4. The DLM representations of ARMA processes, and their extensions to non-stationary noise models based on ARMA models with time-varying coefficients, have a structure similar to the models above when these parameters are uncertain; see Section 9.4. Again the non-linearities involve bilinear terms in evolution and/or observation equations of what would be DLMs if the parameters were known.

EXAMPLE 13.5. A wide range of problems with rather different structure is typified by Example 12.1. There a discount factor was viewed as uncertain; more generally, the evolution variance matrices $\mathbf{W}_t$ may depend on constant but uncertain parameters to be estimated.

Whatever the particular structure of parameter non-linearities is in any application, the basic DLM analysis must be extended to allow for it. We have come a long way with linear models, the associated normal theory being analytically tractable and satisfyingly complete. When extending the framework to allow non-linearities, formal theory defines the analysis as usual, but essentially without exception, the calculation of the required components of analysis becomes burdensome. Implementation of the formally well-defined analysis requires the use of numerical integration to approximate mathematically defined integrals arbitrarily well, and so is in fact impossible. However, a variety of approximation techniques exist, and in practice, much can be done using such techniques. We begin in this chapter by discussing simple mathematical approximations that have been very widely used in the last three decades, followed by some specific developments of direct numerical approximation using quadrature methods. Later, Chapter 15 introduces modern simulation methods that represent the current frontiers of numerical approximation and integration in Bayesian statistics generally.

13.2 LINEARISATION AND RELATED TECHNIQUES

Many models with parameter non-linearities may be written in the following form:

$$\begin{aligned} Y_t &= F_t(\boldsymbol{\theta}_t) + \nu_t, \\ \boldsymbol{\theta}_t &= \mathbf{g}_t(\boldsymbol{\theta}_{t-1}) + \boldsymbol{\omega}_t, \end{aligned} \tag{13.1}$$

where $F_t(.)$ is a known, non-linear regression function mapping the n-vector $\boldsymbol{\theta}_t$ to the real line, $\mathbf{g}_t(.)$ is a known, non-linear vector evolution function, and ν_t and $\boldsymbol{\omega}_t$ are error terms subject to the usual assumptions. Various *linearisation* techniques have been developed for such models, all being based in essence on the use of linear approximations to non-linearities. The most straightforward, and easily interpreted, approach is that based on the use of first order Taylor series approximations to the non-linear regression and evolution functions in (13.1). This requires the assumptions that both $F_t(.)$ and $\mathbf{g}_t(.)$ be (at least once-) differentiable functions of their vector arguments.

Suppose, as usual, that $\nu_t \sim \mathrm{N}[0, V]$ for some constant but unknown variance V. Assume also that at time $t-1$, historical information about the state vector $\boldsymbol{\theta}_{t-1}$ and V is (approximately) summarised in terms of standard posterior distributions:

$$(\boldsymbol{\theta}_{t-1} \mid V, D_{t-1}) \sim \mathrm{N}[\mathbf{m}_{t-1}, \mathbf{C}_{t-1}V/S_{t-1}],$$

and, with $\phi = 1/V$,

$$(\phi \mid D_{t-1}) \sim \mathrm{G}[n_{t-1}/2, d_{t-1}/2],$$

with estimate of V given by $S_{t-1} = d_{t-1}/n_{t-1}$. Also,

$$(\boldsymbol{\omega}_t \mid V, D_{t-1}) \sim \mathrm{N}[\,\mathbf{0}, \mathbf{W}_t V/S_{t-1}],$$

with $\mathbf{W}_t$ known. Then

$$(\boldsymbol{\theta}_{t-1} \mid D_{t-1}) \sim \mathrm{T}_{n_{t-1}}[\mathbf{m}_{t-1}, \mathbf{C}_{t-1}],$$

being uncorrelated with

$$(\boldsymbol{\omega}_t \mid D_{t-1}) \sim \mathrm{T}_{n_{t-1}}[\,\mathbf{0}, \mathbf{W}_t].$$

The mean $\mathbf{m}_{t-1}$ is an estimate of $\boldsymbol{\theta}_{t-1}$, about which a Taylor series expansion of the evolution function gives

$$\begin{aligned} \mathbf{g}_t(\boldsymbol{\theta}_t) = \mathbf{g}_t(\mathbf{m}_{t-1}) &+ \mathbf{G}_t(\boldsymbol{\theta}_{t-1} - \mathbf{m}_{t-1}) \\ &+ \text{quadratic and higher-order terms in } (\boldsymbol{\theta}_{t-1} - \mathbf{m}_{t-1}), \end{aligned}$$

where $\mathbf{G}_t$ is the $n \times n$ matrix derivative of the evolution function evaluated at the estimate $\mathbf{m}_{t-1}$, namely

$$\mathbf{G}_t = \left[\frac{\partial \mathbf{g}_t(\boldsymbol{\theta}_{t-1})}{\partial \boldsymbol{\theta}'_{t-1}}\right]_{\boldsymbol{\theta}_{t-1}=\mathbf{m}_{t-1}};$$

obviously $\mathbf{G}_t$ is known. Assuming that terms other than the linear term are negligible, the evolution equation becomes

$$\boldsymbol{\theta}_t \approx \mathbf{g}_t(\mathbf{m}_{t-1}) + \mathbf{G}_t(\boldsymbol{\theta}_{t-1} - \mathbf{m}_{t-1}) + \boldsymbol{\omega}_t = \mathbf{h}_t + \mathbf{G}_t\boldsymbol{\theta}_{t-1} + \boldsymbol{\omega}_t, \qquad (13.2)$$

where $\mathbf{h}_t = \mathbf{g}_t(\mathbf{m}_{t-1}) - \mathbf{G}_t\mathbf{m}_{t-1}$ is known. Equation (13.2) is a *linearised* version of the evolution equation, linearised about the expected value of $\boldsymbol{\theta}_{t-1}$.

Assuming (13.2) to be an adequate approximation to the model, it follows immediately that the usual DLM evolution applies, with the minor extension to include an additional, known term $\mathbf{h}_t$ in the evolution equation. Thus, the prior for $\boldsymbol{\theta}_t$ is determined by

$$(\boldsymbol{\theta}_t \mid V, D_{t-1}) \sim \mathrm{N}[\mathbf{a}_t, \mathbf{R}_t V/S_{t-1}],$$

so that

$$(\boldsymbol{\theta}_t \mid D_{t-1}) \sim \mathrm{T}_{n_{t-1}}[\mathbf{a}_t, \mathbf{R}_t], \qquad (13.3)$$

with defining quantities

$$\begin{aligned}\mathbf{a}_t &= \mathbf{h}_t + \mathbf{G}_t\mathbf{m}_{t-1} = \mathbf{g}_t(\mathbf{m}_{t-1}),\\ \mathbf{R}_t &= \mathbf{G}_t\mathbf{C}_{t-1}\mathbf{G}'_t + \mathbf{W}_t.\end{aligned} \qquad (13.4)$$

Proceeding to the observation equation, similar ideas apply. Now the non-linear regression function is linearised about the expected value $\mathbf{a}_t$ for $\boldsymbol{\theta}_t$, leading to

$$F_t(\boldsymbol{\theta}_t) = F_t(\mathbf{a}_t) + \mathbf{F}'_t(\boldsymbol{\theta}_t - \mathbf{a}_t) + \text{quadratic and higher order terms in } (\boldsymbol{\theta}_t - \mathbf{a}_t),$$

where $\mathbf{F}_t$ is the n-vector derivative of $F_t(.)$ evaluated at the prior mean $\mathbf{a}_t$, namely

$$\mathbf{F}_t = \left[\frac{\partial F_t(\boldsymbol{\theta}_t)}{\partial \boldsymbol{\theta}_t}\right]_{\boldsymbol{\theta}_t=\mathbf{a}_t},$$

that is known. Assuming that the linear term dominates the expansion, we deduce the linearised observation equation

$$Y_t \approx f_t + \mathbf{F}'_t(\boldsymbol{\theta}_t - \mathbf{a}_t) + \nu_t = (f_t - \mathbf{F}'_t\mathbf{a}_t) + \mathbf{F}'_t\boldsymbol{\theta}_t + \nu_t, \qquad (13.5)$$

where $f_t = F_t(\mathbf{a}_t)$. Combining this with (13.3) and (13.4) leads to a DLM with, in addition to the extra term in the evolution equation, a similar term $f_t - \mathbf{F}'_t\mathbf{a}_t$ in the observation equation. From these equations it follows that for forecasting one-step ahead,

$$(Y_t \mid D_{t-1}) \sim \mathrm{T}_{n_{t-1}}[f_t, Q_t],$$

where $Q_t = \mathbf{F}_t'\mathbf{R}_t\mathbf{F}_t + S_{t-1}$. Also, once Y_t is observed giving forecast error $e_t = Y_t - f_t$, the standard updating equations for $\boldsymbol{\theta}_t$ and V apply directly. Thus, the linearised model is a DLM. Some comments on this analysis are in order.

(1) Firstly, the one-step ahead point forecast $f_t = F_t[\mathbf{g}_t(\mathbf{m}_{t-1})]$ retains precisely the form of the non-linear mean response from the model (13.1) with $\boldsymbol{\theta}_{t-1}$ assumed known and equal to its estimate $\mathbf{m}_{t-1}$. Hence predictions accord with the model; if, for example, either regression or evolution functions impose bounds on the mean response, the predictor accords with these bounds.

(2) Consider forecasting ahead to time $t+k$ from time t. Applying linearisation to the evolution equations successively over time, it follows that

$$(\boldsymbol{\theta}_{t+k}|D_t) \sim \mathrm{T}_{n_t}[\mathbf{a}_t(k), \mathbf{R}_t(k)],$$

with moments successively defined as follows. Setting $\mathbf{a}_t(0) = \mathbf{m}_t$ and $\mathbf{R}_t(0) = \mathbf{C}_t$, we have, for $k = 1, 2, \ldots,$

$$\mathbf{a}_t(k) = \mathbf{g}_{t+k}(\mathbf{a}_t(k-1)),$$
$$\mathbf{R}_t(k) = \mathbf{G}_t(k)\mathbf{R}_t(k-1)\mathbf{G}_t(k)' + \mathbf{W}_{t+k},$$

where

$$\mathbf{G}_t(k) = \left[\frac{\partial \mathbf{g}_{t+k}(\boldsymbol{\theta}_{t+k-1})}{\partial \boldsymbol{\theta}'_{t+k-1}}\right]_{\boldsymbol{\theta}_{t+k-1}=\mathbf{a}_t(k-1)}.$$

Then,

$$(Y_{t+k} \mid D_t) \sim \mathrm{T}_{n_t}[f_t(k), Q_t(k)],$$

where

$$f_t(k) = F_{t+k}(\mathbf{a}_t(k)),$$
$$Q_t(k) = \mathbf{F}_t(k)'\mathbf{R}_t(k)\mathbf{F}_t(k) + S_t,$$

with

$$\mathbf{F}_t(k) = \left[\frac{\partial F_{t+k}(\boldsymbol{\theta}_{t+k})}{\partial \boldsymbol{\theta}_{t+k}}\right]_{\boldsymbol{\theta}_{t+k}=\mathbf{a}_t(k)}.$$

The forecast function $f_t(k)$ can be written as

$$f_t(k) = F_{t+k}[\mathbf{g}_{t+k}\{\mathbf{g}_{t+k-1}(\ldots \mathbf{g}_{t+1}(\mathbf{m}_t)\ldots)\}],$$

retaining the required non-linear form into the future.

(3) In truncating Taylor series expansions, the implicit assumption is that the higher-order terms are negligible relative to the first. To be

specific, consider the linear approximation to the regression function about $\boldsymbol{\theta}_t = \mathbf{a}_t$. When linearising smooth functions, local linearity is evident, so that in a region near the centre $\mathbf{a}_t$ of linearisation, the approximation will tend to be adequate. Far away from this region, it may be that the approximation error increases, often dramatically. However, if most of the probability under $p(\boldsymbol{\theta}_t \mid D_{t-1})$ is concentrated tightly about the estimate $\mathbf{a}_t$, then there is high probability that the approximation is adequate.

(4) The above comments notwithstanding, it should be noted that the linearised model as defined in equations (13.2) and (13.5) defines a valid DLM without reference to its use as an approximation to (13.1). Guided by (13.1), the linearised model and DLM analysis may be accepted for forecasting, producing forecast functions with the desired non-linear features, its appropriateness being judged on the usual basis of forecast accuracy. With this view, the extent to which the DLM approximates (13.1) is irrelevant; note, however, that the resulting posterior distributions cannot be considered as valid for inference about the state vector in (13.1) without such considerations.

(5) As in non-linear modelling generally, non-linear transformations of original parameters in $\boldsymbol{\theta}_t$ may lead to evolution or regression functions that are much closer to linearity in regions of interest than with the original metric.

(6) On approximation error, note that the evolution (and observation) errors can be viewed as implicity accounting for the neglected terms in the Taylor series expansions.

(7) Some or all of the elements of $\boldsymbol{\theta}_t$ may be constant over time. Thus, models such as in Examples 13.2 and 13.3 above lie within this class; the constant parameters are simply incorporated into an extended state vector.

(8) If either $F_t(.)$ or $\mathbf{g}_t(.)$ are linear functions, then the corresponding linearised equations are exact.

(9) Note that the model (13.2) is easily modified to incorporate a stochastic drift in $\boldsymbol{\theta}_{t-1}$ *before* applying the non-linear evolution function, rather than by adding $\boldsymbol{\omega}_t$ *after* transformation. The alternative is simply

$$\boldsymbol{\theta}_t = \mathbf{g}_t(\boldsymbol{\theta}_{t-1} + \boldsymbol{\delta}_t),$$

for some zero-mean error $\boldsymbol{\delta}_t$ with a known variance matrix $\mathbf{U}_t$. Then $\mathbf{a}_t$ and $\mathbf{R}_t$ are as defined in (13.4) with

$$\mathbf{W}_t = \mathbf{G}_t\mathbf{U}_t\mathbf{G}_t'.$$

EXAMPLE 13.1 (continued). Models in which the non-linearities are bilinear abound. Examples 13.1 to 13.4 inclusive are of this form. In Example

13.1, the multiplicative seasonal model, we have state vector $\boldsymbol{\theta}_t' = (\boldsymbol{\theta}_{t1}', \boldsymbol{\theta}_{t2}')$, a linear evolution equation with $\mathbf{g}_t(\boldsymbol{\theta}_{t-1}) = \mathbf{G}_t\boldsymbol{\theta}_{t-1}$ for some known evolution matrix $\mathbf{G}_t$, and a bilinear regression function

$$F_t(\boldsymbol{\theta}_t) = (\mathbf{F}_{t1}'\boldsymbol{\theta}_{t1})[1 + (\mathbf{F}_{t2}'\boldsymbol{\theta}_{t2})].$$

Thus, the model simplifies, linearisation is only used in the observation equation. With $\mathbf{a}_t' = (\mathbf{a}_{t1}', \mathbf{a}_{t2}')$, it follows that the prior estimate for the non-seasonal trend component α_t is $f_{t1} = \mathbf{F}_{t1}'\mathbf{a}_{t1}$, whilst that for the seasonal effect ϕ_t is $f_{t2} = \mathbf{F}_{t2}'\mathbf{a}_{t2}$. Hence the one-step ahead point forecast is $f_t = f_{t1}(1 + f_{t2})$, the product of the estimated components. This model, and form of approximation, is the basis for many practical schemes of analysis of time series exhibiting the forms of behaviour consistent with multiplicative seasonality (Harrison 1965; West, Harrison and Pole 1987).

This approach, along with many variants, has been widely used in various fields of application, particularly in non-linear state space models used in communications and control engineering. Here and elsewhere, non-Bayesian techniques using linearisation combined with classical least squares theory led, in the 1960s and 1970s, to a plethora of related techniques under such names as extended Kalman filtering, generalised Kalman filtering, non-linear filtering, and so forth. The books of Anderson and Moore (1979), Jazwinski (1970) and Sage and Melsa (1971) provide good discussion and references.

One obvious variant on the approach is to extend the Taylor series expansion to include higher-order terms in the relevant state vectors. Considering the regression function, for example, the linear approximation can be refined by including the second-order term to give

$$F_t(\boldsymbol{\theta}_t) = F_t(\mathbf{a}_t) + \mathbf{F}_t'(\boldsymbol{\theta}_t - \mathbf{a}_t) + \frac{1}{2}(\boldsymbol{\theta}_t - \mathbf{a}_t)'\mathbf{H}_t(\boldsymbol{\theta}_t - \mathbf{a}_t)$$

$$+ \text{cubic and higher order terms in } (\boldsymbol{\theta}_t - \mathbf{a}_t),$$

where

$$\mathbf{F}_t = \left[\frac{\partial F_t(\boldsymbol{\theta}_t)}{\partial \boldsymbol{\theta}_t}\right]_{\boldsymbol{\theta}_t = \mathbf{a}_t} \quad \text{and} \quad \mathbf{H}_t = \left[\frac{\partial \mathbf{F}_t(\boldsymbol{\theta}_t)}{\partial \boldsymbol{\theta}_t'}\right]_{\boldsymbol{\theta}_t = \mathbf{a}_t}.$$

With this quadratic approximation to $F_t(\boldsymbol{\theta}_t)$, the distribution implied for Y_t is no longer T or normal. However, based on the assumed T or normal prior for $\boldsymbol{\theta}_t$, moments can be calculated. Assuming $n_{t-1} > 2$, of course, it follows directly that

$$f_t = \mathrm{E}[Y_t \mid D_{t-1}] = F_t(\mathbf{a}_t) + \frac{1}{2} \text{ trace } (\mathbf{H}_t\mathbf{R}_t).$$

Note the additional term here that accounts for some of the uncertainty in $\boldsymbol{\theta}_t$, introducing the *curvature* matrix $\mathbf{H}_t$ at the point $\boldsymbol{\theta}_t = \mathbf{a}_t$. Similarly, if $n_{t-1} > 4$, the variance of Y_t may be calculated. Notice that this involves calculation of mixed, fourth-order moments of the multivariate T distribution. For updating once Y_t is observed, however, this higher-order approximation leads to a difficult analysis.

One other variant of the basic linearisation technique specific to the case of models with bilinearities is suggested, and implemented in a case study, in Migon (1984) and Migon and Harrison (1985). Suppose that $\mathbf{g}_t(\boldsymbol{\theta}_{t-1})$ in (13.1) is a bilinear function of $\boldsymbol{\theta}_{t-1}$. Then given the conditional normal distribution for $\boldsymbol{\theta}_{t-1}$, the exact mean and variance of $\mathbf{g}_t(\boldsymbol{\theta}_{t-1})$ can be calculated using standard normal theory. Thus, the exact prior mean and variance matrix of $\boldsymbol{\theta}_t$ can be calculated. Assuming approximate normality for $\boldsymbol{\theta}_t$, then a linear or bilinear regression function $\mathbf{F}_t(\boldsymbol{\theta}_t)$ implies, by a similar argument, the joint first- and second-order moments of Y_t and $\boldsymbol{\theta}_t$. Thus, again assuming approximate normality of all components, the forecast and posterior distributions are deduced. Given the bilinearity, the approximations here involve the assumptions of approximate normality of products of normal components. In practice, this approach is typically rather similar to direct linearisation up to second-order although it remains to be further investigated from a theoretical standpoint.

13.3 CONSTANT PARAMETER NON-LINEARITIES: MULTI-PROCESS MODELS

A second class of models may be analysed rather more formally within the framework of multi-process, class I models of Chapter 12. These are models whose non-linearities are due to the appearance of constant, but unknown, parameters $\boldsymbol{\alpha}$ in one or more of the components. Examples 13.2 to 13.4 are special cases. Thus, $\boldsymbol{\alpha}$ could comprise elements of a constant $\mathbf{G}$ matrix, transformation parameters, ARMA parameters, discount factors, and so forth. The parameter space for $\boldsymbol{\alpha}$, denoted by $\mathcal{A}$, is usually continuous (at least in part), and often bounded. It is assumed that $\boldsymbol{\alpha}$ is constant over time but uncertain. In such cases, the series follows a standard DLM *conditional* on any chosen value of $\boldsymbol{\alpha} \in \mathcal{A}$. Generally, therefore, write the defining quadruple as

$$\{\mathbf{F}_t(\boldsymbol{\alpha}),\ \mathbf{G}_t(\boldsymbol{\alpha}),\ V_t(\boldsymbol{\alpha}),\ \mathbf{W}_t(\boldsymbol{\alpha})\}. \tag{13.6}$$

The formal theory for analysing such models is developed in Section 12.2 to which the reader is referred. It is made clear there that the theoretically well-specified analysis is practically infeasible due to the abundance of integrals that cannot be analytically calculated. Multi-process, class I models provide a method of numerical integration that allows the approximation of the formal analysis, discussion of a particular example appearing in Section 12.2.3. A summary of the relevant multi-process theory is given here.

Let $\mathbf{X}_t$ be any vector of random quantities of interest at time t. Thus, $\mathbf{X}_t$ may be a function of $\boldsymbol{\theta}_t$ or past values of the state vector, Y_t or future observations, the unknown observational variance, and so forth. Conditional on any value of $\boldsymbol{\alpha}$, the required posterior distribution for inference about $\mathbf{X}_t$ is derived from the conditional DLM (13.6) in standard, analytically manageable form. Denote the corresponding density by $p(\mathbf{X}_t \mid \boldsymbol{\alpha}, D_t)$ as usual. Also, D_t informs on the parameters $\boldsymbol{\alpha}$ in terms of a posterior density $p(\boldsymbol{\alpha} \mid D_t)$. Formally, the marginal posterior for $\mathbf{X}_t$ is the object of interest, given by

$$p(\mathbf{X}_t \mid D_t) = \int_{\mathcal{A}} p(\mathbf{X}_t \mid \boldsymbol{\alpha}, D_t) p(\boldsymbol{\alpha} \mid D_t) d\boldsymbol{\alpha}. \tag{13.7}$$

It is at this point that the analysis becomes difficult, many such integrals of this form are required for posterior inference and prediction, but most will be impossible to evaluate analytically. Numerical integration is called for. The multi-process, class I framework provides approximations to these integrals based on the use of a fixed and finite *grid* of points for the parameters, $\{\boldsymbol{\alpha}_1, \ldots, \boldsymbol{\alpha}_k\}$, as a discrete approximation, in some sense, to the full parameter space $\mathcal{A}$. The analysis is thus based on the use of a finite collection of DLMs, each corresponding to a different choice of the, possibly vector-valued, parameter $\boldsymbol{\alpha}$, these DLMs being analysed in parallel. It is supposed that the collection of k values chosen for $\boldsymbol{\alpha}$ in some sense adequately represents the possibly much larger or continuous true parameter space $\mathcal{A}$. Large spaces $\mathcal{A}$ can often be adequately approximated for some purposes by a fairly small discrete set that somehow spans the larger space, leading to the consideration of a small number of distinct DLMs. When the dimension of $\boldsymbol{\alpha}$ is small, then k can often be chosen large enough so that the points fairly well cover the parameter space. Otherwise, the notion that they be chosen to appropriately span $\mathcal{A}$, representing different regions that may lead to rather different conditional DLM analyses, underlies the use of multi-processes. Obviously, if $\mathcal{A}$ is discrete to begin with, then this approach can be exact if the finite collection of k points coincides with $\mathcal{A}$.

With this in mind, $\boldsymbol{\alpha}$ is a discretised random quantity whose posterior at time t is a mass function rather than a density, with weights

$$p_t(j) = p(\boldsymbol{\alpha}_j \mid D_t) = \Pr[\boldsymbol{\alpha} = \boldsymbol{\alpha}_j \mid D_t], \qquad (j = 1, \ldots, k)$$

for all t. The integral in (13.7) is replaced by the discretised form

$$p(\mathbf{X}_t \mid D_t) = \sum_{j=1}^{k} p(\mathbf{X}_t \mid \boldsymbol{\alpha}, D_t) p_t(j). \tag{13.8}$$

As time progresses and data are obtained, the conditional DLM analyses proceed in parallel. The additional learning about $\boldsymbol{\alpha}$ proceeds via the updating of the posterior masses defined by

$$p_t(j) \propto p_{t-1}(j) p(Y_t \mid \boldsymbol{\alpha}_j, D_{t-1}),$$

where $p(Y_t \mid \boldsymbol{\alpha}_j, D_{t-1})$ is the normal or T one-step forecast density at time t. Thus, with normalising constant c_t defined by $c_t^{-1} = \sum_{j=1}^{k} p_{t-1}(j) p(Y_t \mid$

$\boldsymbol{\alpha}_j, D_{t-1})$, $p_t(j) = c_t p_{t-1}(j)p(Y_t \mid \boldsymbol{\alpha}_j, D_{t-1})$. It follows from (13.8) that all posterior distributions for linear functions of $\boldsymbol{\theta}_t$, and predictive distributions for future observations, are *discrete probability mixtures* of the standard T or normal distributions. Refer to Sections 12.2 and 12.3 for further discussion.

The general approach using mixtures of standard models is rather well known and quite widely used. Apart from the usages discussed in Chapter 12, the main applications of mixtures have been to just this problem of parameter non-linearities. In the control literature, such models are referred to under various names. Analogues of mixtures of normal DLMs (each with known variances) were used by Sorenson and Alspach (1971) and Alspach and Sorenson (1972) under the name of *Gaussian sums*. Anderson and Moore (1979, Chapter 9) discuss this and provide various related references. See also Chapter 10 of Anderson and Moore for further reference to the use of mixtures under the heading of *parallel processing*.

13.4 CONSTANT PARAMETER NON-LINEARITIES: EFFICIENT NUMERICAL INTEGRATION

13.4.1 The need for efficient techniques of integration

The multi-process approach to numerical integration described above and in Chapter 12 attempts to approximate integrals by discretising them into summations, the chosen grid of values for $\boldsymbol{\alpha}$ being fixed for all time. Generally, the quality of approximation increases with k, the size of the grid; a finer grid improves the representation of the space $\mathcal{A}$ and so the posterior probabilities $p_t(j)$ can more accurately follow the variation in the implicitly defined "true" posterior for $\boldsymbol{\alpha}$. However, with more than very few parameters $\boldsymbol{\alpha}$, a reasonably fine grid typically involves the use of many points, and the resulting computational demands can be enormous. Thus, the use of more refined techniques of numerical approximation are called for. In the 1980s, numerical methods suitable for models with low-dimensional, non-linear parameters were developed, based on basic methods of Gaussian quadrature, initiated by Naylor and Smith (1982), and described by Smith et al (1985, 1987). Additional background on quadrature methods appears in Shaw (1988). In the dynamic modelling context, these methods were extended and explored in Pole (1988a) and Pole and West (1988, 1990). These references give full details of the novel numerical problems arising in the sequential context usually adopted for the analysis of dynamic models, and explore various examples. Here we restrict discussion to the basic concepts and issues.

13.4.2 Gaussian quadrature in Bayesian analyses

In a general setting, consider a parameter vector $\boldsymbol{\alpha} \in \mathcal{A}$, assuming that the object of interest is the posterior density $p(\boldsymbol{\alpha})$ from an analysis. Gaussian quadrature is designed to approximate integrals of the form

$$I(q) = \int_{\mathcal{A}} q(\boldsymbol{\alpha})p(\boldsymbol{\alpha})d\boldsymbol{\alpha} \tag{13.9}$$

for functions of interest $q(.)$. The approximations are via sums of the form

$$I(q) \approx \sum_{i=1}^{n} q(\boldsymbol{\alpha}_i)w_i p(\boldsymbol{\alpha}_i), \tag{13.10}$$

where the number n controls the accuracy of the approximation. The integral is effectively "discretised" at *grid points* $\boldsymbol{\alpha}_1, \ldots, \boldsymbol{\alpha}_n$, the density $p(\boldsymbol{\alpha}_i)$ at these points being multiplied in the summation by positive *weights* $w_1, \ldots, w_n$. Basic application of multivariate Gauss-Hermite quadrature (Smith et al 1985, 1987, and supporting material in Davis and Rabinowitz 1984, and Salzer, Zucker and Capuano 1952) provides direct construction of the n quadrature grid points and associated grid weights based on specification of the mean and variance matrix of $\boldsymbol{\alpha}$. The closeness of approximations to integrals improves with n, and depends on the regularity of $h(.)$ and $q(.)$. In particular, if the product $q(\boldsymbol{\alpha})h(\boldsymbol{\alpha})$ is well approximated by a polynomial form of degree not exceeding $2n-1$, then the approximation will tend to be good. If this product is exactly such a polynomial, then the approximating sum is exactly equal to the required integral. The strategy for numerical posterior integration in Smith et al (1985, 1987), applies Gaussian quadrature methods *adaptively,* re-estimating the mean and variance of $p(\boldsymbol{\alpha})$ through several iterations and redefining the grid points and weights at each step. A final iteration delivers a grid that is used as the basis of approximating posterior expectations.

Note that this essentially defines a discrete approximation to the posterior; the grid point $\boldsymbol{\alpha}_i$ have probability masses $w_i p(\boldsymbol{\alpha}_i)$ and the expectation of any function $q(\boldsymbol{\alpha})$ is the usual sum. If, as is often the case, the posterior is known only up to an unknown, positive constant of normalisation c, so that for any $\boldsymbol{\alpha} \in \mathcal{A}$, it is possible to evaluate only $l(\boldsymbol{\alpha}) = c\, p(\boldsymbol{\alpha})$, then the normalisation constant is defined by $c = \int_{\mathcal{A}} l(\boldsymbol{\alpha})d\boldsymbol{\alpha}$ and can be approximately evaluated using (13.10) with $p(.)$ replaced by $l(.)$ and $q(\boldsymbol{\alpha}) = 1$. Then $p(\boldsymbol{\alpha}) = l(\boldsymbol{\alpha})/c$ can be approximately calculated. The earlier references apply this sort of procedure in a variety of applications, and for n relatively small, even in single figures, the approximations for common functions $q(.)$ of interest can be excellent. Variations allow n to change at each iteration. An important issue is the use of reparametrisations of $\boldsymbol{\alpha}$ (Smith et al 1985, 1987). With a final grid approximation, estimates of the posterior density over the range of $\boldsymbol{\alpha}$ generally may be derived via some form of interpolation of the density at values of $\boldsymbol{\alpha}$ between and outside the grid points. The most commonly used such technique is cubic spline interpolation on the log-density scale for univariate posterior margins (Shaw 1987; Smith et al 1985, 1987).

13.4.3 Efficient integration for non-linear parameters

Consider now dynamic models with parametric non-linearities. To be specific, we take the model as specified in (13.6) and suppose that $\boldsymbol{\alpha} = \alpha$ is a scalar, and that the observational variance sequence is known. In Section 13.3, the multi-process model assumes a fixed grid of k values for α that are chosen to represent the full parameter space $\mathcal{A}$. As data are processed, the posterior weights over the grid are updated, providing learning about α. Obvious questions now arise about the accuracy and efficiency of this approach to learning about α. Firstly, concerning accuracy, there is an implicit assumption that the grid will adequately represent regions of the parameter space deemed plausible a priori and supported by the data. However, as time progresses, it may be the case that regions not adequately covered by grid points are supported by the data, in which case the extent to which the discrete, approximate posterior for α appropriately reflects the information available will be questionable. What is needed in such cases is a change in grid values and sometimes an increase in the number of grid values. The second, related point concerns efficiency and is pertinent in the opposite case when the posterior masses concentrate more and more around one or a small number of grid points. Here grid values with little or no posterior probability may be dropped from the grid and more added in the high-probability region. Thus, there is a general need for the grid to evolve over time, i.e., for a *dynamic grid* underlying a multi-process model; the approach using quadrature techniques, referenced above, has such features, as follows.

At time $t-1$, the posterior for the current state vector summarises the history of the analysis. In the current DLM framework, the posterior conditional on any value of α is normal,

$$(\boldsymbol{\theta}_{t-1}|\alpha, D_{t-1}) \sim \mathrm{N}[\mathbf{m}_{t-1}(\alpha), \mathbf{C}_{t-1}(\alpha)], \qquad (13.11)$$

with moments that are theoretically available. Suppose also that the following quantities are available:

- (an approximation to) the full posterior density

$$p(\alpha|D_{t-1}), \qquad (\alpha \in \mathcal{A}); \qquad (13.12)$$

- the associated (approximations to the) posterior mean and variance of $(\alpha|D_{t-1})$, denoted by a_{t-1} and A_{t-1}.

Inferences about $\boldsymbol{\theta}_{t-1}$ (and other quantities of interest) are based on the unconditional posterior

$$p(\boldsymbol{\theta}_{t-1}|D_{t-1}) = \int_{\mathcal{A}} p(\boldsymbol{\theta}_{t-1}|\alpha, D_{t-1})p(\alpha|D_{t-1})d\alpha. \tag{13.13}$$

Of course, this and related integrals for moments of $\boldsymbol{\theta}_{t-1}$ and so forth require that (a) the conditional moments of $(\boldsymbol{\theta}_{t-1}|\alpha, D_{t-1})$ be available for *all* values of $\alpha \in \mathcal{A}$, and (b) that the integration be performed, typically requiring numerical approximation. While (b) can now be entertained, (a) is impossible generally. The conditional density (13.11), summarised by its moments, can only be evaluated for a finite number of values of α. This is where quadrature ideas enter in (Pole and West 1988), as follows.

Suppose that the quadrature-based analysis at time $t-1$ has provided the "current" grid points

$$\mathcal{A}_{t-1} = \{\alpha_{t-1,1}, \ldots, \alpha_{t-1,k_{t-1}}\}, \tag{13.14}$$

of current size k_{t-1}, with associated quadrature weights

$$w_{t-1,1}, \ldots, w_{t-1,k_{t-1}}.$$

Then the implied posterior approximation at $t-1$ is simply

$$p(\boldsymbol{\theta}_{t-1}|D_{t-1}) \approx \sum_{j=1}^{k_{t-1}} p(\boldsymbol{\theta}_{t-1}|\alpha_{t-1,j}, D_{t-1})w_{t-1,j}p(\alpha_{t-1,j}|D_{t-1}). \tag{13.15}$$

Similarly, the posterior mean of $\boldsymbol{\theta}_{t-1}$ is estimated by

$$\mathrm{E}[\boldsymbol{\theta}_{t-1}|D_{t-1}] \approx \sum_{j=1}^{k_{t-1}} \mathbf{m}_{t-1}(\alpha_{t-1,j})w_{t-1,j}p(\alpha_{t-1,j}|D_{t-1}).$$

Then the moments in (13.11) are required only for those values of α in the grid $\mathcal{A}_{t-1}$. The particular points in this grid at time $t-1$ will depend generally on D_{t-1} and the accuracy of approximation upon the size of the grid. Notice that in terms of inference for $\boldsymbol{\theta}_{t-1}$ and related quantities, equations such as (13.15) resemble the marginal posterior distributions derived in standard multi-process modelling as described in Section 13.3. It is as if the posterior for α at time $t-1$ were approximated by a discrete mass function placing masses $p_{t-1}(j) = w_{t-1,j}p(\alpha_{t-1,j}|D_{t-1})$ at the grid values $\alpha_{t-1,j}$. This is essentially the case so far as the calculation of expectations of functions of α is concerned, although inference about α itself is based on the full density (13.12).

In summary, the assumed components summarising the history of the analysis and providing the basis of posterior inferences at time $t-1$ are as follows: For α we have

(i) the density function (13.12), and
(ii) the associated mean and variance a_{t-1} and A_{t-1}.

For given k_{t-1} these moments define the grid (13.14) together with the associated weights. Then for $\boldsymbol{\theta}_{t-1}$ we have

(iii) the defining moments $\mathbf{m}_{t-1}(\alpha)$ and $\mathbf{C}_{t-1}(\alpha)$ for values of $\alpha \in \mathcal{A}_{t-1}$.

For forecasting ahead from time $t-1$, the defining quadruple (13.6) is also obviously required only for those values of $\alpha \in \mathcal{A}_{t-1}$. The one-step ahead forecast density $p(Y_t|D_{t-1}) = \int p(Y_t|\alpha, D_{t-1})p(\alpha|D_{t-1})d\alpha$ is, for each Y_t, an integral with respect to α of standard normal forms. For each α, $(Y_t|\alpha, D_{t-1}) \sim \mathrm{N}[f_t(\alpha), Q_t(\alpha)]$, with moments defined by the usual equations for the particular value of α. The quadrature approximation is then the mixture of normal components

$$p(Y_t|D_{t-1}) = \sum_{j=1}^{k_{t-1}} p(Y_t|\alpha_{t-1,j}, D_{t-1})w_{t-1,j}p(\alpha_{t-1,j}|D_{t-1}).$$

Consider now updating when Y_t is observed. The steps in the analysis are as follows.

(A) As in multi-process models, the k_{t-1} DLMs corresponding to values of $\alpha \in \mathcal{A}_{t-1}$ are updated in parallel. This leads to standard posteriors

$$(\boldsymbol{\theta}_t|\alpha_{t-1,j}, D_t) \sim \mathrm{N}[\mathbf{m}_t(\alpha_{t-1,j}), \mathbf{C}_t(\alpha_{t-1,j})]$$

for $j = 1, \ldots, k_{t-1}$. Obviously, if the observational variance is unknown, the usual extension to normal/gamma distributions for the state vector and the unknown variance parameter is made conditional on each value of α. The above normal posterior is then replaced by the relevant T distribution.

(B) Bayes' Theorem for α leads to $p(\alpha|D_t) \propto p(Y_t|\alpha, D_{t-1})p(\alpha|D_{t-1})$ for all α. Now, whilst the prior density is assumed known from (13.12) for all α, the likelihood component (just the observed value of the one-step ahead forecast density) is only known for $\alpha \in \mathcal{A}_{t-1}$. Fuller information about the posterior is reconstructed in the next two steps.

(C) Bayes' Theorem above provides the values $p(\alpha_{t-1,j}|D_t)$ for $j = 1, \ldots, k_{t-1}$, the ordinates of the posterior density over the existing grid. Using the existing weights $w_{t-1,j}$ on this grid, the posterior is approximately normalised and a first approximation to the posterior mean and variance of α calculated by quadrature. For normalisation, $p(\alpha|D_t) \approx c\, p(Y_t|\alpha, D_{t-1})p(\alpha|D_{t-1})$, where

$$c^{-1} = \sum_{j=1}^{k_{t-1}} p(Y_t|\alpha_{t-1,j})w_{t-1,j}p(\alpha_{t-1,j}|D_{t-1}),$$

the mean and variance being estimated by similar summations. Having calculated these initial estimates, the iterative quadrature strategy referred to in Section 13.4.2 is used to iteratively refine estimates of posterior mean and variance for α and the set of grid points with associated weights. Note that the number of grid points is to be chosen; it may differ from k_{t-1} if required and also change between iterations. Using the revised grid at each stage, revised estimates of the normalising constant c, posterior mean and variance are computed and the cycle continues. Iterations halt when these estimates do not change markedly. At this point, the posterior density is evaluated (including normalising constant) at points on a new grid of chosen size k_t points,

$$\mathcal{A}_t = \{\alpha_{t,1}, \dots, \alpha_{t,k_t}\}.$$

The associated quadrature weights $w_{t,1}, \dots, w_{t,k_t}$ are available, as are estimates of the posterior mean a_t and variance A_t of α.

(D) To reconstruct the full posterior density function $p(\alpha|D_t)$, some form of interpolation is needed based on the k_t evaluations over $\mathcal{A}_t$. Cubic spline interpolation (Shaw 1987) is used in Pole (1988a), and Pole and West (1988, 1990), splines being fitted to the density on the log scale. This ensures positive density everywhere after interpolation, and linear extrapolation outside the grid leads to exponential decay to zero in the tails of the posterior. Fuller details can be found in these references. This follows the use of splines mentioned earlier in Smith et al (1987). In more general models with α vector valued, the basic development is similar but the problems of density reconstruction much more difficult. Following Shaw (1987), Pole and West (1988, 1990) use tensor product splines. Returning to the model here in which α is scalar, the net result is that the posterior density $p(\alpha|D_t)$ may be (approximately) evaluated at any value of α, forming the basis of posterior inference for α. Thus far, then, the updating analysis is completed as far as α is concerned.

(E) In addition, the analysis has identified a new grid $\mathcal{A}_t$ of k_t values of α, that with the associated weights, provides the basis for approximating expectations of functions of α needed for inference about $\boldsymbol{\theta}_t$, etc. At this point, it becomes clear that further work is required to obtain such inferences. The problem lies in the fact that the conditional posteriors for $\boldsymbol{\theta}_t$ in (A) are based on values of α over the original grid $\mathcal{A}_{t-1}$; what is required are posteriors based on the new grid $\mathcal{A}_t$. Formally, the unconditional posterior for $\boldsymbol{\theta}_t$ is defined by the quadrature formula

$$p(\boldsymbol{\theta}_t|D_t) = \sum_{j=1}^{k_t} p(\boldsymbol{\theta}_t|\alpha_{t,j}, D_t) w_{t,j} p(\alpha_{t,j}|D_t),$$

so that the standard form, conditional posteriors $p(\boldsymbol{\theta}_t|\alpha_{t,j}, D_t)$ are required. Again the answer to the problem of their calculation lies in the use of some form of interpolation. The approach in Pole and West (1988, 1990) uses direct linear interpolation. For each value $\alpha_{t,j} \in \mathcal{A}_t$, the required mean vector $\mathbf{m}_t(\alpha_{t,j})$ is calculated by linear interpolation between appropriate values in the set $\mathbf{m}_t(\alpha_{t-1,j})$ from (A). The posterior variance matrix is similarly linearly interpolated. See the above references for further details. Of course, linear interpolation is not possible outside the range of the old grid. To avoid this problem, the referenced works ensure that the two grids always have extreme or boundary points in common so that no interpolation is necessary at the extremes. This is always possible since the choice of the number of grid points is under control and extra points at the extremes can be added as desired. Once this interpolation is complete, the updating is complete with interpolated posteriors over the new grid

$$(\boldsymbol{\theta}_t|\alpha_{t,j}, D_t) \sim \mathrm{N}[\mathbf{m}_t(\alpha_{t,j}), \mathbf{C}_t(\alpha_{t,j})], \qquad (j = 1, \dots, k_t).$$

At this point, the components assumed available at time $t-1$ have all been updated to time t, and so the full updating analysis ends.

13.5 EFFICIENT INTEGRATION FOR GENERAL DYNAMIC MODELS

The ideas underlying the use of quadrature techniques for efficient integration have been extended in Pole and West (1988, 1990) to apply to much more general models than those above, restricted as they are to constant parameter non-linearities. In fact, the strategy presented in this reference is essentially as general as possible within the context of dynamic models having Markovian evolution of state vectors. Some such wider developments are briefly outlined here; the discussion involves widely useful concepts, and leads naturally into more recent and incisive methodological developments using simulation-based methods in the following section.

The primary need is for analyses of models in which parameters that introduce non-linearities are time dependent. If this is the case, then there is no way in which the above ideas of standard analyses conditional on non-linear parameters apply. To be specific, consider the model in (13.1), namely,

$$\text{Observation equation:} \qquad Y_t = F_t(\boldsymbol{\theta}_t) + \nu_t, \qquad (13.16)$$

$$\text{Evolution equation:} \qquad \boldsymbol{\theta}_t = \mathbf{g}_t(\boldsymbol{\theta}_{t-1}) + \boldsymbol{\omega}_t, \qquad (13.17)$$

where $F_t(.)$ is a known, non-linear regression function mapping the n-vector $\boldsymbol{\theta}_t$ to the real line, $\mathbf{g}_t(.)$ a known, non-linear vector evolution function, and ν_t and $\boldsymbol{\omega}_t$ are error terms subject to the usual zero-mean and normality assumptions with known variances V_t and $\mathbf{W}_t$. Suppose also that an initial prior density $p(\boldsymbol{\theta}_0|D_0)$ is specified as usual, though not necessarily in standard, normal form. Note that:

(a) The assumption of known variances is purely for convenience in notation and exposition here. If the observational variance is unknown, for example, then the vector $\boldsymbol{\theta}_t$ can be extended to include it as a component to be estimated along with the original elements of the state vector. In addition, the errors ν_t and $\boldsymbol{\omega}_t$ may, of course, have known, non-zero means.

(b) These equations define a rather general normal model. However, the normality of the errors is not a crucial assumption. Since the analysis is to be via numerical integration, there is no need to restrict to normality any longer. For example, the errors may have heavy-tailed distributions, such as T distributions, in order to permit larger changes in the state vector and outlying observations from time to time (West 1981; Pole and West 1988)

(c) In fact, the model need not be defined directly in terms of these two equations. A rather more general framework will specify the model via observational and evolution densities $p(Y_t|\boldsymbol{\theta}_t, D_{t-1})$ and $p(\boldsymbol{\theta}_t|\boldsymbol{\theta}_{t-1}, D_{t-1})$ respectively. This is very general. The observational distribution may be derived from an equation such as (13.16), but it allows for essentially any continuous or discrete distribution that may depend on the state vector in complicated, non-linear ways. Similarly, the evolution density is general. This too may take the usual DLM form, but the framework allows for any state evolution density, normal or otherwise, with any desired dependence on the state vector at $t-1$. The standard DLM is obviously a special case, as is the DLM with unknown observational variance, since the variance parameter can be incorporated into the state vector as a constant component. In fact, all models considered in this book can be written in the form of these two densities, hence, in principle, all models can be analysed within the framework here. In practice, this is currently not the case; the technical and computational difficulties of analysis have yet to be thoroughly investigated even in the simplest special cases.

Conceptually, the analysis of this model involves the following three steps at each time t, familiar from the usual normal DLM.

(1) Assuming knowledge of the posterior distribution for the state vector at time $t-1$, defined via a density function

$$p(\boldsymbol{\theta}_{t-1}|D_{t-1}), \tag{13.18}$$

compute the implied prior density for $\boldsymbol{\theta}_t$, namely

$$p(\boldsymbol{\theta}_t|D_{t-1}) = \int p(\boldsymbol{\theta}_t|\boldsymbol{\theta}_{t-1}, D_{t-1})p(\boldsymbol{\theta}_{t-1}|D_{t-1})d\boldsymbol{\theta}_{t-1}, \tag{13.19}$$

for all values of $\boldsymbol{\theta}_t$. Typically this implies the need for many numerical integrations over the n-dimensional state parameter space, one for each value of $\boldsymbol{\theta}_t$ required.

(2) Use this to produce forecast distributions one and more steps ahead. The one-step ahead density requires the further integrations implicit in the definition

$$p(Y_t|D_{t-1}) = \int p(Y_t|\boldsymbol{\theta}_t, D_{t-1})p(\boldsymbol{\theta}_t|D_{t-1})d\boldsymbol{\theta}_t. \tag{13.20}$$

Again the implication is that numerical integrations are needed here.

(3) Update the prior (13.19) to the posterior for $\boldsymbol{\theta}_t$ when Y_t is observed to complete the cycle at time t. As usual,

$$p(\boldsymbol{\theta}_t|D_t) \propto p(\boldsymbol{\theta}_t|D_{t-1})p(Y_t|\boldsymbol{\theta}_t, D_{t-1}), \tag{13.21}$$

for all $\boldsymbol{\theta}_t$, the constant of proportionality being the observed value of (13.20). Summarisation of (13.21) involves the calculation, in particular, of moments and posterior probabilities that require further integrations with respect to elements of $\boldsymbol{\theta}_t$. Subsidiary activities include filtering and smoothing, which are not discussed here (see Pole and West 1988).

Kitegawa (1987) describes implementation of the cycle defined in these three steps using the direct approach through numerical integration based on a fixed grid of values of $\boldsymbol{\theta}_t$. This approach requires that the grid of values, fixed for all time, remain appropriate for all time. We have commented earlier on the need for such grids to evolve as $\boldsymbol{\theta}_t$ evolves. Also, the very heavy computational demands of fixed grid methods with very large numbers of grid points in models of even very moderate size can be prohibitive. These points mitigate against any fixed grid-based method being of general applicability, and thus we explore the extension of efficient integration techniques to this context. The ideas underlying the numerical integration in Section 13.4 can be considered to apply to each of the steps summarised in equations (13.18)-(13.21), as follows.

Step 1:
Assume that $p(\boldsymbol{\theta}_{t-1}|D_{t-1})$ has been the subject of Gaussian quadrature, being evaluated on a grid of k_{t-1} values $\mathcal{A}_{t-1} = \{\boldsymbol{\theta}_{t-1,1}, \dots, \boldsymbol{\theta}_{t-1,k_{t-1}}\}$ with associated quadrature weights $w_{t-1,1}, \dots, w_{t-1,k_{t-1}}$. From the evolution equation, this provides an estimate of the prior mean $\mathbf{a}_t = \mathrm{E}[\boldsymbol{\theta}_t|D_{t-1}]$ via the usual quadrature formula

$$\mathbf{a}_t \approx \sum_{j=1}^{k_{t-1}} \mathbf{g}_t(\boldsymbol{\theta}_{t-1,j})w_{t-1,j}p(\boldsymbol{\theta}_{t-1,j}|D_{t-1}).$$

Given this estimate of $\mathbf{a}_t$, the prior variance matrix $\mathbf{R}_t = \mathrm{V}[\boldsymbol{\theta}_t|D_{t-1}]$ is estimated similarly by

$$\mathbf{R}_t \approx \sum_{j=1}^{k_{t-1}} [\mathbf{g}_t(\boldsymbol{\theta}_{t-1,j}) - \mathbf{a}_t][\mathbf{g}_t(\boldsymbol{\theta}_{t-1,j}) - \mathbf{a}_t]' w_{t-1,j} p(\boldsymbol{\theta}_{t-1,j}|D_{t-1}).$$

These initial estimates form the starting point for quadrature applied to the prior (13.19), using the general theory outlined and referenced in Section 13.3. The iterative strategy is applied to $\boldsymbol{\theta}_t$, the prior density at any point being evaluated via

$$p(\boldsymbol{\theta}_t|D_{t-1}) \approx \sum_{j=1}^{k_{t-1}} p(\boldsymbol{\theta}_t|\boldsymbol{\theta}_{t-1,j}, D_{t-1}) w_{t-1,j} p(\boldsymbol{\theta}_{t-1,j}|D_{t-1}). \tag{13.22}$$

Once the iterative strategy converges, the result is final estimates of $\mathbf{a}_t$ and $\mathbf{R}_t$, and a *prior* quadrature grid of some k_t points $\mathcal{A}_t = \{\boldsymbol{\theta}_{t,1}, \ldots, \boldsymbol{\theta}_{t,k_t}\}$ with associated quadrature weights $w_{t,1}, \ldots, w_{t,k_t}$. Note that there is no need for interpolation in this approach, the density actually being evaluated (approximately) at all points of interest.

Step 2:
Forecasting ahead simply repeats this basic procedure to produce approximations to distributions for future state vectors and hence for future observations. Restricting attention to one-step ahead, for any value Y_t the one-step density (13.20) is estimated via

$$p(Y_t|D_{t-1}) = \sum_{j=1}^{k_t} p(Y_t|\boldsymbol{\theta}_{t,j}, D_{t-1}) w_{t,j} p(\boldsymbol{\theta}_{t,j}|D_{t-1}).$$

Moments and predictive probabilities may be similarly evaluated.

Step 3:
Updating is directly achieved again using quadrature. The posterior density is evaluated at all points in $\mathcal{A}_t$ via (13.21), which then leads to initial estimates of posterior moments via the usual quadrature formula. These form the basis of a new set of grid points and weights, the density is evaluated at these points, and the process iterated to convergence. The result is a revision of the prior grid $\mathcal{A}_t$ to a final, posterior grid with associated final weights. Thus, the analysis at time t is complete, the position now being just at at the start of Step 1 with $t-1$ increased to t.

The key problem underlying this sort of analysis, and the basic way in which this application of efficient integration differs from more usual analyses, is that the posterior density for $\boldsymbol{\theta}_t$ at any time is used as an input into the model for future times. Pole and West (1988) discuss some of the problems that arise with its application, and illustrate them in DLMs in which the evolution and observation error terms have non-normal distributions. Some examples include the use of heavy-tailed T distributions for these errors rather than normal distributions, as in Masreliez (1975), Masreliez and Martin (1977), Kitagawa (1988), and West (1981).

13.6 A FIRST SIMULATION METHOD: ADAPTIVE IMPORTANCE SAMPLING

13.6.1 Introduction

The late 1980s and early 1990s saw tremendous growth in the development and use of methods of stochastic simulation for approximating posterior distributions in Bayesian inference quite generally. One class of tools, based on importance sampling ideas, was very much in evidence. This section discusses some application of adaptive importance sampling methods in dynamic models. The outlook is precisely that of Section 13.5, but now the tools are simulation-based rather than those of deterministic (Gaussian) quadrature. As we move towards the end of the twentieth century, simulation methods are becoming dominant tools in statistics, and it is likely that they will provide the way ahead in time series and dynamic modelling, as in other areas. Chapter 15 to follow picks up simulation methods with a focus on more recent developments using Markov chain implementations for analysis of a fixed stretch of data, in contrast to the sequential updating approach here based on importance sampling.

13.6.2 Posterior approximation by importance sampling

West (1992a) introduced an adaptive importance sampling scheme to develop discrete posterior approximations, and methods to provide smooth posterior reconstructions, in general statistical models. Suppose $p(\boldsymbol{\theta})$ is the continuous posterior density function for a continuous parameter vector $\boldsymbol{\theta}$. An approximating density $g(\boldsymbol{\theta})$ is used as an importance sampling function (Geweke 1989) as is now described. Let $\mathcal{T} = \{\boldsymbol{\theta}_j, \quad j = 1, \ldots, n\}$ be a random sample from $g(\boldsymbol{\theta})$, and define *weights* $\mathcal{W} = \{w_j, \quad j = 1, \ldots, n\}$ by $w_j = p(\boldsymbol{\theta}_j)/(kg(\boldsymbol{\theta}_j))$, for each j, where $k = \sum_{j=1}^{n} p(\boldsymbol{\theta}_j)/g(\boldsymbol{\theta}_j)$. The weights are evaluated via $w_j \propto p(\boldsymbol{\theta}_j)/g(\boldsymbol{\theta}_j)$ and then normalised to unit sum. Inference under $p(\boldsymbol{\theta})$ is then approximated using the discrete distribution having masses w_j at $\boldsymbol{\theta}_j$, for each $j = 1, \ldots, n$. To achieve reasonable approximations, we require that g have the same support as p and that the tails of g be heavier than those of p. Multivariate T distributions, and so-called split-T distributions, have become popular (Geweke 1989). In West (1992a) mixtures of T distributions are proposed, one additional reason being that mixtures have the flexibility to represent the possibly quite complex and varied forms of (posterior) densities. This is done using kernel density estimation techniques. With an importance sampling function $g(\boldsymbol{\theta})$ close to the true density $p(\boldsymbol{\theta})$, kernel density estimation (or other smoothing techniques) provides continuous estimates of joint and marginal densities. West (1992a) uses weighted variations on multivariate kernel estimates as importance sampling functions, and with some modification, to more directly estimate marginal densities of $p(\boldsymbol{\theta})$. The basic idea is as follows.

Given a chosen importance sampling density $g_0(\boldsymbol{\theta})$, the sample of size n, $\mathcal{T}$ and associated weights $\mathcal{W}$, the exact density $p(\boldsymbol{\theta})$ may be approximated by a *weighted* kernel estimate of the form

$$g_1(\boldsymbol{\theta}) = \sum_{j=1}^{n} w_j d(\boldsymbol{\theta}|\boldsymbol{\theta}_j, \mathbf{V}h^2), \tag{13.23}$$

where $d(\boldsymbol{\theta}|\mathbf{m}, \mathbf{M})$ denotes a p-variate, elliptically symmetric density function (determining the "kernel"), with mode $\mathbf{m}$ and scale matrix $\mathbf{M}$, $\mathbf{V}$ an estimate of the variance matrix of $p(\boldsymbol{\theta})$ (usually the Monte Carlo estimate based on $\mathcal{T}$ and $\mathcal{W}$), and h a "window-width" smoothing parameter, depending on the Monte Carlo sample size n. A key example has $d(\boldsymbol{\theta}|\mathbf{m}, \mathbf{M})$ as the density of a multivariate T distribution with some $a > 0$ degrees of freedom, whence the density $g_1(\boldsymbol{\theta})$ is a discrete mixture of n component T distributions. Conventional density estimation techniques (Silverman 1986) choose the window width h as a slowly decreasing function of n, so that the kernel components are naturally more concentrated about the locations $\boldsymbol{\theta}_j$ for larger sample sizes. Then $g_1(\boldsymbol{\theta})$ approaches $p(\boldsymbol{\theta})$ as n increases. For moderate sample sizes, $g_1(\boldsymbol{\theta})$ will tend to be overdispersed relative to $p(\boldsymbol{\theta})$ if $\mathbf{V}$ is the usual Monte Carlo estimate of $V[\boldsymbol{\theta}]$ under $p(\boldsymbol{\theta})$. This feature proves useful in developing kernel forms as importance density functions, though the over-dispersion can be counter-balanced using "shrinkage" of kernel locations to provide more direct approximation of the true density and its margins (West 1992a).

Adaptive importance sampling describes any process by which the importance sampling distribution is sequently revised based on information derived from successive Monte Carlo samples. Let $g_0(\boldsymbol{\theta})$ be an initially chosen importance sampling function. For a sample size n, this leads to points $\mathcal{T}_0 = \{\boldsymbol{\theta}_{0,j},\quad j = 1, \ldots, n_0\}$, weights $\mathcal{W}_0 = \{w_{0,j},\quad j = 1, \ldots, n_0\}$ and the summary

$$\mathcal{G}_0 = \{g_0, n_0, \mathcal{T}_0, \mathcal{W}_0\}.$$

Adaptive importance sampling suggests taking n_0 rather small, say several hundreds, and, based on the Monte Carlo outcome $\mathcal{G}_0$, revising $g_0(\boldsymbol{\theta})$ to a "better guess". It is natural to use a mixture such as (13.23) as a second step importance sampling function, and the following adaptive routine arises:

(1) Choose an initial importance sampling distribution with density $g_0(\boldsymbol{\theta})$, draw a fairly small sample n_0 and compute weights, deducing the summary $\mathcal{G}_0 = \{g_0, n_0, \mathcal{T}_0, \mathcal{W}_0\}$. Compute the Monte Carlo estimates $\bar{\boldsymbol{\theta}}_0$ and $\mathbf{V}_0$ of the mean and variance of $p(\boldsymbol{\theta})$.

(2) Construct a revised importance function $g_1(\boldsymbol{\theta})$ using (13.23) with sample size n_0, points $\boldsymbol{\theta}_{0,j}$, weights $w_{0,j}$, and variance matrix $\mathbf{V}_0$.
(3) Draw a larger sample of size n_1 from $g_1(\boldsymbol{\theta})$, and replace $\mathcal{G}_0$ with $\mathcal{G}_1 = \{g_1, n_1, \mathcal{T}_1, \mathcal{W}_1\}$.
(4) Either stop, and base inferences on $\mathcal{G}_1$, or proceed, if desired, to a further revised version $g_2(\boldsymbol{\theta})$, constructed similarly. Continue as desired.

Even though the initial $g_0(\boldsymbol{\theta})$ may poorly represent $p(\boldsymbol{\theta})$, successive refinement through smaller samples can, and usually does, mean that after one or two revisions, the resulting kernel estimate is a much better approximation to $p(\boldsymbol{\theta})$. Hence, once the process of refinement is terminated, a much smaller sample size is necessary for the desired accuracy of approximation. Often just one refinement is sufficient to adjust a very crude approximation, $g_0(\boldsymbol{\theta})$, say a single multivariate T density, to a mixture $g_1(\boldsymbol{\theta})$ of, say, 500 T densities, that much more closely represents the true $p(\boldsymbol{\theta})$. In approximating moments and probabilities, a Monte Carlo sample of two or three thousand draws from $g_1(\boldsymbol{\theta})$ may do as well as, or better than, a sample of two or three times that from the original $g_0(\boldsymbol{\theta})$. Useful diagnostic information is generated in this process at each stage, such as the configuration of points in each dimension of the parameter space and the distributions of weights. This can be used to guide successive choice of sample sizes and possible interventions to adjust successive kernel smoothing parameter values, and also to assess whether further refinement is likely to be additionally effective. Several illuminating examples appear in West (1992a).

13.6.3 Approximating mixtures by mixtures

Suppose the above adaptive strategy has $n_0 = 500$, so that $g_1(\boldsymbol{\theta})$ is a mixture of 500 p-dimensional T distributions, and that the revision process stops here, $g_1(\boldsymbol{\theta})$ being adopted as the "final" importance sampling density to be used for Monte Carlo inference. It is straightforward to sample $\mathcal{T}_1 = \{\boldsymbol{\theta}_{1,i}, \quad i = 1, \ldots, n_1\}$ from $g_1(\boldsymbol{\theta})$; the computational benefit of the components sharing a common scale matrix $\mathbf{V}_0$ is apparent here. It is also straightforward to then evaluate the weights $\mathcal{W}_1 = \{w_{1,i}, \quad i = 1, \ldots, n_1\}$, though the denominator of $w_{1,i}$ requires evaluation of the mixture $g_1(\boldsymbol{\theta}_{1,i})$. The computational burden here clearly increases if further refinement of importance functions with rather larger sample sizes is desired. One way of reducing such computations is to note that quite typically, approximating $p(\boldsymbol{\theta})$ using mixtures of several thousand T distributions is really overkill; even very irregular densities can be adequately matched using mixtures having far fewer components. The Monte Carlo kernel density construction in particular typically leads to a huge redundancy, with many points $\boldsymbol{\theta}_{0,j}$ closely grouped and contributing essentially similar components to the overall mixture density. The discussion of approximating mixtures in Chapter 12 is relevant.

A very basic method of "clustering" mixture components, combining ideas from each of these two references, is used in West (1992a). At the simplest, it involves reducing the number of components by replacing "nearest neighbouring" components with some form of average. The examples below, and those in West (1992a), involve reducing mixtures of n in several hundreds or thousands to around 10% (though sometimes rather less) of the initial value, and performing this reduction using the following simple clustering routine.

(1) Set $r = n$, and starting with the $r = n$ component mixture (13.23), choose $k < n$ as the number of components for the final mixture.
(2) Sort the r values of $\boldsymbol{\theta}_j$ in $\mathcal{T}$ in order of increasing values of weights w_j in $\mathcal{W}$; thus $\boldsymbol{\theta}_1$ corresponds to the component with smallest weight.
(3) Find the index i, $(i = 1, \ldots, r)$, such that $\boldsymbol{\theta}_i$ is the nearest neighbour of $\boldsymbol{\theta}_1$, and reduce the sets $\mathcal{T}$ and $\mathcal{W}$ to sets of size $r-1$ by removing components 1 and i, and inserting "average" values $\boldsymbol{\theta}_* = (w_1\boldsymbol{\theta}_1 + w_i\boldsymbol{\theta}_i)/(w_1 + w_i)$ and $w_* = w_1 + w_i$. Set $r = r - 1$.
(4) Proceed to (2) and repeat the procedure, stopping when $r = k$.
(5) The resulting mixture has the form (13.23) with n reduced to k, the locations based on the final k averaged values, with associated combined weights, the same scale matrix $\mathbf{V}$, but new and larger window-width h based on the current, reduced "sample size" r.

The reduction process can be monitored by evaluating and plotting margins of the mixture over fairly crude grids at stages in the reduction process. If in reducing from, say, n to $80\%n$ components these densities do not appear to change much, then further reduction may proceed, and so forth in later stages of reduction. This is one possible route to approximating mixtures of large numbers of components with mixtures of far fewer.

13.6.4 Sequential updating and dynamic models

Consider now the use of adaptive importance sampling techniques in the dynamic modelling context, where, as we have discussed in earlier sections, we need to allow "grids" of evaluation points in parameter spaces to be changed as time progresses and as data indicate support for different regions in parameter space. Other issues to be addressed include the severe computational demands in problems with more than very few parameters, and the difficulties in reconstructing smooth posterior distributions based on approximate evaluation at only very few points in what may be several dimensions.

We develop the ideas in a rather general context of possibly non-normal and non-linear dynamic models. The time series of (possibly multivariate) observations Y_t has, at each time t, a known sampling distribution with density (or mass function) $p(Y_t|\boldsymbol{\theta}_t)$; here $\boldsymbol{\theta}_t$ is the p-state vector, and conditional on $\boldsymbol{\theta}_t$, Y_t is assumed independent of Y_s and $\boldsymbol{\theta}_s$ for all past and future values of s, as usual in the DLM. The state vector evolves in time according to a known, Markovian process described by an evolution density $p(\boldsymbol{\theta}_t|\boldsymbol{\theta}_{t-1})$; given $\boldsymbol{\theta}_{t-1}$, $\boldsymbol{\theta}_t$ is conditionally independent of past values $\boldsymbol{\theta}_s$ for $s < t-1$. Note that the evolution equation, now defined implicitly, may be quite non-linear and non-normal. Summarising the model, for each $t = 1, 2, \ldots$, the defining equations are

$$\begin{aligned} \text{Observation model:} \quad & (Y_t|\boldsymbol{\theta}_t) \sim p(Y_t|\boldsymbol{\theta}_t), \\ \text{Evolution model:} \quad & (\boldsymbol{\theta}_t|\boldsymbol{\theta}_{t-1}) \sim p(\boldsymbol{\theta}_t|\boldsymbol{\theta}_{t-1}). \end{aligned}$$

Note that these densities may depend also on t and on elements of D_{t-1}, though the notation ignores this for clarity. At time $t-1$ prior to evolution, historical information is summarised through (an approximation to) the posterior $p(\boldsymbol{\theta}_{t-1}|D_{t-1})$. The primary computational problems addressed are

(a) **Evolution step**: compute the current prior for $\boldsymbol{\theta}_t$, defined via

$$p(\boldsymbol{\theta}_t|D_{t-1}) = \int p(\boldsymbol{\theta}_t|\boldsymbol{\theta}_{t-1})p(\boldsymbol{\theta}_{t-1}|D_{t-1})d\boldsymbol{\theta}_{t-1}; \tag{13.24}$$

(b) **Updating step**: on observing Y_t, compute the current posterior

$$p(\boldsymbol{\theta}_t|D_t) \propto p(\boldsymbol{\theta}_t|D_{t-1})p(Y_t|\boldsymbol{\theta}_t). \tag{13.25}$$

Subsidiary calculations include forecasting ahead (discussed below) and filtering problems for retrospection.

Suppose that $p(\boldsymbol{\theta}_{t-1}|D_{t-1})$ has been previously approximated via $\mathcal{G}_{t-1} = \{g_{t-1}, n_{t-1}, \mathcal{T}_{t-1}, \mathcal{W}_{t-1}\}$, where $g_{t-1}(\boldsymbol{\theta}_{t-1})$ is a (final) importance density used for inference about $(\boldsymbol{\theta}_{t-1}|D_{t-1})$; n_{t-1} is the (final) Monte Carlo sample size in that inference; $\mathcal{T}_{t-1} = \{\boldsymbol{\theta}_{t-1,i}, \quad i = 1, \ldots, n_{t-1}\}$ is the sample from $g_{t-1}(\boldsymbol{\theta}_{t-1})$; and $\mathcal{W}_{t-1} = \{w_{t-1,i}, \quad i = 1, \ldots, n_{t-1}\}$ is the set of associated weights. The objective is to perform the evolution, updating and forecasting computations and finally to summarise $p(\boldsymbol{\theta}_t|D_t)$ in terms of $\mathcal{G}_t = \{g_t, n_t, \mathcal{T}_t, \mathcal{W}_t\}$, and this may be done as follows.

Computations: evolution step

The following facts are of use in computations for the evolution step.

(a) Various features of the prior $p(\boldsymbol{\theta}_t|D_{t-1})$ of interest can be computed directly using the Monte Carlo structure $\mathcal{G}_{t-1}$. The prior mean, for example, is computable as

$$\mathrm{E}[\boldsymbol{\theta}_t|D_{t-1}] \approx \sum_{i=1}^{n_{t-1}} w_{t-1,i}\mathrm{E}_e[\boldsymbol{\theta}_t|\boldsymbol{\theta}_{t-1,i}],$$

where E_e stand for expectation under the evolution distribution, if this expectation is available in closed form.

(b) Similarly, the prior density function can be evaluated by Monte Carlo integration at any required point, viz.,

$$p(\boldsymbol{\theta}_t|D_{t-1}) \approx \sum_{i=1}^{n_{t-1}} w_{t-1,i}p(\boldsymbol{\theta}_t|\boldsymbol{\theta}_{t-1,i}). \tag{13.26}$$

(c) An initial Monte Carlo sample of size n_{t-1} may be drawn from the prior by generating one value of $\boldsymbol{\theta}_t$ from $p(\boldsymbol{\theta}_t|\boldsymbol{\theta}_{t-1,i})$, for each $i = 1, \ldots, n_{t-1}$. The resulting sample points, denoted by $\mathcal{T}_t^*$, provide starting values for the evaluation of the prior (13.26).

(d) This prior sample $\mathcal{T}_t^*$ may be used with weights $\mathcal{W}_{t-1}$ to construct a generalised kernel density estimate of the prior. In many models, this is of little interest since the prior may be evaluated as in (b) above, unless the evolution density is extremely complex and difficult to work with. Consider, however, a model in which the evolution equation is degenerate, $\boldsymbol{\theta}_t = G_t(\boldsymbol{\theta}_{t-1})$ with probability one, for some known, possibly non-linear and time-dependent function G_t. This covers many interesting examples. Then prior evaluation as in (b) is vacuous, and so, since values of the prior are required as input to Bayes' theorem in the updating step, some form of interpolation is needed.

(e) Note at this point, that subsidiary computations for forecasting ahead can be performed along these lines. Forecasting $(Y_t|D_{t-1})$, for example, requires computing $\int p(Y_t|\boldsymbol{\theta}_t)p(\boldsymbol{\theta}_t|D_{t-1})d\boldsymbol{\theta}_t$. The observation density typically has a standard form, so that Monte Carlo computations can be performed to approximate forecast moments and probabilities. For example, the forecast mean is evaluated as

$$\mathrm{E}[Y_t|D_{t-1}] \approx \sum_{\boldsymbol{\theta}_t \in \mathcal{T}_t^*} w_{t-1,i}\mathrm{E}_o[Y_t|\boldsymbol{\theta}_t],$$

where E_o stands for expectation under the observation distribution, assuming this expectation is available in closed form. Forecast probabilities are similarly derived. By simulating from the observation density for each value of $\boldsymbol{\theta}_t \in \mathcal{T}_t^*$, and using the associated weights $\mathcal{W}_{t-1}$, regions of interest in the sample space can be identified and the predictive density/mass function evaluated there. Similar comments apply to forecasting more than one step ahead; the process of simulating from the evolution density is repeated into the future, generating samples of future parameter vectors, and proceeding to inference about the future values of the series.

It is important to note the generality of the above strategy for computations. At no stage is it necessary, or interesting, to worry about functional forms of evolution equations, or to cater for many special cases. This simplifies programming the analysis; all that is needed is a collection of general routines for evaluating and sampling from the evolution and observation densities, and if required, generating kernel estimates.

Computations: updating step

Following evolution, the prior $p(\boldsymbol{\theta}_t|D_{t-1})$ to input to Bayes' theorem (13.25) for updating is available in approximate form either via evaluation at step (b) above, or in terms of a generalised kernel form as described in step (d). Updating may then proceed using adaptive Monte Carlo density estimation of the previous section. This begins with an initial "guess" at the posterior $p(\boldsymbol{\theta}_t|D_t)$, denoted $g_{t,0}(\boldsymbol{\theta}_t)$, to be used as an initial importance sampling function. Choice of this density depends on context, though it may be guided by general ideas described in the examples below. This provides the starting point for application of the adaptive strategy. On completion, this results in a final summary of the posterior given by the quadruple $\mathcal{G}_t = \{g_t, n_t, \mathcal{T}_t, \mathcal{W}_t\}$, where $g_t(\boldsymbol{\theta}_t)$ is the final importance sampling density for the posterior $p(\boldsymbol{\theta}_t|D_t)$, n_t is the final Monte Carlo sample size, and $\mathcal{T}_t$ is the set of n_t points in parameter space at which the Monte Carlo weights in $\mathcal{W}_t$ are evaluated.

A variety of examples are discussed in West (1992a, b and c), to which the reader is referred for further details and illustrations. This specific simulation method is based on key general principles for sequential analysis, and it provides a useful strategy for models with fairly low-dimensional state vectors. Beyond a few dimensions, the computational overheads increase substantially, and there are currently no generally useful extensions or alternatives for sequential, dynamic model analysis. However, alternative simulation methods using iterative Markov chain approaches are of immense utility in higher dimensions, so long as the focus on sequential updating is relinquished in favour of retrospective analysis of a fixed stretch of data. This is the topic of Chapter 15, to follow, after a diversion into some specific non-linear models analysed with direct analytic approximations in sequential contexts.

CHAPTER 14

EXPONENTIAL FAMILY DYNAMIC MODELS

14.1 INTRODUCTION

We now turn to a specific class of dynamic models with non-normal sampling distributions and other non-linear components. DLMs have been extended and generalised to various non-normal problems, the largest class being that based on the use of *exponential family models* for observational distributions. This chapter is devoted to these models, the primary references being Migon (1984), Migon and Harrison (1985), and West, Harrison and Migon (1985). Our development follows these references and explores technical issues in analyses that utilise direct analytic approximations. As discussed in the previous chapter, research in more recent years has developed more refined numerical and simulation-based approaches to the analysis of non-linear models, and these kinds of developments are beginning to have practical impact in the class of models discussed here, and in generalisations of them. Though not developed in this chapter, this is a currently active area of dynamic modelling and one that can be expected to grow in coming years. Further discussion of simulation methods appears in Chapter 15, with some recent references to developments with models related to those of this chapter.

A primary need for extension of the class of normal DLMs is to allow for data distributions that are not likely to be adequately modelled using normality, even after transformation. With continuous data plausibly having skewed distributions such as exponential, Weibull, or gamma, it is often the case that a data transformation can be used to achieve approximate symmetry on the transformed scale, and normal models can then be quite useful. However, the meaning and interpretation of model parameters is usually obscured through transformation, the original data scale being natural and interpretable, and it may often be viewed as desirable to develop a model directly for the data on the original scale. See also the related comments in Section 10.6. With discrete data in the form of counts, transformations are often not sensible, and approximation using continuous normal distributions may be radically inappropriate. Discrete data often arise in the form of counts from finite or conceptually infinite distributions and event indicators. Sample surveys, for example, typically result in data that are essentially binomial-like in character. Poisson and compound Poisson data arise in demand and inventory processes through "random" arrival of orders and supplies. At the extreme, binary series arise as indicators of the occurrence, or non-occurrence, of sequences of events, such as rise/fall in a financial index, daily rainfall indicators, and so forth. Thus, we need to consider non-normal observational distributions.

Major progress in non-normal modelling, and hence in practical statistical methodology, has flowed from the development of the theoretical framework of generalised linear models (GLMs), beginning with Nelder and Wedderburn (1972). The ensuing development of the interactive GLIM computer package (Baker and Nelder 1985) presaged an explosion in the application and development of formal, parametric statistical models. The GLM framework provides regression techniques in the context of non-normal models with regression effects on a non-linear scale, one of the most important modelling tools for applied statisticians and researchers. See McCullagh and Nelder (1989) for coverage of theoretical and applied aspects of these and related models, and West (1985a) for Bayesian modelling within this class. The theoretical framework of GLMs provides a starting point for non-normal, dynamic models. Here we describe such models, beginning with an exposition of the basic structure of exponential family analyses.

14.2 EXPONENTIAL FAMILY MODELS

14.2.1 Observational distributions in the EF

Consider the time series of scalar observations Y_t, $(t = 1, 2, \dots)$, continuous or discrete random quantities taking values in the sample space $\mathcal{Y}$. If Y_t is assumed to have a sampling distribution in the *exponential family*, then the density (if discrete, the probability mass function) of Y_t may be described as follows. For some defining quantities η_t and V_t, and three known functions $y_t(Y_t)$, $a(\eta_t)$ and $b(Y_t, V_t)$, the density is

$$p(Y_t|\eta_t, V_t) = \exp\{V_t^{-1}[y_t(Y_t)\eta_t - a(\eta_t)]\}b(Y_t, V_t), \qquad (Y_t \in \mathcal{Y}). \quad (14.1)$$

The primary properties of the distribution may be found in the references above. These properties, some terminology and further points of note are summarised as follows.

(1) η_t is the *natural parameter* of the distribution, a continuous quantity.
(2) $V_t > 0$ is a *scale parameter*; the *precision parameter* of the distribution is defined as $\phi_t = V_t^{-1}$.
(3) As a function of the natural parameter for fixed Y_t, equation (14.1), viewed as a likelihood for η_t, depends on Y_t through the transformed value $y_t(Y_t)$.
(4) The function $a(\eta_t)$ is assumed twice differentiable in η_t. It follows that

$$\mu_t = \mathrm{E}[y_t(Y_t)|\eta_t, V_t] = \frac{da(\eta_t)}{d\eta_t} = \dot{a}(\eta_t).$$

(5) It follows that

$$\mathrm{V}[y_t(Y_t)|\eta_t, V_t] = V_t\ddot{a}(\eta_t).$$

(6) Rather often (in the key practical cases in particular) $y_t(\cdot)$ is the identity function. In such cases,

$$p(Y_t|\eta_t, V_t) = \exp\{V_t^{-1}[Y_t\eta_t - a(\eta_t)]\}b(Y_t, V_t) \tag{14.2}$$

for $Y_t \in \mathcal{Y}$. Also

$$\mathrm{E}[Y_t|\eta_t, V_t] = \mu_t = \dot{a}(\eta_t), \tag{14.3}$$

$$\mathrm{V}[Y_t|\eta_t, V_t] = V_t\ddot{a}(\eta_t). \tag{14.4}$$

Here, in an obvious terminology, $\dot{a}(\cdot)$ is the *mean function* and $\ddot{a}(\cdot)$ the *variance function* of the distribution. Thus, the precision parameter $\phi_t = V_t^{-1}$ appears as a divisor of the variance function to provide the variance in (14.4). The function $a(\cdot)$ is assumed convex, so that $\dot{a}(\cdot)$ is a monotonically increasing function and η_t and μ_t are related via the one-to-one transformation (14.3); the inverse is simply $\eta_t = \dot{a}^{-1}(\mu_t)$.

EXAMPLE 14.1. The usual normal model $(Y_t|\mu_t, V_t) \sim \mathrm{N}[\mu_t, V_t]$ is a key special case. Here $y_t(Y_t) = Y_t$, $a(\eta_t) = \eta_t^2/2$ so that $\mu_t = \eta_t$, and $b(Y_t, V_t) = (2\pi V_t)^{-1/2}\exp(-Y_t^2/2)$.

EXAMPLE 14.2. Consider the binomial model where Y_t is the number of successes in $n_t > 0$ Bernoulli trials with success probability π_t. Here $\mathcal{Y}$ is the positive integers, and the mass function is

$$p(Y_t|\mu_t, n_t) = \begin{cases} \binom{n_t}{Y_t}\mu_t^{Y_t}(1-\mu_t)^{n_t-Y_t}, & (Y_t = 0, 1, \ldots, n_t), \\ 0, & \text{otherwise}. \end{cases}$$

This is a special case of (14.2) in which $y_t(Y_t) = Y_t/n_t$, $\eta_t = \log[\mu_t/(1-\mu_t)]$ (the *log-odds*, or *logistic* transform of the probability μ_t), $V_t^{-1} = \phi_t = n_t$, $a(\eta_t) = \log[1 + \exp(\eta_t)]$, and $b(Y_t, V_t) = \binom{n_t}{Y_t}$.

Several other important distributions, including Poisson and gamma, are also special cases. It is left to the exercises for the reader to verify that in each such case, the density or mass function can be written in exponential family form. Thus, the development for the general model (14.1) or (14.2) applies to a wide and useful variety of distributions.

14.2.2 Conjugate analyses

The development will, with no loss of generality, be in terms of (14.2). In addition, the scale parameter V_t is assumed known for all t. So in modelling the observation, the only unknown quantity in the sampling density (14.2) is the natural parameter η_t, or, equivalently, the conditional mean μ_t of Y_t. At time $t-1$, historical information relevant to forecasting Y_t is denoted, as usual, by D_{t-1}. Note that the density for Y_t may depend in some way on D_{t-1} (in particular, through the assumed value of V_t), so that (14.2) provides $p(Y_t|\eta_t, V_t, D_{t-1})$. For notational convenience V_t is dropped from the list of conditioning arguments as it is assumed known. Thus, the sampling density (14.1) or (14.2) is denoted simply by $p(Y_t|\eta_t)$, $(Y_t \in \mathcal{Y})$, the dependence on D_{t-1} being understood.

Now the only uncertainty about the distribution for Y_t given the past history D_{t-1} is due to the uncertainty about η_t. This will be summarised in terms of a prior (to time t) for η_t, the density denoted by $p(\eta_t|D_{t-1})$ as usual. It then follows, as usual, that the one-step ahead forecast distribution is defined via

$$p(Y_t|D_{t-1}) = \int p(Y_t|\eta_t)p(\eta_t|D_{t-1})d\eta_t, \tag{14.5}$$

the integration being over the full parameter space for η_t, a subset of the real line. Similarly, once Y_t is observed, the prior is updated to a posterior for η_t by Bayes' Theorem,

$$p(\eta_t|D_t) \propto p(\eta_t|D_{t-1})p(Y_t|\eta_t). \tag{14.6}$$

The calculations in (14.5) and (14.6), so familiar and easily performed in the cases of normal models, are analytically tractable in this exponential family framework when the prior belongs to the *conjugate family*. The use of exponential models with conjugate priors is well known in Bayesian statistics generally; see Aitchison and Dunsmore (1976, Chapter 2) for a thorough discussion and technical details.

With reference to (14.2), a prior density from the conjugate family has the form

$$p(\eta_t|D_{t-1}) = c(r_t, s_t)\exp[r_t\eta_t - s_t a(\eta_t)], \tag{14.7}$$

for some defining quantities r_t and s_t (known functions of D_{t-1}), and a known function $c(.\ ,.)$ that provides the normalising constant $c(r_t, s_t)$. Some comments and properties are as follows.

(1) Given the quantities r_t and s_t, the conjugate prior is completely specified. Here $s_t > 0$ and defining $x_t = r_t/s_t$, equation (14.7) can be written as

$$p(\eta_t|D_{t-1}) \propto \exp\{s_t[x_t\eta_t - a(\eta_t)]\}.$$

As a function of η_t, the prior thus has the same form as the likelihood function (14.2) when Y_t is fixed; s_t is analogous to ϕ_t and x_t to Y_t.

(2) All such conjugate priors are unimodal (in fact, *strongly* unimodal) by virtue of the fact that $a(\cdot)$ is a convex function. The prior mode for η_t is defined via $x_t = \dot{a}(\eta_t)$, and is therefore just $\dot{a}^{-1}(x_t)$. Thus, x_t defines the *location* of the prior.

(3) s_t is the *precision parameter* of the prior; larger values of $s_t > 0$ imply a prior increasingly concentrated about the mode.

(4) The roles of x_t and s_t as defining location and precision are further evident in cases when (as is often true) the prior density and its

first derivative decay to zero in the tails. Under such conditions, it is easily verified that $\mathrm{E}[\mu_t|D_{t-1}] = x_t$. It follows that the one-step ahead forecast mean of Y_t is given by $\mathrm{E}[Y_t|D_{t-1}] = x_t$. In addition, $\mathrm{V}[\mu_t|D_{t-1}] = \mathrm{E}[\ddot{a}(\eta_t)|D_{t-1}]/s_t$. See West (1985a, 1986a) for further details.

Assuming that the values r_t and s_t are specified, it easily follows (and the verification is left to the reader) that the predictive density (14.5) and the posterior (14.6) are determined by

$$p(Y_t|D_{t-1}) = \frac{c(r_t, s_t)b(Y_t, V_t)}{c(r_t + \phi_t Y_t, s_t + \phi_t)} \tag{14.8}$$

and

$$p(\eta_t|D_t) = c(r_t + \phi_t Y_t, s_t + \phi_t)\exp[(r_t + \phi_t Y_t)\eta_t - (s_t + \phi_t)a(\eta_t)]. \tag{14.9}$$

The predictive distribution may now be easily calculated through (14.8). The posterior (14.9) has the same form as the conjugate prior, with defining quantities r_t and s_t updated to $r_t + \phi_t Y_t$ and $s_t + \phi_t$ respectively. The posterior precision parameter is thus the sum of the prior precision s_t and the precision parameter $\phi_t = V_t^{-1}$ of the sampling model for Y_t. Hence precision increases with data since $\phi_t > 0$. The prior location parameter $x_t = r_t/s_t$ is updated to

$$\frac{r_t + \phi_t Y_t}{s_t + \phi_t} = \frac{s_t x_t + \phi_t Y_t}{s_t + \phi_t} = (1 - \alpha_t)x_t + \alpha_t Y_t,$$

where $\alpha_t = \phi_t/(s_t + \phi_t)$, so that $0 < \alpha_t < 1$. Thus, the posterior location parameter is a convex linear combination of the prior location parameter x_t and the observation Y_t. The weight α_t on the observation in the combination is an increasing function of the relative precision ϕ_t/s_t.

Full discussion of conjugate analyses can be found in Aitchison and Dunsmore (1976, Chapter 2), De Groot (1970, Chapter 9). Further details of the properties of conjugate priors appear in West (1985a, 1986a). The normal and binomial cases of Examples 14.1 and 14.2 are illustrated here, being special cases of great practical interest. Further special cases are left to the exercises for the reader. See Aitchison and Dunsmore for a full list and a complete summary of prior, posterior and forecast distributions.

EXAMPLE 14.1 (continued). In the normal model the conjugate prior is also normal, $(\eta_t|D_{t-1}) \sim \mathrm{N}[x_t, s_t^{-1}]$. The forecast density in (14.8) is normal, $(Y_t|D_{t-1}) \sim \mathrm{N}[f_t, Q_t]$ with $f_t = x_t$ and $Q_t = V_t + s_t^{-1}$. So, of course, is the posterior (14.9). The posterior mean of $\eta_t = \mu_t$ may be written in the familiar form $x_t + \alpha_t(Y_t - x_t)$, the prior mean plus a weighted correction proportional to the forecast error $Y_t - x_t$. The posterior variance is $(s_t + \phi_t)^{-1} = s_t^{-1} - \alpha_t^2 Q_t$.

EXAMPLE 14.2 (continued). In the binomial model, the conjugate prior $\mu_t = [1 + \exp(-\eta_t)]^{-1}$ is beta, namely $(\mu_t|D_{t-1}) \sim \mathrm{Beta}[r_t, s_t]$. This has density

$$p(\mu_t|D_{t-1}) \propto \mu_t^{r_t-1}(1-\mu_t)^{s_t-1}, \qquad (0 \le \mu_t \le 1),$$

with normalising constant

$$c(r_t, s_t) = \frac{\Gamma(r_t + s_t)}{\Gamma(r_t)\Gamma(s_t)}.$$

Here both r_t and s_t must be positive. Transforming to the natural parameter $\eta_t = \log[\mu_t/(1-\mu_t)]$ provides (14.7), although it is not necessary to work on the η_t scale, and much more convenient to work directly with the conjugate beta on the μ_t scale. The forecast distribution (14.8) is the beta-binomial, and the posterior (14.9) beta, $(\mu_t|D_{t-1}) \sim$ Beta$[r_t + Y_t, s_t + n_t - Y_t]$.

14.3 DYNAMIC GENERALISED LINEAR MODELS

14.3.1 Dynamic regression framework

The class of generalised linear models (Nelder and Wedderburn 1972; Baker and Nelder 1983; West 1985a) assumes data conditionally independently drawn from distributions with common exponential family form, but with parameters η_t and V_t possibly differing. Regression effects of independent variables are modelled by relating the natural parameters for each observation to a linear function of regression variables. In the time series context, the use of time-varying regression type models is appropriate, applying to define the following *dynamic generalised linear model.*

Definition 14.1. Define the following quantities at time t.

- $\boldsymbol{\theta}_t$, an n-dimensional state vector at time t;
- $\mathbf{F}_t$, a known n-dimensional regression vector;
- $\mathbf{G}_t$, a known $n \times n$ evolution matrix;
- $\boldsymbol{\omega}_t$, an n-vector of evolution errors having zero mean and known variance matrix $\mathbf{W}_t$, denoted by $\boldsymbol{\omega}_t \sim [\,\mathbf{0}, \mathbf{W}_t]$;
- $\lambda = \mathbf{F}_t'\boldsymbol{\theta}_t$, a linear function of the state vector parameters;
- $g(\eta_t)$, a known, continuous and monotonic function mapping η_t to the real line.

Then the dynamic generalised linear model (DGLM) for the series Y_t, $(t = 1, 2, \dots)$ is defined by the following components:

Observation model:

$$p(Y_t|\eta_t) \quad \text{and} \quad g(\eta_t) = \lambda_t = \mathbf{F}_t'\boldsymbol{\theta}_t, \tag{14.10}$$

Evolution equation:

$$\boldsymbol{\theta}_t = \mathbf{G}_t\boldsymbol{\theta}_{t-1} + \boldsymbol{\omega}_t \quad \text{with} \quad \boldsymbol{\omega}_t \sim [\,\mathbf{0}, \mathbf{W}_t]. \tag{14.11}$$

This definition is an obvious extension of the standard DLM in the observation model. The sampling distribution is now possibly non-normal, and the linear regression λ_t affects the observational distribution through a possibly non-linear function of the mean response function $\mu_t = \mathrm{E}[Y_t|\eta_t] = \dot{a}(\eta_t)$. The additional component $g(\cdot)$ provides a transformation from the natural parameter space to that of the linear regression. In practice, it is very common that this will be the identity map. This provides, for example, logistic regression in binomial models and log-linear models when the sampling distribution is Poisson. See McCullagh and Nelder (1989) for much further discussion. The evolution equation for the state vector is precisely as in the normal DLM, although normality is not assumed, only the first and second order moments of the evolution errors being so far assumed. As in the DLM, the zero mean assumption for $\boldsymbol{\omega}_t$ may be relaxed to include a known, non-zero mean without any essential change to the structure of the model. These evolution errors are assumed uncorrelated over time, as usual, and conditional on η_t, Y_t is independent of $\boldsymbol{\omega}_t$. The standard, static GLM is a special case in which $\boldsymbol{\theta}_t = \boldsymbol{\theta}$ for all time, given by taking $\mathbf{G}_t$ as the identity and $\mathbf{W}_t$ as the zero matrix.

This class of models provides a generalisation of the DLM to non-normal error models for the time series, and of the GLM to time varying parameters. However, having left the standard, normal/linear framework, analysis becomes difficult, the nice theory of normal models being lost. Thus, there is a need for some form of approximation in analysis. The possibilities, as in the non-linear models of Chapter 13, include (a) analytic approximations, such as the use of transformations of the data to approximate normality if sensible, and normal approximations for prior/posterior distributions; and (b) numerical approximations, using some form of numerical integration technique. An approach of type (a) is described here, that attempts to retain the essential non-normal features of the observational model and forecast distributions by exploiting the conjugate analysis of Section 14.2.2. This analysis follows West, Harrison and Migon (1985). Other possible approaches, such as the use of normal approximations of various forms, are certainly possible, though at the time of writing, these remain largely unexplored and so are not discussed here.

14.3.2 The DLM revisited

The components of the analysis are motivated by first considering a reformulation of the standard, exact analysis in the special case of the normal DLM.

In the special case that $p(Y_t|\eta_t)$ is the normal density, $(Y_t|\eta_t) \sim \mathrm{N}[\mu_t, V_t]$ with $\mu_t = \eta_t$, the DLM is obtained by taking $g(\eta_t) = \eta_t$, so that $\mu_t = \eta_t = \lambda_t = \mathbf{F}_t'\boldsymbol{\theta}_t$. We work in terms of the μ_t notation. In the usual sequential analysis, the model at time t is completed by the posterior for the state vector at $t-1$, namely

$$(\boldsymbol{\theta}_{t-1}|D_{t-1}) \sim \mathrm{N}[\mathbf{m}_{t-1}, \mathbf{C}_{t-1}], \tag{14.12}$$

where the moments are known functions of the past information D_{t-1}. Also, the evolution errors are normal, and so given the assumption of no correlation between $\boldsymbol{\omega}_t$ and $\boldsymbol{\theta}_{t-1}$ in (14.12), the prior for $\boldsymbol{\theta}_t$ is deduced as

$$(\boldsymbol{\theta}_t|D_{t-1}) \sim \mathrm{N}[\mathbf{a}_t, \mathbf{R}_t], \tag{14.13}$$

where $\mathbf{a}_t = \mathbf{G}_t\mathbf{m}_{t-1}$ and $\mathbf{R}_t = \mathbf{G}_t\mathbf{C}_{t-1}\mathbf{G}_t' + \mathbf{W}_t$. So much is standard theory. From (14.13) and the normal sampling model, the one-step ahead forecast distribution for Y_t and the posterior for $\boldsymbol{\theta}_t$ are obtained via the usual equations.

Consider, however, the following alternative derivations of these key components of the analysis.

Step 1: Prior for μ_t.

$\mu_t = \lambda_t = \mathbf{F}_t'\boldsymbol{\theta}_t$ is a linear function of the state vector. Hence, under the prior (14.13), μ_t and $\boldsymbol{\theta}_t$ have a joint (singular) normal prior distribution

$$\left(\begin{matrix}\mu_t \\ \boldsymbol{\theta}_t\end{matrix} \,\middle|\, D_{t-1}\right) \sim \mathrm{N}\left[\begin{pmatrix} f_t \\ \mathbf{a}_t\end{pmatrix}, \begin{pmatrix} q_t & \mathbf{F}_t'\mathbf{R}_t \\ \mathbf{R}_t\mathbf{F}_t & \mathbf{R}_t\end{pmatrix}\right], \tag{14.14}$$

where

$$f_t = \mathbf{F}_t'\mathbf{a}_t \qquad \text{and} \qquad q_t = \mathbf{F}_t'\mathbf{R}_t\mathbf{F}_t. \tag{14.15}$$

Step 2: One-step ahead forecasting.

The sampling distribution of Y_t depends on $\boldsymbol{\theta}_t$ only via the single quantity μ_t, and thus the historical information relevant to forecasting Y_t is completely summarised in the marginal prior $(\mu_t|D_{t-1}) \sim \mathrm{N}[f_t, q_t]$. The one-step ahead forecast distribution is given in Example 14.1 as $(Y_t|D_{t-1}) \sim \mathrm{N}[f_t, Q_t]$, where $Q_t = q_t + V_t$. The reader will verify that this is the usual result.

Step 3: Updating for μ_t.

Observing Y_t, Example 14.1 provides the posterior for μ_t as normal,

$$(\mu_t|D_t) \sim \mathrm{N}[f_t^*, q_t^*],$$

where

$$f_t^* = f_t + (q_t/Q_t)e_t \qquad \text{and} \qquad q_t^* = q_t - q_t^2/Q_t, \tag{14.16}$$

with forecast error $e_t = Y_t - f_t$.

Step 4: Conditional structure for $(\boldsymbol{\theta}_t|\mu_t, D_{t-1})$.

The objective of the updating is to calculate the posterior for $\boldsymbol{\theta}_t$. This can be derived from the *joint* posterior for μ_t and $\boldsymbol{\theta}_t$, namely

$$\begin{aligned} p(\mu_t, \boldsymbol{\theta}_t|D_t) &\propto p(\mu_t, \boldsymbol{\theta}_t|D_{t-1})\ p(Y_t|\mu_t) \\ &\propto [p(\boldsymbol{\theta}_t|\mu_t, D_{t-1})p(\mu_t|D_{t-1})]\ p(Y_t|\mu_t) \\ &\propto p(\boldsymbol{\theta}_t|\mu_t, D_{t-1})\ [p(\mu_t|D_{t-1})p(Y_t|\mu_t)] \\ &\propto p(\boldsymbol{\theta}_t|\mu_t, D_{t-1})\ p(\mu_t|D_t). \end{aligned}$$

Hence, given μ_t, $\boldsymbol{\theta}_t$ is conditionally independent of Y_t, and it follows that

$$p(\boldsymbol{\theta}_t|D_t) = \int p(\boldsymbol{\theta}_t|\mu_t, D_{t-1})p(\mu_t|D_t)d\mu_t. \tag{14.17}$$

The key point is that information about the state vector from Y_t is channelled through the posterior for μ_t. The first component in the integrand is, using standard normal theory in (14.14), just the conditional normal distribution

$$(\boldsymbol{\theta}_t|\mu_t, D_{t-1}) \sim \mathrm{N}[\mathbf{a}_t + \mathbf{R}_t\mathbf{F}_t(\mu_t - f_t)/q_t, \mathbf{R}_t - \mathbf{R}_t\mathbf{F}_t\mathbf{F}_t'\mathbf{R}_t/q_t], \tag{14.18}$$

for all μ_t. The second is that derived in Step 3 above.

Step 5: Updating for $\boldsymbol{\theta}_t$.

Now, since all components are normal, the required posterior (14.17) is normal and is thus defined by the mean and variance matrix. By analogy with (14.17), we can express these as follows. Firstly, the mean is given by

$$\begin{aligned} \mathbf{m}_t &= \mathrm{E}[\boldsymbol{\theta}_t|D_t] \\ &= \mathrm{E}[\mathrm{E}\{\boldsymbol{\theta}_t|\mu_t, D_{t-1}\}|D_t] \\ &= \mathrm{E}[\mathbf{a}_t + \mathbf{R}_t\mathbf{F}_t(\mu_t - f_t)/q_t|D_t] \\ &= \mathbf{a}_t + \mathbf{R}_t\mathbf{F}_t(\mathrm{E}[\mu_t|D_t] - f_t)/q_t \\ &= \mathbf{a}_t + \mathbf{R}_t\mathbf{F}_t(f_t^* - f_t)/q_t. \end{aligned} \tag{14.19}$$

Similarly,

$$\begin{aligned} \mathbf{C}_t &= \mathrm{V}[\boldsymbol{\theta}_t|D_t] \\ &= \mathrm{V}[\mathrm{E}\{\boldsymbol{\theta}_t|\mu_t, D_{t-1}\}|D_t]\ +\ \mathrm{E}[\mathrm{V}\{\boldsymbol{\theta}_t|\mu_t, D_{t-1}\}|D_t] \\ &= \mathrm{V}[\mathbf{a}_t + \mathbf{R}_t\mathbf{F}_t(\mu_t - f_t)/q_t|D_t]\ +\ \mathrm{E}[\mathbf{R}_t - \mathbf{R}_t\mathbf{F}_t\mathbf{F}_t'\mathbf{R}_t/q_t|D_t] \\ &= \mathbf{R}_t\mathbf{F}_t\mathbf{F}_t'\mathbf{R}_t\mathrm{V}[\mu_t|D_t]/q_t^2\ +\ \mathbf{R}_t - \mathbf{R}_t\mathbf{F}_t\mathbf{F}_t'\mathbf{R}_t/q_t \\ &= \mathbf{R}_t - \mathbf{R}_t\mathbf{F}_t\mathbf{F}_t'\mathbf{R}_t(1 - q_t^*/q_t)/q_t. \end{aligned} \tag{14.20}$$

Substituting the values of f_t^* and q_t^* from (14.15) leads to the usual expressions $\mathbf{m}_t = \mathbf{a}_t + \mathbf{A}_t e_t$ and $\mathbf{C}_t = R_t - \mathbf{A}_t\mathbf{A}_t'Q_t$, where $\mathbf{A}_t = \mathbf{R}_t\mathbf{F}_t/Q_t$ is the usual adaptive vector.

Steps 1 to 5 provide an alternative proof of the usual forecasting and updating equations at time t, this route being rather circuitous relative to the usual, direct calculations. However, in non-normal models, an approximate analysis along these lines leads to a parallel analysis for DGLMs.

14.3.3 DGLM updating

Return now to the DGLM of Definition 14.1. Given the non-normality of the observational model and in general the non-linearity of the observation mean μ_t as a function of $\boldsymbol{\theta}_t$, there is no general, exact analysis. The approach in West, Harrison and Migon (1985) develops as follows, paralleling the steps above in the DLM.

Definition 14.1 provides the basic observation and evolution model at time t. To complete the model specification for time t, we need to fully define two more component distributions: (a) that of the evolution error $\boldsymbol{\omega}_t$, as yet only specified in terms of mean and variance matrix; and (b) $p(\boldsymbol{\theta}_{t-1}|D_{t-1})$ that sufficiently summarises the historical information and analysis prior to time t. In the DLM, of course, these are both normal. Whatever forms they may take in the DGLM, prior and posterior distributions for the state vector at any time will not now be normal. Without considerable development of numerical integration-based methodology for the DGLM, there is no way in which, in any generality, such distributions can be adequately calculated and summarised. West, Harrison and Migon (1985) proceed to develop an approximate analysis based on assuming that these required distributions are only partially specified in terms of their first- and second- order moments.

Assumption: In the DGLM of Definition 14.1, suppose that the distribution of $\boldsymbol{\omega}_t$ is unspecified apart from the moments in (14.11). In addition, suppose that the posterior mean and variance of $(\boldsymbol{\theta}_{t-1}|D_{t-1})$ are as in the DLM case (14.12) but without the normality of the distribution.

Thus, the model definition is completed at time t via

$$(\boldsymbol{\theta}_{t-1}|D_{t-1}) \sim [\mathbf{m}_{t-1}, \mathbf{C}_{t-1}], \tag{14.21}$$

the full distributional form being unspecified. It follows that from (14.11), the prior moments of $\boldsymbol{\theta}_t$ are

$$(\boldsymbol{\theta}_t|D_{t-1}) \sim [\mathbf{a}_t, \mathbf{R}_t], \tag{14.22}$$

where $\mathbf{a}_t = \mathbf{G}_t\mathbf{m}_{t-1}$ and $\mathbf{R}_t = \mathbf{G}_t\mathbf{C}_{t-1}\mathbf{G}_t' + \mathbf{W}_t$. This parallels the normal theory; the price paid for the greater generality in not assuming normality is that full distributional information is lost, the prior being only partially specified. The parallel with normal theory is now developed in each of the five steps of the previous section. Now the notation is a little more complex, and worth reiterating. As usual, μ_t is the observation mean, related to the natural parameter η_t via $\mu_t = \dot{a}(\eta_t)$. Similarly, η_t relates to the linear function $\lambda_t = \mathbf{F}_t'\boldsymbol{\theta}_t$ of state parameters via $\lambda_t = g(\eta_t)$. Since $\dot{a}(\cdot)$ and $g(\cdot)$ are bijective, we can (and do) work in terms of μ_t, η_t, or λ_t interchangeably.

Step 1: Prior for λ_t.

$\lambda_t = \mathbf{F}_t'\boldsymbol{\theta}_t$ is a linear function of the state vector. Hence, under (14.22), λ_t and $\boldsymbol{\theta}_t$ have a joint prior distribution that is only partially specified in terms of moments

$$\left(\begin{matrix}\lambda_t \\ \boldsymbol{\theta}_t\end{matrix} \,\middle|\, D_{t-1}\right) \sim \left[\begin{pmatrix} f_t \\ \mathbf{a}_t \end{pmatrix}, \begin{pmatrix} q_t & \mathbf{F}_t'\mathbf{R}_t \\ \mathbf{R}_t\mathbf{F}_t & \mathbf{R}_t \end{pmatrix}\right], \tag{14.23}$$

where

$$f_t = \mathbf{F}_t'\mathbf{a}_t \qquad \text{and} \qquad q_t = \mathbf{F}_t'\mathbf{R}_t\mathbf{F}_t. \tag{14.24}$$

These moments coincide precisely, of course, with those of the special case of normality in (14.14) and (14.15).

Step 2: One-step ahead forecasting.

As in the DLM, the sampling distribution of Y_t depends on $\boldsymbol{\theta}_t$ only via $\eta_t = g^{-1}(\lambda_t)$, and thus the historical information relevant to forecasting Y_t is completely summarised in the marginal prior for $(\eta_t|D_{t-1})$. However, this is now only partially specified through the mean and variance of $\lambda_t = g(\eta)$ from (14.23),

$$(\lambda_t|D_{t-1}) \sim [f_t, q_t]. \tag{14.25}$$

In order to calculate the forecast distribution (and to update to the posterior for η_t), further assumptions about the form of the prior for η_t are necessary. At this point it should be stressed that the form of the prior is arbitrary; apart from (14.25), no further restrictions have been made on the prior. Thus, there is no prior form to be calculated or approximated in any sense, the forecaster may choose any desired form consistent with the mean and variance. The prior may be assumed approximately normal, for example, or to take any other convenient form. In view of the conjugate analysis described in Section 14.2.2, it is clear that the most convenient form is that of the conjugate family, and thus a conjugate prior is supposed. This requires, of course, that such a prior can be found consistent with the mean and variance of λ_t; happily, priors in the conjugate family typically permit a full range of unrelated values for the mean and variance, so that this poses no problem.

Thus, given (14.25), assume that the prior for η_t has the conjugate form in (14.7), namely

$$p(\eta_t|D_{t-1}) = c(r_t, s_t)\exp[r_t\eta_t - s_t a(\eta_t)]. \tag{14.26}$$

The defining parameters r_t and s_t are chosen to be consistent with the moments for λ_t in (14.25), thus implicitly satisfying the equations

$$\mathrm{E}[g(\eta_t)|D_{t-1}] = f_t \qquad \text{and} \qquad \mathrm{V}[g(\eta_t)|D_{t-1}] = q_t. \tag{14.27}$$

The one-step ahead forecast distribution now follows from the density in (14.8). For all $Y_t \in \mathcal{Y}$,

$$p(Y_t|D_{t-1}) = \frac{c(r_t, s_t)b(Y_t, V_t)}{c(r_t + \phi_t Y_t, s_t + \phi_t)}. \tag{14.28}$$

Step 3: Updating for η_t.

Observing Y_t, (14.9) provides the posterior for η_t in the conjugate form

$$p(\eta_t|D_t) = c(r_t + \phi_t Y_t, s_t + \phi_t)\exp[(r_t + \phi_t Y_t)\eta_t - (s_t + \phi_t)a(\eta_t)].$$

By analogy with the prior, denote the posterior mean and variance of $\lambda_t = g(\eta_t)$ by

$$f_t^* = \mathrm{E}[g(\eta_t)|D_t] \qquad \text{and} \qquad q_t^* = \mathrm{V}[g(\eta_t)|D_t]. \tag{14.29}$$

Step 4a: Conditional structure for $(\boldsymbol{\theta}_t|\lambda_t, D_{t-1})$.

As in the normal model, the objective of the updating is to calculate the posterior for $\boldsymbol{\theta}_t$. Again following the normal theory, these can be derived from the *joint* posterior for λ_t and $\boldsymbol{\theta}_t$. The joint density is, by Bayes' Theorem,

$$\begin{aligned} p(\lambda_t, \boldsymbol{\theta}_t|D_t) &\propto p(\lambda_t, \boldsymbol{\theta}_t|D_{t-1})\ p(Y_t|\lambda_t) \\ &\propto [p(\boldsymbol{\theta}_t|\lambda_t, D_{t-1})p(\lambda_t|D_{t-1})]\ p(Y_t|\lambda_t) \\ &\propto p(\boldsymbol{\theta}_t|\lambda_t, D_{t-1})\ [p(\lambda_t|D_{t-1})p(Y_t|\lambda_t)] \\ &\propto p(\boldsymbol{\theta}_t|\lambda_t, D_{t-1})\ p(\lambda_t|D_t). \end{aligned}$$

Hence, given λ_t and D_{t-1}, $\boldsymbol{\theta}_t$ is conditionally independent of Y_t and it follows that

$$p(\boldsymbol{\theta}_t|D_t) = \int p(\boldsymbol{\theta}_t|\lambda_t, D_{t-1})p(\lambda_t|D_t)d\lambda_t. \tag{14.30}$$

As in the DLM, information about the state vector from Y_t is channelled through the posterior for λ_t. The second component in the integrand $p(\lambda_t|D_t)$ may be obtained directly from the conjugate form posterior for η_t in (14.26). The first component, defining the conditional prior for $\boldsymbol{\theta}_t$ given λ_t, is, of course, not fully specified. Note, however, that to complete the updating cycle, we need to calculate only the posterior mean and variance matrix of $\boldsymbol{\theta}_t$, the full posterior remaining unspecified and indeterminate. From (14.30) and in parallel with (14.18) in the normal model, the key ingredients in these calculations are the *prior* mean and variance matrix of $(\boldsymbol{\theta}_t|\lambda_t, D_{t-1})$. Unfortunately, due to the incomplete specification of the joint prior, these conditional moments are unknown, non-linear and *indeterminate* functions of λ_t. They cannot be calculated without imposing further structure. However, given the partial moments specification in (14.23), they can be *estimated* using standard Bayesian techniques as follows.

Step 4b: Linear Bayes' estimation of moments of $(\boldsymbol{\theta}_t|\lambda_t, D_{t-1})$.

From (14.23) we have the joint prior mean and variance matrix of $\boldsymbol{\theta}_t$ and λ_t, and are interested in the conditional moments $\mathrm{E}[\boldsymbol{\theta}_t|, \lambda_t, D_{t-1}]$ and $\mathrm{V}[\boldsymbol{\theta}_t|\lambda_t, D_{t-1}]$. Under our assumptions, these functions of λ_t cannot be calculated, but they can be estimated using methods of linear Bayesian estimation (LBE), as in Section 4.9. Recall that LBE provides a Bayesian decision-theoretic approach to estimation of unknown, non-linear functions by linear "regressions". Section 4.9.2 describes the general theory of LBE, that is applied here. Theorem 4.9 of that section applies here directly as follows. Within the class of *linear* functions of λ_t, and subject only to the prior information in (14.23), the LBE optimal estimate of $\mathrm{E}[\boldsymbol{\theta}_t|, \lambda_t, D_{t-1}]$ is given by

$$\hat{\mathrm{E}}[\boldsymbol{\theta}_t|, \lambda_t, D_{t-1}] = \mathbf{a}_t + \mathbf{R}_t\mathbf{F}_t(\lambda_t - f_t)/q_t, \tag{14.31}$$

for all λ_t. The associated estimate of $\mathrm{V}[\boldsymbol{\theta}_t|\lambda_t, D_{t-1}]$ is

$$\hat{\mathrm{V}}[\boldsymbol{\theta}_t|, \lambda_t, D_{t-1}] = \mathbf{R}_t - \mathbf{R}_t\mathbf{F}_t\mathbf{F}_t'\mathbf{R}_t/q_t, \tag{14.32}$$

for all λ_t. These estimates of conditional moments are adopted. They obviously coincide with the exact values in the normal case.

Step 5: Updating for $\boldsymbol{\theta}_t$.

From (14.30), it follows that (again as in the normal case)

$$\mathrm{E}[\boldsymbol{\theta}_t|D_t] = \mathrm{E}[\mathrm{E}\{\boldsymbol{\theta}_t|\lambda_t, D_{t-1}\}|D_t]$$

and

$$\mathrm{V}[\boldsymbol{\theta}_t|D_t] = \mathrm{V}[\mathrm{E}\{\boldsymbol{\theta}_t|\lambda_t, D_{t-1}\}|D_t] \;+\; \mathrm{E}[\mathrm{V}\{\boldsymbol{\theta}_t|\lambda_t, D_{t-1}\}|D_t].$$

These may be estimated by substituting the LBE estimates for the arguments of the expectations. This leads to the posterior moment

$$(\boldsymbol{\theta}_t|D_t) \sim [\mathbf{m}_t, \mathbf{C}_t],$$

the posterior moments defined as follows. Firstly,

$$\begin{aligned}\mathbf{m}_t &= \mathrm{E}[\hat{\mathrm{E}}\{\boldsymbol{\theta}_t|\lambda_t, D_{t-1}\}|D_t] \\ &= \mathrm{E}[\mathbf{a}_t + \mathbf{R}_t\mathbf{F}_t(\lambda_t - f_t)/q_t|D_t] \\ &= \mathbf{a}_t + \mathbf{R}_t\mathbf{F}_t(\mathrm{E}[\lambda_t|D_t] - f_t)/q_t \\ &= \mathbf{a}_t + \mathbf{R}_t\mathbf{F}_t(f_t^* - f_t)/q_t. \end{aligned} \tag{14.33}$$

Similarly,

$$\begin{aligned}\mathrm{C}_t &= \mathrm{V}[\hat{\mathrm{E}}\{\boldsymbol{\theta}_t|\lambda_t, D_{t-1}\}|D_t] \ + \ \mathrm{E}[\hat{\mathrm{V}}\{\boldsymbol{\theta}_t|\lambda_t, D_{t-1}\}|D_t] \\ &= \mathrm{V}[\mathbf{a}_t + \mathbf{R}_t\mathbf{F}_t(\lambda_t - f_t)/q_t|D_t] \ + \ \mathrm{E}[\mathbf{R}_t - \mathbf{R}_t\mathbf{F}_t\mathbf{F}_t'\mathbf{R}_t/q_t|D_t] \\ &= \mathbf{R}_t\mathbf{F}_t\mathbf{F}_t'\mathbf{R}_t\mathrm{V}[\lambda_t|D_t]/q_t^2 \ + \ \mathbf{R}_t - \mathbf{R}_t\mathbf{F}_t\mathbf{F}_t'\mathbf{R}_t/q_t \\ &= \mathbf{R}_t - \mathbf{R}_t\mathbf{F}_t\mathbf{F}_t'\mathbf{R}_t(1 - q_t^*/q_t)/q_t. \end{aligned} \tag{14.34}$$

Substituting the values of f_t^* and q_t^* from (14.29) completes the updating. Note that these equations are just as in the normal case, equations (14.19) and (14.20), the differences lying in the ways that f_t^* and q_t^* are calculated in the conjugate updating component in Step 3. In particular, in non-normal models the posterior variance matrix $\mathbf{C}_t$ will depend on the data Y_t through q_t^*. This is not the case in the DLM, where $\mathbf{C}_t$ is data independent.

EXAMPLE 14.3. Consider the binomial framework of Example 14.2, Y_t being the number of "successes" in a total of n_t trials. Note that this includes the important special case of binary time series, when $n_t = 1$ for all t. Recall that $\mu_t = \mathrm{E}[Y_t/n_t|\eta_t]$ is the binomial probability, $(0 \le \mu_t \le 1)$, and $\eta_t = \log[\mu_t/(1-\mu_t)]$ is the log-odds or logistic transform of μ_t. Perhaps the most common regression structure with binomial data is *logistic* regression (McCullagh and Nelder 1989, Chapter 4). This corresponds to $g(\cdot)$ being the identity function, so that $\eta_t = \lambda_t = \mathbf{F}_t'\boldsymbol{\theta}_t$ is given by

$$\eta_t = \lambda_t = \log\left[\frac{\mu_t}{1-\mu_t}\right], \qquad (0 \le \mu_t \le 1).$$

Here the conjugate prior (14.26) is beta on the μ_t scale, $(\mu_t|D_{t-1}) \sim \mathrm{Beta}[r_t, s_t]$ having density

$$p(\mu_t|D_{t-1}) = c(r_t, s_t)\mu_t^{r_t-1}(1-\mu_t)^{s_t-1}, \qquad (0 \le \mu_t \le 1).$$

The normalising constant is $c(r_t, s_t) = \mathrm{o}(r_t + s_t)/[\mathrm{o}(r_t)\mathrm{o}(s_t)]$. Given the prior moments f_t and q_t for η_t, this prior is specified by calculating the corresponding values of r_t and s_t. For any positive quantity x, let $\gamma(x) = \dot{\mathrm{o}}(x)/\mathrm{o}(x)$ denote the digamma function at x. It can be verified that under this prior,

$$f_t = \mathrm{E}[\eta_t|D_{t-1}] = \gamma(r_t) - \gamma(s_t)$$

and

$$q_t = \mathrm{V}[\eta_t|D_{t-1}] = \dot{\gamma}(r_t) + \dot{\gamma}(s_t).$$

Given f_t and q_t, these equations implicitly define the beta parameters r_t and s_t, and can be solved numerically by an iterative method.† For moderate and large values of r_t and s_t (e.g., both exceeding 3) the prior for μ_t is reasonably concentrated away from the extremes of 0 and 1, and the digamma function may be adequately approximated using

$$\gamma(x) \approx \log(x) + \frac{1}{2x}.$$

The corresponding approximation for the derivative function is

$$\dot{\gamma}(x) \approx \frac{1}{x}\left(1 - \frac{1}{2x}\right).$$

For even larger values of x, $\gamma(x) \approx \log(x)$ and $\dot{\gamma}(x) \approx x^{-1}$. Thus, for large r_t and s_t, the mean and variance of η_t are approximately given by

$$f_t \approx \log(r_t/s_t) \qquad \text{and} \qquad q_t \approx \frac{1}{r_t} + \frac{1}{s_t}.$$

These may be inverted directly to give

$$r_t = q_t^{-1}[1 + \exp(f_t)] \qquad \text{and} \qquad s_t = q_t^{-1}[1 + \exp(-f_t)].$$

In fact these approximations may be used with satisfactory results even when f_t and q_t are consistent with rather small values of r_t and s_t. West, Harrison and Migon (1985) provide discussion and examples of this.

Given r_t and s_t, the one-step ahead beta-binomial distribution for Y_t is calculable. Updating to the posterior, $(\mu_t|D_t) \sim \text{Beta}[r_t + Y_t, s_t + n_t - Y_t]$, so that

$$f_t^* = \text{E}[\eta_t|D_t] = \gamma(r_t + Y_t) - \gamma(s_t + n_t - Y_t)$$

and

$$q_t^* = \text{V}[\eta_t|D_t] = \dot{\gamma}(r_t + Y_t) + \dot{\gamma}(s_t + n_t - Y_t).$$

These are easily calculated from the digamma function, and updating proceeds via (14.33) and (14.34).

EXAMPLE 14.4. A binomial *linear* regression structure is given by taking $\mu_t = \lambda_t$ for all t, so that $g^{-1}(\cdot)$ is the logistic or log-odds transform. Thus, the expected level of Y_t is just as in the DLM, $\mu_t = \mathbf{F}_t'\boldsymbol{\theta}_t$, but variation about this level is binomial rather than normal. With $(\mu_t|D_{t-1}) \sim \text{Beta}[r_t, s_t]$, $\mu_t = \lambda_t$ implies directly that

$$f_t = \text{E}[\mu_t|D_{t-1}] = \frac{r_t}{(r_t + s_t)} \qquad \text{and} \qquad q_t = \text{V}[\mu_t|D_{t-1}] = \frac{f_t(1 - f_t)}{(r_t + s_t + 1)}.$$

†See Abramowitz and Stegun 1965, for full discussion of the digamma function. The recursions $\gamma(x) = \gamma(x+1) - x^{-1}$ and $\dot{\gamma}(x) = \dot{\gamma}(x+1) + x^{-2}$ are useful for evaluations. They also provide useful tables of numerical values and various analytic approximations.

These equations invert to define the beta prior via

$$r_t = f_t[q_t^{-1} f_t(1-f_t) - 1] \quad \text{and} \quad s_t = (1-f_t)[q_t^{-1} f_t(1-f_t) - 1].$$

After updating,

$$f_t^* = \mathrm{E}[\mu_t|D_t] = \frac{r_t + Y_t}{(r_t + s_t + n_t)}$$

and

$$q_t^* = \mathrm{V}[\mu_t|D_{t-1}] = \frac{f_t^*(1-f_t^*)}{(r_t + s_t + n_t + 1)}.$$

14.3.4 Step ahead forecasting and filtering

Forecasting ahead to time $t+k$ from time t, Steps 1 and 2 of the one-step analysis extend directly. At time t, the posterior moments of $\boldsymbol{\theta}_t$ are available,

$$(\boldsymbol{\theta}_t|D_t) \sim [\mathbf{m}_t, \mathbf{C}_t].$$

From the evolution equation (14.11) applied at times $t+1, \dots, t+k$, it follows that

$$(\boldsymbol{\theta}_{t+k}|D_t) \sim [\mathbf{a}_t(k), \mathbf{R}_t(k)],$$

with moments defined sequentially into the future via, for $k = 1, 2, \dots,$

$$\mathbf{a}_t(k) = \mathbf{G}_{t+k}\mathbf{a}_t(k-1)$$

and

$$\mathbf{R}_t(k) = \mathbf{G}_{t+k}\mathbf{R}_t(k-1)\mathbf{G}'_{t+k} + \mathbf{W}_{t+k},$$

with $\mathbf{a}_t(0) = \mathbf{m}_t$ and $\mathbf{R}_t(0) = \mathbf{C}_t$. These equations are, of course, just as in the DLM (see Theorem 4.2). Now, $\lambda_{t+k} = \mathbf{F}'_{t+k}\boldsymbol{\theta}_{t+k}$ has moments

$$(\lambda_{t+k}|D_t) \sim [f_t(k), q_t(k)],$$

where

$$f_t(k) = \mathbf{F}'_{t+k}\mathbf{a}_t(k) \quad \text{and} \quad q_t(k) = \mathbf{F}'_{t+k}\mathbf{R}_t(k)\mathbf{F}_{t+k}.$$

The conjugate prior for $\eta_{t+k} = g^{-1}(\lambda_{t+k})$ consistent with these moments is now determined; denote the corresponding defining parameters by $r_t(k)$ and $s_t(k)$, so that, by analogy with (14.28), the required forecast density is just

$$p(Y_{t+k}|D_t) = \frac{c(r_t(k), s_t(k))b(Y_{t+k}, V_{t+k})}{c(r_t(k) + \phi_{t+k}Y_{t+k}, s_t(k) + \phi_{t+k})}.$$

Note that if using discount factors to construct $\mathbf{W}_t$ at each time t, then as usual, $\mathbf{W}_{t+k}$ in the above equations should be replaced by $\mathbf{W}_t(k) = \mathbf{W}_t$ for extrapolating.

Filtering backwards over time from time t, the ideas underlying the use of linear Bayes' methods in updating may be applied to derive optimal (in the LBE sense) estimates of the posterior moments of $\boldsymbol{\theta}_{t-k}$ given D_t, for all t and $k > 0$. It is clear from the results in Section 4.9 that the linear Bayes' estimates of filtered moments are derived just as in the case of normality. Thus, the estimated moments of the filtered distributions are given by Theorem 4.4.

14.3.5 Linearisation in the DGLM

Some practical models will introduce parameter nonlinearities into either of the model equations (14.10) or (14.11). Within the approximate analysis here, such nonlinearities can be treated as in Section 13.2, i.e., using some form of linearisation technique. Such an approach underlies the applications in Migon and Harrison (1985). To parallel the use of linearisation applied to the normal non-linear model (13.1) consider the following non-normal, non-linear version of the DGLM of Definition 14.1. Suppose that the sampling model is just as described in (14.10), but now that

- for some known, non-linear regression function $F_t(\cdot)$,

$$\lambda = F_t(\boldsymbol{\theta}_t); \tag{14.35}$$

- for some known, non-linear evolution function $\mathbf{g}_t(\cdot)$,

$$\boldsymbol{\theta}_t = \mathbf{g}_t(\boldsymbol{\theta}_{t-1}) + \boldsymbol{\omega}_t. \tag{14.36}$$

Based on the use of linearisation in Section 13.2 the above DGLM analysis can be applied to this non-linear model at time t as follows.

(1) Linearise the evolution equation to give

$$\boldsymbol{\theta}_t \approx \mathbf{h}_t + \mathbf{G}_t\boldsymbol{\theta}_{t-1} + \boldsymbol{\omega}_t,$$

where $\mathbf{G}_t$ is the known evolution matrix

$$\mathbf{G}_t = \left[\frac{\partial \mathbf{g}_t(\boldsymbol{\theta}_{t-1})}{\partial \boldsymbol{\theta}'_{t-1}}\right]_{\boldsymbol{\theta}_{t-1}=\mathbf{m}_{t-1}},$$

and

$$\mathbf{h}_t = \mathbf{g}_t(\mathbf{m}_{t-1}) - \mathbf{G}_t\mathbf{m}_{t-1}$$

is also known. Then the usual DLM evolution applies, with the minor extension to include $\mathbf{h}_t$ in the evolution equation. The prior moments for $\boldsymbol{\theta}_t$ are

$$\begin{aligned}\mathbf{a}_t &= \mathbf{h}_t + \mathbf{G}_t\mathbf{m}_{t-1} = \mathbf{g}_t(\mathbf{m}_{t-1}),\\ \mathbf{R}_t &= \mathbf{G}_t\mathbf{C}_{t-1}\mathbf{G}'_t + \mathbf{W}_t.\end{aligned}$$

(2) Proceeding to the observation equation, the non-linear regression function is linearised to give

$$\lambda_t = f_t + \mathbf{F}_t'(\boldsymbol{\theta}_t - \mathbf{a}_t),$$

where

$$f_t = F_t(\mathbf{a}_t),$$

and $\mathbf{F}_t$ is the known regression vector

$$\mathbf{F}_t = \left[\frac{\partial F_t(\boldsymbol{\theta}_t)}{\partial \boldsymbol{\theta}_t}\right]_{\boldsymbol{\theta}_t=\mathbf{a}_t}.$$

Assuming an adequate approximation, the prior moments of λ_t and $\boldsymbol{\theta}_t$ in (14.23) are thus given by a modification of the usual equations.

$\mathbf{R}_t$ and q_t are as usual, with regression vector $\mathbf{F}_t$ and evolution matrix $\mathbf{G}_t$ derived here via linearisation of possibly non-linear functions. The means $\mathbf{a}_t$ and f_t retain the fundamental non-linear characteristics of (14.35) and (14.36), given by

$$\mathbf{a}_t = \mathbf{g}_t(\mathbf{m}_{t-1})$$

and

$$f_t = F_t(\mathbf{a}_t).$$

As noted in comment (9) in Section 13.2 the model is easily modified to incorporate a stochastic drift in $\boldsymbol{\theta}_{t-1}$ *before* applying the non-linear evolution function in (14.36), rather than by adding $\boldsymbol{\omega}_t$ *after* transformation. The alternative to (14.36) is simply

$$\boldsymbol{\theta}_t = \mathbf{g}_t(\boldsymbol{\theta}_{t-1} + \boldsymbol{\delta}_t), \tag{14.37}$$

for some zero-mean error $\boldsymbol{\delta}_t$, with a known variance matrix $\mathbf{U}_t$. Then $\mathbf{a}_t$ and $\mathbf{R}_t$ are as defined above, with

$$\mathbf{W}_t = \mathbf{G}_t\mathbf{U}_t\mathbf{G}_t'. \tag{14.38}$$

This method is applied in the case study reported in the following section. Note that the comments in Section 13.2 concerning the use of this and alternative techniques of linearisation in normal models apply equally here in the DGLM. Also, step ahead forecast and filtered moments and distributions may be derived in an obvious manner, as the interested reader may easily verify.

14.4 CASE STUDY IN ADVERTISING AWARENESS

14.4.1 Background

An illustration of the above type of analysis is described here. Extensive application of a particular class of non-normal, non-linear dynamic models has been made in the assessment of the impact on consumer awareness of the advertising of various products. Some of this work is reported in Migon and Harrison (1985), Migon (1984), and for less technical discussion, Colman and Brown (1983). Related models are discussed in West, Harrison and Migon (1985), and implemented in West and Harrison (1986b).

Analysts of consumer markets define and attempt to measure many variables in studies of the effectiveness of advertising. The *awareness* in a consumer population of a particular advertisement is one such quantity, the subject of the above-referenced studies. Attempts to measure consumer awareness may take one of several forms, though a common approach in the UK is to simply question sampled members of the TV-viewing population as to whether or not they have seen TV commercials for any of a list of branded items during a recent, fixed time interval such as a week. The proportion of positive respondents for any particular brand then provides a measure of population awareness for the advertisement of that brand. In assessing advertising effectiveness, such awareness measures can be taken as defining the response variable to be related to independent variables that describe the nature of the advertising campaign being assessed. The extent of television advertising of a product, or a group of products, of a company is often measured in standardised units known as TVRs, for television ratings. TVRs are based on several factors, including the length of the TV commercial in the TV areas in which the sampled awareness is measured. Broadbent (1979) discusses the construction of TVR measurement. Here the key point is that TVRs provide standardised measurements of the expenditure of the company on TV commercials. The main issues in the area now concern the assessment of the relationship between TVRs and awareness, and the comparison of this relationship at different times and across different advertising campaigns.

The basic information available in this study is as follows. Suppose that weekly surveys sample a given number of individuals from the population of TV viewers, counting the number "aware" of the current or recent TV commercial for a particular branded item marketed by a company. Suppose that the number of sampled individuals in week t is n_t, and that Y_t of them respond positively. Finally, let X_t denote the weekly TVR measurement for week t.[†] The problem now concerns the modelling of the effect of X_s, $(s = t, t-1, \dots)$ on Y_t.

[†]In fact, due to uncertainties about the precise timing of surveys, the TVR measurements used in the referenced studies are usually formed from a weighted average of that in week t and week $t-1$.

Migon and Harrison (1985) discuss elements of the modelling problem and previous modelling efforts, notably that of Broadbent (1979), and develop some initial models based essentially on extensions of normal DLMs. Practically more appropriate and useful models are then developed by moving to non-normal and non-linear structures that are consistent with the nature of the problem, and more directly geared to answering the major questions of interest and importance to the company. Several key features to be modelled are as follows.

(1) The data collection mechanism leads to a binomial-like sampling structure for the observations; the data count the number of positive responses out of a total number sampled. Thus normal models are generally inappropriate (except as approximations).
(2) The level of awareness in the population is expected to be bounded below by a minimum, usually non-zero lower threshold value. This base level of awareness represents a (small) proportion of the population who would tend to respond positively in the sample survey even in the absence of any historical advertising of the product (due to the effects of advertisements for other, related products, misunderstanding and misrepresentation).
(3) Awareness is similarly expected to be bounded above, the maximum population proportion aware being less than some upper threshold level that will typically be less than unity.
(4) Given no advertising into the future, the effect of past advertising on awareness is expected to decay towards the lower threshold.
(5) The effect of TVR on awareness is expected to be non-linear, exhibiting a *diminishing returns* form. That is, at higher values of X_t, the marginal effect of additional TVR is likely to be smaller than at lower values of X_t.
(6) The thresholds and rates of change with current and past TVR will tend to be relatively stable over time for a given advertising campaign, but can be expected to change to some small degree due to factors not included in the model (as is usual in regression models in time series; see Chapters 3 and 9). In addition, such characteristics can be expected to change markedly in response to a major change in the advertising campaign.

Points (1) to (6) are incorporated in a model described in the next section, along with several other features of the practical data collection and measurement scheme that bear comment. The primary objectives in modelling are as follows.

(a) To incorporate these features and provide monitoring of the awareness/TVR relationship through sequential estimation of model parameters relating to these features.

(b) To allow the incorporation of subjective information. Initially this is desired to produce forecasts before data are collected. Initial priors for model parameters may be based, for example, on experience with other, similar branded items in the past. Such inputs are also desired when external information is available. In particular, information about changing campaigns should be fed-forward in the form of increased uncertainty about (some) model parameters to allow for immediate adaptation to observed change.
(c) To cater for weeks with no sample, i.e., missing data.
(d) To produce forecasts of future awareness levels based on projected TVR values, i.e., so-called *What-if?* analyses.
(e) To produce filtered estimates of past TVR effects, i.e., so-called *What-Happened?* analyses.

Clearly, the sequential analysis within a dynamic model provides for each of these goals once an appropriate model is constructed.

14.4.2 The non-linear binomial model

The non-linear binomial model is defined as follows, following points (1) to (6) itemised in the previous section.

(1) Sampling model
At time t, the number Y_t of positive respondents out of the known total sample size n_t is assumed binomially distributed. Thus, if μ_t represents the population proportion aware in week t, $(0 \leq \mu_t \leq 1)$, the sampling model is defined by

$$p(Y_t|\mu_t) = \binom{n_t}{Y_t}\mu_t^{Y_t}(1-\mu_t)^{n_t-Y_t}, \qquad (Y_t = 0, 1, \dots, n_t).$$

Let E_t denote the *effect* of current and past weekly TVR on the current expected awareness μ_t. Then, following points (2) to (6), introduce the following components.

(2) Lower threshold for awareness
The minimum level of awareness expected at time t is denoted by the lower threshold parameter α_t. Thus, $\mu_t \geq \alpha_t \geq 0$ for all time.

(3) Upper threshold for awareness
The maximum level of awareness expected at time t is denoted by the upper threshold parameter β_t. Thus, $\mu_t \leq \beta_t \leq 1$ for all time

(4) TVR effect: decay of awareness
Suppose that $X_t = 0$. Then the effect on awareness level is expected to decay between times $t-1$ and t by a factor of $100(1-\rho_t)\%$, moving from E_{t-1} to $E_t = \rho_t E_{t-1}$, for some ρ_t between 0 and 1, typically closer to 1.

ρ_t measures the persistence, or memory effect, of the advertising campaign between weeks $t-1$ and t, and is typically expected to be rather stable, even constant, over time. Larger values are obviously associated with more effective (in terms of retention of the effect) advertising. If $\rho_t = \rho$ for all time, then a period of s weeks with no TVR implies an exponential decay from E_{t-1} to $\rho^s E_{t-1}$. The *half-life* σ associated with any particular value of ρ is the number of weeks taken for the effect to decay to exactly half any initial value with no further input TVR. Thus, $\rho^\sigma = 0.5$, so that

$$\sigma = -\frac{\log(2)}{\log(\rho)},$$

and obviously, $\sigma > 0$. With time-varying ρ_t, the associated half-life parameter is denoted by σ_t. Clearly, $\rho_t = 2^{-1/\sigma_t}$.

(5) **TVR effect: diminishing returns**

At time t with current TVR input X_t, the input of X_t is expected to instantaneously increase the advertising effect by a fraction of the remaining available awareness. To model diminishing returns at higher values of TVR, this fraction should decay with increasing X_t. Specifically, suppose that for some $\kappa_t > 0$, this fraction is determined by the penetration function $\exp(-\kappa_t X_t)$, decaying exponentially as TVR increases. Then before including the instantaneous effect of X_t, the expected level of awareness is the base level α_t from (2) plus the effect E_{t-1} from the previous week discounted by the memory decay parameter ρ_t from (3), giving $\alpha_t + \rho_t E_{t-1}$. With upper threshold level β_t from (3), the remaining possible awareness is $\beta_t - (\alpha_t + \rho_t E_{t-1})$ so that the new expected effect at time t is given by

$$\begin{aligned} E_t =& \rho_t E_{t-1} + [1 - \exp(-\kappa_t X_t)][\beta_t - (\alpha_t + \rho_t E_{t-1})] \\ =& (\beta_t - \alpha_t) - (\beta_t - \alpha_t - \rho_t E_{t-1})\exp(-\kappa_t X_t). \end{aligned} \tag{14.39}$$

The expected level is then

$$\mu_t = \alpha_t + E_t. \tag{14.40}$$

As with ρ_t, the penetration parameter κ_t measures advertising effectiveness, here in terms of the immediate impact of expected consumer awareness of weekly TVR. Also, by analogy with σ_t, the *half-penetration* parameter $\xi_t > 0$ is defined as the TVR needed to raise the effect (from any initial level) by exactly half the remaining possible. Thus, $1 - \exp(-\kappa_t \xi_t) = 0.5$, so that $\xi_t = \log(2)/\kappa_t$ and $\kappa_t = \log(2)/\xi_t$.

(6) **Time-varying parameters**

Define the 5-dimensional state vector $\boldsymbol{\theta}_t$ by

$$\boldsymbol{\theta}_t' = (\alpha_t, \beta_t, \rho_t, \kappa_t, E_t) \tag{14.40}$$

for all t. All five parameters are viewed as possibly subject to minor stochastic variation over time, due to factors not included in the model. Some pertinent comments are as follows.

(i) Changes in the composition of the sampled population can be expected to lead to changes in the awareness/TVR relationship.
(ii) In a stable population, the maturity of the product is a guide to stability of the model. Early in the life of a product, base level awareness tends to rise to a stable level, with an associated rise in the upper threshold for awareness. With a mature product, threshold levels will tend to be stable.
(iii) The memory decay rate ρ_t will tend to be stable, even constant, over time and across similar products and styles of advertisement.
(iv) The penetration parameter κ_t will also tend to be stable over time, though differing more widely across advertising campaigns.
(v) A point related to (i) concerns the possibility of additional, week-to-week variation in the data due to surveys being made in different regions of the country. Such regional variation is best modelled, not by changes in model parameters, but as unpredictable, "extra-binomial" variation.
(vi) A degree of stochastic change in all parameters can be viewed as a means of accounting for model misspecification generally (as in all dynamic models), in particular in the definition of X_t and the chosen functional forms of the model above.

14.4.3 Formulation as a dynamic model

The model formulation is as follows. The observation model is binomial with probability parameter $\mu_t = \lambda_t$. With $\boldsymbol{\theta}_t$ given by (14.40) we have

$$\mu_t = \alpha_t + E_t = \mathbf{F}'\boldsymbol{\theta}_t,$$

where $\mathbf{F}' = (1, 0, 0, 0, 1)$ for all t. For the evolution, suppose that $\boldsymbol{\theta}_{t-1}$ changes by the addition of a zero-mean evolution error $\boldsymbol{\delta}_t$ between $t-1$ and t. The only element of the state vector that is expected to change systematically is the cumulative effect of TVR i.e., the variable E_t evolving according to (14.39). Thus, the evolution is a special case of the general, non-linear model (14.37),

$$\boldsymbol{\theta}_t = \mathbf{g}_t(\boldsymbol{\theta}_{t-1} + \boldsymbol{\delta}_t),$$

where

- $\boldsymbol{\delta}_t$ is an evolution error, with

$$\boldsymbol{\delta}_t \sim [\,\mathbf{0}, \mathbf{U}_t]$$

  for some known variance matrix $\mathbf{U}_t$;
- For any 5-vector $\mathbf{z} = (z_1, z_2, z_3, z_4, z_5)'$,

$$\mathbf{g}_t(\mathbf{z})' = [z_1, z_2, z_3, z_4, (z_2 - z_1) - (z_2 - z_1 - z_3 z_5)\exp(-z_4 X_t)].$$

The linearisation of this evolution equation follows Section 14.3.5. The derivative matrix

$$\mathbf{G}_t(\mathbf{z}) = \frac{\partial \mathbf{g}_t(\mathbf{z})}{\partial \mathbf{z}'}$$

is given in transposed form as

$$\mathbf{G}_t(\mathbf{z})' = \begin{pmatrix} 1 & 0 & 0 & 0 & \mathrm{e}^{-z_4 X_t} - 1 \\ 0 & 1 & 0 & 0 & 1 - \mathrm{e}^{-z_4 X_t} \\ 0 & 0 & 1 & 0 & z_5 \mathrm{e}^{-z_4 X_t} \\ 0 & 0 & 0 & 1 & (z_2 - z_1 - z_3 z_5) X_t \mathrm{e}^{-z_4 X_t} \\ 0 & 0 & 0 & 0 & z_3 \mathrm{e}^{-z_4 X_t} \end{pmatrix}.$$

Then the linearised evolution equation at time t is specified as in the previous section. With

$$(\boldsymbol{\theta}_{t-1} | D_{t-1}) \sim [\mathbf{m}_{t-1}, \mathbf{C}_{t-1}],$$

we have

$$(\boldsymbol{\theta}_t | D_{t-1}) \sim [\mathbf{a}_t, \mathbf{R}_t],$$

with

$$\mathbf{a}_t = \mathbf{g}_t(\mathbf{m}_{t-1}), \quad \mathbf{R}_t = \mathbf{G}_t \mathbf{C}_{t-1} \mathbf{G}_t' + \mathbf{W}_t \quad \text{and} \quad \mathbf{W}_t = \mathbf{G}_t \mathbf{U}_t \mathbf{G}_t',$$

where

$$\mathbf{G}_t = \mathbf{G}_t(\mathbf{m}_{t-1}).$$

For calculation, note that $\mathbf{R}_t = \mathbf{G}_t(\mathbf{C}_{t-1} + \mathbf{U}_t)\mathbf{G}_t'$.

Some points of detail are as follows.

(i) The error term $\boldsymbol{\delta}_t$ controls the extent and nature of stochastic change in the state vector through the variance matrix $\mathbf{U}_t$. Following the earlier comments, it is clear that with a stable product, such changes are likely to be relatively small. Supposing this to be defined via discount factors, all such factors should be close to unity. In Migon and Harrison (1985), various possible cases are considered. In some of these, subsets of $\boldsymbol{\theta}_t$ are assumed constant over time, so that the corresponding rows and columns of $\mathbf{U}_t$ are zero, and the comments in Section 14.4.2 about the relative stability of the elements indicate that generally, there is a need to construct $\mathbf{U}_t$ from components with different discount factors. For the illustration here, however, a single discount factor is applied to the five-element vector as a single block, permitting a very small amount of variation over time in all parameters (so long as the initial prior does not constrain any to be fixed for all time; a point reconsidered below).

(ii) As always with discount models, in forecasting ahead, a fixed evolution variance matrix is used. Thus, with $\mathbf{W}_t$ as defined at time $t-1$

via discount factors based on the current $\mathbf{C}_{t-1}$, forecasting ahead from time t to time $t+k$ we use $\mathbf{W}_{t+k} = \mathbf{W}_t$ for extrapolation.

(iii) There may be a need to consider complications arising due to the bounded ranges of the state parameters; all are positive and some lie in the unit interval. The model assumes, at any time, only prior means, variances and covariances for these quantities. If the model is generally appropriate, then it will be found that the data processing naturally restricts the means, as point estimates, to the relevant parameter ranges. Also, if initial priors are reasonably precise within the ranges, then the restrictions should not cause problems. However, with less precise priors, posterior estimates may stray outside the bounds if correction is not made. One possibility, used in West and Harrison (1986b), is simply to constrain the individual prior and posterior means to lie within the bounds, adjusting them to a point near but inside the boundary if they stray outside. An alternative approach, not explored further here, is to reparametrise the model in terms of unbounded parameters. If this is done, the evolution equation can be re-expressed in terms of the new parameters and linearised in that parametrisation.

(iv) Migon and Harrison (1985) note the possibility of extra-binomial variation due to surveys taken in different regions from week-to-week (point 6(vi) of Section 14.4.2). This can easily be incorporated within the dynamic model by taking $\mu_t = \mathbf{F}'\boldsymbol{\theta}_t + \partial\mu_t$, where $\partial\mu_t$ is an additional zero-mean error. With $\partial\mu_t = 0$, the original model is obtained. Otherwise, the variance of this error term introduces extra variation in the binomial level that caters for such region-to-region changes (and also for other sources of extra-binomial variation). This sort of consideration is often important in practice; the inclusion of such a term protects the estimation of the state parameters from possible corruption due to purely random, extraneous variation in the data not adequately modelled through the basic binomial distribution. See Migon and Harrison (1985) for further comments, and Migon (1985) for discussion of similar points in different models.

14.4.4 Initial priors and predictions

Consider application to the monitoring of advertising effectiveness of a forthcoming TV commercial campaign for a single product, to begin at $t = 1$. Suppose that the product has been previously advertised and that the forecaster concerned has some information from the previous campaigns (as well as from experience with other products). In fact, the product here is confectionary, the data to be analysed below taken from the studies of Migon and Harrison (1985). The levels of TVR in the forthcoming campaign are expected to be consistent with past advertising, lying between zero and ten units on a standardised scale. In the past, similar campaigns have been associated with awareness levels that rarely exceeded 0.5. In forming an initial view about the effectiveness of a forthcoming advertising campaign, this background information D_0 is summarised as follows to provide the initial prior for forecasting future awareness.

(1) The base level for awareness is currently thought to be almost certainly less that 20%, and probably close to 0.1. The initial prior for α_0 has mean 0.1 and standard deviation 0.025.

(2) In the past, observed awareness levels have never exceeded 0.5, so that there has been little opportunity to obtain precise information about the upper threshold. It is the case, of course, that if levels remain low in the future, then the model analysis should not be unduly affected by uncertainty about the upper threshold. Based on general experience with consumer awareness, the upper threshold is expected to lie between 0.75 and 0.95, the uncertainty being relatively high. The initial prior is specified in terms of the mean and variance for the *available* awareness $\gamma_0 = \beta_0 - \alpha_0$ rather then for β_0 directly; the mean is 0.75 and standard deviation 0.2. Assuming, in addition, that γ_0 and α_0 are uncorrelated, this implies that β_0 has mean 0.85 and standard deviation approximately 0.202, and that the correlation between α_0 and β_0 is 0.124.

(3) The memory decay rate parameter ρ_t is viewed as stable over time and across products, with value near 0.9 for weekly awareness decay (anticipating approximately 10% decay of the advertising effect each week). The initial prior for ρ_0 has mean 0.9 and standard deviation 0.01. In terms of the memory half-life σ_0, a value of 0.9 for ρ_0 corresponds to a half-life of 6.6 weeks, 0.87 to a half-life of 5 weeks, and 0.93 to a half-life of 9.6 weeks.

(4) The penetrative effect of the advertising on awareness is expected to be fairly low, consistent with previous advertising. The previously low TVRs have never achieved marked penetration, so the previous maximum value with TVR near 10 units per week is considered to be well below the half-penetration level. Thus, ξ_0 is expected to be much greater than 10, the forecaster assessing the most likely value as about 35 units. Converting to κ_0, this provides the initial mean as $\log(2)/35 \approx 0.02$. The prior standard deviation of κ_0 is chosen as 0.015. Note that κ_0 values of 0.01 and 0.04 correspond approximately to ξ_0 at 69 and 17 respectively.

(5) The previous campaign has only very recently terminated, and had been running at reasonably high TVR levels. Thus, the retained effect E_0 of past advertising is considered to be fairly high, although there is a reasonable degree of uncertainty. The prior mean is taken as 0.3, the standard deviation is 0.1.

(6) For convenience and in the absence of further information, the covariances that remain to be specified are set to zero. Thus,

$$\mathbf{m}_0 = (0.10, 0.85, 0.90, 0.02, 0.30)'$$

and

$$\mathbf{C}_0 = 0.0001 \begin{pmatrix} 6.25 & 6.25 & 0 & 0 & 0 \\ 6.25 & 406.25 & 0 & 0 & 0 \\ 0 & 0 & 1 & 0 & 0 \\ 0 & 0 & 0 & 2.25 & 0 \\ 0 & 0 & 0 & 0 & 100 \end{pmatrix}. \tag{14.41}$$

(7) Week-to-week variation in model parameters is expected to be fairly low and modelled using a single discount factor of 0.97. Thus, the variance matrix of $\boldsymbol{\delta}_t$ in (14.37) is defined as

$$\mathbf{U}_t = (0.97^{-1} - 1)\mathbf{C}_{t-1} \approx 0.03\mathbf{C}_{t-1}.$$

The initial prior, based on previous experience with this, and similar, campaigns, represents a reasonable degree of precision about the model parameters. It is therefore directly useful in forecasting the forthcoming campaign without further data, such forecasting providing an assessment of model implications, and the effects of the specified priors. Note that the models allow the forecaster the facility to impose constraints on parameters by assuming the relevant prior variances and covariances to be zero. The applied models in Migon and Harrison (1985), and West and Harrison (1986b), make some use of such constraints. In initial forecasting, this allows an assessment of the implications of prior plus model assumptions for each of the component parameters, free from the effects of the uncertainties about the constrained components. This is done here, temporarily constraining E_0.

First, consider the implied decay of awareness into the future in the absence of any advertising, $X_k = 0$ for $k = 1, \ldots, 25$. Take the prior in (14.41) but modified, so that E_0 has prior mean 0.75 and zero variance. Thus, initially, the effect of past advertising is assumed to be saturated at the maximum level. Looking ahead, awareness is expected to decay towards the base level, initially expected to be 0.1 but with some uncertainty. For forecasting ahead, the theory in Sections 14.3.4 and 14.3.5 applies with $t = 0$; for $k = 1, \ldots, 25$, $\mu_k = \lambda_k \sim [f_0(k), q_0(k)]$. The conjugate distribution consistent with these moments is $(\mu_k | D_0) \sim \text{Beta}[r_0(k), s_0(k)]$ with parameters defined as in Example 14.4. Figures 14.1 and 14.2 provide some features of forecasts based on this initial distribution. Figure 14.1 provides a plot of $f_0(k)$ against k, clearly illustrating the expected exponential decay in the effect of awareness in the absence of further advertising. To give an indication of uncertainty, the dotted lines provide standard deviation

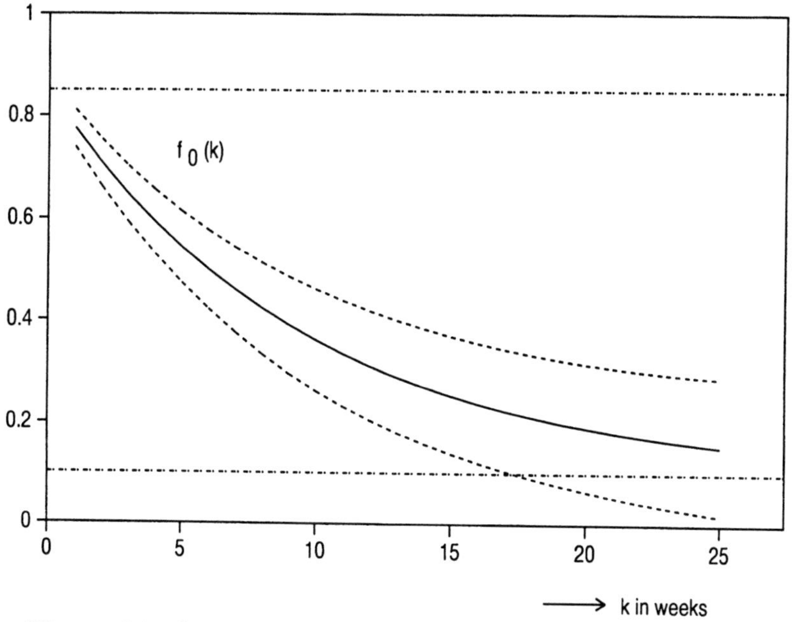

Figure 14.1 Initial forecasts of expected memory decay

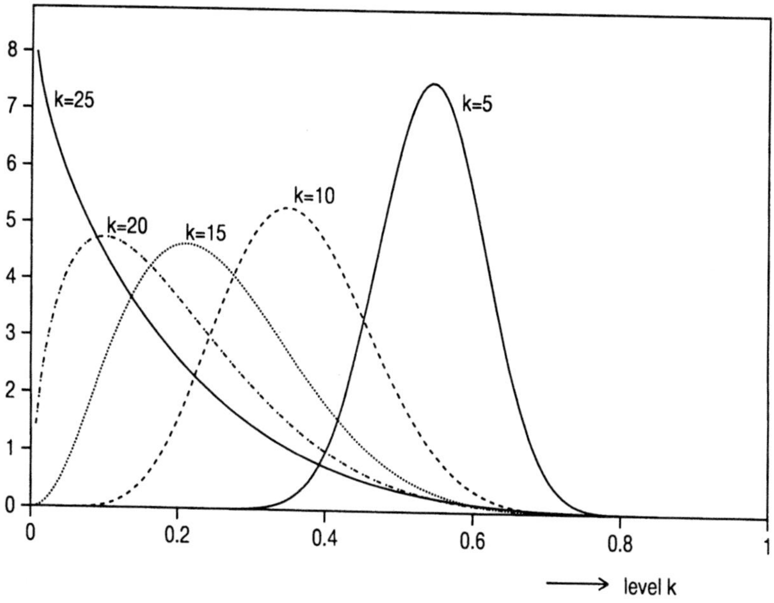

Figure 14.2 Initial forecast densities for decay in awareness

limits above and below the forecast mean, $f_0(k) \pm \sqrt{q_0(k)}$. From the graph, awareness is forecast as decaying to roughly half the initial value (whatever this initial value may be) after roughly 7 weeks. To examine the uncertainty in forecasting more closely, Figure 14.2 plots the beta distributions of $(\mu_k|D_0)$ at values $k = 5, 10, 15, 20$ and 25. Note the highly skewed forms as the expected level moves towards the extreme at 0 and the variance increases.

We can similarly explore the implications for expected penetration due to various initial levels of TVR. Take the prior (14.41) modified to have zero mean and variance for E_0. Thus, there is no retained effect of past advertising. For any given TVR X_1 in the first week of the campaign, awareness is expected to increase towards the threshold level with expected value 0.85. The one-step ahead forecast distribution for μ_1 has moments $f_0(1)$ and $q_0(1)$ that depend on X_1. Figure 14.3 plots $f_0(1)$ with one standard deviation limits for $X_1 = 1, \ldots, 25$. The increase in uncertainty about the level as X_1 increases is marked, as is the fact that penetration is assumed to be rather low even for TVR values as high as 25; recall that the prior estimate of the penetration parameter is consistent with an estimated half-penetration of $X_1 \approx 35$.. Since the forthcoming campaign will have TVR values of less than 10 units per week, it is apparent that penetration is expected to be rather low. Figure 14.4 provides the beta densities of $(\mu_1|D_0)$ for $X_1 = 5, 10, 25$.

14.4.5 Data analysis

The campaign date now becomes available, week by week. As the data are observed, any available external information will also be incorporated in the model via subjective intervention as in Chapter 11, and typically applied models will also be subject to continuous monitoring. Suppose here that intervention and monitoring are not considered, and that the weekly observations are simply processed as normal via the analysis of Sections 14.3.3 and 14.3.4.

The sample surveys in this dataset count the number of positive respondents out of a nominal total of $n_t = 66$ for all t. The full set of observations becoming available during the first 75 weeks of the campaign are given in Table 14.1. The awareness measurements are given in terms of proportions out of a nominal 66 (rounded to 2 decimal places). These are also plotted in the upper frame of Figure 14.5, the TVR measurements appearing in the lower frame of that figure. Note that three observations during the campaign are missing, corresponding to weeks in which no sample was taken. Of course these are routinely handled by the sequential analysis, the posterior moments for state vectors following a missing observation being just the prior moments. Note also that advertising over the 75 weeks comes in essentially two bursts. The first, very short spell during weeks 3 to 7 is followed by no TVR for a period of weeks, the second, long spell beginning

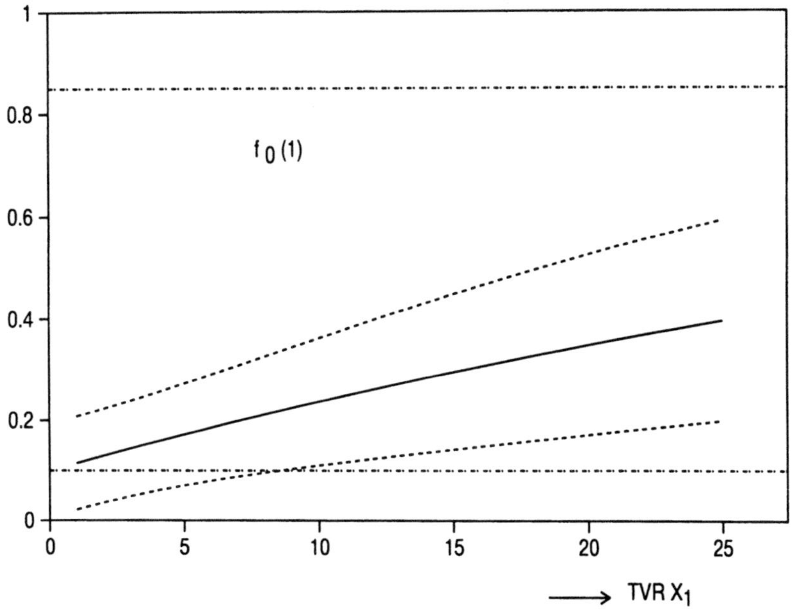

Figure 14.3 Initial forecasts of expected TVR penetration

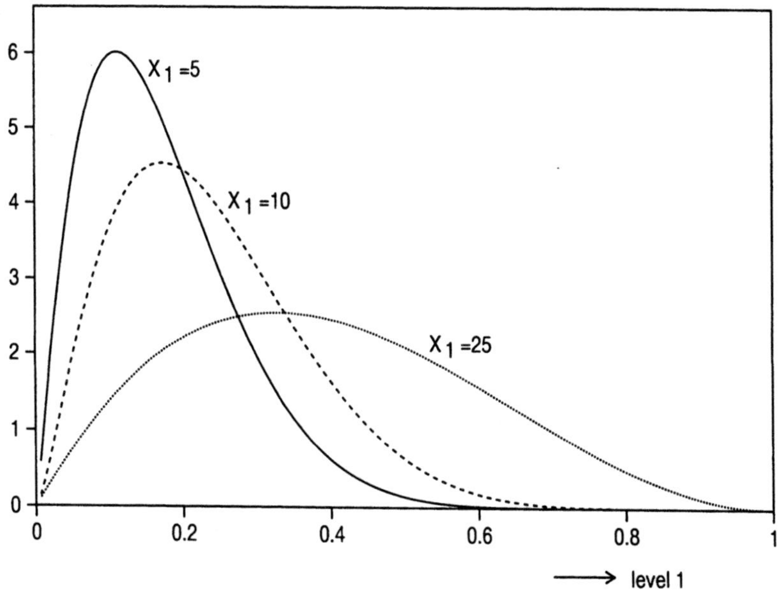

Figure 14.4 Initial forecast densities for TVR penetration

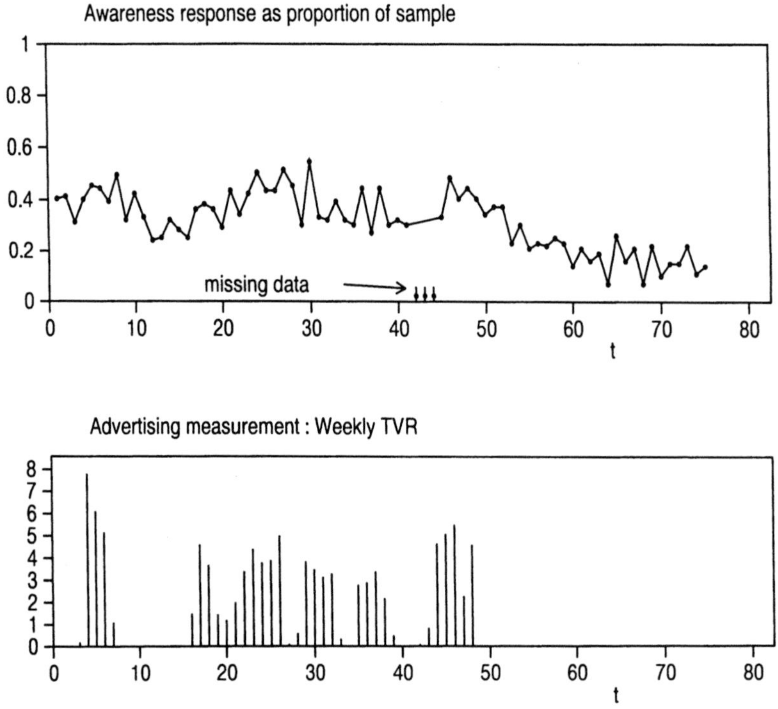

Figure 14.5 Awareness response and TVR time series

Table 14.1. Awareness response proportions and TVR

Weekly TVR measurements (to be read across rows)

0.05	0.00	0.20	7.80	6.10	5.15	1.10	0.00	0.00	0.00	0.00	0.00	0.00	0.00	0.00
1.50	4.60	3.70	1.45	1.20	2.00	3.40	4.40	3.80	3.90	5.00	0.10	0.60	3.85	3.50
3.15	3.30	0.35	0.00	2.80	2.90	3.40	2.20	0.50	0.00	0.00	0.10	0.85	4.65	5.10
5.50	2.30	4.60	0.00	0.00	0.00	0.00	0.00	0.00	0.00	0.00	0.00	0.00	0.00	0.00
0.00	0.00	0.00	0.00	0.00	0.00	0.00	0.00	0.00	0.00	0.00	0.00	0.00	0.00	0.00

Awareness proportions (read across rows, * denotes missing values)

0.40	0.41	0.31	0.40	0.45	0.44	0.39	0.50	0.32	0.42	0.33	0.24	0.25	0.32	0.28
0.25	0.36	0.38	0.36	0.29	0.43	0.34	0.42	0.50	0.43	0.43	0.52	0.45	0.30	0.55
0.33	0.32	0.39	0.32	0.30	0.44	0.27	0.44	0.30	0.32	0.30	*	*	*	0.33
0.48	0.40	0.44	0.40	0.34	0.37	0.37	0.23	0.30	0.21	0.23	0.22	0.25	0.23	0.14
0.21	0.16	0.19	0.07	0.26	0.16	0.21	0.07	0.22	0.10	0.15	0.15	0.22	0.11	0.14

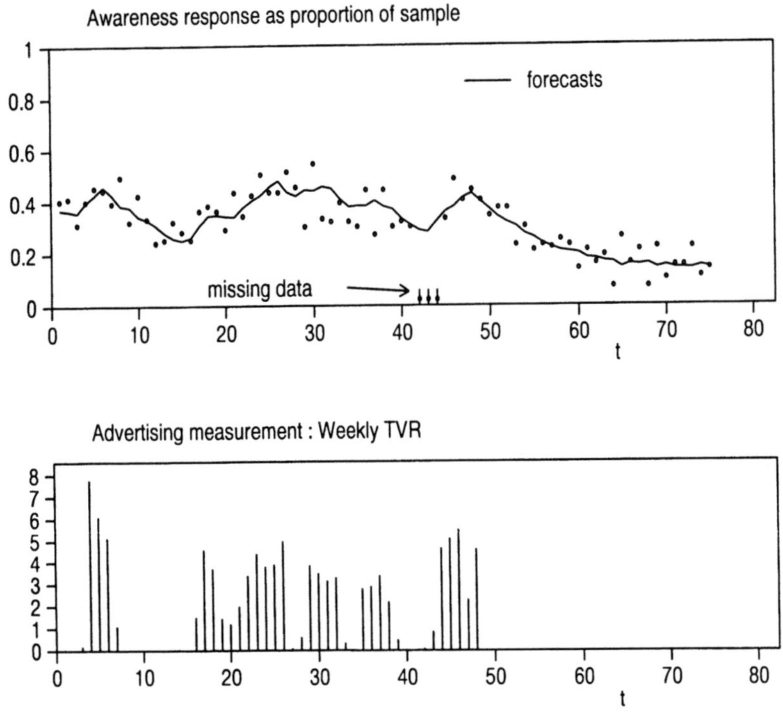

Figure 14.6 Awareness data and one-step ahead point forecasts

in week 16 and ending in week 48. The decay of awareness in the final period of no advertising is clear from the graph.

Based on the prior moments in (14.41), the sequential one-step ahead forecasting and updating equations apply directly. Figure 14.6 provides the data with a line plot of the one-step point forecasts f_t, $(t = 1, \dots, 75)$, as they are calculated over time. These are simply the one-step forecast means, since under the binomial model for Y_t, we have $\mathrm{E}[Y_t/n_t|\mu_t] = \mu_t$, and so $\mathrm{E}[Y_t/n_t|D_{t-1}] = \mathrm{E}[\mu_t|D_{t-1}] = f_t$. Note the response to renewed advertising after the periods of no TVR and the smooth decay of forecast awareness during these periods. For clarity, forecast uncertainties are not indicated on the graph (see below). However, under the binomial model, $\mathrm{V}[Y_t/n_t|\mu_t] = \mu_t(1-\mu_t)/n_t$, so that

$$\begin{aligned}\mathrm{V}[Y_t/n_t|D_{t-1}] &= \mathrm{E}[\mathrm{V}\{Y_t/n_t|\mu_t\}|D_{t-1}] + \mathrm{V}[\mathrm{E}\{Y_t/n_t|\mu_t\}|D_{t-1}]\\ &= \mathrm{E}[\mu_t(1-\mu_t)/n_t|D_{t-1}] + \mathrm{V}[\mu_t|D_{t-1}]\\ &= f_t(1-f_t)/n_t + q_t(1-1/n_t).\end{aligned}$$

Thus, the variation in the one-step forecast (beta/binomial) distribution naturally exceeds that of a standard binomial distribution with μ_t estimated by the value f_t. This variance decreases as μ_t moves towards 0 or 1. At the lowest values of f_t in the graph, $f_t \approx 0.1$ corresponds to $f_t(1 - f_t)/66 \approx 0.035^2$, so that the forecast standard deviation for all t exceeds 0.035. With this in mind, the variation in the observed proportions about their forecast means appears to be reasonably attributable to unpredictable sampling variation.†

Figures 14.7a to 14.10b provide graphs of the on-line and retrospectively filtered trajectories of the four state parameters α_t, β_t, ρ_t, κ_t. The on-line trajectories are, as usual, just the posterior means from $\mathbf{m}_t$ over time, together with one standard deviation limits using the relevant posterior variances from $\mathbf{C}_t$. The smoothed trajectories are similar, but now the estimates are based on all the data, derived through the standard filtering algorithms (Section 4.9; see comments in Section 14.3.4 above). The following features are apparent.

(a) The on-line trajectory of α_t in Figure 14.7a is stable, the estimate remaining near the prior mean of 0.1 until the last 10 or 15 observations. Until this time, little information is obtained from the data about the lower threshold, since awareness does not decay sufficiently. At the end, the TVR is zero for a long period and awareness decays, the data thus informing about the lower threshold. The graph then indicates a slight increase to values nearer 0.12 at the end. The smoothed trajectory in Figure 14.7b, re-estimating the threshold at each time based on all the data, confirms this.

(b) The initial variance for β_0 is rather large, resulting in the possibility of marked adaptation to the data in learning about the upper threshold. This is apparent in Figure 14.8a. The smoothed version in Figure 14.8b confirms that the upper threshold is fairly stable over time, though estimated as rather lower than the initial prior, around 0.8 at the end of the series. Note, however, the wide intervals about the estimates; there is really very little information in the data about the upper threshold levels.

(c) Figures 14.9a and 14.9b illustrate the stability of the decay parameter ρ_t at around 0.9.

(d) There is a fair degree of movement in the on-line trajectory for the penetration parameter κ_t in Figure 14.10a. In part this is due to conflict between the information in the data and that in the initial prior. This is confirmed in the smoothed version, Figure 14.10b, that indicates that κ_t is really fairly stable, taking values nearer 0.03 than the initial mean 0.02. In terms of the half-penetration effect, 0.02 corresponds to TVR levels of 35, whereas the more appropriate value of 0.03 corresponds to TVR at around 24. These

†One point of detail concerns the last few observations, that at low awareness levels, are somewhat more erratic than perhaps expected. Here there is a degree of evidence for extra-binomial variation at low levels that could be incorporated in the model for forecasting ahead, as previously mentioned.

points are further illustrated in Figures 14.11 and 14.12. These are the posterior versions of Figures 14.1 and 14.3, constructed in precisely the same way but now at $t = 75$ rather than at $t = 0$. Hence they represent forecasts based on the final posterior moments $\mathbf{m}_{75}$ and $\mathbf{C}_{75}$ at $t = 75$, modified as follows.

(e) E_{75} is constrained to the expected maximum value $\mathrm{E}[\beta_{75} - \alpha_{75}|D_{75}]$. Figure 14.11 provides point forecasts $f_{75}(k)$, with limits $f_{75}(k) \pm \sqrt{q_{75}(k)}$, for the next 25 weeks, assuming no further TVR, $X_{75+k} = 0$ for $k = 1, \ldots, 75$. The decay of awareness forecast is similar in expectation to that initially (Figure 14.1), although the forecast distributions are more precise due to the processing of the 75 observations.

(f) Constraining E_{75} to be zero, Figure 14.12 provides one-step point forecasts $f_{75}(1)$ with standard deviation limits. The expected penetration is rather larger than that initially in Figure 14.2, consistent with the posterior view expressed in (d) above that κ_{75} is very likely to be larger than the initial estimate of κ_0.

Finally, model validation *What-if?* forecasts are represented in Figure 14.13. At $t = 75$, consider forecasting ahead to times $75+k$, $(k = 1, \ldots, 75)$, based on hypothesised values of TVR over the next 75 weeks. In particular, suppose that the company considers repeating the advertising campaign, reproducing the original TVR series precisely. Starting with the full posterior moments $\mathbf{m}_{75}$ and $\mathbf{C}_{75}$, the expected value of E_{75} is altered to 0.3, consistent with initial expectations. Figure 14.13 then provides step ahead forecasts of awareness proportions over the coming weeks. The forecasts are given in terms of means and standard deviation limits from the beta/binomial predictive distribution. These moments are defined, for all k and with $t = 75$, by

$$\mathrm{E}[Y_{t+k}/n_{t+k}|D_t] = f_t(k)$$

and

$$\mathrm{V}[Y_{t+k}/n_{t+k}|D_t] = f_t(k)(1 - f_t(k))/n_{t+k} + q_t(k)(1 - 1/n_{t+k}).$$

Given the past stability of the model parameters, these predictions of course resemble those in the past. The historical data are also plotted in this figure for comparison. This sort of exercise can be viewed as a form of model validation, examining the question of just how well the model performs in forecasting a further data series that happens to coincide precisely with the past data. Any systematic discrepancies will show up most sharply here, although in this case the forecasts are adequate.

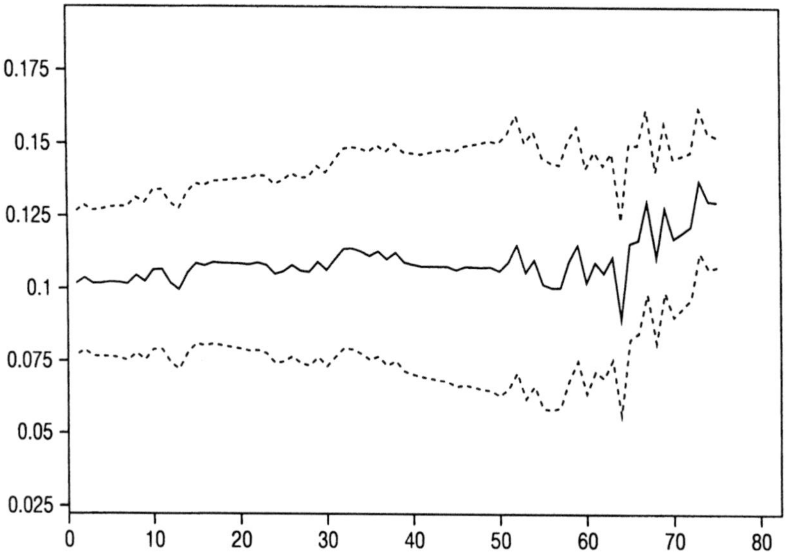

Figure 14.7a On-line trajectory of lower threshold α_t

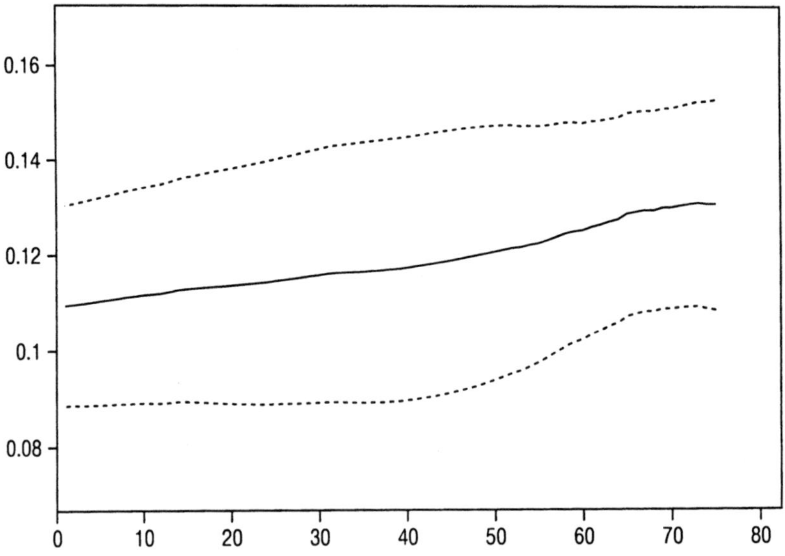

Figure 14.7b Smoothed trajectory of lower threshold α_t

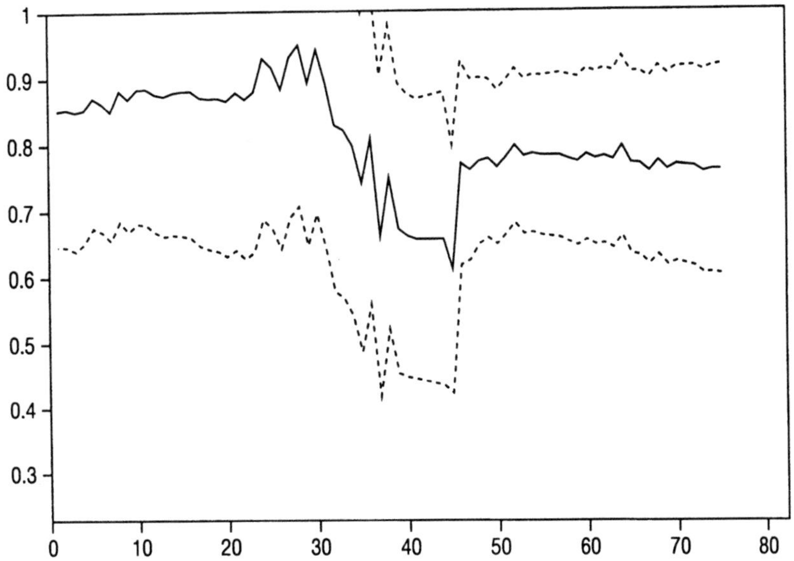

Figure 14.8a On-line trajectory of upper threshold β_t

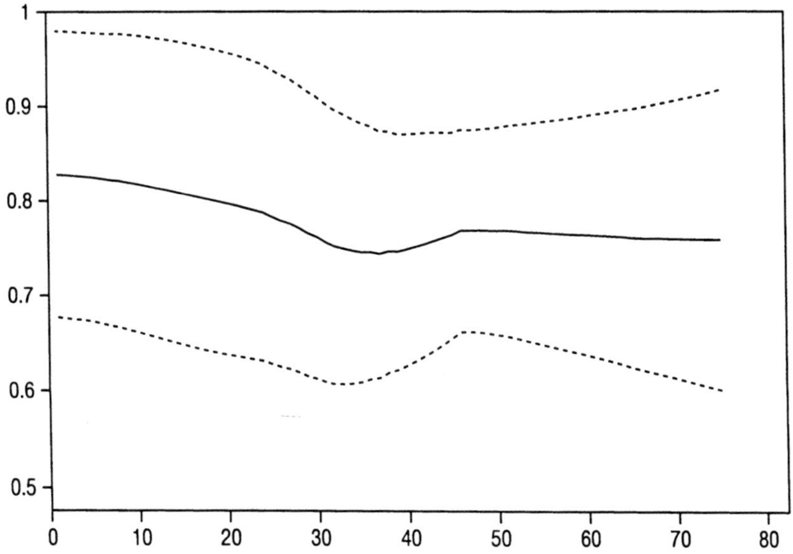

Figure 14.8b Smoothed trajectory of upper threshold β_t

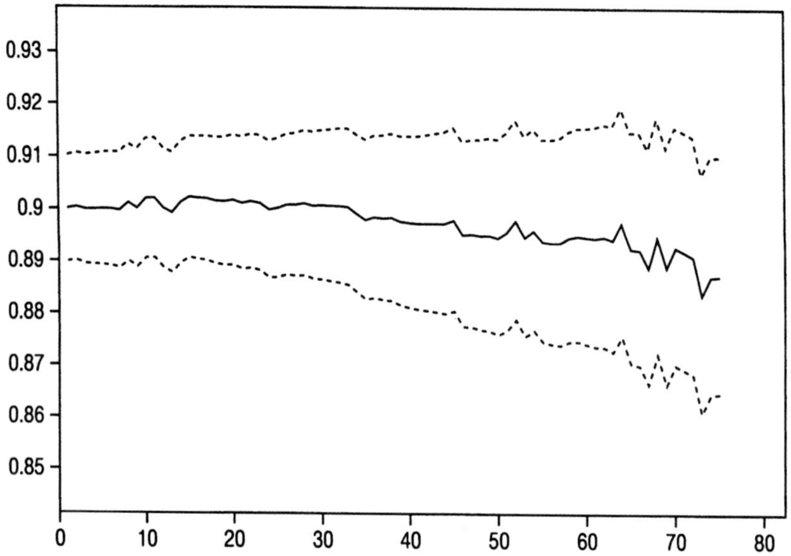

Figure 14.9a On-line trajectory of memory decay ρ_t

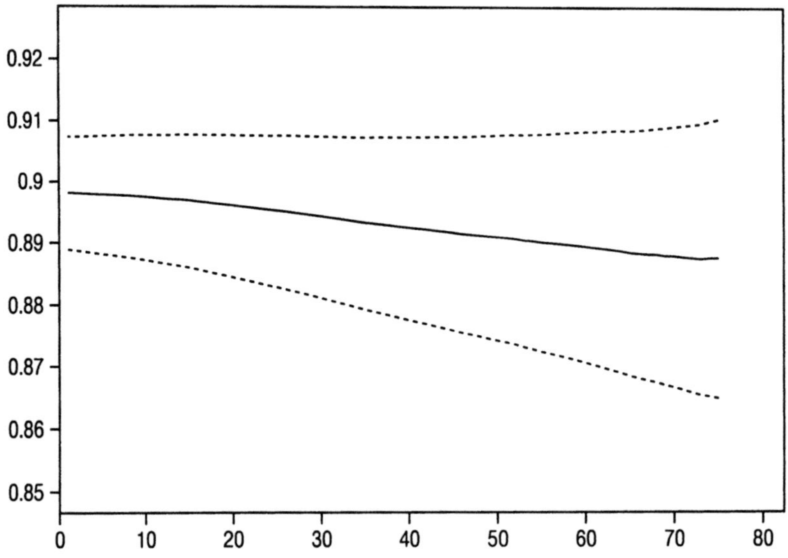

Figure 14.9b Smoothed trajectory of memory decay ρ_t

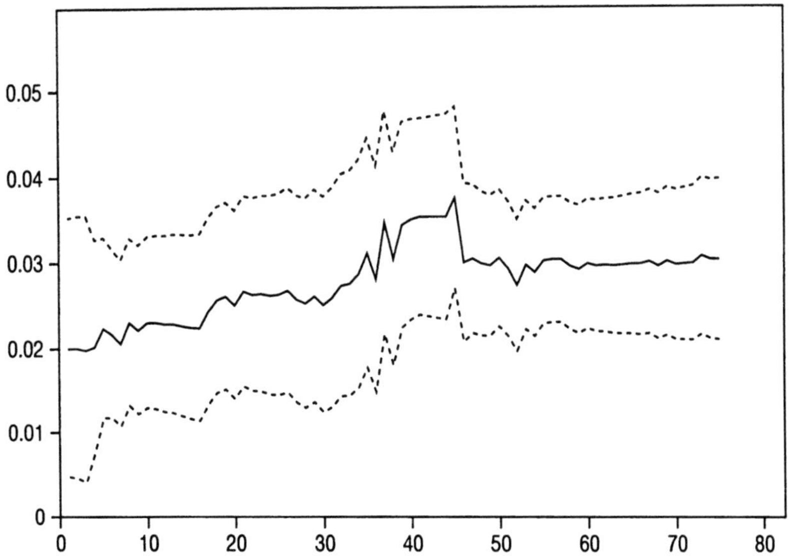

Figure 14.10a On-line trajectory of penetration κ_t

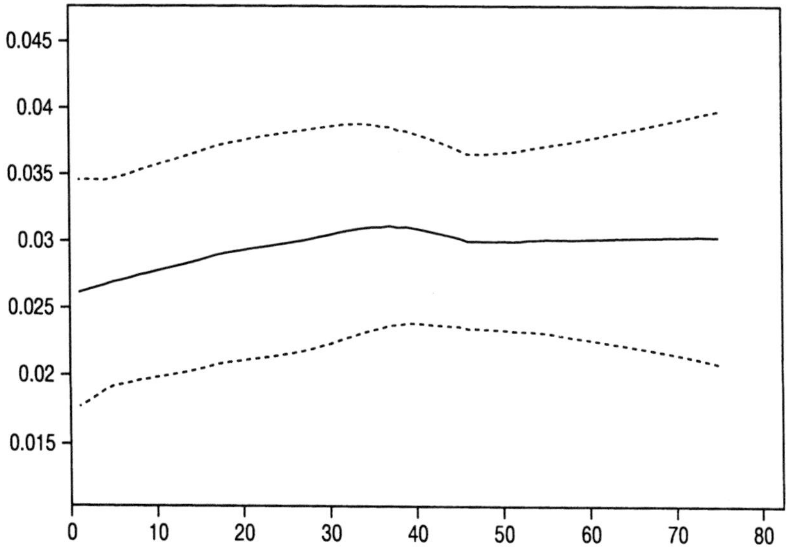

Figure 14.10b Smoothed trajectory of penetration κ_t

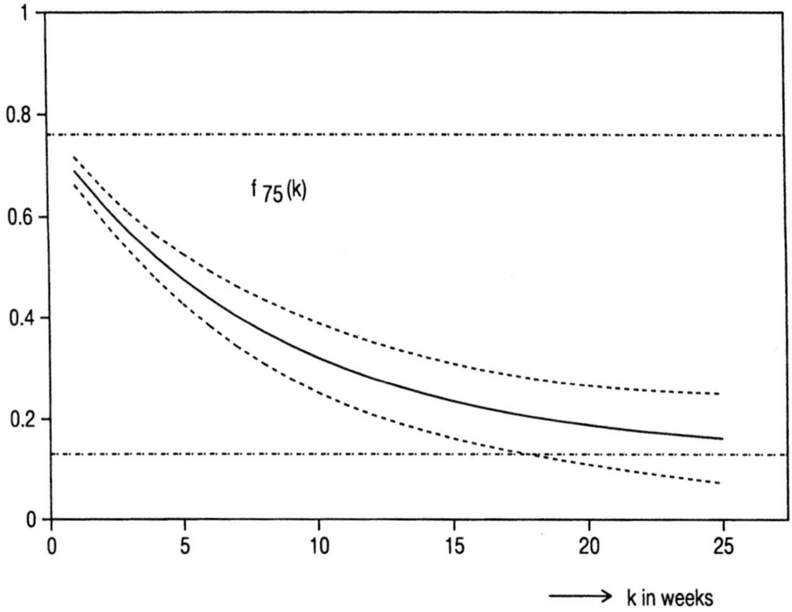

Figure 14.11 Posterior forecasts of expected memory decay

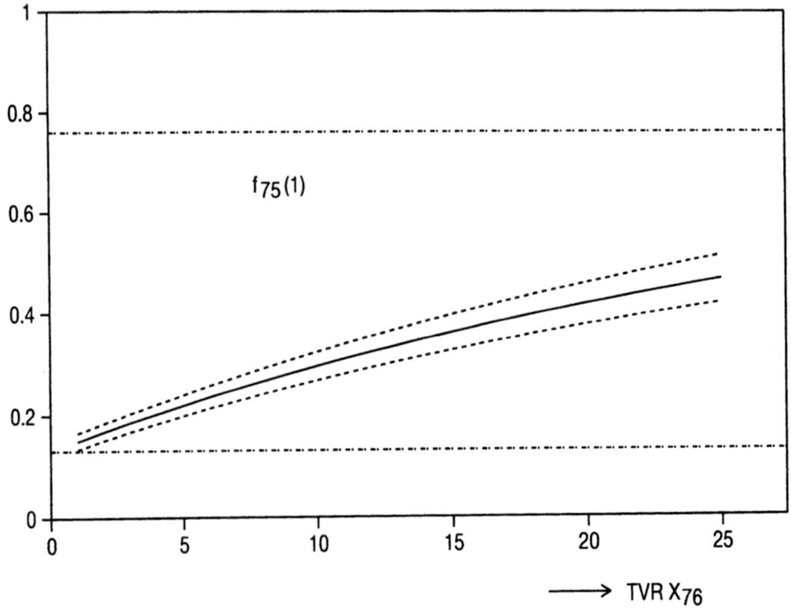

Figure 14.12 Posterior forecasts of expected TVR penetration

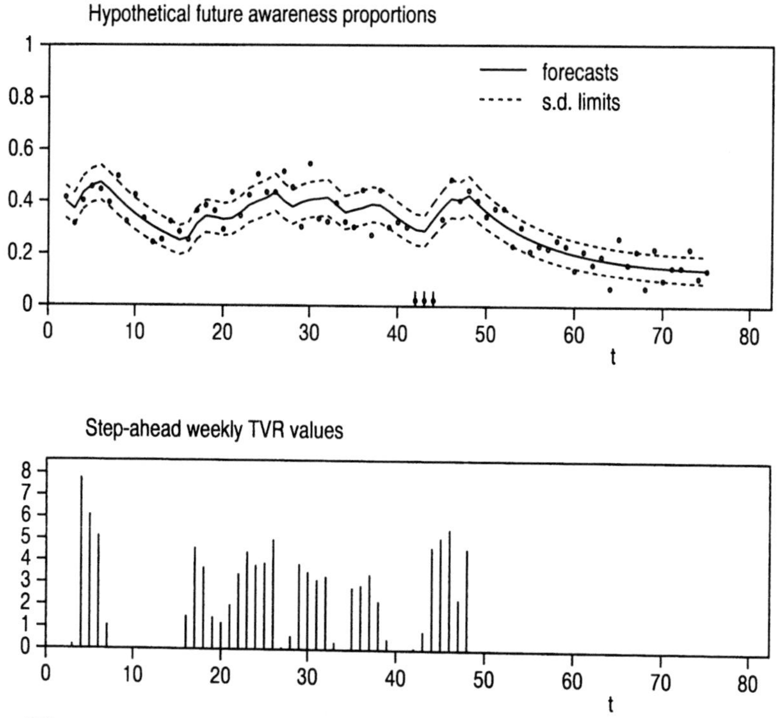

Figure 14.13 'What-if?' step-ahead forecast awareness

14.5 FURTHER COMMENTS AND EXTENSIONS

14.5.1 Monitoring, change-points and outliers

The panoply of techniques for subjective intervention of Chapter 11 apply to non-normal models directly. Interventions may be made on the prior and posterior moments of the state vector to incorporate external information routinely. Thus, for example, abrupt changes in parameters can be modelled through additional evolution error terms, as in the normal DLM.

Concerning automatic detection of outliers and monitoring for change-points, West (1986a) develops a framework that provides non-normal analogues of the feedback intervention schemes in Sections 11.4 and 11.5. Recall from that chapter that the central component of any model testing scheme is the use of one or more alternative models to provide relative assessment of the predictive ability of the original model. Here consider the use of a single alternative model in the framework of Section 11.5 to provide an automatic monitor on the predictive performance of the basic model. As in Section 11.4, comparison between the original model, denoted by M_0, and any single alternative M_1 involves the data at time t through the Bayes' factor $H_t = p_0(Y_t|D_{t-1})/p_1(Y_t|D_{t-1})$, where the subscripts denote the relevant model. Sequential model monitoring is based on the sequence of Bayes' factors and the cumulative versions in Section 11.4.2, and all that is required to extend to the non-normal case is a suitable class of alternative predictive densities $p_1(Y_t|D_{t-1})$. The ideas in Section 11.4.3, in particular Example 11.3, suggest the following alternatives, based on West (1986a). At time t in M_0, the predictive distribution is defined by (14.28), with defining quantities r_t and s_t given in (14.26) as functions of f_t and q_t in (14.23). In particular, the precision of the forecast distribution is a decreasing function of q_t, equivalently an increasing function of the prior precision parameter s_t. In assessing the predictive density $p_0(Y_t|D_{t-1})$ with a view to identifying observations that are poorly predicted, a suitable alternative will give greater probability to regions not heavily favoured, whilst being more diffuse in regions of high density. The idea, as in Example 11.3, is that a *flatter* version of the standard forecast density is appropriate. A class of such alternatives is easily constructed by adapting the standard model to have smaller prior precision parameters. In modelling terms, the standard equation $\lambda_t = \mathbf{F}_t'\boldsymbol{\theta}_t$ is revised to

$$\lambda_t = \mathbf{F}_t'\boldsymbol{\theta}_t + \partial\lambda_t,$$

where $\partial\lambda_t$ is an additional zero-mean error term with known variance Δq_t, say. With this defining the alternative model at time t, the analogues of equations (14.23) to (14.28) define the components of the alternative model analysis, the only difference being that q_t is replaced by $q_t + \Delta q_t$ throughout. Thus, the resulting conjugate prior distribution under M_1 is more diffuse than under M_0, and this carries over to the predictive density $p_1(Y_t|D_{t-1})$. Full details about this construction, in particular concerning structuring and quantifying Δq_t, are given in West (1986a).

West (1986b) shows how the use of the above ideas provides non-normal extension of the multi-process models for change-point estimation and outlier accommodation. It is clear that so far as modelling change-points is concerned, the multi-process framework extends directly, abrupt changes in $\boldsymbol{\theta}_t$ being ascribed to evolution errors with appropriately large variances. One difference in the non-normal case is, of course, that the predictive densities forming the components of mixtures in a multi-process model are not normal or T, but this is a technicality; the calculations proceed just as in Chapter 12. The main point of difference requiring thought concerns multi-process models with an outlier modelling component. West (1986b) uses the ideas above here, modelling outliers with a component having a diffuse predictive distribution.

Alternative approaches to monitoring and model assessment are found in West, Harrison and Migon (1985), based on developments in West (1985a, 1986a). These have more in common with standard statistical approaches to outlier detection and accommodation, and related methods of influence assessment, but are not discussed further here.

14.5.2 Applications, extensions and related approaches

Several further illustrations and applications of the DGLM in forecasting and smoothing time series appear in West, Harrison and Migon (1985). Various Poisson and binomial models are covered. An application in reliability growth analysis and forecasting is reported in Mazzuchi and Soyer (1992). Also, related models are considered from different viewpoints by Smith (1979), Souza (1981), and Smith and Miller (1986). Multivariate versions and extensions appear in Fahrmeir (1992); see also Grunwald, Raftery and Guttorp (1993), and Cargnoni, Müller and West (1996) for related models for discrete multinomial time series.

The special case of binary time series is discussed in West and Mortera (1987). The observational model here is a special case of the binomial model in Example 14.2, binary time series in which $n_t = 1$ for all t, the observations being simple event indicators, 0 or 1. A variety of applications are discussed in this paper, and related topics in the area of subjective probability assessment, calibration and combination are discussed in West (1985b and 1985c). For example, simple Markov models can be written as special static cases of the DGLM for binary data. In the framework of Example 14.2 with $n_t = 1$, suppose the simplest case of a time-homogeneous, one-step Markov process. Thus, the "success" probability μ_t at time t actually depends on D_{t-1}, but only through the previous observation Y_{t-1}. This may be written in many forms, the most obvious being

$$\mu_t = \alpha Y_{t-1} + \beta(1 - Y_{t-1}),$$

where α and β are the probabilities *conditional* on $Y_{t-1} = 1$ and $Y_{t-1} = 0$ respectively. If $\boldsymbol{\theta} = (\alpha, \beta)'$ and $\mathbf{F}_t = (Y_{t-1}, 1 - Y_{t-1})'$, then the model is a special, static case of the DGLM. Various similar models appear in the above references. Easy extensions cover Markov depedencies to order higher than the first, dynamic versions in which α and β vary stochastically over time to provide non-homogenous processes and similar models, but with μ_t undergoing a non-linear transformation, such as logistic, before being related to the state vector.

A major extension and application of the DGLM approach has been made in Gamerman (1985, 1987a) to models for the analysis of survival data. Discussion of this important area is far beyond the scope of this book, although the underlying modelling ideas are essentially those of dynamic modelling of time series. Interested readers should consult the above references for full theoretical details and illustrations. Applications in medicine and economics, along with further theoretical and practical details, can be found in Gamerman (1987b), Gamerman and West (1987a and b), Gamerman, West and Pole (1987) and West (1992d).

Finally, we note that new approaches to analysis of non-linear and non-normal models based on simulation methods are beginning to impact on practical applications, as was mooted in the previous chapter. The following chapter takes up simulation methods in various models, and the potential for such tools to apply in various non-linear contexts, such as the models in this chapter, will be apparent from that development. Indeed, recent work by several authors has begun the development of simulation-based analyses in various dynamic generalised linear models as well as in other, "custom" non-linear and non-normal frameworks. This is a new and growing field, and one sure to develop rapidly in coming years. Some key early works of interest include Cargnoni, Müller and West (1996), Carlin, Polson and Stoffer (1992), Gamerman (1996), Jaquier, Polson and Rossi (1994), Scipione and Berliner (1993), Shephard (1994), and Shephard and Pitt (1995).

14.6 EXERCISES

(1) Suppose a Poisson sampling model, $(Y_t|\mu_t) \sim \mathrm{P}[\mu_t]$, where $\mu_t > 0$ is the Poisson mean, the density being given by

$$p(Y_t|\mu_t) = \mu_t^{Y_t} \mathrm{e}^{-\mu_t}/Y_t!, \qquad (Y_t = 0, 1, \ldots).$$

(a) Verify that this is the density of an exponential family distribution, identifying the natural parameter η_t and the defining functions $y_t(\cdot)$, $a(\cdot)$ and $b(\cdot\ ,\cdot)$ in (14.1).

(b) Verify that μ_t is both the mean and the variance of the distribution.

(c) Identify the conjugate prior family defined through (14.7). Verify that the conjugate distributions are such that μ_t is gamma distributed.

(d) For given prior quantities r_t and s_t in (14.7), calculate the one-step forecast density $p(Y_t|D_{t-1})$.

(2) Suppose Y_t is gamma distributed, $(Y_t|\mu_t, n_t) \sim \mathrm{G}[n_t, \mu_t]$ for some n_t and μ_t, both positive quantities, with n_t known. The density is

$$p(Y_t|\mu_t, n_t) = \mu_t^{n_t} Y_t^{n_t - 1} \mathrm{e}^{-\mu_t Y_t}/\Gamma(n_t), \qquad (Y_t > 0).$$

(a) Verify that this is the density of an exponential family distribution, identifying the natural parameter η_t and the defining functions $y_t(\cdot)$, $a(\cdot)$ and $b(\cdot\ ,\cdot)$ in (14.1).

(b) Calculate the mean and the variance of the distribution as functions of n_t and μ_t, and verify that the distribution has a quadratic variance function, $\mathrm{V}[Y_t|\mu_t, n_t] \propto \mathrm{E}[Y_t|\mu_t, n_t]^2$.

(c) Identify the conjugate prior family defined through (14.7). Verify that the conjugate distributions are such that μ_t is also gamma distributed, and identify the prior mean and variance as functions of r_t and s_t.

(d) Calculate the one-step forecast density $p(Y_t|D_{t-1})$.

(3) Verify the form of the predictive and posterior densities given in (14.8) and (14.9).

(4) Verify the results used in Example 14.3 that in a model in which $(\mu_t|D_{t-1}) \sim \text{Beta}[r_t, s_t]$ and $\eta_t = \log[\mu_t/(1-\mu_t)]$, then

$$\text{E}[\eta_t|D_{t-1}] = \gamma(r_t) - \gamma(s_t)$$

and

$$\text{V}[\eta_t|D_{t-1}] = \dot{\gamma}(r_t) - \dot{\gamma}(s_t).$$

(5) Consider Poisson or gamma models above in which the conjugate priors are gamma, $(\mu_t|D_{t-1}) \sim \text{G}[r_t, s_t]$. Suppose as usual that $(\lambda_t|D_{t-1}) \sim [f_t, q_t]$.
   (a) In a linear regression model, $\text{E}[Y_t|\mu_t] = \lambda_t$, show that the conjugate gamma prior is defined via $r_t = f_t^2/q_t$ and $s_t = f_t/q_t$.
   (b) In a log-linear regression model, $\text{E}[Y_t|\mu_t] = \log(\lambda_t)$, show that the conjugate gamma prior has r_t and s_t implicitly defined by $f_t = \gamma(r_t) - \log(s_t)$ and $q_t = \dot{\gamma}(s_t)$.

(6) Verify that given the moments (14.16), the equations (14.19) and (14.20) reduce to the usual DLM updating recurrences.

(7) Verify that equations (14.19) and (14.20) reduce to the standard DLM updating equations.

(8) This exercise and that following concern the use of the "power discount" procedure for non-normal, first-order polynomial dynamic models, as developed in Smith (1979). See also Exercise 13 of Section 6.5 for a related example in the normal DLM.

A non-normal discount model appropriate for time-varying rates of Poisson distributed time series is defined, at each time t, as follows:

$$\text{observation:} \quad (Y_t|\mu_t, D_{t-1}) \sim \text{Poisson}[\mu_t],$$

$$\text{system:} \quad p(\mu_t = \mu|D_{t-1}) \propto \{p(\mu_{t-1} = \mu|D_{t-1})\}^{\delta},$$

for some discount factor δ in $(0, 1]$. The system equation is interpreted as follows: writing $f_{t-1}(\cdot)$ for the density $p(\mu_{t-1}|D_{t-1})$, the implied prior density function $p(\mu_t|D_{t-1})$ is explicitly given by $c_t f_{t-1}(\cdot)^{\delta}$ for an appropriate normalising constant c_t.
   (i) Show that $(\mu_t|D_{t-1}) \sim \text{G}[n_t^*, \alpha_t^*]$, where

$$\alpha_t^* = \alpha_{t-1}\delta \quad \text{and} \quad n_t^* = n_{t-1}\delta + 1 - \delta.$$

   (ii) With $D_t = \{Y_t, D_{t-1}\}$, prove that $(\mu_t|D_t) \sim \text{G}[n_t, \alpha_t]$, expressing α_t and n_t in terms of α_t^* and n_t^*.
   (iii) Show that the forecast distribution $(Y_t|D_{t-1})$ is negative binomial; explicitly,

$$p(Y_t|D_{t-1}) = \binom{n_t^* + Y_t - 1}{Y_t} \left(\frac{\alpha_t^*}{1+\alpha_t^*}\right)^{n_t^*} \left(\frac{1}{1+\alpha_t^*}\right)^{Y_t}$$

   for $Y_t \geq 0$.

(9) In analysing brand switching in market share analysis, a non-normal discount model describes market share proportions $\boldsymbol{\pi}_t = (\pi_{1t}, \ldots, \pi_{nt})'$ of a set of n competing products as slowly varying in time. Data arise from weekly sample survey results $\mathbf{Y}_t = (Y_{1t}, \ldots, Y_{nt})'$, where Y_{it} is the number of purchases of product i and $s_t = \sum_{i=1}^n Y_{it}$ the total purchases in the sample in week t. The standard multinomial sampling model has

$$p(\mathbf{Y}_t|\boldsymbol{\pi}_t) = s_t! \prod_{i=1}^n \pi_{it}^{Y_{it}}/Y_{it}!.$$

At $t-1$, suppose the posterior for $\boldsymbol{\pi}_{t-1}$ is a Dirichlet distribution, namely

$$(\boldsymbol{\pi}_{t-1}|D_{t-1}) \sim \text{Dirichlet}(\mathbf{r}_t),$$

where $\mathbf{r}_t = (r_{1t}, \ldots, r_{nt})'$ is known from past data analyses. This has density function

$$p(\boldsymbol{\pi}_{t-1}|D_{t-1}) = K(\mathbf{r}_t) \prod_{i=1}^n \pi_{i,t-1}^{r_{it}}$$

over the n-simplex such that $0 \le \pi_{i,t-1} \le 1$ for $i = 1, \ldots, n$ and $\sum_{i=1}^n \pi_{i,t-1} = 1$. The factor $K(\mathbf{r}_t)$ is a known normalising constant. The system equation provides the prior

$$p(\boldsymbol{\pi}_t = \boldsymbol{\pi}|D_{t-1}) \propto \{p(\boldsymbol{\pi}_{t-1} = \boldsymbol{\pi}|D_{t-1})\}^\delta,$$

for some discount factor δ, via the same kind of power discount construction as in the previous question.

(a) Obtain the forecast distribution $p(\mathbf{Y}_t|s_t, D_{t-1})$, represented in terms of $K(\cdot)$.

(b) Obtain the posterior distribution $p(\boldsymbol{\pi}_t|\mathbf{Y}_t, D_{t-1})$.

CHAPTER 15

SIMULATION-BASED METHODS IN DYNAMIC MODELS

15.1 INTRODUCTION

Chapters 13 and 14 describe computational problems arising in models with non-linear components and non-normal structure. Many such problems are now amenable to analysis, at least in part, based on computational approaches using stochastic simulation methods. The material in Section 13.6 demonstrates this and is prelude, both historically and thematically, to approaches now to be presented. In the early 1990s, developments in simulation methods of computation have had dramatic impact on the entire field of Bayesian statistics, and dynamic models are no exception. This chapter describes some such methods, including examples of direct posterior simulation in interesting, non-stationary DLMs and iterative methods based on *Markov chain Monte Carlo* approaches, referred to by the acronym MCMC. As we approach the end of the twentieth century, it seems quite clear that routine statistical analysis in more complex and realistic models will continue to become more and more accessible as such tools are refined and extended.

In Section 15.1 we briefly review elements of simulation in Bayesian analyses. Section 15.2 discusses iterative, Markov chain Monte Carlo approaches in dynamic models, with various examples and illustrations. A final section concerns a specific class of time-varying autoregressive component DLMs in which we use MCMC methods for analysis.

15.1.1 Elements of posterior simulation

Some concepts and elements of Bayesian inference based on stochastic simulation are reviewed first. Useful background on posterior simulation appears in Bernardo and Smith (1994, Section 5.5), and Gelman, Carlin, Stern and Rubin (1995), Chapters 10 and 11. For general issues and methods of simulation, see Ripley (1987).

In a specified statistical model, suppose that $\mathbf{X}$ represents a collection of quantities of interest, and suppose that an analysis leads to a posterior distribution with density $p(\mathbf{X})$, based on available prior information and observed data. Direct posterior sampling refers to any process by which random samples may be drawn from $p(\mathbf{X})$. If $\mathbf{X}$ is multivariate normal, for example, then the posterior is trivially sampled directly. Suppose a simulation study draws a direct sample of Monte Carlo size m, namely $\mathbf{X}_1, \ldots, \mathbf{X}_m$. Then approximate posterior inference is often based on simple sample summaries, such as sample histograms to approximate marginal densities for elements of $\mathbf{X}$, sample means to approximate posterior means, and so forth. This is based simply on the law of large numbers; under very generally applicable conditions, sample averages converge to corresponding "exact" posterior expectations, i.e.,

$$m^{-1}\sum_{i=1}^{m} g(\mathbf{X}_i) \to \int g(\mathbf{X})p(\mathbf{X})d\mathbf{X} = \mathrm{E}[g(\mathbf{X})]$$

in probability as $m \to \infty$. Sample histograms represent approximations to posterior densities, with the underlying assurance that the empirical cumulative distribution converges to the true posterior marginal distribution in this sense. Often, direct posterior simulation samples with m in the several or tens of thousands are easily generated, and resulting posterior approximations are highly accurate.

Refined approximations to some characteristics of posterior distributions are often available via the device of so-called "Rao-Blackwellisation" (Gelfand and Smith 1990). For example, suppose $\mathbf{X}$ is partitioned into $\mathbf{X} = (X_1, \mathbf{X}_2)$ where X_1 is scalar, and that the conditional density $p(X_1|\mathbf{X}_2)$ is a "standard" distribution of known form, such as a conditional normal. Partition the sampled values similarly, so that $\mathbf{X}_i = (X_{1,i}, \mathbf{X}_{2,i})$ for each i. Then

$$p(X_1) \approx m^{-1}\sum_{i=1}^{m} p(X_1|\mathbf{X}_{2,i})$$

is an approximation to the univariate margin $p(X_1)$ that utilises some of the known structure in $p(\mathbf{X})$ as well as the Monte Carlo sample. This "mixture" approximation is always a more accurate estimate of $p(X_1)$ than the raw histogram of $X_{1,i}$ values, in the sense of being subject to a lesser degree of sampling variability in repeated Monte Carlo samples. In some specific applications, it is very easy to compute such refined approximations, and then they can be compared with raw histograms. Similarly, posterior characteristics, such as probabilities and means, may be more accurately approximated by sample averages of conditionally exact values, based on

$$\mathrm{E}[g(X_1)] \approx m^{-1}\sum_{i=1}^{m} \mathrm{E}[g(X_1)|\mathbf{X}_{2,i}],$$

for specified functions $g(\cdot)$ when the exact conditional mean $\mathrm{E}[g(X_1)|\mathbf{X}_2]$ is analytically available for all $\mathbf{X}_2$.

Monte Carlo integration and posterior approximation via importance sampling involves direct simulation from an importance sampling distribution, usually viewed as an approximation to the true density $p(\mathbf{X})$; we refer to the brief development in Section 13.6.1, and reference therein, and do not discuss this topic further here. There are, however, direct connections between importance sampling and various MCMC methods, noted below. Useful discussions in the context of Bayesian computation more generally appear in Gelman et al (1995, Chapters 10 and 11); see also West (1992a,c).

15.1.2 Markov chain Monte Carlo methods

Direct simulation is often impossible due to the complicated mathematical form of posterior distributions in many applied modelling contexts. In the 1990s, the Bayesian statistics field has developed a range of simulation schemes based on iterative, Markov chain Monte Carlo methods, referred to by the acronym MCMC, to address problems of posterior simulation when direct methods are infeasible. Some brief review of specific methods is given here, following some notation and basic concepts.

MCMC methods are designed to successively simulate values of the $\mathbf{X}$ vector based on a strategy designed to eventually draw these values from the desired, or so-called target, posterior distribution $p(\mathbf{X})$. They are Markov chain methods, so an ordered sequence of simulated values, $\mathbf{X}_1, \mathbf{X}_2, \ldots$, is generated by

(1) specifying a starting value $\mathbf{X}_1$, then
(2) sampling successive values from a specified transition distribution with density $f(\mathbf{X}_i|\mathbf{X}_{i-1})$, for $i = 2, 3, \ldots$; $\mathbf{X}_i$ is generated conditionally independently of $\mathbf{X}_{i-2,}\mathbf{X}_{i-3,}, \ldots$.

Methods differ in the choice of transition densities $f(\cdot|\cdot)$. Variants of this basic idea have transition distributions that may depend on i, being modified or updated based on both the number of sampled values and the actual past $\mathbf{X}_i$ values generated, but these are not considered here. MCMC sequences of $\mathbf{X}$ values are thus first-order dependent, not random samples. Typically, dependencies between $\mathbf{X}_i$ and $\mathbf{X}_j$ are positive and decay exponentially with $|i-j|$. Under quite weak conditions (Tierney 1994), the Markov chain "converges" in the sense that ultimately, the sequence of sampled values appears stationary and the individual $\mathbf{X}_i$ are marginally distributed according to a unique stationary distribution. MCMC designs choose transition densities $f(\cdot|\cdot)$ in order to ensure that this stationary distribution is the true target posterior $p(\mathbf{X})$. Then simulations are run for some number of iterations until the effect of the chosen starting value is assumedly negligible and the process is approximately stationary. Thereafter, successively sampled values may be assumed to be approximately drawn from the posterior and though dependent appropriately summarised to produce approximate posterior inferences.

The literature on MCMC methods, covering issues of design and choice of methods, convergence theory and practical methods of diagnosing convergence, numerous applications and other importance topics, is simply huge and growing dramatically. Some recent useful references include Gelman et al (1994), and Gilks, Richardson and Spiegelhalter (1996); in addition to containing excellent discussions of many aspects of MCMC methods, in theory and in practice, these books provide access to the literature through their many up-to-date references.

Gibbs sampling

For the p-vector $\mathbf{X}$, define the notation $X_j|X_{(-j)}$ where

$$X_{(-j)} = \{X_1, \dots, X_{j-1}, X_{j+1}, \dots, X_p\}$$

for each j. The posterior $p(\mathbf{X})$ is then (usually) uniquely characterised by the full set of p "complete conditional" densities $p(X_j|X_{(-j)})$, for $j = 1, \dots, p$. Gibbs sampling MCMC methods are built around sequences of simulations from such conditionals, based on this characterisation. In moving one step in the chain, the current sampled value $\mathbf{X}_{i-1}$ is updated to $\mathbf{X}_i$ by sequentially updating the elements one by one: at each stage $j = 1, 2, \dots, p$, a new value of X_j is simulated from the true conditional $p(X_j|X_{(-j)})$, where the $p-1$ elements of $X_{(-j)}$ are each set at their most recently sampled values. In this way, the full p-vector is updated element by element, and a transition distribution is implicitly defined that delivers the full posterior as the stationary distribution of the resulting Markov Chain (Gelfand and Smith 1990).

Generalisations update sub-vectors rather than just individual scalar elements of $\mathbf{X}$. Thus, the above story applies, but now with each of the elements X_j replaced by a sub-vector of $\mathbf{X}$. In applications, this is more common than the scalar element-wise approach, as is illustrated in dynamic modelling contexts below.

Other MCMC algorithms and practical issues

Gibbs sampling is attractive in its simplicity and because the use of exact conditional distributions for simulation requires no other choices to be made in connection with defining a transition density. Other MCMC methods require choices. Variants of the wide and very general class of Metropolis-Hastings methods (Tierney 1994) have been developed and are growing in application in dynamic models. Though further development is beyond our scope in this book, this is a very interesting and likely critical area for the future.

Any application of the MCMC method requires care and expertise in setting up algorithms, in specifying and assessing sensitivity to starting values for the Markov chain, and in related studies of convergence of the chain; convergence means that the successively generated parameter values are approximately marginally distributed according to the required stationary distribution, i.e., the posterior distribution of interest. Theoretical and practical issues relevant to determining convergence criteria and assessing convergence of specific analyses has been a very significant research field in recent years, and continues to be. Useful discussions appear in the references in the previous section.

Finally, we note that there have recently been quite varied developments in simulation methods in ranges of time series models outside the DLM framework. Some interesting and relevant references include Albert and Chib (1993), McCulloch and Tsay (1994), and Müller, West and MacEachern (1996). Additional references more directly related to dynamic modelling are mentioned in context in the following sections.

15.2 BASIC MCMC IN DYNAMIC MODELS

15.2.1 Posterior sampling state-by-state in general models

In a rather general framework, a non-linear and non-normal dynamic model is defined by the sequence of observation and evolution distributions

$$p(Y_t|\boldsymbol{\theta}_t, D_{t-1}) \quad \text{and} \quad p(\boldsymbol{\theta}_t|\boldsymbol{\theta}_{t-1}, D_{t-1}), \tag{15.1}$$

where each density may depend on both t and independent variables, but is assumed known apart from the values of the state vectors. Model specification is completed through the addition of an initial prior density $p(\boldsymbol{\theta}_0|D_0)$, as usual. Note that this is essentially the framework of Section 13.6.

Based on a series of observations $Y_1, \ldots, Y_n$, we are interested in sampling the full posterior $p(\boldsymbol{\theta}_1, \ldots, \boldsymbol{\theta}_n|D_n)$. Applying the Gibbs sampling concept directly, the apparently natural approach is sampling state-by-state, as follows. A version of Gibbs sampling proposes that we iteratively resample conditional posteriors

$$p(\boldsymbol{\theta}_t|\boldsymbol{\theta}_{(-t)}, D_n), \tag{15.2}$$

sequencing through $t = 1, 2, \ldots, n$ and updating the conditioning values to the most recently sampled value at each step. Directly via Bayes' theorem, we see that

$$p(\boldsymbol{\theta}_t|\boldsymbol{\theta}_{(-t)}, D_n) \propto p(\boldsymbol{\theta}_t|\boldsymbol{\theta}_{t-1}, D_{t-1})p(\boldsymbol{\theta}_{t+1}|\boldsymbol{\theta}_t, D_{t-1})p(Y_t|\boldsymbol{\theta}_t, D_{t-1}), \tag{15.3}$$

as a function of $\boldsymbol{\theta}_t$.

Rarely will the model be tractable enough so that (15.3) may be directly simulated, so that this is a context in which Metropolis-Hastings methods will come into play, nested within the Gibbs iterations. Carlin et al (1992) discuss this in detail in various models; see also Jaquier et al (1994) for interesting special cases in modelling stochastic volatility in financial time series. Related ideas and approaches are discussed in Scipione and Berliner (1993) in rather different but related modelling contexts. The generality of this approach is evident from the broad class of structures encompassed by (15.1), and at time of writing, this approach is being explored and extended by researchers in various fields. A technical issue arising is that as in other applications of Gibbs sampling with many parameters and uncertain variables, convergence to sampling stationary distributions may be very slow, partly due to the "state-by-state" approach in which each $\boldsymbol{\theta}_t$ is sampled conditional on its "neighbours". Other examples of this arise, and with more severe computational consequences, in spatial modelling with Markov random field prior distributions; it should be remarked that all prior structures in DLMs, and non-linear dynamic models, discussed throughout this book are specific cases of the general class of Markov random field models.

We move away from this general framework here to discuss conditionally linear/normal models and rather more specific MCMC methods in some important model classes. In this development, rather more efficient MCMC approaches arise and partly overcome some of the computational issues of slow convergence associated with the state-by-state simulation approach. This is possible due to the conditionally linear/normal structure of the models considered, and does not, unfortunately, generalise to other classes of inherently non-linear/non-normal models; the development of more efficient and effective MCMC methods in other models raises open research question.

15.2.2 *Inference on parameters in DLMs*

A variety of "non-linear" learning problems may be characterised as problems of inference for defining parameters in otherwise standard normal DLMs. We have seen several examples throughout this book, including DLMs with autoregressive noise components (Section 9.4), DLMs with uncertain observational and evolution variances, or discount factors (Section 13.3), and multi-process mixtures with normal-mixture error distributions in various components (throughout Chapter 12 and Section 13.3). In such models, Gibbs sampling often provides a very natural and easily implemented method for sampling posterior distributions of these model parameters and the DLM state vectors over a fixed time interval. In fact, specific multi-process mixture DLMs provided the initiating context for some of the very first developments of MCMC methods in dynamic models (Carter and Kohn 1994, Frühwirth-Schnatter 1994). The basic idea is discussed here, with an example or two, prior to more extensive discussion of efficient simulation in conditionally normal linear models in the following sections.

Suppose that across the time interval $t = 1, \ldots, n$, the Y_t series is modelled via a DLM with defining parameters collected in a vector $\boldsymbol{\alpha}_n$. This structure is essentially as in Section 12.1, e.g., equation (12.1) and the following discussion, with the explicit recognition that the required defining parameter vector may depend on n. Typically $\boldsymbol{\alpha}_n$ will include several constant parameters, such as defining entries in $\mathbf{G} = \mathbf{G}(\boldsymbol{\alpha}_n)$ and variance components, but may also include latent variables whose number increases with sample size. Some examples remind us of important special cases.

EXAMPLE 15.1. In a standard univariate DLM, write $\boldsymbol{\alpha}_n = \boldsymbol{\alpha} = \{V, \mathbf{W}\}$ for the assumedly constant observation and evolution variances. Obviously, the analysis is standard conditional on these variance components. Thus, for any specified $\boldsymbol{\alpha}$, posterior distributions $p(\boldsymbol{\theta}_1, \ldots, \boldsymbol{\theta}_n | \boldsymbol{\alpha}, D_n)$ may be computed; they may also be simulated.

EXAMPLE 15.2. Consider the normal mixture observational error distribution

$$\nu_t \sim 0.95\text{N}[0, V] + 0.05\text{N}[0, k^2 V],$$

where $k > 1$ is a specified scale inflation factor. As in Sections 12.3 and 12.4, this models occasional "outliers" in the observations, with extreme observational errors coming from the inflated variance component of the mixture around 5% of the time. We can write this as a conditionally normal error model, $(\nu_t|\lambda_t) \sim \text{N}[0, V\lambda_t]$, where for $t = 1, 2, \ldots,$ λ_t is a latent variable taking values 1 or k^2 with corresponding probabilities 0.95 and 0.05. For the fixed series of observation times $t = 1, \ldots, n$, include the latent variables $\{\lambda_1, \ldots, \lambda_n\}$ in the "parameter" vector $\boldsymbol{\alpha}_n$, together with variance components V and possibly evolution variances. Conditional on $\boldsymbol{\alpha}_n$, the data model is a standard normal DLM; for any set of λ_t values (together with any other elements of $\boldsymbol{\alpha}_n$), posterior distributions $p(\boldsymbol{\theta}_1, \ldots, \boldsymbol{\theta}_n|\boldsymbol{\alpha}_n, D_n)$ may be computed and simulated.

EXAMPLE 15.3. The framework of Example 15.2 extends to allow essentially any "mixing" distribution for the normal scale factors λ_t. In particular, assuming that the λ_t follow inverse gamma distributions of the form $(\lambda_t^{-1}|D_0) \sim \text{G}[r/2, r/2]$ implies that the ν_t are marginally Student-T distributed with r degrees of freedom. This is a particular case of a "heavy-tailed" observational error distribution constructed as a continuous scale mixture of normals, in contrast to the above discrete mixture. This has uses in modelling outlying observations, and falls into the conditionally normal context of this section, i.e., analysis may proceed using the MCMC simulation approach.

The importance and utility of this structure is evident in considering MCMC based on Gibbs sampling. Write $\boldsymbol{\Theta}_n = \{\boldsymbol{\theta}_0, \boldsymbol{\theta}_1, \ldots, \boldsymbol{\theta}_n\}$ for the n state vectors in the DLM for observations, $Y_1, \ldots, Y_n$, together with the initial vector $\boldsymbol{\theta}_0$ (by convention). Suppose the model is a standard DLM conditional on parameters and latent variables in $\boldsymbol{\alpha}_n$. Gibbs sampling suggests that the full posterior distribution

$$p(\boldsymbol{\Theta}_n, \boldsymbol{\alpha}_n|D_n)$$

may be simulated by iterating between the two conditional posteriors

$$p(\boldsymbol{\Theta}_n|\boldsymbol{\alpha}_n, D_n) \longleftrightarrow p(\boldsymbol{\alpha}_n|\boldsymbol{\Theta}_n, D_n). \tag{15.4}$$

Note that this involves sampling $\boldsymbol{\Theta}_n$, and separately $\boldsymbol{\alpha}_n$, from their full, multivariate conditional posteriors, rather than sequencing through individual scalar elements as in the original definition of Gibbs sampling. Based on the observation that the analysis is standard conditional on $\boldsymbol{\alpha}_n$, simulation of the first component of (15.4) will be accessible as a (complicated) multivariate normal distribution; efficient approachs to sampling this are discussed in the following section. The second component will depend very much on model form, as an example indicates.

EXAMPLE 15.1 (continued). In the model $\{\mathbf{F}_t, \mathbf{G}_t, V, \mathbf{W}\}$ with $\boldsymbol{\alpha}_n = \boldsymbol{\alpha} = \{V, \mathbf{W}\}$, assume independent priors on the variances, i.e., $p(V, \mathbf{W}|D_0) = p(V|D_0)p(\mathbf{W}|D_0)$. Then the conditional posterior for $\boldsymbol{\alpha}$ is

$$p(\boldsymbol{\alpha}|\boldsymbol{\Theta}_n, D_n) = p(V|\boldsymbol{\Theta}_n, D_n)p(\mathbf{W}|\boldsymbol{\Theta}_n, D_n), \tag{15.5}$$

with

$$p(V|\boldsymbol{\Theta}_n, D_n) \propto p(V|D_0)V^{-n/2}\exp(-n\hat{V}_n/2V)$$

and

$$p(\mathbf{W}|\boldsymbol{\Theta}_n, D_n) \propto p(\mathbf{W}|D_0)|\mathbf{W}|^{-n/2}\exp(-n\text{trace}(\hat{\mathbf{W}}_n\mathbf{W}^{-1})/2),$$

where based on conditionally known values of $\nu_t = Y_t - \mathbf{F}_t'\boldsymbol{\theta}_t$ and $\boldsymbol{\omega}_t = \boldsymbol{\theta}_t - \mathbf{G}_t\boldsymbol{\theta}_{t-1}$ for each t, $\hat{V}_n = n^{-1}\sum_{t=1}^n \nu_t^2$ and $\hat{\mathbf{W}}_n = n^{-1}\sum_{t=1}^n \boldsymbol{\omega}_t\boldsymbol{\omega}_t'$. Note, for example, that conditionally conjugate priors are inverse gamma for V, and inverse Wishart (see Section 16.4) for $\mathbf{W}$. Under such priors, the conditional posterior for variance components may be easily simulated. Often, as in further and more elaborate examples below, the evolution variance matrix $\mathbf{W}$ is structured in terms of just a few uncertain parameters, and the computations simplify further.

In some cases, direct sampling of $p(\boldsymbol{\alpha}_n|\boldsymbol{\Theta}_n, D_n)$ is infeasible, but writing in terms of further sub-vectors of $\boldsymbol{\alpha}_n$ leads to an alternative form of Gibbs sampling that is easily implemented. Generally, partition $\boldsymbol{\alpha}_n$ into k sub-vectors, so that $\boldsymbol{\alpha}_n = \{\boldsymbol{\alpha}_{n,1}, \ldots, \boldsymbol{\alpha}_{n,k}\}$. Suppose that each of the (possibly multivariate) conditionals for the $\boldsymbol{\alpha}_{n,j}$ may be routinely simulated. In an extension of our earlier notation, let $\boldsymbol{\alpha}_{n,-j}$ be all elements of $\boldsymbol{\alpha}_n$ but $\boldsymbol{\alpha}_{n,j}$. Then the revised form of iterative simulation involves sequencing through draws of $\boldsymbol{\Theta}_n$ given $\boldsymbol{\alpha}_n$, followed by samples of each of the sub-vectors in turn, viz,

$$\begin{aligned} &p(\boldsymbol{\Theta}_n|\boldsymbol{\alpha}_n, D_n) \\ &\qquad \leftrightarrow\ p(\boldsymbol{\alpha}_{n,1}|\boldsymbol{\alpha}_{n,-1}, \boldsymbol{\Theta}_n, D_n) \\ &\qquad\qquad \ddots \\ &\qquad\qquad\quad \leftrightarrow\ p(\boldsymbol{\alpha}_{n,k}|\boldsymbol{\alpha}_{n,-k}, \boldsymbol{\Theta}_n, D_n). \end{aligned} \tag{15.6}$$

EXAMPLE 15.2 (continued). With the discrete normal mixture structure for observational errors, identify $\boldsymbol{\alpha}_{n,1} = \{V, \mathbf{W}\}$ and $\boldsymbol{\alpha}_{n,2} = \{\lambda_1, \ldots, \lambda_n\}$. Suppose prior independence between V and $\mathbf{W}$ and the λ_t, so that

$$p(\boldsymbol{\alpha}_n|D_0) = p(V|D_0)p(\mathbf{W}|D_0)\prod_{t=1}^n p(\lambda_t|D_0).$$

Here $p(\lambda_t|D_0)$ is discrete with masses 0.95 and 0.05 on the values 1 and k^2, respectively. It follows easily that $p(\boldsymbol{\alpha}_{n,1}|\boldsymbol{\alpha}_{n,-1},\boldsymbol{\Theta}_n,D_n)$ has the form of (15.5) with the simple modification that the observational errors ν_t are weighted by the appropriate values of the scale factors, i.e., $\hat{V}_n = n^{-1}\sum_{t=1}^n \nu_t^2/\lambda_t$ for each t. For the scale factors in $\boldsymbol{\alpha}_{n,2}$, it is clear that

$$p(\boldsymbol{\alpha}_{n,2}|\boldsymbol{\alpha}_{n,-2},\boldsymbol{\Theta}_n,D_n) = \prod_{t=1}^{n} p(\lambda_t|V,\mathbf{W},\boldsymbol{\Theta}_n,D_n),$$

with component posteriors defined in terms of odds ratios as

$$\frac{\Pr(\lambda_t=1|V,\mathbf{W},\boldsymbol{\Theta}_n,D_n)}{\Pr(\lambda_t=k^2|V,\mathbf{W},\boldsymbol{\Theta}_n,D_n)} = \frac{0.95}{0.05}\,k\exp(-\nu_t^2(1-k^{-2})/2V)$$

with $\nu_t = Y_t - \mathbf{F}_t'\boldsymbol{\theta}_t$ as before. Thus, λ_t is simulated by choosing 1 or k^2 with the resulting probability.

EXAMPLE 15.3 (continued). In the case of the continuous normal mixture, the discussion of the above example may be followed with the modification that the conditional posteriors for scale factors λ_t are continuous, i.e.,

$$p(\lambda_t|V,\mathbf{W},\boldsymbol{\Theta}_n,D_n) \propto p(\lambda_t|D_0)\lambda_t^{-1/2}\exp(-\nu_t^2/2V\lambda_t).$$

In the case of the inverse gamma model, the prior $\mathrm{G}[r/2,r/2]$ distribution for λ_t^{-1} is updated to the conditional gamma $\mathrm{G}[(r+1)/2,(r+\nu_t^2/V)/2]$, for each t. Simulation is then direct.

In other cases when direct sampling of $p(\boldsymbol{\alpha}_n|\boldsymbol{\Theta}_n,D_n)$ is not possible, alternatives to the above "nested" Gibbs approach may be based on embedding some form of Metropolis "step" to sample $\boldsymbol{\alpha}_n$ at each stage of the iterations in (15.4). This can involve use of a direct approximation to $p(\boldsymbol{\alpha}_n|\boldsymbol{\Theta}_n,D_n)$ to be used as a proposal distribution or a random walk Metropolis step simulating candidate values of $\boldsymbol{\alpha}_n$ from a symmetric proposal centred at the "current" value, or others.

These examples give some idea of the scope of MCMC methods in extending DLM analyses to include uncertain parameters and latent variables. For much more extensive development in mixture modelling and related contexts, see Carter and Kohn (1994), Frühwirth-Schnatter (1994), and West (1996c), for example.

We now turn to more specific developments in DLM contexts, based essentially on the variants of Gibbs sampling mentioned above. As this field develops in research and in practice, we will surely see refined and generalised approaches utilising other forms of Metropolis-Hastings algorithms, though we do not develop these further here.

15.2.3 *Efficient MCMC in normal DLMs*

Before proceeding to illustrate in DLMs with autoregressive components, we need to attend to the technical issues arising in sampling the key conditional distribution $p(\boldsymbol{\Theta}_n|\boldsymbol{\alpha}, D_n)$ of (15.6). Carter and Kohn (1994) and Frühwirth-Schnatter (1994) provide the basic and original development. The latter article introduced the prototype simulation method under the very descriptive name of the *forward filtering, backward sampling* algorithm. In a general context, this is structured as follows.

We have the model $\{\mathbf{F}_t, \mathbf{G}_t, V_t, \mathbf{W}_t\}$, in general, suppressing the notational dependence on the conditioning value of $\boldsymbol{\alpha}$ for clarity. We wish to sample a full set of state vectors $\boldsymbol{\Theta}_n = \{\boldsymbol{\theta}_0, \boldsymbol{\theta}_1, \ldots, \boldsymbol{\theta}_n\}$ from the full, multivariate normal posterior $p(\boldsymbol{\Theta}_n|D_n)$. Note that exploiting the Markov structure of the evolution equation of the DLM, we may write

$$p(\boldsymbol{\Theta}_n|D_n) = p(\boldsymbol{\theta}_n|D_n)p(\boldsymbol{\theta}_{n-1}|\boldsymbol{\theta}_n, D_{n-1})\ldots p(\boldsymbol{\theta}_1|\boldsymbol{\theta}_2, D_1)p(\boldsymbol{\theta}_0|\boldsymbol{\theta}_1, D_0). \tag{15.7}$$

As a result, we may sample the entire $\boldsymbol{\Theta}_n$ by sequentially simulating the individual state vectors $\boldsymbol{\theta}_n, \boldsymbol{\theta}_{n-1}, \ldots, \boldsymbol{\theta}_0$, as follows:

(1) Sample $\boldsymbol{\theta}_n$ from $(\boldsymbol{\theta}_n|D_n) \sim \mathrm{N}[\mathbf{m}_n, \mathbf{C}_n]$, then
(2) for each $t = n-1, n-2, \ldots, 1, 0$, sample $\boldsymbol{\theta}_t$ from $p(\boldsymbol{\theta}_t|\boldsymbol{\theta}_{t+1}, D_t)$, where the conditioning value of $\boldsymbol{\theta}_{t+1}$ is the value just sampled.

The required conditional distributions in the second item here are obtained as described in developing filtering recurrences in Section 4.7. Explicitly from equation (4.5) of that chapter,

$$(\boldsymbol{\theta}_t|\boldsymbol{\theta}_{t+1}, D_t) \sim \mathrm{N}[\mathbf{h}_t, \mathbf{H}_t], \tag{15.8}$$

where

$$\mathbf{h}_t = \mathbf{m}_t + \mathbf{B}_t(\boldsymbol{\theta}_{t+1} - \mathbf{a}_{t+1})$$

and

$$\mathbf{H}_t = \mathbf{C}_t - \mathbf{B}_t\mathbf{R}_{t+1}\mathbf{B}_t',$$

with

$$\mathbf{B}_t = \mathbf{C}_t\mathbf{G}_{t+1}'\mathbf{R}_{t+1}^{-1},$$

for each t.

Hence, the process of sampling $\boldsymbol{\Theta}_n$ starts by running the standard analysis forward from $t = 0$ up to $t = n$, computing and saving the summaries $\mathbf{m}_t, \mathbf{C}_t$, and the concomitant quantities $\mathbf{a}_t, \mathbf{R}_t$ and $\mathbf{B}_t$ at each stage. At $t = n$, a vector $\boldsymbol{\theta}_n$ is sampled, then we sequence backwards through time, computing the elements $\mathbf{h}_t$ and $\mathbf{H}_t$ at each step and generating a value of $\boldsymbol{\theta}_t$. Hence the the forward filtering (or forward updating), backward sampling terminology arises.

On computing, note that sampling multivariate normal distributions involves decomposing the relevant variance matrices, and this is often done via Cholesky decomposition. In many cases, correlations in these conditional posteriors may be high, so that singular-value decomposition methods are preferable for numerical stability.

The algorithm outlined here is very general. Though it therefore applies to many different DLM contexts, it may tend to be computationally inefficient relative to more specialised, or "customised," variants in specific model frameworks. We elaborate on this in developing a modified algorithm for a specific model class in the following section. Note further that at the frontiers of current research in the area at the time of writing, investigators are active in developing and extending these kinds of MCMC algorithms, with a view to furthering applicability as well as computational efficiency and effectiveness (e.g., Shephard 1994; Cargnoni et al 1996, among many others). Future developments are likely to have major impact on applied work in time series and forecasting using dynamic models.

15.3 STATE-SPACE AUTOREGRESSION

15.3.1 A specific AR component DLM

More detailed development and illustration of MCMC simulation-based analysis is given in the context of the specific autoregressive component DLM discussed in Section 9.4.7. Much of the material here follows West (1996a,c, 1997), where further details and examples may be found. See also West (1995 and 1996b) for similar developments in a closely related class of dynamic models.

Consider the model

$$Y_t = \mu_t + X_t + \nu_t,$$

where μ_t represents an underlying, first-order polynomial trend, X_t is an autoregressive noise component, and ν_t is the usual observational noise term. Specifically, for all t,

(1) $\mu_t = \mu_{t-1} + \omega_t$, with stochastic level changes $\omega_t \sim \text{N}[0, W]$,
(2) $X_t = \sum_{j=1}^{p} \phi_j X_{t-j} + \epsilon_t$, with stochastic innovations $\epsilon_t \sim \text{N}[0, U]$, for each t, and
(3) $\nu_t \sim \text{N}[0, V\lambda_t]$, with fixed and known variance weights λ_t.

Dealing generally with weighted observational errors leads into extended analyses to allow for errors following normal mixture distributions, as in Example 15.2, to be discussed further below. We note that similar weightings could be used for the variances of the innovations of the AR component, though this is not discussed further here.

We make the additional assumptions that the three sequences of error components are independent and mutually independent. We are interested in inference about the underlying, latent trend and noise processes, together with the uncertain variance components, W and U, and the autoregressive parameters $\boldsymbol{\phi} = (\phi_1, \ldots, \phi_p)'$. We have discussed such models in Section 9.4 and now turn to their analyses using MCMC methods. Note that the trend component μ_t is of the simplest DLM form here; the development can be pursued with alternative trend terms, combined trend and regression terms, and so forth, with direct modification to the details below.

As in Section 9.4.7, we can represent this model in DLM form as

$$Y_t = \mathbf{F}'\boldsymbol{\theta}_t + \nu_t \quad \text{and} \quad \boldsymbol{\theta}_t = \mathbf{G}\boldsymbol{\theta}_{t-1} + \boldsymbol{\omega}_t,$$

where

(1) $\boldsymbol{\theta}_t = (\mu_t, X_t, X_{t-1}, \ldots, X_{t-p+1})'$, the state vector at time t,
(2) $\mathbf{F} = (1, 1, 0, \ldots, 0)'$,
(3) $\mathbf{G}$ is the $(p+1) \times (p+1)$ matrix

$$\mathbf{G} = \begin{pmatrix} 1 & 0 & 0 & 0 & \cdots & 0 \\ 0 & \phi_1 & \phi_2 & \phi_3 & \cdots & \phi_p \\ 0 & 1 & 0 & 0 & \cdots & 0 \\ 0 & 0 & 1 & 0 & \cdots & 0 \\ \vdots & \vdots & & \ddots & \cdots & \vdots \\ 0 & 0 & 0 & \cdots & 1 & 0 \end{pmatrix},$$

and
(4) $\boldsymbol{\omega}_t = (\omega_t, \epsilon_t, 0, \ldots, 0)'$ with (singular) variance matrix

$$\mathbf{W} = \begin{pmatrix} W & 0 & 0 & \cdots & 0 \\ 0 & U & 0 & \cdots & 0 \\ 0 & 0 & 0 & \cdots & 0 \\ \vdots & \vdots & \vdots & \cdots & \vdots \\ 0 & 0 & 0 & \cdots & 0 \end{pmatrix}.$$

In the context of (15.6), we identify model parameters $\boldsymbol{\alpha}_1 = \boldsymbol{\alpha}_{n,1} = \boldsymbol{\phi} = (\phi_1, \ldots, \phi_p)'$ and $\boldsymbol{\alpha}_2 = \boldsymbol{\alpha}_{n,2} = \{V, W, U\}$. Section 15.3.3 develops the conditional posterior $p(\boldsymbol{\alpha}_2|\boldsymbol{\phi}, \boldsymbol{\Theta}_n, D_n)$ under specific prior assumptions.

For each t, write $\mathbf{Z}_t = (X_t, X_{t-1}, \ldots, X_{t-p+1})'$, the final p elements of $\boldsymbol{\theta}_t$ corresponding to the previous p values of the AR process. In the posterior computations below, we will encounter the term $p(\mathbf{Z}_0|\boldsymbol{\alpha}, D_0)$ at various places. This is the prior distribution for the p initial values $X_0, X_{-1}, \ldots, X_{-p+1}$ of the AR process, and the algorithm simulates from the corresponding posterior. We need to specify this prior. Were we to assume the X_t process to be stationary, this prior would be theoretically determined and depend, in complicated ways, on parameters U and $\boldsymbol{\phi}$ of $\boldsymbol{\alpha}$. In such a case, the analysis reported below should be modified to account for the dependence of this function on these parameters (among other things, this would involve the use of Metropolis-Hastings simulation steps rather than the direct Gibbs sampling developed here). We do not do this here, assuming directly that $p(\mathbf{Z}_0|\boldsymbol{\alpha}, D_0)$ is a specified normal prior distribution whose moments are independent of $\boldsymbol{\alpha}$. This simplifies the analysis somewhat. Moreover, if stationarity is in fact a valid assumption, then it will be supported under the conditional posterior distribution in any case.

Indeed, modelling without the restriction permits posterior assessment of the stationarity assumption.

15.3.2 Sampling the state vectors

Conditional on the model parameters $\boldsymbol{\alpha} = \{\phi, V, W, U\}$, we require the simulation of $p(\boldsymbol{\Theta}_n|\boldsymbol{\alpha}, D_n)$, where $\boldsymbol{\Theta}_n = \{\boldsymbol{\theta}_0, \boldsymbol{\theta}_1, \ldots, \boldsymbol{\theta}_n\}$. Though theoretically correct, the general algorithm of Section 15.2.3 degenerates in models with AR components, as it does in other models in which consecutive state vectors contain common components. In such cases, the central conditional distributions (15.8) are singular, and so need examining in detail to sample. The general recipe for simulation given there may be redeveloped to provide a more direct and efficient form in such specific models. In the current model, this is done as follows.

To sample (15.7), proceed as described in the general algorithm:

(1) Sample a value of $\boldsymbol{\theta}_n$ from $(\boldsymbol{\theta}_n|D_n) \sim \mathrm{N}[\mathbf{m}_n, \mathbf{C}_n]$, then
(2) for each $t = n-1, n-2, \ldots, 1, 0$, compute new values of $\boldsymbol{\theta}_t$ conditional on the values of the previous $\boldsymbol{\theta}_{t+1}$ just sampled.

The computations at the second step are now simplified and specific to the model structure, based on the fact that elements $2, \ldots, p-1$ of the state vector $\boldsymbol{\theta}_t$ are known if $\boldsymbol{\theta}_{t+1}$ is known; they are simply the elements $3, \ldots, p$ of the latter, i.e.,

$$\begin{array}{llllll} \boldsymbol{\theta}'_{t+1} = (\mu_{t+1}, X_{t+1}, & X_t, & X_{t-1}, & \cdots, & X_{t-p+2}) & \\ & \updownarrow & \updownarrow & \cdots & \updownarrow & \\ \quad\quad \boldsymbol{\theta}'_t = (\mu_t, & X_t, & X_{t-1}, & \cdots, & X_{t-p+2}, & X_{t-p+1}). \end{array}$$

Hence, given $\boldsymbol{\theta}_{t+1}$, replace entries $2, \ldots, p-1$ of $\boldsymbol{\theta}_t$ accordingly. Sample the remaining two elements $(\mu_t, X_{t-p+1})'$ as follows.

First, compute the moments of $p(\mu_t, X_{t-p+1}|X_t, X_{t-1}, \ldots, X_{t-p+2}, D_t)$; this is just the bivariate conditional for the first and final elements of $\boldsymbol{\theta}_t$ from the full joint distribution $(\boldsymbol{\theta}_t|D_t) \sim \mathrm{N}[\mathbf{m}_t, \mathbf{C}_t]$. This is most efficiently done by sequentially conditioning on the elements X_{t-i} for $i = 0, \ldots, p-2$ in turn, reducing the dimension of the normal distribution by 1 at each stage, eventually reducing from the full $p+1$ dimensions to 2.

Next, compute the partial residual $e_{t+1} = X_{t+1} - \sum_{j=1}^{p-1} \phi_j X_{t+1-j}$ based on the previously sampled values of the elements of $\boldsymbol{\theta}_{t+1}$. We then have a pair of independent "observations"

$$\mu_{t+1} \sim \mathrm{N}[\mu_t, W] \text{ and } e_{t+1} \sim \mathrm{N}[\phi_p X_{t-p+1}, U]$$

on the two parameters $(\mu_t, X_{t-p+1})'$. Use the corresponding likelihood to update the bivariate normal "prior" already computed, i.e., compute the bivariate normal posterior proportional to

$$p(\mu_t, X_{t-p+1}|X_t, X_{t-1}, \ldots, X_{t-p+2}, D_t)$$
$$\times \exp(-(\mu_{t+1} - \mu_t)^2/2W) \exp(-(e_{t+1} - \phi_p X_{t-p+1})^2/2U);$$

this is the appropriate bivariate margin of the first and last elements of $\boldsymbol{\theta}_t$ under the distribution $p(\boldsymbol{\theta}_t|\boldsymbol{\theta}_{t+1}, D_n)$. Sample this bivariate normal distribution, and so fill in these two elements of $\boldsymbol{\theta}_t$ to complete this step.

Sequencing through this process results in a complete set of sampled state vectors $\boldsymbol{\Theta}_n$ that represents a sample from $p(\boldsymbol{\Theta}_n|\boldsymbol{\alpha}, D_n)$, as required.

15.3.3 Sampling the DLM parameters

We now explore the required conditional distributions for model parameters $\boldsymbol{\alpha}_1 = \boldsymbol{\phi}$ and $\boldsymbol{\alpha}_2 = \{V, W, U\}$, as required in (15.6). We assume a joint prior with independent components,

$$p(\boldsymbol{\alpha}|D_0) = p(\boldsymbol{\phi}|D_0)p(V|D_0)p(W|D_0)p(U|D_0),$$

implicitly assuming prior independence of $\boldsymbol{\alpha}$ and $\boldsymbol{\Theta}_n$. Among other things, this implies, as we shall see, that the conditional posterior for the variance components factors into three independent components. Hence sampling $p(\boldsymbol{\alpha}_2|\boldsymbol{\phi}, \boldsymbol{\Theta}_n, D_n)$ reduces to three independent draws from $p(V|\boldsymbol{\phi}, \boldsymbol{\Theta}_n, D_n)$, $p(W|\boldsymbol{\phi}, \boldsymbol{\Theta}_n, D_n)$ and $p(U|\boldsymbol{\phi}, \boldsymbol{\Theta}_n, D_n)$. These conditional posteriors are described, following that for $\boldsymbol{\phi}$.

Sampling $p(\boldsymbol{\phi}|V, W, U, \boldsymbol{\Theta}_n, D_n)$

Conditioning on $\boldsymbol{\Theta}_n$ provides values for the entire AR process X_t for $t = -(p-1), \ldots, -1, 0, 1, \ldots, n$ (the initial values entering as elements of $\boldsymbol{\theta}_0$.) With $\mathbf{Z}_t = (X_t, X_{t-1}, \ldots, X_{t-p+1})'$ as above, the required conditional posterior for $\boldsymbol{\phi}$ is proportional to

$$p(\boldsymbol{\phi}|D_0)p(\mathbf{Z}_0|\boldsymbol{\alpha}, D_0) \prod_{t=1}^{n} \exp(-(X_t - \mathbf{Z}_t'\boldsymbol{\phi})^2/2U).$$

The third component here, the conditional likelihood, has the normal form in $\boldsymbol{\phi}$ arising in the now standard AR model for the n values of the process X_t and with known initial values in $\mathbf{Z}_0$. We have already assumed that $p(\mathbf{Z}_0|\boldsymbol{\alpha}, D_0)$ does not depend on $\boldsymbol{\alpha}$. Hence a normal prior $p(\boldsymbol{\phi}|D_0)$ leads to a posterior as in the standard AR regression analysis of Section 9.4.6, resulting in a normal conditional posterior. We make this assumption here; hence the required conditional posterior is determined by application of the standard formulae in equation (9.11) or, in the case of a reference initial prior, (9.12).

As an aside, we note that an explicit assumption of stationarity of the AR process would involve modification of the normal prior (in addition to making explicit the dependence of $p(\mathbf{Z}_0|\boldsymbol{\alpha}, D_0)$ on $\boldsymbol{\phi}$). In such cases, the analysis from here on would be modified, using Metropolis sampling methods for $p(\boldsymbol{\phi}|\boldsymbol{\alpha}_2, \boldsymbol{\Theta}_n, D_n)$.

Sampling $p(V|\boldsymbol{\phi}, \boldsymbol{\Theta}_n, D_n)$

Based on the conditioning values of $\boldsymbol{\theta}_t$, compute the imputed residuals $\nu_t = Y_t - \mathbf{F}'\boldsymbol{\theta}_t$ for each $t = 1, \ldots, n$, and then sample the posterior proportional to

$$p(V|D_0)V^{-n/2} \exp\Big(-\sum_{t=1}^{n} \nu_t^2/2\lambda_t V\Big).$$

An inverse gamma prior for V is conjugate; under such a prior, the conditional posterior is also inverse gamma, and easily sampled. Alternative priors, such as proper uniform priors for functions of V, lead to modified and truncated inverse gamma posteriors for V, that may also be easily sampled, sometimes using rejection methods.

Sampling $p(W|\boldsymbol{\phi}, \boldsymbol{\Theta}_n, D_n)$

Similar to the above, compute the imputed level changes $\omega_t = \mu_t - \mu_{t-1}$ for each t, and sample the posterior proportional to

$$p(W|D_0)W^{-n/2} \exp\Big(-\sum_{t=1}^{n} \omega_t^2/2W\Big).$$

Similar comments about prior and posterior inverse gamma forms are relevant.

Sampling $p(U|\boldsymbol{\phi}, \boldsymbol{\Theta}_n, D_n)$

Again as above, compute the imputed values of the AR innovations $\epsilon_t = X_t - \sum_{j=1}^{p} \phi_j X_{t-j}$, and sample the posterior proportional to

$$p(U|D_0)U^{-n/2} \exp\Big(-\sum_{t=1}^{n} \epsilon_t^2/2U\Big).$$

Inverse gamma posteriors, or truncated or otherwise modified versions, are again deduced with typical prior forms.

Before proceeding to some illustrations, we mention an extension of the above analysis to include inference on uncertain observational variance scale factors. As introduced in Examples 15.2 and 15.3, treating the λ_t as random weights with assigned prior distributions is a way of inducing non-normal distributions for the ν_t in order to accommodate outlying errors, among other things. In a general context, suppose the λ_t are initially independent and identically distributed according to some common, specified prior $p(\lambda_t|D_0)$. The above development of the MCMC algorithm still holds conditional on any set of values $\Lambda_n = \{\lambda_1, \ldots, \lambda_n\}$; formally, the notation should now include Λ_n as a third component of the model parameter $\boldsymbol{\alpha}$, and all the conditional distributions above should include λ_n in the conditioning.

Then these scale factors may be included in the analysis by linking with a further simulation step to sample their values from appropriate conditional posteriors, as discussed in the examples, as follows.

Sampling $p(\Lambda_n|\phi, V, W, U, \Theta_n, D_n)$

Following Examples 15.2 and 15.3, the λ_t are conditionally independent, with for each t,

$$p(\lambda_t|\phi, V, W, U, \Theta_n, D_n) = p(\lambda_t|V, \boldsymbol{\theta}_t, Y_t)$$
$$\propto p(\lambda_t|D_0)\lambda_t^{-1/2}\exp(-\nu_t^2/2V\lambda_t).$$

With a discrete or continuous mixing prior distribution $p(\lambda_t|D_0)$, these conditionals are often easily sampled to produce a set Λ_n, as required, so extending the overall iterations to provide learning on scale factors along with the other model parameters. We note that with discrete prior distributions, this leads to models with multi-process, class I structure. Extensions to include normal mixture distributions for evolution error terms are direct, though are not pursued further here (see, for example, Carter and Kohn 1994, Frühwirth-Schnatter 1994).

15.3.4 Illustration

Geological time variations in oxygen, and other, isotope measurements from deep ocean cores relate to patterns of variation in global ice volume and ocean temperature (Shackleton and Hall 1989; Park and Maasch 1993). The single series graphed in Figure 15.1 is representative of several oxygen isotope series from cores from various geographical locations, derived from original δ^{18}O Site 677 measurements presented in Shackleton and Hall (1989); the data are courtesy of J. Park of Yale University. The values estimate relative abundance of δ^{18}O and are timed at equal spacings of 3,000 years (or 3 kyears). The time scale calibration is based on that of Ruddiman, McIntyre and Raymo (1989), and discussed in Park (1992) and Park and Maasch (1993). This latter reference also discusses the process of interpolation of original, unequally spaced measurements to this equally spaced scale. This section of 400 observations stretches back roughly 1.2 million years, and is plotted with a reverse of sign, by convention, so that the apparent increase in levels in modern times reflects generally warmer average global temperatures and smaller average ice masses. Time-varying periodicities are evident; the nature and structure of these periodicities is of some geological importance, and a focus for analysis here.

Useful discussion of the relationships between climatic indicators, such as the δ^{18}O measures here, and cyclical patterns of changes in the earth's orbital dynamics appears in Park (1992). The precession and obliquity of the earth's orbit impact on insolation received by the earth and so induce substantial variations in climate characteristics. Periodicities in eccentricity are generally believed to be associated with periods of around 95-100

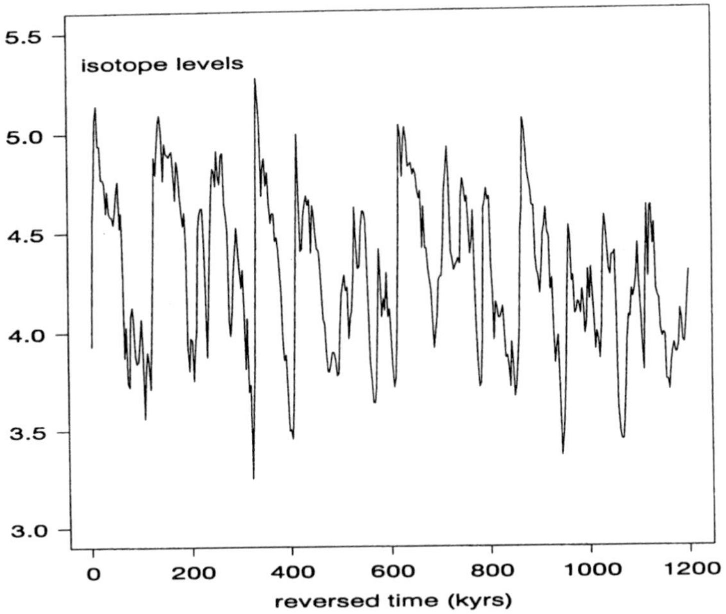

Figure 15.1 Oxygen isotope series

kyears and 120-130 kyears; each of these figures is subject to uncertainty. Shorter-term cycles include those associated with precession, with period around 19-23 kyears, and obliquity of the earth's orbit, with period around 40-42 kyears. The "100,000-year ice-age cycle", so-called, is of major interest and has been the subject of intensive investigation in recent years. Identifying the nature and structure of quasi-periodic components of period around 100 kyears is of importance in contributing to debates over the genesis of the ice-age cycles, roughly a million years ago, and particularly to questions of whether or not the onset was gradual and inherent or the result of a significant structural climatic change. (e.g., Ruddiman et al 1989; Park 1992, and references therein).

We report an analysis allowing for routine measurement, sampling and laboratory errors contaminating the oxygen recordings, in addition to possible occasional gross errors, or outliers. Note that there are surely errors in the timing of the observations, due to the inherent difficulty in estimating true calendar times of geochemical data, and also due to the process of calibrating times to the nearest unit (3,000 years here). We make no further attempt to explore this; our analysis is therefore consistent with previous analysis by Park and Maasch (1993). Approaches to dealing with timing errors generally are a current topic of active research; see West (1996b) for developments in a very similar context. Here we are accepting the imputed, equally spaced times as accurate, though the use of a measurement error model may, incidentally, play a role in accounting for timing errors too.

The model assumed is as described in Section 15.3.1 above, with a first-order polynomial trend superimposed on a latent AR(20) process X_t. The model permits heavy-tailed observational errors through a normal mixture of precisely the form described in Example 15.2, with a scale inflation factor of $k = 10$. This admits a background level of routine measurement error, with variance V, together with occasional extremes generated by the inflated variance component with probability 0.05. The implied prior for each of the variance weights λ_t is thus discrete, and the conditional posteriors required in posterior sampling, detailed at the end of Section 15.3.2, are simply computed as in the continuation of Example 15.2. Initial prior specification is as follows: a reference uniform prior for the AR parameter $\boldsymbol{\phi}$; a vague prior for the initial state vector $\boldsymbol{\theta}_0$ with zero mean and a diagonal variance matrix with elements 10^3; and independent uniform priors for the square roots of all three variance components V, W and U. A Gibbs sampling analysis was run exactly as described in the previous section; an initial 500 iterations were judged sufficient for the simulations to "burn-in," after which samples of all state vectors and parameters were saved and summarised to give approximate posterior inferences. Convergence checks include repeat analysis, with short runs of 500 iterations, at a small number of different starting values to ensure that the burn-in was adequate and that successive samples thereafter are essentially independent of the starting values. Also, plots of sampled parameters, including variance components and elements of $\boldsymbol{\phi}$, against MCMC iteration number were explored to provide additional support for the view that the sampling process was running "cleanly". Following the burn-in period, a posterior sample of 5,000 draws was saved and used to compute the approximate posterior inferences now summarised. First, the reciprocals of the characteristic roots of the AR process were computed by solving the AR polynomial at each of the 5,000 sampled values of $\boldsymbol{\phi}$. In all 5,000 cases, the moduli of the reciprocal roots are all less than unity, indicating stationarity of the latent X_t process. Further, in each sampled set of roots, the three pairs of complex roots of largest period have periods (after multiplication by 3,000 years) around 26, 42 and 100 kyears, respectively, and tend to be the most "persistent" roots in terms of having moduli closer to unity than the rest. Recalling the decomposition of AR models into latent components detailed in Section 9.4.8, it is seen that these three (pairs of) roots correspond to quasi-periodic components of X_t, i.e., underlying damped cycles with stochastically time-varying amplitudes and phases but fixed periods. As mentioned above, the earth's orbital dynamics are expected to show up in these data, and the periods of these three leading components of the estimated X_t process correspond closely to the ranges mentioned above.

More formal summaries of the Monte Carlo samples of the roots confirm this, as follows. Approximate posterior 25% quartile, median (mean) and 75% quartiles for each of the three longest periods are, respectively, 95, 102 (104), 112 kyears for the longest period, 39.4, 41.7 (45.3), 45 kyears for the second, and 22.9, 24.1 (25.3), 25.9 kyears for the third. Evidently, these posteriors support the view that the ~100 kyears ice-age cycle is dominant, followed by cycles related to the obliquity (~ 41+ kyears) and precession (~23-24 kyears).

Further exploration of these components can be derived by evaluating the actual decomposition of the estimated X_t process at an estimate of ϕ, following the development in Section 9.4.8 leading to the decomposition in equations (9.17) and (9.18). We do this using the Monte Carlo estimate of the posterior mean of ϕ. This leads to estimates over time t of the three dominant quasi-cyclical components $z_{t,j}$, $j = 1, 2, 3$, and the remaining components that sum to give X_t in (9.18). Also, we have posterior means of the smooth trend μ_t over all t, and as a result, posterior means of the observational errors ν_t by subtraction, i.e., $\mathrm{E}[\nu_t|D_n] = Y_t - \mathrm{E}[\mu_t|D_n] - \mathrm{E}[X_t|D_n]$. We graph some of the estimated components of Y_t in Figure 15.2; the upper time series plot is the data series, followed by the posterior mean trajectory of the trend, followed by quasi-cyclical components $z_{t,j}$ for $j = 1, 2$ and 3, followed by the sum of the remaining components of X_t, and finally, the estimated observation error sequence, or residuals. The several series here are plotted on the same vertical scales to enable direct assessment of their relative contributions to explaining the observed data. The data series is the direct sum of all components graphed. We note that the estimated trend is smooth and indicates increasing levels in more recent times (recall the reversed time scale). The three components driven by the earth's orbital dynamics are apparently ordered in terms of decreasing amplitudes as well as periods; each clearly shows periodicity in the relevant frequency range, as well as time-variation in amplitude and phase characteristics. The sum of remaining components, while necessary to give an adequate model fit, has amplitude comparable with the third component.

Observational errors are represented by the estimated residuals in the figure. Posterior inferences about the controlling variance components are summarised in terms of approximate posterior median and quartiles from the Monte Carlo samples. Those for $\sqrt{V}$ (the observational error s.d.) are approximately $(0.127, 0.145, 0.165)$, those for $\sqrt{W}$ (controlling variation in the trend μ_t) are $(0.014, 0.018, 0.023)$, and those for $\sqrt{U}$ (variation in X_t) are $(0.247, 0.323, 0.362)$. A small fraction of the measurement errors are apparently relatively large, but their impact on inferences is apparently slight.

Related studies and further details of this and associated models can be found in West (1995, 1996a,b,c and 1997) and in references cited. These simulation-based methods are nowadays accessible computationally and will become routine research tools in coming years.

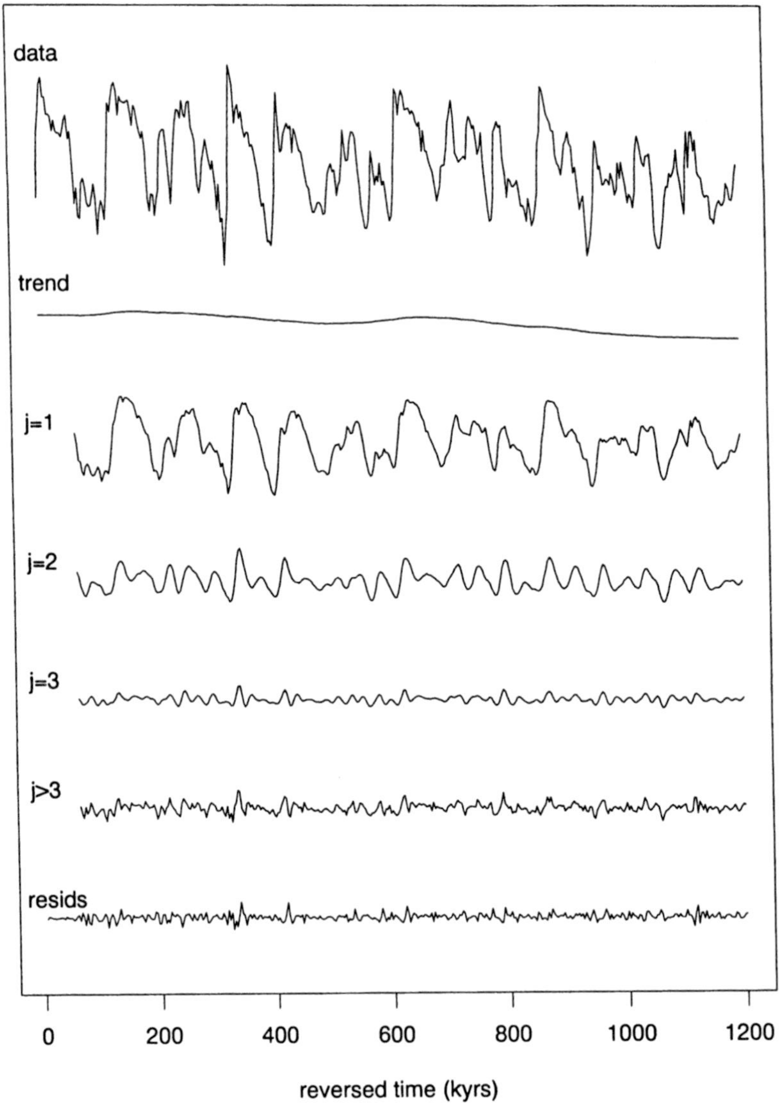

Figure 15.2 Decomposition of oxygen isotope series into estimated trend, latent quasi-cyclical components and residual observational error

CHAPTER 16

MULTIVARIATE MODELLING AND FORECASTING

16.1 INTRODUCTION

In this final chapter we return to the linear/normal framework to explore models for multivariate time series. Univariate DLMs can be extended in an obvious way to multivariate problems simply by taking the observations at each time as vectors rather than scalars. In fact, such models have already been defined in Definition 4.1 and developed somewhat in later sections of that chapter. The basic theory of the univariate DLM as developed in Chapter 4 extends directly to such models. The observational errors are now vectors, so that model specification requires observational variance *matrices* defining the joint stochastic structure of the observations conditional on state parameters. In DLMs having vector observations with observational errors following a multivariate normal distribution, the univariate theory extends easily only when it is assumed that the observational error variance matrices are known for all time. However, as soon as uncertainties about observational variance matrices are admitted, the tractability of analysis is lost. In general, there is no neat, conjugate analysis of multivariate DLMs whose observational errors are multivariate normal with unknown (constant or otherwise) variance matrices. The theory for models with known variance matrices is developed in Section 16.2.

Concerning unknown covariance structures, there are many models that are amenable to fully conjugate analyses, and these are described in Section 16.4. These models are extensions of the basic multivariate DLM in which the state parameters are naturally involved through a state *matrix*, rather than the usual vector, and the analysis most easily developed in terms of *matrix normal*, rather than multivariate normal, distribution theory. The relevant theory is developed below, as is the associated theory for learning about variance matrices within a matrix-normal framework. Some applications illustrate the scope for these models in assessing *cross-sectional* structure of several, possibly many, similar time series.

Section 16.3 is concerned with practical issues arising in multivariate forecasting of hierarchically related time series. Here we identify and discuss problems that arise when forecasting at different levels of aggregation within such hierarchies. Related issues also arise concerning the combination of forecasts made at different levels of aggregation and also possibly by different forecasters or models.

The chapter ends with mention of additional multivariate modelling developments and provides one or two additional references to work in both modelling and applications.

16.2 THE GENERAL MULTIVARIATE DLM

16.2.1 General framework

The general, multivariate normal DLM for a vector time series of observations $\mathbf{Y}_t$ is given in Definition 4.1. Suppose, for $t = 1, \ldots$, that $\mathbf{Y}_t$ is a (column) vector of r observations on the series following a multivariate DLM as in Definition 4.1. The model is defined via a quadruple

$$\{\mathbf{F}, \mathbf{G}, \mathbf{V}, \mathbf{W}\}_t = \{\mathbf{F}_t, \mathbf{G}_t, \mathbf{V}_t, \mathbf{W}_t\}$$

for each time t, where

(a) $\mathbf{F}_t$ is a known $(n \times r)$ dynamic regression matrix;
(b) $\mathbf{G}_t$ is a known $(n \times n)$ state evolution matrix;
(c) $\mathbf{V}_t$ is a known $(r \times r)$ observational variance matrix;
(d) $\mathbf{W}_t$ is a known $(n \times n)$ evolution variance matrix.

The corresponding model equations are

$$\begin{aligned} \mathbf{Y}_t &= \mathbf{F}_t'\boldsymbol{\theta}_t + \boldsymbol{\nu}_t, \qquad & \boldsymbol{\nu}_t &\sim \mathrm{N}[\mathbf{0}, \mathbf{V}_t], \\ \boldsymbol{\theta}_t &= \mathbf{G}_t\boldsymbol{\theta}_{t-1} + \boldsymbol{\omega}_t, \qquad & \boldsymbol{\omega}_t &\sim \mathrm{N}[\mathbf{0}, \mathbf{W}_t], \end{aligned} \tag{16.1}$$

where the error sequences $\boldsymbol{\nu}_t$ and $\boldsymbol{\omega}_t$ are independent and mutually independent. As usual in univariate DLMs, $\boldsymbol{\theta}_t$ is the n-dimensional state vector. With all components of the defining quadruple known, the following results are immediate extensions of the standard updating, forecasting and filtering results in the univariate DLM.

16.2.2 Updating, forecasting and filtering

Suppose the model to be closed to inputs of external information, so that given initial prior information D_0 at $t = 0$, the information set available at any time t is simply $D_t = \{\mathbf{Y}_t, D_{t-1}\}$. Suppose also that the initial prior at $t = 0$ is the usual multivariate normal,

$$(\boldsymbol{\theta}_0 | D_0) \sim \mathrm{N}[\mathbf{m}_0, \mathbf{C}_0], \tag{16.2}$$

for some known moments $\mathbf{m}_0$ and $\mathbf{C}_0$. The following results parallel those in the univariate DLM.

Theorem 16.1. *One-step forecast and posterior distributions in the model just defined are given, for each t, as follows.*

(a) Posterior at $t-1$:
For some mean $\mathbf{m}_{t-1}$ and variance matrix $\mathbf{C}_{t-1}$,

$$(\boldsymbol{\theta}_{t-1} \mid D_{t-1}) \sim \mathrm{N}[\mathbf{m}_{t-1}, \mathbf{C}_{t-1}].$$

(b) Prior at t:

$$(\boldsymbol{\theta}_t \mid D_{t-1}) \sim \mathrm{N}[\mathbf{a}_t, \mathbf{R}_t],$$

where

$$\mathbf{a}_t = \mathbf{G}_t\mathbf{m}_{t-1} \qquad \text{and} \qquad \mathbf{R}_t = \mathbf{G}_t\mathbf{C}_{t-1}\mathbf{G}_t' + \mathbf{W}_t.$$

(c) One-step forecast:

$$(\mathbf{Y}_t \mid D_{t-1}) \sim \mathrm{N}[\mathbf{f}_t, \mathbf{Q}_t],$$

where

$$\mathbf{f}_t = \mathbf{F}_t'\mathbf{a}_t \qquad \text{and} \qquad \mathbf{Q}_t = \mathbf{F}_t'\mathbf{R}_t\mathbf{F}_t + \mathbf{V}_t.$$

(d) Posterior at t:

$$(\boldsymbol{\theta}_t \mid D_t) \sim \mathrm{N}[\mathbf{m}_t, \mathbf{C}_t],$$

with

$$\mathbf{m}_t = \mathbf{a}_t + \mathbf{A}_t\mathbf{e}_t \qquad \text{and} \qquad \mathbf{C}_t = \mathbf{R}_t - \mathbf{A}_t\mathbf{Q}_t\mathbf{A}_t',$$

where

$$\mathbf{A}_t = \mathbf{R}_t\mathbf{F}_t\mathbf{Q}_t^{-1} \qquad \text{and} \qquad \mathbf{e}_t = \mathbf{Y}_t - \mathbf{f}_t.$$

Proof. Proof is by induction, completely paralleling that for the univariate DLM in Theorem 4.1. Suppose (a) to hold. Then (b) follows directly as in the univariate case, the evolution equation being no different here. Using (b) together with the evolution equation implies that $\mathbf{Y}_t$ and $\boldsymbol{\theta}_t$ are jointly normally distributed conditional on D_{t-1}, with covariance matrix

$$\mathrm{C}[\mathbf{Y}_t, \boldsymbol{\theta}_t | D_{t-1}] = \mathrm{C}[\mathbf{F}_t'\boldsymbol{\theta}_t + \boldsymbol{\nu}_t, \boldsymbol{\theta}_t \mid D_{t-1}] = \mathbf{F}_t'\mathrm{V}[\boldsymbol{\theta}_t \mid D_{t-1}] = \mathbf{F}_t'\mathbf{R}_t = \mathbf{Q}_t\mathbf{A}_t',$$

where $\mathbf{A}_t = \mathbf{R}_t\mathbf{F}_t\mathbf{Q}_t^{-1}$. The mean $\mathbf{f}_t$ and variance $\mathbf{Q}_t$ of $\mathbf{Y}_t$ follow easily from the observation equation and (b), establishing (c). It is also clear that

$$\left(\begin{matrix}\mathbf{Y}_t \\ \boldsymbol{\theta}_t\end{matrix}\middle| D_{t-1}\right) \sim \mathrm{N}\left[\begin{pmatrix}\mathbf{f}_t \\ \mathbf{a}_t\end{pmatrix}, \begin{pmatrix}\mathbf{Q}_t & \mathbf{Q}_t\mathbf{A}_t' \\ \mathbf{A}_t\mathbf{Q}_t & \mathbf{R}_t\end{pmatrix}\right].$$

Hence, using normal theory from Section 17.2, the conditional distribution of $\boldsymbol{\theta}_t$ given $D_t = \{Y_t, D_{t-1}\}$ is directly obtained. The $n \times r$ matrix of regression coefficients in the regression of $\boldsymbol{\theta}_t$ on $\mathbf{Y}_t$ is just $\mathbf{A}_t$, and (d) follows.

◇

$\mathbf{e}_t$ is the r-vector of one-step forecast errors and $\mathbf{A}_t$ the $n \times r$ matrix of adaptive coefficients. Clearly, the standard results for the univariate case are given when $r = 1$. Note also that the model definition may be marginally extended to incorporate known, non-zero means for the observational or evolution noise terms. Thus, if $\mathrm{E}[\boldsymbol{\nu}_t]$ and/or $\mathrm{E}[\boldsymbol{\omega}_t]$ are known and non-zero, the above results apply with the modifications that $\mathbf{a}_t = \mathbf{G}_t\mathbf{m}_{t-1} + \mathrm{E}[\boldsymbol{\omega}_t]$ and $\mathbf{f}_t = \mathbf{F}_t'\mathbf{a}_t + \mathrm{E}[\boldsymbol{\nu}_t]$.

Theorem 16.2. *For each time t and $k \geq 0$, the k-step ahead distributions for $\boldsymbol{\theta}_{t+k}$ and $\mathbf{Y}_{t+k}$ given D_t are given by*

$$(a)\quad \textit{State distribution:} \qquad (\boldsymbol{\theta}_{t+k} \mid D_t) \sim \mathrm{N}[\mathbf{a}_t(k), \mathbf{R}_t(k)],$$

$$(b)\quad \textit{Forecast distribution:} \qquad (\mathbf{Y}_{t+k} \mid D_t) \sim \mathrm{N}[\mathbf{f}_t(k), \mathbf{Q}_t(k)],$$

with moments defined recursively by

$$\mathbf{f}_t(k) = \mathbf{F}_t'\mathbf{a}_t(k) \qquad \text{and} \qquad \mathbf{Q}_t(k) = \mathbf{F}_t'\mathbf{R}_t(k)\mathbf{F}_t + \mathbf{V}_{t+k},$$

where

$$\mathbf{a}_t(k) = \mathbf{G}_{t+k}\mathbf{a}_t(k-1) \qquad \text{and} \qquad \mathbf{R}_t(k) = \mathbf{G}_{t+k}\mathbf{R}_t(k-1)\mathbf{G}_{t+k}' + \mathbf{W}_{t+k},$$

with starting values $\mathbf{a}_t(0) = \mathbf{m}_t$ and $\mathbf{R}_t(0) = \mathbf{C}_t$.

Proof. The state forecast distributions in (a) are directly deduced from Theorem 4.2 since the evolution into the future is exactly as in the univariate DLM. The forecast distribution is deduced using the observation equation at time $t+k$; the details are left as an exercise for the reader.

◇

Theorem 16.3. *The filtered distributions $p(\boldsymbol{\theta}_{t-k} \mid D_t)$, $(k > 1)$, are defined as in the univariate DLM in Theorem 4.4.*

Proof. Also left to the reader.

◇

16.2.3 Comments

The above results may be applied in any context where the defining components are known. In particular, they are based on the availability of known or estimated values of the observational variance matrices $\mathbf{V}_t$ for all t. This is clearly a limiting assumptions in practice, such variance matrices will often be uncertain, at least in part. In general, there is no neat, conjugate analysis available to enable sequential learning about unknown variance matrices in the model of (16.1).

In principle, the Bayesian analysis with unknown variance matrix is formally well-defined in cases when $\mathbf{V}_t = \boldsymbol{\Sigma}$, constant for all time t. The unknown parameters in $\boldsymbol{\Sigma}$ introduce complications that can, in principle, be handled using some form of approximate analysis. Analytic approximations are developed in West (1982, Chapter 4) and Barbosa and Harrison (1992). Numerical approximation techniques include multi-process, class I models, and more efficient numerical approaches as described in Chapter 13. In these a prior for $\boldsymbol{\Sigma}$ is specified over a discrete set of values at $t = 0$, this prior (and possibly the set of values) being sequentially updated over time to give an approximate posterior at each time t. Though well-defined in principle, this approach has been little developed to date. As simulation methods of analysis, especially those based on MCMC approaches as discussed in Chapter 15, become more and more prevalent, this is very likely to be a growth area. Some features to be aware of in developing an analysis along these lines are as follows.

First, note that the variance matrix has $r(r+1)/2$ free parameters, so that the dimension of the problem grows rapidly with the dimension r of the time series observation vector. Numerical approximations involving grids of points in the parameter space are therefore likely to be highly computationally demanding unless r is fairly small.

Second, there are restrictions on the parameter space in order that the variance matrix be non-negative definite. Unless structure is imposed on $\boldsymbol{\Sigma}$, it can be rather difficult to identify the relevant subspace in $r(r+1)/2$ dimensions over which to determine a prior distribution for $\boldsymbol{\Sigma}$. Even then, the problems of assessing prior distributions for variance matrices are hard (Dickey, Dawid and Kadane 1986).

Third, we note that some applications will require that structure be imposed on the elements of $\boldsymbol{\Sigma}$, relating them to a small number of unknown quantities. This structures the covariances across the elements of $\mathbf{Y}_t$ and also reduces computational burdens. For example, an assumption of equi-correlated observational errors implies that for some variance σ^2 and correlation ρ,

$$\boldsymbol{\Sigma} = \sigma^2 \begin{pmatrix} 1 & \rho & \rho & \dots & \rho \\ \rho & 1 & \rho & \dots & \rho \\ \vdots & \vdots & \vdots & \ddots & \vdots \\ \rho & \rho & \rho & \dots & 1 \end{pmatrix}.$$

Here the dimension of the parameter space for $\boldsymbol{\Sigma}$ is effectively reduced to 2, that of σ and ρ.

Finally, we note that much can be done in practice using off-line estimated values for $\boldsymbol{\Sigma}$, and more generally, a possibly time-varying variance matrix $\mathbf{V}_t$, possibly updated sequentially over time externally and rather less than formally. Substituting estimated values will always lead to the uncertainties in forecasting and posterior inferences being understated, however, and this must be taken into account in decisions based on such inferences. Harrison, Leonard and Gazard (1977) describe an application to hierarchical forecasting. The particular context is in forecasting industrial demand/sales series that are hierarchically disaggregated into sub-series classified by geographic region, although the approach is very widely applicable. The models developed in this reference involve vector time series of compositional data. Thus, a *multinomial* type of covariance structure is postulated, the DLM observation equation forming a normal approximation to the multinomial. In more detail, write the elements of $\mathbf{Y}_t$ as $\mathbf{Y}_t = (Y_{t1}, \ldots, Y_{tr})'$ supposing that $Y_{tj} > 0$ for all t and j, being compositions of the total

$$N_t = \mathbf{1}'\mathbf{Y}_t = \sum_{j=1}^{r} Y_{tj},$$

where $\mathbf{1} = (1, \ldots, 1)'$. The mean response is given by

$$\mu_t = \mathbf{F}_t'\boldsymbol{\theta}_t = (\mu_{t1}, \ldots, \mu_{tr})',$$

and with an underlying multinomial model,

$$\mu_{tj} = N_t p_{tj}, \qquad (j = 1, \ldots, r),$$

where the proportions p_{tj} sum to unity for all t. Also, the multinomial structure implies that given N_t,

$$\mathrm{V}[Y_{tj}|\boldsymbol{\theta}_t] = N_t p_{tj}(1 - p_{tj}), \qquad (j = 1, \ldots, r),$$

and

$$\mathrm{C}[Y_{tj}, Y_{ti}|\boldsymbol{\theta}_t] = -N_t p_{tj} p_{ti}, \qquad (i, j = 1, \ldots, r;\ i \neq j).$$

Thus, the observational variance matrix $\mathbf{V}_t$ is time-varying and depends on $\boldsymbol{\theta}_t$ (similar in general terms to univariate models with variance laws; see Section 10.7), given by

$$\mathbf{V}_t = N_t[\mathrm{diag}(\mathbf{p}_t) - \mathbf{p}_t\mathbf{p}_t'],$$

where $\mathbf{p}_t = (p_{t1}, \ldots, p_{tr})'$. The form of approximation in Harrison, Leonard and Gazard (1977) is to approximate this matrix by substituting current estimates of the elements of $\mathbf{p}_t$. For example, given D_{t-1} the estimate of $\boldsymbol{\theta}_t$ is the prior mean $\mathbf{a}_t$, so that $\mathbf{p}_t$ is estimated by $\mathbf{a}_t/N_t$ for known N_t. The latter may be estimated externally from a different model for the total or by combining such an estimate with the forecast total $\mathbf{1}'\mathbf{f}_t$, where $\mathbf{f}_t$ is the one-step forecast mean for $\mathbf{Y}_t$; see the relevant comments on top-up and bottom-down forecasting and forecast combination in the next section. For further discussion and details, see Harrison, Leonard and Gazard (1977).

16.3 AGGREGATE FORECASTING

16.3.1 Bottom-up and top-down forecasting

> *"Just as processes cannot be predicted upward from a lower level, they can never be completely analysed downward into their components. To analyse means to isolate parts from the whole, and the functioning of a part in isolation is not the same as its functioning in the whole."*
>
> Arthur Koestler, *The Yogi and the Commissar*

Koestler's words accord entirely with the fundamental tenet of forecasting several or many series in a system or hierarchy, namely that in contributing to the movement and variation in series, different factors dominate at different levels of aggregation. Unfortunately, this is an often neglected fact. It is still very common practice, particularly in commercial forecasting, for forecasts at an aggregate level to be made by aggregating the forecasts of the constituents. In manufacturing industries, for example, stock control systems often produce forecasts and control on individual, component items of a product or process with little or no control at the product level. As a result it is commonly found that stocks build up in slump periods and that shortages are encountered in boom periods. Planning at the macro level based on economic and market factors is rarely adequately used to feed-down information relevant to control at the micro level of the individual products. These factors, that dominate variations at the macro level, often have relatively little *apparent* effect at the disaggregate level and so are ignored.

Now, whilst this neglect of factors that are apparently insignificant at the component level is perfectly sensible when forecasting individual components, the mistake is often made of combining such individual forecasts to produce overall forecasts of the aggregate. Following this "bottom-up" strategy can lead to disastrous results, since moving to the aggregate level means that the previously unimportant factors are now critical. Poor forecasts often result, and the resulting inefficiency and consequent losses can be great.

A simple example illustrates this. Here, in this and the next section, we drop the time subscript t for simplicity in notation and since it is not central to the discussion.

EXAMPLE 16.1. The following simple illustration appears in Green and Harrison (1972). Consider a company selling 1,000 individual products. In one month, let Y_i be the sales of the ith product. Suppose that Y_i may be expressed as

$$Y_i = f_i + X + \epsilon_i, \qquad (i = 1, \dots, 1{,}000),$$

where f_i is the expected value of Y_i, assumed known, X and ϵ_i are zero-mean, uncorrelated random quantities with variances

$$\mathrm{V}[X] = 1 \qquad \text{and} \qquad \mathrm{V}[\epsilon_i] = 99.$$

Assume that ϵ_i represents variation affecting only the ith product, so that ϵ_i and ϵ_j are uncorrelated for $i \neq j$. X is a factor common to all products, related, for example, to changes in consumer disposable income through taxation changes, interest rates, a seasonal effect of a business cycle, etc.

It is clear that for each individual product, $\mathrm{V}[Y_i] = 100$ and that only 1% of this variance is contributed by the common factor X. Variation in individual sales is dominated, to the tune of 99%, by individual, uncorrelated factors. Even if X were precisely known, little improvement in forecasting accuracy would result, the forecast variance reducing by only 1%. Hence it is justifiable, from the point of view of forecasting the individual Y_i, to omit consideration of X and simply adopt the model $Y_i = f_i + \epsilon_i^*$, where the ϵ_i^* are zero-mean, uncorrelated with common variance 100.

Consider now total sales

$$T = \sum_{i=1}^{1000} Y_i.$$

It is commonly assumed that X, having essentially no effect on any of the individual products, is of little or no importance when forecasting the total T. This is generally not the case; surprising things happen when aggregating or disaggregating processes. To see this, write

$$f = \sum_{i=1}^{1000} f_i \qquad \text{and} \qquad \epsilon = \sum_{i=1}^{1000} \epsilon_i.$$

Then

$$T = \sum_{i=1}^{1000} (f_i + X + \epsilon_i) = f + 1000X + \epsilon.$$

Now, $\mathrm{E}[T] = f$, so that the point forecasts of individual sales are simply summed to give the point forecast of the total, but

$$\mathrm{V}[T] = V[1000X] + \mathrm{V}[\epsilon] = 10^6 + 1000 \times 99 = 1,099,000,$$

and X, unimportant at the disaggregated level, is now of paramount importance. The term $1,000X$ accounts for over 91% of the variance of the total T, the aggregate of the individual ϵ_i terms contributing the remaining 9%. At this level, precise knowledge of X reduces the forecast variance dramatically, leading to

$$\mathrm{E}[T|X] = f + 1000X \qquad \text{and} \qquad \mathrm{V}[T|X] = 99,000.$$

The key points arising are summarised as follows.

(a) Suppose that X remains unknown but the fact that X is common to the 1,000 individual series is recognised. Then the total as forecast from the model for the individuals has mean f and variance 1,099,000. Although relative to knowing X, the point forecast has a bias of $1,000X$ units, the high degree of uncertainty in the forecast distribution is appropriately recognised. This is so since the proper joint distribution for the Y_i has been used, with the relevant positive correlation amongst the Y_i induced by X becoming critically evident following aggregation. Formally, writing

$$\mathbf{Y} = (Y_1, \ldots, Y_{1000}),'$$

$$\mathbf{f} = (f_1, \ldots, f_{1000})',$$

and

$$\mathbf{Q} = \begin{pmatrix} 100 & 1 & 1 & \ldots & 1 \\ 1 & 100 & 1 & \ldots & 1 \\ 1 & 1 & 100 & \ldots & 1 \\ \vdots & \vdots & \vdots & \ddots & \vdots \\ 1 & 1 & 1 & \ldots & 100 \end{pmatrix},$$

we have $\mathrm{E}[\mathbf{Y}] = \mathbf{f}$ and $\mathrm{V}[\mathbf{Y}] = \mathbf{Q}$, with $\mathrm{C}[Y_i, Y_j] = 1$. In aggregating, $\mathrm{V}[T] = \mathbf{1}'\mathbf{Q}\mathbf{1}$ is appropriately large due to the large number of positive covariance terms.

(b) A forecaster who unwittingly follows the argument that X is irrelevant will end up with the representation

$$T = f + \sum_{i=1}^{1000} \epsilon_i^*,$$

so that

$$\mathrm{E}[T] = f \qquad \text{and} \qquad \mathrm{V}[T] = 100,000,$$

dramatically under-estimating the uncertainty about T.

(c) It is very worthwhile devoting effort to learning further about X, possibly using a macro model, since this corrects the bias and dramatically reduces the uncertainty about T.

The related procedure of decomposing a point forecast into parts, often referred to as "top-down" forecasting, must also be carefully considered, since it can lead to poor individual predictions. Typically, such procedures convert a total forecast to a set of individuals using a set of forecast proportions. Thus, in the previous example, a forecast mean f and variance V for T will provide a forecast mean $p_i f$ and variance $p_i^2 V$ for Y_i, where p_i is the assumed or estimated proportion that Y_i contributes to total sales. This will sometimes be refined to include uncertainties about the proportions. If the proportions are indeed well determined and stable, then such a strategy can perform well even though the proportions are themselves forecasts. However, it is clear that if major events occur on individual products, the directly disaggregated forecasts, not accounting for such events, may be dismal. Related discussion of these ideas is found in Green and Harrison (1972), and Harrison (1985c).

16.3.2 Combination of forecasts

If consistent and efficient forecasts are to be obtained in a hierarchy at all levels, the previous section underpins the need for a framework in which forecasting at any level is primarily based on consideration of issues relevant at that level. It is also clear, however, that information relevant at any level will tend to be of relevance, though typically of much reduced importance, at other levels. Hence there is a need to consider how forecasts made at different levels, by different forecasters based on different, though possibly related information sets, may be appropriately combined.

Combination of forecast information more generally is a many-faceted subject about which much has been written. Many authors have considered the problems of combining point forecasts of a single random quantity, or random vector, such forecasts being obtained from several forecasters or models. Common approaches have been to form some kind of weighted average of point forecasts, attempting to correct for biases and information overlap amongst the forecasters. Key references include Bates and Granger (1969), Bordley (1982), Bunn (1975), Dickinson (1975), and Granger and Ramanathan (1984). Such simple rules of forecast combination can sometimes be justified within conceptually sound frameworks that have been developed for more general consideration of forecasts made in terms of full or partially specified probability forecast distributions. Such approaches are developed in Lindley, Tversky and Brown (1979), Lindley (1983, 1985, and with review and references, 1988), Harrison (1985c), Morris (1983), West (1984b, 1985b,c, 1988, 1992e), West and Crosse (1992), Winkler (1981), and by other authors. Importantly, such approaches allow consideration of problems of interdependencies amongst forecasters and models, coherence, calibration and time variation in such characteristics.

Forecast combination, and more generally, the synthesis of inferences from different, possibly subjective sources, is a wide-ranging subject, a full discussion being beyond the scope of this book. Attention is restricted to combination of forecasts within the above aggregation framework, with forecasts of linear combinations of a random vector being made in terms of forecast means and variances. Normality is assumed throughout although this is not central to the development.

Consider forecasting a vector $\mathbf{Y} = (Y_1, \dots, Y_n)'$. It is instructive to bear in mind the archetypical example in which all observations are on sales of items produced by a company. The vector $\mathbf{Y}$ may then represent sales across a collection of possibly a large number n of individual products or product lines, sales disaggregated according to market sector or geographical region, and so forth. Interest lies in forecasting the individual sales Y_i, $(i = 1, \dots, n)$, total sales $T = \sum_{i=1}^{n} Y_i$, and also in subtotals of subsets of the elements of $\mathbf{Y}$. Assume that the forecaster has a current, joint forecast distribution, assumed normal though this is a side detail, given by

$\mathbf{Y} \sim \mathrm{N}[\mathbf{f}, \mathbf{Q}]$, with moments

$$\mathbf{f} = \begin{pmatrix} f_1 \\ \vdots \\ f_n \end{pmatrix} \quad \text{and} \quad \mathbf{Q} = \begin{pmatrix} q_{1,1} & q_{1,2} & q_{1,3} & \cdots & q_{1,n} \\ q_{1,2} & q_{2,2} & q_{2,3} & \cdots & q_{2,n} \\ q_{1,3} & q_{2,3} & q_{3,3} & \cdots & q_{3,n} \\ \vdots & \vdots & \vdots & \ddots & \vdots \\ q_{1,n} & q_{2,n} & q_{3,n} & \cdots & q_{n,n} \end{pmatrix}.$$

Several considerations arise in aggregate forecasting, typified by the following questions that the forecaster may wish to answer.

(A) Based only on the forecast information embodied in $p(\mathbf{Y})$, what is the forecast of the total T? More generally, what is the forecast of any linear function $\mathbf{X} = \mathbf{LY}$ where $\mathbf{L}$ is a matrix of dimension $k \times n$ with k an integer, $1 \leq k < n$. $\mathbf{X}$, for example, may represent sales in k product groups or market sectors.

(B) Suppose that in a *What-if?* analysis, the total T is assumed fixed. What is the revised distribution for $\mathbf{Y}$? More generally, what is the forecast distribution $p(\mathbf{Y}|\mathbf{X})$ where $\mathbf{X} = \mathbf{LY}$ with $\mathbf{L}$ as in (A)? This sort of analysis is often required in considering forecasts of $\mathbf{Y}$ subject to various aggregates being constrained to nominal, "target" values.

(C) An additional forecast, in terms of a mean and variance for T, is provided from some other source, such as a macro model for the total sales involving factors not considered in forecasting the individual Y_i. What is the revised forecast for T?

(D) Under the circumstances in (C), what is the implied, revised forecast for the vector of individuals $\mathbf{Y}$?

(E) More generally, an additional forecast of a vector of subaggregates $\mathbf{X} = \mathbf{LY}$ is obtained, with $\mathbf{L}$ as in (A). What is the revised forecast for $\mathbf{Y}$?

These points are considered in turn under the assumption that all forecast distributions are assumed normal.

(A) Forecasting linear functions of Y

This is straightforward; $T \sim \mathrm{N}[f, Q]$, with

$$f = \mathbf{1}'\mathbf{f} = \sum_{i=1}^{n} f_i$$

and

$$Q = \mathbf{1}'\mathbf{Q}\mathbf{1} = \sum_{i=1}^{n} \sum_{j=1}^{n} q_{ij}.$$

Note that the comments in Example 16.1 apply here; sensible results are obtained if it can be assumed that the appropriate correlation structure amongst the individual Y_i is embodied in $\mathbf{Q}$. More generally, it is clear that $\mathbf{X} = \mathbf{LY}$ may be forecast using

$$\mathbf{X} \sim \mathrm{N}[\mathbf{Lf}, \mathbf{LQL}'],$$

for any $k \times n$ matrix $\mathbf{L}$.

(B) Conditioning on subaggregates

For any $k \times n$ matrix $\mathbf{L}$ with $\mathbf{X} = \mathbf{LY}$, we have the joint (singular) distribution

$$\begin{pmatrix} \mathbf{Y} \\ \mathbf{X} \end{pmatrix} \sim \mathrm{N}\left[\begin{pmatrix} \mathbf{f} \\ \mathbf{Lf} \end{pmatrix}, \begin{pmatrix} \mathbf{Q} & \mathbf{QL}' \\ \mathbf{LQ} & \mathbf{LQL}' \end{pmatrix}\right].$$

Then, using standard multivariate normal theory from Section 17.2, the conditional distribution for $\mathbf{Y}$ when $\mathbf{X}$ is known is

$$(\mathbf{Y}|\mathbf{X}) \sim \mathrm{N}[\mathbf{f} + \mathbf{A}(\mathbf{X} - \mathbf{Lf}), \mathbf{Q} - \mathbf{ALQL}'\mathbf{A}'],$$

where

$$\mathbf{A} = \mathbf{QL}'(\mathbf{LQL}')^{-1}.$$

Consider the special case $k = 1$ and $\mathbf{L} = \mathbf{1}'$, so that $\mathbf{X} = T$, the total. Write

$$\mathbf{q} = \mathbf{Q1} = (q_1, \ldots, q_n)',$$

where

$$q_i = \sum_{j=1}^{n} q_{ij}, \qquad (i = 1, \ldots, n).$$

We then have

$$(\mathbf{Y}|T) \sim \mathrm{N}[\mathbf{f} + \mathbf{A}(T - f), \mathbf{Q} - \mathbf{AA}'Q],$$

with $\mathbf{A} = \mathbf{q}/Q$. For the individual Y_i, the marginal forecasts conditional on the total are

$$(Y_i|T) \sim \mathrm{N}[f_i + (q_i/Q)(T - f), q_{ii} - q_i^2/Q].$$

It may be remarked that this is just the approach used to constrain seasonal factors to zero-sum, as in Theorem 8.2, and is obviously applicable generally when precise, linear constraints are to be imposed.

(C) Revising $p(T)$ based on an additional forecast

Suppose that an additional forecast of T is made in terms of a forecast mean m and variance M. Consider how this should be used to revise the prior forecast distribution for the total, namely $T \sim \mathrm{N}[f, Q]$, from (A). Let $H = \{m, M\}$ denote the information provided. Prior to learning the values of m and M, the information H is a random vector, and the objective is to calculate the posterior distribution for T when H is known, identifying the density $p(T|H)$. Formally, via Bayes' Theorem,

$$p(T|H) \propto p(T)p(H|T) \propto p(T)p(m, M|T).$$

The problem is thus to identify an appropriate *model* for the distribution of m and M conditional on T. This special case is an example of a wider class of problems in which T is a very general random quantity and H may represent a variety of information sets. The general approach here is based on the foundational works of Lindley (1983, 1985 and 1988). There are many possible models for $p(m, M|T)$, the choice of an appropriate form depending on the features and circumstances of the application. Some important possibilities are as follows.

(1) Various models are developed in Lindley (1983, 1988). Harrison (1985c) develops similar ideas, both authors discussing the following important special case. Write

$$p(m, M|T) = p(m|M, T)p(M|T),$$

and suppose that these two densities are given as follows.

Firstly, the conditional density of the forecast mean m when both M and T are known is normal,

$$(m|M, T) \sim \mathrm{N}[T, M].$$

Thus, the point forecast m is assumed to be unbiased and distributed about the correct value T with variance equal to that stated. Secondly, the component, $p(M|T)$, is assumed not to depend on T, so it contributes nothing to the likelihood for T. Full discussion of such assumptions is given by Lindley (1988, Section 17).

This results in

$$p(H|T) = p(m, M|T) \propto p(m|M, T),$$

as a function of T, and the revised, posterior forecast density for the total is given through Bayes' Theorem by

$$p(T|H) \propto p(T)p(H|T) \propto p(T)p(m|M, T).$$

From the normal prior $p(T)$ and likelihood $p(m|M, T)$ here, standard normal theory leads to a normal posterior

$$(T|H) \sim \mathrm{N}[f + \rho(m - f), (1 - \rho^2)Q - \rho^2 M],$$

where $\rho = Q/(Q+M)$. The posterior variance here may be alternatively written as ρM; in terms of precision, $\mathrm{V}[T|H]^{-1} = Q^{-1}+M^{-1}$.

(2) The above approach is a special case of more general models in Lindley (1983, 1988), an important class of such allowing for anticipated

biases in the forecast data m and M. Such biases can be modelled by writing, for some known quantities a, b and $c > 0$,

$$(m|M,T) \sim \mathrm{N}[a + bT, cM].$$

This generalises (1) to allow for systematic biases a and/or b in the point forecast m, and also, importantly, to allow for over- or under- optimism in the forecast variance M through the scale factor c. With this replacing the original assumption in (1) above, (but retaining the assumption that $p(M|T)$ does not involve T), Bayes' Theorem easily leads to the revised forecast distribution

$$(T|H) \sim \mathrm{N}[f + \rho(m - a - bf), (1 - \rho^2 b^2)Q - \rho^2 cM],$$

where $\rho = bQ/(b^2Q + cM)$. The posterior variance here may be alternatively written as $\rho(cM/b)$; in terms of precision, $\mathrm{V}[T|H]^{-1} = Q^{-1} + b^2(cM)^{-1}$. The result in (1) is obtained when the forecast information is assumed unbiased, so that $a = 0$ and $b = c = 1$.

Further extensions of this model are also considered by Lindley. One such allows for the possibilities of uncertainties about some or all of the quantities a, b and c, for example. In a time series context where $\mathbf{Y} = \mathbf{Y}_t$, $(t = 1, 2, \ldots,)$, and independent forecast information $H_t = \{m_t, M_t\}$ for the total T_t is sequentially obtained, these biases can be estimated, and also allowed to vary, over time.

(3) In the above, additional forecast information is viewed essentially as data informing about T, whether corrected for anticipated biases as in (2) or not, such data to be combined with the information summarised in the original forecast distribution in more or less standard ways. In some circumstances, this direct combination may be viewed as inappropriate. One of its consequences, for example, concerns the resulting forecast precision. In (1), ρ decreases towards unity as the stated variance M of the additional forecast decreases, with $p(T|H)$ concentrating about m. Often this will lead to spuriously precise inferences that may also be seriously biased. Approach (2) can adequately allow for spurious precision and biases through the use of, possibly uncertain, bias parameters a, b and c. Obviously, some form of correction is desirable if the forecast information H is derived from external sources that may suffer from biases and over- or under-optimism.

This problem is clearly evident when dealing with additional forecasts provided from other forecasters, models, or agencies. In such cases, the concepts of forecasting *expertise* and *calibration* must be considered. West and Crosse (1992), and West (1992e), directly address these issues in a very general framework, and they develop models for the synthesis of various types of forecast information deriving from several different, possibly subjective, sources. Within the specific, normal and linear framework here, and with a single additional forecast, this approach leads to results similar to those of (1) and (2), though with important differences. Refer to the source of the forecast information as an *agent*, whether this be an individual, a group of individuals, or an alternative, possibly related, forecasting model. The essence of the approach is to put a numerical limit on the extent to which the agent's forecast information affects the revised distribution for T. Consider the extreme case in which the forecast provided is perfectly precise, given by taking $M = 0$, so that the information is $H_0 = \{X, 0\}$, the agent's model indicating that T takes the forecast value X. Starting with $T \sim \mathrm{N}[f, Q]$, the question now concerns just how reliable this "perfect" forecast is. One special case of these models leads to $p(T|H_0) = p(T|X)$ with

$$(T|X) \sim \mathrm{N}[f + \rho(X - f), (1 - \rho^2)Q]$$

for some quantity ρ, $(0 \leq \rho \leq 1)$. The quantity ρ represents a measure of assessed expertise of the agent in forecasting. A value near unity is consistent with the view that the agent is rather good in forecasting, $\rho = 1$ leading to direct acceptance of the stated value X. For $\rho < 1$, there is an implied limit on the posterior precision, unlike the approaches in (1) and (2). Hence this case of hypothetical, perfect forecast information H_0 serves to provide the expertise measure ρ. The effect of the received information $H = \{m, M\}$ is modelled by substituting the moments m and M as those of X, taking $(X|H) \sim \mathrm{N}[m, M]$. It should be noted that bias corrections, as in (2), may be used to correct for mis-calibration in forecasts made by other individuals or models, writing $(X|H) \sim \mathrm{N}[a + bm, cM]$, for example. For clarity suppose here that such biases are not deemed necessary, the agent forecast being assumed calibrated. Then, putting together the components

$$(T|X) \sim \mathrm{N}[f + \rho(X - f), (1 - \rho^2)Q]$$

and

$$(X|H) \sim \mathrm{N}[m, M],$$

it follows that the revised forecast for T is given by the posterior

$$(T|H) \sim \mathrm{N}[f + \rho(m - f), (1 - \rho^2)Q + \rho^2 M].$$

The posterior mean here is similar in form to that of the earlier approaches, although now the extent ρ of the correction made to the point forecast m directly relates to the assessed expertise of the agent. The posterior variance, unlike the earlier approaches, will exceed the prior variance Q if $M > Q$, never being less than the value $(1 - \rho^2)Q$ obtained when $M = 0$, providing a rather more conservative synthesis of the two information sources.

Variations of this model, and other, more general models, appear in West and Crosse (1992) and West (1992e). Forms of distribution other than normal lead to results with practically important differences, as do variations in which ρ is uncertain. Details are beyond the scope of the current discussion, however, and the interested reader is referred to the above sources.

(D) Revising $p(\mathbf{Y})$ using an additional forecast of T

Whatever approach is used in (C) above, the result is a revised forecast distribution $(T|H) \sim \mathrm{N}[f^*, Q^*]$ for some moments f^* and Q^*. Given that the additional information H is relevant only to forecasting the total T, it follows that

$$p(\mathbf{Y}|T, H) = p(\mathbf{Y}|T),$$

i.e., given T, $\mathbf{Y}$ is conditionally independent of H. This distribution is given in (B) by

$$(\mathbf{Y}|T) \sim \mathrm{N}[\mathbf{f} + \mathbf{q}(T - f)/Q, \mathbf{Q} - \mathbf{q}\mathbf{q}'/Q],$$

with $\mathbf{q} = \mathbf{Q}\mathbf{1}$.

Hence the implications of H for forecasting $\mathbf{Y}$ are derived through the posterior

$$p(\mathbf{Y}|H) = \int p(\mathbf{Y}|T, H)p(T|H)dT = \int p(\mathbf{Y}|T)p(T|H)dT.$$

It easily follows (the proof left as an exercise for the reader) that

$$(\mathbf{Y}|H) \sim \mathrm{N}[\mathbf{f} + \mathbf{q}(f^* - f)/Q, \mathbf{Q} - \mathbf{q}\mathbf{q}'(Q - Q^*)/Q^2].$$

(E) Additional forecasts of aggregates

Models in (D) generalise directly to the case of independent forecasts of subaggregates, and linear combinations generally. Let $\mathbf{X} = \mathbf{L}\mathbf{Y}$ as above, so that from (A),

$$\mathbf{X} \sim \mathrm{N}[\mathbf{s}, \mathbf{S}],$$

with moments $\mathbf{s} = \mathbf{L}\mathbf{f}$ and $\mathbf{S} = \mathbf{L}\mathbf{Q}\mathbf{L}'$. Independent forecast information is obtained, providing a forecast mean $\mathbf{m}$ and variance matrix $\mathbf{M}$. The models of (D), in the special case that $\mathbf{L} = \mathbf{1}'$, extend directly. Under model (1) of part (C), the additional forecast information is taken at face value. The direct generalisation of the results there has $p(H|T)$ defined via the two components $p(\mathbf{m}|\mathbf{M}, \mathbf{X})$ and $p(\mathbf{M}|\mathbf{X})$ given as follows. First,

$$(\mathbf{m}|\mathbf{M}, \mathbf{X}) \sim \mathrm{N}[\mathbf{X}, \mathbf{M}];$$

second, $p(\mathbf{M}|\mathbf{X})$ does not depend on $\mathbf{X}$. Under these assumptions, Bayes' Theorem leads to the posterior

$$(\mathbf{X}|H) \sim \mathrm{N}[\mathbf{s}^*, \mathbf{S}^*],$$

where the revised moments are given by

$$\mathbf{s}^* = \mathbf{s} + \mathbf{B}(\mathbf{m} - \mathbf{s})$$

and

$$\mathbf{S}^* = \mathbf{S} - \mathbf{B}(\mathbf{S} + \mathbf{M})\mathbf{B}',$$

where

$$\mathbf{B} = \mathbf{S}(\mathbf{S} + \mathbf{M})^{-1}.$$

Biases can be modelled, generalising (2) of part (C), the details being left as an exercise for the reader. Similarly, using approaches in West (1992), additional information from an agent may be combined through models generalising that described in part (3) of (C) above.

Given any revised moments $\mathbf{s}^*$ and $\mathbf{S}^*$, and again assuming that the additional information H is relevant only to forecasting $\mathbf{X}$, it follows that

$$p(\mathbf{Y}|\mathbf{X}, H) = p(\mathbf{Y}|\mathbf{X}),$$

i.e., given $\mathbf{X}$, $\mathbf{Y}$ is conditionally independent of H. Also, from (B),

$$(\mathbf{Y}|\mathbf{X}) \sim \mathrm{N}[\mathbf{f} + \mathbf{A}(\mathbf{X} - \mathbf{s}), \mathbf{Q} - \mathbf{A}\mathbf{S}\mathbf{A}'],$$

where $\mathbf{A} = \mathbf{Q}\mathbf{L}'\mathbf{S}^{-1}$.

Hence $p(\mathbf{Y})$ is revised in the light of H to

$$p(\mathbf{Y}|H) = \int p(\mathbf{Y}|\mathbf{X}, H)p(\mathbf{X}|H)d\mathbf{X} = \int p(\mathbf{Y}|\mathbf{X})p(\mathbf{X}|H)d\mathbf{X},$$

that (as can be verified by the reader) is given by

$$(\mathbf{Y}|H) \sim \mathrm{N}[\mathbf{f}^*, \mathbf{Q}^*],$$

with moments given by

$$\mathbf{f}^* = \mathbf{f} + \mathbf{A}(\mathbf{s}^* - \mathbf{s})$$

and

$$\mathbf{Q} - \mathbf{A}(\mathbf{S} - \mathbf{S}^*)\mathbf{A}'.$$

Finally, the revised forecasts for any other aggregates $\mathbf{K}\mathbf{Y}$, say, can be deduced from part (A).

16.4 MATRIX NORMAL DLMS

16.4.1 Introduction and general framework

A general framework for multivariate time series analysis when the covariance structure across series is unknown is presented in Quintana (1985, 1987), and developed and applied in Quintana and West (1987, 1988). The resulting models are extensions of the basic DLM that allow fully conjugate, closed-form analyses of covariance structure when it may be assumed that the scalar component time series follow univariate DLMs with common $\mathbf{F}_t$ and $\mathbf{G}_t$. Thus, the models are appropriate in applications when several similar series are to be analysed. Such a setup is common in many application areas. In economic and financial modelling, such series arise as measurements of similar financial indicators, share prices or exchange rates, and compositional series such as energy consumption classified by energy sources. In medical monitoring, collections of similar time series are often recorded, such as with measurements on each of several, related biochemical indicators in post-operative patient care.

The framework developed in the above references is as follows. Suppose that we have q univariate series Y_{tj} with, for each $j = 1, \ldots, q$, Y_{tj} following a standard, univariate DLM with defining quadruple

$$\{\mathbf{F}_t, \mathbf{G}_t, V_t\sigma_j^2, \mathbf{W}_t\sigma_j^2\}.$$

The model is n-dimensional, and the defining quantities of the quadruple are assumed known apart from the scale factors σ_j^2, $(j = 1, \ldots, q)$. In terms of observation and evolution equations, the univariate series Y_{tj} is given by

$$\text{Observation:} \quad Y_{tj} = \mathbf{F}_t'\boldsymbol{\theta}_{tj} + \nu_{tj}, \qquad \nu_{tj} \sim \mathrm{N}[0, V_t\sigma_j^2], \qquad (16.3a)$$

$$\text{Evolution:} \quad \boldsymbol{\theta}_{tj} = \mathbf{G}_t\boldsymbol{\theta}_{t-1,j} + \boldsymbol{\omega}_{tj}, \quad \boldsymbol{\omega}_{tj} \sim \mathrm{N}[\mathbf{0}, \mathbf{W}_t\sigma_j^2]. \quad (16.3b)$$

Note the key feature of the model here: the defining components $\mathbf{F}_t$, $\mathbf{G}_t$ and $\mathbf{W}_t$ are the same for each of the q series. In addition, the model permits a common, known, observational scale factor V_t across all q series, to allow, for example, for common measurement scales, common sampling variation, common occurrence of missing values, outliers, and so forth. Otherwise, the series have individual state vectors $\boldsymbol{\theta}_{tj}$ that are typically different. They also possibly vary through the scales of measurement, defined via individual variances σ_j^2, that are assumed uncertain. Note also that as usual σ_j^2 appears as a multiplier of the known evolution variance matrix $\mathbf{W}_t$. Of course the model equations above are defined conditional upon these variances. Finally, the usual conditional independence assumptions are made: given all defining parameters, we assume, for all j, that the errors ν_{tj} are independent over time, the evolution errors $\boldsymbol{\omega}_{tj}$ are independent over time, and that the two sequences are mutually independent.

The joint, *cross-sectional* structure across series at time t comes in via the covariances between the observational errors of each of the q series, and also between evolution errors. Introduce a $q \times q$ covariance matrix $\boldsymbol{\Sigma}$ given by

$$\boldsymbol{\Sigma} = \begin{pmatrix} \sigma_1^2 & \sigma_{1,2} & \cdots & \sigma_{1,q} \\ \sigma_{1,2} & \sigma_2^2 & \cdots & \sigma_{2,q} \\ \vdots & \vdots & \ddots & \vdots \\ \sigma_{1,q} & \sigma_{2,q} & \cdots & \sigma_q^2 \end{pmatrix},$$

where for all i and j, $(i = 1, \dots, q;\ j = 1, \dots, q;\ i \neq j)$ σ_{ij} determines the covariance between series Y_{ti} and Y_{tj}. Then the model equations (16.3a and b) are supplemented by the cross-sectional assumptions that conditional on $\boldsymbol{\Sigma}$,

$$\mathrm{C}[\nu_{ti}, \nu_{tj}] = V_t \sigma_{ij}, \tag{16.4a}$$

$$\mathrm{C}[\boldsymbol{\omega}_{ti}, \boldsymbol{\omega}_{tj}] = \mathbf{W}_t \sigma_{ij}, \tag{16.4b}$$

for $i \neq j$.

The individual series Y_{tj}, $(j = 1, \dots, q)$, follow the same *form* of DLM, the model parameters $\boldsymbol{\theta}_{tj}$ being different across series. Correlation structure induced by $\boldsymbol{\Sigma}$ affects both the observational errors through (16.4a) and the evolution errors $\boldsymbol{\omega}_{tj}$ through (16.4b). Thus, if, for example, σ_{ij} is large and positive, series i and j will tend to follow similar patterns of behaviour in both the underlying movements in their defining state parameters and in the purely random, observational variation about their levels. Of course the scales σ_i and σ_j may differ.

The model equations may be written in matrix notation. For notation, define the following quantities for each t:

- $\mathbf{Y}_t = (Y_{t1}, \dots, Y_{tq})'$, the q-vector of observations at time t;
- $\boldsymbol{\nu}_t = (\nu_{t1}, \dots, \nu_{tq})'$, the q-vector of observational errors at time t;
- $\boldsymbol{\Theta}_t = [\boldsymbol{\theta}_{t1}, \dots, \boldsymbol{\theta}_{tq}]$, the $n \times q$ matrix whose columns are the state vectors of the individual DLMs (16.3a);
- $\boldsymbol{\Omega}_t = [\boldsymbol{\omega}_{t1}, \dots, \boldsymbol{\omega}_{tq}]$, the $n \times q$ matrix whose columns are the evolution errors of the individual DLMs in (16.3b).

With these definitions, (16.3a and b) are re-expressible as

$$\begin{aligned} \mathbf{Y}_t' &= \mathbf{F}_t' \boldsymbol{\Theta}_t + \boldsymbol{\nu}_t', \\ \boldsymbol{\Theta}_t &= \mathbf{G}_t \boldsymbol{\Theta}_{t-1} + \boldsymbol{\Omega}_t. \end{aligned} \tag{16.5}$$

Note here that the observation is a *row* vector and the state parameters are in the form of a *matrix*. The fact that $\mathbf{F}_t$ and $\mathbf{G}_t$ are common to each of the q univariate DLMs is fundamental to this new representation. To proceed we need to identify the distributions of the observational error vector $\boldsymbol{\nu}_t$ and the evolution error *matrix* $\boldsymbol{\Omega}_t$, all, of course, conditional on $\boldsymbol{\Sigma}$ (as well as V_t and $\mathbf{W}_t$ for all t). The former is obviously multivariate normal,

$$\boldsymbol{\nu}_t \sim \mathrm{N}[\mathbf{0}, V_t \boldsymbol{\Sigma}],$$

independently over time, where $\boldsymbol{\Sigma}$ defines the cross-sectional covariance structure for the multivariate model. The latter is a *matrix-variate normal distribution* (Dawid 1981; see also Press 1985), described as follows.

16.4.2 The matrix normal distribution for $\boldsymbol{\Omega}_t$

Clearly, any collection of the qn elements of $\boldsymbol{\Omega}_t$ are multivariate normally distributed, as is the distribution of any linear function of the elements.

Dawid (1981) describes the class of matrix-variate (or just matrix) normal distributions that provides a concise mathematical representation and notation for such matrices of jointly normal quantities. Full theoretical details of the structure and properties of such distributions, and their uses in Bayesian analyses, are given in Dawid's paper, to which the interested reader may refer. See also Quintana (1987, Chapter 3). Some of these features are described here.

The random matrix $\boldsymbol{\Omega}_t$ has a matrix normal distribution with **mean matrix 0, left variance matrix $\mathbf{W}_t$** and **right variance matrix $\boldsymbol{\Sigma}$.** The density function is given by

$$p(\boldsymbol{\Omega}_t) = k(\mathbf{W}_t, \boldsymbol{\Sigma})\exp\{-\frac{1}{2}\text{trace}[\boldsymbol{\Omega}_t'\mathbf{W}_t^{-1}\boldsymbol{\Omega}_t\boldsymbol{\Sigma}^{-1}]\},$$

where

$$k(\mathbf{W}_t, \boldsymbol{\Sigma}) = (2\pi)^{-qn/2}|\mathbf{W}_t|^{-q/2}|\boldsymbol{\Sigma}|^{-n/2}.$$

Some important properties are that (i) all marginal and conditional distributions of elements of $\boldsymbol{\Omega}_t$, and linear functions of them, are uni-, multi-, or matrix- variate normal; (ii) the definition of the distribution remains valid when either or both of the variance matrices is non-negative definite; (iii) the distribution is non-singular if and only if each variance matrix is positive definite; and (iv) if either of $\mathbf{W}_t$ and $\boldsymbol{\Sigma}$ is the zero matrix then $\boldsymbol{\Omega}_t$ is zero with probability one.

Concerning the model (16.5), note that $q = 1$ leads to a standard, univariate DLM with unknown observational scale factor $\boldsymbol{\Sigma} = \sigma_1^2$. Also, for any q, if $\boldsymbol{\Sigma}$ is diagonal then the q series Y_{tj} are unrelated. The distribution for $\boldsymbol{\Omega}_t$ is denoted, in line with the notation in the above references, by

$$\boldsymbol{\Omega}_t \sim \text{N}[\mathbf{0}, \mathbf{W}_t, \boldsymbol{\Sigma}]. \tag{16.6}$$

A simple extension of (16.5) to include a known mean vector for each of the evolution error vectors poses no problem. Suppose that $\text{E}[\boldsymbol{\omega}_{tj}] = \mathbf{h}_{tj}$ is known at time t for each j, and let $\mathbf{H}_t$ be the $n \times q$ mean matrix $\mathbf{H}_t = [\mathbf{h}_{t1}, \ldots, \mathbf{h}_{tq}]$. Then $\boldsymbol{\Omega}_t - \mathbf{H}_t$ has the distribution (16.6); equivalently, $\boldsymbol{\Omega}_t$ has a matrix normal distribution with the same variance matrices, but now with mean matrix $\mathbf{H}_t$, the notation being $\boldsymbol{\Omega}_t \sim \text{N}[\mathbf{H}_t, \mathbf{W}_t, \boldsymbol{\Sigma}]$.

16.4.3 The matrix normal/inverse Wishart distribution

Having introduced the matrix normal distribution for the matrix of evolution errors, it is no surprise that the same form of distribution provides the basis of prior and posterior distributions for the state matrices $\boldsymbol{\Theta}_t$ for all t. This is directly parallel to the use of multivariate normals in standard DLMs when the state parameters form a vector. Analysis conditional on $\boldsymbol{\Sigma}$ is completely within the class of matrix normal distributions, but, in line with the focus of this chapter, the results are developed below in the more general framework in which $\boldsymbol{\Sigma}$ is unknown. Thus, we consider the class of **matrix normal/inverse Wishart** distributions suitable for learning jointly about $\boldsymbol{\Theta}_t$ and $\boldsymbol{\Sigma}$. Before proceeding to the model analysis in the next section, the structure of this class of distributions is summarised here (free from notational complications due to the t subscripts).

(A) The inverse Wishart distribution

Consider a $q \times q$ positive definite random matrix $\boldsymbol{\Sigma}$. Following Dawid (1981), $\boldsymbol{\Sigma}$ has an **inverse (or inverted) Wishart** distribution if and only if the density of the distribution of $\boldsymbol{\Sigma}$ is given (up to a constant of normalisation) by

$$p(\boldsymbol{\Sigma}) \propto |\boldsymbol{\Sigma}|^{-(q+n/2)} \exp[-\frac{1}{2}\text{trace}(n\mathbf{S}\boldsymbol{\Sigma}^{-1})],$$

where $n > 0$ is a known scalar *degrees of freedom* parameter, and $\mathbf{S}$ a known $q \times q$ positive definite matrix. Full discussion and properties appear in Box and Tiao (1973, Section 8.5), and Press (1985, Chapter 5), some features of interest being as follows.

- $\boldsymbol{\Phi} = \boldsymbol{\Sigma}^{-1}$ has a Wishart distribution with n degrees of freedom and mean $\text{E}[\boldsymbol{\Phi}] = \mathbf{S}^{-1}$. Thus, $\mathbf{S}$ is an estimate of $\boldsymbol{\Sigma}$, the harmonic mean under the inverse Wishart distribution. If $\mathbf{S}$ has elements

$$\mathbf{S} = \begin{pmatrix} S_1 & S_{1,2} & S_{1,3} & \cdots & S_{1,q} \\ S_{1,2} & S_2 & S_{2,3} & \cdots & S_{2,q} \\ \vdots & \vdots & \vdots & \ddots & \vdots \\ S_{1,q} & S_{2,q} & S_{3,q} & \cdots & S_q \end{pmatrix},$$

  then S_j is an estimate of the variance σ_j^2, and S_{jk} an estimate of the covariance σ_{jk}, for all j and k, $(j \neq k)$.
- If $q = 1$, so that $\boldsymbol{\Sigma}$ is scalar, then so are $\boldsymbol{\Phi} = \phi$ and $\mathbf{S} = S$. Now ϕ has the usual gamma distribution, $\phi \sim \text{G}[n/2, nS/2]$ with mean $\text{E}[\phi] = 1/S$.
- For $n > 2$, $\text{E}[\boldsymbol{\Sigma}] = \mathbf{S}n/(n-2)$.
- The marginal distributions of the variances σ_j^2 on the diagonal of $\boldsymbol{\Sigma}$ are inverse gamma; with precisions $\phi_j = \sigma_j^{-2}$, $\phi_j \sim \text{G}[n/2, nS_j/2]$.
- As $n \to \infty$, the distribution concentrates around $\mathbf{S}$, ultimately degenerating there.

By way of notation here, the distribution is denoted by

$$\boldsymbol{\Sigma} \sim \text{W}_n^{-1}[\mathbf{S}]. \tag{16.7}$$

(B) The matrix normal/inverse Wishart distribution

Suppose that (16.7) holds, and introduce a further random matrix $\boldsymbol{\Theta}$ of dimension $p \times q$. Suppose that conditional on $\boldsymbol{\Sigma}$, $\boldsymbol{\Theta}$ follows a matrix-normal distribution with $p \times q$ mean matrix $\mathbf{m}$, $p \times p$ left variance matrix $\mathbf{C}$, and right variance matrix $\boldsymbol{\Sigma}$. The defining quantities $\mathbf{m}$ and $\mathbf{C}$ are assumed known. Thus, following (16.6),

$$(\boldsymbol{\Theta}|\boldsymbol{\Sigma}) \sim \mathrm{N}[\mathbf{m}, \mathbf{C}, \boldsymbol{\Sigma}]. \tag{16.8}$$

Equations (16.7) and (16.8) define a joint distribution for $\boldsymbol{\Theta}$ and $\boldsymbol{\Sigma}$ that is referred to as a **matrix normal/inverse Wishart** distribution. The special case of scalar $\boldsymbol{\Sigma}$, when $q = 1$, corresponds to the usual normal/inverse gamma distribution (Section 16.3) in variance learning in DLMs in earlier chapters. Dawid (1981) details this class of distributions, deriving many important results. By way of notation, if (16.7) and (16.8) hold, then the joint distribution is denoted by

$$(\boldsymbol{\Theta}, \boldsymbol{\Sigma}) \sim \mathrm{NW}_n^{-1}[\mathbf{m}, \mathbf{C}, \mathbf{S}]. \tag{16.9}$$

Equation (16.9) is now understood to imply both (16.7) and (16.8). In addition, the marginal distribution of $\boldsymbol{\Theta}$ takes the following form.

(C) The matrix T distribution

Under (16.9) the marginal distribution of the matrix $\boldsymbol{\Theta}$ is a matrix-variate analogue of the multivariate T distribution (Dawid 1981). This is completely analogous to the matrix normal, the differences lying in the tail weight of the marginal distributions of elements of $\boldsymbol{\Theta}$. As with the matrix normal, the component columns of $\boldsymbol{\Theta}$ themselves follow p-dimensional multivariate T distributions with n degrees of freedom. Write

$$\boldsymbol{\Theta} = [\boldsymbol{\theta}_1, \dots, \boldsymbol{\theta}_q]$$

and

$$\mathbf{m} = [\mathbf{m}_1, \dots, \mathbf{m}_q].$$

It follows (Dawid 1981) that

$$\boldsymbol{\theta}_j \sim \mathrm{T}_n[\mathbf{m}_j, \mathbf{C}S_j], \qquad (j = 1, \dots, q).$$

If $n > 1$, $\mathrm{E}[\boldsymbol{\theta}_j] = \mathbf{m}_j$. If $n > 2$, $\mathrm{V}[\boldsymbol{\theta}_j] = \mathbf{C}S_j n/(n-2)$ and the covariance structure between the columns is given by

$$\mathrm{C}[\boldsymbol{\theta}_j, \boldsymbol{\theta}_k] = \mathbf{C}S_{jk}n/(n-2), \qquad (j, k = 1, \dots, p;\ j \neq k).$$

As with the matrix normal notation in (16.8), the matrix T distribution of $\boldsymbol{\Theta}$ is denoted by

$$\boldsymbol{\Theta} \sim \mathrm{T}_n[\mathbf{m}, \mathbf{C}, \mathbf{S}]. \tag{16.10}$$

16.4.4 Updating and forecasting equations

The model (16.5) is amenable to a conjugate, sequential analysis that generalises the standard analysis for univariate series with variance learning. The analysis is based on the use of matrix normal/inverse Wishart prior and posterior distributions for the $n \times q$ state matrix $\boldsymbol{\Theta}_t$ and the $q \times q$ variance matrix $\boldsymbol{\Sigma}$ at all times t. The theory is essentially based on the Bayesian theory in Dawid (1981). Full details are given in Quintana (1985, 1987), the results being stated here without proof and illustrated in following sections.

Suppose $\mathbf{Y}_t$ follows the model defined in equations (16.3-16.6), summarised as

$$\begin{aligned} \mathbf{Y}_t' &= \mathbf{F}_t'\boldsymbol{\Theta}_t + \boldsymbol{\nu}_t', & \boldsymbol{\nu}_t &\sim \mathrm{N}[\mathbf{0}, V_t\boldsymbol{\Sigma}], \\ \boldsymbol{\Theta}_t &= \mathbf{G}_t\boldsymbol{\Theta}_{t-1} + \boldsymbol{\Omega}_t, & \boldsymbol{\Omega}_t &\sim \mathrm{N}[\mathbf{0}, \mathbf{W}_t, \boldsymbol{\Sigma}]. \end{aligned} \tag{16.11}$$

Here $\boldsymbol{\nu}_t$ are independent over time, $\boldsymbol{\Omega}_t$ are independent over time, and the two series are mutually independent. Suppose also that the initial prior for $\boldsymbol{\Theta}_0$ and $\boldsymbol{\Sigma}$ is matrix normal/inverse Wishart, as in (16.9), namely

$$(\boldsymbol{\Theta}_0, \boldsymbol{\Sigma}|D_0) \sim \mathrm{NW}_{n_0}^{-1}[\mathbf{m}_0, \mathbf{C}_0, \mathbf{S}_0], \tag{16.12}$$

for some known defining parameters $\mathbf{m}_0$, $\mathbf{C}_0$, $\mathbf{S}_0$ and n_0. Then, for all times $t > 1$, the following results apply.

Theorem 16.4. *One-step forecast and posterior distributions in the model (16.11) and (16.12) are given, for each t, as follows.*

(a) Posteriors at $t-1$:
For some $\mathbf{m}_{t-1}$, $\mathbf{C}_{t-1}$, $\mathbf{S}_{t-1}$ and n_{t-1},

$$(\boldsymbol{\Theta}_{t-1}, \boldsymbol{\Sigma}|D_{t-1}) \sim \mathrm{NW}_{n_{t-1}}^{-1}[\mathbf{m}_{t-1}, \mathbf{C}_{t-1}, \mathbf{S}_{t-1}].$$

(b) Priors at t:

$$(\boldsymbol{\Theta}_t, \boldsymbol{\Sigma}|D_{t-1}) \sim \mathrm{NW}_{n_{t-1}}^{-1}[\mathbf{a}_t, \mathbf{R}_t, \mathbf{S}_{t-1}],$$

where

$$\mathbf{a}_t = \mathbf{G}_t\mathbf{m}_{t-1} \quad \text{and} \quad \mathbf{R}_t = \mathbf{G}_t\mathbf{C}_{t-1}\mathbf{G}_t' + \mathbf{W}_t.$$

(c) One-step forecast:

$$(\mathbf{Y}_t|\boldsymbol{\Sigma}, D_{t-1}) \sim \mathrm{N}[\mathbf{f}_t, Q_t\boldsymbol{\Sigma}]$$

with marginal

$$(\mathbf{Y}_t|D_{t-1}) \sim \mathrm{T}_{n_{t-1}}[\mathbf{f}_t, Q_t\mathbf{S}_{t-1}],$$

where

$$\mathbf{f}_t' = \mathbf{F}_t'\mathbf{a}_t \quad \text{and} \quad Q_t = V_t + \mathbf{F}_t'\mathbf{R}_t\mathbf{F}_t.$$

(d) Posteriors at t:

$$(\boldsymbol{\Theta}_t, \boldsymbol{\Sigma}|D_t) \sim \mathrm{NW}_{n_t}^{-1}[\mathbf{m}_t, \mathbf{C}_t, \mathbf{S}_t],$$

with

$$\mathbf{m}_t = \mathbf{a}_t + \mathbf{A}_t\mathbf{e}_t' \qquad \textit{and} \qquad \mathbf{C}_t = \mathbf{R}_t - \mathbf{A}_t\mathbf{A}_t'Q_t,$$

$$n_t = n_{t-1} + 1 \qquad \textit{and} \qquad \mathbf{S}_t = n_t^{-1}[n_{t-1}\mathbf{S}_{t-1} + \mathbf{e}_t\mathbf{e}_t'/Q_t],$$

where

$$\mathbf{A}_t = \mathbf{R}_t\mathbf{F}_t/Q_t \qquad \textit{and} \qquad \mathbf{e}_t = \mathbf{Y}_t - \mathbf{f}_t.$$

Proof. Omitted, an exercise for the reader. Full details appear in Quintana (1985, 1987).

◇

As mentioned, this is a direct extension of the univariate theory. Theorem 4.3 is the special case $q = 1$ (and with no real loss of generality, $V_t = 1$).† Here, generally, $\boldsymbol{\Theta}_t$ is a matrix, the prior and posterior distributions in (a), (b) and (d) being matrix normal. Thus, the prior and posterior means $\mathbf{a}_t$ and $\mathbf{m}_t$ are both $n \times q$ matrices, their columns providing the means of the q state vectors of the individual DLMs (16.3a and b). Notice the key features that since $\mathbf{F}_t$, $\mathbf{G}_t$ and $\mathbf{W}_t$ are common to these q DLMs, the elements $\mathbf{R}_t$, $\mathbf{C}_t$, $\mathbf{A}_t$ and Q_t are common too. Thus, although there are q series analysed, the calculation of these elements need only be done once. Their dimensions are as in the univariate case, $\mathbf{R}_t$ and $\mathbf{C}_t$ are both $n \times n$, the common adaptive vector $\mathbf{A}_t$ is $n \times 1$, and the common variance Q_t is a scalar.

The analysis essentially duplicates that for the individual univariate DLMs, this being seen in detail by decomposing the vector/matrix results as follows.

(a) For each of the q models, the posterior in (a) of the Theorem has the following marginals:

$$(\boldsymbol{\theta}_{t-1,j}|\sigma_j^2, D_{t-1}) \sim \mathrm{N}[\mathbf{m}_{t-1,j}, \mathbf{C}_{t-1}\sigma_j^2]$$

and

$$(\sigma_j^{-2}|D_{t-1}) \sim \mathrm{G}[n_{t-1}/2, d_{t-1,j}/2],$$

†Note that results are given here in the original form with the variance matrices $\mathbf{C}_t$, $\mathbf{W}_t$, etc. all being subject to multiplication by scale factors from $\boldsymbol{\Sigma}$, as in the univariate case in Theorem 4.4. The alternative, and preferred, representation summarised in the table in Section 4.6 and used throughout the book, has no direct analogue here in the multivariate framework.

where $d_{t-1,j} = n_{t-1}S_{t-1,j}$. Unconditional on the variance σ_j^2,

$$(\boldsymbol{\theta}_{t-1,j}|D_{t-1}) \sim \mathrm{T}_{n_{t-1}}[\mathbf{m}_{t-1,j}, \mathbf{C}_{t-1}S_{t-1,j}].$$

Thus, for each j, the prior is the standard normal/gamma. Note again that the scale-free variance matrix $\mathbf{C}_{t-1}$ is common across the q series.

(b) Evolving to time t, similar comments apply to the prior distribution in (b). Writing $\mathbf{a}_t$ in terms of its columns

$$\mathbf{a}_t = [\mathbf{a}_{t1}, \dots, \mathbf{a}_{tq}],$$

it is clear that

$$\mathbf{a}_{tj} = \mathbf{G}_t\mathbf{m}_{t-1,j}, \qquad (j = 1, \dots, q).$$

Thus, for each of the q univariate DLMs, the evolution is standard, $(\boldsymbol{\theta}_{tj}|D_{t-1}) \sim \mathrm{T}_{n_{t-1}}[\mathbf{a}_{tj}, \mathbf{R}_tS_{t-1,j}]$, where $\mathbf{a}_{tj}$ and $\mathbf{R}_t$ are calculated as usual. Again note that $\mathbf{R}_t$ is common to the q series and so need only be calculated once.

(c) In the one-step ahead forecasting result (c), the q-vector of forecast means $\mathbf{f}_t = (f_{t1}, \dots, f_{tq})'$ has elements $f_{tj} = \mathbf{F}_t'\mathbf{a}_{tj}$, $(j = 1, \dots, q)$, that are as usual for the univariate series. So is the one-step ahead, scale-free variance Q_t. For the jth series, the forecast distribution is Student T,

$$(Y_{tj}|D_{t-1}) \sim \mathrm{T}_{n_{t-1}}[f_{tj}, Q_tS_{t-1,j}],$$

as usual.

(d) In updating in part (d) of the theorem, the equation for the common variance matrix $\mathbf{C}_t$ is as usual, applying to each of the q series. The mean matrix update can be written column by column, as

$$\mathbf{m}_{tj} = \mathbf{a}_{tj} + \mathbf{A}_te_{tj}, \qquad (j = 1, \dots, q),$$

where for each j, $e_{tj} = Y_{tj} - f_{tj}$ is the usual forecast error and $\mathbf{A}_t$ the usual, and common, adaptive vector. For $\boldsymbol{\Sigma}$, n_t increases by one degree of freedom, being common to each series. The update for the estimate $\mathbf{S}_t$ can be decomposed into elements as follows. For the variances on the diagonal,

$$S_{tj} = n_t^{-1}(n_{t-1}S_{t-1,j} + e_{tj}^2/Q_t).$$

With $d_{tj} = n_tS_{tj}$, then $d_{tj} = d_{t-1,j} + e_{tj}^2/Q_t$ and $(\sigma_j^{-2}|D_t) \sim \mathrm{G}[n_t/2, d_{tj}/2]$. This is again exactly the standard updating, with a common value of n_t across the q series.

Thus, although we are analysing several related series together, there is no effect on the posterior and forecast distributions generated. The univariate theory applies separately to each of the q series, although the calculations are reduced since some of the components are common. The difference lies in the fact that the covariance structure across series is also identified through part (d) of the theorem. Here the full posterior of $\boldsymbol{\Sigma}$ is given by

$$(\boldsymbol{\Sigma}|D_t) \sim \mathrm{W}_{n_t}^{-1}[\mathbf{S}_t],$$

having mean $\mathbf{S}_t n_t/(n_t-2)$ if $n_t > 2$, and harmonic mean $\mathbf{S}_t$. The covariance terms are updated via

$$S_{tjk} = n_t^{-1}(n_{t-1}S_{t-1,jk} + e_{tj}e_{tk}/Q_t), \qquad (j,k = 1,\ldots,q;\ j \neq k).$$

It follows that

$$\begin{aligned} S_{tjk} &=(n_0 + t)^{-1}(n_0 S_{0jk} + \sum_{r=1}^{t} e_{rj}e_{rk}/Q_r) \\ &=(1-\alpha_t)S_{0,jk} + \alpha_t c_{jk}(t), \end{aligned}$$

where $\alpha_t = t/(n_0+t)$ is a weight lying between 0 and 1, and

$$c_{jk}(t) = \frac{1}{t}\sum_{r=1}^{t} e_{rj}e_{rk}/Q_r.$$

This is just the *sample correlation* of the standardised, one-step ahead forecast errors observed up to time t. Thus, S_{tjk} is a weighted average of the prior estimate $S_{0,jk}$ and the sample estimate $c_{jk}(t)$.

16.4.5 Further comments and extensions

Further theoretical and practical features of the analysis, and some extensions, deserve mention.

(a) Step ahead forecasting and filtering

Forecasting ahead to time $t+k$, the standard results apply to each of the q univariate series without alteration, giving Student T distributions

$$(Y_{t+k,j}|D_t) \sim \mathrm{T}_{n_t}[f_{tj}(k), Q_t(k)S_{tj}].$$

The joint forecast distribution is multivariate T,

$$(\mathbf{Y}_{t+k}|D_t) \sim \mathrm{T}_{n_t}[\mathbf{f}_t(k), Q_t(k)\mathbf{S}_t],$$

where $\mathbf{f}_t(k) = [f_{t1}(k),\ldots,f_{tq}(k)]'$.

Similarly, retrospective time series analysis is based on filtered distributions for the state vectors, the usual results applying for each series. Thus, for $1 < k \leq t$,

$$(\boldsymbol{\theta}_{t-k,j}|D_t) \sim \mathrm{T}_{n_t}[\mathbf{a}_{tj}(-k), \mathbf{R}_t(-k)S_{tj}],$$

with the filtered mean vector $\mathbf{a}_{tj}(-k)$ and the common filtered variance matrix $\mathbf{R}_t(-k)$ calculated as in Theorem 4.4. The filtered distribution of the state matrix $\boldsymbol{\Theta}_{t-k}$ is matrix-variate T, being defined from the filtered joint distribution

$$(\boldsymbol{\Theta}_{t-k}, \boldsymbol{\Sigma}|D_t) \sim \mathrm{NW}_{n_t}^{-1}[\mathbf{a}_t(-k), \mathbf{R}_t(-k), \mathbf{S}_t],$$

where $\mathbf{a}_t(-k) = [\mathbf{a}_{t1}(-k), \ldots, \mathbf{a}_{tq}(-k)]$. In full matrix notation, the filtering equation for the matrix mean is analogous to that in Theorem 4.4 for each of its columns, given by

$$\mathbf{a}_t(-k) = \mathbf{m}_{t-k} + \mathbf{C}_{t-k}\mathbf{G}'_{t-k+1}\mathbf{R}^{-1}_{t-k+1}[\mathbf{a}_t(-k+1) - \mathbf{a}_{t-k+1}].$$

(b) Correlation structure and principal components analysis

The covariance structure across the series can be explored by considering the inverse Wishart posterior for $\boldsymbol{\Sigma}$ at any time t. The matrices $\mathbf{S}_t$ provide estimates of the variances and covariances of the series. These provide obvious estimates of the correlations between series, that of the correlation σ_{jk}, $(j, k = 1, \ldots, q;\ j \neq k)$, being given by $S_{tjk}/(S_{tj}S_{tk})^{1/2}$. These estimates can be derived as optimal Bayesian estimates of the actual correlations in a variety of ways, one such being developed in Quintana (1987), and noted in Quintana and West (1987).

One common way of exploring joint structure is to subject an estimate of the covariance matrix to a principal components decomposition (Mardia, Kent and Bibby 1979, Chapter 8; and Press 1985, Chapter 9). $\boldsymbol{\Sigma}$ is non-negative definite, and usually positive definite, so it has q real-valued, distinct and non-negative eigenvalues with corresponding real-valued and orthogonal eigenvectors. The orthonormalised eigenvectors define the principal components of the matrix. Denote the eigenvalues of $\boldsymbol{\Sigma}$ by λ_j, $(j = 1, \ldots, q)$, the corresponding orthonormal eigenvectors by $\boldsymbol{\eta}_j$, $(j = 1, \ldots, q)$, satisfying $\boldsymbol{\eta}'_j\boldsymbol{\eta}_j = 1$ and $\boldsymbol{\eta}'_i\boldsymbol{\eta}_j = 0$ for $i \neq j$. Suppose, without loss of generality, that they are given in order of decreasing eigenvalues, so that $\boldsymbol{\eta}_1$ is the eigenvector corresponding to the largest eigenvalue λ_1, and so forth. Then the covariation of the elements of any random vector $\mathbf{Y}$ having variance matrix $\boldsymbol{\Sigma}$ is explained through the random quantities $X_j = \boldsymbol{\eta}'_j\mathbf{Y}$, $(j = 1, \ldots, q)$, the principal components of $\mathbf{Y}$. These X variates are uncorrelated and have variances $\mathrm{V}[X_j] = \lambda_j$, decreasing as j increases. Total variation in $\mathbf{Y}$ is measured by $\lambda = \mathrm{trace}(\boldsymbol{\Sigma}) = \sum_{j=1}^{q} \lambda_j$, and so the jth principal component explains a proportion λ_j/λ of this total. Interpretation of such a principal components decomposition rests upon (i) identifying those components that contribute markedly to the variation, often just the first one or two; and (ii) interpreting the vectors $\boldsymbol{\eta}_j$ defining these important components.

As in estimating correlations, Quintana (1987) shows that the eigenvalues and eigenvectors of the estimate $\mathbf{S}_t$ are, at time t, optimal Bayesian estimates of those of $\boldsymbol{\Sigma}$, and so may be considered as summarising the covariance structure based on the posterior at t. The use of such estimated principal components is described in the next section in the context of an application. For further applications and much further discussion, see Press (1985) and Mardia, Kent and Bibby (1979) as referenced above.

(c) Discount models

The use of discount factors to structure evolution variance matrices applies here directly as in the univariate DLM. $\mathbf{W}_t$ is common to each of the q univariate series, as are $\mathbf{C}_{t-1}$, $\mathbf{R}_t$, etc. Thus, $\mathbf{W}_t$ may be based on $\mathbf{C}_{t-1}$ using discount factors that are common to each of the q univariate models.

(d) Time-varying $\boldsymbol{\Sigma}$

In Section 10.8 consideration was given to the possibility that observational variances may vary stochastically over time, and to several arguments in favour of models at least allowing for such variation. Those arguments are applied and extended here. One way of modelling such variation is via an important extension of the stochastic model for time-varying variances, and the resulting use of variance discounting in Section 10.8.2. The results of such model extensions are summarised here and applied in the following sections. The basic issue is that of appropriately modelling a time-varying sequence of variance matrices $\boldsymbol{\Sigma}_t$, and attempting to retain the nice conjugate structure of prior and posterior Wishart distributions. The basic technique was developed originally via direct extension of variance discounting ideas to variance matrix discounting, in Quintana (1985, 1987), and developed and applied in Quintana and West (1987, 1988). A formal theoretical foundation is provided by a multivariate extension of the scalar model in equation (10.4) of Section 10.8.2, based on results about combining Wishart and specific versions of matrix-variate beta distributions in Uhlig (1994, 1997). Further details appear in Quintana, Chopra and Putnam (1995), and in as yet unpublished results in joint work of J.M. Quintana and F. Li; the essentials are as follows.

Write $\boldsymbol{\Sigma}_{t-1}$ for the observational variance matrix at time $t-1$ and assume that as in the constant variance case, the time $t-1$ posterior distribution is inverse Wishart with n_{t-1} degrees of freedom and variance estimate $\mathbf{S}_{t-1}$, i.e., $(\boldsymbol{\Sigma}_{t-1}|D_{t-1}) \sim \mathrm{W}^{-1}_{n_{t-1}}[\mathbf{S}_{t-1}]$. Let $\mathbf{U}_{t-1}$ be the upper triangular matrix in the Cholesky decomposition of $\boldsymbol{\Sigma}_{t-1}$, so that $\boldsymbol{\Sigma}_{t-1} = \mathbf{U}'_{t-1}\mathbf{U}_{t-1}$. For any specified $n \times n$ matrix $\mathbf{B}_t$, consider the resulting matrix $\mathbf{U}'_{t-1}\mathbf{B}_t\mathbf{U}_{t-1}$. As outlined in Quintana et al (1995), we can find an appropriate matrix-variate distribution such that for appropriate choices of defining parameters, a matrix $\mathbf{B}_t$ generated from such a distribution leads to the product $\mathbf{U}'_{t-1}\mathbf{B}_t\mathbf{U}_{t-1}$ having an inverse Wishart distribution too. In particular, for any specified discount factor β, $(0 < \beta \leq 1)$, such a matrix $\mathbf{B}_t$ can be found such that

$$\boldsymbol{\Sigma}_t = \mathbf{U}'_{t-1}\mathbf{B}_t\mathbf{U}_{t-1}$$

has the implied prior distribution

$$(\boldsymbol{\Sigma}_t|D_{t-1}) \sim \mathrm{W}^{-1}_{\beta n_{t-1}}[\mathbf{S}_{t-1}].$$

The relevant distribution for $\mathbf{B}_t$ is a scaled version of a matrix-variate "inverse beta" distribution. As a result, the evolution from $\boldsymbol{\Sigma}_{t-1}$ to $\boldsymbol{\Sigma}_t$ implied by this multiplicative model results in a loss of precision about the variance matrix, evident simply in the reduction of the degrees of freedom from n_{t-1} to βn_{t-1}. The analogy with the development in the case of a scalar variance in Section 10.8 is immediate; here we have a direct multivariate extension, and models that formally justify the use of discounting in learning a time-varying variance matrix as originally developed in Quintana and West (1987). Using this model, updating at time t now takes the form

$$(\boldsymbol{\Sigma}_t|D_t) \sim \mathrm{W}_{n_t}^{-1}[\mathbf{S}_t],$$

where

$$n_t = \beta n_{t-1} + 1 \quad and \quad \mathbf{S}_t = n_t^{-1}(\beta n_{t-1}\mathbf{S}_{t-1} + \mathbf{e}_t\mathbf{e}_t'/Q_t).$$

It follows that for $0 < \beta < 1$, the posterior will not degenerate as t increases. Following Section 10.8.3, n_t converges to the finite limit $(1-\beta)^{-1}$. Also, for large t, the posterior estimate $\mathbf{S}_t$ is given by

$$\mathbf{S}_t \approx (1-\beta)\sum_{r=0}^{t-1} \beta^r \mathbf{e}_{t-r}\mathbf{e}_{t-r}'/Q_{t-r}.$$

Hence the estimates of variances and covariances across series are exponentially weighted moving averages of past sample variances and sample covariances.

In filtering backwards in time, the univariate theory of filtered posteriors of Section 10.8.4 can be redeveloped for $\boldsymbol{\Sigma}_{t-k}$. Thus, for $k > 1$, the filtered estimate of $\boldsymbol{\Sigma}_{t-k}$ at time t is given by $\mathbf{S}_t(-k)$, recursively calculated using

$$\mathbf{S}_t(-k)^{-1} = (1-\beta)\mathbf{S}_{t-k}^{-1} + \beta\mathbf{S}_t(-k+1)^{-1},$$

and with associated degrees of freedom parameters $n_t(-k)$ evaluated as

$$n_t(-k) = (1-\beta)n_{t-k} + \beta n_t(-k+1).$$

Of course the static $\boldsymbol{\Sigma}$ model obtains when $\beta = 1$. Otherwise, a value of β slightly less than unity at least allows the flexibility to explore a minor deviation away from constancy of $\boldsymbol{\Sigma}_t$ that may be evidence of true changes in covariance structure, model misspecification, and so forth. See Section 10.8 for further discussion.

(e) Matrix observations

The matrix model framework extends directly to matrix-valued time series, applying when $\mathbf{Y}_t$ is an $r \times q$ matrix of observations, comprising q separate time series of r-vector valued observations, in which the q series follow multivariate DLMs as in Section 16.2 but having the same $\mathbf{F}_t$ and $\mathbf{G}_t$ elements. Readers interested in further theoretical details of this, the other topics in this section, and further related material, should refer to the earlier-referenced works by Dawid, Quintana and West, and also the book by Press for general background reading. Related models are discussed, from different perspectives, in Harvey (1986), and Highfield (1984).

16.4.6 Application to exchange rate modelling

The model developed above is illustrated here, following Quintana and West (1987). The interest is in analysing the joint variation over a period of years in monthly series of observations on international exchange rates. The data series considered are the exchange rates for pounds sterling of the U.S. dollar, the Canadian dollar, the Japanese yen, the Deutschmark, the French franc and the Italian lira. The original analysis in Quintana and West (1987) concerned data from January 1975 up to and including August 1984. Figure 16.1 displays plots of these six exchange rates from the same starting point but including data up to the end of 1986. This full data set is given in Table 16.1.[†]

From the graphs of the series, it can be seen that the trends may be approximately modelled as linear over rather short periods of time, the magnitudes and directions of the trends subject to frequent and often very marked change. After logarithmic transformation, the local linearity is still strongly evident, the marked changes no less important but the general appearance less volatile.

The model is based on the use of logarithmic transformations of each of the series, and variation in each of these transformed series is described as *locally linear*, using a second-order polynomial DLM, $\{\mathbf{E}_2, \mathbf{J}_2(1), ., .\}$. Note that this is *not* assumed to be a useful forecasting model for any other than the very short term (one-step ahead), but is used as a flexible, local trend description that has the ability to adapt to the series as time progresses, so long as the evolution variances are appropriately large. The focus is on identifying the covariance structure across the series, and the basic DLM form serves mainly to adapt appropriately to changes in trends over time. Various analyses are reported by Quintana and West. Here slightly different models are considered, the evolution variance sequence being based on the use of discount factors unlike those originally considered in by Quintana and West. The general results are rather similar, however.

The model requires the specification of the observational variance factors V_t, the evolution variances $\mathbf{W}_t$, the variance matrix discount factor β and the initial prior (16.12). We report results from a model in which $V_t = 1$ for all t, as is usually the case in the absence of information about additional sampling variation or recording errors, etc. The marked and frequent abrupt changes in trend would most adequately be handled using

[†]Data source: U.K. Central Statistical Office data bank, by permission of the Controller of H.M. Stationery Office

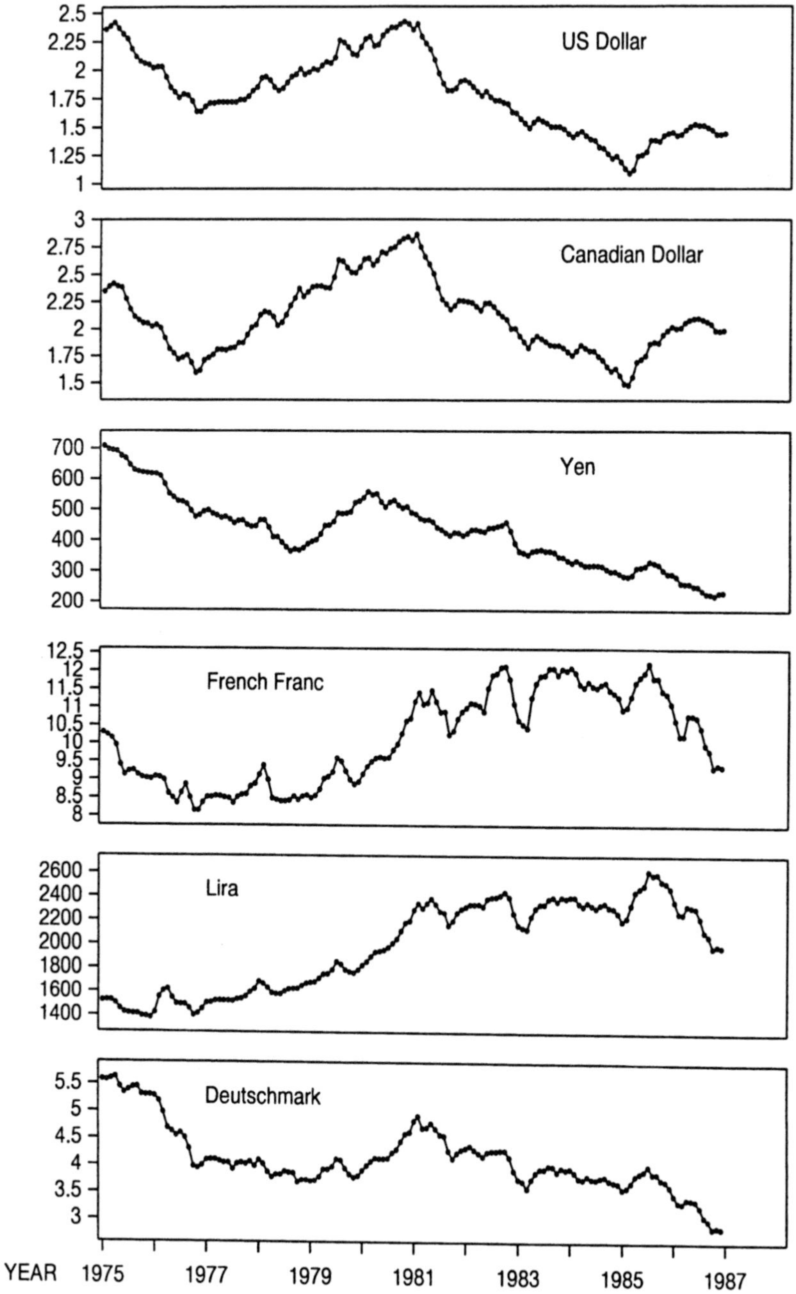

Figure 16.1 International exchange rate time series

Table 16.1. International exchange rate time series
(Source: C.S.O. Data Bank)

Year	Month	$ U.S.A.	$ Canada	Yen	Franc	Lira	Mark
1975	1	2.36	2.35	707.8	10.31	1522	5.58
	2	2.39	2.40	698.3	10.24	1527	5.57
	3	2.42	2.42	695.1	10.15	1526	5.60
	4	2.37	2.40	692.4	9.95	1502	5.63
	5	2.32	2.39	676.3	9.41	1456	5.45
	6	2.28	2.28	668.9	9.13	1426	5.34
	7	2.19	2.19	647.5	9.23	1419	5.39
	8	2.12	2.12	630.1	9.25	1413	5.44
	9	2.08	2.09	624.8	9.13	1413	5.45
	10	2.06	2.06	621.9	9.06	1395	5.31
	11	2.05	2.05	620.2	9.03	1391	5.30
	12	2.02	2.02	618.0	9.01	1381	5.30
1976	1	2.03	2.04	617.8	9.08	1424	5.28
	2	2.03	2.01	611.0	9.06	1554	5.19
	3	1.94	1.92	584.2	8.99	1605	4.98
	4	1.85	1.82	552.3	8.62	1622	4.69
	5	1.81	1.77	540.5	8.50	1549	4.63
	6	1.76	1.72	527.9	8.36	1498	4.55
	7	1.79	1.74	526.4	8.64	1495	4.60
	8	1.78	1.76	518.6	8.86	1493	4.51
	9	1.73	1.69	497.1	8.51	1460	4.31
	10	1.64	1.60	477.1	8.16	1401	3.98
	11	1.64	1.61	483.0	8.16	1416	3.95
	12	1.68	1.71	494.3	8.37	1455	4.00
1977	1	1.71	1.73	498.3	8.52	1506	4.10
	2	1.71	1.76	487.2	8.52	1509	4.11
	3	1.72	1.81	481.7	8.55	1523	4.11
	4	1.72	1.81	473.5	8.54	1525	4.08
	5	1.72	1.80	476.7	8.51	1523	4.05
	6	1.72	1.82	468.6	8.49	1522	4.05
	7	1.72	1.83	456.1	8.36	1520	3.93
	8	1.74	1.87	463.8	8.52	1535	4.03
	9	1.74	1.87	465.2	8.58	1540	4.05
	10	1.77	1.95	450.9	8.60	1559	4.03
	11	1.82	2.02	444.8	8.81	1596	4.07
	12	1.85	2.04	446.6	8.88	1623	3.99

Table 16.1. (continued)

Year	Month	$ U.S.A.	$ Canada	Yen	Franc	Lira	Mark
1978	1	1.93	2.13	466.1	9.13	1686	4.10
	2	1.94	2.16	466.0	9.39	1668	4.03
	3	1.91	2.15	442.3	8.99	1633	3.88
	4	1.85	2.11	410.4	8.49	1589	3.78
	5	1.82	2.03	411.0	8.45	1582	3.83
	6	1.84	2.06	393.2	8.41	1580	3.83
	7	1.89	2.13	378.2	8.41	1604	3.89
	8	1.94	2.21	365.7	8.43	1623	3.87
	9	1.96	2.28	372.1	8.54	1625	3.86
	10	2.01	2.37	368.6	8.45	1628	3.69
	11	1.96	2.30	375.8	8.54	1651	3.73
	12	1.98	2.34	388.8	8.57	1671	3.73
1979	1	2.01	2.39	396.4	8.50	1677	3.71
	2	2.00	2.40	401.6	8.56	1683	3.72
	3	2.04	2.39	420.5	8.73	1714	3.79
	4	2.07	2.38	447.9	9.02	1748	3.93
	5	2.06	2.38	449.6	9.08	1752	3.93
	6	2.11	2.48	461.3	9.21	1785	3.98
	7	2.26	2.63	489.2	9.60	1854	4.12
	8	2.24	2.62	487.4	9.52	1833	4.10
	9	2.20	2.56	489.2	9.24	1785	3.95
	10	2.14	2.52	493.3	9.00	1770	3.84
	11	2.13	2.52	522.5	8.87	1760	3.78
	12	2.20	2.57	528.0	8.94	1785	3.81
1980	1	2.27	2.64	538.8	9.15	1823	3.91
	2	2.29	2.65	558.9	9.37	1854	4.00
	3	2.21	2.59	548.2	9.51	1896	4.08
	4	2.22	2.63	552.2	9.60	1934	4.14
	5	2.30	2.70	525.4	9.63	1942	4.13
	6	2.34	2.69	509.2	9.60	1951	4.13
	7	2.37	2.73	524.3	9.62	1973	4.14
	8	2.37	2.75	530.8	9.83	2008	4.24
	9	2.40	2.80	515.3	9.99	2045	4.30
	10	2.42	2.83	505.6	10.28	2111	4.45
	11	2.40	2.84	510.1	10.63	2177	4.59
	12	2.35	2.81	491.1	10.70	2194	4.62

Table 16.1. (continued)

Year	Month	\$ U.S.A.	\$ Canada	Yen	Franc	Lira	Mark
1981	1	2.40	2.86	485.9	11.16	2289	4.83
	2	2.29	2.75	471.8	11.41	2343	4.92
	3	2.23	2.66	466.6	11.09	2303	4.70
	4	2.18	2.59	467.9	11.13	2345	4.71
	5	2.09	2.51	461.0	11.47	2384	4.79
	6	1.97	2.38	442.3	11.16	2339	4.69
	7	1.88	2.27	435.1	10.87	2278	4.58
	8	1.82	2.23	425.2	10.89	2267	4.56
	9	1.82	2.18	416.8	10.25	2159	4.28
	10	1.84	2.22	426.4	10.36	2196	4.14
	11	1.90	2.26	424.8	10.69	2267	4.24
	12	1.91	2.26	416.5	10.89	2300	4.30
1982	1	1.89	2.25	423.8	10.99	2317	4.33
	2	1.85	2.24	435.1	11.12	2338	4.37
	3	1.81	2.21	435.6	11.10	2337	4.30
	4	1.77	2.17	431.9	11.05	2339	4.24
	5	1.81	2.23	428.8	10.89	2321	4.18
	6	1.76	2.24	441.2	11.54	2385	4.26
	7	1.73	2.20	442.1	11.89	2397	4.28
	8	1.73	2.15	446.6	11.94	2400	4.28
	9	1.71	2.12	450.4	12.11	2415	4.29
	10	1.70	2.09	460.1	12.13	2442	4.29
	11	1.63	2.00	431.9	11.79	2400	4.17
	12	1.62	2.00	392.4	11.11	2265	3.92
1983	1	1.57	1.93	366.0	10.66	2162	3.76
	2	1.53	1.88	361.5	10.54	2142	3.72
	3	1.49	1.83	354.9	10.44	2129	3.59
	4	1.54	1.90	366.2	11.28	2239	3.76
	5	1.57	1.93	369.2	11.67	2308	3.88
	6	1.55	1.91	371.8	11.87	2340	3.95
	7	1.53	1.88	367.3	11.89	2340	3.95
	8	1.50	1.85	367.0	12.08	2386	4.01
	9	1.50	1.85	363.3	12.08	2400	4.00
	10	1.50	1.85	348.6	11.90	2368	3.90
	11	1.48	1.83	347.3	12.06	2401	3.97
	12	1.44	1.79	336.4	12.02	2391	3.94

Table 16.1. (continued)

Year	Month	$ U.S.A.	$ Canada	Yen	Franc	Lira	Mark
1984	1	1.41	1.76	329.1	12.10	2403	3.96
	2	1.44	1.80	336.4	11.97	2402	3.89
	3	1.46	1.85	327.9	11.65	2350	3.78
	4	1.42	1.82	320.3	11.56	2327	3.76
	5	1.39	1.80	320.3	11.71	2357	3.82
	6	1.38	1.80	321.4	11.59	2333	3.77
	7	1.32	1.75	320.8	11.55	2311	3.76
	8	1.31	1.71	318.3	11.64	2338	3.79
	9	1.26	1.65	308.8	11.69	2352	3.81
	10	1.22	1.61	301.1	11.48	2317	3.74
	11	1.24	1.63	302.2	11.40	2309	3.71
	12	1.19	1.57	294.3	11.29	2270	3.69
1985	1	1.13	1.49	286.8	10.95	2199	3.58
	2	1.09	1.48	284.7	11.02	2229	3.61
	3	1.12	1.55	289.8	11.31	2336	3.70
	4	1.24	1.70	312.3	11.69	2446	3.83
	5	1.25	1.72	314.6	11.84	2476	3.88
	6	1.28	1.75	318.7	11.96	2502	3.92
	7	1.38	1.86	332.6	12.21	2620	4.01
	8	1.38	1.88	328.4	11.81	2591	3.87
	9	1.37	1.87	322.8	11.81	2597	3.87
	10	1.42	1.94	305.2	11.46	2537	3.76
	11	1.44	1.98	293.6	11.38	2523	3.73
	12	1.45	2.02	293.2	11.11	2478	3.63
1986	1	1.42	2.00	284.7	10.66	2368	3.47
	2	1.43	2.01	263.8	10.23	2269	3.34
	3	1.47	2.06	262.1	10.23	2263	3.32
	4	1.50	2.08	262.2	10.79	2332	3.40
	5	1.52	2.09	253.8	10.79	2322	3.39
	6	1.51	2.10	252.8	10.74	2312	3.37
	7	1.51	2.08	239.4	10.45	2229	3.25
	8	1.49	2.06	229.2	9.99	2111	3.07
	9	1.47	2.04	227.6	9.83	2075	3.00
	10	1.43	1.98	223.1	9.36	1980	2.86
	11	1.43	1.98	232.0	9.43	1996	2.88
	12	1.44	1.98	233.2	9.39	1984	2.86

some form of monitoring and intervention technique. Alternatively, a simpler approach to tracking the changes in the series is to use a model allowing a reasonably large degree of variation in the trend component over all time. The model reported here is structured using a single discount factor $\delta = 0.7$ for the trend component, so that $\mathbf{W}_t = \mathbf{J}_2(1)\mathbf{C}_{t-1}\mathbf{J}_2(1)'(0.7^{-1} - 1)$, for all t. This very low discount factor is consistent with the view that there is to be a fair degree of change over time, particularly in the growth components of the series. The initial prior for the trend and growth parameters at $t = 0$ (January 1975) is very vague, being specified via $\mathbf{m}_0 = \mathbf{0}$, the 2×6 zero matrix, and

$$\mathbf{C}_0 = \begin{pmatrix} 100 & 0 \\ 0 & 4 \end{pmatrix}.$$

Recall that we are working on the logarithmic scale for each of the series.

For $\boldsymbol{\Sigma}$, the initial prior is also rather vague, based on $n_0 = 1$ and $\mathbf{S}_0 = \mathbf{I}$, the 6×6 identity matrix. Consistent with the volatility of the individual trends, it is apparent that the relationships amongst the series can be expected to vary noticeably over time. This is reinforced through consideration of changes in economic trading relationships between the various countries, and between the EEC, Japanese and North American sectors, over the long period of 12 years. Such changes are also evident from the figures, particularly, for example, in the changes from concordance to discordance, and vice versa, between the trends of the yen and each of the North American currencies. Thus, variance discounting is used to allow for adaptation to changes in $\boldsymbol{\Sigma}$ over time, and again consistent with the marked nature of changes anticipated, a fairly low variance discount factor $\beta = 0.9$ is used. Note from the previous section that the limiting value of the degrees of freedom parameter n_t in this case is $(1 - \beta)^{-1} = 10$, and that this limit is very rapidly approached (in this case, from below since $n_0 = 1$). At any time t, the estimate $\mathbf{S}_t$ is less heavily dependent on data in the past than in the static model, and therefore it more closely estimates local, or current, cross-sectional structure.

The estimate $\mathbf{S}_t$ may be analysed using a principal components decomposition as described in the previous section. This provides inferences about $\boldsymbol{\Sigma}$ based on the data up to that time. On the basis of the model, the one-step ahead forecast variance matrix at t is simply proportional to $\mathbf{S}_t$, so that this is an analysis of the covariance structure of the forecast distribution. Having proceeded to the end of the analysis at $t = 144$, the revised, filtered estimates $\mathbf{S}_{144}(-k)$ provide the raw material for such analyses.

Table 16.2 provides information from the principal components analysis of the estimated covariance matrix $\mathbf{S}_{144}(-36)$, the filtered estimate of $\boldsymbol{\Sigma}$ at $t = 108$, corresponding to December 1983; this provides comparison with Quintana and West (1987) who consider time $t = 116$, albeit using a different model. Of the six eigenvalues of the estimate $\mathbf{S}_{144}(-36)$, the first three account for over 97% of the total estimated variation in $\boldsymbol{\Sigma}$. The

Table 16.2. Weights of currencies in first three components (December 1983)

	λ_1	λ_2	λ_3
\$ U.S.A.	0.27	−0.64	−0.13
\$ Canada	0.27	−0.65	−0.09
Yen	0.50	0.06	0.86
Franc	0.49	0.25	−0.28
Lira	0.43	0.15	−0.23
Mark	0.44	0.29	−0.31
% Variation	73	18	6

proportions of total variation contributed by the first three components are approximately 73%, 18% and 6% respectively. The first principal component clearly dominates, explaining almost three quarters of total variation, although the second is also important, the third rather less so. Table 16.2 gives the corresponding three eigenvectors, defining the weightings of the elements of principal components $X_j = \boldsymbol{\lambda}_j' \mathbf{Y}_{108}$ (for notational convenience, the earlier notation $\boldsymbol{\lambda}_j$ is used for the eigenvalues, although recall that they are estimates based on $\mathbf{S}_{144}(-36)$).

The interpretation of the eigenvectors, and of the first three principal components, is relatively straightforward and in line with the economic background. Bearing in mind that the vectors in the tables are uncertain estimates, the first vector implies that the first principal component is a weighted average of the six currencies, providing an average measure of the international value of the pound (as estimated for December 1983). In fact the compound measure defined by this component is very similar to U.K. official measures relative to "baskets" of currencies. Similar conclusions are found using a range of similar models. See Quintana (1987), Chapter 6, especially Figure 6.3, for further details. This average measure, and its development over time, provides an assessment of the buoyancy of sterling on the international exchanges, and is obviously (and unsurprisingly) the major factor underlying the joint variation of the six exchange rates. The second principal component explains about 18% of the variation through a contrast of an average of the North American rates with an average of those of the EEC countries. The yen receives little weight here. The contrast arises through the opposite signs of the weights between the two sectors, and reflects the major effects on the value of sterling of trading relationships between the two sectors. Note also the magnitudes of the weights. The U.S. and Canadian dollars are roughly equally weighted, being extremely highly correlated and so essentially indistinguishable. In the contrasting sector, the deutschmark and franc are also roughly equally weighted, the lira somewhat less so, reflecting the dominance in the EEC of the effect of

Table 16.3. Weights of currencies in first three components (December 1986)

	λ_1	λ_2	λ_3
$ U.S.A.	0.37	−0.51	0.29
$ Canada	0.33	−0.52	0.34
Yen	0.42	0.67	0.59
Franc	0.42	0.04	−0.53
Lira	0.47	0.06	−0.20
Mark	0.42	0.10	−0.35
% Variation	71	20	9

the German and French currencies on sterling. Finally, the third component contrasts the yen with the remaining currencies, though giving relatively little weight to the North American currencies. This therefore essentially reflects the strength of the yen relative to an aggregate measure of EEC countries.

Proceeding to the end of the series, $t = 144$ at December 1986, the estimate $\mathbf{S}_{144} = \mathbf{S}_{144}(0)$ may be similarly analysed. The conclusions are broadly similar to those above, although the differences are of interest. Table 16.3 provides details of the principal components analysis of $\mathbf{S}_{144}$ similar to those in Table 16.2.

The amount of variation explained by each of the components is similar to that at $t = 108$. The first component dominates at over 70%, the second is significant at 20%, the third less so at 9% and the other three negligible. The differences between this and the previous analysis lie in a more appropriate reflection of the covariance structure in late 1986, the estimate of $\boldsymbol{\Sigma}$ more appropriately reflecting local conditions due to the discounting through $\beta = 0.9$ of past data. The first component is a similar basket of currencies, though slightly more weight is given to the North American currencies, and very slightly less to the yen, in determining this weighted average. The second component is now seen to contribute rather more to total variation, accounting for over 20% in late 1986. This contrasts the North American currencies with the yen and the EEC, although the EEC currencies are only very marginally weighted. The third component then explains a further 9% of total variation through a contrast between the EEC and non-EEC currencies. The dominance of the effect of the yen is evident in a higher weighting relative to the North American currencies.

16.4.7 Application to compositional data analysis

A natural setting for the application of this class of models is in the analysis of time series of compositions, similar to the application in hierarchical forecasting mentioned in Section 16.2.3. One such application appears in

Quintana (1987), and Quintana and West (1987), the series of interest there being U.K. monthly energy consumption figures according to energy generating source (gas, electricity, petroleum and nuclear). Although the models applied there do not directly address the compositional nature of the data, the application typifies the type of problem for which these models are suited. A common, univariate DLM form is appropriate for each of the series, and the focus is on analysis of covariance structure rather than forecasting.

With compositional (and other) series, the analysis of *relative* behaviour of series may often be simplified by converting from the original data scale to the corresponding time series of proportions. The idea is that if general environmental conditions (such as national economic policies, recessions, and so forth) can be assumed to affect each of the series through a common multiplicative factor at each time, then converting to proportions removes such effects and leads to a simpler analysis on the transformed scale. Quintana and West (1988) describe the analysis of such series, the models introduced there being developed as follows.

Suppose that $\mathbf{z}_t$, $(t = 1, \dots)$, is a q-vector valued time series of *positive* quantities of similar nature, naturally defining q compositions of a total $\mathbf{1}'\mathbf{z}_t$ at time t. The time series of proportions is defined simply via

$$\mathbf{p}_t = (\mathbf{1}'\mathbf{z}_t)^{-1}\mathbf{z}_t = \mathbf{z}_t / \sum_{i=1}^{q} z_{ti}.$$

Having used the transformation to proportions in an attempt to remove the common effects of unidentified factors influencing the time series, we now face the technical problems of analysing the proportions themselves. A first problem is the restriction on the ranges of the elements of $\mathbf{p}_t$; our normal models of course assume each of the components to be essentially unrestricted, and so further transformation may be needed. Aitchison (1986) describes the use of the logistic/log ratio transformation in the analysis of proportions, in a variety of application areas. Here this transformation may be applied to map the vector of proportions $\mathbf{p}_t$ into a vector of real-valued quantities $\mathbf{Y}_t$. A particular, symmetric version of the log ratio transformation is given by taking

$$Y_{tj} = \log(p_{tj}/\hat{p}_t) = \log(p_{tj}) - \log(\hat{p}_t), \qquad (j = 1, \dots, q),$$

where

$$\hat{p}_t = \prod_{j=1}^{q} p_{tj}^{1/q}$$

is the geometric mean of the p_{tj}. The inverse of this log ratio transformation is the logistic transformation

$$p_{ti} = \frac{\exp(Y_{ti})}{\sum_{j=1}^{q} \exp(Y_{tj})}, \qquad (i = 1, \ldots, q).$$

Modelling $\mathbf{Y}_t$ with the matrix model of (16.11) implies a conditional multivariate normal structure. Thus, the observational distribution of the proportions $\mathbf{p}_t$ is the multivariate logistic-normal distribution as defined in Aitchison and Shen (1980); see also Aitchison (1986). A sufficient condition for $\mathbf{p}_t$ to have a logistic-normal distribution is that the series $\mathbf{z}_t$ be distributed as a multivariate log-normal, although the converse does not hold.

Suppose then that $\mathbf{Y}_t$ follows the model (16.11). Having proceeded from the proportions to the real-valued elements on the $\mathbf{Y}_t$ scale, we now face a second problem. It follows from the definition of Y_{tj} that the elements sum to zero, $\mathbf{1}'\mathbf{Y}_t = 0$ for all t. This implies that $\boldsymbol{\Sigma}$ is singular, having rank $q-1$. This and other singularities can be handled as follows. Suppose that we model $\mathbf{Y}_t$ ignoring the constraint, assuming as above that $\boldsymbol{\Sigma}$ is non-singular, its elements being unrestricted. Then the constraint can be imposed directly by transforming the series to $\mathbf{KY}_t$, where

$$\mathbf{K} = \mathbf{K}' = \mathbf{I} - q^{-1}\mathbf{1}\mathbf{1}', \qquad \mathbf{1} = (1, \ldots, 1)'.$$

Then from equations (16.11), it follows that

$$\begin{aligned} \mathbf{Y}_t'\mathbf{K} &= \mathbf{F}_t'\boldsymbol{\Psi}_t + (\mathbf{K}\boldsymbol{\nu}_t)', \qquad & \mathbf{K}\boldsymbol{\nu}_t &\sim \mathrm{N}[\mathbf{0}, V_t\boldsymbol{\Xi}^{-1}], \\ \boldsymbol{\Psi}_t &= \mathrm{G}_t\boldsymbol{\Psi}_{t-1} + \boldsymbol{\Omega}_t\mathbf{K}, \qquad & \boldsymbol{\Omega}_t\mathbf{K} &\sim \mathrm{N}[\mathbf{0}, \mathbf{W}_t, \boldsymbol{\Xi}^{-1}], \end{aligned}$$

where $\boldsymbol{\Psi}_t = \boldsymbol{\Theta}_t\mathbf{K}$ and $\boldsymbol{\Xi} = \mathbf{K}\boldsymbol{\Sigma}\mathbf{K}$. Similarly, from (16.12) it follows that

$$(\boldsymbol{\Psi}_0, \boldsymbol{\Xi}|D_0) \sim \mathrm{NW}_{n_0}^{-1}[\mathbf{m}_0\mathbf{K}, \mathbf{C}_0, \mathbf{K}\mathbf{S}_0\mathbf{K}].$$

Thus, the constrained series follows a matrix model, and the results in Theorem 16.4 apply. For a derivation of this and related results see Quintana (1987). Note that the quantities $\mathbf{F}_t$, V_t, $\mathbf{G}_t$, $\mathbf{W}_t$ and $\mathbf{C}_t$ are unaffected by such linear transforms, as is the use of discount factors to define $\mathbf{W}_t$.

With this in mind, initial priors for $\boldsymbol{\Theta}_0$ and $\boldsymbol{\Sigma}$ at time $t = 0$ can be specified in non-singular forms and then transformed as above to conform to the constraint.† Moreover, the unconstrained model is defined for the logarithms of $\mathbf{z}_t$ rather than the log-ratios, so that the unconstrained priors are those of meaningful quantities. The transformation changes the correlation structure in obvious ways since it imposes a constraint. Whatever the correlation structure may be in the unconstrained $\boldsymbol{\Sigma}$, transformation

†This procedure is similar to that for setting a prior for a form-free seasonal DLM component subject to zero-sum constraints (Section 8.4)

will lead to negative correlations afterwards. This is most easily seen in the case $q = 2$, when the constraint leads to perfect negative correlation in $\boldsymbol{\Xi}$. Hence the joint structure may be interpreted through relative values of the resulting constrained correlations.

Illustration of this model appears in Quintana and West (1988). The data series are several years of monthly estimates of values of imports into the Mexican economy. There are three series, determining the composition of total imports according to end use, classified as Consumer, Intermediate and Capital. The data, running from January 1980 to December 1983 inclusive, appear in Table 16.4. The series are labelled as A=Consumer, B=Intermediate and C=Capital imports.

The time span of these series includes periods of marked change in world oil prices that seriously affected the Mexican economy, notably during 1981 and 1982. Rather marked changes are apparent in each of the three series. The log-ratio transformed series are plotted in Figure 16.2, where the compositional features are clearly apparent.

Note the relative increase in imports of intermediate-use goods at the expense of consumer goods during the latter two years. An important feature of the transformation is that it removes some of the common variation over time evident on the original scale, such as inflation effects and national economic cycles.

The model used for the transformed series is a second-order polynomial form as in the example of the previous section. Again, as in that application, this is *not* a model to be seriously entertained for *forecasting* the series other than in the very short term. However, as in many other applications, it serves here as an extremely flexible *local smoothing* model that can track the changes in trend in the data and provide robust retrospective estimates of the time-varying trend and cross-sectional correlation structure. Thus, for our purposes here, this simple, local trend description is adequate, the basic, *locally linear* form of the model being supported by the graphs in Figure 16.2.

Quintana and West (1988) use a relatively uninformative initial prior that is used here for illustration (the problems involved in specifying an informative prior for $\boldsymbol{\Sigma}$ are well discussed by Dickey, Dawid and Kadane 1986). For the unconstrained series,

$$\mathbf{m}_0 = \begin{pmatrix} -1 & 1 & 0 \\ 0 & 0 & 0 \end{pmatrix}, \qquad \mathbf{C}_0 = \begin{pmatrix} 1 & 0 \\ 0 & 1 \end{pmatrix}.$$

The prior for $\boldsymbol{\Sigma}$ has $\mathbf{S}_0 = 10^{-5}\mathbf{I}$ and $n_0 = 10^{-3}$, a very vague specification. These values are used to initialise the analysis, the corresponding posterior and filtered values for each time t then being transformed as noted above to impose the linear constraint. The evolution variance sequence is based on a single discount factor δ, so that $\mathbf{W}_t = \mathbf{G}\mathbf{C}_{t-1}\mathbf{G}'(\delta^{-1} - 1)$, with the standard value of $\delta = 0.9$. In addition, the possibility of time variation in $\boldsymbol{\Sigma}$ is allowed for through the use of a variance discount factor $\beta = 0.98$,

Table 16.4. Monthly Mexican import data ($\$10^6$)
(Source: Bank of Mexico)

		J	F	M	A	M	J
1980	A:	98.1	112.1	123.1	150.2	148.0	180.2
	B:	736.3	684.1	793.9	856.2	1020.9	986.9
	C:	301.2	360.7	4290.9	358.7	354.6	419.8
1981	A:	246.5	135.4	183.4	178.2	194.4	189.2
	B:	1046.9	1051.6	1342.2	1208.0	1213.6	1291.1
	C:	575.3	565.4	699.2	598.3	622.6	709.7
1982	A:	157.2	175.4	191.5	167.9	123.5	143.9
	B:	933.4	808.9	1029.2	894.3	811.8	866.1
	C:	514.5	684.7	568.4	436.0	312.6	388.7
1983	A:	52.3	44.5	42.9	39.2	55.7	56.6
	B:	257.0	318.6	457.2	440.3	506.4	529.6
	C:	85.2	83.2	126.7	131.3	144.4	165.5

Table 16.4. (continued)

		J	A	S	O	N	D
1980	A:	244.9	225.0	253.1	321.5	262.4	331.1
	B:	1111.8	962.9	994.0	1049.2	954.0	1058.6
	C:	438.4	445.2	470.3	514.8	483.4	606.1
1981	A:	226.6	254.8	298.9	362.2	270.2	273.0
	B:	1229.6	994.4	999.3	1120.4	977.1	1067.2
	C:	689.6	588.8	510.9	773.9	588.3	653.3
1982	A:	156.8	115.0	70.0	90.6	42.6	82.3
	B:	762.0	631.0	476.6	470.0	314.9	419.5
	C:	367.1	330.1	232.9	270.1	220.8	176.5
1983	A:	41.9	60.5	40.2	33.1	39.5	47.4
	B:	488.1	548.3	502.2	474.9	392.4	431.7
	C:	159.5	137.4	128.2	190.7	268.8	198.1

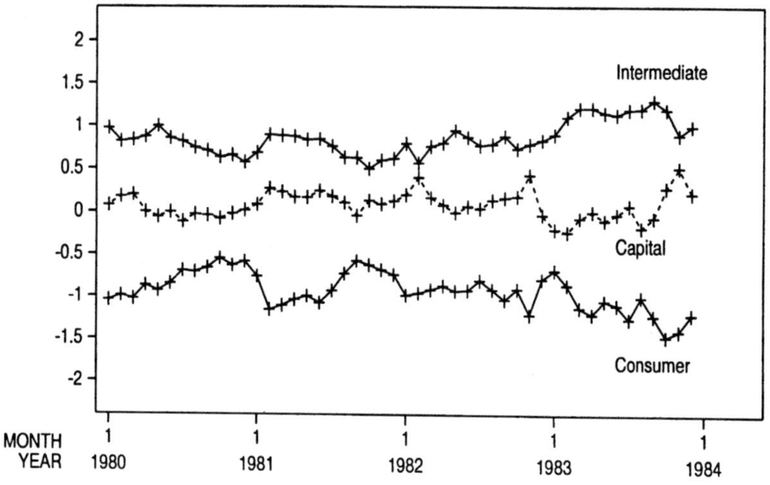

Figure 16.2 Log-ratio compositions of Mexican imports by use

small, though sufficient to permit some degree of change.† In addition to this, one further modification is made to the model (and to each of those explored) to allow by intervention for the marked changes in trend of the series that are essentially responses to major events in the international markets, particularly the plunge into recession in early 1981, for example. Interventions are made to allow for changes in model parameters that are greater than those anticipated through the standard discount factors δ and β. The points of intervention are January 1981, September 1981 and January 1983. At these points only, the standard discount factors are replaced by lower values $\delta = 0.1$ and $\beta = 0.9$ to model the possibility of greater *random* variation in both $\boldsymbol{\Theta}_t$ and $\boldsymbol{\Sigma}$, though not anticipating the direction of such changes.

Figures 16.3, 16.4 and 16.5 display the filtered estimates of the trends (the *fitted* trends) in the three imports series, with approximate 90% posterior intervals; these should not be confused with the associated *predictive* intervals that though not displayed here, are roughly twice as wide. The times of intervention are indicated by arrows on the time axes, and the response of the model is apparent at points of abrupt change. Though

†Of several models with values in the a priori plausible ranges $0.8 \leq \delta \leq 1$ and $0.95 \leq \beta \leq 1$, that illustrated has essentially the highest likelihood. In particular, the static $\boldsymbol{\Sigma}$ models with $\beta = 1$ are very poor by comparison, indicating support for time variation in $\boldsymbol{\Sigma}$. The inferences illustrated are, of course, representative of inferences from models with discount factors near the chosen values.

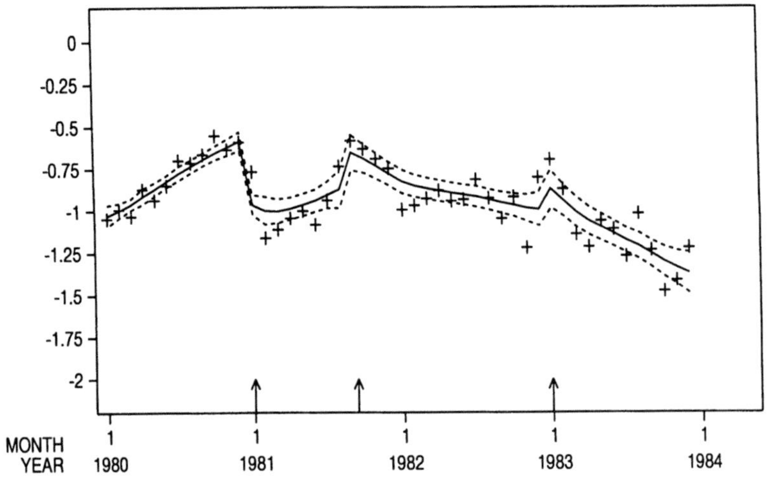

Figure 16.3 Filtered trend in Consumer series

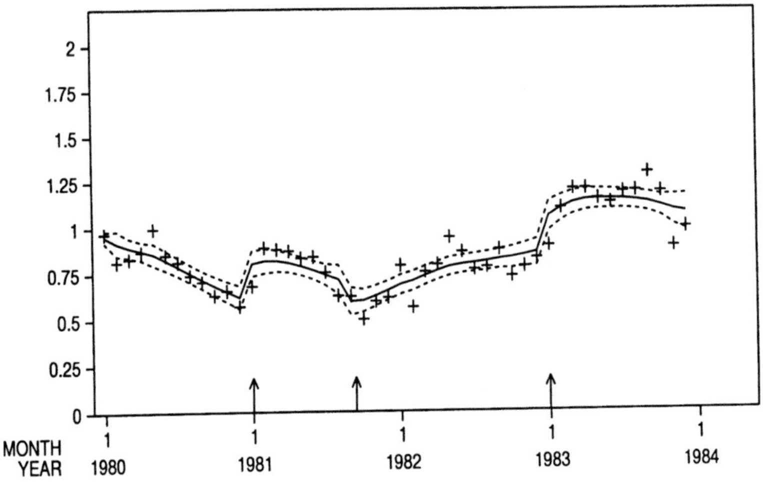

Figure 16.4 Filtered trend in Intermediate series

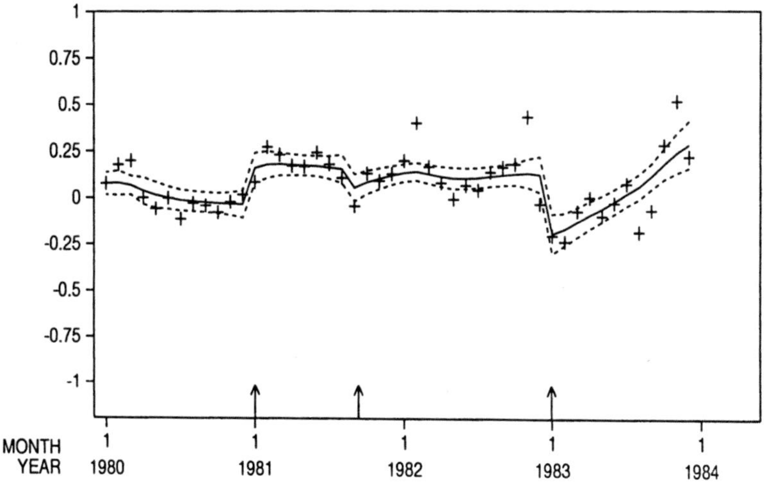

Figure 16.5 Filtered trend in Capital series

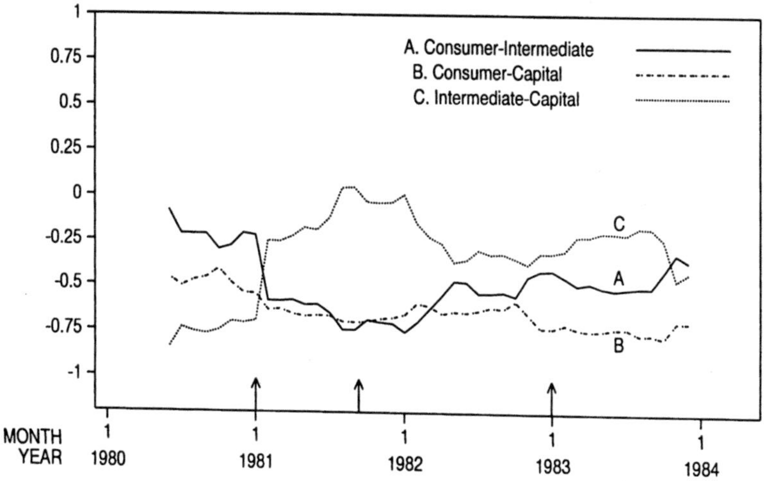

Figure 16.6 Filtered estimates of correlations

the model could be refined, improvements are likely to be very minor, the fitted trends *smooth* the series rather well. The main interest is in correlation structure across series, and Figure 16.6 clearly identifies the time variation in $\boldsymbol{\Xi}$. The graph is of the correlations taken from the smoothed estimates as defined in Section 16.4.5. Note again that though the series may move together at times suggesting positive values of the correlations in $\boldsymbol{\Sigma}$, corresponding elements in $\boldsymbol{\Xi}$ will be negative due to the constraint. The estimates in the figure bear this out. Slow, random changes are apparent here, accounting for changing economic conditions that affect the three import areas in different ways. A more marked change occurs at intervention prior to the onset of recession in early 1981. Cross-sectional inferences made in early 1981 based on this model would reflect these changes, that are quite marked, whereas those based on a static $\boldsymbol{\Sigma}$ model would be much more heavily based on the pre-1981 estimates of $\boldsymbol{\Sigma}$. Thus, for example, this model would suggest a correlation near -0.25 between the Intermediate and Capital series, whereas the pre-1981 based value is near -0.75.

Note that we can also analyse the principal components of $\boldsymbol{\Xi}$ over time. The fact that $\boldsymbol{\Xi}$ has rank $q-1$ rather than full rank q simply implies that one of the eigenvalues is zero, the corresponding eigenvector being undefined and of no interest. Variation is explained by the remaining $q-1$ principal components.

16.5 OTHER MODELS AND RELATED WORK

There has been much work on Bayesian models and methods in problems of multivariate time series and forecasting not discussed above, but of relevance and interest to readers in various fields. Recent extensions and applications of the models of Section 16.4, especially in connection with financial time series and portfolio management, are reported in Quintana (1992) and Quintana et al (1995), and this is a growing area. Additional developments on variance matrix estimation appear in Barbosa and Harrison (1992). Modifications of our multivariate models, as well as quite distinct, non-normal dynamic models, have been developed for problems in which data series represent proportions, or compositions. These include a variety of multinomial models, including those in Grunwald, Raftery and Guttorp (1993), and Cargnoni, Müller and West (1996). One area that has seen some exploration to date, and is likely to be a growth area in future, is the development of multivariate dynamic models using traditional Bayesian hierarchical modelling concepts and methods; some relevant works include Zellner and Hong (1989), Zellner, Hong and Min (1991), Quintana et al (1995), Gamerman and Migon (1993) and Prado and West (1996). Finally, though not explicitly based in dynamic linear models, the Bayesian forecasting approaches using vector autoregression and structured prior distributions (Litterman 1986a,b) represent a significant area of development and an important branch of the wider Bayesian forecasting arena.

16.6 EXERCISES

(1) Consider the general multivariate DLM of Section 16.2. In Theorem 16.1, assume the prior distribution for $(\boldsymbol{\theta}_t|D_{t-1})$ of part (b). Using Bayes' Theorem, show directly that $(\boldsymbol{\theta}_t|D_t)$ is normally distributed with moments given by the alternative expressions

$$\mathbf{m}_t = \mathbf{C}_t[\mathbf{R}_t^{-1}\mathbf{a}_t + \mathbf{F}_t\mathbf{V}_t^{-1}\mathbf{Y}_t],$$

with

$$\mathbf{C}_t^{-1} = \mathbf{R}_t^{-1} + \mathbf{F}_t\mathbf{V}_t^{-1}\mathbf{F}_t'.$$

(2) Use the standard updating equations of Theorem 16.1 and the results of Exercise 1 above to deduce the following identities:

(a) $\mathbf{A}_t = \mathbf{R}_t\mathbf{F}_t\mathbf{Q}_t^{-1} = \mathbf{C}_t\mathbf{F}_t\mathbf{V}_t^{-1}$;

(b) $\mathbf{C}_t = \mathbf{R}_t - \mathbf{A}_t\mathbf{Q}_t\mathbf{A}_t' = \mathbf{R}_t(\mathbf{I} - \mathbf{F}_t\mathbf{A}_t')$;

(c) $\mathbf{Q}_t = (\mathbf{I} - \mathbf{F}_t'\mathbf{A}_t)^{-1}\mathbf{V}_t$;

(d) $\mathbf{F}_t'\mathbf{A}_t = \mathbf{I} - \mathbf{V}_t\mathbf{Q}_t^{-1}$.

(3) In the multivariate DLM of Section 16.2, consider the vector mean response function $\boldsymbol{\mu}_t = \mathbf{F}_t'\boldsymbol{\theta}_t$ at time t. The defining moments of the posterior for the mean response at time t,

$$(\boldsymbol{\mu}_t|D_t) \sim \mathrm{N}[\mathbf{f}_t(0), \mathbf{Q}_t(0)],$$

are easily seen to be given by

$$\mathrm{E}[\boldsymbol{\mu}_t|D_t] = \mathbf{f}_t(0) = \mathbf{F}_t'\mathbf{m}_t$$

and

$$\mathrm{V}[\boldsymbol{\mu}_t|D_t] = \mathbf{Q}_t(0) = \mathbf{F}_t'\mathbf{C}_t\mathbf{F}_t.$$

Use the updating equations for $\mathbf{m}_t$ and $\mathbf{C}_t$ to derive the following results.

(a) Show that the mean can be updated via

$$\mathrm{E}[\boldsymbol{\mu}_t|D_t] = \mathrm{E}[\boldsymbol{\mu}_t|D_{t-1}] + (\mathbf{I} - \mathbf{B}_t)\mathbf{e}_t,$$

or via the similar expression

$$\mathbf{f}_t(0) = (\mathbf{I} - \mathbf{B}_t)\mathbf{Y}_t + \mathbf{B}_t\mathbf{f}_t,$$

where $\mathbf{B}_t = \mathbf{V}_t\mathbf{Q}_t^{-1}$. Interpret the second expression here as a weighted average of two estimates of $\boldsymbol{\mu}_t$.

(b) Show that the variance can be similarly updated using

$$\mathrm{V}[\boldsymbol{\mu}_t|D_t] = \mathbf{B}_t\mathrm{V}[\boldsymbol{\mu}_t|D_{t-1}],$$

or equivalently, using

$$\mathbf{Q}_t(0) = (\mathbf{I} - \mathbf{B}_t)\mathbf{V}_t.$$

(4) Prove the general step ahead forecasting results of Theorem 16.2 for the multivariate DLM.

(5) Prove the general filtering results of Theorem 16.3 for the multivariate DLM.

(6) $\mathbf{Y}$ is an n-vector of sales of a large number n of competitive products. The forecast distribution is given by $\mathbf{Y} \sim \mathrm{N}[\mathbf{f}, \mathbf{Q}]$, where

$$\mathbf{Q} = Q \begin{pmatrix} 1 & -\epsilon & -\epsilon & \dots & -\epsilon \\ -\epsilon & 1 & -\epsilon & \dots & -\epsilon \\ -\epsilon & -\epsilon & 1 & \dots & -\epsilon \\ \vdots & \vdots & \vdots & \ddots & \vdots \\ -\epsilon & -\epsilon & -\epsilon & \dots & 1 \end{pmatrix},$$

for some variance $Q > 0$ and scalar $\epsilon > 0$.

(a) Calculate the implied forecast distribution for total sales $T = \mathbf{1}'\mathbf{Y}$, where $\mathbf{1} = (1, \dots, 1)'$.

(b) Verify that $\mathrm{V}[T] \geq 0$ if and only if $\epsilon \leq (n-1)^{-1}$, with $\mathrm{V}[T] = 0$ when $\epsilon = (n-1)^{-1}$.

(c) With n large, comment on the implications for aggregate forecasting.

(7) A series of n product sales $\mathbf{Y}_t$ is forecast using the constant model $\{\mathbf{I}, \mathbf{I}, \mathbf{V}, \mathbf{W}\}$, where all four known matrices are $n \times n$. At time t, the prior for the state vector $\boldsymbol{\theta}_t$ is $(\boldsymbol{\theta}_t | D_{t-1}) \sim \mathrm{N}[\mathbf{a}_t, \mathbf{R}_t]$ as usual. Let T_s be the total sales at any time s, $T_s = \mathbf{1}'\mathbf{Y}_s$.

(a) Show, directly from normal theory or otherwise that

$$\mathrm{V}[T_{t+1} | \mathbf{Y}_t, D_{t-1}] = \mathrm{V}[T_{t+1} | \mathbf{Y}_t, T_t, D_{t-1}] \leq \mathrm{V}[T_{t+1} | T_t, D_{t-1}],$$

with equality if and only if

$$\mathrm{C}[\mathbf{Y}_t, T_{t+1} | T_t, D_{t-1}] = \mathbf{0}.$$

(b) Calculate the joint normal distribution of T_{t+1}, T_t and $\mathbf{Y}_t$ given D_{t-1}.

(c) Using (b), prove that the required covariance in (a) is given by

$$\mathbf{r}_t - (\mathbf{r}_t + \mathbf{v}) R_t / (R_t + W + V),$$

where $\mathbf{r}_t = \mathbf{R}_t \mathbf{1}$, $R_t = \mathbf{1}'\mathbf{r}_t$, $W = \mathbf{1}'\mathbf{W}\mathbf{1}$, $\mathbf{v} = \mathbf{V}\mathbf{1}$ and $V = \mathbf{1}'\mathbf{v}$. Deduce that the equality in (a) holds if and only if

$$\mathbf{r}_t = R_t \mathbf{v} / V.$$

(d) Comment on the relevance of the above results to two forecasters interested in T_{t+1} when $\mathbf{Y}_t$ becomes known. The first proposes to forecast T_{t+1} using $p(T_{t+1} | \mathbf{Y}_t, D_{t-1})$; the second proposes to use $p(T_{t+1} | T_t, D_{t-1})$,

(8) Verify the general results in parts (D) and (E) of Section 16.3.2. In particular, suppose that $\mathbf{Y} \sim \mathrm{N}[\mathbf{f}, \mathbf{Q}]$ and $T = \mathbf{1}'\mathbf{Y}$. If the marginal distribution is revised to $(T | H) \sim \mathrm{N}[f^*, Q^*]$ and $\mathbf{Y}$ is conditionally

independent of H when T is known, show that $p(Y|H)$ is as stated in (D), namely

$$(\mathbf{Y}|H) \sim \mathrm{N}[\mathbf{f} + \mathbf{q}(f^* - f)/Q, \mathbf{Q} - \mathbf{q}\mathbf{q}'(Q - Q^*)/Q^2].$$

Extend this proof to verify the more general result in (E) when independent information H is used to revise the marginal forecast distribution of a vector of subaggregates, or more general linear combinations of the elements of $\mathbf{Y}$.

(9) In the framework of Section 16.3.2 (C), part (2), suppose that $a = 0$, $b = 1$ and $c = 1/\phi$ is uncertain, so that

$$(m|M, T, \phi) \sim \mathrm{N}[T, M/\phi].$$

Suppose that uncertainty about the scale bias ϕ is represented by a gamma distribution with mean $\mathrm{E}[\phi] = 1$ and n degrees of freedom,

$$\phi \sim \mathrm{G}[n/2, n/2],$$

and that ϕ is considered to be independent of M and T.

(a) Show that unconditional on ϕ,

$$(m|M, T) \sim \mathrm{T}_n[T, M].$$

(b) Given the prior forecast distribution $T \sim \mathrm{N}[f, Q]$, show that the posterior density $p(T|H)$ is proportional to

$$\exp\{-0.5(T - f)^2/Q\}\{n + (T - m)^2/M\}^{-(n+1)/2}.$$

(c) Suppose that $f = 0$, $M = Q = 1$ and $n = 5$. For each value $m = 0, 2, 4$ and 6 plot the (unnormalised) posterior density in (b) as a function of T over the interval $-2 \leq T \leq m + 2$. Discuss the behaviour of the posterior in each cases, making comparison with the normal results in the case of known scale bias $c = 1$ (cf. Lindley 1988).

(10) Consider the 3-dimensional DLM

$$\{\mathbf{I}, \mathbf{I}, \mathrm{diag}(150, 300, 600), \mathrm{diag}(25, 75, 150)\}$$

applied to forecasting demand for three products. At time t, information D_t is summarised by

$$\mathbf{m}_t' = (100, 400, 600)$$

and

$$\mathbf{C}_t = \mathrm{diag}(75, 125, 250).$$

(a) Calculate the one-step ahead forecast distribution $p(\mathbf{Y}_{t+1}|D_t)$.

(b) The total sales at time $t + 1$, $T_{t+1} = \mathbf{1}'\mathbf{Y}_{t+1}$, is targeted at $T_{t+1} = 960$. Calculate $p(\mathbf{Y}_{t+1}|T_{t+1}, D_t)$ for any T_{t+1} and deduce the joint forecast distribution for the three products conditional on the target being met.

(c) An independent forecasting method produces a point forecast m of the total. This is modelled as a random quantity such that

$$(m|T_{t+1}, D_t) \sim \mathrm{N}[T_{t+1}, 750].$$

Calculate the revised forecast distribution $p(\mathbf{Y}_{t+1}|m, D_t)$ for any value of m, and evaluate the moments at $m = 960$. Verify also that the revised forecast of the total is

$$(T_{t+1}|m, D_t) \sim \mathrm{N}[1002, 525].$$

CHAPTER 17

DISTRIBUTION THEORY AND LINEAR ALGEBRA

17.1 DISTRIBUTION THEORY

17.1.1 Introduction

This chapter contains a review and discussion of some of the basic results in distribution theory used throughout the book. Many of these results are stated without proof and the reader is referred to a standard text on multivariate analysis, such as Mardia, Kent and Bibby (1979), for further details. The exceptions to this general rule are some of the results relating to Bayesian prior to posterior calculations in multivariate normal and joint normal/gamma models. Press (1985) is an excellent reference for such results in addition to standard, non-Bayesian multivariate theory. These results are discussed in detail as they are of some importance and should be clearly understood. For a comprehensive reference to all the distribution theory, see Johnson and Kotz (1972). For distribution theory specifically within the contexts of Bayesian analyses, see Aitchison and Dunsmore (1976), Box and Tiao (1973), and De Groot (1971).

It is assumed that the reader is familiar with standard univariate distribution theory. Throughout this chapter we shall be concerned with sets of scalar random variables such as $X_1, X_2, \dots$ $(-\infty < X_i < \infty$, for each $i)$, that may be either discrete or continuous. For each i, the density of the distribution of X_i exists and is defined with respect to the natural counting or Lebesgue measure in the discrete and continuous cases respectively. In either case, the distribution function of X_i is denoted by $P_i(x)$, and the density by $p_i(x)$, $(-\infty < x < \infty)$. The density is only non-zero when x belongs to the range of X_i. In the discrete case,

$$p_i(x) = \Pr(X_i = x) \qquad \text{and} \qquad P_i(x) = \sum_{y \le x} p_i(y), \qquad (-\infty < x < \infty),$$

so that $p_i(x)$ is obtained from the differences of the distribution at consecutive values of x. In the continuous case,

$$p_i(x) = \frac{dP_i(x)}{dx} \qquad \text{and} \qquad P_i(x) = \int_{-\infty}^{x} p_i(y)\,dy.$$

From here on, the derivative and integral expressions for the continuous case will be applied throughout, with the understanding that differences and summations are implied if X_i is discrete. This simplifies the exposition and unifies notation of the multivariate theory to follow.

17.1.2 Random vectors and distributions

$\mathbf{X} = (X_1, X_2, \ldots, X_n)'$ is a random n-vector for any positive integer n. For each $\mathbf{x} = (x_1, x_2, \ldots, x_n)'$, the distribution function of $\mathbf{x}$ is given by

$$P(\mathbf{x}) = \Pr(X_1 \leq x_1 \cap \ldots \cap X_n \leq x_n).$$

$P(.)$ is also referred to as the joint distribution function of the random variables $X_1, \ldots, X_n$ and is defined for all combinations of discrete and continuous variables. As in the univariate case we have

(1) $P(\mathbf{x}) \geq 0$ and is non-decreasing in each x_i;
(2) as $x_i \to -\infty$ for all i, $P(\mathbf{x}) \to 0$;
(3) as $x_i \to \infty$ for all i, $P(\mathbf{x}) \to 1$.

MARGINAL DISTRIBUTIONS.
If $\mathbf{X}'$ is partitioned into

$$\mathbf{X}_1' = (X_1, \ldots, X_m) \qquad \text{and} \qquad \mathbf{X}_2' = (X_{m+1}, \ldots, X_n),$$

the marginal distribution of $\mathbf{X}_1$ is simply the joint distribution of its elements $X_1, \ldots, X_m$. Similarly, the marginal distribution of any subset of m of the X_i in $\mathbf{X}$ is defined by permuting these elements into the first m-subvector $\mathbf{X}_1$. Denote this marginal distribution by $P_1(\mathbf{x}_1)$ for any $\mathbf{x}_1$. Then

(1) $P_1(\mathbf{x}_1) = \lim P(\mathbf{x})$, where the limit is that obtained as $x_i \to \infty$ for each $i = m+1, \ldots, n$;
(2) for $m = 1$, $P_1(x_1)$ is the univariate distribution of the scalar random variable X_1.

INDEPENDENCE.
Given $P(\mathbf{x})$ and the marginal distributions $P_1(\mathbf{x}_1)$ and $P_2(\mathbf{x}_2)$, the vectors $\mathbf{X}_1$ and $\mathbf{X}_2$ are independent if and only if

$$P(\mathbf{x}) = P_1(\mathbf{x}_1)P_2(\mathbf{x}_2)$$

for all $\mathbf{x}$. More generally, for a partition of $\mathbf{X}'$ into $\mathbf{X}_1', \ldots, \mathbf{X}_k'$, mutual independence of the subvectors is defined by the factorisation

$$P(\mathbf{x}) = \prod_{i=1}^{k} P_i(\mathbf{x}_i),$$

where $P_i(\mathbf{x}_i)$ is the distribution function of $\mathbf{X}_i$.

DENSITY FUNCTIONS.
Joint density functions are now defined, and all further theory is based

on density rather than distribution functions. $\mathbf{X}$ is a continuous random vector if and only if the joint distribution function has the form

$$P(\mathbf{x}) = \int_{-\infty}^{x_1} \cdots \int_{-\infty}^{x_n} p(\mathbf{y})\, d\mathbf{y}$$

for some real- and positive-valued function $p(\mathbf{y})$. The function $p(\cdot)$ is the joint density of the X_i, $i = 1, \ldots, n$. The notation will be taken as including the cases that some, or all, of the X_i are discrete when the integrals are replaced by summations accordingly. In general, the density is given by

$$p(\mathbf{x}) = \frac{\partial^n P(\mathbf{x})}{\partial x_1 \ldots \partial x_n},$$

with derivatives being replaced by differences for discrete X_i. Note also that $p(\mathbf{x})$ is normalised, i.e.,

$$\int_{-\infty}^{\infty} \cdots \int_{-\infty}^{\infty} p(\mathbf{x})\, d\mathbf{x} = 1.$$

MARGINAL DENSITIES.
Given the above partition of $\mathbf{X}'$ into $\mathbf{X}_1'$ and $\mathbf{X}_2'$, the marginal density of the marginal distribution of $\mathbf{X}_1$ is given by

$$p_1(\mathbf{x}_1) = \int_{-\infty}^{\infty} \cdots \int_{-\infty}^{\infty} p(\mathbf{x})\, d\mathbf{x}_2,$$

the integration being taken over all $(n-m)$ elements of $\mathbf{X}_2 = \mathbf{x}_2$.

INDEPENDENCE AGAIN.
If $\mathbf{X}_1, \ldots \mathbf{X}_k$ are (mutually) independent, then

$$p(\mathbf{x}) = \prod_{i=1}^{k} p_i(\mathbf{x}_i).$$

CONDITIONAL DENSITIES.
Conditional distributions are defined via conditional densities. The density of the m-vector $\mathbf{X}_1$ conditional on a fixed value of the $(n-m)$-vector $\mathbf{X}_2$ is defined, for all $\mathbf{x}_1$ and $\mathbf{x}_2$, by

$$p_1(\mathbf{x}_1 \mid \mathbf{x}_2) \propto p(\mathbf{x}).$$

Since densities are normalised, $p_1(\mathbf{x}_1 \mid \mathbf{x}_2) = k(\mathbf{x}_2)p(\mathbf{x})$, where the normalisation constant is simply $k(\mathbf{x}_2) = p_2(\mathbf{x}_2)^{-1}$, assumed to exist. If $p_2(\mathbf{x}_2) = 0$ then $p_1(\mathbf{x}_1 \mid \mathbf{x}_2)$ is undefined. Thus, the conditional density of $\mathbf{X}_1$ given

$\mathbf{X}_2$ is the ratio of the joint density of $\mathbf{X}_1$ and $\mathbf{X}_2$ to the marginal density of $\mathbf{X}_2$. Clearly then

(1) $p(\mathbf{x}) = p_1(\mathbf{x}_1 \mid \mathbf{x}_2)p_2(\mathbf{x}_2) = p_1(\mathbf{x}_1)p_2(\mathbf{x}_2 \mid \mathbf{x}_1)$;
(2) $p_1(\mathbf{x}_1 \mid \mathbf{x}_2) = p_1(\mathbf{x}_1)p_2(\mathbf{x}_2 \mid \mathbf{x}_1)/p(\mathbf{x}_2)$. This is the multivariate form of Bayes' Theorem;
(3) $p_1(\mathbf{x}_1 \mid \mathbf{x}_2) \equiv p_1(\mathbf{x}_1)$ and $p_2(\mathbf{x}_2 \mid \mathbf{x}_1) \equiv p_2(\mathbf{x}_2)$ if and only if $\mathbf{X}_1$ and $\mathbf{X}_2$ are independent.

17.1.3 Expectations

The expectation, or mean value, of a scalar function $g(\mathbf{X})$ is defined as

$$\mathrm{E}[g(\mathbf{X})] = \int_{-\infty}^{\infty} \cdots \int_{-\infty}^{\infty} g(\mathbf{x})p(\mathbf{x})\, d\mathbf{x},$$

when the integral exists. A vector function $\mathbf{g}(\mathbf{X})$ with i^{th} element $g_i(\mathbf{X})$ has expectation, or mean vector, $\mathrm{E}[\mathbf{g}(\mathbf{X})]$ whose i^{th} element is $\mathrm{E}[g_i(\mathbf{X})]$. Similarly, the random matrix $\mathbf{G}(\mathbf{X})$ with ij^{th} element $G_{ij}(\mathbf{X})$ has expectation given by the matrix with ij^{th} element $\mathrm{E}[G_{ij}(\mathbf{X})]$.

MEAN VECTORS.
The mean vector (or simply mean) of $\mathbf{X}$ is $\mathrm{E}[\mathbf{X}] = (\mathrm{E}[\mathbf{X}_1], \ldots, \mathrm{E}[\mathbf{X}_n])'$.

VARIANCE MATRICES.
The variance (or variance-covariance) matrix of $\mathbf{X}$ is the $(n \times n)$ symmetric, positive semi-definite matrix

$$\mathrm{V}[\mathbf{X}] = \mathrm{E}[(\mathbf{X} - \mathrm{E}[\mathbf{X}])(\mathbf{X} - \mathrm{E}[\mathbf{X}])'],$$

whose ij^{th} element is $\mathrm{E}[(X_i - \mathrm{E}[X_i])(X_j - \mathrm{E}[X_j])]$. Thus, the ith diagonal element is the variance of X_i, and the ij^{th} off-diagonal element is the covariance between X_i and X_j.

Given the above partition of $\mathbf{X}'$ into the m-vector $\mathbf{X}_1'$ and the $(m-n)$-vector $\mathbf{X}_2'$, the corresponding partitions of the mean and variance matrix are

$$\mathrm{E}[\mathbf{X}] = \begin{pmatrix} \mathrm{E}[\mathbf{X}_1] \\ \mathrm{E}[\mathbf{X}_2] \end{pmatrix} \quad \text{and} \quad \mathrm{V}[\mathbf{X}] = \begin{pmatrix} \mathrm{V}[\mathbf{X}_1] & \mathrm{C}[\mathbf{X}_1, \mathbf{X}_2] \\ \mathrm{C}[\mathbf{X}_2, \mathbf{X}_1] & \mathrm{V}[\mathbf{X}_2] \end{pmatrix}$$

where $\mathrm{C}[\mathbf{X}_1, \mathbf{X}_2] = \mathrm{E}[(\mathbf{X}_1 - \mathrm{E}[\mathbf{X}_1])(\mathbf{X}_2 - \mathrm{E}[\mathbf{X}_2])'] = \mathrm{C}[\mathbf{X}_2, \mathbf{X}_1]'$ is the $m \times (n-m)$ covariance matrix between $\mathbf{X}_1$ and $\mathbf{X}_2$. More generally, if $\mathbf{X}' = (\mathbf{X}_1', \ldots, \mathbf{X}_k')$, then $\mathrm{E}[\mathbf{X}]' = (\mathrm{E}[\mathbf{X}_1]', \ldots, \mathrm{E}[\mathbf{X}_k]')$, $\mathrm{V}[\mathbf{X}]$ has diagonal blocks $\mathrm{V}[\mathbf{X}_i]$ and off-diagonal blocks $\mathrm{C}[\mathbf{X}_i, \mathbf{X}_j] = \mathrm{C}[\mathbf{X}_j, \mathbf{X}_i]'$, $(i = 1, \ldots, k;\ j = 1, \ldots, k)$. If $\mathrm{C}[\mathbf{X}_i, \mathbf{X}_j] = \mathbf{0}$, then $\mathbf{X}_i$ and $\mathbf{X}_j$ are uncorrelated.

CONDITIONAL MEAN AND VARIANCE MATRIX.
In general, the expectation of any function $g(\mathbf{X})$ may be calculated iteratively via

$$\mathrm{E}[g(\mathbf{X})] = \mathrm{E}[\mathrm{E}[g(\mathbf{X}) \mid \mathbf{X}_2]].$$

The inner expectation is with respect to the conditional distribution of $(\mathbf{X}_1 \mid \mathbf{X}_2)$; the outer with respect to the marginal for $\mathbf{X}_2$. In particular, this implies that

(1) $\mathrm{E}[\mathbf{X}_1] = \mathrm{E}[\mathbf{m}_1(\mathbf{X}_2)]$, where $\mathbf{m}_1(\mathbf{X}_2) = \mathrm{E}[\mathbf{X}_1 \mid \mathbf{X}_2]$;
(2) $\mathrm{V}[\mathbf{X}_1] = \mathrm{V}[\mathbf{m}_1(\mathbf{X}_2)] + \mathrm{E}[\mathrm{V}_1[\mathbf{X}_2]]$, where $\mathrm{V}_1[\mathbf{X}_2] = \mathrm{V}[\mathbf{X}_1 \mid \mathbf{X}_2]$.

INDEPENDENCE.
If $\mathbf{X}_1$ and $\mathbf{X}_2$ are independent, then they are uncorrelated, so that

$$\mathrm{V}[\mathbf{X}] = \begin{pmatrix} \mathrm{V}[\mathbf{X}_1] & \mathbf{0} \\ \mathbf{0} & \mathrm{V}[\mathbf{X}_2] \end{pmatrix},$$

a block diagonal variance matrix.

17.1.4 Linear functions of random vectors

LINEAR TRANSFORMATIONS.
For constant matrices $\mathbf{A}$ and vectors $\mathbf{b}$ of suitable dimensions, if $\mathbf{Y} = \mathbf{AX} + \mathbf{b}$, then

$$\mathrm{E}[\mathbf{Y}] = \mathbf{A}\mathrm{E}[\mathbf{X}] + \mathbf{b} \qquad \text{and} \qquad \mathrm{V}[\mathbf{Y}] = \mathbf{A}\mathrm{V}[\mathbf{X}]\mathbf{A}'.$$

LINEAR FORMS.
More generally, for suitable constant matrices $\mathbf{A}_1, \ldots, \mathbf{A}_k$ and vectors $\mathbf{b}$, if $\mathbf{Y} = \sum_{i=1}^k \mathbf{A}_i\mathbf{X}_i + \mathbf{b}$, then

$$\mathrm{E}[\mathbf{Y}] = \sum_{i=1}^{k} \mathbf{A}_i \mathrm{E}[\mathbf{X}_i] + \mathbf{b}$$

and

$$\mathrm{V}[\mathbf{Y}] = \sum_{i=1}^{k} \mathbf{A}_i \mathrm{V}[\mathbf{X}_i]\mathbf{A}_i' + \sum_{i=1}^{k}\sum_{j\neq i}^{k} \mathbf{A}_i \mathrm{C}[\mathbf{X}_i, \mathbf{X}_j]\mathbf{A}_j'.$$

In particular, if $\mathbf{Y} = \sum_{i=1}^k \mathbf{X}_i$, then

$$\mathrm{E}[\mathbf{Y}] = \sum_{i=1}^{k} \mathrm{E}[\mathbf{X}_i]$$

and

$$\mathrm{V}[\mathbf{Y}] = \sum_{i=1}^{k} \mathrm{V}[\mathbf{X}_i] + 2\sum_{i=1}^{k}\sum_{j=1}^{i-1} \mathrm{C}[\mathbf{X}_i, \mathbf{X}_j].$$

Finally, if $\mathbf{Y} = \sum_{i=1}^{k} \mathbf{A}_i\mathbf{X}_i + \mathbf{b}$ and $\mathbf{Z} = \sum_{j=1}^{h} \mathbf{B}_j\mathbf{X}_j + \mathbf{c}$, then

$$\mathrm{C}[\mathbf{Y}, \mathbf{Z}] = \sum_{i=1}^{k}\sum_{j=1}^{h} \mathbf{A}_i \mathrm{C}[\mathbf{X}_i, \mathbf{X}_j]\mathbf{B}_j'.$$

17.2 THE MULTIVARIATE NORMAL DISTRIBUTION

17.2.1 The univariate normal

A real random variable X has a normal (or Gaussian) distribution with mean (mode and median) m, and variance V if and only if

$$p(X) = (2\pi V)^{-1/2} \exp\left[-(X-m)^2/2V\right], \qquad (-\infty < X < \infty).$$

In this case we write $X \sim \mathrm{N}[m, V]$.

SUMS OF NORMAL RANDOM VARIABLES.
If $X_i \sim \mathrm{N}[m_i, V_i]$, $(i = 1, \dots, n)$, have covariances $\mathrm{C}[X_i, X_j] = c_{ij} = c_{ji}$, then $Y = \sum_{i=1}^{n} a_i X_i + b$ has a normal distribution with

$$\mathrm{E}[Y] = \sum_{i=1}^{n} a_i m_i + b \qquad \text{and} \qquad \mathrm{V}[Y] = \sum_{i=1}^{n} a_i^2 V_i + 2\sum_{i=1}^{n}\sum_{j=1}^{i-1} a_i a_j c_{ij},$$

where a_i and b are constants. In particular, $c_{ij} = 0$ for all $i \neq j$ if and only if the X_i are independent normal, in which case $\mathrm{V}[Y] = \sum_{i=1}^{n} a_i^2 V_i$.

17.2.2 The multivariate normal

A random n-vector $\mathbf{X}$ has a multivariate normal distribution in n dimensions if and only if $Y = \sum_{i=1}^{n} a_i X_i$ is normal for all constant, non-zero vectors $\mathbf{a} = (a_1, \dots, a_n)'$.

If $\mathbf{X}$ is multivariate normal, then $\mathrm{E}[\mathbf{X}] = \mathbf{m}$ and $\mathrm{V}[\mathbf{X}] = \mathbf{V}$ exist, and we use the notation $\mathbf{X} \sim \mathrm{N}[\mathbf{m}, \mathbf{V}]$. The moments $\mathbf{m}$ and $\mathbf{V}$ completely define the distribution whose density is

$$p(\mathbf{X}) = \{(2\pi)^n |\mathbf{V}|\}^{-1/2} \exp\left[-(\mathbf{X}-\mathbf{m})'\mathbf{V}^{-1}(\mathbf{X}-\mathbf{m})/2\right].$$

The subvectors of $\mathbf{X}$ are independent if and only if they are uncorrelated. In particular, if $\mathbf{V}$ is block diagonal, then the corresponding subvectors of $\mathbf{X}$ are mutually independent.

LINEAR TRANSFORMATIONS.
For any constant $\mathbf{A}$ and $\mathbf{b}$ of suitable dimensions, if $\mathbf{Y} = \mathbf{AX} + \mathbf{b}$, then $\mathbf{Y} \sim \mathrm{N}[\mathbf{Am} + \mathbf{b}, \mathbf{AVA}']$. If $\mathbf{AVA}'$ is diagonal, then the elements of $\mathbf{Y}$ are independent normal.

LINEAR FORMS.
Suppose $\mathbf{X}_i \sim \mathrm{N}[\mathbf{m}_i, \mathbf{V}_i]$ independently, $i = 1, \ldots, k$, and consider constant matrices and vectors $\mathbf{A}_1, \ldots, \mathbf{A}_k$ and $\mathbf{b}$ of suitable dimensions; $\mathbf{Y} = \sum_{i=1}^k \mathbf{A}_i \mathbf{X}_i + \mathbf{b}$ is multivariate normal with mean $\sum_{i=1}^k \mathbf{A}_i \mathbf{m}_i + \mathbf{b}$ and variance matrix $\sum_{i=1}^k \mathbf{A}_i \mathbf{V}_i \mathbf{A}_i'$.

MARGINAL DISTRIBUTIONS.
Suppose that we have conformable partitions

$$\mathbf{X} = \begin{pmatrix} \mathbf{X}_1 \\ \mathbf{X}_2 \end{pmatrix}, \qquad \mathbf{m} = \begin{pmatrix} \mathbf{m}_1 \\ \mathbf{m}_2 \end{pmatrix}, \qquad \text{and} \quad \mathbf{V} = \begin{pmatrix} \mathbf{V}_{11} & \mathbf{V}_{12} \\ \mathbf{V}_{21} & \mathbf{V}_{22} \end{pmatrix}.$$

Then $\mathbf{X}_i \sim \mathrm{N}[\mathbf{m}_i, \mathbf{V}_{ii}]$, $i = 1, 2$. In particular, if X_i is univariate normal, $X_i \sim \mathrm{N}[m_i, V_{ii}]$ for $i = 1, \ldots, n$.

BIVARIATE NORMAL.
Any two elements X_i and X_j of $\mathbf{X}$ are bivariate normal with joint density

$$p_{ij}(x_i, x_j) = \left[(2\pi)\sqrt{V_{ii}V_{jj}(1 - \rho_{ij}^2)}\right]^{-1} \exp[-Q(x_i, x_j)/2],$$

where $Q(x_i, x_j) = A(x_i, x_j)/\sqrt{(1 - \rho_{ij}^2)}$ and

$$A(x_i, x_j) = \frac{(x_i - m_i)^2}{V_{ii}} + \frac{(x_j - m_j)^2}{V_{jj}} - 2\rho_{ij}\frac{(x_i - m_i)}{\sqrt{V_{ii}}}\frac{(x_j - m_j)}{\sqrt{V_{jj}}},$$

with $\rho_{ij} = \mathrm{corr}(X_i, X_j) = V_{ij}/\sqrt{V_{ii}V_{jj}}$.

CONDITIONAL DISTRIBUTIONS.
For the partition of $\mathbf{X}'$ into $\mathbf{X}_1'$ and $\mathbf{X}_2'$, we have

(1) $(\mathbf{X}_1 \mid \mathbf{X}_2) \sim \mathrm{N}[\mathbf{m}_1(\mathbf{X}_2), \mathbf{V}_{11}(\mathbf{X}_2)]$, where

$$\mathbf{m}_1(\mathbf{X}_2) = \mathbf{m}_1 + \mathbf{V}_{12}\mathbf{V}_{22}^{-1}(\mathbf{X}_2 - \mathbf{m}_2)$$

and

$$\mathbf{V}_{11}(\mathbf{X}_2) = \mathbf{V}_{11} - \mathbf{V}_{12}\mathbf{V}_{22}^{-1}\mathbf{V}_{21}.$$

The matrix $\mathbf{A}_1 = \mathbf{V}_{12}\mathbf{V}_{22}^{-1}$ is called the *regression matrix* of $\mathbf{X}_1$ on $\mathbf{X}_2$. The conditional moments are given in terms of the regression matrix by

$$\mathbf{m}_1(\mathbf{X}_2) = \mathbf{m}_1 + \mathbf{A}_1(\mathbf{X}_2 - \mathbf{m}_2)$$

and

$$\mathbf{V}_{11}(\mathbf{X}_2) = \mathbf{V}_{11} - \mathbf{A}_1\mathbf{V}_{22}\mathbf{A}_1'.$$

(2) $(\mathbf{X}_2 \mid \mathbf{X}_1) \sim \mathrm{N}[\mathbf{m}_2(\mathbf{X}_1), \mathbf{V}_{22}(\mathbf{X}_1)]$, where

$$\mathbf{m}_2(\mathbf{X}_1) = \mathbf{m}_2 + \mathbf{V}_{21}\mathbf{V}_{11}^{-1}(\mathbf{X}_1 - \mathbf{m}_1)$$

and

$$\mathbf{V}_{22}(\mathbf{X}_1) = \mathbf{V}_{22} - \mathbf{V}_{21}\mathbf{V}_{11}^{-1}\mathbf{V}_{12}.$$

The matrix $\mathbf{A}_2 = \mathbf{V}_{21}\mathbf{V}_{11}^{-1}$ is called the *regression matrix* of $\mathbf{X}_2$ on $\mathbf{X}_1$. The conditional moments are given in terms of the regression matrix by

$$\mathbf{m}_2(\mathbf{X}_1) = \mathbf{m}_2 + \mathbf{A}_2(\mathbf{X}_1 - \mathbf{m}_1)$$

and

$$\mathbf{V}_{22}(\mathbf{X}_1) = \mathbf{V}_{22} - \mathbf{A}_2\mathbf{V}_{11}\mathbf{A}_2'.$$

(3) In the special case of the bivariate normal in (1) above, the moments are all scalars, and the correlation between X_1 and X_2 is $\rho = \rho_{12} = V_{12}/\sqrt{V_{11}V_{22}}$. The regressions are determined by regression coefficients A_1 and A_2 given by

$$A_1 = \rho\sqrt{V_{11}/V_{22}} \qquad \text{and} \qquad A_2 = \rho\sqrt{V_{22}/V_{11}}.$$

Also

$$\mathrm{V}[X_1 \mid X_2] = (1-\rho^2)V_{11} \quad \text{and} \quad \mathrm{V}[X_2 \mid X_1] = (1-\rho^2)V_{22}.$$

17.2.3 Conditional normals and linear regression

Many of the important results in this book may be derived directly from the multivariate normal theory reviewed above. A particular regression model is reviewed here to provide the setting for those results.

Suppose that the p-vector $\mathbf{Y}$ and the n-vector $\boldsymbol{\theta}$ are related via the conditional distribution

$$(\mathbf{Y} \mid \boldsymbol{\theta}) \sim \mathrm{N}[\mathbf{F}'\boldsymbol{\theta}, \mathbf{V}],$$

where the $(n \times p)$ matrix $\mathbf{F}$ and the $(p \times p)$ positive definite symmetric matrix $\mathbf{V}$ are constant. An equivalent statement is

$$\mathbf{Y} = \mathbf{F}'\boldsymbol{\theta} + \boldsymbol{\nu},$$

where $\boldsymbol{\nu} \sim \mathrm{N}[\mathbf{0}, \mathbf{V}]$. Suppose further that the marginal distribution of $\boldsymbol{\theta}$ is given by

$$\boldsymbol{\theta} \sim \mathrm{N}[\mathbf{a}, \mathbf{R}],$$

where both $\mathbf{a}$ and $\mathbf{R}$ are constant, and that θ is independent of ν. Equivalently,

$$\boldsymbol{\theta} = \mathbf{a} + \boldsymbol{\omega},$$

where $\boldsymbol{\omega} \sim \mathrm{N}[\mathbf{0}, \mathbf{R}]$ independently of $\boldsymbol{\nu}$.

From these distributions it is possible to construct the joint distribution for $\mathbf{Y}$ and $\boldsymbol{\theta}$ and hence both the marginal for $\mathbf{Y}$ and the conditional for $(\boldsymbol{\theta} \mid \mathbf{Y})$.

MULTIVARIATE JOINT NORMAL DISTRIBUTION.
Since $\boldsymbol{\theta} = \mathbf{a} + \boldsymbol{\omega}$ and $\mathbf{Y} = \mathbf{F}'\boldsymbol{\theta} + \boldsymbol{\nu} = \mathbf{F}'\mathbf{a} + \mathbf{F}'\boldsymbol{\omega} + \boldsymbol{\nu}$, then the vector $(\mathbf{Y}', \boldsymbol{\theta}')'$ is a linear transformation of $(\boldsymbol{\nu}', \boldsymbol{\omega}')'$. By construction the latter has a multivariate normal distribution, so that $\mathbf{Y}$ and $\boldsymbol{\theta}$ are jointly normal. Further

(1) $\mathrm{E}[\boldsymbol{\theta}] = \mathbf{a}$ and $\mathrm{V}[\boldsymbol{\theta}] = \mathbf{R}$;
(2) $\mathrm{E}[\mathbf{Y}] = \mathrm{E}[\mathbf{F}'\boldsymbol{\theta} + \boldsymbol{\nu}] = \mathbf{F}'\mathrm{E}[\boldsymbol{\theta}] + \mathrm{E}[\boldsymbol{\nu}] = \mathbf{F}'\mathbf{a}$ and

$$\mathrm{V}[\mathbf{Y}] = \mathrm{V}[\mathbf{F}'\boldsymbol{\theta} + \boldsymbol{\nu}] = \mathbf{F}'\mathrm{V}[\boldsymbol{\theta}]\mathbf{F} + \mathrm{V}[\boldsymbol{\nu}] = \mathbf{F}'\mathbf{R}\mathbf{F} + \mathbf{V};$$

(3) $\mathrm{C}[\mathbf{Y}, \boldsymbol{\theta}] = \mathrm{C}[\mathbf{F}'\boldsymbol{\theta} + \boldsymbol{\nu}, \boldsymbol{\theta}] = \mathbf{F}'\mathrm{C}[\boldsymbol{\theta}, \boldsymbol{\theta}] + \mathrm{C}[\boldsymbol{\nu}, \boldsymbol{\theta}] = \mathbf{F}'\mathbf{R}$.

It follows that

$$\begin{pmatrix} \mathbf{Y} \\ \boldsymbol{\theta} \end{pmatrix} \sim \mathrm{N}\left[\begin{pmatrix} \mathbf{F}'\mathbf{a} \\ \mathbf{a} \end{pmatrix}, \begin{pmatrix} \mathbf{F}'\mathbf{R}\mathbf{F} + \mathbf{V} & \mathbf{F}'\mathbf{R} \\ \mathbf{R}\mathbf{F} & \mathbf{R} \end{pmatrix}\right].$$

Therefore, identifying $\mathbf{Y}$ with $\mathbf{X}_1$ and $\boldsymbol{\theta}$ with $\mathbf{X}_2$ in the partition of $\mathbf{X}$ in 17.2.2, we have

(4) $\mathbf{Y} \sim \mathrm{N}[\mathbf{F}'\mathbf{a}, \mathbf{F}'\mathbf{R}\mathbf{F} + \mathbf{V}]$;
(5) $(\boldsymbol{\theta} \mid \mathbf{Y}) \sim \mathrm{N}[\mathbf{m}, \mathbf{C}]$, where

$$\mathbf{m} = \mathbf{a} + \mathbf{R}\mathbf{F}[\mathbf{F}'\mathbf{R}\mathbf{F} + \mathbf{V}]^{-1}[\mathbf{Y} - \mathbf{F}'\mathbf{a}]$$

and

$$\mathbf{C} = \mathbf{R} - \mathbf{R}\mathbf{F}[\mathbf{F}'\mathbf{R}\mathbf{F} + \mathbf{V}]^{-1}\mathbf{F}'\mathbf{R}.$$

By defining $\mathbf{e} = \mathbf{Y} - \mathbf{F}'\mathbf{a}$, $\mathbf{Q} = \mathbf{F}'\mathbf{R}\mathbf{F} + \mathbf{V}$ and $\mathbf{A} = \mathbf{R}\mathbf{F}\mathbf{Q}^{-1}$, these equations become

$$\mathbf{m} = \mathbf{a} + \mathbf{A}\mathbf{e} \qquad \text{and} \qquad \mathbf{C} = \mathbf{R} - \mathbf{A}\mathbf{Q}\mathbf{A}'.$$

MULTIVARIATE BAYES' THEOREM.
An alternative derivation of the conditional distribution for $(\boldsymbol{\theta} \mid \mathbf{Y})$ via Bayes' Theorem provides alternative expressions for $\mathbf{m}$ and $\mathbf{C}$. Note that

$$p(\boldsymbol{\theta} \mid \mathbf{Y}) \propto p(\mathbf{Y} \mid \boldsymbol{\theta})p(\boldsymbol{\theta})$$

as a function of $\boldsymbol{\theta}$, so that

$$\ln[p(\boldsymbol{\theta} \mid \mathbf{Y})] = k + \ln[p(\mathbf{Y} \mid \boldsymbol{\theta})] + \ln[p(\boldsymbol{\theta})],$$

where k depends on $\mathbf{Y}$ but not on $\boldsymbol{\theta}$. Therefore

$$\ln[p(\boldsymbol{\theta} \mid \mathbf{Y})] = k - \frac{1}{2}\left[(\mathbf{Y} - \mathbf{F}'\boldsymbol{\theta})'\mathbf{V}^{-1}(\mathbf{Y} - \mathbf{F}'\boldsymbol{\theta}) + (\boldsymbol{\theta} - \mathbf{a})'\mathbf{R}^{-1}(\boldsymbol{\theta} - \mathbf{a})\right].$$

The bracketed term here is simply

$$\begin{aligned}\mathbf{Y}'\mathbf{V}^{-1}\mathbf{Y} - 2\mathbf{Y}'\mathbf{V}^{-1}\mathbf{F}'\boldsymbol{\theta} + \boldsymbol{\theta}'\mathbf{F}\mathbf{V}^{-1}\mathbf{F}'\boldsymbol{\theta} + \boldsymbol{\theta}'\mathbf{R}^{-1}\boldsymbol{\theta} - 2\mathbf{a}'\mathbf{R}^{-1}\boldsymbol{\theta} + \mathbf{a}'\mathbf{R}^{-1}\mathbf{a}\\ = \boldsymbol{\theta}'\left\{\mathbf{F}\mathbf{V}^{-1}\mathbf{F}' + \mathbf{R}^{-1}\right\}\boldsymbol{\theta} - 2\left\{\mathbf{Y}'\mathbf{V}^{-1}\mathbf{F}' + \mathbf{a}'\mathbf{R}^{-1}\right\}\boldsymbol{\theta} + h\end{aligned}$$

where h depends on $\mathbf{Y}$ but not on $\boldsymbol{\theta}$. Completing the quadratic form gives

$$(\boldsymbol{\theta} - \mathbf{m})'\,\mathbf{C}^{-1}(\boldsymbol{\theta} - \mathbf{m}) + h^*,$$

where again h^* does not involve $\boldsymbol{\theta}$,

$$\mathbf{C}^{-1} = \mathbf{R}^{-1} + \mathbf{F}\mathbf{V}^{-1}\mathbf{F}',$$

and

$$\mathbf{m} = \mathbf{C}\left\{\mathbf{F}\mathbf{V}^{-1}\mathbf{Y} + \mathbf{R}^{-1}\mathbf{a}\right\}.$$

Hence

$$p(\boldsymbol{\theta} \mid \mathbf{Y}) \propto \exp\left[-\frac{1}{2}(\boldsymbol{\theta} - \mathbf{m})'\,\mathbf{C}^{-1}(\boldsymbol{\theta} - \mathbf{m})\right]$$

as a function of $\boldsymbol{\theta}$, so that $(\boldsymbol{\theta} \mid \mathbf{Y}) \sim \mathrm{N}[\mathbf{m}, \mathbf{C}]$, just as derived earlier.

Note that the two derivations give different expressions for $\mathbf{m}$ and $\mathbf{C}$ that provide, in particular, the matrix identity for $\mathbf{C}$ given by

$$\mathbf{C} = \left[\mathbf{R}^{-1} + \mathbf{F}\mathbf{V}^{-1}\mathbf{F}'\right]^{-1} = \mathbf{R} - \mathbf{R}\mathbf{F}\left[\mathbf{F}'\mathbf{R}\mathbf{F} + \mathbf{V}\right]^{-1}\mathbf{F}'\mathbf{R},$$

that is easily verified once stated.

17.3 JOINT NORMAL/GAMMA DISTRIBUTIONS

17.3.1 The gamma distribution

A random variable $\phi > 0$ has a gamma distribution with parameters $n > 0$ and $d > 0$, denoted by $\phi \sim \mathrm{G}[n, d]$, if and only if

$$p(\phi) \propto \phi^{n-1}\exp(-\phi d), \qquad (\phi > 0).$$

Normalisation leads to $p(\phi) = d^n\Gamma(n)^{-1}\phi^{n-1}\exp(-\phi d)$, where $\Gamma(n)$ is the gamma function. Note that $\mathrm{E}[\phi] = n/d$ and $\mathrm{V}[\phi] = \mathrm{E}[\phi]^2/n$.

Two special cases of interest are

(1) $n = 1$, when ϕ has a (negative) exponential distribution with mean $1/d$;

(2) $\phi \sim \mathrm{G}[n/2, d/2]$ when n is a positive integer. In this case $d\phi \sim \chi^2_n$, a chi-squared distribution with n degrees of freedom.

17.3.2 Univariate normal/gamma distribution

Let $\phi \sim \mathrm{G}[n/2, d/2]$ for any $n > 0$ and $d > 0$, and suppose that the conditional distribution of a further random variable X given ϕ is normal $(X \mid \phi) \sim \mathrm{N}[m, C\phi^{-1}]$, for some m and C. Note that $\mathrm{E}[X] \equiv \mathrm{E}[X \mid \phi] = m$ does not depend on ϕ. However $\mathrm{V}[X \mid \phi] = C\phi^{-1}$. The joint distribution of X and ϕ is called (univariate) normal/gamma. Note that

(1)

$$p(X, \phi) = \left(\frac{\phi}{2\pi C}\right)^{1/2} \exp\left[-\frac{\phi(X-m)^2}{2C}\right] \times \frac{d^{n/2}}{2^{n/2}\Gamma(\frac{n}{2})}\phi^{\frac{n}{2}-1}\exp\left[-\frac{\phi d}{2}\right] \propto \phi^{(\frac{n+1}{2})-1}\exp\left[-\frac{\phi}{2}\left\{\frac{(X-m)^2}{C}+d\right\}\right],$$

as a function of ϕ and X.

(2)

$$p(\phi \mid X) \propto \phi^{(\frac{n+1}{2})-1}\exp\left[-\frac{\phi}{2}\left\{\frac{(X-m)^2}{C}+d\right\}\right],$$

so that

$$(\phi \mid X) \sim \mathrm{G}\left[\frac{n^*}{2}, \frac{d^*}{2}\right],$$

where $n^* = n + 1$ and $d^* = d + (X - m)^2/C$.

(3)

$$p(X) = p(X, \phi)/p(\phi \mid X) \qquad (-\infty < X < \infty)$$
$$\propto [n + (X-m)^2/R]^{-(n+1)/2},$$

where $R = C(d/n) = C/\mathrm{E}[\phi]$. This is proportional to the density of the Student T distribution with n degrees of freedom, mode m and scale R. Hence $X \sim \mathrm{T}_n[m, R]$, or $(X - m)/R^{1/2} \sim \mathrm{T}_n[0, 1]$, a standard Student T distribution with n degrees of freedom.

17.3.3 Multivariate normal/gamma distribution

As an important generalisation, suppose that $\phi \sim \mathrm{G}[n/2, d/2]$ and that the p-vector $\mathbf{X}$ is normally distributed conditional on ϕ, as $\mathbf{X} \sim \mathrm{N}[\mathbf{m}, \mathbf{C}\phi^{-1}]$. Here the p-vector $\mathbf{m}$ and the $(p \times p)$ symmetric positive definite matrix $\mathbf{C}$

are known. Thus, each element of V[$\mathbf{X}$] is scaled by the common factor ϕ. The basic results are similar to 17.3.2 in that

(1) $(\phi \mid \mathbf{X}) \sim \mathrm{G}[n^*/2, d^*/2]$, where $n^* = n + p$ and $d^* = d + (\mathbf{X} - \mathbf{m})'\mathbf{C}^{-1}(\mathbf{X} - \mathbf{m})/2$ (notice the degrees of freedom increases by p).

(2) $\mathbf{X}$ has a (marginal) multivariate T distribution in p dimensions with n degrees of freedom, mode $\mathbf{m}$ and scale matrix $\mathbf{R} = \mathbf{C}(d/n) = \mathbf{C}/\mathrm{E}[\phi]$, denoted by $\mathbf{X} \sim \mathrm{T}_n[\mathbf{m}, \mathbf{R}]$, with density

$$p(\mathbf{X}) \propto [n + (\mathbf{X} - \mathbf{m})'\mathbf{R}^{-1}(\mathbf{X} - \mathbf{m})]^{-(n+p)/2}.$$

In particular, if X_i is the ith element of $\mathbf{X}$, m_i and C_{ii} the corresponding mean and diagonal element of $\mathbf{C}$, then

$$X_i \sim \mathrm{T}_n[m_i, R_{ii}],$$

where $R_{ii} = C_{ii}(d/n)$.

17.3.4 Simple regression model

The normal/gamma distribution plays a key role in providing closed form Bayesian analyses of linear models with unknown scale parameters. Details may be found in De Groot (1971) and Press (1985), for example. A particular regression setting is reviewed here for reference. The details follow from the above joint normal/gamma theory. Suppose that a scalar variable Y is related to the p-vector $\boldsymbol{\theta}$ and the scalar ϕ via

$$(Y \mid \boldsymbol{\theta}, \phi) \sim \mathrm{N}[\mathbf{F}'\boldsymbol{\theta}, k\phi^{-1}],$$

where the p-vector $\mathbf{F}$ and the variance multiple k are fixed constants. Suppose also that $(\boldsymbol{\theta}, \phi)$ have a joint normal/gamma distribution, namely

$$(\boldsymbol{\theta} \mid \phi) \sim \mathrm{N}[\mathbf{a}, \mathbf{R}\phi^{-1}]$$

and

$$\phi \sim \mathrm{G}[n/2, d/2]$$

for fixed scalars $n > 0$, $d > 0$, p-vector $\mathbf{a}$ and $(p \times p)$ variance matrix $\mathbf{R}$, and let $S = d/n = 1/\mathrm{E}[\phi]$. Then

(1) $(Y \mid \phi) \sim \mathrm{N}[f, Q\phi^{-1}]$, where $f = \mathbf{F}'\mathbf{a}$ and $Q = \mathbf{F}'\mathbf{R}\mathbf{F} + k$;

(2) $Y \sim \mathrm{T}_n[f, QS]$, unconditional on $\boldsymbol{\theta}$ or ϕ;

(3) $(\boldsymbol{\theta}, \phi \mid Y)$ have a joint normal/gamma distribution. Specifically,

$$(\boldsymbol{\theta} \mid \phi, Y) \sim \mathrm{N}[\mathbf{m}, \mathbf{C}\phi^{-1}],$$

where from 17.2.3, $\mathbf{m} = \mathbf{a} + \mathbf{R}\mathbf{F}(Y - f)/Q$, $\mathbf{C} = \mathbf{R} - \mathbf{R}\mathbf{F}\mathbf{F}'\mathbf{R}/Q$ and

$$(\phi \mid Y) \sim \mathrm{G}[(n + 1)/2, \{d + (Y - f)^2/Q\}/2];$$

(4) $(\boldsymbol{\theta} \mid Y)$ has a multivariate T distribution in p dimensions with $n+1$ degrees of freedom, mode $\mathbf{m}$ and scale matrix $\mathbf{C}S$, denoted by $(\boldsymbol{\theta} \mid Y) \sim \mathrm{T}_{n+1}[\mathbf{m}, \mathbf{C}S]$. In particular, for the ith element $\boldsymbol{\theta}_i$ of $\boldsymbol{\theta}$,

$$\theta_i \sim \mathrm{T}_{n+1}[m_i, C_{ii}S], \qquad \text{or} \qquad \frac{(\theta_i - m_i)}{\sqrt{C_{ii}S}} \sim \mathrm{T}_{n+1}[0,1].$$

17.3.5 HPD regions

Consider any random vector $\mathbf{X}$ with (posterior) density $p(\mathbf{X})$. Density measures relative support for values of $\mathbf{X}$ in the sense that $\mathbf{X} = \mathbf{x}_0$ is better supported than $\mathbf{X} = \mathbf{x}_1$ if $p(\mathbf{x}_0) > p(\mathbf{x}_1)$. **Highest posterior density** regions (HPD regions) form the bases for inferences about $\mathbf{X}$ based on sets of values with high density. The basic notion is that the posterior probability of the set or region of values of $\mathbf{X}$ that have higher density than any chosen value $\mathbf{x}_0$ measures the support for or against $\mathbf{X} = \mathbf{x}_0$. Formally, the HPD region generated by the value $\mathbf{X} = \mathbf{x}_0$ is

$$\{\mathbf{X} \ : \ p(\mathbf{X}) \geq p(\mathbf{x}_0)\}.$$

The probability that $\mathbf{X}$ lies in this region is simply

$$\Pr[p(\mathbf{X}) \geq p(\mathbf{x}_0)].$$

If this probability is high, then $\mathbf{x}_0$ is poorly supported. In normal linear models, HPD regions are always connected regions, or intervals, and these provide interval-based inferences and tests of hypotheses through the use of posterior normal, T and F distributions, described below. Fuller theoretical details are provided by Box and Tiao (1973) and De Groot (1971).

MULTIVARIATE NORMAL POSTERIOR.

(1) Suppose that $\mathbf{X} = X$, a scalar, with posterior $X \sim \mathrm{N}[m, C]$. Then, as is always the case with symmetric distributions, HPD regions are intervals symmetrically located about the median (here also the mode and mean) m. For any $k > 0$, the equal-tails interval

$$m - kC^{1/2} \leq X \leq m + kC^{1/2}$$

is the HPD region with posterior probability

$$\Pr[|X - m|/C^{1/2} \leq k] = 2\Phi(k) - 1,$$

where $\Phi(.)$ is the standard normal cumulative distribution function. With $k = 1.645$, so that $\Phi(k) = 0.95$, this gives the 90% region $m \pm 1.645C^{1/2}$. With $k = 1.96$, $\Phi(k) = 0.975$ and the 95% region is $m \pm 1.96C^{1/2}$. For any $k > 0$, the $100[2\Phi(k) - 1]\%$ HPD region for X is simply $m \pm kC^{1/2}$.

(2) Suppose that $\mathbf{X}$ is n-dimensional for some $n > 1$,

$$\mathbf{X} \sim \mathrm{N}[\mathbf{m}, \mathbf{C}]$$

for some mean vector $\mathbf{m}$ and covariance matrix $\mathbf{C}$. Then HPD regions are defined by elliptical shells centred at the mode $\mathbf{m}$, defined by the points $\mathbf{X}$ that lead to common values of the quadratic form in the density, namely

$$Q(\mathbf{X}) = (\mathbf{X} - \mathbf{m})'\mathbf{C}^{-1}(\mathbf{X} - \mathbf{m}).$$

For any $k > 0$, the region

$$\{\mathbf{X} \ : \ Q(\mathbf{X}) \leq k\}$$

is the HPD region with posterior probability

$$\Pr[Q(\mathbf{X}) \leq k] = \Pr[\kappa \leq k],$$

where κ is a gamma distributed random quantity,

$$\kappa \sim \mathrm{G}[n/2, 1/2].$$

When n is an integer, this gamma distribution is chi-squared with n degrees of freedom, and so

$$\Pr[Q(\mathbf{X}) \leq k] = \Pr[\chi^2_n \leq k].$$

MULTIVARIATE T POSTERIORS.

The results for T distribution parallel those for the normal, T distributions replace normal distributions and F distributions replace gamma.

(1) Suppose that $\mathbf{X} = X$, a scalar, with posterior $X \sim \mathrm{T}_r[m, C]$ for some degrees of freedom $r > 0$. Again, HPD regions are intervals symmetrically located about the mode m. For any $k > 0$, the equal-tails interval

$$m - kC^{1/2} \leq X \leq m + kC^{1/2}$$

is the HPD region with posterior probability

$$\Pr[|X - m|/C^{1/2} \leq k] = 2\Psi_r(k) - 1,$$

where $\Psi_r(.)$ is the cumulative distribution function of the standard Student T distribution on r degrees of freedom. For any $k > 0$, the $100[2\Psi_r(k) - 1]\%$ HPD region for X is simply $m \pm kC^{1/2}$.

(2) Suppose that $\mathbf{X}$ is n-dimensional for some $n > 1$, $\mathbf{X} \sim \mathrm{T}_r[\mathbf{m}, \mathbf{C}]$ for some mean vector $\mathbf{m}$, covariance matrix $\mathbf{C}$, and degrees of freedom $r > 0$. HPD regions are again defined by elliptical shells centred at the mode $\mathbf{m}$, identified by values of $\mathbf{X}$ having a common value of the quadratic form

$$Q(\mathbf{X}) = (\mathbf{X} - \mathbf{m})'\mathbf{C}^{-1}(\mathbf{X} - \mathbf{m}).$$

For any $k > 0$, the region

$$\{\mathbf{X} \,:\, Q(\mathbf{X}) \leq k\}$$

is the HPD region with posterior probability

$$\Pr[Q(\mathbf{X}) \leq k] = \Pr[\xi \leq k/n],$$

where ξ is an F distributed random quantity with n and r degrees of freedom,

$$\xi \sim \mathrm{F}_{n,r}$$

(tabulated in Lindley and Scott 1984, pages 50-55). Note that when r is large, this is approximately a χ^2_n distribution.

17.4 ELEMENTS OF MATRIX ALGEBRA

The reader is assumed to be familiar with basic concepts and results of matrix algebra including, for example, numerical operations on matrices; trace; determinant; transposition; linear and quadratic forms; singularity and inversion; symmetry; positive definiteness and positive semi-definiteness (or non-negative definiteness); orthogonality; vector and matrix derivatives. Here we review some simple results concerning functions of square matrices and then explore in depth the eigenstructure and diagonalisability of such matrices. Foundational material and more advanced theory may be found in Bellman (1970) and Graybill (1969).

Throughout the Appendix we consider square $(n \times n)$ matrices, such as $\mathbf{G}$, whose elements G_{ij}, $(i = 1, \ldots n;\ j = 1, \ldots, n)$, are real-valued.

17.4.1 Powers and polynomials

POWERS.
For any $k = 1, 2, \ldots$, $\mathbf{G}^k = \mathbf{GG}\ldots\mathbf{G} = \mathbf{GG}^{k-1} = \mathbf{G}^{k-1}\mathbf{G}$. If $\mathbf{G}$ is non-singular, then $\mathbf{G}^{-k} = \left(\mathbf{G}^{-1}\right)^k$. Also, for all integers h and k, $\mathbf{G}^k\mathbf{G}^h = \mathbf{G}^{k+h}$ and $\left(\mathbf{G}^k\right)^h = \mathbf{G}^{kh}$.

POLYNOMIALS.
If $p(\alpha)$ is a polynomial of degree m in α, $p(\alpha) = \sum_{r=0}^{m} p_r\alpha^r$, the matrix polynomial in $\mathbf{G}$, $p(\mathbf{G})$ is the $(n \times n)$ matrix defined by $p(\mathbf{G}) = \sum_{r=0}^{m} p_r\mathbf{G}^r$.

INFINITE SERIES.
Suppose that for all α with $|\alpha| \leq \max\{|G_{ij}|\}$, $p(\alpha)$ is a convergent power series $p(\alpha) = \sum_{r=0}^{\infty} p_r\alpha^r$. Then the matrix $p(\mathbf{G})$ defined as $\sum_{r=0}^{\infty} p_r\mathbf{G}^r$ exists with finite elements.

17.4.2 Eigenstructure of square matrices

The $(n \times n)$ matrix $\mathbf{G}$ has n eigenvalues $\lambda_1, \ldots, \lambda_n$ that are the roots of the polynomial of degree n in λ given by the determinant

$$p(\lambda) = |\mathbf{G} - \lambda\mathbf{I}|,$$

where $\mathbf{I}$ is the $(n \times n)$ identity matrix. $p(\lambda)$ is the characteristic polynomial of $\mathbf{G}$ and the roots may be real or complex, occurring in pairs of complex conjugates in the latter case. Eigenvalues are sometimes referred to as the *characteristic values*, or *roots*, of the matrix.

The eigenvectors of $\mathbf{G}$ are n-vectors $\boldsymbol{\eta}$ satisfying

$$\mathbf{G}\boldsymbol{\eta} = \lambda_i\boldsymbol{\eta}, \qquad (i = 1, \ldots, n).$$

Eigenvectors are sometimes referred to as the *characteristic vectors* of the matrix.

This equation has at least one solution for each i. Various important properties are as follows:

(1) $|\mathbf{G}| = \prod_{i=1}^n \lambda_i$ and trace $(\mathbf{G}) = \sum_{i=1}^n \lambda_i$.
(2) The elements of $\boldsymbol{\eta}$ corresponding to λ_i are real valued if and only if λ_i is real valued.
(3) If the λ_i are distinct, then $\mathbf{G}\boldsymbol{\eta} = \lambda_i\boldsymbol{\eta}$ has a unique solution $\boldsymbol{\eta}_i$ (up to a constant scale factor of course) and the $\boldsymbol{\eta}_i$ are themselves distinct.
(4) If $\mathbf{G}$ is diagonal, then $\lambda_1, \ldots, \lambda_n$ are the diagonal elements.
(5) If $\mathbf{G}$ is symmetric then the λ_i are real. If, in addition, $\mathbf{G}$ is positive definite (or non-negative definite) then the λ_i are positive (or non-negative).
(6) If $\mathbf{G}$ has rank $p \leq n$, then $n - p$ of the λ_i are zero.
(7) The eigenvalues of a polynomial or power series function $p(\mathbf{G})$ are given by $p(\lambda_i)$, $i = 1, \ldots, n$.
(8) If $p(\lambda) = |\mathbf{G} - \lambda\mathbf{I}|$ is the characteristic polynomial of $\mathbf{G}$, then

$$p(\mathbf{G}) = \mathbf{0}.$$

Since $p(\lambda) = \prod_{i=1}^n (\lambda - \lambda_i)$, then $\prod_{i=1}^n (\mathbf{G} - \lambda_i\mathbf{I}) = \mathbf{0}$. This result is known as the *Cayley-Hamilton Theorem*. It follows that if $p(\lambda) = \sum_{i=0}^n p_i\lambda^i$, then $\sum_{i=0}^n p_i\mathbf{G}^i = 0$, so that

$$\mathbf{G}^n = -p_n^{-1}\sum_{i=0}^{n-1} p_i\mathbf{G}^i = q_o\mathbf{I} + q_1\mathbf{G} + \ldots + q_{n-1}\mathbf{G}^{n-1}, \text{ say.}$$

EIGENSTRUCTURE OF SPD MATRICES.

The eigenstructure of a symmetric positive definite (or SPD) matrix has particularly special features. Suppose that the $n \times n$ matrix $\mathbf{V}$ is symmetric and positive definite, i.e., a variance matrix. Then the n eigenvalues of $\mathbf{V}$ are real and positive. Without loss of generality, order the eigenvalues so that $\lambda_1 \geq \lambda_2 \geq \ldots \geq \lambda_n$. The corresponding eigenvectors are real-valued and orthogonal so that $\boldsymbol{\eta}_i'\boldsymbol{\eta}_j = 0$, $(i,j = 1,\ldots,n;\ i \neq j)$ (Graybill 1969, Chapter 3). The *orthonormalised* eigenvectors are defined by normalising the eigenvectors to have unit norm, so that in addition to orthogonality, $\boldsymbol{\eta}_j'\boldsymbol{\eta}_j = 1$, $(j = 1,\ldots,n)$. These define the **principal components** of the matrix $\mathbf{V}$. Suppose that $\mathbf{Y}$ is an $n \times 1$ random vector with variance matrix $\mathbf{V}$, so that $\mathrm{V}[\mathbf{Y}] = \mathbf{V}$. Covariation of the elements of any random vector $\mathbf{Y}$ is explained through the random quantities $X_j = \boldsymbol{\eta}_j'\mathbf{Y}$, $(j = 1,\ldots,n)$, the principal components of $\mathbf{Y}$. These X variates are uncorrelated and have variances $\mathrm{V}[X_j] = \lambda_j$, decreasing as j increases. Total variation in $\mathbf{Y}$ is measured by $\lambda = \mathrm{trace}(\mathbf{V}) = \sum_{j=1}^{n} \lambda_j$, and so the jth principal component explains a proportion λ_j/λ of this total.

17.4.3 Similarity of square matrices

Let $\mathbf{G}$ and $\mathbf{L}$ be $(n \times n)$ matrices and the eigenvalues of $\mathbf{G}$ be $\lambda_1,\ldots,\lambda_n$. Then $\mathbf{G}$ and $\mathbf{L}$ are similar (or $\mathbf{G}$ is similar to $\mathbf{L}$, $\mathbf{L}$ is similar to $\mathbf{G}$) if there exists a non-singular $(n \times n)$ matrix $\mathbf{H}$ such that

$$\mathbf{H}\mathbf{G}\mathbf{H}^{-1} = \mathbf{L}.$$

$\mathbf{H}$ is called the similarity matrix (or similarity transformation) and may be complex valued. Note that

(1) $\mathbf{G} = \mathbf{H}^{-1}\mathbf{L}\mathbf{H}$.
(2) $|\mathbf{G} - \lambda\mathbf{I}| = |\mathbf{H}^{-1}(\mathbf{L} - \lambda\mathbf{I})\mathbf{H}| = |\mathbf{H}|^{-1}|\mathbf{L} - \lambda\mathbf{I}||\mathbf{H}| = |\mathbf{L} - \lambda\mathbf{I}|$, so that the eigenvalues of $\mathbf{L}$ are also $\lambda_1,\ldots,\lambda_n$. Hence the traces of $\mathbf{G}$ and $\mathbf{L}$ coincide, as do their determinants.
(3) If $\mathbf{G}$ is similar to $\mathbf{L}$ and $\mathbf{L}$ is similar to $\mathbf{K}$, then $\mathbf{G}$ is similar to $\mathbf{K}$.
(4) Suppose that $\mathbf{G}_1,\ldots,\mathbf{G}_k$ are any k square matrices with the $(n_i \times n_i)$ matrix $\mathbf{G}_i$ similar to $\mathbf{L}_i$, $(i = 1,\ldots,k)$. Let $n = \sum_{i=1}^{k} n_i$. Then if $\mathbf{G}$ and $\mathbf{L}$ are the block diagonal matrices

$$\mathbf{G} = \text{block diag}\,[\mathbf{G}_1,\ldots,\mathbf{G}_k]$$

and

$$\mathbf{L} = \text{block diag}\,[\mathbf{L}_1,\ldots,\mathbf{L}_k],$$

it follows that $\mathbf{G}$ and $\mathbf{L}$ are similar. In particular, if the similarity matrix for $\mathbf{G}_i$ and $\mathbf{L}_i$ is $\mathbf{H}_i$, so that $\mathbf{H}_i\mathbf{G}_i\mathbf{H}_i^{-1} = \mathbf{L}_i$, then the similarity matrix for $\mathbf{G}$ and $\mathbf{L}$ is $\mathbf{H} = \text{block diag}\,[\mathbf{H}_1,\ldots,\mathbf{H}_k]$.

By way of terminology, $\mathbf{G}$, $\mathbf{L}$ and $\mathbf{H}$ are said to be formed from the **superposition** of the matrices $\mathbf{G}_i$, $\mathbf{L}_i$ and $\mathbf{H}_i$, $(i = 1,\ldots,k)$, respectively.

DIAGONALIZATION.

If $\mathbf{G}$ is similar to a diagonal matrix $\mathbf{L}$, then $\mathbf{G}$ is diagonalisable. In particular, if $\mathbf{L} = \boldsymbol{\Lambda} = \mathrm{diag}(\lambda_1, \ldots, \lambda_n)$, then, since

$$\mathbf{G}\mathbf{H}^{-1} = \mathbf{H}^{-1}\boldsymbol{\Lambda},$$

the columns of $\mathbf{H}^{-1}$ are the eigenvectors of $\mathbf{G}$. It follows that

(1) $\mathbf{G}$ is similar to $\boldsymbol{\Lambda}$ if the eigenvectors of $\mathbf{G}$ are linearly independent.
(2) If $\lambda_1, \ldots, \lambda_n$ are distinct, then $\mathbf{G}$ is diagonalisable.
(3) If $\lambda_1, \ldots, \lambda_n$ are not distinct, then $\mathbf{G}$ may not be diagonalisable. If $\mathbf{G}$ is symmetric or skew-symmetric, or orthogonal, then $\mathbf{G}$ is diagonalisable. In general, a different similar form for $\mathbf{G}$ is appropriate, as discussed in the next section.

DISTINCT EIGENVALUES.
If $\lambda_1, \ldots, \lambda_n$ are distinct, then the eigenvectors are unique (up to a scalar constant) and linearly independent; they form the columns of $\mathbf{H}^{-1}$ in $\mathbf{H}\mathbf{G}\mathbf{H}^{-1} = \boldsymbol{\Lambda} = \mathrm{diag}\,(\lambda_1, \ldots, \lambda_n)$. In such a case it follows that

(1) $\mathbf{G}^k = \left(\mathbf{H}^{-1}\boldsymbol{\Lambda}\mathbf{H}\right)^k = \mathbf{H}^{-1}\boldsymbol{\Lambda}^k\mathbf{H}$ for any integer k. Thus, $\mathbf{G}^k$ is similar to $\boldsymbol{\Lambda}^k$, for each k, and the similarity matrix is $\mathbf{H}$. Note that $\boldsymbol{\Lambda}^k = \mathrm{diag}\left(\lambda_1^k, \ldots, \lambda_n^k\right)$, so that the eigenvalues of $\mathbf{G}^k$ are λ_i^k.
(2) For any n-vectors $\mathbf{a}$ and $\mathbf{b}$ not depending on k,

$$p(k) = \mathbf{a}'\mathbf{G}^k\mathbf{b} = \left(\mathbf{a}'\mathbf{H}^{-1}\right)\boldsymbol{\Lambda}^k\left(\mathbf{H}\mathbf{b}\right) = \sum_{i=1}^{n} c_i\lambda_i^k$$

for some constants $c_1, \ldots, c_n$ not depending on k.

In general, the eigenvalues of $\mathbf{G}$ may be real or complex, with the latter occurring in conjugate pairs. Suppose that $\mathbf{G}$ has p, $1 \leq p \leq n/2$, pairs of distinct complex conjugate eigenvalues ordered in pairs on the lower diagonal of $\boldsymbol{\Lambda}$. Thus, $\lambda_1, \ldots, \lambda_{n-2p}$ are real and distinct, and

$$\left.\begin{aligned} \lambda_j &= r_j \exp\left(i\omega_j\right) \\ \lambda_{j+1} &= r_j \exp\left(-i\omega_j\right) \end{aligned}\right\}, \qquad \begin{aligned} j &= n - 2(p-h) - 1 \\ \text{for } h &= 1, \ldots, p, \end{aligned}$$

for some real, non-zero r_j and ω_j, (ω_j not an integer multiple of π), for each j.

Define the $(n \times n)$ matrix F as the block diagonal form

$$\mathbf{F} = \text{block diag}\left[\mathbf{I}, \begin{pmatrix} 1 & 1 \\ i & -i \end{pmatrix}, \ldots, \begin{pmatrix} 1 & 1 \\ i & -i \end{pmatrix}\right],$$

where $\mathbf{I}$ is the $(n-2p)$ square identity matrix. By noting that

$$\begin{pmatrix} 1 & 1 \\ i & -i \end{pmatrix}\begin{pmatrix} e^{i\omega_j} & 0 \\ 0 & e^{-i\omega_j} \end{pmatrix}\begin{pmatrix} 1 & 1 \\ i & -i \end{pmatrix}^{-1} = \begin{pmatrix} \cos(\omega_j) & \sin(\omega_j) \\ -\sin(\omega_j) & \cos(\omega_j) \end{pmatrix}$$

for each ω_j it is clear, by superposition, that $\mathbf{F}\boldsymbol{\Lambda}\mathbf{F}^{-1} = \boldsymbol{\Phi}$ where $\boldsymbol{\Phi}$ is the block diagonal matrix with lead diagonal elements $\lambda_1, \ldots, \lambda_{n-2p}$, followed by p separate 2×2 diagonal block components

$$\begin{pmatrix} r_j \cos(\omega_j) & r_j \sin(\omega_j) \\ -r_j \sin(\omega_j) & r_j \cos(\omega_j) \end{pmatrix}$$

for $j = 1, \ldots, p$. Thus, $\boldsymbol{\Lambda}$, and so $\mathbf{G}$, is similar to $\boldsymbol{\Phi}$. By contrast to $\boldsymbol{\Lambda}$, the block diagonal matrix $\boldsymbol{\Phi}$ is real-valued; $\boldsymbol{\Phi}$ is called the real canonical form of $\mathbf{G}$. Reducing $\mathbf{G}$ via the similarity transformation $\mathbf{H}$ to $\boldsymbol{\Lambda}$ simplifies the structure of the matrix but introduces complex entries. By contrast, the reduction to the "almost" diagonal $\boldsymbol{\Phi}$ remains real valued.

COMMON EIGENVALUES AND JORDAN FORMS.

We now consider the case of replicated eigenvalues of $\mathbf{G}$ in an observable DLM. In such cases, $\mathbf{G}$ must must be similar to the precise Jordan form $\mathbf{J}$ given below; (see also Theorem 5.2, p155). For any positive integer r and any complex number λ, the Jordan block $\mathbf{J}_r(\lambda)$ is the $(r \times r)$ upper-triangular matrix

$$\mathbf{J}_r(\lambda) = \begin{pmatrix} \lambda & 1 & 0 & \ldots & 0 \\ 0 & \lambda & 1 & \ldots & 0 \\ 0 & 0 & \lambda & \ldots & 0 \\ \vdots & \vdots & \vdots & \ddots & \vdots \\ 0 & 0 & 0 & \ldots & \lambda \end{pmatrix}.$$

Thus, the diagonal elements are all equal to λ, the super-diagonal elements are unity, and all other entries are zero.

Suppose that in general, the $(n \times n)$ matrix $\mathbf{G}$ has s distinct eigenvalues $\lambda_1, \ldots, \lambda_s$ such that λ_i has multiplicity $r_i \geq 1$, $\sum_{i=1}^{s} r_i = n$. Thus, the characteristic polynomial of $\mathbf{G}$ is simply $p(\lambda) = \prod_{i=1}^{s} (\lambda - \lambda_i)^{r_i}$. It can be shown that $\mathbf{G}$ is similar to the block diagonal Jordan form matrix $\mathbf{J}$ given by the superposition of the Jordan blocks

$$\mathbf{J} = \text{block diag}\left[\mathbf{J}_{r_1}(\lambda_1), \ldots, \mathbf{J}_{r_s}(\lambda_s)\right].$$

$\mathbf{J}$ is sometimes called the Jordan canonical form of $\mathbf{G}$. In this case it follows that since $\mathbf{G} = \mathbf{H}^{-1}\mathbf{J}\mathbf{H}$ for some $\mathbf{H}$, then

$$\mathbf{G}^k = \left(\mathbf{H}^{-1}\mathbf{J}\mathbf{H}\right)^k = \left(\mathbf{H}^{-1}\mathbf{J}\mathbf{H}\right)\left(\mathbf{H}^{-1}\mathbf{J}\mathbf{H}\right)\ldots\left(\mathbf{H}^{-1}\mathbf{J}\mathbf{H}\right) = \mathbf{H}^{-1}\mathbf{J}^k\mathbf{H},$$

so that for any integer k, $\mathbf{G}^k$ is similar to $\mathbf{J}^k$ and the similarity matrix is $\mathbf{H}$, for each k. The structure of $\mathbf{J}^k$ may be explored as follows:

(1) From the block diagonal form of $\mathbf{J}$,

$$\mathbf{J}^k = \text{block diag}\left[\mathbf{J}_{r_1}(\lambda_1)^k, \ldots, \mathbf{J}_{r_s}(\lambda_s)^k\right],$$

for each integer k.

(2) For a general Jordan block $\mathbf{J}_r(\lambda)$, let $\mathbf{J}_r(\lambda)^k$ have elements $m_{i,j}(k)$, $(i = 1, \ldots, r;\ j = 1, \ldots, r)$, for each integer k. Then for $k \geq 1$,

$$m_{i,i+j}(k) = \begin{cases} \binom{k}{j}\lambda^{k-j} & \text{for } 0 \leq j \leq \min(k, r-i), \\ 0 & \text{otherwise.} \end{cases}$$

This may be proved by induction as follows. For $k = 1$, $m_{i,i}(1) = \lambda$ and $m_{i,i+1}(1) = 1$ are the only non-zero elements of $\mathbf{J}_r(\lambda)^1 = \mathbf{J}_r(\lambda)$, so the result holds. Assuming the result to hold for a general k, we have $\mathbf{J}_r(\lambda)^{k+1} = \mathbf{J}_r(\lambda)^k \mathbf{J}_r(\lambda)$, so that

$$\begin{aligned} m_{i,i+j}(k+1) &= \sum_{h=1}^{r} m_{i,h}(k) m_{h,i+j}(1) \\ &= \begin{cases} \lambda m_{i,i+j}(k) + m_{i,i+j-1}(k), & j \neq 0; \\ \lambda m_{i,i}(k), & j = 0. \end{cases} \end{aligned}$$

Hence $m_{i,i}(k+1) = \lambda m_{i,i}(k) = \lambda\lambda^k = \lambda^{k+1}$ as required. Otherwise, it follows for $0 \leq j \leq k+1$ that

$$\begin{aligned} m_{i,i+j}(k+1) &= \lambda\binom{k}{j}\lambda^{k-j} + \binom{k}{j-1}\lambda^{k-j+1} \\ &= \binom{k+1}{j}\lambda^{k+1-j}, \end{aligned}$$

as required. Hence assuming the result true for any k implies it is true for $k+1$. Since the result holds for $k = 1$, then it is true in general by induction.

(3) From (2) it follows that for any fixed r-vectors $\mathbf{a}_r$ and $\mathbf{b}_r$,

$$\mathbf{a}_r \mathbf{J}_r(\lambda)^k \mathbf{b}_r = \lambda^k p_r(k),$$

where $p_r(k) = p_{0r} + p_{1r}k + \ldots + p_{r-1\,r}k^{r-1}$ is a polynomial of degree $r-1$ in k; the coefficients p_{ir} depend on $\mathbf{a}_r$, $\mathbf{b}_r$ and λ but not on k.

(4) From (1) and (2) it follows that for any fixed n-vectors $\mathbf{a}$ and $\mathbf{b}$ and integer $k \geq 1$,

$$\begin{aligned} f(k) &= \mathbf{a}'\mathbf{G}^k\mathbf{b} = \mathbf{a}'\mathbf{H}^{-1}\mathbf{J}^k\mathbf{H}\mathbf{b} \\ &= \sum_{i=1}^{s} p_i(k)\lambda_i^k, \end{aligned}$$

where $p_i(k)$ is a polynomial function of k of degree $r_i - 1$, for each $i = 1, \ldots, s$.

When some of the eigenvalues of $\mathbf{G}$ are complex, a similar real canonical form is usually preferred to the complex-valued Jordan form. In general, order the eigenvalues of $\mathbf{G}$ so that the complex eigenvalues occur in conjugate pairs. Thus, if $\mathbf{G}$ has p distinct pairs of complex eigenvalues, then $\lambda_1, \dots, \lambda_{s-2p}$ are real and distinct, and

$$\left.\begin{aligned} \lambda_j &= r_j \exp(i\omega_j) \\ \lambda_{j+1} &= r_j \exp(-i\omega_j) \end{aligned}\right\}, \qquad \begin{aligned} j &= s - 2(p-h) - 1, \\ &\text{for } h = 1, \dots, p. \end{aligned}$$

Note that both λ_j and λ_{j+1} have multiplicity r_j $(= r_{j+1})$, for each such j. Now for each j, the $(2r_j \times 2r_j)$ Jordan form matrix

$$\text{block diag}[\mathbf{J}_{r_j}(\lambda_j), \mathbf{J}_{r_j}(\lambda_{j+1})]$$

may be shown to be similar to the real-valued matrix shown below. Define the (2×2) real matrix $\mathbf{G}_j$, $(j = s - 2p + 1, \dots, s-1)$, by

$$\mathbf{G}_j = \begin{pmatrix} r\cos(\omega_j) & r\sin(\omega_j) \\ -r\sin(\omega_j) & r\cos(\omega_j) \end{pmatrix},$$

and let $\mathbf{I}$ be the (2×2) identity matrix. Then the similar matrix required is the $(2r_j \times 2r_j)$ (Jordan-form) matrix

$$\mathbf{L}_j = \begin{pmatrix} \mathbf{G}_j & \mathbf{I} & 0 & \dots & 0 \\ 0 & \mathbf{G}_j & \mathbf{I} & \dots & 0 \\ 0 & 0 & \mathbf{G}_j & \dots & 0 \\ \vdots & \vdots & & \ddots & \vdots \\ 0 & 0 & 0 & \dots & \mathbf{G}_j \end{pmatrix}.$$

Using this result it can be shown that the general $\mathbf{G}$ matrix is similar to the $(n \times n)$ block diagonal matrix whose upper-diagonal block is the Jordan form corresponding to real eigenvalues and whose lower-diagonal block is given by block diag$[\mathbf{L}_1, \dots, \mathbf{L}_p]$.

BIBLIOGRAPHY

Abraham, B., and Ledolter, A., 1983. *Statistical Methods for Forecasting.* Wiley, New York.

Abramowitz, M., and Stegun, I.A., 1965. *Handbook of Mathematical Functions.* Dover, New York.

Aguilar, O., and West, M., 1998a. Analysis of hospital quality monitors using hierarchical time series models. In *Case Studies in Bayesian Statistics, Volume 4*, C. Gatsonis, R.E. Kass, B. Carlin, A. Carriquiry, A. Gelman, I. Verdinelli and M. West (Eds.). Springer-Verlag, New York.

Aguilar, O., and West, M., 1998b. Bayesian dynamic factor models and variance matrix discounting for portfolio allocation. *ISDS Discussion Paper* 98-03, Duke University.

Aguilar, O., Huerta, G., Prado, R., and West, M., 1999. Bayesian inference on latent structure in time series (with discussion). In *Bayesian Statistics 6*, J.O. Berger, J.M. Bernardo, A.P. Dawid, and A.F.M. Smith (Eds.). Oxford University Press.

Aitchison, J., 1986. *The Statistical Analysis of Compositional Data.* Chapman-Hall, London.

Aitchison, J., and Brown, J.A.C., 1957. *The Lognormal Distribution.* Cambridge University Press, Cambridge.

Aitchison, J., and Dunsmore, I.R., 1975. *Statistical Prediction Analysis.* Cambridge University Press, Cambridge.

Aitchison, J., and Shen, S.M., 1980. Logistic-normal distributions: some properties and uses. *Biometrika* **67**, 261-272.

Albert, J.H., and Chib, S., 1993. Bayesian inference via Gibbs sampling of autoregressive time series subject to Markov mean and variance shifts. *J. Econ. and Bus. Stat.* **11**, 1-15.

Alspach, D.L., and Sorenson, H.W., 1972. Nonlinear Bayesian estimation using Gaussian sum approximations. *IEEE Trans. Automatic Control* **17**, 439-448.

Amaral, M.A., and Dunsmore, I.R., 1980. Optimal estimates of predictive distributions. *Biometrika* **67**, 685-689.

Ameen, J.R.M., 1984. *Discount Bayesian models and forecasting.* Unpublished Ph.D. thesis, University of Warwick.

Ameen, J.R.M., and Harrison, P.J., 1984. Discount weighted estimation. *J. Forecasting* **3**, 285-296.

Ameen, J.R.M., and Harrison, P.J., 1985a. Normal discount Bayesian models. In *Bayesian Statistics 2*, J.M. Bernardo, M.H. DeGroot, D.V. Lindley, and A.F.M. Smith (Eds.). North-Holland, Amsterdam, and Valencia University Press.

Ameen, J.R.M., and Harrison, P.J., 1985b. Discount Bayesian multiprocess modelling with cusums. In *Time Series Analysis: Theory and Practice 5*, O.D. Anderson (Ed.). North-Holland, Amsterdam.

Anderson, B.D.O., and Moore, J.B., 1979. *Optimal Filtering.* Prentice-Hall, New Jersey.

Ansley, C.F., 1979. An algorithm for the exact likelihood of a mixed autoregressive moving average process. *Biometrika* **66**, 59-65.

Azzalini, A., 1983. Approximate filtering of parameter driven processes. *J. Time Series Analysis* **3**, 219-224.

Baker, R.J., and Nelder, J.A., 1985. *GLIM Release 3.77.* Oxford University, Numerical Algorithms Group.

Barbosa, E., and Harrison, P.J., 1992. Variance estimation for multivariate DLMs. *J. Forecasting* **11**, 621-628.

Barnard, G.A., 1959. Control charts and stochastic processes (with discussion). *J. Roy. Statist. Soc.* (Ser. B) **21**, 239-271.

Barnett, V., and Lewis, T., 1978. *Outliers in Statistical Data.* Wiley, Chichester.

Bates, J.M., and Granger, C.W.J., 1969. The combination of forecasts. *Oper. Res. Quart.* **20**, 451-468.

Bellman, R., 1970. *Introduction to Matrix Analysis.* McGraw-Hill, New York.

Berger, J.O., 1985. *Statistical Decision Theory and Bayesian Analysis* (2nd ed.). Springer-Verlag, New York.

Berk, R.H., 1970. Consistency a posteriori. *Ann. Math. Stat.* **41**, 894-906.

Bernardo, J.M., 1979. Reference posterior distributions for Bayesian inference (with discussion). *J. Roy. Statist. Soc.* (Ser. B) **41**, 113-148.

Bernardo, J.M., and Smith, A.F.M., 1994. *Bayesian Theory.* Wiley, Chichester.

Bordley, R.F., 1982. The combination of forecasts: a Bayesian approach. *J. Oper. Res. Soc.* **33**, 171-174.

Box, G.E.P., 1980. Sampling and Bayes' inference in scientific modelling and robustness. *J. Roy. Statist. Soc.* (Ser. A) **143**, 383-430.

Box, G.E.P., and Cox, D.R., 1964. An analysis of transformations (with discussion). *J. Roy. Statist. Soc.* (Ser. B) **26**, 241-252.

Box, G.E.P., and Draper, N.R., 1969. *Evolutionary Operation.* Wiley, New York.

Box, G.E.P., and Draper, N.R., 1987. *Empirical Model-Building and Response Surfaces.* Wiley, New York.

Box, G.E.P., and Jenkins, G.M., 1976. *Time Series Analysis: Forecasting and Control,* (2nd ed.). Holden-Day, San Francisco.

Box, G.E.P., and Tiao G.C., 1968. A Bayesian approach to some outlier problems. *Biometrika* **55**, 119-129.

Box, G.E.P., and Tiao G.C., 1973. *Bayesian Inference in Statistical Analysis.* Addison-Wesley, Massachusetts.

Broadbent, S., 1979. One way T.V. advertisements work. *J. Mkt. Res. Soc.* **21**, 139-165.

Broemling, L.D., 1985. *Bayesian Analysis of Linear Models.* Marcel Dekker, New York.

Broemling, L.D., and Tsurumi, H., 1987. *Econometrics and Structural Change.* Marcel Dekker, New York.

Brown, R.L., Durbin, J., and Evans, J.M., 1975. Techniques for testing the constancy of regression relationships over time. *J. Roy. Statist. Soc.* (Ser. B) **37**, 149-192.

Brown, R.G., 1959. *Statistical Forecasting for Inventory Control.* McGraw-Hill, New York.

Brown, R.G., 1962. *Smoothing, Forecasting and Prediction of Discrete Time Series.* Prentice-Hall, Englewood Cliffs, NJ.

Bunn, D.W., 1975. A Bayesian approach to the combination of forecasts. *J. Oper. Res. Soc.* **26**, 325-330.

Carlin, B.P., Polson, N.G., and Stoffer, D.S., 1992. A Monte Carlo approach to nonnormal and nonlinear state-space modelling. *J. Amer. Statist. Assoc.* **87**, 493-500.

Cargnoni, C., Müller, P., and West, M., 1996. Bayesian forecasting of multinomial time series through conditionally Gaussian dynamic models. *J. Amer. Statist. Assoc.* (to appear).

Carter, C.K., and Kohn, R., 1994. On Gibbs sampling for state space models. *Biometrika* **81**, 541-553.

Cleveland, W.S., 1974. Estimation of parameters in distributed lag models. In *Studies in Bayesian Econometrics and Statistics*, S.E. Fienberg and A. Zellner (Eds.). North-Holland, Amsterdam.

Colman, S., and Brown, G., 1983. Advertising tracking studies and sales effects. *J. Mkt. Res. Soc.* **25**, 165-183.

Cooper, J.D., and Harrison, P.J., 1997. A Bayesian approach to modelling the observed bovine spongiform encephalopathy epidemic. *J. Forecasting* **16**, 355-374.

Davis, P.J., and Rabinowitz, P. 1984 (2nd Edition). *Methods of Numerical Integration.* Academic Press, Orlando, Florida.

Dawid, A.P., 1981. Some matrix-variate distribution theory: notational considerations and a Bayesian application. *Biometrika* **68**, 265-274.

de Finetti, B., 1974, 1975. *Theory of Probability* (Vols. 1 and 2). Wiley, Chichester.

DeGroot, M.H., 1971. *Optimal Statistical Decisions.* McGraw-Hill, New York.

Dickey, J.M., Dawid, A.P., and Kadane, J.B., 1986. Subjective-probability assessment methods for multivariate-t and matrix-t models. In *Bayesian Inference and Decision Techniques: Essays in Honor of Bruno de Finetti*, P.K. Goel and A. Zellner (Eds.). North-Holland, Amsterdam.

Dickinson, J.P., 1975. Some statistical results in the combination of forecasts. *Oper. Res. Quart.* **26**, 253-260.

Duncan, D.B., and Horne, S.D., 1972. Linear dynamic regression from the viewpoint of regression analysis. *J. Amer. Statist. Assoc.* **67**, 815-821.

Dyro, F.M., 1989. *The EEG Handbook.* Little, Brown and Co., Boston.

Ewan, W.D., and Kemp, K.W., 1960. Sampling inspection of continuous processes with no autocorrelation between successive results. *Biometrika* **47**, 363-380.

Fahrmeir, L., 1992. Posterior mode estimation by extended Kalman filtering for multivariate dynamic GLM. *J. Amer. Statist. Assoc.* **87**, 501-509.

Frühwirth-Schnatter, S., 1994. Data augmentation and dynamic linear models. *J. Time Series Analysis* **15**, 183-102.

Fox, A.J., 1972. Outliers in time series. *J. Roy. Statist. Soc.* (Ser. B) **34**, 350-363.

Fuller, W.A., 1976. *Introduction to Statistical Time Series.* Wiley, New York.

Gamerman, D., 1985. Dynamic Bayesian models for survival data. *Research Report* 75, Department of Statistics, University of Warwick.

Gamerman, D., 1987a. *Dynamic analysis of survival models and related processes.* Unpublished Ph.D. thesis, University of Warwick.

Gamerman, D., 1987b. Dynamic inference on survival functions. In *Probability and Bayesian Statistics*, R. Viertl (Ed.). Plenum, New York and London.

Gamerman, D., 1996. MCMC in dynamic generalised linear models. *Technical Report* 90, Universidade Federal do Rio de Janeiro, Brazil.

Gamerman, D., and Migon H.S., 1993. Dynamic hierarchical models. *J. Roy. Statist. Soc.* (Ser. B) **55**, 629-642.

Gamerman, D., and West, M., 1987a. A time series application of dynamic survival models in unemployment studies. *The Statistician* **36**, 269-174.

Gamerman, D., and West, M., 1987b. Dynamic survival models in action. *Research Report* 111, Department of Statistics, University of Warwick.

Gamerman, D., West, M., and Pole, A., 1987. A guide to SURVIVAL: Bayesian analysis of survival data. *Research Report* 121, Department of Statistics, University of Warwick.

Gardner, G., Harvey, A.C., and Phillips, G.D.A., 1980. An algorithm for exact maximum likelihood estimation by means of Kalman filtering. *Applied Statistics* **29**, 311-322.

Gelfand, A.E., and Smith A.F.M., 1990. Sampling based approaches to calculating marginal densities. *J. Amer. Statist. Assoc.* **85**, 398-409.

Gelman, A., Carlin, B., Stern, H.S., and Rubin, D.B., 1995. *Bayesian Data Analysis.* Chapman & Hall, London.

Geweke, J.F., 1989. Bayesian inference in econometrics using Monte Carlo integration. *Econometrica* **23**, 156-163.

Gilchrist, W., 1976. *Statistical Forecasting.* Wiley, New York.

Gilks, W.R., Richardson, S., and Spiegelhalter, D.J., 1996. *Markov Chain Monte Carlo in Practice.* Chapman & Hall, London.

Godolphin, E.J., and Harrison, P.J., 1975. Equivalence theorems for polynomial projecting predictors. *J. Roy. Statist. Soc.* (Ser. B) **37**, 205-215.

Goldstein, M., 1976. Bayesian analysis of regression problems. *Biometrika* **63**, 51-58.

Good, I.J., 1985. Weights of evidence: a critical survey. In *Bayesian Statistics 2*, J.M. Bernardo, M.H. DeGroot, D.V. Lindley, and A.F.M. Smith (Eds.). North-Holland, Amsterdam, and Valencia University Press.

Granger, C.W.J., and Newbold, P., 1977. *Forecasting Economic Time Series.* Academic Press, New York.

Granger, C.W.J., and Ramanathan, R., 1984. Improved methods of combining forecasts. *J. Forecasting* **3**, 197-204.

Graybill, F.A., 1969. *Introduction to Matrices with Applications in Statistics.* Wadsworth, California.

Green, M., and Harrison, P.J., 1972. On aggregate forecasting. *Research Report* 2, Department of Statistics, University of Warwick.

Green, M., and Harrison, P.J., 1973. Fashion forecasting for a mail order company. *Oper. Res. Quart.* **24**, 193-205.

Grunwald, G.K., Raftery, A.E., and Guttorp, P., 1993. Time series of continuous proportions. *J. Roy. Statist. Soc.* (Ser. B) **55**, 103-116.

Harrison, P.J., 1965. Short-term sales forecasting. *Applied Statistics* **15**, 102-139.

Harrison, P.J., 1967. Exponential smoothing and short-term forecasting. *Man. Sci.* **13**, 821-842.

Harrison, P.J., 1973. Discussion of Box-Jenkins seasonal forecasting: a case study. *J. Roy. Statist. Soc.* (Ser. A) **136**, 319-324.

Harrison, P.J., 1985a. First-order constant dynamic models. *Research Report* 66, Department of Statistics, University of Warwick.

Harrison, P.J., 1985b. Convergence for dynamic linear models. *Research Report* 67, Department of Statistics, University of Warwick.

Harrison, P.J., 1985c. The aggregation and combination of forecasts. *Research Report* 68, Department of Statistics, University of Warwick.

Harrison, P.J., 1988. Bayesian forecasting in O.R.. In *Developments in Operational Research 1988*, N.B. Cook and A.M. Johnson (Eds.). Pergamon Press, Oxford.

Harrison, P.J., 1996. Weak probability. *Research Report*, Department of Statistics, University of Warwick.

Harrison, P.J., 1997. Convergence and the constant dynamic linear model. *J. Forecasting* **16**, 287-291.

Harrison, P.J., and Akram, M., 1983. Generalised exponentially weighted regression and parsimonious dynamic linear modelling. In *Time Series Analysis: Theory and Practice 3*, O.D. Anderson (Ed.). North-Holland, Amsterdam.

Harrison, P.J., and Davies, O.L., 1964. The use of cumulative sum (CUSUM) techniques for the control of routine forecasts of produce demand. *Oper. Res.* **12**, 325-33.

Harrison, P.J., and Lai, I.C.H., 1999. Statistical process control and model monitoring. *J. App. Stat.* **26**, 273-292.

Harrison, P.J., Leonard, T., and Gazard, T.N., 1977. An application of multivariate hierarchical forecasting. *Research Report* 15, Department of Statistics, University of Warwick.

Harrison, P.J., and Pearce, S.F., 1972. The use of trend curves as an aid to market forecasting. *Industrial Marketing Management* **2**, 149-170.

Harrison, P.J., and Quinn, M.P., 1978. A brief description of part of the work done to date concerning a view of agricultural and related systems. In *Econometric Models*, presented to the Beef-Milk symposium, Commission of European Communities, Agriculture EUR6101, March 1977, Brussels.

Harrison P.J., and Reed, R.J., 1996. Dynamic linear modelling with S-Plus. *Research Report* 310, Department of Statistics, University of Warwick.

Harrison P.J., and Scott, F.A., 1965. A development system for use in short-term forecasting (original draft 1965, re-issued 1982). *Research Report* 26, Department of Statistics, University of Warwick.

Harrison, P.J., and Smith, J.Q., 1980. Discontinuity, decision and conflict (with discussion). In *Bayesian Statistics*, J.M. Bernardo, M.H. De Groot, D.V. Lindley, and A.F.M. Smith (Eds.). University Press, Valencia.

Harrison P.J., and Stevens, C., 1971. A Bayesian approach to short-term forecasting. *Oper. Res. Quart.* **22**, 341-362.

Harrison P.J., and Stevens, C., 1976a. Bayesian forecasting (with discussion). *J. Roy. Statist. Soc.* (Ser. B) **38**, 205-247.

Harrison P.J., and Stevens, C., 1976b. Bayes forecasting in action: case studies. *Warwick Research Report* 14, Department of Statistics, University of Warwick.

Harrison, P.J., and Veerapen, P.P., 1993. Incorporating and deleting information in dynamic models. In *Developments in Time Series Analysis: in Honour of Maurice B. Priestley*, T. Subba Rao (Ed.). Chapman & Hall, London.

Harrison, P.J., and Veerapen, P.P., 1994. A Bayesian decision approach to model monitoring and Cusums. *J. Forecasting* **13**, 29-36.

Harrison, P.J., and West, M., 1986. Bayesian forecasting in practice. *Bayesian Statistics Study Year Report* 13, University of Warwick.

Harrison, P.J., and West, M., 1987. Practical Bayesian forecasting. *The Statistician* **36**, 115-125.

Harrison, P.J., and West, M., 1991. Dynamic linear model diagnostics. *Biometrika* **78**, 797-808.

Harrison, P.J., West, M., and Pole, A., 1987. FAB: A training package for Bayesian forecasting. *Warwick Research Report* 122, Department of Statistics, University of Warwick.

Hartigan, J.A., 1969. Linear Bayesian methods. *J. Roy. Statist. Soc.* (Ser. B) **31**, 446-454.

Harvey, A.C., 1981. *Time Series Models.* Philip Allan, Oxford.

Harvey, A.C., 1984. A unified view of statistical forecasting procedures. *J. Forecasting* **3**, 245-275.

Harvey, A.C., 1986. Analysis and generalisation of a multivariate exponential smoothing model. *Man. Sci.* **32**, 374-380.

Harvey, A.C., and Durbin, J., 1986. The effects of seat belt legislation on British road casualties: a case study in structural time series modelling (with discussion). *J. Roy. Statist. Soc.* (Ser. A) **149**, 187-227.

Heyde, C.C., and Johnstone, I.M., 1979. On asymptotic posterior normality for stochastic processes. *J. Roy. Statist. Soc.* (Ser. B) **41**, 184-189.

Highfield, R., 1984. Forecasting with Bayesian state space models. *Technical Report*, Graduate School of Business, University of Chicago.

Holt, C.C., 1957. Forecasting seasonals and trends by exponentially weighted moving averages. *O.N.R. Research Memo.* 52, Carnegie Institute of Technology.

Jaquier, E., Polson, N.G., and Rossi, P.E., 1994. Bayesian analysis of stochastic volatility. *J. Bus. and Econ. Stat.* **12**, 371-417.

Jazwinski, A.H., 1970. *Stochastic Processes and Filtering Theory.* Academic Press, New York.

Jeffreys, H., 1961. *Theory of Probability* (3rd ed.). Oxford University Press, London.

Johnston, F.R., and Harrison, P.J., 1980. An application of forecasting in the alcoholic drinks industry. *J. Oper. Res. Soc.* **31**, 699-709.

Johnston, F.R., Harrison, P.J., Marshall, A.S., and France, K.M., 1986. Modelling and the estimation of changing relationships. *The Statistician* **35**, 229-235.

Johnson, N.L., and Kotz, S., 1972. *Distributions in Statistics: Continuous Multivariate Distributions.* Wiley, New York.

Kalman, R.E., 1960. A new approach to linear filtering and prediction problems. *J. Basic Engineering* **82**, 35-45.

Kalman, R.E., 1963. New methods in Wiener filtering theory. In *Proceedings of the First Symposium of Engineering Applications of Random Function Theory and Probability*, J.L. Bogdanoff and F. Kozin (Eds.). Wiley, New York.

Kitegawa, G., 1987. Non-Gaussian state-space modelling of nonstationary time series (with disussion). *J. Amer. Statist. Assoc.* **82**, 1032-1063.

Kitegawa, G., and Gersch, W., 1985. A smoothness priors time varying AR coefficient modeling of nonstationary time series. *IEEE Trans. Automatic Control* **30**, 48-56.

Kleiner, B., Martin, R.D., and Thompson, D.J., 1979. Robust estimation of power spectra (with discussion). *J. Roy. Statist. Soc.* (Ser. B) **3**, 313-351.

Kullback, S., 1983. Kullback information. In *Encyclopedia of Statistical Sciences* (Vol. 4), S. Kotz and N.L. Johnston (Eds.). Wiley, New York.

Kullback, S., and Leibler, R.A., 1951. On information and sufficiency. *Ann. Math. Statist.* **22**, 79-86.

Leamer, E.E., 1972. A class of information priors and distributed lag analysis. *Econometrica* **40**, 1059-1081.

Lindley, D.V., 1965. *Introduction to Probability and Statistics from a Bayesian Viewpoint* (Parts 1 and 2). Cambridge University Press, Cambridge.

Lindley, D.V., 1983. Reconciliation of probability distributions. *Oper. Res.* **31**, 806-886.

Lindley, D.V., 1985. Reconciliation of discrete probability distributions. In *Bayesian Statistics 2*, J.M. Bernardo, M.H. DeGroot, D.V. Lindley, and A.F.M. Smith (Eds.). North-Holland, Amsterdam, and Valencia University Press.

Lindley, D.V., 1988. The use of probability statements. In *Accelerated life tests and experts' opinion in reliability*, C.A. Clarotti and D.V. Lindley (Eds.). North-Holland, Amsterdam.

Lindley, D.V., and Scott, W.F., 1984. *New Cambridge Elementary Statistical Tables.* Cambridge University Press, Cambridge.

Lindley, D.V., and Smith, A.F.M., 1972. Bayes' estimates for the linear model. *J . Roy. Statist. Soc.* (Ser. B) **34**, 1-41.

Lindley, D.V., Tversky, A., and Brown, R.V., 1979. On the reconciliation of probability assessments (with discussion). *J. Roy. Statist. Soc.* (Ser. A) **142**, 146-180.

Litterman, R.B., 1986a. Specifying vector autoregressions for macro-economic forecasting. In *Bayesian Inference and Decision Techniques: Essays in Honor of Bruno de Finetti*, P.K. Goel and A. Zellner (Eds.). North-Holland, Amsterdam.

Litterman, R.B., 1986b. Forecasting with Bayesian vector autoregressions: Five years of experience. *J. Bus. and Econ. Stat.* **4**, 25-38.

Mardia, K.V., Kent, J.T., and Bibby, J.M., 1979. *Multivariate Analysis.* Academic Press, London.

Marriot, J., 1987. Bayesian numerical integration and graphical methods for Box-Jenkins time series. *The Statistician* **36**, 265-268.

Masreliez, C.J., 1975. Approximate non-Gaussian filtering with linear state and observation relations. *IEEE Trans. Aut. Con.* **20**, 107-110.

Masreliez, C.J., and Martin, R.D., 1977. Robust Bayesian estimation for the linear model and robustifying the Kalman filter. *IEEE Trans. Aut. Con.* **22**, 361-371.

Mazzuchi, T.A., and Soyer, R., 1992. A dynamic general linear model for inference from accelerated life tests. *Naval Research Logistics* **39**, 757-773.

McCullagh, P., and Nelder, J.A., 1989. *Generalised Linear Models (2nd edition).* Chapman Hall, London and New York.

McCulloch, R., and Tsay, R., 1994. Bayesian analysis of autoregressive time series via the Gibbs sampler. *J. Time Series Anal.* **15**, 235-250.

McKenzie, E., 1974. A comparison of standard forecasting systems with the Box-Jenkins approach. *The Statistician* **23**, 107-116.

McKenzie, E., 1976. An analysis of general exponential smoothing. *Oper. Res.* **24**, 131-140.

Migon, H.S., 1984. *An approach to non-linear Bayesian forecasting problems with applications.* Unpublished Ph.D. thesis, Department of Statistics, University of Warwick.

Migon, H.S., and Harrison, P.J., 1985. An application of non-linear Bayesian forecasting to television advertising. In *Bayesian Statistics 2*, J.M. Bernardo, M.H. DeGroot, D.V. Lindley, and A.F.M. Smith (Eds.). North-Holland, Amsterdam, and Valencia University Press.

Morris, C.N., 1983. Natural exponential families with quadratic variance functions: statistical theory. *Ann. Statist.* **11**, 515-519.

Morris, P.A., 1983. An axiomatic approach to expert resolution. *Man. Sci.* **29**, 24-32.

Müller, P., West, M., and MacEachern, S.N., 1997. Bayesian models for non-linear auto-regressions. *J. Time Series Anal.* **18**, ,. 593-614.

Muth, J.F., 1960. Optimal properties of exponentially weighted forecasts. *J. Amer. Statist. Assoc.* **55**, 299-306.

Naylor, J.C., and Smith, A.F.M., 1982. Applications of a method for the efficient computation of posterior distributions. *Applied Statistics* **31**, 214-225.

Nelder, J.A., and Wedderburn, R.W.M., 1972. Generalised linear models. *J. Roy. Statist. Soc.* (Ser. A) **135**, 370-384.

Nerig, E.D., 1969. *Linear Algebra and Matrix Theory.* Wiley, New York.

Nerlove, M., and Wage, S., 1964. On the optimality of adaptive forecasting. *Man. Sci.* **10**, 207-229.

O'Hagan, A., 1994. *Kendall's Advanced Theory of Statistics, Volume 2B: Bayesian Inference.* Edward Arnold, London.

Page, E.S., 1954. Continuous inspection schemes. *Biometrika* **41**, 100-114.

Park, J., 1992. Envelope estimation for quasi-periodic geophysical signals in noise: a multitaper approach. In *Statistics in the Environmental and Earth Sciences*, A.T. Walden and P. Guttorp (Eds.). Edward Arnold.

Park, J., and Maasch, K.A., 1993. Plio-Pleistocene time evolution of the 100-kyr cycle in marine paleoclimate records. *J. Geophys. Res.* **98**, 447-461.

Pole, A., 1988a. Transfer response models: a numerical approach. In *Bayesian Statistics 3*, J.M. Bernardo, M.H. DeGroot, D.V. Lindley, and A.F.M. Smith (Eds.). Oxford University Press.

Pole, A., 1988b. NOBS: Non-linear Bayesian forecasting software. *Warwick Research Report* 154, Department of Statistics, University of Warwick.

Pole, A., and Smith, A.F.M., 1983. A Bayesian analysis of some threshold switching models. *J. Econometrics* **29**, 97-119.

Pole, A., and West, M., 1988. Efficient numerical integration in dynamic models. *Warwick Research Report* 136, Department of Statistics, University of Warwick.

Pole, A., and West, M., 1989. Reference analysis of the DLM. *J. Time Series Analysis* **10**, 131-147.

Pole, A., and West, M., 1990. Efficient Bayesian learning in non-linear dynamic models. *J. Forecasting* **9**, 119-136.

Pole, A., West, M., Harrison, P.J., 1988. Non-normal and non-linear dynamic Bayesian modelling. In *Bayesian Analysis of Time Series and Dynamic Models*, J.C. Spall (Ed.). Marcel Dekker, New York.

Pole, A., West, M., Harrison, P.J., 1994. *Applied Bayesian Forecasting and Time Series Analysis.* Chapman & Hall, New York.

Prado, R., Krystal, A.D., and West, M., 1999. Evaluation and comparison of EEG traces: Latent structure in non-stationary time series. *J. Amer. Statist. Assoc.* **94**, -.

Prado, R., and West, M., 1997. Bayesian analysis and decomposition of multiple non-stationary time series. In *Modelling Longitudinal and Spatially Correlated Data*, T. Gregoire (Ed.). Springer-Verlag, New York.

Press, S.J., 1985. *Applied Multivariate Analysis: Using Bayesian and Frequentist Methods of Inference.* Krieger, California.

Priestley, M.B., 1980. System identification, Kalman filtering, and stochastic control. In *Directions in Time Series*, D.R. Brillinger and G.C.Tiao (Eds.). Institute of Mathematical Statistics.

Quintana, J.M., 1985. A dynamic linear matrix-variate regression model. *Research Report* 83, Department of Statistics, University of Warwick.

Quintana, J.M., 1987. *Multivariate Bayesian forecasting models.* Unpublished Ph.D. thesis, University of Warwick.

Quintana, J.M., 1992. Optimal portfolios of forward currency contracts. In *Bayesian Statistics 4*, J.O. Berger, J.M. Bernardo, A.P. Dawid, and A.F.M. Smith (Eds.). Oxford University Press.

Quintana, J.M., Chopra, V.K., and Putnam, B.H., 1995. Global asset allocation: Stretching returns by shrinking forecasts. In *Proceedings of the ASA Section on Bayesian Statistical Science, 1995 Joint Statistical Meetings*, American Statistical Association.

Quintana, J.M., and West, M., 1987. Multivariate time series analysis: new techniques applied to international exchange rate data. *The Statistician* **36**, 275-281.

Quintana, J.M., and West, M., 1988. Time series analysis of compositional data. In *Bayesian Statistics 3*, J.M. Bernardo, M.H. DeGroot, D.V. Lindley, and A.F.M. Smith (Eds.). Oxford University Press.

Ripley, B.D., 1987. *Stochastic Simulation.* Wiley, New York.

Roberts, S.A., and Harrison, P.J., 1984. Parsimonious modelling and forecasting of seasonal time series. *European J. Oper. Res.* **16**, 365-377.

Ruddiman W.F., McIntyre, A., and Raymo, M., 1986. Paleoenvironmental results from North Atlantic Sites 607 and 609. In *Init. Reports DSDP: 94*, R.B. Kidd et al (Eds.). U.S. Govt. Printing Office, Washington.

Russell, B., 1921. *The Analysis of Mind.* Allen and Unwin.

Salzer, H.E., Zucker, R., and Capuano, R., 1952. Tables of the zeroes and weight factors of the first twenty Hermite polynomials. *J. Research of the National Bureau of Standards* 48, 111-116.

Sage, A.P., and Melsa, J.L., 1971. *Estimation Theory with Applications to Communications and Control.* McGraw-Hill, New York.

Savage, L.J., 1954. *The Foundations of Inference.* Wiley, New York.

Schnatter, S., 1988. Bayesian forecasting of time series by Gaussian sum approximation. In *Bayesian Statistics 3*, J.M. Bernardo, M.H. De Groot, D.V. Lindley, and A.F.M. Smith (Eds.). Oxford University Press.

Schervish, M.J., and Tsay, R., 1988. Bayesian modelling and forecasting in autoregressive models. In *Bayesian Analysis of Time Series and Dynamic Models*, J.C. Spall (Ed.). Marcel Dekker, New York.

Scipione, C.M., and Berliner, L.M., 1993. Bayesian statistical inference in nonlinear dynamical systems. In *1993 Proceedings of the Bayesian Statistical Science Section of the ASA*, American Statistical Association.

Shackleton, N.J., and Hall, M.A., 1989. Stable isotope history of the Pleistocene at ODP Site 677. In *Proc. ODP, Sci. Results*, K. Becker et al (Eds.). College Station, Texas.

Shaw, J.E.H., 1987. A strategy for reconstructing multivariate probability distributions. *Research Report* 123, Department of Statistics, University of Warwick.

Shaw, J.E.H., 1988. Aspects of numerical integration and summarisation. In *Bayesian Statistics 3*, J.M. Bernardo, M.H. DeGroot, D.V. Lindley, and A.F.M. Smith (Eds.). Oxford University Press.

Shephard, N., 1994. Partial non-Gaussian state space models. *Biometrika* **81**, 115-131.

Shephard, N., and Pitt, M.K., 1995. Parameter-driven, exponential family models. *Technical Report*, University of Oxford.

Silverman, B.W., 1986. *Density Estimation for Statistics and Data Analysis.* London, Chapman-Hall.

Smith, A.F.M., 1975. A Bayesian approach to inference about a change point in a sequence of random variables. *Biometrika* **63**, 407-416.

Smith, A.F.M., 1980. Change point problems: approaches and applications. In *Bayesian Statistics*, J.M. Bernardo, M.H. De Groot, D.V. Lindley, and A.F.M. Smith (Eds.). University Press, Valencia.

Smith, A.F.M., and Cook, D.G., 1980. Switching straight lines: a Bayesian analysis of some renal transplant data. *Applied Statistics* **29**, 180-189.

Smith, A.F.M., and Pettit, L.I., 1985. Outliers and influential observations in linear models. In *Bayesian Statistics 2*, J.M. Bernardo, M.H. DeGroot, D.V. Lindley, and A.F.M. Smith (Eds.). North-Holland, Amsterdam, and Valencia University Press.

Smith, A.F.M., Skene, A.M., Shaw, J.E.H., Naylor, J.C., and Dransfield, M., 1985. The implementation of the Bayesian paradigm. *Commun. Statist.: Theor. Meth.* **14**, 1079-1102.

Smith, A.F.M., Skene, A.M., Shaw, J.E.H., and Naylor, J.C., 1987. Progress with numerical and graphical methods for practical Bayesian statistics. *The Statistician* **36**, 75-82.

Smith, A.F.M., and West, M., 1983. Monitoring renal transplants: an application of the multi-process Kalman filter. *Biometrics* **39**, 867-878.

Smith, A.F.M., West, M., Gordan, K., Knapp, M.S., and Trimble, I., 1983. Monitoring kidney transplant patients. *The Statistician* **32**, 46-54.

Smith, J.Q., 1979. A generalisation of the Bayesian steady forecasting model. *J. Roy. Statist. Soc.* (Ser. B) **41**, 378-387.

Smith, R.L., and Miller, J.E., 1986. Predictive records. *J. Roy. Statist. Soc.* (Ser. B) **48**, 79-88.

Sorenson, H.W., and Alspach, D.L., 1971. Recursive Bayesian estimation using Gaussian sums. *Automatica* **7**, 465-479.

Souza, R.C., 1981. A Bayesian-entropy approach to forecasting: the multi-state model. In *Time Series Analysis*, O.D. Anderson (Ed.). North-Holland, Houston, Texas.

Spiegelhalter, D.J., Thomas, A., Best, N.G., and Gilks, W.R., 1994. BUGS: Bayesian Inference Using Gibbs Sampling. *Medical Research Council Biostatistics Units*, Cambridge.

Stevens, C.F., 1974. On the variability of demand for families of items. *Oper. Res. Quart.* **25**, 411-420.

Sweeting, T.J., and Adekola, A.O., 1987. Asymptotic posterior normality for stochastic processes revisited. *J. Roy. Statist. Soc.* (Ser. B) **49**, 215-222.

Tierney, L., 1994. Markov chains for exploring posterior distributions (with discussion). *Ann. Statist.* **22**, 1701-1762.

Theil, H., 1971. *Principles of Econometrics.* Wiley, New York.

Thiel, H., and Wage, S., 1964. Some observations on adaptive forecasting. *Man. Sci.* **10**, 198-206.

Titterington, D.M., Smith, A.F.M., and Makov, U.E., 1985. *Statistical Analysis of Finite Mixture Distributions.* Wiley, Chichester.

Tong, H., 1983. *Threshold Models in Non-Linear Time Series Analysis.* Springer-Verlag, New York.

Tong, H., and Lim, K.S., 1980. Threshold autoregression, limit cycles and cyclical data (with discussion). *J. Roy. Statist. Soc.* (Ser. B) **42**, 245-292.

Trimble, I., West, M., Knapp, M.S., Pownall, R., and Smith, A.F.M., 1983. Detection of renal allograft rejection by computer. *British Medical Journal* **286**, 1695-1699.

Uhlig, H., 1994. On singular Wishart and singular multivariate beta distributions. *Ann. Statist.* **22**, 395-405.

Uhlig, H., 1997. Bayesian vector-autoregressions with stochastic volatility. *Econometrica* (to appear).

Van Dijk, H.K., Hop, J.P., and Louter, A.S., 1987. Some algorithms for the computation of posterior moments and densities using Monte Carlo integration. *The Statistician* **36**, 83-90.

Van Duijn, J.J., 1983. *The Long Wave in Economic Life.* George Allen and Unwin, London.

Vasconcellos, K.L.P., 1992. *Aspects of forecasting aggregate and discrete data.* Unpublished Ph.D. thesis, University of Warwick.

West, M., 1981. Robust sequential approximate Bayesian estimation. *J. Roy. Statist. Soc.* (Ser. B) **43**, 157-166.

West, M., 1982. *Aspects of Recursive Bayesian Estimation.* Unpublished Ph.D. thesis, University of Nottingham.

West, M., 1984a. Outlier models and prior distributions in Bayesian linear regression. *J. Roy. Statist. Soc.* (Ser. B) **46**, 431-439.

West, M. 1984b. Bayesian aggregation. *J. Roy. Statist. Soc.* (Ser. A) **147**, 600-607.

West, M., 1985a. Generalised linear models: outlier accomodation, scale parameters and prior distributions. In *Bayesian Statistics 2*, J.M. Bernardo, M.H. DeGroot, D.V. Lindley, and A.F.M. Smith (Eds.). North-Holland, Amsterdam, and Valencia University Press.

West, M., 1985b. Combining probability forecasts. *Research Report* 67, Department of Statistics, University of Warwick.

West, M., 1985c. Assessment and control of probability forecasts. *Research Report* 69, Department of Statistics, University of Warwick.

West, M., 1986a. Bayesian model monitoring. *J. Roy. Statist. Soc.* (Ser. B) **48**, 70-78.

West, M., 1986b. Non-normal multi-process models. *Research Report* 81, Department of Statistics, University of Warwick.

West, M., 1988. Modelling expert opinion (with discussion). In *Bayesian Statistics 3*, J.M. Bernardo, M.H. De Groot, D.V. Lindley, and A.F.M. Smith (Eds.). Oxford University Press.

West, M., 1992a. Approximating posterior distributions with mixtures. *J. Roy. Statist. Soc.* (Ser. B) **55**, 409-422.

West, M., 1992b. Mixture models, Monte Carlo, Bayesian updating and dynamic models. *Computing Science and Statistics* **24**, 325-333.

West, M., 1992c. Modelling with mixtures (with discussion). In *Bayesian Statistics 4*, J.O. Berger, J.M. Bernardo, A.P. Dawid, and A.F.M. Smith (Eds.). Oxford University Press.

West, M., 1992d. Modelling time-varying hazards and covariate effects (with discussion). In *Survival Analysis: State of the Art*, J.P. Klein and P.K. Goel (Eds.). Kluwer.

West, M., 1992e. Modelling agent forecast distributions. *J. Roy. Statist. Soc.* (Ser. B) **54**, 553-567.

West, M., 1995. Bayesian inference in cyclical component dynamic linear models. *J. Amer. Statist. Assoc.* **90**, 1301-1312.

West, M., 1996a. Bayesian time series: Models and computations for the analysis of time series in the physical sciences. In *Maximum Entropy and Bayesian Methods 15*, K. Hanson and R. Silver (Eds.). Kluwer.

West, M., 1996b. Some statistical issues in Palæoclimatology (with discussion). In *Bayesian Statistics 5*, J.O. Berger, J.M. Bernardo, A.P. Dawid and A.F.M. Smith (Eds.). Oxford University Press.

West, M., 1996c. Modelling and robustness issues in Bayesian time series analysis. In *Bayesian Robustness 2*, J.O. Berger, F. Ruggeri, and L. Wasserman (Eds.). IMS Monographs.

West, M., 1997. Time series decomposition. *Biometrika* **84**, 489-494.

West, M., and Crosse, J. 1992. Modelling of probabilistic agent opinion. *J. Roy. Statist. Soc.* (Ser. B) **54**, 285-299.

West, M., and Harrison, P.J., 1986a. Monitoring and adaptation in Bayesian forecasting models. *J. Amer. Statist. Assoc.* **81**, 741-750.

West, M., and Harrison, P.J., 1986b. Advertising awareness response model: micro-computer APL*PLUS/PC implementation. *Warwick Research Report* 118, Department of Statistics, University of Warwick.

West, M., and Harrison, P.J., 1989. Subjective intervention in formal models. *J. Forecasting* **8**, 33-53.

West, M., Harrison, P.J., and Migon, H.S., 1985. Dynamic generalised linear models and Bayesian forecasting (with discussion). *J. Amer. Statist. Assoc.* **80**, 73-97.

West, M., Harrison, P.J., and Pole, A., 1987a. BATS : Bayesian Analysis of Time Series. *The Professional Statistician* **6**, 43-46. (This work concerns the early version of the BATS software, written in APL for PC implementation. The more recent software is available with the 1994 book by Pole, West and Harrison, in this bibliography)

West, M., Harrison, P.J., and Pole, A., 1987b. BATS : A user guide. *Warwick Research Report* 114, Department of Statistics, University of Warwick. (This is the guide to the early version of the BATS software, written in APL for PC implementation. The more recent software is available with the 1994 book by Pole, West and Harrison, in this bibliography)

West, M., and Mortera, J., 1987. Bayesian models and methods for binary time series. In *Probability and Bayesian Statistics*, R. Viertl (Ed.). Plenum, New York and London.

Whittle, P., 1965. Recursive relations for predictors of non-stationary processes. *J. Roy. Statist. Soc.* (Ser. B) **27**, 523-532.

Winkler, R.L., 1981. Combining probability distributions from dependent information sources. *Man. Sci.* **27**, 479-488.

Winters, P.R., 1960. Forecasting sales by exponentially weighted moving averages. *Man. Sci.* **6**, 324-342.

Woodward, R.H., and Goldsmith, P.L., 1964. *I.C.I. Monograph No. 3: Cumulative Sum Techniques.* Mathematical and Statistical Techniques for Industry, Oliver and Boyd, Edinburgh.

Young, A.S., 1977. A Bayesian approach to prediction using polynomials. *Biometrika* **64**, 309-317.

Young, A.S., 1983. A comparative analysis of prior families for distributed lags. *Empirical Econonomics* **8**, 215-227.

Young, P.C., 1984. *Recursive Estimation and Time Series Analysis.* Springer-Verlag, Berlin.

Zellner, A., 1971. *An Introduction to Bayesian Inference in Econometrics.* Wiley, New York.

Zellner, A., and Hong, C., 1989. Forecasting international growth rates using Bayesian shrinkage and other procedures. *J. Econometrics* **40**, 183-202.

Zellner, A., Hong, C., and Min, C., 1991. Forecasting turning points in international output growth rates using Bayesian exponentially weighted autoregression, time-varying parameter, and pooling techniques. *J. Econometrics* **49**, 275-304.

AUTHOR INDEX

SUBJECT INDEX

Springer Series in Statistics

(continued from p. ii)

Kotz/Johnson (Eds.): Breakthroughs in Statistics Volume II.
Kotz/Johnson (Eds.): Breakthroughs in Statistics Volume III.
Kres: Statistical Tables for Multivariate Analysis.
Küchler/Sørensen: Exponential Families of Stochastic Processes.
Le Cam: Asymptotic Methods in Statistical Decision Theory.
Le Cam/Yang: Asymptotics in Statistics: Some Basic Concepts.
Longford: Models for Uncertainty in Educational Testing.
Manoukian: Modern Concepts and Theorems of Mathematical Statistics.
Miller, Jr.: Simultaneous Statistical Inference, 2nd edition.
Mosteller/Wallace: Applied Bayesian and Classical Inference: The Case of the Federalist Papers.
Parzen/Tanabe/Kitagawa: Selected Papers of Hirotugu Akaike.
Pollard: Convergence of Stochastic Processes.
Pratt/Gibbons: Concepts of Nonparametric Theory.
Ramsay/Silverman: Functional Data Analysis.
Read/Cressie: Goodness-of-Fit Statistics for Discrete Multivariate Data.
Reinsel: Elements of Multivariate Time Series Analysis, 2nd edition.
Reiss: A Course on Point Processes.
Reiss: Approximate Distributions of Order Statistics: With Applications to Non-parametric Statistics.
Rieder: Robust Asymptotic Statistics.
Rosenbaum: Observational Studies.
Ross: Nonlinear Estimation.
Sachs: Applied Statistics: A Handbook of Techniques, 2nd edition.
Särndal/Swensson/Wretman: Model Assisted Survey Sampling.
Schervish: Theory of Statistics.
Seneta: Non-Negative Matrices and Markov Chains, 2nd edition.
Shao/Tu: The Jackknife and Bootstrap.
Siegmund: Sequential Analysis: Tests and Confidence Intervals.
Simonoff: Smoothing Methods in Statistics.
Small: The Statistical Theory of Shape.
Tanner: Tools for Statistical Inference: Methods for the Exploration of Posterior Distributions and Likelihood Functions, 3rd edition.
Tong: The Multivariate Normal Distribution.
van der Vaart/Wellner: Weak Convergence and Empirical Processes: With Applications to Statistics.
Vapnik: Estimation of Dependences Based on Empirical Data.
Weerahandi: Exact Statistical Methods for Data Analysis.
West/Harrison: Bayesian Forecasting and Dynamic Models, 2nd edition.
Wolter: Introduction to Variance Estimation.
Yaglom: Correlation Theory of Stationary and Related Random Functions I: Basic Results.
Yaglom: Correlation Theory of Stationary and Related Random Functions II: Supplementary Notes and References.